T0135239

Advances in Intelligent Systems and Computing

Volume 689

About this Series

The series "Advances in Intelligent Systems and Computing" contains publications on theory, applications, and design methods of Intelligent Systems and Intelligent Computing. Virtually all disciplines such as engineering, natural sciences, computer and information science, ICT, economics, business, e-commerce, environment, healthcare, life science are covered. The list of topics spans all the areas of modern intelligent systems and computing.

The publications within "Advances in Intelligent Systems and Computing" are primarily textbooks and proceedings of important conferences, symposia and congresses. They cover significant recent developments in the field, both of a foundational and applicable character. An important characteristic feature of the series is the short publication time and world-wide distribution. This permits a rapid and broad dissemination of research results.

More information about this series at http://www.springer.com/series/11156

Natalia Shakhovska · Volodymyr Stepashko
Editors

Advances in Intelligent Systems and Computing II

Selected Papers from the International Conference on Computer Science and Information Technologies, CSIT 2017, September 5–8, Lviv, Ukraine

Editors
Natalia Shakhovska
Lviv Polytechnic National University
Lviv
Ukraine

Volodymyr Stepashko
IRTC ITS NASU
Kiev
Ukraine

ISSN 2194-5357 ISSN 2194-5365 (electronic)
Advances in Intelligent Systems and Computing
ISBN 978-3-319-70580-4 ISBN 978-3-319-70581-1 (eBook)
https://doi.org/10.1007/978-3-319-70581-1

Library of Congress Control Number: 2016950408

Printed on acid-free paper

This Springer imprint is published by Springer Nature
The registered company is Springer International Publishing AG
The registered company address is: Gewerbestrasse 11, 6330 Cham, Switzerland

Preface

This book is designed for scientists and junior- or senior-level computer science students with a basic background in intelligence systems, data mining, inductive modeling, and computing. The international conference *Computer Sciences and Information Technologies* is annually organized with the principal aim to discuss modern trends in computer sciences, information technologies, applied linguistics, and others related areas. To achieve this goal, various aspects of computer science will be presented in 11 major topics: artificial intelligence; computational intelligence; computer vision; information modeling of database and knowledge systems; big data and cloud computing; intelligence control systems and technologies; computational linguistics; cyber-physical systems; project management; software engineering; IT education.

September 2017

Natalia Shakhovska
Program Committee Member

Contents

GMDH-Based Learning System for Mobile Robot Navigation in Heterogeneous Environment

Anatoliy Andrakhanov(✉) and Alexander Belyaev

Department of Control Systems and Mechatronics, National Research Tomsk Polytechnic University, Lenin Avenue, 30, Tomsk 634050, Russian Federation
riml282a@gmail.com, belyaewas@mail.ru

Abstract. One of the key tasks of mobile robotics is navigation, which for Outdoor-type robots is exacerbated by the functioning in an environment with a priori of unknown characteristics of underlying surfaces. In this paper, for the first time, the learning navigation system for mobile robot based on the group method of data handling (GMDH) is presented. The paper presents the results of training of models both for evaluating the robot's pose (coordinates and angular orientation) in heterogeneous environment and classification of the type of underlying surfaces. In addition to the direct readings of the on-board sensors, additional parameters (reflecting how the robot perceives the surface terramechanics) were introduced to train the models. The results of testing of the obtained models demonstrate their performance in an essentially heterogeneous environment, when areas of the underlying surfaces are comparable with the robot's dimensions. This testifies the operability of developed GMDH-based learning system for mobile robot navigation.

Keywords: Mobile robot · Heterogeneous environment · Underlying surface Testing ground · Navigation · Coordinates evaluation · Machine learning Inductive modeling · GMDH · Active neuron · Festo Robotino

1 Introduction

Despite the rapid development of mobile robotics, the development of an intelligent control system remains one of the main challenges in the creation of autonomous robotic systems.

One of the key tasks is the navigation, which can be divided into 2 parts: the estimation of the current position (coordinates and angular orientation) of the robot in the working space and the development of control actions on the actuators to sequentially achieve all the intermediate robot positions along the planned trajectory. In this case, the solution of the second part of the problem is impossible without solving of the first. In some cases, the evaluation of the robot position in the environment can be carried out only by means of on-board inertial system, because global positioning systems (GPS) may be not exist (planetary rovers), may be inaccessible (fire fighting robots, autonomous mining vehicles) or suppressed by electronic countermeasures equipment (combat robots).

N. Shakhovska and V. Stepashko (eds.), *Advances in Intelligent Systems and Computing II*, Advances in Intelligent Systems and Computing 689,
https://doi.org/10.1007/978-3-319-70581-1_1

For outdoor-type mobile robots, this problem is exacerbated by the natural conditions of the environment which is characterized by the a priori unknown of an environment model, the heterogeneous characteristics of surfaces to be traversed, and the difficulty of determining the features of the robot-terrain interaction based on on-board sensor readings.

There are a great number of papers on this subject [1–9]. In a first approximation the four main approaches can be identified for solving navigation problem in a heterogeneous environment (Table 1).

Table 1. The man ways to solve the problem of a mobile robot's navigation in a heterogeneous environment

Approach to models construction	Classification of the underlying surface type	
	Used	Not used
Detection and analysis of a physical patterns of robot-terrain interaction	S. Khaleghian and S. Taheri [9] Sensors: 3-axis and 1-exis accelerometers, encoders Method: Fuzzy logic	L. Ojeda et al. [7] Iagnemma K. et al. [8] Sensors: encoders, gyroscopes, accelerometers [8] + current sensors [7] Method: Wheel slip analysis
Construction of non-physical models by means of machine learning	DuPont E.M. et al. [6] Sensors: 3-axis gyroscope, 3-axis accelerometer Method: Probabilistic neural network	A.A. Andrakhanov [10, 11] Sensors: encoders, motor current sensors Method: Twice Multilayered Modified Polynomial Neural Network with active neurons

There is currently no generally accepted dominant methodology at present, and each research group is trying to solve the problem in its own way. The authors of this work believe that one of the most promising paths is to construct non-physical models using the advantages of the inductive modeling approach.

The basic method of the inductive modeling approach is Group Method of Data Handling (GMDH). To date, the most complete overview of the use of GMDH in robotics is shown in the work [12]. This method was already used by the authors to solve the problem of evaluating the robot's pose in homogeneous and heterogeneous environments and demonstrated an acceptable result [10, 11].

2 The Advantages of GMDH for Finding Dependencies Based on the Analysis of Sensor Readings

A number of field experiments were carried out in order to determine the interaction between the robot's propulsion system and the underlying surface of various types based on on-board sensor readings.

A serial-produced mobile platform Festo Robotino was used as the mobile robot. Experiments were carried out at a specially designed testing ground consisting of 28 modules with different terramechanical characteristics (Fig. 1a). The testing ground was designed in such a way that the terramechanical interaction of its areas with Festo Robotino's wheel system was correspond in terms of quality with the level of terramechanical interaction of the outdoor robot with the areas of the natural environment. Figure 1b shows the robot's test motions (along a square and a triangle path) with the same motor speed setpoints under conditions of a homogeneous (an ideal flat surface) and heterogeneous (Testing Ground) environment. Holonomic character of robot movements is provided due to 3 wheels of omnidirectional type, located at an angle of 120° with respect to each other (Fig. 1c).

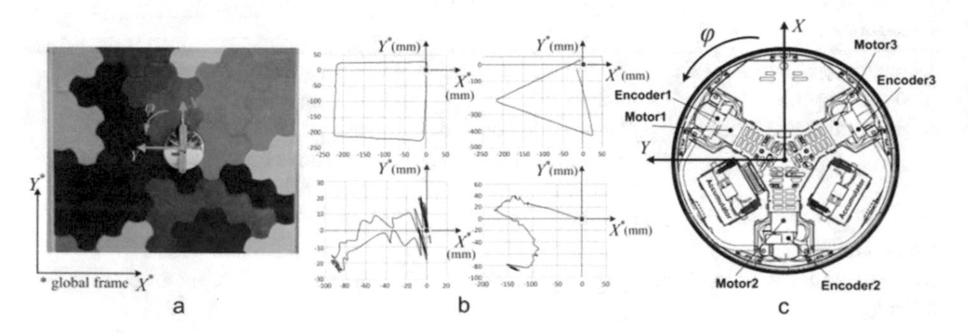

Fig. 1. (a) Testing Ground appearance and the robot; (b) Trajectories for test motions (along a square and a triangle path) in homogeneous (upper) and heterogeneous (lower) environments; (c) arrangement of wheels in Robotino's omni-drive system [13] (*X*, *Y*, φ – Robotino's local coordinate system)

It is clear from Fig. 1b that the areas of the testing ground have a significant impact on the nature of the robot's movement. In order to maximize the transparency between sensor readings and the nature of the robot's motion, we decided to assign only the simplest robot motions and only the homogeneous areas of the testing ground consisting of modules of the same type. Using the simplest driving setpoints makes it possible to eliminate the features of the robot's kinematics and propulsion system, which in turn allows it to make complex motions (curvilinear motion with rotational motion) without being affected by the features of the environment. Moving in a homogeneous local area eliminates the comlex influence of different local areas on the robot behavior because on the testing ground the each wheel interacts with its local area (Fig. 1a).

Five types of local surfaces, which were different in terms of terramechanics, were selected by expertise, as well as four of the simplest movements setpoints: three translational motion, without rotational component (along the X-axis, along the Y-axis, in the XY plane at the same X- and Y-axis speeds) and rotational motion without translational component, all motions mentioned in relation to the robot's local coordinate system (Fig. 1c). The speed of the translational motion was set to be 100 mm/s (for both the X- and Y-axes), and that of rotational motion was 24, 48 and 96 (deg/s).

Sensor data subject to analysis included: $\{N\} = \{N_1, N_2, N_3\}$ – a set of incremental encoders values, $\{\omega\} = \{\omega_1, \omega_2, \omega_3\}$ – values of the speeds of motors, $\{I\} = \{I_1, I_2, I_3\}$ – values of the motors consumption currents, $\{g\} = \{g_x, g_y, g_z\}$ – angular velocity values in relation to the X, Y and Z axes (Z-axis is perpendicular to XY-plane), $\{a\} = \{a_x, a_y, a_z\}$ – a set of acceleration values along the X, Y and Z axes.

Figure 2 shows the readings of three sensors when the robot traverses over five different underlying surfaces.

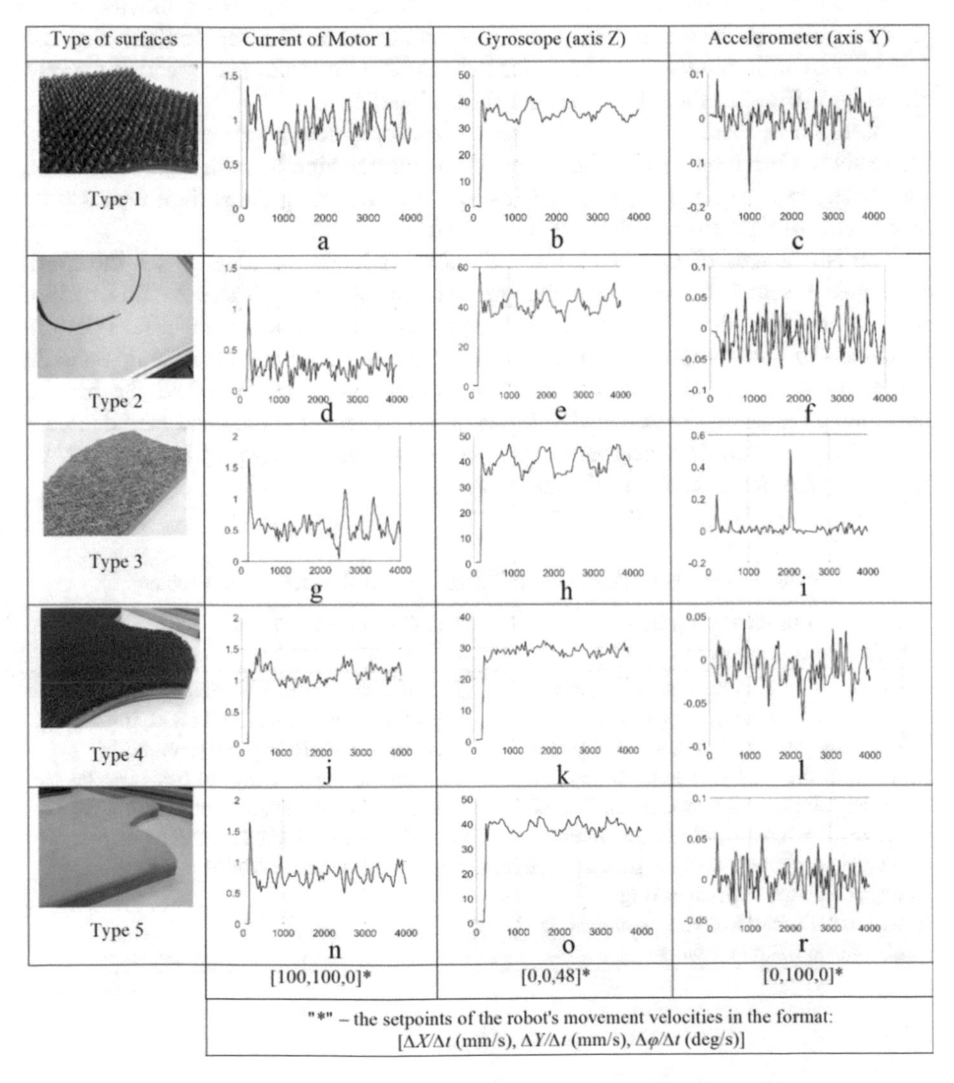

Fig. 2. Sensor readings during the robot's movement over different surfaces (horizontal axis for all graphics is the time axis in milliseconds)

The analysis shows that, on the one hand, the sensor data are correlated to the type of underlying surface and the nature of the robot movement on this surface. On the other hand, the correlation is ambiguous.

For example, the average values of currents in the first motor (I_1) demonstrate the laboriousness of overcoming the area (Fig. 2a and g), on the one hand, while on the other hand, these values may be close despite the fact that the surfaces may have different terramechanical characteristics (Fig. 2a and j). In addition, there are other features in the sensor data, which reflect the interaction of the robot with a surface of a particular type. In particular, the gyroscope data along the Z-axis (g_z), shown in Fig. 2b and h, clearly show the periodic behavior and the curve shape when moving over a particular surface. However, these may not be shown in the sensor data (Fig. 2k). A similar trend is observed in accelerometer readings (a_y): the features may manifest themselves (Fig. 2i), though not always (Fig. 2l and f).

Inasmuch as there are unique data properties for different subsets of sensors, it is necessary to derive models on their basis by using machine learning and data mining techniques. In our opinion, one promising tool for evaluating the surface type and the coordinates using sensor readings is the inductive approach.

The advantages of GMDH (as a method of inductive modeling) for the aforementioned research topics and developments are shown in Table 2. This method provides maximal flexibility at the stage of model construction both in handling the parameters of data sample (the method of dividing the sample into the training and the test parts, sorting by the dispersion of the output variable, etc.) and the training algorithm parameters (for neural algorithms, it means selecting the maximum degree of a partial description of a neuron, the number of selectable neurons in the layer, the maximum number of layers in the network, etc.).

Table 2. The benefits of GMDH in addressing the navigation problem

The problem of navigation in a heterogeneous environment	The GMDH advantages
A variety of tools is necessary to generate methods for evaluating coordinates and determining the underlying surface type. For instance, in the well-known paper [6], the technical solution contained operations of extracting the most significant features (principal component analysis), interpolation (fundamental splines), clustering (Eigen-transformation), and classification (probabilistic neural network)	The method includes a wide range of algorithms for predicting, classifying, clustering, identifying and data mining The unified methodological basis for the aforementioned spectrum of algorithms contributes to the standardization of the system's program modules

(*continued*)

Table 2. (*continued*)

The problem of navigation in a heterogeneous environment	The GMDH advantages
There are no simple and obvious correlations between sensor readings and the terramechanical properties of the underlying surfaces	The most effective input variables (with respect to some quality criterion) are selected automatically from the set of variables available to the system, and relationships in data are interpreted The resulting dependencies have an analytical form (it is also typical for the GMDH-type neural networks), which enhances the capabilities for analysis and makes it possible to interpret the results
The number of local areas that affect the robot motion in different ways can be arbitrary large. Therefore, it is necessary that the functional dependences derived by the onboard computing system should be generalizable for other areas that have not yet been traversed	The resulting dependencies have a generalizing ability because an external criterion of model quality is used (evaluation of model parameters and selection of model structure are performed using independent data subsamples)
Considering that the size of local areas may be relatively small, it is important that the methods used to derive models be able to work with short data samples In some cases (for instance, the time limit for making a decision, limited learning time, as well as energy costs, and so on), it makes sense to collect a relatively small number of samples even if the size of areas is significant	For short, inaccurate, or noisy data, an optimal nonphysical model can be found, whose accuracy is higher and structure is simpler than the structure of a complete physical model [14] Finding a solution within a limited training time is guaranteed. The system can calculate the training time before training algorithm run

In addition, GMDH provides great opportunities in analyzing dependencies found during the training phase: which sensor readings are used for the model's output as the input variables; with what degrees and/or coefficients these variables are included in the model; how often these variables are chosen by the neural network algorithm when constructing a neural network structure from layer to layer sequentially, etc.

In order to ensure maximum access to these features and the advantages of GMDH, the authors have decided to develop a navigational training system on its basis.

3 Navigation Learning System

3.1 Description of the System

The system architecture is represented in Fig. 3 (only the basic connections between the modules are shown).

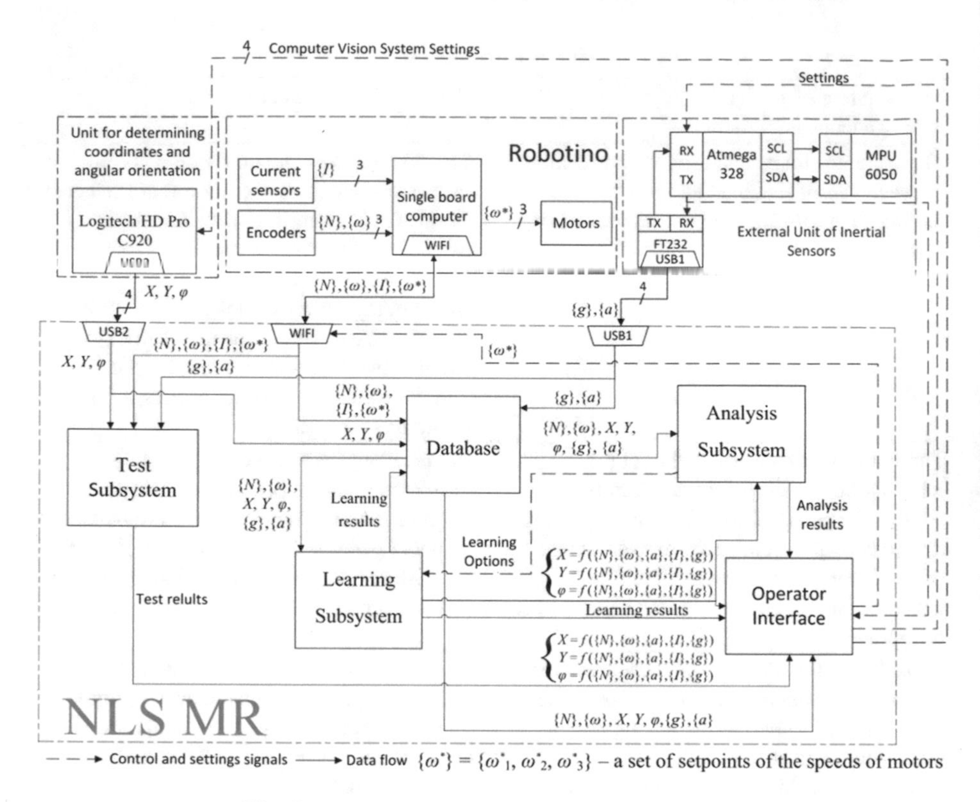

Fig. 3. The navigation learning system architecture

The navigation learning system (NLS) consists of the following modules:

- The Learning Subsystem that implements computational intelligence algorithms within a framework of the inductive modeling approach.
- The Database which is necessary for collecting and storing sensor data as well as learning results.
- The Test subsystem which is required to test the hardware of the system (sensors and actuators of the robot, unit for determining coordinates and angular orientation) and the learning subsystem. The hardware test is based on comparing the stored data (sensor readings, the coordinates and angular orientation values) and the data obtained as a result of the robot's test movements. Testing of the learning subsystem is done by models training on test data samples and comparing the obtained results with the stored results.
- The Analysis subsystem which analyzes the previous results of models training in order to determine the influence of different subsets of input variables, the variants of splitting a data sample, the quality criteria and the algorithm parameters on the quality of obtained models.
- The Operator interface that provides the operator's access to data samples, as well as to the results of testing, training and analysis.

The hardware which is external in relation to NLS includes the on-board sensors and actuators of the Festo Robotino platform, the additional unit of inertial sensors (three-axis gyroscope and three-axis accelerometer) and the Full HD camera to obtain the coordinates and angular orientation of the robot.

This system was used for training and testing models in all the experiments mentioned below.

3.2 The Models Training Algorithm

Twice-Multilayered Modified Polynomial Neural Network with Active Neurons (TMMPNN) algorithm makes it possible to find the optimal network structure (from the view of the external criterion) and partial descriptions of neurons automatically (in the self-organization mode). The concept of twice-multilayered polynomial neural network algorithm was first proposed by A.G. Ivakhnenko and J.A. Muller [15].

The modification is that the generation of partial descriptions on each layer (starting from the 2nd) involves not only the outputs of neurons of the previous layer, but also the input variables. Thus, such modification provides an opportunity to avoid losing important input variables on the first and subsequent layers of the network. The structure of the modified polynomial neural network is shown in Fig. 4 [12]:

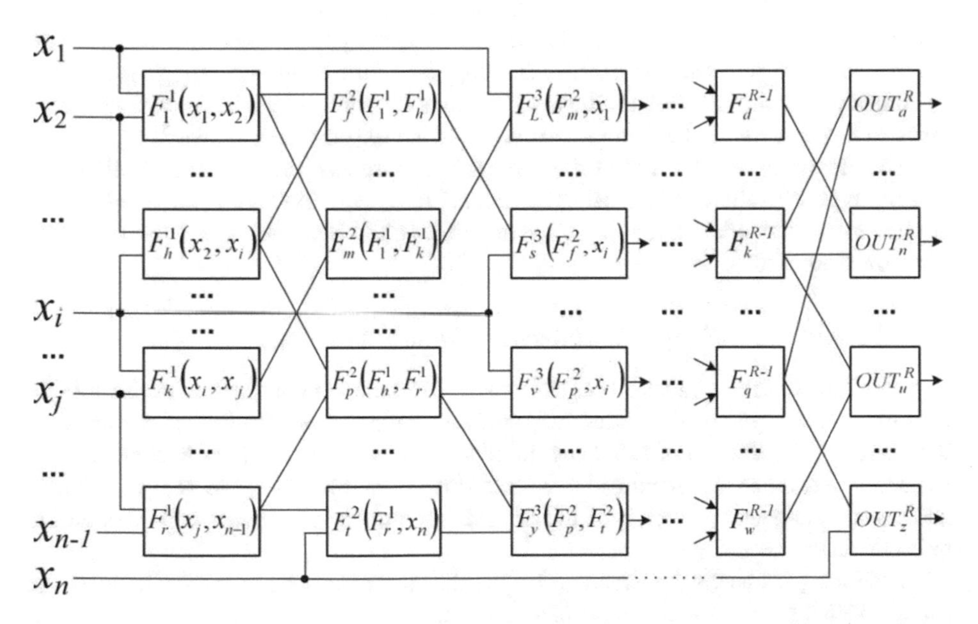

Fig. 4. Twice-multilayered modified GMDH-type polynomial neural network ($x_1, x_2, x_i, x_j, x_{n-1}, x_n$ are input variables; $F_k^1(x_i, x_j)$ are partial descriptions of k-th selected neuron of the 1st layer; $F_d^{R-1}, F_k^{R-1}, F_q^{R-1}, F_w^{R-1}$ are partial descriptions of neurons of the layer (R–1); $OUT_a^R, OUT_n^R, OUT_u^R, OUT_z^R$ are partial descriptions of selected neurons of the output layer)

Since this algorithm was described in detail in earlier papers [10, 12] and its software implementation was published on the CD to the book [12], we will discuss only the main points related to the settings of this algorithm in the following experiments.

The classic combinatorial algorithm of GMDH [16] is used to search for partial descriptions of neurons. In this case, two-input neurons were used limited by the maximum polynomial degree 2 of the partial description:

$$F_k^l(x_i, x_j) = a_0 + a_1 \cdot x_i + a_2 \cdot x_j + a_3 \cdot x_i^2 + a_4 \cdot x_j^2 + a_5 \cdot x_i \cdot x_j \quad (1)$$

The regularity criterion was used as an external criterion for the selection of partial descriptions of neurons:

$$CR = \frac{1}{N_B} \sum_{i=1}^{N_B} (f_i - y_i)^2 \quad (2)$$

where N_B is the number of rows of the testing data sample; f_i is the output of the model for row i; y_i is the output value for row i of the data sample.

In case of network structure construction the regularity criterion was also used as an external criterion for the selection of the best neurons of a layer. The external layer criterion (the arithmetic mean value of the regularity criteria of the best neurons in a given layer), the limit of maximal network capacity (maximal number of layers $\times$ number of selected best neurons in a layer) and the additional stopping criterion (an improvement in the value for the external layer criterion should be more than ε from layer to layer) were used as stopping criteria of the expansion of network layers.

The algorithm, criteria and settings described were also used for the training of a surface type classifier. When using a trained classifier, the threshold condition is applied to the network output: «1» if the network output is greater than or equal to 0.5, and «0» otherwise.

3.3 Forming Sets of Input Variables for Models Training

In Sect. 2, it was shown that it was difficult to estimate the coordinates of the robot's position and the type of the underlying surfaces directly from the sensor readings. In the paper [17], parameters reflecting how the robot sense the terramechanics of a surface based on sensor readings were introduced. In this work, we also introduced additional parameters of such type to increase the number of relevant variables on the training stage.

The three parameters that characterize the displacement in a given local area are the robot's velocities in its local coordinate system:

$$\begin{pmatrix} V_x \\ V_y \\ \Omega \end{pmatrix} = R \cdot \begin{pmatrix} -\frac{2}{3}\cos(\alpha-\theta) & \frac{2}{3}\sin(\alpha) & \frac{2}{3}\cos(\alpha+\theta) \\ -\frac{2}{3}\sin(\alpha-\theta) & -\frac{2}{3}\cos(\alpha) & \frac{2}{3}\sin(\alpha+\theta) \\ \frac{1}{3 \cdot L} & \frac{1}{3 \cdot L} & \frac{1}{3 \cdot L} \end{pmatrix} \cdot \begin{pmatrix} \omega_1 \\ \omega_2 \\ \omega_3 \end{pmatrix} \quad (3)$$

where V_x, V_y is the velocity along the X and Y axes of the robot's local coordinate system; Ω is the angular rotation velocity of the robot in the local coordinate system; ω_1, ω_2, ω_3 are the angular speeds of wheels (associated with the speeds of the motors through the 1:16 gear ratio); L is the distance from the center of the robot to the wheel (125 mm); R is the wheel radius (40 mm); α is the robot orientation angle; θ is the wheel orientation angle (30°).

The kinematics Eq. (3) was also used by the authors to obtain the parameters of the laboriousness of translational and rotational motion of the robot:

$$\begin{pmatrix} I_x \\ I_y \\ I_\varphi \end{pmatrix} = R \cdot \begin{pmatrix} -\frac{2}{3}\cos(\alpha-\theta) & \frac{2}{3}\sin(\alpha) & \frac{2}{3}\cos(\alpha+\theta) \\ -\frac{2}{3}\sin(\alpha-\theta) & -\frac{2}{3}\cos(\alpha) & \frac{2}{3}\sin(\alpha+\theta) \\ \frac{1}{3} & \frac{1}{3} & \frac{1}{3} \end{pmatrix} \cdot \begin{pmatrix} I_1 \\ I_2 \\ I_3 \end{pmatrix} \tag{4}$$

where I_x, I_y are the values of the currents that characterize the laboriousness of the robot's movement along the X and Y axes of the local coordinate system; I_φ is the value of the current which characterizes the laboriousness of the robot's angular rotation in the local coordinate system; I_1, I_2, I_3 are the consumption currents of motors.

In (4), the consumption currents of motors have the same sign, which is determined by the direction of the wheel rotation, as for the speeds of motors in (3). Unlike (3), R and L values are not used in (4) because, firstly, they are not related to a geometric transformation of the current vectors. Secondly, the dimension of the output quantities and their physical meaning will be inconsistent with each other, which is unacceptable. Thirdly, these values influence only the amplitude of the output values, which is not important for GMDH, since the TMMPNN algorithm independently selects necessary weighting coefficients. The preliminary analysis showed an appropriate separability for all five types of surfaces used in the experiments in case of use of parameters – I_x, I_y, I_φ (Fig. 5b, c and d).

Another parameter used, which characterizes the interaction of the robot with the surface, is I_Σ – the total consumption current of motors:

$$I_\Sigma = |I_1| + |I_2| + |I_3| \tag{5}$$

As can be seen in Fig. 5a, the mean value of this parameter varies for different types of surfaces, which makes it a useful variable both for classifying the surface type and for estimating the coordinates and angular orientation. Since the robot arrives to different coordinates on different surfaces with the same motor velocities setpoints, the coordinate estimation can be related to this parameter.

In addition to the aforementioned absolute parameters, the following relative parameters were also introduced:

$$\begin{gathered} T_x = \frac{V_x}{I_x}; T_y = \frac{V_y}{I_y}; T_\varphi = \frac{\Omega}{I_\varphi}; T_z = \frac{g_z}{I_\varphi}; T_x^* = \frac{V_x^* - V_x}{I_x}; T_y^* = \frac{V_y^* - V_y}{I_y}; \\ T_\varphi^* = \frac{\Omega^* - \Omega}{I_\varphi}; T_z^* = \frac{\Omega^* - g_z}{I_\varphi} \end{gathered} \tag{6}$$

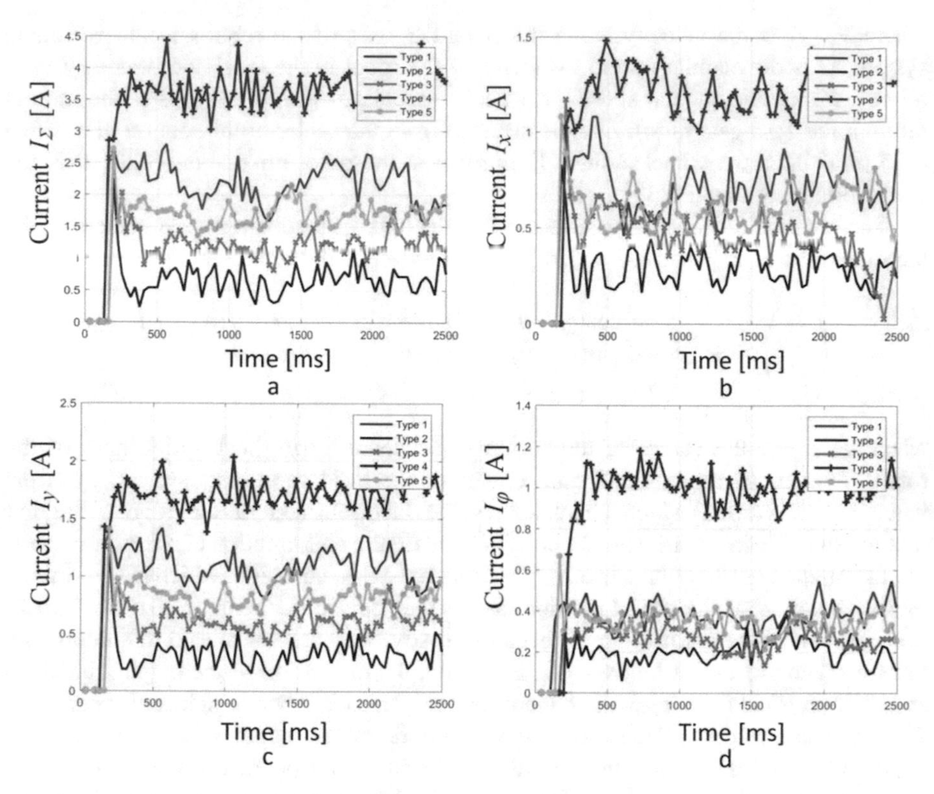

Fig. 5. Analysis of the relevance of the parameters (the setpoints of the robot's movement velocities in the format – [V_x^* (mm/s), V_y^* (mm/s), Ω^* (deg/s)]: (a) [0; 100; 0]; (b) [100; 100; 0]; (c) [0; 100; 0]; (d) [0; 0; 24])

The values V_x, V_y, and Ω are calculated using actual wheel speeds values, based on (3), while I_x, I_y, and I_φ are calculated using the current sensors values based on (4). The values V_x^*, V_y^*, and Ω^* are the setpoints of the robot's movement velocities in the local coordinate system.

It should be noted that the dimension of relative parameters has the physical interpretation as a unit of the translational/rotational movement on a particular surface for the expended current impulse, which is normalized to the same type and direction of movement. In the case of a difference in the numerator between the setpoint and the real velocity, the physical interpretation changes into: by how many millimeters/ degrees the actual displacement/rotation of the robot on the surface will differ from the setpoint value after one current impulse.

Thus, while implementing inductive modeling, three sets of input variables were used:

- $\{V_1\} = \{\{N\},\{\omega\},\{I\},\{g\},\{a\}\}$ are values obtained directly from the robot's sensors;
- $\{V_2\} = \{V_x, V_y, \Omega, I_x, I_y, I_\varphi, I_\Sigma\}$ are absolute parameters obtained by means of mathematical transformations of values measured by sensors;

- $\{V_3\} = \left\{T_x, T_y, T_\varphi, T_z, T_x^*, T_y^*, T_\varphi^*, T_z^*\right\}$ are relative parameters obtained by means of algebraic relations between values of the second and the first sets.

The purpose of the experiments series was:

- Determination of the obtained models accuracy for robot pose evaluation taking into account three sets of parameters;
- Estimation of the relevance of each set of parameters for the constructing of robot pose estimation models;
- Determination of the accuracy of the underlying surface type classification taking into account three sets of parameters;
- Estimation of the relevance of each set of parameters for constructing classifiers of the surface type;
- Testing the obtained models during robot movement in an essentially heterogeneous environment, when dimensions of both the surfaces and the robot are comparable.

4 Results of Experiments

4.1 Results of Models Training

All experiments were carried out on five selected types (see Fig. 2) of surfaces using the Festo Robotino mobile platform. There were 30 robot launches lasting 4 s with the following combinations of robot movement setpoints (in the format [$\Delta X/\Delta t$ (mm/s), $\Delta Y/\Delta t$ (mm/s), $\Delta\varphi/\Delta t$ (deg/s)]): [100,0,0], [0,100,0], [100,100,0], [0,0,24], [0,0,48], [0,0,96]. Data sample was formed by dividing of the sensor readings into half-second intervals, sensor values was averaged for these intervals (except values of $\{N\}$, the increment of values per half-second intervals were calculated for this case). Thus, for each robot launch four examples were included into the training data sample, four – into the testing data sample.

All models were obtained with help on software that was published on the CD to the book «GMDH-Methodology and Implementation in C» [12].

In the all experiments, the constraints were used to both the maximum power of neuron (power – 2) and the network capacity (10 layers x 10 neurons per layer). The choice of these parameters is due to the experience of our previous experiments, including in [10, 11]. In particular, it was found that such a limitation on the network capacity makes it possible to obtain the most stable (in terms of accuracy and bias) models.

Since the absolute error in the determination of the coordinates used by the computer vision unit is 1 mm, an additional criterion for stopping the network construction $\varepsilon = 0.1$ was given. Data sample was divided into the two equal parts (training and testing).

The results of the experiments are shown in Table 3 ("[Avr]" is the arithmetic mean error, "Max" is the maximum error, "GM" are denoted ("General Model") the training results of model on the combined data sample for all types of the surfaces) and Table 4.

The best trained models with minimum error are highlighted in bold in Tables 3 and 4. The average values ("[Avr]") of the coordinates and angular orientation in Table 3 are less 1 (mm or deg) on the all types of underlying surfaces for the all subsets of input variables.

Table 3. Results of models training for robot pose estimation

Value	Type	Input variable set						
		$\{V_1\}$	$\{V_2\}$	$\{V_3\}$	$\{V_1\}$, $\{V_2\}$	$\{V_1\}$, $\{V_3\}$	$\{V_2\}$, $\{V_3\}$	All
		Max [Avr]	Max [Avr]	Max [Avr]	Max [Avr]	Max [Avr]	Max [Avr]	Max [Avr]
X, mm	1	**3.6**	6.0	6.8	4.6	3.6	4.9	4.6
	2	10.5	**9.3**	9.4	10.9	10.5	9.7	10.9
	3	**5.9**	8.1	6.4	6.3	5.9	6.6	6.1
	4	2.2	2.1	2.8	2.0	2.3	**1.8**	2.0
	5	7.2	7.2	6.8	**6.1**	7.2	6.3	6.4
	GM	**8.7 [2.02]**	9.9 [1.7]	12.6 [2.16]	10.5 [1.7]	9.3 [1.8]	10.0 [1.8]	10.3 [1.7]
Y, mm	1	5.1	**4.8**	6.6	4.8	7.3	4.8	4.8
	2	9.0	8.9	8.1	7.7	8.1	8.5	**6.5**
	3	**6.1**	9.7	29.3	12.2	6.2	8.6	12.2
	4	**1.5**	2.2	2.0	1.9	1.7	2.8	2.0
	5	**5.5**	8.1	9.2	5.8	5.8	9.2	9.2
	GM	9.9 [1.7]	15.1 [1.5]	243 [6.6]	**9.0 [1.4]**	9.9 [1.7]	15.1 [1.5]	9.0 [1.4]
φ, deg	1	**5.3**	6.8	65.2	5.5	5.3	6.8	5.5
	2	4.3	5.8	9.0	4.9	**4.0**	5.8	4.9
	3	6.4	5.7	7.6	6.0	6.1	**4.9**	6.0
	4	4.8	9.2	**3.3**	4.8	4.8	9.2	4.8
	5	8.2	7.7	**3.6**	4.5	5.4	6.5	5.4
	GM	14.6 [1.4]	**10.3 [1.74]**	305 [13.8]	10.6 [1.6]	14.6 [1.4]	11.1 [1.6]	10.6 [1.6]

Table 4. Percentage of correct classification for trained classifiers

Type	Input variable set						
	$\{V_1\}$	$\{V_2\}$	$\{V_3\}$	$\{V_1\}$, $\{V_2\}$	$\{V_1\}$, $\{V_3\}$	$\{V_2\}$, $\{V_3\}$	All
1	81.2	85.0	81.2	86.3	82.5	**88.0**	82.1
2	96.6	95.7	83.3	**97.0**	96.2	95.7	97.0
3	86.3	85.0	81.6	84.6	85.0	81.6	**86.8**
4	97.9	**98.7**	79.5	97.8	97.9	98.7	98.3
5	78.6	**82.1**	79.9	79.5	78.2	79.5	80.8

Insomuch as this neural network is based on the inductive principles of self-organization of models, the very process of the self-organization of its structure serves not only as a means of obtaining the final model but also as a tool for analysis. Thus, based on the selection of appropriate input variables on each layer of the network by active neurons, we can estimate the contribution of the sensor data to the overall dependency. The received structures of the GMDH-type neural networks for the best models of robot pose estimation are shown in Fig. 6.

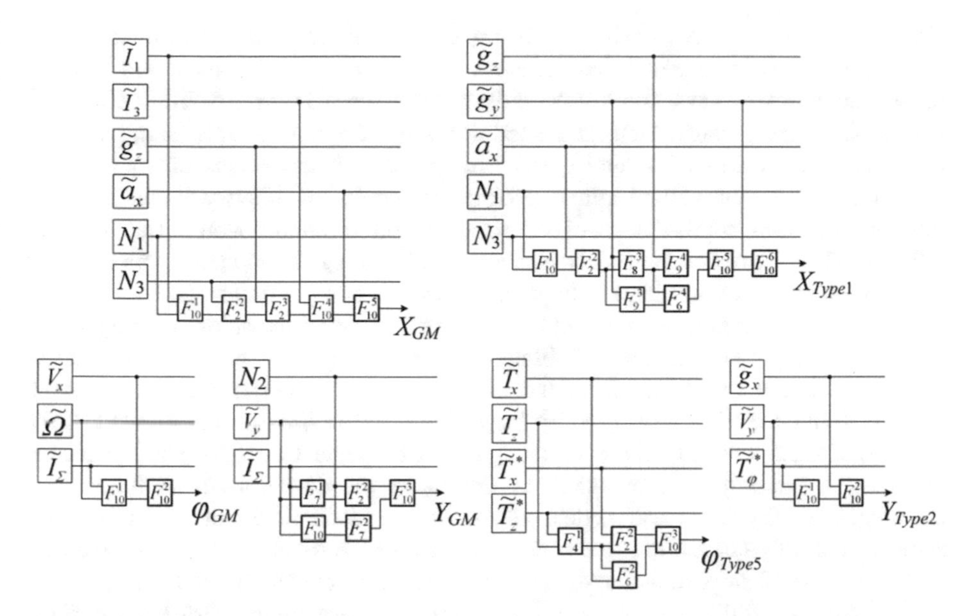

Fig. 6. Structures of twice-multilayered modified polynomial neural networks for the some best trained models ("$\sim$" – average value per half-second interval, "Type" – the type of underlying surface (see Table 3), the networks evaluates X, Y and φ in 0.5 s of robot's movement)

Active neurons first of all choose input parameters taking into account their direct physical correspondence with the output variable, although the GMDH algorithm constructs non-physical models. For example:

- In case X_{GM} in Fig. 6 the neurons chosen the values of I_1, I_3 and N_1, N_3, which are directly related to X-coordinate (see Fig. 1c). The a_x is also directly related to the output variable.
- In case Y_{GM} the neurons chosen the V_y and N_2, which are directly related to Y-coordinate (see Fig. 1c). The choice of I_Σ is also associated with the Y-variable evaluation, since, as mentioned above, for each type of surface this parameter has different values (see Fig. 5a).

At the same time, the choice of some input parameters is not obvious, because the features of robot-terrain interaction for specific types of surfaces are taken into account (for example, g_y and g_z for X_{Type1}, g_x and $T^*{}_{\varphi}$ for Y_{Type2}).

Analysis of classifiers shows that in order to construct better models, active neurons in all cases use the variables of $\{V_2\}$.

4.2 Results of Testing the Trained Models During the Robot Movement in a Heterogeneous Environment

This section shows some results of testing of the trained models in an essentially heterogeneous environment, when the areas of the surfaces are comparable with the robot's dimensions. Thus, the test conditions for navigation are much more difficult than at the training stage. First, at the training stage the robot moved along separate homogeneous surfaces of several types. But in this test, the effects of not only the influence but also the mutual influence of the properties of different surfaces on the robot's movement appear. Secondly, such transition zones between surfaces (when different wheels are simultaneously located at different surfaces) appear often during the movement, which accelerates the accumulation of navigation errors.

The task was to movement along the triangle (lengths of the sides are assigned by operator) through the surfaces of different types, using only the on-board sensor readings (without the signal from the global positioning system) and the trained models. Trained models were used by the robot to determine the achievement of the vertexes of triangle (with the aim of changing the movement direction) and to correct the trajectory during the movement. To determine the deviations from the desired trajectory the outputs of these models were used as a feedback signals instead of GPS signals. With these deviations, the control signals for the robot motions are generated by the method of proportional regulation well-known in the automatic control theory.

The purpose of the series of experiments was to determine the performance of the best trained models (see Table 3) for two cases:

- movement for mentioned above conditions by means of models general to all types of surfaces (denoted "GM" in Table 3);
- movement under the same conditions using coordinates evaluation models specialized for a specific type of surface (we denote this models as "MT" ("Model for Type")). These models are selected by signal from the corresponding classifier. If signals from all classifiers are absent or there are signals from several classifiers, then the coordinates and orientation angle are evaluated by the "GM"-models.

In Fig. 7 shows the final trajectory of the robot along the specified sides of triangle.

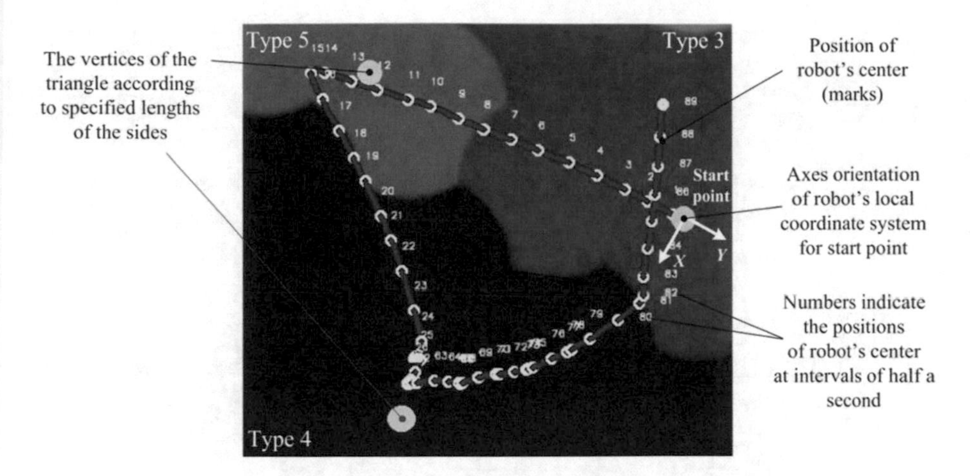

Fig. 7. Trajectory of robot movement obtained by means of "GM"-models

The achievement of the coordinates of the vertices of the triangle was estimated by means of the best "GM" models for *X*, *Y* and φ. Also these models were used for trajectory correction during the robot movement. In this experiment, the length of the side of triangle was set at 0.5 m. In combination with three different types of surfaces and the robot's diameter of 37 cm provides specified conditions for testing the models, because about half of path the robot moves through the transition zones. For example, already at points 7 and 12 the robot's wheels are simultaneously located on two surfaces: for the first point – both on 3 and 5 types of surfaces, for the second point – both on 4 and 5 types of surfaces.

In Fig. 8b and d shows the evaluation of the *X* and *Y* coordinates by means of both the specified models (X_{Type3}, X_{Type4}, X_{Type5}, Y_{Type3}, Y_{Type4}, Y_{Type5}) for three types of surfaces (see Fig. 7) and best general models for all types of surfaces (X_{GM}, Y_{GM}). As X_{CSV} and Y_{CSV} are denoted real coordinates of robot movement detected by computer vision system (see Fig. 3). As can be seen from the curves of X_{CSV} and Y_{CSV}, the movement of the robot is complex (for example, deviations of coordinates for a period of 15-23 robot's steps and pause in the movement for a period of 30–57 steps) and is not rectilinear, which indicates a significant influence of the surfaces on it. It can also be seen that on separate time intervals, different "MT"-models are more accurate than general models. At the same time, general models (X_{GM}, Y_{GM}) show an acceptable (in average) result during the overall time of the movement.

Figure 8a and c show the errors in determining both *X* and *Y* coordinates by the best "GM"-models (X_Error_{GM}, Y_Error_{GM}) and the set of specialized models (X_Error_{MT}, Y_Error_{MT}) that are selected at the moment the classifier of surface type is triggered (output is equal "1") in accordance with the above rule. In general, the trained classification models demonstrate their operability. However, in addition to its own classification errors (for example, at the interval 55–75 steps in Fig. 8f), there are errors

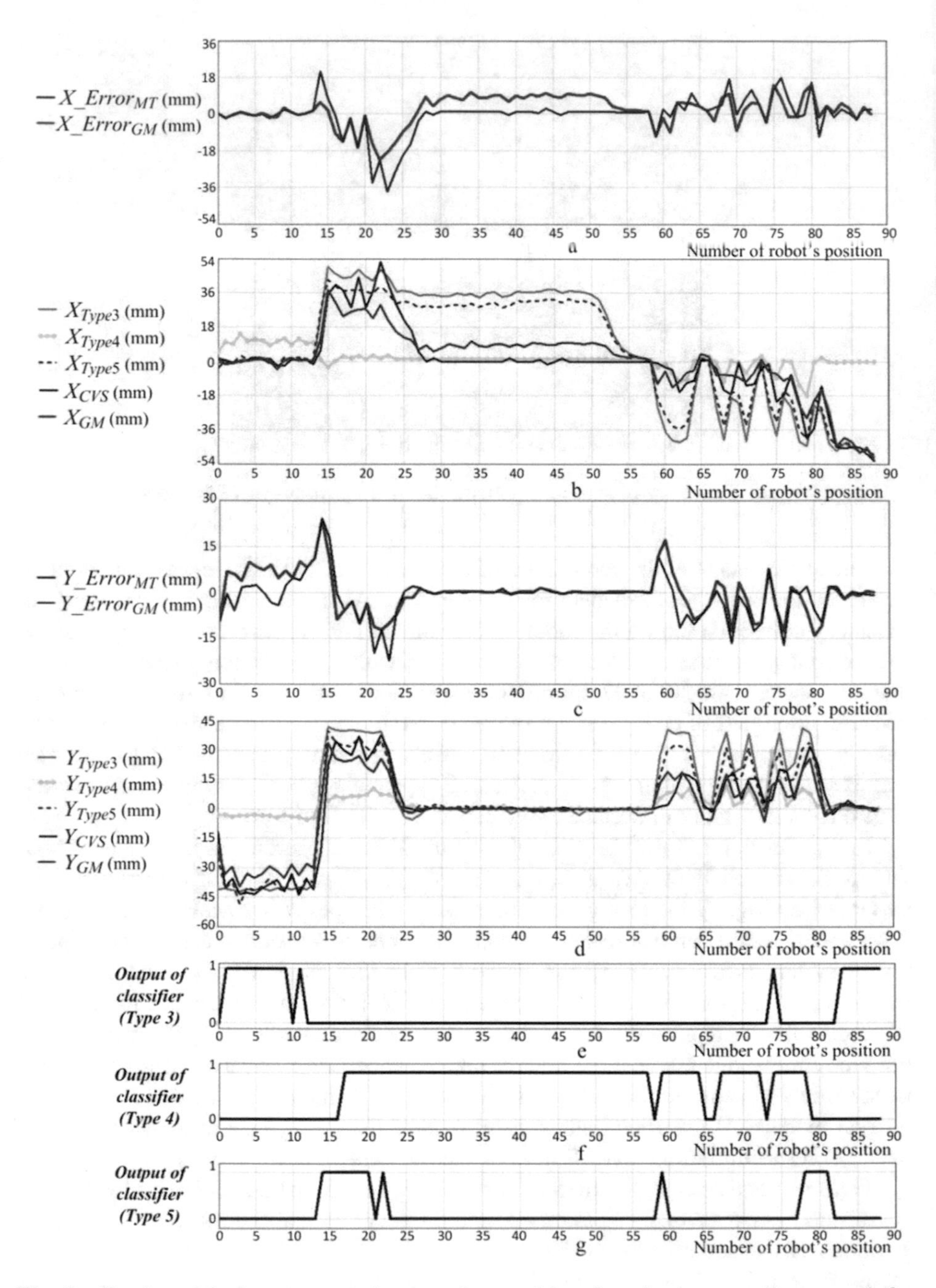

Fig. 8. Graphs of both outputs and errors for models of evaluating coordinates and for classification models during the robot movement

caused by the transition of the robot from one surface to another (for example, at the interval of 15–20 steps in Fig. 8f and g, two classifiers are triggered simultaneously).

In Fig. 8a, b , c and d all values of output variables for each robot's step are shown in robot's local coordinate system (Fig. 1c).

Based on the results of testing the models, we can conclude the following:

- With the simultaneous contact of the robot wheels with surfaces of different types, there are classification errors (i.e. errors of selection of "MT"-models) and errors of "MT"-models. This leads to the fact that the use of general (i.e., averaged for all types of surfaces) and specified models gives a comparable result.
- In general, all models (both pose estimation and classifications) demonstrate their performance by an example much more complex than the conditions for their training. This indicates the operability of developed learning navigation system (in point of view quality of obtained models) and practical applicability of GMDH to solving the problem of local navigation of a robot.

It should be noted that the purpose of this section was to test the obtained models, and not to solve complex questions of developing a system for local navigation. The focus of this study is the synthesis of models, but to increase the accuracy of the local navigation system, it is necessary to consider a wider range of issues related more to the stage of using the obtained models, rather than to the stage of their training. In order to improve the quality of the obtained trajectory and the accuracy of movement to the given coordinates, it is necessary to solve the following tasks: synthesis of a better regulator for robot motion along the desired trajectory; the development of methods to use the classifiers and "MT"-models according to considered conditions and etc.

In Fig. 8b and d the accuracy both "GM"- and "MT"-models during test motion correlates with its accuracy on training stage (see Table 3).

5 Conclusion

We have obtained higher accuracy (arithmetic mean error is less) of models for evaluating the coordinates and angular orientation than in our previous research [10, 11] due to the extension of the input parameters set (only $\{N\}$, $\{\omega\}$ and $\{I\}$ were used in past).

In general case, it is not sufficient to use a certain subset of input parameters to obtain better models for different outputs (X, Y and φ) and underlying surface types. We recommend using parameters derived from sensor readings (for example, $\{V_2\}$ and $\{V_3\}$) to improve the quality of the models. For example, it is interesting to note that the variables of $\{V_2\}$ were selected by active neurons to train classifiers for each of the five types of surfaces (also see best training results (highlighted in bold) in Table 4). Also it should be noted that the physical meaning of the set of parameters $\{V_2\}$ and $\{V_3\}$ is not associated to this testing ground and this robot, which makes it possible to use them in other projects on the same subject.

The results of testing of the trained models (both pose estimation and classification) demonstrate their performance in an essentially heterogeneous environment (the areas of the surfaces are comparable with the robot's dimensions). It is important to note that the conditions for testing of models were much more complicated than the conditions for their training. First, during the training there were no transition zones, when different wheels are simultaneously located at different surfaces. Secondly, the movement in these zones was half of the path. Thirdly, the control actions on the motors were changed dynamically (during the robot movement in Fig. 7) due to correction of both the trajectory and the orientation angle (at the training stage, the control actions were statically assigned from a specified set of motor velocities). As result, the parameters of robot movement and, as a consequence, the readings of the sensors were significantly different from those observed at the training stage. Fourthly, the models (both for pose estimation and classification) were used as a feedback signal for motors control, which could disturb stability of robot movement along the trajectory. Thus, testing was a very serious test for bias of models. From this point of view, the results shown in Fig. 8 testify the practical possibility of models training based on GMDH for task of local navigation in heterogeneous environment.

Future work is the development of methods and algorithms for applying both the trained models and learning navigation system to construct the local navigation system based on the readings of on-board sensors (in the absence of a GPS-signals) in heterogeneous environment.

References

1. Rogers-Marcovitz, F., George, M., Seegmiller, N., Kelly, A.: Aiding off-road inertial navigation with high performance models of wheel slip. In: Proceedings of the IEEE International Conference on Intelligent Robots and Systems, pp. 215–222. IEEE, Vilamoura, Portugal (2012)
2. Madhavan, R., Nettleton, E., Nebot, E., Dissanayake, G., Cunningham, J., Durrant-Whyte, H., Corke, P., Roberts, J.: Evaluation of internal navigation sensor suites for underground mining vehicle navigation. In: Proceedings of the IEEE International Conference on Robotics and Automation, pp. 999–1004. IEEE, Detroit, MI, USA (1999)
3. Koch, J., Hillenbrand, C., Bems, K.: Inertial navigation for wheeled robots in outdoor terrain. In: Fifth International Workshop on Robot Motion and Control, pp. 169–174. IEEE, Dymaczewo, Poland (2005)
4. Bingbing, L., Adams, M., Ibañez-Guzmán, J.: Multi-aided inertial navigation for ground vehicles in outdoor uneven environments. In: Proceedings of the IEEE International Conference on Robotics and Automation, pp. 4703–4708. IEEE, Barcelona, Spain (2005)
5. Liu, Y., Xiong, R., Wang, Y., Huang, H., Xie, X., Liu, X., Zhang, G.: Stereo visual-inertial odometry with multiple kalman filters ensemble. IEEE Trans. Indus. Electron. **63**(10), 6205–6216 (2016)
6. Dupont, E., Collins, E., Coyle, E., Roberts, R.: Terrain classification using vibration sensors: theory and methods. In: New Research on Mobile Robotics, pp. 1–41 (2010)
7. Ojeda, L., Cruz, D., Reina, G., Borenstein, J.: Current-based slippage detection and odometry correction for mobile robots and planetary rovers. IEEE Trans. Robot. **22**(2), 366–378 (2006)
8. Iagnemma, K., Ward, C.: Classification-based wheel slip detection and detector fusion for mobile robots on outdoor terrain. Auton. Robots **26**(1), 33–46 (2009)
9. Khaleghian, S., Taheri, S.: Terrain classification using intelligent tire. J. Terramech. **71**, 15–24 (2017)

10. Andrakhanov, A.: Technology of autonomous mobile robot control based on the inductive method of self-organization of models. In: Proceedings of the 7th International Symposium “Robotics for Risky Environment – Extreme Robotics”, pp. 361–368. Saint-Petersburg, Russia (2013)
11. Andrakhanov, A.: Navigation of autonomous mobile robot in homogeneous and heterogeneous environments on basis of GMDH neural networks. In: Proceedings of the 4th International Conference on Inductive Modelling, pp. 133–138. Kiev, Ukraine (2013)
12. Tyryshkin, A., Andrakhanov, A., Orlov, A.: GMDH-based modified polynomial neural network algorithm.In: GMDH-methodology and Implementation in C (With CD-ROM), Imperial College Press, World Scientific, London (2015). ISBN: 978-1-84816-610-3
13. Robotino Manual (Order No. 544305). http://www.festo-didactic.com/ov3/media/customers/1100/544305_robotino_deen2.pdf. Accessed 05 June 2017
14. Ivakhnenko, A., Ivakhnenko, G.: The review of problems solvable by algorithms of the group method of data handling. Pattern Recogn. Image Anal. Adv. Math. Theory Appl. **5**(4), 527–535 (1994)
15. Ivakhnenko, A., Ivakhnenko, G., Mueller, J.: Self-organization of neuronets with active neurons. Int. J. Pattern Recogn. Image Anal. Adv. Math. Theory Appl. **4**(4), 177–188 (1994)
16. Madala, H.R., Ivakhnenko, A.G.: Inductive Learning Algorithms for Complex System Modeling. CRC Press, Boca Raton (1994). ISBN 0-8493-4438-7
17. Martin, S., Murphy, L., Corke, P.: Building large scale traversability maps using vehicle experience. In: Desai, J.P., Dudek, G., Khatib, O., Kumar, V. (eds.) 13th International Symposium on Experimental Robotics 2012, STAR, vol. 88, pp. 891–905. Springer, Heidelberg (2013)

Model of the Objective Clustering Inductive Technology of Gene Expression Profiles Based on SOTA and DBSCAN Clustering Algorithms

Sergii Babichev[1(✉)], Volodymyr Lytvynenko[2], Jiri Skvor[1], and Jiri Fiser[1]

[1] Jan Evangelista Purkyne University, Usti nad Labem, Czech Republic
sergii.babichev@ujep.cz, jskvor@physics.ujep.cz, ithil@jf.cz
[2] Kherson National Technical University, Kherson, Ukraine
immun56@gmail.com
http://www.sci.ujep.cz

Abstract. The paper presents the hybrid model of the objective clustering inductive technology based on complex using of the self-organizing SOTA and the density DBSCAN clustering algorithms. The inductive methods of complex systems analysis were used as the basis to implement the objective clustering inductive technology of gene expression profiles. To estimate the clustering quality for equal power subsets (include the same quantity of pairwise similar objects) the complex multiplicative criterion was calculated as the combination of the Calinski-Harabasz criterion and WB-index. The external clustering quality criterion is calculated as the normalized difference of the internal clustering quality criteria for the equal power subsets. The final decision concerning the determination of the optimal parameters of the clustering algorithm operation is done based on the maximum value of the Harrington desirability function that takes into account both the character of the objects and the clusters distribution in various clustering and the difference between clustering, which are implemented on the equal power subsets. The studied data grouping within the framework of the objective clustering inductive technology was performed in two stages. Firstly, the studied gene expression profiles were grouped with the use DBSCAN clustering algorithm. Then, the obtained set of gene expression profiles was divided into two clusters using SOTA clustering algorithm. This step-by-step procedure of the data clustering crates the conditions to save more useful information for following data processing.

Keywords: Objective clustering · Inductive modeling
Gene expression profiles · Clustering quality criteria
SOTA clustering algorithm · DBSCAN clustering algorithm

1 Introduction

The gene regulatory network creation based on the gene expression profiles is one of the current problems of the modern bioinformatics. The gene regulatory

N. Shakhovska and V. Stepashko (eds.), *Advances in Intelligent Systems and Computing II*, Advances in Intelligent Systems and Computing 689,
https://doi.org/10.1007/978-3-319-70581-1_2

network is a set of genes, which interact with each other to control the specific cells functions. Qualitatively constructed gene regulatory network allows us to study the influence of the corresponding group of genes or the individual genes on functional possibilities of the biology objects in order to correct this process. The gene expression profiles, which are obtained by DNA microarray experiments or by RNA sequences technology are the basis to construct the gene regulatory networks. High dimension of feature space is one of the distinctive peculiarities of the studied profiles. About tens of thousands genes are contained in the gene expression profiles. The creation of gene regulatory network based on the whole dataset of the gene expression profiles is very difficult problem because: it requests large computer resources; it needs large time expenses to process the information; the complexity of the obtained network complicates the interpretation of results. In this context, it is necessary firstly to divide the gene expression profiles into subsets, each of which includes a group of genes that performs similar functions in the studied biological object. Biclustering technology is actual to solve this problem nowadays. Implementation of this technology allows grouping the objects and the features according to their mutual correlation. So, in the paper [14,17] authors provide a review of a large quantity of biclustering approaches existing in the literature with analysis their advantages and disadvantages. In [7] authors have proposed and implemented the convex biclustering method using gene expression profiles of the lung cancer patient. The authors have shown the efficiency of the proposed method during simulation process. However, it should be noted that one of the significant problem of this technology qualitative implementation is selection of the biclustering level during the objects and the genes grouping. The qualitative validation of the obtained model is another task, which has no solution nowadays. High dimension of the features space promotes to the large quantity of the obtained biclusters. Limitation of their quantity by removing of small biclusters leads to the loss of some useful information. To solve this problem we propose the cluster-bicluster technology the implementation of which involves two stage: clustering of the gene expression profiles at the first step and biclustering of the obtained clusters at the second step. The reproducibility error is one of the current problems of the existing clustering algorithms, in other words, successful clustering results obtained on one dataset do not repeat while using another similar dataset. Reduction of this error can be achieved by careful verification of the obtained model using "fresh information", which was not used during the model making. A higher degree of coincidence between the clustering results on the similar data corresponds to a higher degree of the obtained model objectivity. This idea is the basis of the objective clustering inductive technology, the main conception of which was presented in [15] and further developed in [16,18,19]. The practical implementation of the objective clustering inductive technology is possible using various clustering algorithms. The choice of the clustering algorithm is determined by the structure and character of the studied data. The practical implementation of this technology based on the agglomerative hierarchical and self-organizing SOTA clustering algorithms were presented in [2,3]. One of the key conditions of successful implementation of this technology is careful determination of the internal, the external and the complex balance clustering quality criteria, which should take into account both the character of the objects grouping within the clusters and the character of the clusters distribution in the features space. This paper presents the research concerning the complex using of the density-based DBSCAN (Density Based Spatial Clustering of Application with Noise)[9] and self-organizing SOTA (Self Organizing Tree Algorithm)[8,10] clustering algorithms within the framework of the objective clustering inductive technology. The implementation of the proposed step-by-step procedure of the gene expression profiles grouping allows us to save more useful information of following data processing.

The aim of the paper is development of the hybrid model of the objective clustering inductive technology of gene expression profiles based on DBSCAN and SOTA clustering algorithms.

2 Problem Statement

Let the initial dataset of the gene expression profiles is a matrix: $A = \{x_{ij}\}$, $i = 1, \ldots, n; j = 1, \ldots, m$, where n – is the quantity of genes observed, m – is the quantity of the studied objects. The aim of the clustering process is a partition of the genes expression profiles into non empty subsets of pairwise non-intersecting clusters in accordance with the clustering quality criteria taking into account the properties of the studied profiles:

$$K = \{K_s\}, s = 1, \ldots, k; K_1 \bigcup K_2 \bigcup \cdots \bigcup K_k = A; K_i \bigcap K_j = \emptyset, i \neq j,$$

where k – is the quantity of clusters, $i, j = 1, \ldots, k$. The objective clustering technology is based on the inductive methods of complex systems analysis, which involves sequential enumeration of the clustering within a given range in order to select from them the best variants. Let W – is a set of all admissible clustering for given set A. The clustering is the best (an optimal) in terms of clustering quality criteria QC(K) is the following condition is performed:

$$K_{opt} = \arg \min_{K \subseteq W} CQ(K) or\ K_{opt} = \arg \max_{K \subseteq W} CQ(K)$$

The clustering $K_{opt} \subseteq W$ is the objective if the difference of the objects and clusters distribution in different clustering for equal power subsets is minimal:

$$K_{obj} = \arg \min_{K \subseteq W} (QC(K_{opt})^A - QC(K_{opt})^B)$$

The architecture of the objective clustering inductive technology is presented in Fig. 1. Implementation of the technology involves the following steps:

1. Problem statement. Clustering aim formation according to the stated task. Studied data preprocessing and their formation as a matrix.

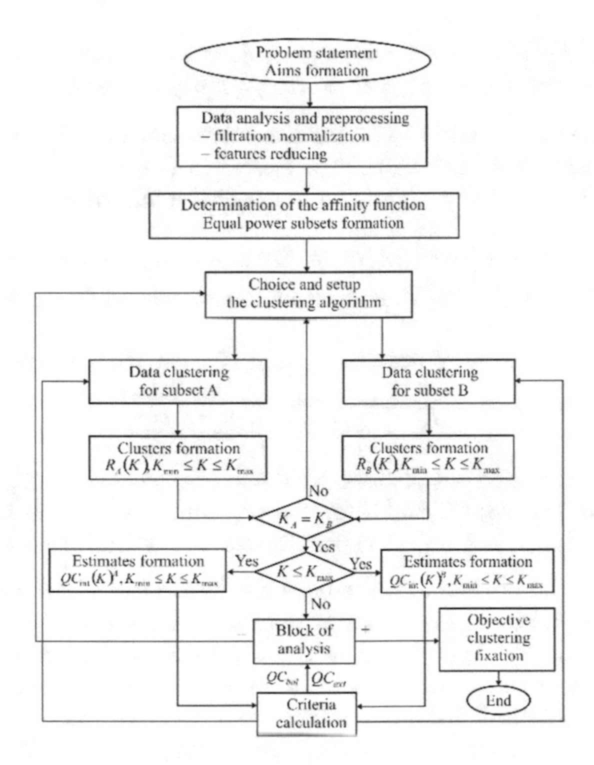

Fig. 1. Architecture of the objective clustering inductive technology

2. Determination of the affinity function of the studied data. Division of the initial dataset into two equal power subsets A and B using chosen affinity function. The equal power subsets include the same quantity of the pairwise similar objects.
3. Choice of the clustering algorithm. Setup of its initial parameters. These parameters are changed during the algorithm operation to obtain the different variants of the studied data clustering.
4. Data clustering on subsets A and B concurrently and clusters formation within the range of the algorithm's parameters change. If the clusters quantity in various clustering differs, it is necessary to change the setup of the algorithm or to use another admissible clustering algorithm and to repeat the step 5.
5. Calculation of the internal QC_{int}, the external QC_{ext} and the complex balance QC_{bal} clustering quality criteria for the current clustering on equal power subsets A and B.
6. Analysis of the complex balance clustering quality criterion values. In case of absence of this criterion extremums or if their values are less than admissible standards, choose another clustering algorithm and repeat the steps 4–7 of this procedure.

7. Fixation of the objective clustering in correspondents with the maximum values of the complex balance clustering quality criterion.

The idea of the algorithm to divide the initial dataset of the objects Ω into two equal power subsets Ω^A and Ω^B is stated in [15] and further developed in [18]. Implementation of this algorithm involves the following steps:

1. Calculation of $\frac{n \times (n-1)}{2}$ pairwise distances between the gene expression profiles of the initial data. The result of this step is a triangular matrix of the distances.
2. Allocation of the pairs of objects X_s and X_p, the distance between which is minimal:
$$d(X_s, X_p) = \min_{i,j} d(X_i, X_j);$$
3. Distribution of the object X_s to subset Ω^A, and the object X_p to subset Ω^B.
4. Repetition of the steps 2 and 3 for the remaining objects. If the quantity of objects is odd, the last object is distributed to the both subsets.

The example of the objects and the clusters distribution in the objective clustering inductive technology is shown in the Fig. 2. Obviously, that the best clustering corresponds to the higher density of the objects distribution relative to the mass centers of the clusters where these objects are and less density of the clusters' mass centers distribution in the feature space. Moreover, it is necessary that the difference of the clustering results which are obtained on the equal power subsets was minimal. Thus, to implement this technology it is necessary to determine the gene expression profiles proximity metric, the internal, the external and the complex balance clustering quality criteria.

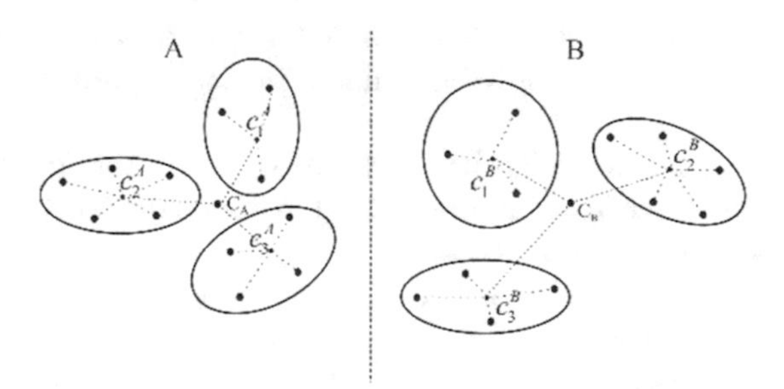

Fig. 2. The example of the objects and the clusters distribution in the objective clustering inductive technology in case of three clusters structure

3 Criteria to Estimate the Gene Expression Sequences Proximity and Clustering Quality

It is obvious that the qualitative clustering corresponds to the high division ability of different clusters and high density of the objects concentration inside the clusters. Thus, it is necessary firstly to determine the proximity metric of the gene expression profiles. The [4] presents the results of the research concerning comparison of the three well know metrics efficiency to estimate the proximity level of numeric vectors: Manhattan, Euclidean and Correlation distances. Evaluation of the effectiveness of the studied metrics was performed using the model data representing the gene expression profiles of the objects in two different clusters. Centers of the corresponding clusters are calculated by the formula:

$$C_S = \frac{1}{N_S} \sum_{i=1}^{N_S} x_i^S,$$

where N_S is the quantity of gene expression profiles in cluster S, x_i^S is i–th sequence in cluster S. The research technique consists the next steps:

- calculation of the average distance d_{int} from the profiles to the clusters' centers, where these profiles are:

$$d_{int}(X^{S,P}, C_{S,P}) = \frac{1}{N} (\sum_{i=1}^{N_S} d(x_i^S, C_S) + \sum_{j=1}^{N_P} d(x_j^P, C_P));$$

- calculation the average distance d_{ext} from the profiles to the centers of the neighbouring clusters:

$$d_{ext}(X^{S,P}, C_{S,P}) = \frac{1}{N} (\sum_{i=1}^{N_S} d(x_i^S, C_P) + \sum_{j=1}^{N_P} d(x_j^P, C_S));$$

- calculation the relative coefficient:

$$d_{rel}(X^{S,P}, C_{S,P}) = \frac{d_{ext}(X^{S,P}, C_{S,P})}{d_{int}(X^{S,P}, C_{S,P})};$$

It is obvious the higher value of the relative coefficient corresponds to the higher separating ability of the used proximity metric. In order to estimate the effectiveness of the metrics we used the data of the lung cancer patients of the database Array Express [5], which includes the gene expression profiles of 96 patients, 10 of which were healthy and 86 patients were divided by the degree of the health severity into three groups (Well, Moderate and Poor). Each of the profiles includes 7129 genes expressions. Data preprocessing in order of gene expression matrix formation was carried out accordingly to the technique, which is presented in [1]. To choose the metrics of the gene expression profiles similarity class of the health patient (10 profiles) and class of patients with Poor state of health (21 profiles) were used. The results of the relative criteria values distribution while using different metrics to estimate the level of the gene expression profiles similarity are shown in Fig. 3. The analysis of the Fig. 3 allows us to conclude that in case of the gene expression profiles the correlation metric has higher separating ability in comparison with Euclid and Manhattan metrics because the values of the relative criterion which is calculated basing on the correlation distance, are higher in comparison with the use of Euclid and Manhattan distances.

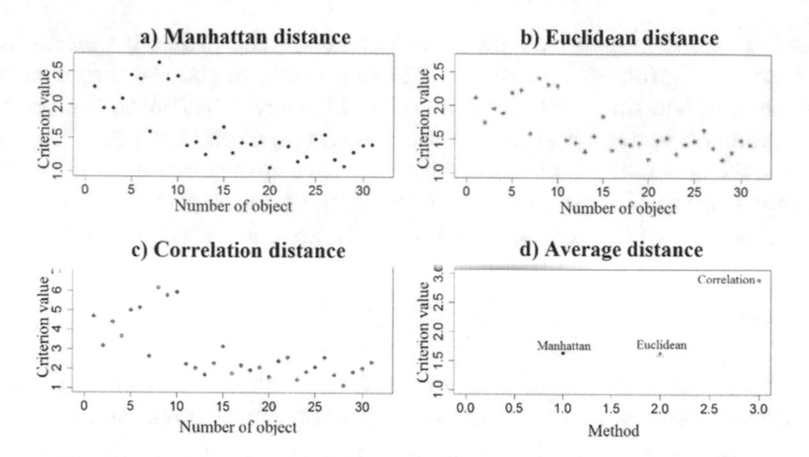

Fig. 3. The distribution of the relative criteria values using different metrics to estimate the gene expression profiles of the lung cancer patients: (a) Manhattan distance; (b) Euclidean distance; (c) Correlation distance; (d) Average of all distances

3.1 Internal and External Clustering Quality Criteria

As it was noted herein before, it is obvious that the qualitative clustering corresponds to the high division ability of different clusters and high density of the objects concentration inside the clusters. Thus, the internal clustering quality criterion should be complex and takes into account both the objects distribution inside different clusters and the clusters distribution in the features space. The first component of the complex internal criterion is calculated as average distance from the objects to the mass centers of the clusters, where these objects are:

$$QCW = \frac{1}{N}\sum_{s=1}^{K}\sum_{i=1}^{N_s} d(x_i^s, C_s)$$

The second component of this criterion, which takes into account the singularity of the clusters distribution in the feature space, is calculated as an average distance between the mass centers of the clusters:

$$QCB = \frac{2}{K(K-1)}\sum_{i=1}^{K-1}\sum_{j=i+1}^{K} d(C_i, C_j)$$

where K – is the quantity of clusters, N – is the general quantity of objects, N_s – is the quantity of the objects in cluster s, x_i^s – is the i-th vector in S cluster, C_i, C_j and C_s – are the mass centers of the clusters i, j and S concurrently, $d(\cdot)$ – is the metric used to estimate the proximity level of the studied vectors. Various combinations of these components allow obtaining the clustering quality criteria for studied data subsets. During the simulation process the following internal criteria to estimate the data grouping quality were used:

– Calinski-Harabasz [6]:

$$Q_{CH} = \frac{QCB(N-K)}{QCW(K-1)};$$

– WB index [20]:

$$Q_{WB} = \frac{KQCW}{QCB};$$

– Hartigan [12]:

$$Q_H = \log_2(\frac{QCB}{QCW}).$$

In order to obtain more complete information about the effectiveness of these criteria operation the complex multiplicative criteria were calculated as follow:

$$QC_{CX1} = \frac{QC_{WB}}{QC_{CH}} = \frac{K(K-1)QCW^2}{(N-K)QCB^2};$$

$$QC_{CX2} = \frac{QC_H}{QC_{CH}} = \frac{log_2(\frac{QCB}{QCW})(K-1)QCW}{(N-K)QCB};$$

$$QC_{CX3} = QC_{WB}QC_H = \frac{QCB(N-K)}{QCW(K-1)}log_2(\frac{QCB}{QCW});$$

$$QC_{CX4} = \frac{QC_{WB}QC_H}{QC_{CH}} = \frac{K(K-1)QCW^2}{(N-K)QCB^2}log_2(\frac{QCB}{QCW}).$$

The external clustering quality criterion is calculated as the normalized difference of the internal clustering quality criteria for the equal power subsets A and B:

$$QC_{ext}(A,B) = \frac{|QC_{int}(A) - QC_{int}(B)|}{QC_{int}(A) + QC_{int}(B)}.$$

To estimate the effectiveness of the internal and the external clustering quality criteria within the framework of the objective clustering inductive technology the gene expression profiles of the lung cancer patients were used [5]. Firstly, the data were divided into two equal power subsets with the use of the algorithm that had been presented in [15,18]. Then, each of the subsets was sequentially divided into clusters from Kmin = 2 to Kmax = 5. In case of two-cluster structure in first cluster there were the gene expression profiles of the healthy patients (NORM) and gene expression of the patients with good state of health (WELL), second cluster included the gene expression of the patients with poor (POOR) and moderate (MODERATE) states. In case of three-cluster structure the first cluster contained the data of the healthy patients, the second – the data of the patients with good state, the third cluster included the gene expression of the patients with poor and moderate states. In case of four-cluster structure the first cluster contained the data of the healthy patients, the second – the data of the patients with good state, the third cluster included the gene expression of the patients with poor state and the fourth cluster contained the gene expression of the patients with moderate state. To obtain the five-cluster structure the gene expression profiles of the patients with moderate state were divided into two groups randomly. Objective clustering in this case corresponds to four-cluster structure. To estimate the proximity level of the appropriate vectors the correlation metric was used. Figure 4 shows the charts of the internal clustering quality criteria for equal power subsets A and B versus the clusters quantity. Figure 5 presents the charts of the complex multiplicative internal criteria versus the clusters quantity. Figure 6 shows the charts of the external clustering quality criteria, which were calculated based on the internal criteria versus the clusters quantity. Analysis of the charts which are shown in Fig. 4 and Fig. 5 allows us to conclude that the internal clustering quality criteria give the same results in terms of the local extremums existence. They have local extremums, which corresponds to the objects division into 4 clusters, however, it should be noted that in case of the QC_H, QC_{CX2}, QC_{CX3} and QC_{CX4} criteria use, the clustering, which correspond to the objects division into 4 and 5 clusters are badly distinguished. Analysis of the external criteria values, which are shown in Fig. 6, allows concluding that in terms of the clustering objectivity (proximity level of the results, which have been obtained on equal power subsets A and B) the QC_H Hartigan criterion and the QC_{CX2}, QC_{CX3} complex criteria are ineffective, because they have not a local minimums corresponding to the objects division into 4 clusters (the objective clustering). The QC_{CX1} and QC_{CX4} criteria are the most informative to select the objective clustering, however, the QC_{CX1} criterion is more preferable because it has more expressed local minimum, which corresponds to four clusters existence in the obtained clustering.

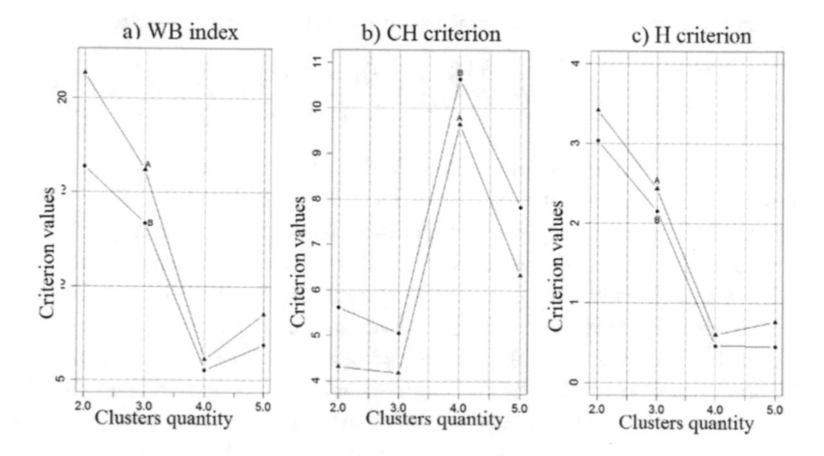

Fig. 4. Charts of the internal clustering quality criteria versus the clusters quantity: (a) WB index; (b) Calinski-Harabasz criterion; (c) Hartigan criterion

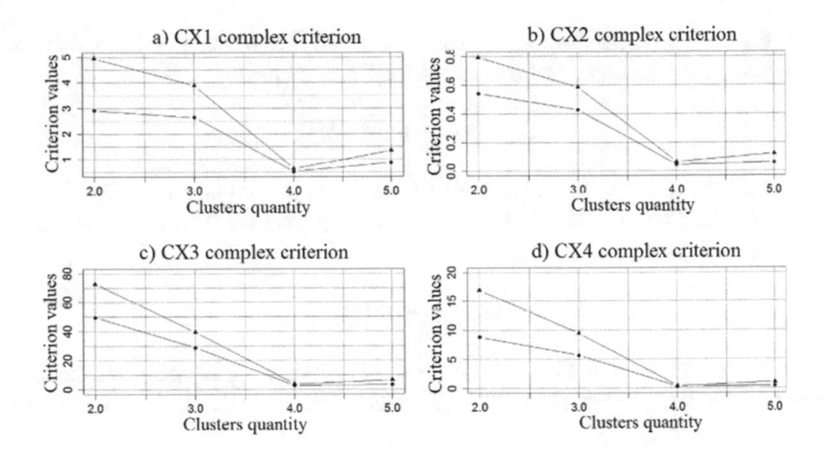

Fig. 5. Charts of the complex multiplicative internal clustering quality criteria versus the clusters quantity

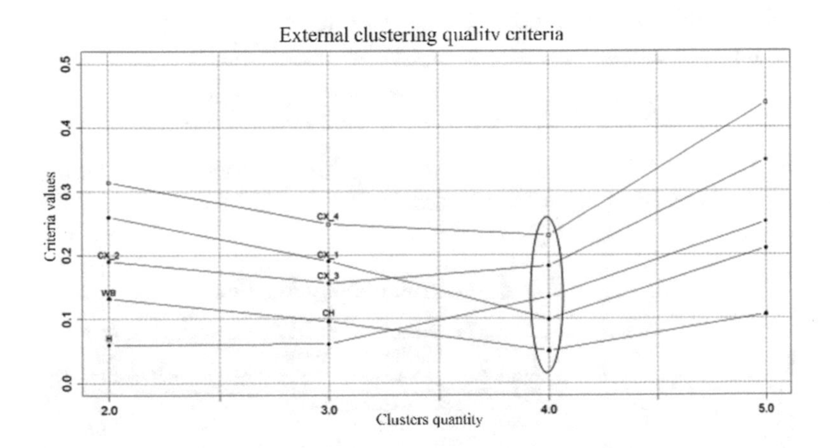

Fig. 6. Charts of the external clustering quality criteria versus the clusters quantity

3.2 Complex Balance Clustering Quality Criterion

It is obvious that the objective clustering corresponds to the minimum values of the internal and the external clustering quality criteria. However, it is possible that the extremums of these criteria correspond to different clustering. Thus, it is necessary to determine the complex balance clustering quality criterion that takes into account both the character of the objects and the clusters distribution in various clustering and the difference between clustering, which are implemented on the two equal power subsets. To calculate the complex balance clustering quality criterion the Harrington desirability function [11] was used. The implementation of this function involves transformation of the scales of the internal and the external criteria into reaction scale the values of which are changed linearly within the range from −2 to 5. Then the private desirabilities of the appropriate criteria are calculated by the formula:

$$d = \exp(-\exp(-Y))$$

The chart of the Harrington desirability function versus the reaction index Y is shown in Fig. 7. The transformation of the criteria scales into the reaction scales were performed by linear equation:

$$Y = a - b \cdot QC$$

The parameters a and b are determined empirically. The general desirability index value is calculated as geometric average of the private desirabilities indexes:

$$D = \sqrt[n]{\prod_{i=1}^{n} d_i}$$

In case of the objective clustering inductive technology the general Harrington desirability index was used as the complex balance criterion:

$$QC_{bal} = \sqrt[3]{QC_{int}(1) + QC_{int}(2) + QC_{ext}}$$

The largest value of the complex balance criterion corresponds to the best parameters of the clustering algorithm operation.

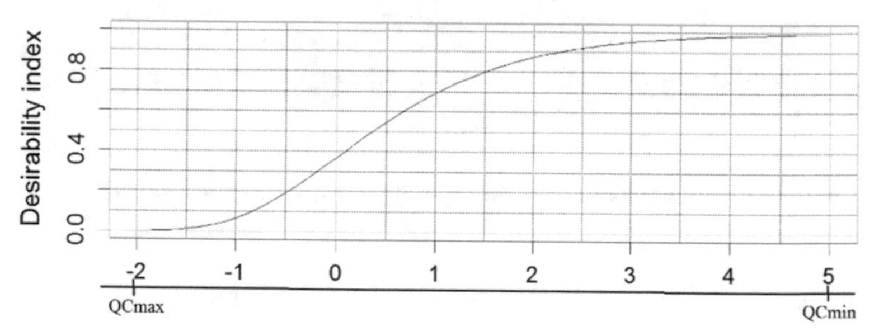

Fig. 7. Chart of Harrington desirability function

4 Implementation of SOTA Clustering Algorithm Within the Framework of the Objective Clustering Inductive Technology

The SOTA clustering algorithm (Self-Organizing Tree Algorithm) [8] represents a type of self-organizing neural networks based on the Kohonen maps and the

Fritzke algorithm of the spatial cell structure growing [10]. Opposed to the Kohonen maps that reflect a set of high dimensional input data on the elements of the two-dimensional array of small dimension, the SOTA algorithm generates a binary topological tree. The Fritzke algorithm performs self-organization of output nodes of the network in such a way that the quantity of the nodes increases in the field of the higher density of objects concentration and decreases in the field of the lower density. Two parameter are determined the effectiveness of the SOTA clustering algorithm operation: weight coefficient of the sister's cell (scell) and maximum divergence coefficient value. The weight coefficients of the parent's and winner' cells are determined automatically: $pcell = scell \cdot 5$; $wcell = pcell \cdot 2$. This ratio is recommended by the algorithm's authors. The block-scheme of the objective clustering model based on the SOTA clustering algorithm is shown in Fig. 8. Implementation of this model involves the following steps:

1. Presentation of the studied data as a matrix $n \times m$, where n – is the quantity of the studied profiles or the quantity of the rows and m – is the quantity of the objects or the quantity of the columns.
2. Division of the initial dataset into two equal power subsets.
3. Setup of the SOTA clustering algorithm. Setting of the initial value of scell weight coefficient, the interval and the step of its change.
4. Data clustering on the equal power subsets A and B concurrently. Clusters formation and the internal, the external and the balance clustering quality criteria calculation within a range of the interval of the algorithm's parameter change.
5. Fixation of the optimal scell parameter corresponding to the maximum value of the balance criterion.
6. Setting of the initial value of the maximum divergence parameter, the interval and the step of its change. Repetition of the step 4 of this algorithm. Fixation of the optimal maximum divergence parameter.
7. Full data clustering by the SOTA clustering algorithm using the optimal parameters of the algorithm operation.

5 Implementation of DBSCAN Clustering Algorithm Within the Framework of the Objective Clustering Inductive Technology

DBSCAN clustering algorithm (Density Based Spatial Clustering of Application with Noise Algorithm) [9] initially needs two parameters: EPS-neighborhood of points (EPS) and the least quantity of the points within EPS-neighborhood (MinPts). Choice of these parameters determines the character of the studied objects grouping during the algorithm operation. In [9] authors proposed the technology based on the sorted 4-dist graph. However, the implementation of this technology does not allow determination of EPS and MinPts values exactly and this fact influences the quality of the algorithm operation. To determine

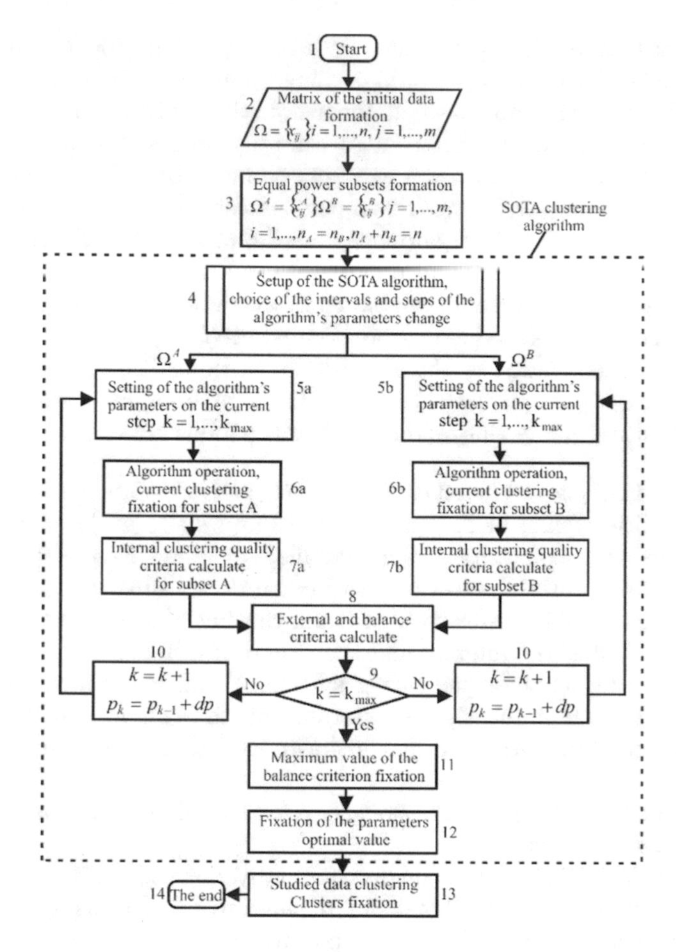

Fig. 8. Block-scheme of the objective clustering model based on the SOTA clustering algorithm

EPS and MinPts values we propose to use the objective clustering inductive technology. The structural scheme of the objective clustering model based on DBSCAN clustering algorithm is presented in Fig. 9. Implementation of this model involves the following steps:

1. The matrix of the studied data formation. The matrix contains n rows or studied profiles and m columns or the objects.
2. Division of the initial dataset into two equal power subsets.
3. The distance matrix between studied profiles for both subsets is calculated using correlation distance. This distance matrix is the input matrix for the next step of the algorithm operation.
4. Setup of DBSCAN clustering algorithm, choice of the intervals and steps of EPS and MinPts change.
5. Fixation of MinPts value (MinPts = 3). Initialization of EPS = EPSmin.

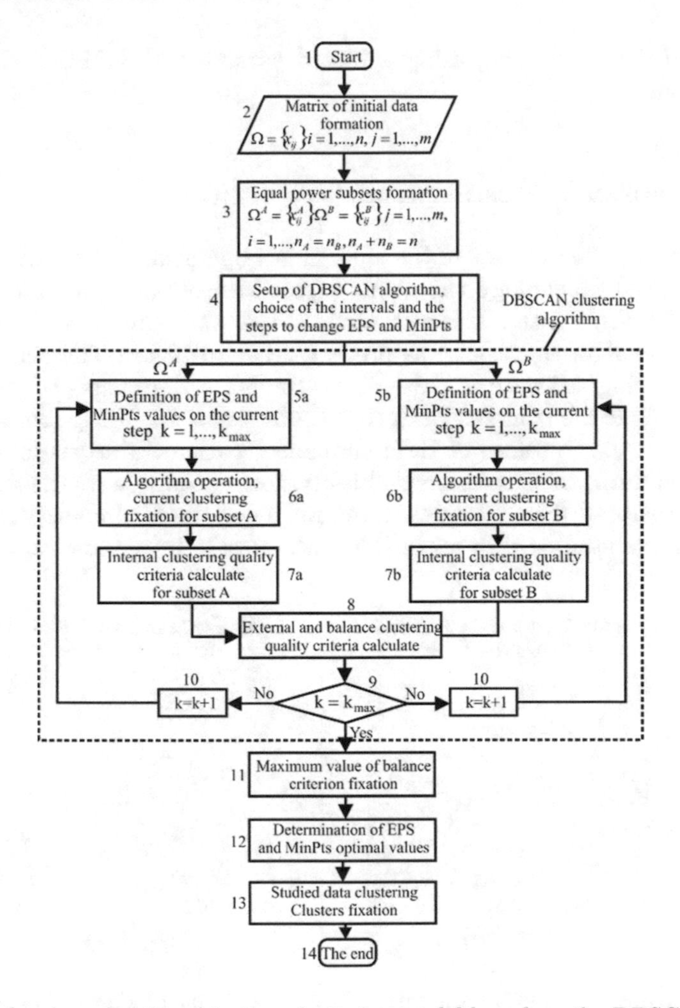

Fig. 9. Block-scheme of the objective clustering model based on the DBSCAN clustering algorithm

6. Data clustering on the two subsets A and B using DBSCAN algorithm in range from EPSmin to EPSmax. Clustering fixation at each step.
7. The internal, the external and the complex balance clustering quality criteria is calculated at each step of the algorithm operation.
8. Analysis of the balance criterion values. Fixation of the optimal value EPS, which corresponds to the maximum value of the balance clustering quality criterion.
9. Data clustering on the two equal power subsets A and B in the range from MinPtsmin to MinPtsmax. Clustering fixation at each step.
10. Repetition of the steps 7 and 8 of this algorithm for MinPts values. Fixation of EPS and MinPts optimal values which correspond to the maximum of the complex balance clustering quality criterion.

11. Studied data clustering using obtained parameters of DBSCAN algorithm operation.

6 Experiment, Results and Discussion

To estimate the effectiveness of the algorithm's operation within the framework of the proposed technology the genes expressions of the lung cancer patients [5] were used. Firstly, the data were divided into two equal power subsets with the use of the algorithm that was presented herein before The simulation was carried out using software R [13]. Figure 10 shows the charts of the internal, the external and the complex balance criteria versus EPS-neighborhood values for gene expression profiles of the lung cancer patient. Two thousand profiles were studied during the experiment. Firstly, these profiles were divided into two equal power subsets using correlation metric. Then the dissimilarity matrices for all pairs of the studied objects of the both subsets using correlation distance

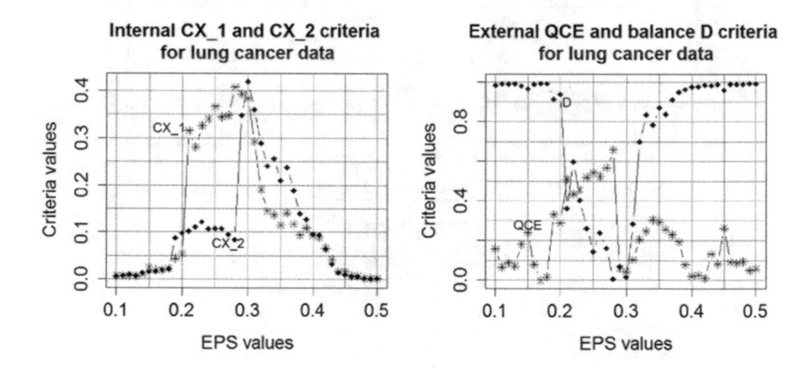

Fig. 10. Charts of the internal, the external and the complex balance clustering quality criteria versus EPS-neighborhood values for gene expression profiles of lung cancer

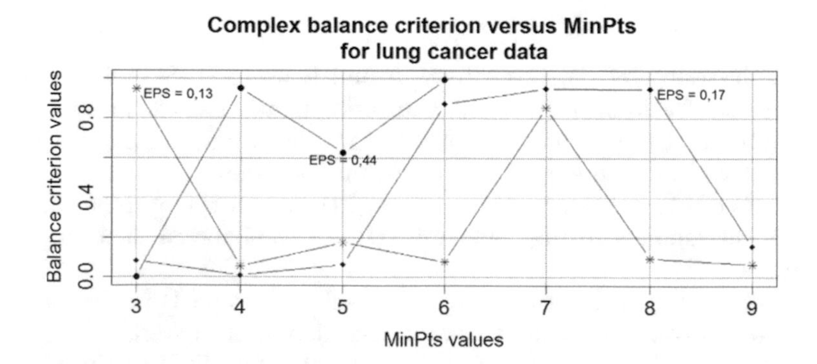

Fig. 11. Charts of the complex balance clustering quality criterion versus MinPts values for gene expression profiles of lung cancer

was calculated. These dissimilarity matrices were used as the input data for next steps of DBSCAN algorithm operation. Three values of EPS-neighborhood were selected based on the maximum values of the complex balance criterion which is shown in Fig. 10: $EPS_1 = 0{,}13$; $EPS_2 = 0{,}17$; $EPS_3 = 0{,}44$. Figure 11 shows the charts of the complex balance criteria for selected EPS versus MinPts values. The analysis of the charts allows concluding that the best clustering in terms of maximum value of the complex balance clustering quality criterion is achieved using the following parameters of DBSCAN algorithm: (a) EPS = 0,13, MinPts = 3; (b) EPS = 0,17, MinPts = 8; (c) EPS = 0,44, MinPts = 6. However, the detail analysis of the obtained results has shown what in the first and in the second cases there were differ clusters quantity in the obtained clustering. Only in case of EPS = 0,44 and MinPts = 6 both clustering contained the same quantity of clusters. The initial dataset contained 2000 gene expression profiles. In this case the studied data were divided in such a way: the first cluster contained 1663 profiles, in the second cluster there were 16 profiles, there were 321 profiles in the third cluster. The objects in the third were identified as the noise component. The results of the simulation have shown also that the largest quantity of the gene expression profiles are concentrated in the first cluster. This fact can be explained by the fact that these genes define the main processes, which are carried out in biological organisms, therefore they have more correlation between each other to compare with genes in other clusters or genes, which are identified as the noise. The results of the internal criteria for the equal power subsets A and B, the external and the balance criteria versus the weight parameter of the sisters cell using SOTA clustering algorithm are presented in Fig. 12. The maximum divergence value in this case E = 0,001 was taken. As it can be seen from Fig. 12, the internal clustering quality criteria CX_1 and CX_2, which have been calculated on equal power subsets A and B do not allow determining the optimal scell value corresponding the objective clustering of the studied data. The external clustering quality criterion CQE has several local minimums corresponding to the successful grouping of the studied vectors. However, the analysis of the complex balance criterion values, which takes into account both the internal and the external criteria, allows us to conclude that the best clustering corresponds to the scell = 0,001. In this case the 6659 profiles were divided into two clusters. The first cluster contained 4276 profiles and the second – 2383 ones. Variation of the maximum divergence value in the range from 0,001 to 1 has not changed the obtained results. The obtained results create the conditions to create the step-by-step technology of gene expression profiles grouping at early stage of the gene regulatory network construction. Objective clustering based on DBSCAN algorithm allows us to select the genes with higher level of their mutual correlation. The noise component also is removed at this step. Then at the second step of the profiles grouping the selected profiles are divided into two group using SOTA clustering algorithm. At the third step of the gene expression profiles grouping the biclustering technology is implemented on the obtained clusters. To our mind the implementation of the proposed technology allows saving more useful information to follow create the gene regulatory network.

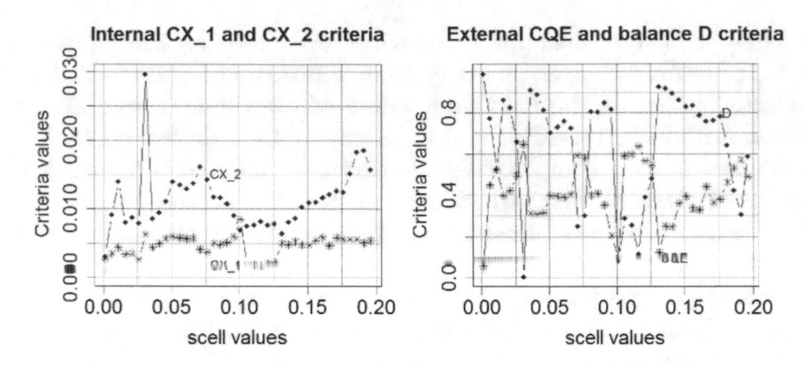

Fig. 12. The internal, the external and the balance criteria versus the weight coefficient of the sisters cell value of the SOTA clustering algorithm

7 Conclusion

The paper presents the model of the objective clustering inductive technology of gene expression profiles based on DBSCAN and SOTA clustering algorithm. The implementation of this technology involves the concurrent data clustering on the two equal power subsets which include the same quantity of the pairwise similar objects. The correlation metric was used as the proximity metric of the gene expression profiles. The internal clustering quality criteria take into account both the character of objects distribution within clusters relative to the mass center of the appropriate cluster and the character of the clusters distribution in the features space. The external clustering quality criteria were calculated as a normalized difference of the internal clustering quality criteria, which were calculated on the equal power subsets A and B. The simulation process involved sequential evaluation of the internal and external criteria for clustering during increase the clusters quantity from Kmin to Kmax. The objective clustering corresponded to the global minimum of the external clustering quality criterion. The gene expression sequences of the patients of the database Array Express, which were investigated on the lung cancer, were used as the experimental data. The quantity of the clusters was changed from 2 to 5 during clustering process. The objective clustering corresponded to the four-cluster structure. The results of the simulation have shown that the complex multiplicative criterion, which is the combination of the WB-index and Calinski-Harabasz criterion is the most effective to determine the objective clustering. This criterion has the clearly expressed minimum corresponding to the four-cluster structure both in case of estimation of the character of the objects and the clusters distribution in the equal power subsets and in case of estimation of the result of clustering difference on these subsets. The external criterion was calculated as the normalized difference of the internal clustering quality criteria which are calculated on the two equal power subsets. The general Harrington desirability index based on the internal and external criteria was used as the complex balance clustering quality criterion. Determination of optimal parameters of the used algorithm operation has been performed based on the maximum value of the complex balance clustering quality criterion during the algorithm operation. The results of the simulation have shown the high efficiency of the proposed technology. In case of DBSCAN clustering algorithm using the noise component in terms of density of the objects distribution was selected during algorithm operation. Implementation of the proposed technology also allows us to group the gene expression sequences based on the similarity of their profiles. The gene expression sequences with high correlation coefficient were distributed into one cluster. This fact allows us to select the groups of the gene expression sequences, which determine the main processes in the biological organisms in order to study and to correct these processes. In case of SOTA clustering algorithm using the studied gene expression profiles were divided into two groups. This fact create the conditions to create the step-by-step technology of gene expression profiles grouping at early stage of the gene regulatory network construction. Objective clustering based on DBSCAN algorithm allows selecting the genes with higher level of their mutual correlation. Then the selected profiles are divided into two group using SOTA clustering algorithm. The further perspective of the authors' research is the development of the hybrid technology of the step-by-step gene expression profiles grouping based on the complex use of the objective clustering and biclustering technologies.

References

1. Babichev, S., Kornelyuk, A., Lytvynenko, V., Osypenko, V.: Computational analysis of gene expression profiles of lung cancer. Biopolymers Cell **32**(1), 70–79 (2016). http://biopolymers.org.ua/content/32/1/070/
2. Babichev, S., Taif, M.A., Lytvynenko, V.: Inductive model of data clustering based on the agglomerative hierarchical algorithm. In: Proceeding of the 2016 IEEE First International Conference on Data Stream Mining and Processing (DSMP), pp. 19–22 (2016). http://ieeexplore.ieee.org/document/7583499/
3. Babichev, S., Taif, M.A., Lytvynenko, V., Korobchynskyi, M., Taif, M.A.: Objective clustering inductive technology of gene expression sequences features. In: Proceeding of the 13th International Conference Beyond Databases, Architectures and Structures. Communication in Computer and Information Science, Ustron, Poland, pp. 359–372 (2017). https://link.springer.com/content/pdf/10.1007/978-3-319-58274-0_29.pdf
4. Babichev, S., Taif, M.A., Lytvynenko, V., Osypenko, V.: Criterial analysis of gene expression sequences to create the objective clustering inductive technology. In: Proceeding of the 2017 IEEE 37th International Conference on Electronics and Nanotechnology (ELNANO), pp. 244–249 (2017). http://apps.webofknowledge.com/full_record.do?product=WOS&search_mode=GeneralSearch&qid=3&SID=U2bB7H8kqTrSyZ2eAKs&page=1&doc=2
5. Beer, D., Kardia, S., et al.: Gene-expression profiles predict survival of patients with lung adenocarcinoma. Nat. Med. **8**(8), 216–224 (2002). https://www.ncbi.nlm.nih.gov/pubmed/12118244
6. Calinski, T., Harabasz, J.: A dendrite method for cluster analysis. Commun. Stat. **3**, 1–27 (1974)

7. Chi, E., Allen, G., Baraniuk, R.: Convex biclustering. Biometrics **73**, 10–19 (2016). http://onlinelibrary.wiley.com/doi/10.1111/biom.12540/full
8. Dorazo, J., Corazo, J.: Phylogenetic reconstruction using an unsupervised growing neural network that adopts the topology of a phylogenetic tree. J. Mol. Evol. **44**(2), 226–259 (1997). https://www.ncbi.nlm.nih.gov/pubmed/9069183
9. Ester, M., Kriegel, H., Sander, J., Xu, X.: A density-based algorithm for discovering clusters in large spatial datasets with noise. In: Proceedings of the Second International Conference on Knowledge Discovery and Data Mining, Portland, pp. 226–231 (1996). http://dl.acm.org/citation.cfm?id=3001507
10. Fritzke, B.: Growing cell structures a self-organizing network for unsupervised and supervised learning. Neural Netw. **7**(9), 1441–1460 (1994). http://www.sciencedirect.com/science/article/pii/0893608094900914
11. Harrington, J.: The desirability function. Ind. Qual. Control **21**(10), 494–498 (1965). http://asq.org/qic/display-item/?item=4860
12. Hartigan, J.: Clustering Algorithms. Wiley, New York (1975). http://dl.acm.org/citation.cfm?id=540298
13. Ihaka, R., Gentleman, R.: R: a linguage for data analysis and graphics. J. Comput. Graph. Stat. **5**(3), 299–314 (1996). http://www.tandfonline.com/doi/abs/10.1080/10618600.1996.10474713
14. Kaiser, S.: Biclustering: Methods, Software and Application (2011). https://edoc.ub.uni-muenchen.de/13073/
15. Madala, H., Ivakhnenko, A.: Inductive Learning Algorithms for Complex Systems Modeling, pp. 26–51. CRC Press, Boca Raton (1994). http://www.gmdh.net/articles/theory/ch2.pdf
16. Osypenko, V.V., Reshetjuk, V.M.: The methodology of inductive system analysis as a tool of engineering researches analytical planning. Agric. Forest Eng. **58**, 67–71 (2011). http://annals-wuls.sggw.pl/?q=node/234
17. Pontes, B., Giraldez, R., Aguilar-Ruiz, J.S.: Biclustering on expression data: a review. J. Biomed. Inf. **57**, 163–180 (2015). https://www.ncbi.nlm.nih.gov/pubmed/26160444
18. Sarycheva, L.: Objective cluster analysis of data based on the group method of data handling. Problems of Control and Automatics **2**, 86–104 (2008)
19. Stepashko, V.: Elements of the Inductive Modeling Theory, State and Prospects of Informatics Development in Ukraine, pp. 471–486. Scientific Thought, Kiev (2010). Monograph/Team of autors
20. Zhao, Q., Xu, M., Frnti, P.: Sum-of-squares based cluster validity index and significance analysis. In: Proceeding of International Conference on Adaptive and Natural Computing Algorithms, pp. 313–322 (2009). https://link.springer.com/chapter/10.1007/978-3-642-04921-7_32

From Close the Door to Do not Click and Back. Security by Design for Older Adults

Bartlomiej Balcerzak(✉), Wieslaw Kopec, Radoslaw Nielek, Kamil Warpechowski, and Agnieszka Czajka

Polish-Japanese Institute of Information Technology, ul. Koszykowa 86, 02-008 Warsaw, Poland
b.balcerzak@pjwstk.edu.pl

Abstract. With the growing number of older adults who adopt mobile technology in their life, a new form of challenge faces both them, as well as the software engineering communities. This challenge is the issue of safety, not only in the context of risk older adults already face online, but also, due to the mobile nature of the used applications, real life safety issues raising from the use of on-line solutions. In this paper, we wish to use a case study they conducted in order to address this issue of interrelating on-line and real life threats. We describe how the observation from the case study relate to the collected body off knowledge in the relevant topic, as well as propose a set of suggestion for improving the design of applications in regards to addressing the issue of older adults safety.

1 Introduction

The trend of population aging is currently an undisputed fact, observed throughout the globe, in developed and developing countries alike [23]. By 2050 over 20% of the population of the USA is projected to be 65+ [27], other developed countries, such as Japan are also expected to become aging populations by that time. Although this trend has been deemed a negative trajectory for population development [11], and a giant challenge for social security frameworks, and in turn a major burden for the state, research shows that this trend may bear out other results. For instance, Spijker et al. argument that although demographic data are true in reality number of older adults requiring assistance in UK and other countries actually been falling in recent years [36]. This point of view is supported by Sanderson et al. [35] who point out that metrical age qualifying to being older adult will increase in coming years.

In parallel to the mentioned trend, an increase in smartphone and on-line tool usage among older adults can also be observed [43]. This trend of increasing usage of mobile ICT can be observed worldwide, and already has inspired many researchers and designers to introduce new and creative solutions tailored towards the needs of the older adults [18]. When considered in relation to the findings suggesting that in future aging societies, older adults may be more affluent, healthy and tech savvy than their current counterparts. This increase in use of mobile technology, opens up new possibilities for addressing the issue of population aging, and creating solutions tailored to a new

N. Shakhovska and V. Stepashko (eds.), *Advances in Intelligent Systems and Computing II*, Advances in Intelligent Systems and Computing 689,
https://doi.org/10.1007/978-3-319-70581-1_3

growing customer base. With these new possibilities however, also come new challenges. The demographics increase in the share of mobile solutions users leads to new, and previously not addressed issues such as those related to the older adults safety.

Research shows that older adult are less likely to be tech savvy [9], and often as a result, are more prone to suffer from threats related to their safety. Moreover, older adults tend to be more aware of the risks they face on-line [1]. Due to the fact that on-line solutions such as social media, communication and financial services or other mobile based apps increasingly blur the line between the online and real life situation, it is understandable that, when addressing the safety issues of older adults, one must consider both aspects of safety. Discussing this topic is our main aim in this paper.

For the sake of this paper, we made an explicit distinction between two aspects of older adults safety; their on-line security, and real life safety. The former, as the name suggest, refers to risk the older adult may face when interacting with ICT (i.e. identity theft, scamming, phishing, spread of malware and spam), the latter on the other hand deals with risks connected with interactions made in the real world, such as theft, assault, burglary, traffic accidents on so on. Although these two aspects were deeply reviewed by researchers (with work by [7, 9] serving as good examples), there has been no attempt, to the best of our knowledge, the analyze the interaction between the two aspects, let alone to analyze how developers can address them in the design process.

Therefore we decided, based also on previous endeavours and activities with older adults in our LivingLab, to use a case study of an intergenerational design team during a Hackathon event conducted at the Polish-Japanese Academy of Information Technology in Warsaw, Poland (PJAIT). In this study we addressed the issue of the connection between on-line security and real life safety in the design process by using participatory design approach. By describing the design process and group dynamics, as well as the final product, we wish to open further discussion, and provide some deeper insight into a topic that, although important in lieu of the aforementioned trend in stronger connection between the on-line and off-line, was not yet fully addressed by the research community. In this paper we provide more comprehensive approach to the security of older adults based on the concept described in case study section which we previously briefly outlined in our conference report [3].

The rest of the paper is organized as follows. In the Related work section, we provide a comprehensive description of findings in regards to on-line security and real life safety of older adults, as well as security design in software engineering, as well as participatory design, and older adults motivation for taking part in design processes and similar tasks. In the section LivingLab insight we describe a broader context of their empirical studies and research activities with older adults in PJAIT LivingLab relevant to the above mentioned topics. The Case study section present in detail the design session conducted during the Hackathon, the specific features of the application created by the intergenerational team, as well as the design process itself. The Conclusions and Future Work section details how the observations made during the design session interact with the findings related to the topics of on-line security and real life safety.

2 Related Works

2.1 Older Adults and On-line Security

According to studies done by [9] older adults are, on average less knowledgeable and aware of on-line risks than younger adults. Also, as shown by [1], older adults show limited trust to on-line technology, however this effect diminishes with usage of ICT. Another dimension crucial with on-line security is trust. Studies done by [19] as well as [9] show that trust among older adults is not affected by age, and relies more on experience that comes with the use of ICT. Our previous study on location-based game with mobile ICT technology [14] which we present briefly in next section also showed the connection between direct ICT usage and self-confidence of older adults in context of mobile technologies. Therefore, the design of applications that address the issue of trust is important. The topic of safety, also plays a role in research related to applying smart house technologies as tools for enabling older adult independence [28]. A topic also related to online safety is the issue using data and text mining techniques for the detection f scamming and the credibility of on-line material [13, 39].

2.2 Older Adults and Real Life Safety

For the sake of this paper, when reviewing the real-life hazards faced by older adults, the authors will focus on a specific aspects of safety, which in the literature is refereed to as neighborhood safety. This refers to risk related to social interaction in the nearest environment of the older adult. As the comprehensive literature review conducted by [7] suggests, five main themes constitute the overall framework of neighborhood safety. These are: general neighborhood safety; crime-related safety; traffic-related safety; fall-related safety; and proxies for safety (e.g., vandalism, graffiti). This literature review provides also a deep and comprehensive review of what main themes are the object of focus among researchers interested in real life hazards and their perception by older adults. It is noteworthy, that in the studies used in the review, main focus was put on health issues and their relation with the feeling of safety. This is a theme which appears in works such as those done by [40], where the authors show that physical health is a correlate with the older adults sense of safety. Another interesting field of study is related to older adults emotion in relation with surrounding environment. We conducted a preliminary study on mapping senior citizens' emotions with urban space [24], which is also connected with our previous work, namely the mentioned mobile game [14] which was related to the topics of wellbeing and happiness of older adults. This issue is a vital subject of many other interesting research, i.e. a longitudinal study on a sample of over 10,000 older adults that pointed out that perception of one's psychological well being affects the older adults feeling of safety [5]. To the best of the authors knowledge, little focus was placed on interaction between thus defined levels and perceptions of neighborhood safety and use of mobile devices and apps. Therefore, besides our previous endeavours mentioned above, we decided also to pursue this topic in exploratory in-depth interviews with older adults form our LivingLab, which are elaborated in the next section.

2.3 User Participation and Interaction

Fundamental idea relevant for this case is connected with a concept of usercenter design, as a part of general idea of human focused approach related to another important idea: participatory design, sometimes called co-design. While there are different origins for the two latter terms they actually refer to the same idea of bottom-up approach which is widely used besides software engineering from architecture and landscape design to healthcare industry [37]. All those concepts put human in the center of designing process, however, there is a small, but significant distinction between user-center design and participatory design: the former refers the process of designing FOR users, while the latter WITH users [33, 34]. From the point of view of this study especially important are concepts that are related to user competences and empowerment provided e.g. by Ladner [16].

Another case is the work done by [41], describing a cooperation between seniors and preschool children in a design task. An interesting observation made in this research was the fact that both groups needed equally time together, and time in separation in order to function properly. The broader context for these topics is covered by the contact theory, a widely recognized psychological concept coined by Allport [2] and developed for many years by other, e.g. Pettigrew [29]. According to the theory the problem of intergroup stereotypes can be faced by intergroup contact. However, there is several condition for optimal intergroup contact, but many studies proved that intergroup contact is worthwhile approach since it typically reduces intergroup prejudice [30]. We had also explored the topic of intergroup interaction in our various studies including mentioned intergenerational location-based game and the hackathon based on previous works and tools i.e. by Rosencranz [31].

2.4 Volunteering of Older Adults

The effects of volunteering has on older adults is a developing field of study within various disciplines. The consensus that can be reached throughout various studies, is that participation in volunteer activity has many positive effects on the elderly. [17] claim that older adults who frequently volunteer in various activities, tend to have improved physical and mental health, compared to those who do not participate in volunteering. The work of [22] extends this correlation to wellbeing (with volunteering being correlated with higher levels of well-being) this is also reinforced by [8, 10]. This is crucial since studies also show that in some regards, elders are more likely to be engaged in volunteer activity [21].

Motivation for volunteering among older adults is also important. In research done by [12] involving the comparison of older and younger adults when participating in a crowd sourcing task of proof reading texts in Japanese, it was shown that older adults where successfully motivated by the use of gamification techniques within the task, which means according to the most widespread definition of gamification by Deterding *the use of game design elements in non-game contexts* [6]. However, a similar task, conducted by [4] on a group of American seniors showed a slightly different pattern, where older adults were bored by the task, and did not comprehend its' significance.

Nevertheless gamification is strictly connected with motivation and therefore inevitably leads to the psychological context, e.g. Zichermann claims that *gamification is 75% of psychology and 25% of technology* [42]. At this point it is worth mentioning that according to the latest reviews more and more solutions are based on solid psychological theories and frameworks [20]. One of the most important theoretical approach in the field of motivation is self-determination theory (SDT) developed by Ryan and Deci [32]. It is based on subtheories formerly developed by the authors of SDT: cognitive evaluation theory (CET) related to the intrinsic motivation and organismic integration theory (OIT) related to extrinsic motivation. The theory was proven to be effective to the elderly as well, e.g. by Vallerand [38]. We had explored the topic of older adults volunteering and motivation in our research i.e. Wikipedia content co-creation [25].

3 LivingLab PJAIT Insights

In this section we provide some insights from our experience with older adults within Polish-Japanese Academy of Information Technology LivingLab. Further details of our LivingLab, it's origin and design as well as older adults activities are provided in separate description [15].

3.1 About LivingLab PJAIT

The term *LivingLab* was coined by William Mitchell from MIT [26] and was used to refer to the real environment, like a home or urban space, where routines and everyday life interactions of users and new technology can be observed and recorded to foster the process of designing new useful and acceptable products and services. The idea of LivingLab is therefore inherently coupled with broad concept human-centered approach described in previous section.

LivingLab at the Polish-Japanese Academy of Information Technology is a longterm framework project, whose goals are related to social inclusion and active engagement of the elderly in social life by facilitating the development of ICT literacy among them as well as creating an active community of stakeholders who are both the beneficiaries and enablers of research into their problems. This framework project has been established in a long-term partnership with the City of Warsaw.

Throughout recent years we have organized a number of activities for older adults focused on various research areas relevant to the topics of this article as we mentioned in related work section. In particular during an intergenerational location based game older adults performed various everyday mobile ICT tasks, such as connecting to Wi-Fi, browsing the information, scanning QR codes or taking panoramic photos with the assistance of their teammates. On the other hand, the younger participants benefited from the background knowledge about the city and its history of the elderly. Thus a positive bi-directional intergenerational interaction was observed alongside positive older adults self-awareness of the technology in context of physical activities and well-being. In other LivingLab activities, including on-line courses and crowdsourcing tasks as well as real life workshops and activities i.e. devoted to both application and content co-creation we also noticed that security issues are important but a bit vague area for older adults. Based on the literature review we decided to conduct a more in-depth qualitative research described in next subsection.

3.2 Older Adults and Security

Based on literature and observations from LivingLab activities, workshops and consultations we decided to obtain a more in-depth insight in the topic, including both relevant perspectives: on-line security and real life safety.

We used individual in-depth interviews in order to extract additional security insights. In total, we conducted four such interviews with older adults from our LivingLab aged 65+ , two less technology advanced (female participants P1 and P2) and a pair of more advanced older adults (female P3 and male P4). To obtain more in-depth information from them we decided to conduct individual semistructured interviews related to several topics: Internet and mobile application usage, endeavours towards on-line security and real life safety alongside with perception of potential interaction between those two realms.

Based on interviews we figured three different strategies toward security: caution, separation and awareness. While awareness is related to the more technologically advanced older adults, the two former strategies are interesting since they are represented by older adults with lower ICT literacy. Generally speaking the caution strategy represented by P1 was based on carefulness in both real life and on-line activities while separation strategy represented by P2 was based on strong conviction on non-transition between virtual and real realm. In particular P2 claimed that *these two worlds should not be comined.*

First we asked participants about their Internet and mobile experience. There are three major areas of their on-line interests:

– doing everyday tasks, like paying the bills –
keeping in touch with family and friends –
source of information.

The latter was the most extensive category and included various topics from health and medical issues (i.e. drugs and food ingredients, dietary information), through transportation to cultural and political news. In this context P2 claimed that *The world has gone so far that it is difficult to live without the Internet these days.*

In order to obtain insight about on-line security we asked the participants about securing themselves in virtual space. In this context participants were aware of anti-virus protection as well trust issues, identity theft and identity verification (the need of verification the identity of on-line entities i.e. shops and companies). However, here we observed the major difference in separate-world approach. P2 explicitly stated, that there is no direct transition between the virtual and real realms. This was directly connected with careless websurfing. Moreover P2 wasn't afraid of her identity theft based on the claim *I am no one special, I am an ordinary Smith.*

We also asked about the safeness endeavor made by the participant in their everyday life. Besides personal physical safety measures like traffic safety, observation of the surroundings and other persons they pointed out two major areas connected with virtual space: finance and health. However they cannot establish the connection between the two realms. In particular, besides techniques of safe cash carrying (P1: *one can deposit the cash in various piece of garment*) they stated that in general they use credit cards instead of cash and they deposit money in bank. In reference to health, besides physical activity they pointed out healthy diet and food and medical ingredients verification.

As we can see some factors from virtual space activities can be directly mapped and connected with the real world. However, it was difficult to the participants to find the connection by themselves. More technologically advanced older adults (P3 and P4), were more aware of two realms i.e. bank account protection and identity theft alongside with interface as a vital concept between virtual and real world. On the other hand older adults with lower ICT literacy did not found themselves the connection between real life safety and increasing on-line security and vice versa. Surprisingly even though they provided a number of proper examples and behaviors from both realms they failed to find themselves spontaneously the connection. This leads to the conclusion that inevitably there is a room for designers to employ a participatory design approach in order to obtain insights from older adults not only to better understand their need but also to foster the process of idea development in order to find better connection between on-line and real life habits and safety. Because usually there is a generation gap between software development teams and end-users in case of older adults application we also decided to obtain a deeper insight into the dynamics of such collaboration, described in next section, based on our experience in the field of intergenerational interaction.

4 Case Study

4.1 Case Study Context

The case study, involving the intergenerational developer team, was observed during a DEVmuster Hackathon organized in March 2016 in The Polish-Japanese Academy for Information Technology in Warsaw Poland, during which older adults and students of the academy had an opportunity to cooperate in designing application that would address the needs of older adults. The team consisted of 4 males, all in their twenties, who were students of the academy, and two older adults, a male, and a female, who were participants of the PJAIT LivingLab, a framework presented in previous section, which involves volunteers wishing to take part in various project aimed at activizing older adults with the use of ICT.

The team formed during the first hours of the event, during which ideas for the potential app were discussed. Upon reviewing the opinions of the older adults, the students decided to change their original idea of an application, in to an app designed for exchange of favor between volunteers and older adults. The name 'F1' was chosen, based on the function key for calling the help menu.

4.2 Platform Architecture

The F1 platform was designed as client-server model and requires access to the Internet for the proper functioning. It might be a serious limitation for less developed and less populated countries but as the platform is intended to be deployed in Poland we decided to sacrifice versatility for simplicity.

Access to the system is possible either through a web site or a dedicated mobile application. Although both ways provide the same functionality there are also substantial differences. Mobile application was designed to be most convenient for people offering support. Web site is more focused on posting requests for help. The reason for this differentiation is that mobile application will be more frequently used by younger volunteers and web browsers are better suited for older adults. In the case described in this paper senior participants were more familiar with traditional desktop or mobile computers than with smartphones, mainly due to their professional background and prior LivingLab activities i.e. computer workshop organized by the City of Warsaw. Moreover computer web browsers can be more suitable for people with certain disabilities – e.g. visually impaired or with limited hand dexterity. On the other hand mobile devices are becoming more and more widespread and the adoption of touch interface by the elderly can be faster and more effective than traditional computer interfaces (e.g. observed in our previous research studies). Thus in final product the application mode (web site or mobile app) is intended to be freely interchangeable at any time by the user.

4.3 Functionality

As presented in Fig. 1 The web based application contains all the key information and functions important for the user searching for the help of a volunteer.

The screen informs the user about his or her previous favor requests, as well as the location of other users in the area. There is also an S.O.S button which can be used in case of emergency.

The mobile application view, used by the potential volunteer is shown in Fig. 2c. When using the mobile view, the user can view a map of the nearest area where favor requests are marked. When the user selects a favor a brief description of the favor and the requesting user is provided.

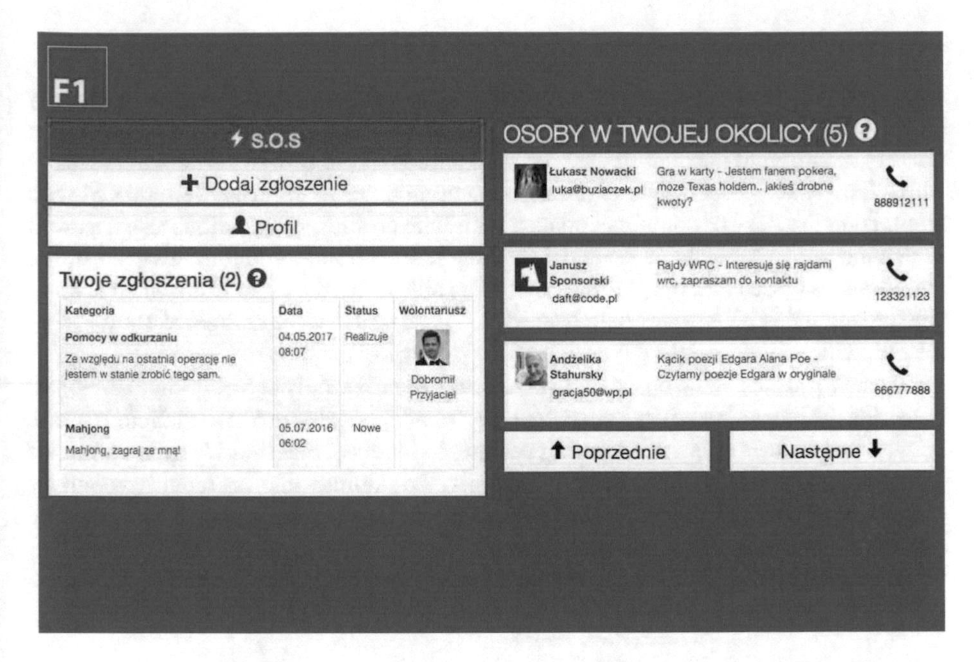

Fig. 1. Main screen of web-based application focused mostly on people requiring assistance.

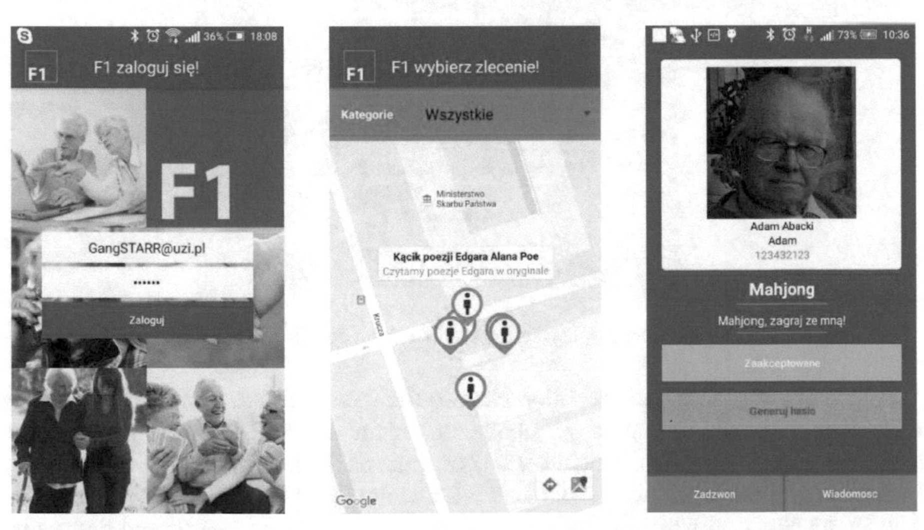

(a) Welcome and login (b) Screen presents re- (c) Detailed information screen. quests for assists in neigh- about request for assist. borhood on the map.

Fig. 2. Mobile application dedicated for those who offer assistance.

4.4 Security Related Features

Another direct benefit from applying participatory design approach refers to security issues. During the pre-design phase, older adults voiced their concerns related to user security, taking into account threats for both parties of the process. The main areas in which they stated older adults require aid when it comes to security were related with minimizing the risks of coming in contact with fake profiles, or malicious users, as well as dealing with problems of potential identity theft. The older adults involved in the project also stressed that the platform should be able to aid the older users when dealing with emergencies when swift help is needed. These issues were addressed by applying the following solutions into the design of the platform.

Trusted profiles Sign up is free for all and requires only a valid email account. Lowering the barrier makes system more user-friendly but also prone to malicious users. Discussions with prospect users during design phase reveal that the elderly are afraid of letting unknown people visit their apartments. To address this problem a voluntary procedure for confirming profiles was implemented. User profile might be confirmed by external organizations that are trustworthy: e.g. schools or local NGOs. Confirmation of the verified status is visible for everyone next to profile picture – see Fig. 3.

Fig. 3. Pop-up window displaying the confirmation details for selected volunteer

Challenge-response authentication Next to the threat of fake or malicious profiles mentioned in previous section yet another problem was identified by participatory approach. Even if identity of volunteer is confirmed on the platform still we need a way to confirm it in real world when volunteer is knocking to the door of senior's apartment. This is the situation when digital system should face analogue world and bottom-up approach proved to be helpful once again. Therefore, a standard challenge-response authentication has been adapted and implemented. The platform generates two keywords for both users. Elderly should ask about the right password before letting someone in. Passwords are randomly selected from a subset of polish words to make them easy to remember and dictation by entry phone.

Reputation score Limiting the amount of frauds is crucial for assuring wide acceptance of the platform but it is not enough. Next to deliberate and planed frauds we can also see a bad quality service. Therefore, the platform contains a reputation management system. Every agreed and conducted service can be evaluated on Likert-type scale. To make scale easier to understand for users first two grades are red, neutral score is gray and the two positive levels are green. Sum of all evaluation for give user are displayed next to the picture – see Fig. 3.

Emergency button In real life exhaustive list of risks and threats is impossible to complete. Therefore, an emergency button has been added to the F1 platform.

5 Conclusions and Future Work

As it was presented in the Related Work section and in our study, the issues of on-line security and real life safety faced by older adults are quite diverse. The span a wide set of threats ranging from health issues, to crime related issues, and problems connected with the spread of malware and spam, and even though older adults can point out a variety of those issues, usually it is difficult for them to spontaneously find the connection between the two realms. However, in our case study we showed that participatory design approach can benefit both younger tech-minded developers and older adults as end-users.

The observations made during the presented case study show, that the key for creating an on-line solution, that would not only aid them in their daily lives, but also but be resilient to safety and security issues mentioned in the literature, is not necessarily to address all potential dangerous scenarios. From the description of the app created through a participatory design within an intergenerational team, where older adults, were not only final users, but also active team members, one can see that the focal point for addressing issues of safety, both on-line and in real life, is modeling of trust within the user base of the application. Use of such elements like a reputation system, two step password verification between users who decided to exchange favors, as well as external confirmation of users provides a wide array of possibilities for encouraging the development of trust between older adults and younger volunteers. This corroborates the general ramifications of the intergroup contact theory, however it extends its scope beyond intergroup stereotyping into the field of limiting the feeling of insecurity among the group of older adults.

It is also important to notice that the participatory approach utilized by the team, allowed to overcome the aforementioned issue of understanding the link between real life safety and on-line security. In light of the results from the interviews conducted within the framework of the LivingLab, where older adults had difficulties with connecting the two aspects of safety, the outcome of the team design process yields great promise.

The case study presented here, of course, serves mostly as a jumping-off point for further considerations in a topic that, although important, is not, in the authors opinion, amply researched. The findings made in this paper show, that by use of a participatory design framework, it is possible not only to address general issues of on-line security, but also, to create new methods of thinking about user safety outside of the paradigm of software engineering. With the increasing role of mobile technologies in real life situation, these initial exploratory findings offer interesting option for improving the initial design process of mobile applications.

In their future work, we plan to further explore this interesting emerging field of study. With the observation made in this paper being mostly exploratory in nature, it seems fitting to conduct a set of more methodologically rigid tests with the aim of verifying, what form of participatory design can further improve the process of addressing the threat on-line and real life threats faced by end users of an application.

Acknowledgments. This project has received funding from the European Union's Horizon 2020 research and innovation programme under the Marie Skłodowska-Curie grant agreement No 690962.

References

1. Adams, N., Stubbs, D., Woods, V.: Psychological barriers to internet usage among older adults in the UK. Med. Inf. Internet Med. **30**(1), 3–17 (2005)
2. Allport, G.W.: The Nature of Prejudice. Basic Books, New York (1979)
3. Balcerzak, B., Kopeć, W., Nielek, R., Kruk, S., Warpechowski, K., Wasik, M., Węgrzyn, M.: Press F1 for help: participatory design for dealing with on-line and real life security of older adults. arXiv preprint arXiv:1706.10223 (2017)
4. Brewer, R., Morris, M.R., Piper, A.M.: "Why would anybody do this?": Older adults' understanding of and experiences with crowd work
5. Choi, Y.J., Matz-Costa, C.: Perceived neighborhood safety, social cohesion, and psychological health of older adults. Gerontol. **1**, gnw187 (2017)
6. Deterding, S., Dixon, D., Khaled, R., Nacke, L.: From game design elements to gamefulness: defining gamification. In: Proceedings of the 15th International Academic MindTrek Conference: Envisioning Future Media Environments, pp. 9–15. ACM (2011)
7. Forjuoh, S.N., Won, J., Ory, M.G., Lee, C.: 172 neighbourhood safety and injury prevention among older adults: a systematic literature review (2016)
8. Greenfield, E.A., Marks, N.F.: Formal volunteering as a protective factor for older adults' psychological well-being. J. Gerontol. Ser. B: Psychol. Sci. Soc. Sci. **59**(5), S258–S264 (2004)
9. Grimes, G.A., Hough, M.G., Mazur, E., Signorella, M.L.: Older adults' knowledge of internet hazards. Educ. Gerontol. **36**(3), 173–192 (2010)
10. Hao, Y.: Productive activities and psychological well-being among older adults. J. Gerontol. Ser. B: Psychol. Sci. Soc. Sci. **63**(2), S64–S72 (2008)
11. Harper, S.: Economic and social implications of aging societies. Science **346**(6209), 587–591 (2014)
12. Itoko, T., Arita, S., Kobayashi, M., Takagi, H.: Involving senior workers in crowdsourced proofreading. In: Universal Access in Human-Computer Interaction. Aging and Assistive Environments, pp. 106–117. Springer (2014)
13. Jankowski-Lorek, M., Nielek, R., Wierzbicki, A., Zieliński, K.: Predicting controversy of Wikipedia articles using the article feedback tool. In: Proceedings of the 2014 International Conference on Social Computing, SocialCom 2014, New York, NY, USA, pp. 22:1–22:7. ACM (2014)

14. Kopeć, W., Abramczuk, K., Balcerzak, B., Juźwin, M., Gniadzik, K., Kowalik, G., Nielek, R.: A location-based game for two generations: teaching mobile technology to the elderly with the support of young volunteers. In: eHealth 360°, pp. 84–91. Springer (2017)
15. Kopeć, W., Skorupska, K., Jaskulska, A., Abramczuk, K., Nielek, R., Wierzbicki, A.: LivingLab PJAIT: towards better urban participation of seniors. arXiv preprint arXiv: 1707.00030 (2017)
16. Ladner, R.E.: Design for user empowerment. Interactions **22**(2), 24–29 (2015)
17. Lum, T.Y., Lightfoot, E.: The effects of volunteering on the physical and mental health of older people. Res. Aging **27**(1), 31–55 (2005)
18. Massimi, M., Baecker, R.M., Wu, M.: Using participatory activities with seniors to critique, build, and evaluate mobile phones. In: Proceedings of the 9th International ACM SIGACCESS Conference on Computers and Accessibility, pp. 155–162. ACM (2007)
19. McCloskey, D.W.: The importance of ease of use, usefulness, and trust to online consumers: an examination of the technology acceptance model with older consumers. J. Organ. End User Comput. **18**(3), 47 (2006)
20. Mora, A., Riera, D., González, C., Arnedo-Moreno, J.: A literature review of gamification design frameworks. In: 2015 7th International Conference on Games and Virtual Worlds for Serious Applications (VS-Games), pp. 1–8. IEEE (2015)
21. Morrow-Howell, N.: Volunteering in later life: Research frontiers. J. Gerontol. Ser. B: Psychol. Sci. Soc. Sci. **65**(4), 461–469 (2010)
22. Morrow-Howell, N., Hinterlong, J., Rozario, P.A., Tang, F.: Effects of volunteering on the well-being of older adults. J. Gerontol. Ser. B: Psychol. Sci. Soc. Sci. **58**(3), S137–S145 (2003)
23. United Nations: World Population Ageing 2015. Department of Economic and Social Affairs (2015)
24. Nielek, R., Ciastek, M., Kopec, W.: Emotions make cities live. Towards mapping emotions of older adults on urban space. arXiv preprint arXiv:1706.10063 (2017)
25. Nielek, R., Lutostanska, M., Kopec, W., Wierzbicki, A.: Turned 70? It is time to start editing Wikipedia. arXiv preprint arXiv:1706.10060 (2017)
26. Niitamo, V.-P., Kulkki, S., Eriksson, M., Hribernik, K.A.: State-of-the-art and good practice in the field of living labs. In: Technology Management Conference (ICE), 2006 IEEE International, pp. 1–8. IEEE (2006)
27. Ortman, J.M., Velkoff, V.A., Hogan, H., et al.: An Aging Nation: The Older Population in the United States. US Census Bureau, Washington, DC (2014)
28. Peek, S.T., Aarts, S., Wouters, E.J.: Can smart home technology deliver on the promise of independent living? A critical reflection based on the perspectives of older adults. In: Handbook of Smart Homes, Health Care and Well-Being, pp. 203–214 (2017)
29. Pettigrew, T.F.: Intergroup contact theory. Annu. Rev. Psychol. **49**(1), 65–85 (1998)
30. Pettigrew, T.F., Tropp, L.R.: A meta-analytic test of intergroup contact theory. J. Pers. Soc. Psychol. **90**(5), 751 (2006)
31. Rosencranz, H.A., McNevin, T.E.: A factor analysis of attitudes toward the aged. Gerontol. **9**(1), 55–59 (1969)
32. Ryan, R.M., Deci, E.L.: Self-determination theory and the facilitation of intrinsic motivation, social development, and well-being. Am. Psychol. **55**(1), 68 (2000)
33. Sanders, E.B.-N.: From user-centered to participatory design approaches. In: Design and the Social Sciences: Making Connections, pp. 1–8 (2002)
34. Sanders, E.B.-N., Stappers, P.J.: Co-creation and the new landscapes of design. Co-design **4**(1), 5–18 (2008)

35. Sanderson, W.C., Scherbov, S.: Faster increases in human life expectancy could lead to slower population aging. PLoS ONE **10**(4), e0121922 (2015)
36. Spijker, J., MacInnes, J.: Population ageing: the timebomb that isn't. BMJ **347**(6598), 1–5 (2013)
37. Szebeko, D., Tan, L.: Co-designing for society. Australas. Med. J. **3**(9), 580–590 (2010)
38. Vallerand, R.J., O'Connor, B.P.: Motivation in the elderly: a theoretical framework and some promising findings. Can. Psychol./Psychol. Can. **30**(3), 538 (1989)
39. Wawer, A., Nielek, R., Wierzbicki, A.: Predicting webpage credibility using linguistic features. In: Proceedings of the 23rd International Conference on World Wide Web, WWW 2014 Companion, New York, NY, USA, pp. 1135–1140. ACM (2014)
40. Won, J., Lee, C., Forjuoh, S.N., Ory, M.G.: Neighborhood safety factors associated with older adults' health-related outcomes: a systematic literature review. Soc. Sci. Med. **165**, 177–186 (2016)
41. Xie, B., Druin, A., Fails, J., Massey, S., Golub, E., Franckel, S., Schneider, K.: Connecting generations: developing co-design methods for older adults and children. Behav. Inf. Technol. **31**(4), 413–423 (2012)
42. Zichermann, G., Cunningham, C.: Gamification by design: implementing game mechanics in web and mobile apps. O'Reilly Media, Inc., Sebastopol (2011)
43. Zickuhr, K., Madden, M.: Older adults and internet use. Pew Internet & American Life Project, 6 June 2012

The Popularization Problem of Websites and Analysis of Competitors

Taras Basyuk(✉)

Information Systems and Networks Department,
Lviv Polytechnic National University, Lviv, Ukraine
Taras.M.Basyuk@lpnu.ua

Abstract. The features of popularizing websites in the global Internet and the main factors that influence this process are analysed in the article. The analysis of existing researches is conducted and implementation of search engines that made it possible to identify the most important factors in the competitive promotion of the resource is described. The methods of assessing backlinks, anchors, resource and content logical structure are developed. The possibilities to apply thematic backlinks of competitors in the promoting website are determined. The scheme, which reflects features of website structure design depending on types of keywords, is developed.

Keywords: Popularization · Ranking · Search engine · Website · Backlinks · Anchor · Structure · Keywords

1 Introduction

The success of any website directly depends on its popularity among search engines. Moreover, the position of Internet resources as a result of search request is an important factor that directly affects the number of visitors [1, 2]. Effective search engine optimization contributes to its display in the top positions of search engine ranking. Achievement of this goal is only possible in case of thorough knowledge of ranking algorithms as well as skilful application of technology of search engines optimization, which is impossible without detailed consideration of problems of this sphere, the analysis of the main methods and work principles [3].

As for search engines, their role is constantly increasing, and it is impossible to imagine modern digital society without them. As is evidenced by short statistics of GlobalStat: 12 billion of search operations were carried out in the world monthly in 2011, which comprised about 400 million of search requests a day, and in 2016 the figure rose to 16.8 billion only on the US territory for desktop devices. This means that averagely over 6500 search operations is performed every second, while Google Inc. owns approximately 64% of the search market, and its search technology enables processing more than 4,000 operations per second [4].

Today, only a few seconds are spent for searching the information whereas twenty years ago we had to visit the library and spend two or three hours. This trend contributed

N. Shakhovska and V. Stepashko (eds.), *Advances in Intelligent Systems and Computing II*, Advances in Intelligent Systems and Computing 689,
https://doi.org/10.1007/978-3-319-70581-1_4

to the fact that the set of information resources, banking operations, social activities or purchases is performed via global network. Given that, the presence of high ranking in the search engines for business website is the key to the survival of the business, because people are constantly searching services and products in the global network. The achievement of such ranking is a great challenge, and basically relies on search engine optimization (SEO) [5, 6].

As the popularity of any search engine depends on the relevance of search results, so in applied algorithms and staff of developers the main attention is paid to the relevance of the search results. In turn it entrusts with huge amount of responsibilities on seo-specialists, as they should be aware of the search engines that the techniques used by them were not considered to be black optimization by search engines which can lead to the imposition of sanctions or deletion of the search engine from directory. Given that, the promotion of websites aimed at increasing their visibility in search engines is the extremely urgent problem.

1.1 Relation of Highlighted Issue to Scientific and Practical Tasks

Solving of the problem of effective websites promoting is an important task within the development of both the IT industry and industry in general. Since there is no company or entity, which in some way is not represented in the Global Internet that in its turn promotes the appropriate mathematical and algorithmic apparatus for implementing this process. Thus, the task of popularizing website can be divided into two types: the first – the actual analysis of search queries and formation of effective resource structure; second – the analysis of competitor websites that exist in this market segment. Understanding of the advancing mechanism provides answer to questions about why two similar-looking sites, with the same topic, take so different places in the search results. For example, one is on the top of search results [7, 8], and the other – on the third or fourth page of search result. In view of this, there are following answers to questions: how is the popularization of websites performed, which of the re-sources are preferred, based on which features the "best" relevance is defined - are becoming increasingly important in today's information society.

The evaluation of search engine by the user is the other side of the study. Specifically, the main purpose of the latter is to obtain information relevant to its request. However, requests may take a plurality of different forms, considering this an important task is to create a marketing strategy for the website (based on optimization mechanisms and search rankings). All this will contribute to the study and understanding of the psychology of target audience and thus provide the means of establishing contacts with new users [9].

As a result of globality of assigned task there is no general solution for it, as for the different types of resources, internal and external optimization factors as well as for different structure of websites. Complexity of research lies also in the fact that web-sites promotion is a complex and continuous process, in the course of which it is required to apply advanced mathematical and algorithmic apparatus [10, 11]. The solution of this task will help to determine the main factors influencing the analysis of competitor resources and thus contribute to construction of optimized websites and provide means of adapting known methods and algorithms to the tasks of construction of intelligent information systems in this subject area.

1.2 Analysis of Recent Researches and Publications

The era of information society significantly changed the concept of information resource by which increasingly the website is understood that can be presented as follows: the ordinary business cards (small website consisting of one or more pages, which contains basic information about the organization, individuals, products or services, contact information), powerful information portal (website containing a comprehensive amount of information on any subject area or several areas) or online store of the global scope (the site for selecting products from the catalog and selling them via the Internet). Thus, the availability of the resource on the Internet is not only prestigious but also profitable. By analysing the situation, we can conclude that the rating of website is comparable with the rating of its operations success. Considering conducted analysis of the literature [5, 11, 12] as well as the review of relevant Internet resources [13, 14] today search requests is often related to: research, search and purchase of goods and services. That is why, global e-commerce market is the mostly expanded, which already imposes financial factors as on the stage of advancement of the information website for complaints and product promotion as well as popularization, including website store spaces in the global network. According to the conducted analysis, the popularization of this area is determined by such factors as [11, 15]:

- the increase in number of users and improvement of the behavior culture in a global network that promotes the domination of users aged 21–35 years, for whom the possibility of purchase via network ranges from 52–63%;
- the introduction of high-speed wireless channels and performance of transactions by means of mobile devices is at the level of 30%;
- preservation of the competitive advantage of goods by providing lower prices, wide range of products and convenience of matching vendors' offers in comparison with traditional trade;
- usability, due to the presence of perfect delivery system at the expense of increasing number of goods outlets, the use of alternative delivery formats (automatic parcel terminal).

It should be highlighted that the study, conducted by Yahoo company! showed certain correspondence between the search in a global network and behavior beyond it. Specifically, the main findings were as follows [16, 17]: every dollar spent on advertisement in the global Internet networks brings nine dollars of income in ordinary stores; search marketing has a significant impact on sales increase in the stores than in demonstrating advertisements (almost three times). In view of above analysis, the popularization of websites has a set of advantages for both information search and online commerce. With regard to methodological research in this field, the following works should be admitted: the work of N.V. Yevdokimova on the basics of optimizing information content; A. Peleschyshyna on positioning websites in the global information environment; A. Yakovlyeva and V. Tkachyeva on the possibilities of promoting websites, I. Ashmanova and A. Ivanova on optimizing and promoting websites. The authors in these works consider general methods of improvement the websites position, but with the diversity of works the lack of thorough research on the positioning of websites, the increase of their popularity and competitive power is noticeable. In addition, there are no studies on the allocation of main reasons that motivate web optimizers to promote resources in top positions and analyze websites of competitors which are located at the same market segment. This, in turn, determines the relevance of research on the development of new methods and means of promoting websites in the global network that would have an appropriate justification and predictable performance.

2 The Main Objectives of the Research and Their Meaning

The main objectives of the research are as follows: the analysis of the functioning features of search engines with further display of search results on user's screen; methodological review of the structure and formulation of recommendations for the analysis of websites-competitors.

According to indicated research tasks, for their solutions it is necessary to: define the overall scheme of search and analyse the need for getting your website in the top of search results system; develop a methodology for analysing websites of competitors under the proposed factors. The solution of these problems will allow developing intelligent system, which will provide the following means: the accumulation of data as regards parameters of optimization and will provide opportunities for their systematization; analysis, implementation and support of new services of search engines; provision of recommendations on the best options for promoting websites.

2.1 Features of Search Mechanisms

The use of search engines has been increased and improved significantly in recent years, but the basic principles of the search procedure remain the same. In general, most search operations are of triple form: requirement, request and execution of request. The main components of which are as follows:

- requirement - involves the emergence of the need for a response or solution of the problem (in particular, the user can make navigation search (website search), transactional search (purchase) or information search (search of information);
- request - specifies the need to implement a search request in form of line of words or phrases;
- execution of request - defines the search results verification, if the result is inaccurate the specification is made, the ultimate goal of which is to meet the needs of the user.

We know that the most relevant pages in the top of search results are displayed in the upper left-hand corner of the screen. The above mentioned location is explained by the fact that while working at a computer, the focus of person is concentrated on the left-upper corner of the screen. As it can be noted the number of advertising as well as search results is displayed elsewhere. Research in this area ascertains the fact that such location is caused by the use of F-like pattern of eye movement (initially the focus is on the left-upper corner of the screen, then moves vertically down the first two or three search results). Then, the view moves across the page on the first result of paid page, then down another few results, and then back to the other results of paid page. As it can be noted in those areas the most relevant search pages and windows of contextual advertising are displayed.

The study shows that the top position of website directly affects the traffic that the resource receives, approximately 62% of people, who go to one of the results of the first page of search results, are satisfied with it. By summarizing the research [16], conducted during 2006–2016, it can be observed how the trend of clicks quantity has been changed depending on the position in the search results (Fig. 1).

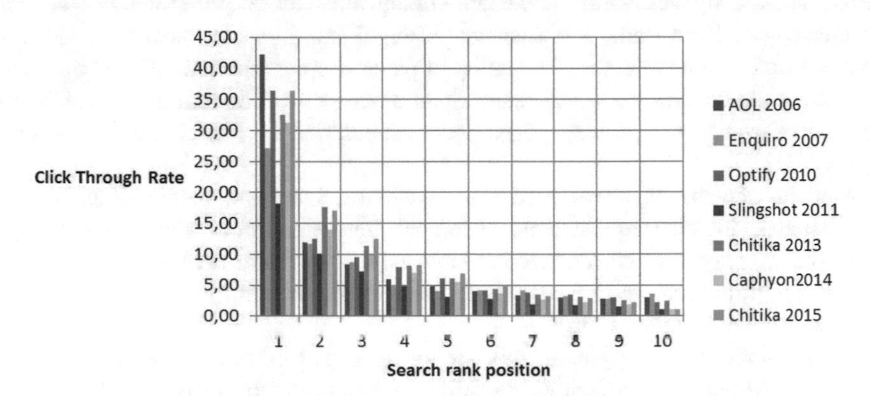

Fig. 1. Click through rate by search position

Consequently, it can be said that the absence of a website in the top search engine significantly reduces the number of entries on it and the number of potential clients on its pages.

2.2 Analysis of Competitors' Resources

Often there is a situation that execution of optimization measures as well as taking into account the recommendations of filtering algorithms (Google Panda, Google Penguin, Google Calibri [18]) website is located on the second or third page of search results. While the resource of competitor with a similar design and functionality occupies the top positions. This situation is typical and quite common on the World Wide Web spaces. In view of this, the task of analysing resource of competitor is important. Analysis of competitors' website firstly is necessary to understand what they do, how they move, what methods are used, how actively the process of popularizing is performed, which strategy and direction is chosen. As a result of conducted research the set, which includes the main factors that are necessary to be considered while analysing competitor's resources, is conducted:

$$Resources\ of\ competitors = \{Rl, Al, C, S\} \tag{1}$$

where Rl is a returning links, Al is a anchor links, C is a content, S – is a structure.

Returning Links

Returning links can be analysed by using various services, among the most influential the following are distinguished: SEOGadget (http://www.seogadget.ru/ backlinks), Majesticseo (https://majestic.com/) XSeo (http://xseo.in/smz), Ahrefs.com (https://ahrefs.com/). Mentioned services reflect the link that contain the specified competitor's domain link. By March 2016, besides this, resources reflect PR index of the resource [19]. Page Rank (PR) displayed information about the theoretical attendance of the page, which was calculated as the attendance probability of given web page by the user; at that the amount of probabilities on all web pages of a network was equal to one, because the user necessarily was visiting any page. Though the module operation for determining the meaning of Page Rank was suspended, PageRank algorithm still are used in the Google search engine to assess the quality of pages despite criticism of SEO optimizers concerning the creation of search spam Spamdexing [20]. The feature of which is the creation of websites and web pages for manipulation of search results in search engines [21].

Methods of webspam are divided into two classes: Spam of content and spam of link. The first class includes replacement of logical representation of how a search engine perceives the page content, the second is based on the features of the functioning of ranking algorithms, created based on links, such as resource rating which is higher than more ranked site is linked to it.

Most notable technologies at this are as follows: mirroring of sites (hosting of multiple websites with conceptually similar content but with the use of different URL - address); URL redirect (transferring of the user to another page without its intervention, for instance, with the help of meta updating of tags, flash, JavaScript, Java, etc.); masking (analysis of request variables in which the site content is transmitted by search engine different from what the user sees); keywords stuffing (includes location of keywords within a page, with the purpose of raising its importance for job search); wiki spam (spammers use open for editing wiki system with the aim to place links from the wiki site to the spam site).

Regarding the analysis of returning links of competitors, it is important to understand what kind of links they use, and which ones are most important to promote resource because uncontrolled increase in links mass can cause the opposite effect - the fall of rating.

Determination of the importance of the returning link is proposed to carry out in accordance with the developed method. Let Returning links – be the set of backlinks competitor's resource in the global network:

$$returning\ links = \{returning\ links_i\}_{i=1}^{N^{Rl}} \quad (2)$$

where, N^{Rl} – is the number of returning links. Each of returning links is described by corresponding topics of the resource to which they refer:

$$theme\ resource = \{theme\ resource_i\}_{i=1}^{N^{Tr}} \quad (3)$$

where, N^{Tr} – is the number of diverse topics of resource with returning links of competitor's website. Thus, the rate of the site importance is determined by the weight of returning links and is used by search engines to provide a higher rating for the resource. Search engine provides higher status to the website, to which the user addresses the maximum number of "quality" external links with significant weight. Given that let us define returning links as a set of resource topics and reference weight:

$$returning\ links_i = \{(theme\ resource_p, weight\ backlinks_{ip})\}_{p=1}^{N^{TrRLi}} \tag{4}$$

where, N^{TrRLi} – is the number of returning links in *thematic resource*;

weight backlinks$_{ip}$ – is the weight of reference in thematic resource of *returning links*.

The number of possible thematic resources to which the references can be done, creates thematic catalog of returning links, which can be determined by:

$$theme\ resource\ catague_i = \{(theme\ resource_p, n_{ip})\}_{i=1}^{N^{Trci}} \tag{5}$$

where, n_{ip} – is the number of pages that matches a resource theme; N^{Trci} – is the number of thematic resources of catalog.

Given that the feasibility of usage of competitor's separate returning link in the projected resource can be defined by the ratio:

$$\delta(analyzed\ site_i, competitor's\ site_z) = \frac{\sum_{p=1}^{N^{Trci}} weight\ backlinks_{zi} n_{ip}}{\sum_{p=1}^{N^{Trz}} weight\ backlinks_{zi}} \tag{6}$$

where,

$weight\ backlinks_{zi} = weight\ backlinks\ (theme\ resource\ catague_z, theme\ resource_i)$ – compliance of returning link weight and thematic resource, catalog.

The determination of actual weights of each separate returning links out of a plurality of thematic resources is an important objective of the analysis of this parameter for defining the feasibility of their use in projected resource.

Anchor Links (Texts of Links)

Besides determining links to competitors' websites it is necessary to understand by which phrase (or text of link) are they addressed. For this, it is necessary to apply advanced statistical service SemRush (https://www.semrush.com), which provides detailed information on these parameters (Fig. 2).

This service displays a list of link texts, by which one can link to the appropriate resource (in this case the National University of Lviv Polytechnic). As it can be seen from Fig. 2, the total number of link texts is equal to 1564, and this is the total number of unique Anchor on the website. Thus, statistics is available for every resource by which phrases they move and how they are distributed on the website.

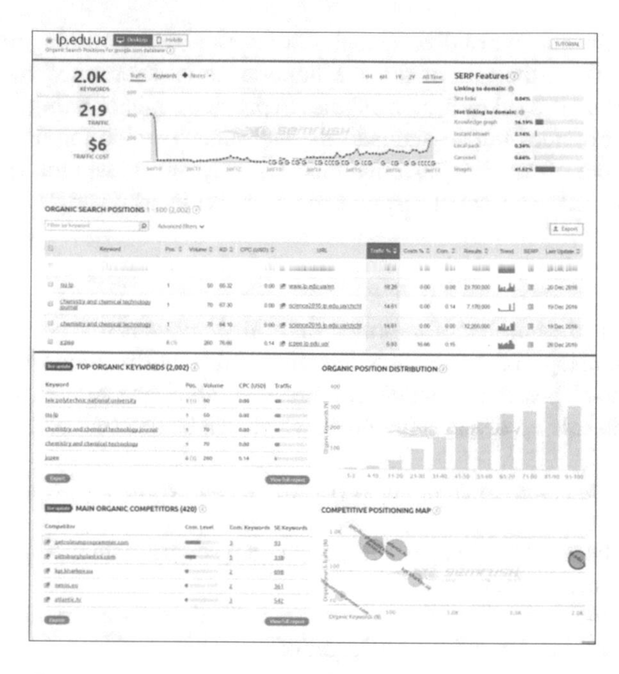

Fig. 2. Statistics of anchor links for the lp.edu.ua resource

Content

The website content is another important factor which should be analysed, namely its textual and graphical components. First of all, it is necessary to assess optimized text and its authenticity as well as originality of the used images. After analysing the set of resources, the following sources for filling site may be allocated:

- an independent content of website (author copywriting) - the best method of filling that promotes high uniqueness, both in terms of technical and semantic aspects, but the most labour-consuming in terms of writing time and the complexity of use;
- independent rewriting of texts from other sources - the most popular method because of the relative simplicity and accessibility. Advantage - the processing of finished material is easier than writing text from scratch, but the main drawback - by using rewrite website risks to lose some people who have already met such information;
- independent translation of information from "foreign" resources - is often used for the promotion of foreign news and popular media, the sufficiently high level of obtaining unique materials that are equally well rated by both users and search engines is the advantage, drawback - it requires considerable time to obtain permission from the text authors;
- independent processing of materials from paper sources - consists in scanning and reprinting of articles from various publications that are not yet published on the Internet. Advantage - the availability of a plurality of such sources that do not require the possession of significant knowledge on the topic of resource, drawback - the process complexity in typing and checking the availability of the resource in a global network;

- formation of content via visitors of website - this opportunity came with the advent of Web 2.0 and lies in the text filling with content of individual sections of the website (forums, comments, social networks) by users. Advantage - the automatic content filling of popular resource is the advantage, drawback - it is necessary constantly to maintain the popularity of the resource and perform moderator functions for the timely removal of false information.

There is no perfect method of filling the website content and each method has its advantages and disadvantages, so while designing the content of the website following basic principles should be observed: application of defined terms (consist in the use of expressions that are commonly-accepted and equally understandable to all community of the website); the feasibility of use (foresees the availability of corresponding elements and information relevant to thematic section); sustainability (determines the application of the same navigation elements on a website with the overall graphic design); division into parts (introduced information should be logically separated into components with its subsequent deployment in different subsections of the designed resource).

Structure

While analysing content resources of competitor it is very important to analyse the structure of the website, its code and contents. Since the structure of each Internet resource is closely connected with links, their clever location is the essential factor. Two aspects of assessment are considered here, developer and user. From the developer's side website structure is divided into two parts: logical (set of pages, united by a common design) and physical (an ordered set of files that form the backbone of the resource). Websites with failed logical structure not only complicate their viewing, but also contribute to the emergence of the idea that there is no stated information on the specified website.

The determination of competition level in the topic of key elements (words, phrases, etc.) directly influences the change of website structure. If the competition is considerable, it is necessary to perform the expansion of the structure of designed website at the expense of pages number increase, which is designed according to corresponding requests. Besides, it is necessary to increase statistic weight of pages data by performing relink operation. General method of website structure designing depending on keywords varieties is represented on Fig. 3.

Summing up the information (Fig. 3) it can be emphasized that in the case of highly competitive key phrases usage in designed website, it is appropriate to add some pages to the website structure with further promotion and use of internal and external links, in other cases – it is unnecessary to make any changes to the structure of the website.

Analysis of competitors' websites is an important step in the popularization of individual Internet resource that allows us to understand how competitors behave, and what should be done in order to go to the top position of search results.

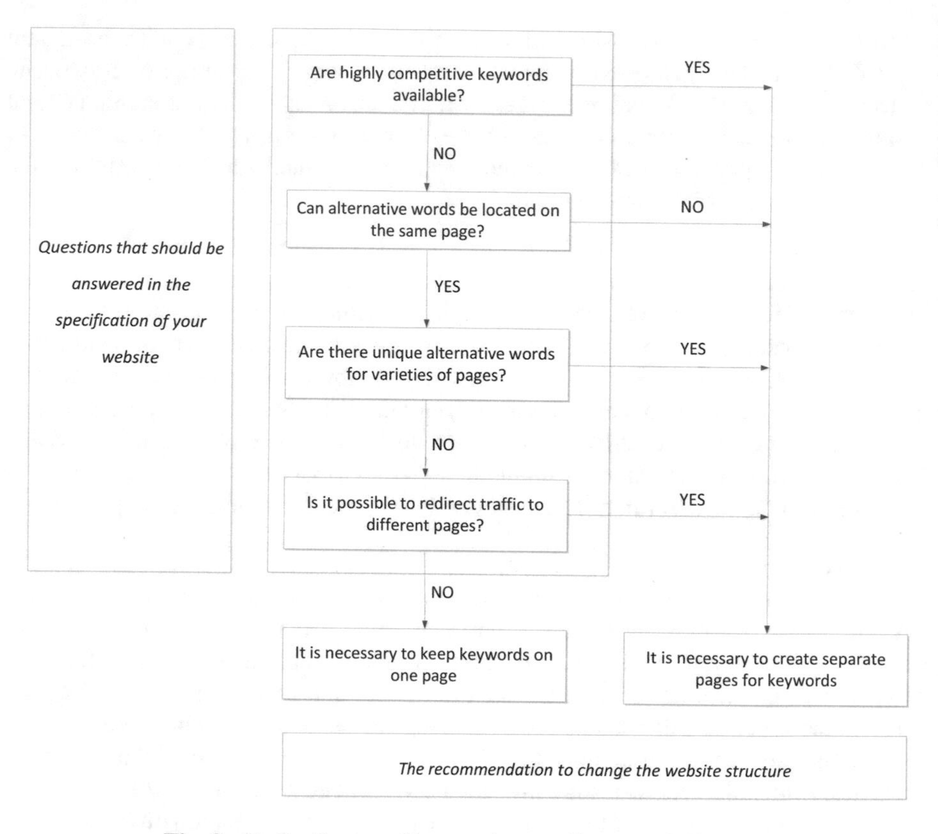

Fig. 3. Redistribution of keywords according to website pages

3 Conclusion

The analysis of important methodological reworks in the field of information content optimization as well as positioning of websites that showed practical lack of research process of identification the main factors of competition between them, is carried out herein. In view of this, the existent problems in the sphere of popularizing websites are considered, the features of search engines functionality are shown and feasibility of resources display in the top of search results is reasoned. The main factors (Returning links, Anchor links, Content, Structure), which should be taken into account during the analysis of websites of competitors are determined. The method of evaluating feasibility of applying some returning links depending on the theme of resource was developed. The general scheme of distribution of keywords on the pages of Internet resources is developed on the basis of analysis of known structures of websites and features of website structure construction are provided. The use of these measures promotes the popularization of website and its promotion in the top of search engine.

Further research will be aimed at designing individual modules of intelligent system of websites popularization and evaluating of their operation.

References

1. Grappone, J.: Search Engine Optimization (SEO): An Hour a Day. Wiley Publishing, Inc., New York (2013). 435 pages
2. Basyuk, T.M., Vasylyuk, A.S.: Ranking of websites on the Internet. Information systems and networks. Visnyk of the Lviv Polytechnic National University, Lviv № 770, pp. 3–12 (2013). (in Ukrainian)
3. Basyuk, T.: Influence of external factors on the position rankings website. In: Proceedings of the V International Scientific-Practical Conference on Information Control Systems and Technologies (ICST-ODESSA 2014), pp. 241–242 (2014)
4. Lella, A.: ComScore Releases February 2016 U.S. Desktop Search Engine Rankings. http://www.comscore.com/Insights/Rankings/comScore-Releases-February-2016-US-Desktop-Search-Engine-Rankings
5. Kennedy, G.: Seo: Marketing Strategies to Dominate the First Page (SEO, Social Media Marketing). CreateSpace Independent Publishing Platform (2016). 122 pages
6. Basyuk, T.: Web site promotion on Google. In: Proceedings of the XVIII International Scientific and Technical Conference on System Analysis and Information Technology, SAIT 2016, pp. 301–302 (2016)
7. Yakovlev, A., Tkachev, V.: Site Promotion. Fundamentals, secrets, tricks - SPb: BHV-SPb (2015). 352 pages. (in Russian)
8. Basyuk, T.: Popularization of website and without anchor promotion. In: Proceedings of the XI International Scientific and Technical Conference on Computer Science and Information Technologies, CSIT-2016, pp. 193–195. IEEE (2016). https://doi.org/10.1109/STC-CSIT.2016.7589904
9. Alpar, A., Koczy, M., Metzen, M.: SEO - Strategie, Taktik und Technik. Springer Gabler (2015). 538 pages. (in German)
10. Peleschishin, A.: Positioning Sites in the Global Information Environment. Publisher Lviv Polytechnic, Lviv (2007). 260 pages. (in Ukrainian)
11. Enge, E., Spencer, S., Stricchiola, J., Fishkin, R.: Mastering Search Engine Optimization. O'Reilly Media, Sebastopol (2015). 994 pages
12. Semenyshyn, V., Oleksiv, I.: Theoretical approaches to communications management in IT industry of Ukraine. Econtechmod Int. Q. J. **2**(4), 67–72 (2013). New technologies and modelling processes. Lublin, Rzeszow
13. Fedasyuk, D., Yakovyna, V., Serdyuk, P., Nytrebych, O.: Variables state-based software usage model. Econtechmod Int. Q. J. Econ. Technol. **3**(2), 15–20 (2014). New technologies and modelling processes. Lublin, Rzeszow
14. Basyuk, T.: The main reasons of attendance falling of internet resource. In: International Scientific and Technical Conference on Computer Science and Information Technologies CSIT-2015, pp. 91–93. IEEE (2015). https://doi.org/10.1109/STC-CSIT.2015.7325440
15. Nielsen: Electronic commerce in the world. http://www.nielsen.com/ua/uk/insights/reports/2016/E-commerce_worldwide.html
16. Petrescu, P.: Google organic click-through rates in 2014. https://moz.com/blog/google-organic-click-through-rates-in-2014
17. Searchengineland: Yahoo's ROBO study: search has big impact on offline purchases. http://searchengineland.com/yahoos-robo-study-search-has-big-impact-on-offline-purchases-11832
18. Slegg, J.: A complete guide to Panda, Penguin, and Hummingbird. https://www.searchenginejournal.com/seo-guide/google-penguin-panda-hummingbird

19. Schwartz, B.: Google has confirmed it is removing Toolbar PageRank. Search Engine Land. http://searchengineland.com/google-has-confirmed-they-are-removing-toolbar-pagerank-244230
20. Sullivan, D.: What is search engine spam? The video edition. http://searchengineland.com/what-is-search-engine-spam-the-video-edition-15202
21. Zoltan, G., Hector, G.-M.: Web spam taxonomy. In: First International Workshop on Adversarial Information Retrieval on the Web at the 14th International World Wide Web Conference, Chiba, Japan (2005)

Analysis of Uncertainty Types for Model Building and Forecasting Dynamic Processes

Peter Bidyuk[1(✉)], Aleksandr Gozhyj[2], Irina Kalinina[2], and Victor Gozhyj[2]

[1] National Technical University of Ukraine "Igor Sikorsky Kyiv Polytechnik Institute", Kiev, Ukraine
pbidyke@gmail.com

[2] Black Sea National University named after Petro Mohyla, Nikolaev, Ukraine
alex.gozhyj@gmail.com, irina.kalinina1612@gmail.com, gozhyi.v@gmail.com

Abstract. The article deals with the issues of handling uncertainties in the problems of modeling and forecasting dynamic systems within the framework of the dynamic planning methodology. To analyze and take into account possible structural, statistical and parametric uncertainties, the Kalman filter, various methods for calculating missing data, numerous methods for estimating the model parameters and the Bayesian approach to programming are used. The questions of an estimation of quality of predicted decisions are considered.

Keywords: Dynamic planning · Stochastic uncertainty · Kalman filter Uncertainties of model parameters · Uncertainties of a level (amplitude) Probabilistic uncertainties · Bayesian networks

1 Introduction

Analysis of dynamic processes in the planning and decision-making procedures is an urgent problem not only for financial organizations and companies but for all industrial enterprises, small and medium business, investment and insurance companies etc. This problem is solved by using dynamic programming methodology. Dynamic planning (DP) could be defined as the process of estimation by an enterprise of its current state on the market in comparison with other competing enterprises, and determining the further goals as well as sequences of actions and resources that are necessary for reaching the goals stated. This process of planning is performed continuously (or quasi-continuously) with acquiring new information (knowledge) about market, technologies, forecast estimates of necessary variables and situations. All this knowledge is used for correcting of actions and activities of the enterprise and supporting its competitiveness with flow of time.

Formally DP could be represented in the form:

$$DP = \{\mathbf{X}_0, \mathbf{G}, \mathbf{R}, \mathbf{D}(t), \mathbf{K}, \mathbf{T}, \mathbf{F}, \Delta\mathbf{D}(t), \Delta\mathbf{R}(t)\},$$

Where $\mathbf{X}_0$ is initial state of an enterprise; $\mathbf{G}$ are the goals stated by the enterprise management; $\mathbf{R}$ are resources that are necessary for reaching the goals stated. $\mathbf{D}(t)$ is a

N. Shakhovska and V. Stepashko (eds.), *Advances in Intelligent Systems and Computing II*, Advances in Intelligent Systems and Computing 689,
https://doi.org/10.1007/978-3-319-70581-1_5

sequence of actions that should be performed on the interval of planning; **K** is a new knowledge about environment; **T** are new technologies. **F** are results of forecasting and foresight; $\Delta\mathbf{D}(t)$ are corrections that are to be performed for reaching the goals; $\Delta\mathbf{R}(t)$ are necessary extra resources. One of the main problems that are to be solved within the DP paradigm is high quality forecasting of relevant processes.

Adequate models of the process and the forecasts generated with them help to take into consideration a set of various influencing factors and make objective planning managerial decisions. Another purpose of the studies is in estimating possible risks using forecasts of volatility. There are several types of processes that could be described with mathematical models in the form of appropriately constructed equations or probability distributions. Among them are the processes with deterministic and stochastic trends, and heteroscedastic processes. As of today the following mathematical models are widely used for describing nonlinear dynamics of processes relevant to planning: linear and nonlinear regression (logit and probit, polynomials, splines), autoregressive integrated moving average (ARIMA) models, autoregressive conditionally heteroscedastic models (ARCH), generalized ARCH (GARCH), dynamic Bayesian networks, support vector machine (SVM) approach, neural networks and neuro-fuzzy techniques as well as combinations of the approaches mentioned [1–5].

All types of mathematical modeling usually need to cope with various kinds of uncertainties related to statistical data, structure of the process under study and its model, parameter uncertainty, and uncertainties relevant to the models and forecasts quality. Reasoning and decision making are very often performed with leaving many facts unknown or rather vaguely represented in processing of data and expert estimates. To avoid or to take into consideration the uncertainties and improve this way quality of the final result (processes forecasts and the planning decisions based upon them) it is necessary to construct appropriate computer based decision support systems (DSS) for solving multiple specific problems.

Selection and application of a specific model for process description and forecasts estimation depends on application area, availability of statistical data, qualification of personnel, who work on the data analysis problems, and availability of appropriate applied software. Better results for estimation of processes forecasts is usually achieved with application of ideologically different techniques combined in the frames of one computer system. Such approach to solving the problems of quality forecasts estimation can be implemented in the frames of modern decision support systems (DSS). DSS today is a powerful instrument for supporting user's (managerial) decision making as far as it combines a set of appropriately selected data and expert estimates processing procedures aiming to reach final result of high quality: objective high quality alternatives for a decision making person (DMP). Development of a DSS is based on modern theory and techniques of system analysis, data processing systems, estimation and optimization theories, mathematical and statistical modeling and forecasting, decision making theory as well as many other results of theory and practice of processing data and expert estimates [6–8].

The paper considers the problem analysis, accounting and handling of uncertainties for solving the problems of modeling and estimating forecasts for selected types of dynamic processes with the possibility for application of alternative data processing techniques, modeling and estimation of parameters and states for the processes under study.

2 Problem Formulation

The purpose of the study is as follows: (1) analysis of uncertainty types characteristic for model building and forecasting dynamic processes; (2) selection of techniques for taking into consideration of the uncertainties; (3) selection of mathematical modeling and forecasting techniques for nonstationary heteroscedastic processes.

3 Coping with Uncertainties

All types of mathematical modeling with the use of statistical experimental data usually need to consider various kinds of uncertainties caused by data, informational structure of a process under study and its model, parameter uncertainty, and uncertainties relevant to the quality of models and forecasts. In many cases a researcher has to cope with the following basic types of uncertainties: structural, statistical and parametric. Structural uncertainties are encountered in the cases when structure of the process under study (and respectively its model) is unknown or not clearly enough defined (known partially). For example, when the functional approach to model constructing is applied usually we do not know object (or a process) structure, it is estimated with appropriate model structure estimation techniques: correlation analysis, estimation of mutual information, estimation of lags, testing for nonlinearity and nonstationarity, identification of external disturbances etc. Uncertainty could also be introduced by an expert who is studying the process and provides its estimates for model structure, parameter restrictions, selection of computational procedures etc. The sequence of actions necessary for identification, processing and taking into consideration of uncertainties could be formulated as follows: – identification and reduction of data uncertainty; – model structure and parameters estimation; – reduction of uncertainties related to the model structure and parameters estimation; – reduction of uncertainties relevant to expert estimates; – estimation of forecasts and reduction of respective uncertainties; – selection of the best final result. All the tasks mentioned above are usually solved sequentially (in an adaptive loop) with appropriately designed and implemented DSS.

We consider uncertainties as the factors that influence negatively the whole process of mathematical model constructing, forecasts and possible risk estimating and generating of alternative decisions. They are inherent to the process being studied due to incomplete or noise corrupted data, complex stochastic external influences, incompleteness or inexactness of our knowledge regarding the objects (systems) structure, incorrect application of computational procedures etc. The uncertainties very often appear due to incompleteness of data, noisy measurements or they are invoked by sophisticated stochastic external disturbances with complex unknown probability distributions, poor estimates of model structure or by a wrong selection of parameter estimation procedure. The problem of uncertainties identification is solved with application of special statistical tests and visual studying of available data.

As far as we usually work with stochastic data, correct application of existing statistical techniques provides a possibility for approximate estimation of a system (and its model) structure. To find "the best" model structure it is recommended to apply adaptive estimation schemes that provide automatic search in a pre-defined range of model structures and parameters (model order, time lags, and possible nonlinearities). It is often possible to perform the search in the class of regression type models with the use of information criterion of the following type [2]:

$$N \log(FPE) = N \log(V_N(\hat{\theta})) + N \log\left(\frac{N+p}{N-p}\right), \tag{1}$$

where $\hat{\theta}$ is a vector of model parameters estimates; N is a power of time series used; FPE is final prediction error term; $V_N(\hat{\theta})$ is determined by the sum of squared errors; p is a number of model parameters. The value of the criteria (1) is asymptotically equivalent to the Akaike information criterion with $N \to \infty$. As the amount of data, N, may be limited, then an alternative, the minimum description length (MDL) criterion

$$MDL = \log(V_N(\hat{\theta})) + p\frac{\log(N)}{N}$$

could be hired to find the model that adequately represents available data with the minimum amount of available information.

There are several possibilities for adaptive model structure estimation: (1) application of statistical criteria for detecting possible nonlinearities and the type of nonstationarity (integrated or heteroskedastic process); (2) analysis of partial autocorrelation for determining autoregression order; (3) automatic estimation of the exogeneous variable lag (detection of leading indicators); (4) automatic analysis of residual properties; (5) analysis of data distribution type and its use for selecting correct model estimation method; (6) adaptive model parameter estimation with hiring extra data; (7) optimal selection of weighting coefficients for exponential smoothing, nearest neighbor and other techniques. The development and use of a specific adaptation scheme depends on volume and quality of data, specific problem statement, requirements to forecast estimates etc.

The adaptive estimation schemes also help to cope with the model parameters uncertainties. New data are used to re-compute model parameter estimates that correspond to possible changes in the object under study. In the cases when model is nonlinear alternative parameter estimation techniques (say, MCMC) could be hired to compute alternative (though admissible) sets of parameters and to select the most suitable of them using statistical quality criteria.

3.1 Processing Some Types of Stochastic Uncertainties

While performing practical modeling very often statistical characteristics (covariance matrix) of stochastic external disturbances and measurement noise (errors) are unknown. To eliminate this uncertainty optimal filtering algorithms are usually applied that provide for a possibility of simultaneous estimation of object (system) states and the covariance matrices. One of the possibilities to solve the problem is optimal Kalman filter. Kalman filter is used to find optimal estimates of system states on the bases of the system model represented in a convenient state space form as follows:

$$\mathbf{x}(k) = \mathbf{A}(k, k-1)\mathbf{x}(k-1) + \mathbf{B}(k, k-1)\mathbf{u}(k-1) + \mathbf{w}(k) \tag{2}$$

where $\mathbf{x}$(k) is n-dimensional vector of system states; $k = 0,1,2,\ldots$ is discrete time; $\mathbf{u}(k-1)$ is m – dimensional vector of deterministic control variables; $\mathbf{w}(k)$ is n - dimensional vector of external random disturbances; $\boldsymbol{A}(k,\ k-1)$ is $(n \times m)$ - matrix of system dynamics; is $\boldsymbol{B}(k,\ k-1)$ $(n \times m)$ matrix of control coefficients. The double argument $(k,\ k-1)$ means that the variable or parameter is used at the moment k, but its value is based on the former (earlier) data processing including moment $(k-1)$. Usually the matrices $\boldsymbol{A}$ and $\boldsymbol{B}$ are written with onc argument like, $\boldsymbol{A}(k)$ and $\boldsymbol{B}(k)$, to simplify the text. Besides the main task, optimal state estimation, Kalman filter can be used to solve the following problems: computing of short-term forecasts, estimation of unknown model parameters including statistics of external disturbances and measurement errors (adaptive extended Kalman filter), estimation of state vector components that cannot be measured directly, and fusion of data coming from various external sources.

Obviously stationary system model is described with constant parameters like $\boldsymbol{A}$, and $\boldsymbol{B}$. As far as matrix $\boldsymbol{A}$ is a link between two consequent system states, it is also called state transition matrix. Discrete time k and continuous time t are linked to each other via data sampling time: $\boldsymbol{T}_s : t = k\,\boldsymbol{T}_s$. In the classic problem statement for optimal filtering the vector sequence of external disturbances $\mathbf{w}(k)$ is supposed to be zero mean white Gaussian noise with covariance matrix $\mathbf{Q}$, i.e. the noise statistics are as follows:

$$E[\mathbf{w}(k)] = 0, \quad \forall k; \quad E[\mathbf{w}(k)\mathbf{w}^T(j)] = \mathbf{Q}(k)\delta_{kj},$$

where δ_{kj} is Kronecker delta-function: $\delta_{kj} = \begin{cases} 0, & k \neq j \\ 1, & k = j \end{cases}$; $\mathbf{Q}(k)$ is positively defined covariance $(n \times n)$ matrix. The diagonal elements of the matrix are variances for the components of disturbance vector $\mathbf{w}(k)$. Initial system state $\mathbf{x}_0$ is supposed to be known and the measurement equation for vector $\mathbf{z}(k)$ of output variables is described by the equation:

$$\mathbf{z}(k) = \mathbf{H}(k)\mathbf{x}(k) + \mathbf{v}(k), \tag{3}$$

where $\mathbf{H}(k)$ is $(r \times n)$ observation (coefficients) matrix; $\mathbf{v}(k)$ is r-dimensional vector of measurement noise with statistics: $E[\mathbf{v}(k)] = 0, E[\mathbf{v}(k)\mathbf{v}^T(j)] = \mathbf{R}(k)\delta_{kj}$,

where $R(k)$ is $(r \times r)$ positively defined measurement noise covariance matrix, the diagonal elements of which represent variances of additive noise for each measurable variable. The noise of measurements is also supposed to be zero mean white noise sequence that is not correlated with external disturbance $\mathbf{w}(k)$ and initial system state. For the system (2), (3) with state vector $\mathbf{x}(k)$ it is necessary to find optimal state estimate $\hat{\mathbf{x}}(k)$ at arbitrary moment k as a linear combination of estimate $\hat{\mathbf{x}}(k-1)$ at the previous moment $(k-1)$ and the last measurement available, $\mathbf{z}(k)$. The estimate of state vector $\hat{\mathbf{x}}(k)$ is computed as optimal one with minimizing the expectation of the sum of squared errors, i.e.:

$$E[(\hat{\mathbf{x}}(k) - \mathbf{x}(k))^T(\hat{\mathbf{x}}(k) - \mathbf{x}(k))] = \min_K, \tag{4}$$

where $\mathbf{x}(k)$ is an exact value of state vector that can be found as deterministic part of the state Eq. (2); $\boldsymbol{K}$ is optimal matrix gain that is determined as a result of minimizing quadratic criterion (4).

Thus, the filter is constructed to compute optimal state vector $\hat{\mathbf{x}}(k)$ in conditions of influence of external random system disturbances and measurement noise. Here one of possible uncertainties arises when we don't know estimates of covariance matrices $\mathbf{Q}$ and $\mathbf{R}$. To solve the problem an adaptive Kalman filter is to be constructed that allows to compute estimates of $\hat{\mathbf{Q}}$ and $\hat{\mathbf{R}}$ simultaneously with the state vector $\hat{\mathbf{x}}(k)$. Another choice is in constructing separate algorithm for computing the values of $\hat{\mathbf{Q}}$ and $\hat{\mathbf{R}}$. A convenient statistical algorithm for estimating the covariance matrices was proposed [11]:

$$\hat{\mathbf{R}} = \frac{1}{2}\left[\hat{\mathbf{B}}_1 + \mathbf{A}^{-1}(\hat{\mathbf{B}}_1 - \hat{\mathbf{B}}_2)(\mathbf{A}^{-1})^T\right],$$
$$\hat{\mathbf{Q}} = \hat{\mathbf{B}}_1 - \hat{\mathbf{R}} - \mathbf{A}\hat{\mathbf{R}}\mathbf{A}^T,$$

where $\hat{\mathbf{B}}_1 = E\{[\mathbf{z}(k) - \mathbf{A}\,\mathbf{z}(k-1)]\,[\mathbf{z}(k) - \mathbf{A}\,\mathbf{z}(k-1)]^T\}$; $\hat{\mathbf{B}}_2 = E\{[\mathbf{z}(k) - \mathbf{A}^2\,\mathbf{z}(k-2)]\,[\mathbf{z}(k) - \mathbf{A}^2\,\mathbf{z}(k-2)]^T\}$.

The matrices $\hat{\mathbf{Q}}$ and $\hat{\mathbf{R}}$ are used in the optimal filtering procedure as follows:

$$\mathbf{S}(k) = \mathbf{A}\mathbf{P}(k-1)\mathbf{A}^T + \hat{\mathbf{Q}};$$
$$\Delta(k) = \mathbf{S}(k)[\mathbf{S}(k) + \hat{\mathbf{R}}]^{\#};$$
$$\mathbf{P}(k) = [\mathbf{I} - \Delta(k)]\mathbf{S}(k), \quad k = 0, 1, 2, \cdots,$$

where $\mathbf{S}(k)$ and $\mathbf{P}(k)$ are prior and posterior covariance matrices of estimates errors respectively; the symbol "#" denotes pseudo-inverse; $\mathbf{A}^T$ means matrix transposition; $\Delta(k)$ is a matrix of intermediate covariance results. The algorithm was successfully applied to the covariance estimating in many practical applications. The computation experiments showed that the values of $\Delta(k)$ become stationary after about 20–25 periods of time (sampling periods) in a scalar case, though this figure is growing substantially with the growth of dimensionality of the system under study. It was also determined that the parameter estimators are very sensitive to the initial conditions of the system. The initial conditions should differ from zero enough to provide stability for the estimates generated.

Other appropriate instruments for taking into consideration the uncertainties are fuzzy logic, neuro-fuzzy models, Bayesian networks, appropriate types of distributions etc. Some of statistical data uncertainties such as missing measurements, extreme values and high level jumps of stochastic origin could be processed with appropriately selected statistical procedures. There exists a number of data imputation schemes that help to complete the sets of the data collected. For example, very often missing measurements for time series could be generated with appropriately selected distributions or in the form of short term forecasts. Appropriate processing of jumps and extreme values helps with adjusting data nonstationarity and to estimate correctly the probability distribution for the stochastic processes under study.

3.2 Processing Data with Missing Observations (Data Are in the Form of Time Series)

As of today for the data in the time series form the most suitable imputation techniques are as follows: simple averaging when it is possible (when only a few values are missing); generation of forecast estimates with the model constructed using available measurements; generation of missing estimates from distributions the form and parameters of which are again determined using available part of data and expert estimates; the use of optimization techniques, say appropriate forms of EM-algorithms (expectation maximization); exponential smoothing etc. It should also be mentioned that optimal Kalman filter can also be used for imputation of missing data because it contains "internal" forecasting function that provides a possibility for generating quality short-term forecasts [12]. Besides, it has a feature of fusion the data coming from various external sources and improving this way the quality of state vector and its forecasts.

Further reduction of this uncertainty is possible thanks to application of several forecasting techniques to the same problem with subsequent combining of separate forecasts using appropriate weighting coefficients. The best results of combining the forecasts is achieved when variances of forecasting errors for different forecasting techniques do not differ substantially (at any rate the orders of the variances should be the same).

3.3 Coping with Uncertainties of Model Parameters Estimates

Usually uncertainties of model parameter estimates such as bias and inconsistency result from low informative data, or data do not correspond to normal distribution, what is required in a case of LS application for parameter estimation. This situation may also take place in a case of multicollinearity of independent variables and substantial influence of process nonlinearity that for some reason has not been taken into account when model was constructed. When power of the data sample is not satisfactory for model construction it could be expanded by applying special techniques, or simulation is hired, or special model building techniques, such as group method for data handling (GMDH), are applied. Very often GMDH produces results of acceptable quality with rather short samples. If data do not correspond to normal distribution, then ML technique could be used or appropriate Monte Carlo procedures for Markov Chains (MCMC) [13]. The last techniques could be applied with quite acceptable computational expenses when the number of parameters is not large.

3.4 Dealing with Model Structure Uncertainties

When considering mathematical models it is convenient to use proposed here a unified notion of a model structure which we define as follows: $S = \{r, p, m, n, d, w, l\}$, where r is model dimensionality (number of equations); p is model order (maximum order of differential or difference equation in a model); m is a number of independent variables in the right hand side of a model; n is a nonlinearity and its type; d is a lag or output reaction delay time; is stochastic external disturbance and its type; l are possible restrictions for the variables and/or parameters. When using DSS, the model structure should practically always be estimated using data. It means that elements of the model structure accept almost always only approximate values. When a model is constructed for forecasting we build several candidates and select the best one of them with a set model quality statistics. Generally we could define the following techniques to fight structural uncertainties: gradual improvement of model order (AR(p) or ARMA(p, q)) applying adaptive approach to modeling and automatic search for the "best" structure using complex statistical quality criteria; adaptive estimation (improvement) of input delay time (lag) and data distribution type with its parameters; describing detected process nonlinearities with alternative analytical forms with subsequent estimation of model adequacy and forecast quality. As another example of complex statistical model adequacy and forecast quality criterion could be the following:

$$J = \left|1 - R^2\right| + \alpha \ln\left[\sum_{k=1}^{N} e^2(k)\right] + |2 - DW| + \beta \ln(1 + MAPE) + U \rightarrow \min_{\hat{\theta}_i},$$

where R^2 is a determination coefficient; DW is Durbin-Watson statistic; $MAPE$ is mean absolute percentage error for forecasts;

$\sum_{k=1}^{N} e^2(k) = \sum_{k=1}^{N} [y(k) - \hat{y}(k)]^2$ is sum of squared model errors; U is Theil coefficient that measures forecasting characteristic of a model; α, β are appropriately selected weighting coefficients; $\hat{\theta}_i$ is parameter vector for the *i-th* candidate model. A criterion of this type is used for automatic selection of the best candidate model. The criterion also allows operation of DSS in the automatic adaptive mode. Obviously, other forms of the complex criteria are possible. While constructing the criterion it is important not to overweigh separate members in the right hand side.

3.5 Coping with Uncertainties of a Level (Amplitude) Type

The use of random (i.e. with random amplitude or a level) and/or non-measurable variables leads to necessity of hiring fuzzy sets for describing such situations. The variable with random amplitude can be described with some probability distribution if the measurements are available or they come for analysis in acceptable time span. However, some variables cannot be measured (registered) in principle, say amount of shadow capital that "disappears" every month in offshore, or amount of shadow salaries paid at some company, or a technology parameter that cannot be measures on-line due to absence of appropriate gauge. In such situations we could assign to the variable a set of possible values in the linguistic form as follows: *capital amount* = {*very low*, *low*, *medium*, *high*, *very high*}. There exists a complete necessary set of mathematical operations to be applied to such fuzzy variables. Finally fuzzy value could be transformed into usual exact form using known techniques.

3.6 Processing Probabilistic Uncertainties

To fight probabilistic uncertainties it is possible to hire Bayesian approach that helps to construct models in the form of conditional distributions for the sets of random variables. Usually such models represent the process (under study) variables themselves, stochastic disturbances and measurement errors or noise. The problem of distribution type identification also arises in regression modeling. Each probability distribution is characterized by a set of specific values that random variable could take and the probabilities for these values. The problem is in the distribution type identification and estimating its parameters. The probabilistic uncertainty (will some event happen or not) could be solved with various models of Bayesian type. This approach is known as Bayesian programming or paradigm. The generalized structure of the Bayesian program includes the following steps: (1) problem description and statement with putting the question regarding estimation of conditional probability in the form: $p\,(X_i|D,\,Kn)$, where X_i - is the main (goal) variable or event; the probability p should be found as a result of application of some probabilistic inference procedure; (2) statistical (experimental) data D and knowledge K_n are to be used for estimating model and parameters of specific type; (3) selected and applied probabilistic inference technique should give an answer to the question put above; (4) analysis of quality of the final result. The steps given above are to some extent "standard" regarding model constructing and computing probabilistic inference using statistical data available. This sequence of actions is naturally consistent with the methods of cyclic structural and parametric model adaptation to the new data and operating modes (and possibly expert estimates).

One of the most popular Bayesian approaches today is created by the models in the form of static and dynamic Bayesian networks (BN). Bayesian networks are probabilistic and statistical models represented in the form of directed acyclic graphs (DAG) with vertices as variables of an object (system) under study, and the arcs showing existing causal relations between the variables. Each variable of BN is characterized with complete finite set of mutually excluding states. Formally BN could be represented with the four following components: $\mathbf{N} = <\mathbf{V}, \mathbf{G}, \mathbf{P}, \mathbf{T}>$, where $\mathbf{V}$ stands for the set of model variables; $\mathbf{G}$ represents directed acyclic graph; $\mathbf{P}$ is joint distribution of probabilities for the graph variables (vertices), $\mathbf{V} = \{X_1, \ldots, X_n\}$; and $\mathbf{T}$ denotes conditional and unconditional probability tables for the graphical model variables [14, 15]. The relations between the variables are established via expert estimates or applying special statistical and probabilistic tests to statistical data (when available) characterizing dynamics of the variables.

The process of constructing BN is generally the same as for models of other types, say regression models. The set of the model variables should satisfy the Markov condition that each variable of the network does not depend on all other variables but for the variable's parents. In the process of BN constructing first the problem is solved of computing mutual information values between all variables of the net. Then an optimal BN structure is searched using acceptable quality criterion, say well-known minimum description length (MDL) that allows analyzing and improving the graph (model) structure at each iteration of the learning algorithm applied. Bayesian networks provide the following advantages for modeling: the model may include qualitative and quantitative variables simultaneously as well as discrete and continuous ones; number of the variables could be very large (thousands); the values for conditional probability tables could be computed with the use of statistical data and expert estimates; the methodology of BN constructing is directed towards identification of actual causal relations between the variables hired what results in high adequacy of the model; the model is also operable in conditions of missing data.

To reduce an influence of probabilistic and statistical uncertainties on models quality and the forecasts based upon them it is also possible to use the models in the form of Bayesian regression based on analysis of actual distributions of model variables and parameters. Consider a simple two variables regression

$$y(k)|x(k) = \beta_1 + \beta_2 x(k) + u(k), \quad k = 0, 1, \ldots, n.$$

It is supposed that of random values $u_1, \ldots, u_n$ are independent and belong, for example, to normal distribution, $\{u(k)\} \sim N(0, \sigma_u^2)$; here vector of unknown parameters includes three elements, $\theta = (\beta_1, \beta_2, \sigma_u^2)^T$. The likelihood function for dependent variable $\mathbf{y} = (y_1, \ldots, y_n)^T$ and predictor $\mathbf{x} = (x_1, \ldots, x_n)^T$ without proportion coefficient is determined as follows:

$$L(\mathbf{y}|\mathbf{x}, \beta_1, \beta_2, \sigma_u) = \frac{1}{\sigma_u^N} \exp\left\{ -\frac{1}{2\sigma_u^2} \sum_{k=1}^{N} [y(k) - \beta_1 - \beta_2 x(k)]^2 \right\}$$

Using simplified (non-informative) distributions for the model parameters:

$$g(\beta_1, \beta_2, \sigma_u) = g_1(\beta_1) g_2(\beta_2) g_3(\sigma_u),$$

$$g_1(\beta_1) \propto const,$$

$$g_2(\beta_{21}) \propto const,$$

$$g_3(\sigma_u) \propto 1/\sigma_u,$$

and Bayes theorem it is possible to find joint posterior distribution for the parameters in the form [16]:

$$h(\beta_1, \beta_2, \sigma_u | x, y) \propto \frac{1}{\sigma} \frac{1}{\sigma^N} \exp\left[-\frac{1}{2\sigma^2} \sum_{k=1}^{N} (y(k) - \beta_1 - \beta_2 x(k))^2 \right],$$

$$-\infty < \beta_1, \beta_2 < +\infty, \quad 0 < \sigma_u < \infty$$

Maximum likelihood estimates for the model parameters are determined as follows:

$$\hat{\beta}_1 = \bar{y} - \hat{\beta}_2 \bar{x}; \quad \hat{\beta}_2 = \frac{\sum_{k=1}^{N} [x(k) - \bar{x}][y(k) - \bar{y}]}{\sum_{k=1}^{N} [x(k) - \bar{x}] \sum_{k=1}^{N} [y(k) - \bar{y}]},$$

where $\bar{x} = N^{-1}\sum_{k=1}^{N} x(k)$, $\bar{y} = N^{-1}\sum_{k=1}^{N} y(k)$, with unbiased sample estimate of variance:

$$\hat{\sigma}_u^2 = s^2 = \frac{1}{N-2}\sum\nolimits_{k=1}^{N} [y(k) - \hat{\beta}_1 - \hat{\beta}_2 x(k)]$$

Joint posterior density for the model parameters corresponds to two dimensional Student distribution:

$$h_1(\beta_1, \beta_2|\mathbf{y}, \mathbf{x}) \propto \left\{(N-2)s^2 + N(\beta_1 - \hat{\beta}_1)^2 + (\beta_2 - \hat{\beta}_2)^2 \sum\nolimits_{k=1}^{N} x(k)^2 + 2(\beta_1 - \hat{\beta}_1)(\beta_2 - \hat{\beta}_2)\sum\nolimits_{k=1}^{N} x(k)\right\}^{-0,5N}.$$

This way we get a possibility for using more exact distributions of models variables and parameters what helps to enhance model quality. Using new observation x^* and prior information regarding particular model it is possible to determine the forecast interval for the dependent variable, y^*, as follows:

$$p(y^*|x^*) = \iiint L(y^*|x^*, \beta_1, \beta_2, \sigma)h(\beta_1, \beta_2, \sigma)|\mathbf{x}, \mathbf{y})d\beta_1, d\beta_2, d\sigma.$$

Another useful Bayesian approach is in hierarchical modeling that is based on a set of simple conditional distributions comprising one model. The approach is naturally combined with the theory of computing Bayesian probabilistic inference using modern computational procedures [17]. The hierarchical models belong to the class of marginal models where the final result is provided in the form of a distribution ***P*(y)**, where **y** is available data vector. The models are formed from the sequence of conditional distributions for selected variables including the hidden ones. The hierarchical representation of parameters usually supposes that data, **y**, is situated at the lower (first) level, model parameters (second level) $\theta = (\theta_i, i = 1, 2, \ldots, n)$, $\theta_i \sim N(\mu, \tau^2)$, determine distributions of dependent variables $y_i \sim N(\theta_i, \sigma^2), i = 1, 2, \ldots, n$, and parameters $\{\theta_i\}$ are determined by the pair, (μ, τ^2), of the third level. Supposing the parameters σ^2 and τ^2 accept known finite values, and parameter μ is unknown with the prior π_μ, then joint prior density for (θ, μ) could be presented in the form: $\pi_\mu(\mu)\prod_i \pi_\theta(\theta_i|\mu)$, and the prior for parameter vector θ will be defined by the integral: $p(\theta) = \int \pi_\mu(\mu)\prod\limits_i \pi_\theta(\theta_i|\mu)d\mu$.

4 Data, Model and Forecasts Quality Criteria

To achieve reliable high quality final result of risk estimation and forecasting at each stage of computational hierarchy separate sets of statistical quality criteria have been used. Data quality control is performed with the following criteria:

- database analysis for missing values using developed logical rules, and imputation of missed values with appropriately selected techniques;

- analysis of data for availability of outliers with special statistical tests, and processing of outliers to reduce their negative influence on statistical properties of the data available;
- normalizing of data in the selected range in a case of necessity;
- application of low-order digital filters (usually low-pass filters) for separation of observations from measurement noise;
- application of optimal (usually Kalman) filters for optimal state estimation and fighting stochastic uncertainties;
- application of principal component method to achieve desirable level of orthogonalization between the variables selected;
- computing of extra indicators for the use in regression and other models (say, moving average processes based upon measurements of dependent variables).

It is also useful to test how informative is the data collected. Very formal indicator for the data being informative is its sample variance. It is considered formally that the higher is the variance the richer is the data with information. Another criterion is based on computing derivatives with a polynomial that describes data in the form of a time series. For example, the equation given below can describe rather complex process with nonlinear trend and short-term variations imposed on the trend curve:

$$y(k) = a_0 + \sum_{i=1}^{p} a_i y(k-i) + c_1 k + c_2 k^2 + \ldots + c_m k^m + \varepsilon(k),$$

Where $y(k)$ is basic dependent variable; a_i, c_i are model parameters; $k = 0, 1, 2, \ldots$ is discrete time; $\varepsilon(k)$ is a random process that integrates the influence of external disturbances to the process being modeled as well as model structure and parameters errors. Autoregressive part of model (1) describes the deviations that are imposed on a trend, and the trend itself is described with the m-th order polynomial of discrete time k. In this case maximum number of derivatives could be m, though in practice actual number of derivatives is defined by the largest number i of parameter c_i, that is statistically significant. To select the best model constructed the following statistical criteria are used: determination coefficient (R^2); Durbin-Watson statistic (DW); Fisher F-statistic; Akaike information criterion (AIC), and residual sum of squares (SSE). The forecasts quality is estimated with hiring the criteria mentioned above in (1) and (2). To perform automatic model selection the above mentioned combined criteria (1) could be hired. The power of the criterion was tested experimentally and proved with a wide set of models and statistical data. Thus, the three sets of quality criteria are used to insure high quality of final result.

5 Conclusions

The general methodology was proposed for mathematical modeling and forecasting dynamics of economic and financial processes that is based on the system analysis principles. As instrumentation for fighting possible structural, statistic and parametric uncertainties the following techniques are used: Kalman filter, various missing data imputation techniques, multiple methods for model parameter estimation, and Bayesian programming approach. The issues of estimating the quality of forecasted solutions are considered.

References

1. Tsay, R.S.: Analysis of Financial Time Series. Wiley, Chicago (2010). 715 pages
2. Harris, L., Hong, X., Gan, Q.: Adaptive Modeling, Estimation and Fusion from Data. Springer, Heidelberg (2002). 323 pages
3. Congdon, P.: Applied Bayesian Modeling. Wiley, Chichester (2003). 472 pages
4. De Lurgio, S.M.: Forecasting Principles and Applications. McGraw-Hill, Boston (1998). 802 pages
5. Taylor, S.J.: Modeling stochastic volatility: a review and comparative study. Math. Fin. **2**, 183–204 (1994)
6. Burstein, F., Holsapple, C.W.: Handbook of Decision Support Systems. Springer-Verlag, Heidelberg (2008). 908 pages
7. Hollsapple, C.W., Winston, A.B.: Decision Support Systems. West Publishing Company, Saint Paul (1996). 860 pages
8. Bidyuk, P.I., Gozhyj, A.P.: Computer Decision Support Systems. Black Sea State University named after Petro Mohyla, Mykolaiv (2012). 380 pages
9. Xekalaki, E., Degiannakis, S.: ARCH Models for Financial Applications. Wiley, Chichester (2010). 550 pages
10. Chatfield, C.: Time Series Forecasting. Chapman & Hall, CRC, Boca Raton (2000). 267 pages
11. Anderson, W.N., Kleindorfer, G.B., Kleindorfer, P.R., Woodroofe, M.B.: Consistent estimates of the parameters of a linear system. Ann. Math. Stat. **40**, 2064–2075 (1969)
12. Gibbs, B.P.: Advanced Kalman Filtering, Least-Squares and Modeling, Wiley, Hoboken (2011). 267 pages
13. Gilks, W.R., Richardson, S., Spiegelhalter, D.J.: Markov Chain Monte Carlo in Practice. Chapman & Hall, CRC, New York (2000). 486 pages
14. Jensen, F.V., Nielsen, T.D.: Bayesian Networks and Decision Graphs. Springer, New York (2007). 457 pages
15. Zgurovsky, M.Z., Bidyuk, P.I., Terentyev, O.M., Prosyankina-Zharova, T.I.: Bayesian Networks in Decision Support Systems. Edelweis, Kyiv (2015). 300 pages
16. Bernardo, J.M., Smith, A.F.M.: Bayesian Theory. Wiley, New York (2000). 586 pages
17. Bolstad, W.M.: Understanding Computational Bayesian Statistics. Wiley, Hoboken (2010). 334 pages

Dynamic Inertia Weight in Particle Swarm Optimization

Bożena Borowska(✉)

Institute of Information Technology, Lodz University of Technology,
Wólczańska 215, 90-924 Łódź, Poland
bozena.borowska@p.lodz.pl

Abstract. This paper proposes a particle swarm optimization method with a novel strategy for inertia weight. Instead of a commonly used linear inertia weight, a nonlinear, dynamic changing inertia weight is applied. The new presented weight is a function of the worst and the best fitness of individuals of a population. In order to investigate the effectiveness of the proposed strategy tests on a set of benchmark function were conducted. The results were compared with those obtained through the LDW-PSO method, EWPSO and the RNW-PSO methods.

Keywords: Inertia weight · Particle swarm optimization
Swarm intelligence · Optimization

1 Introduction

Particle Swarm Optimization (PSO) is an optimization method modeled on the social behavior of the group of organisms in their natural environment [1]. Proposed for the first time by Kennedy and Eberhart in 1995 [2, 3], now belongs to the most frequently used evolutionary methods. Because of its many advantages such as simplicity, few parameters to adjust and easy implementation, it has been applied in almost all fields of science and engineering, where optimization is required [4–9]. A parameter that significantly influences the effectiveness of the particle swarm optimization method is inertia weight. Its role is to control deviation of the particles from their original direction and keep balance between local and global explorations. An incorrectly selected value of inertia weight maintains a balance between local and global exploration and can negatively affect the algorithm performance.

A factor of the inertia weight was first proposed and introduced to the PSO method by Shi and Eberhart [10, 11]. In the subsequent years, a lot of research on inertia weight have been undertaken. Clerc [12] and Trelea [13] suggested that the inertia weight should rather be a constant value. She and Eberhart [11, 14] recommended a linear decreasing inertia weight (LDW). A flexible inertia weight, that can be a positive or negative real number, was used by Han et al. [15]. In order to avoid some troubles of the LDW strategy connected with a poor local search ability at the beginning of the method, and the lack of global search ability at the end of the method, Zhang et al. [16] proposed a random inertia weight (RNW). PSO with random inertia weight was also proposed by Niu et al. [17] and Eberhart and She [18]. Three different concepts of

N. Shakhovska and V. Stepashko (eds.), *Advances in Intelligent Systems and Computing II*, Advances in Intelligent Systems and Computing 689,
https://doi.org/10.1007/978-3-319-70581-1_6

inertia weight were investigated by Arumugan and Rao [19, 20] and Umapathy et al. [21] They considered a constant inertia weight (CIW), time-varying inertia weight (TVIW) and a global-local best inertia weight (GLbestIW) and reported that GLbestIW outperforms the other methods in terms of high quality solution, consistency, faster convergence and accuracy. Another approach was developed by Yang et al. [22]. In their study, the inertia weight depends on two parameters: an aggregation degree factor and an evolution speed factor, and is different for each individual of the swarm. A modified inertia weight was also considered by Miao et al. [23]. In this case, weight is updated by dispersion degree and advance degree factors. Performance of PSO with nonlinear strategies was examined by Chauhan et al. [24]. The authors examined an exponential self adaptive (DESIWPSO) and a dynamic (DEDIVPSO) inertia weight based on Gompertz function and a fine grained inertia weight (FGIWPSO). Another exponential inertia weight was also developed by Ememipour et al. [25], Borowska [26] and Ghali et al. [27]. A fuzzy adaptive inertia weight was proposed by Shi and Eberhart [28]. An approach based on the fuzzy systems was also described in [29–33].

This paper presents a modified PSO method named DWPSO in which a new strategy for inertia weight was developed. In DWPSO, the values of inertia weight are dynamically changing and are determined based on fitness function. The proposed weight is a function of the best and the worst fitness of particles. Moreover, the new presented method was tested on a set of benchmark function. Then the results were compared to the RNW-PSO method with a random inertia weight [16], the LDW-PSO method with a linear decreasing inertia weight, and the nonlinear EWPSO method [26].

2 The PSO Method

The PSO algorithm belongs to the group of optimization methods based on the population. In case of PSO this population is called swarm and consists of individuals named particles. Each particle is a point of the space of the feasible solutions. The movement of the particle in this space enables the velocity vector. Initial location and velocity of the particle are randomly generated at the beginning of the algorithm. The quality of the particles is evaluated according to the fitness function of the optimization problem. In each iteration, particles update information about their own best position (named pbest) found so far. The knowledge about the particle with the best fitness among all the particles in the whole swarm (named gbest) is also remembered and updated in every iteration. The change of the velocity and location of the particles is carried out according to the formula:

$$V_i = wV_i + c_1 r_1 (pbest_i - X_i) + c_2 r_2 (gbest - X_i) \tag{1}$$

$$X_i = X_i + V_i \tag{2}$$

where $V_i = (v_{i1}, v_{i2}, \ldots, v_{iD})$ is a velocity vector of the particle i in the D-dimensional search space. Vector $X_i = (x_{i1}, x_{i2}, \ldots, x_{iD})$ represents a location of the particle i. Factor w is the inertia weight. Vector $pbest_i = (pbest_{i1}, pbest_{i2}, \ldots, pbest_{iD})$ means a personal best location of the particle i and $gbest = (gbest_1, gbest_2, \ldots, gbest_D)$ denotes a location of the particle with the best fitness function among all the particles in the whole swarm. The variables c_1 and c_2 are acceleration coefficients. They decide how strong the particle is influenced by its knowledge about its pbest and gbest value. Parameters r_1 and r_2 represent randomly generated numbers between 0 and 1 to maintain diversity of the population.

3 The Proposed DWPSO Algorithm

The proposed DWPSO algorithm is a variant of the particle swarm optimization method in which the new strategy for determination of inertia weight was introduced. In DWPSO a commonly used linear weight was omitted and replaced with an exponential inertia weight. In the proposed approach, the inertia weight is changing dynamically based on a fitness of the particles in the swarm. In each iteration, the particles of the swarm move in the search space according to Eqs. 1–2. After evaluating the quality of the new location of the particles, the individuals with the best and the worst fitness are found and recorded. Then, on their basis, the new value of the weight is counted. The new weight is represented by a nonlinear function of the best and the worst fitness of the particles. In each iteration, a different inertia weight is calculated and applied for the whole swarm. The proposed strategy has been defined as follows:

$$dw = ((gbest - f_{max}/2)^{-1}/((-\ln[f_{\min}])))/10^5 fh \tag{3}$$

$$w(t+1) = w(t) - dw(t) \tag{4}$$

where f_{max} and f_{min} are the values of maximal and minimal fitness in the current iteration, respectively. Factor fh is a randomly generated number in the range [0, 1].

4 Results

The simulation tests of the DWPSO method with proposed strategy was carried out on a set of nonlinear benchmark functions depicted in Table 1.

Table 1. Optimization test functions

Function	Formula	Minimum	Range of x
Sphere	$f_1 = \sum_{i=1}^{n} x_i^2$	0	(−100, 100)
Rosenbrock	$f_3 = \sum_{i=1}^{n-1} [100(x_{i+1} - x_i^2)^2 + (x_i - 1)^2$	0	(−30, 30)
Ackley	$f_4 = -20\exp\left(-0.2\sqrt{\frac{1}{n}\sum_{i=1}^{n} x_i^2}\right) - \exp\left(\frac{1}{n}\sum_{i=1}^{n}\cos(2\pi x_i)\right) + 20 + e$	0	(−32, 32)
Rastrigin	$f_5 = \sum_{i=1}^{n} (x_i^2 - 10\cos(2\pi x_i) + 10)$	0	(−5.12, 5.12)
Griewank	$f_6 = \frac{1}{4000}\sum_{i=1}^{n} x_i^2 - \prod_{i=1}^{n}\cos\left(\frac{x_i}{\sqrt{i}}\right) + 1$	0	(−600, 600)

The results of the tests were compared with the performance of PSO with a random number inertia weight (RNWPSO), LDW-PSO with a linear decreasing weight as well as EWPSO with an exponential inertia weight.

The DWPSO algorithm started with the inertia weight of 0.7 and was changing according to the formula 3–4. The acceleration coefficients c_1, c_2 used for executed computation were set to 1.6. The simulations were performed with four dimension sizes D = 10, 20 and 30 for N = 20, 40, 60 and 80 particles in the swarm respectively. Each experiment was run 50 times. In all cases, the iteration number was 1000.

The exemplary results of the tests performed for 20, 40 and 80 particles of the swarm are illustrated in Tables 2, 3 and 4. The presented values were averaged over 50 trials.

Table 2. Performance of the LDW-PSO, RNW-PSO, EWPSO and DWPSO algorithms for Rosenbrock function

Population size	Dimension	Algorithm	LDW-PSO	RNW-PSO	EWPSO	DWPSO
20	10	Mean	4.0933e+001	1.8401e+001	1.5489e+001	4.9787e+000
		St. Dev.	2.8127e+001	2.3554e+001	2.0147e+001	6.0743e+000
		Min	4.3607e+000	8.0913e−001	2.2132e−001	1.5060e−001
	20	Mean	8.3192e+001	1.9984e+002	7.2504e+001	5.0429e+001
		St. Dev.	4.5308e+001	2.8759e+002	1.3315e+002	4.1276e+001
		Min	1.5216e+001	1.5318e+001	9.0325e−001	2.1551e+000
	30	Mean	1.3507e+002	4.2856e+002	1.7210e+002	7.2359e+001
		St. Dev.	1.5563e+002	5.3721e+002	1.5334e+002	5.5085e+001
		Min	2.3142e+001	2.9608e+001	2.3512e+001	9.3401e+000
40	10	Mean	2.5379e+001	1.7240e+001	1.4231e+001	4.1631e+000
		St. Dev.	1.7455e+001	3.2178e+001	5.3622e+000	9.1435e+000
		Min	2.8927e−002	1.0806e+000	1.4523e−002	4.8498e−002
	20	Mean	5.7344e+001	5.6102e+001	4.7376e+001	4.0605e+001
		St. Dev.	5.3637e+001	4.9630e+001	4.8744e+001	6.4665e+001
		Min	6.1431+000	5.7073e+000	1.9198e+000	1.5801e−001
	30	Mean	7.1917e+001	6.9343e+001	6.7008e+001	5.2569e+001
		St. Dev.	1.4425e+002	3.5032e+001	7.5692e+001	2.9028e+001
		Min	1.1633e+001	1.0225e+001	6.8376e+000	2.5110e+000
80	10	Mean	1.8261e+001	1.5728e+001	1.2415e+001	5.5479e+000
		St. Dev.	2.9027e+001	2.3569e+001	2.2178e+001	1.8712e+001
		Min	1.2415e−001	2.8725e−003	9.4684e−003	2.2647e−002
	20	Mean	4.4502e+001	3.3485e+001	2.2570e+001	2.0819e+001
		St. Dev.	3.2417e+001	2.7279e+001	2.9342e+001	3.0246e+001
		Min	1.5655e−001	5.2635e−001	1.5904e−005	1.5185e+000
	30	Mean	6.6138e+001	4.2504e+001	3.9453e+001	3.8589e+001
		St. Dev.	1.3050e+002	3.2997e+001	2.6149e+001	3.4396e+001
		Min	9.5624e−001	1.1682e+001	1.2736e+001	1.3776e−002

Table 3. Performance of the LDW-PSO, RNW-PSO, EWPSO and DWPSO algorithms for Rastrigin function

Population size	Dimension	Algorithm	LDW-PSO	RNW-PSO	EWPSO	DWPSO
20	10	Mean	6.4812e+000	6.3174e+000	6.1075e+000	7.7109e+000
		St. Dev.	4.5160e+000	7.2541e+000	4.1338e+000	4.8600e+000
		Min	3.3097e+000	3.0705e+000	9.9775e−001	9.9496e−001
	20	Mean	4.5202e+001	5.3877e+001	4.3706e+001	3.7609e+001
		St. Dev.	1.6925e+001	1.6786e+001	1.4537e+001	1.2697e+001
		Min	2.2130e+001	3.1973e+001	2.5166e+001	1.9899e+001
	30	Mean	8.1753e+001	8.8445e+001	8.3524e+001	7.1985e+001
		St. Dev.	2.2488e+001	2.9007e+001	2.4530e+001	2.1340e+001
		Min	4.5357e+001	6.2319e+001	3.8712e+001	4.5768e+001
40	10	Mean	4.3519e+000	4.3854e+000	4.1303e+000	4.7657e+000
		St. Dev.	3.9689e+000	2.2077e+000	3.6194e+000	3.1695e+000
		Min	1.0526e+000	2.8343e+000	1.2376e+000	1.9899e+000
	20	Mean	3.6697e+001	3.4463e+001	3.1793e+001	2.3779e+001
		St. Dev.	1.1455e+001	9.0821e+000	1.3508e+001	7.9528e+000
		Min	1.1243e+001	1.5354e+001	1.9487e+001	2.9849e+000
	30	Mean	7.3137e+001	7.2862e+001	6.7972e+001	6.3329e+001
		St. Dev.	3.3256e+001	2.0617e+001	2.5146e+001	1.8331e+001
		Min	3.6972e+001	4.1565e+001	3.3887e+001	2.7859e+001
80	10	Mean	2.7314e+000	2.5931e+000	2.4145e+000	2.5801e+000
		St. Dev.	4.3898e+000	2.8769e+000	1.8860e+000	2.2603e+000
		Min	9.7515e−001	9.9004e−001	9.7073e−001	0.0000e+000
	20	Mean	3.1305e+001	3.0798e+001	2.8712e+001	2.1292e+001
		St. Dev.	9.4789e+000	9.7543e+000	1.1676e+001	8.4036e+000
		Min	1.6002e+001	1.6200e+001	1.0841e+001	8.9546e+000
	30	Mean	5.7351e+001	5.5917e+001	5.5712e+001	5.3379e+001
		St. Dev.	1.6879e+001	1.7884e+001	1.6234e+001	1.2905e+001
		Min	4.3125e+001	3.3266e+001	3.6258e+001	2.9849e+001

Table 4. Performance of the LDW-PSO, RNW-PSO, EWPSO and DWPSO algorithms for Griewank function

Population size	Dimension	Algorithm	LDW-PSO	RNW-PSO	EWPSO	DWPSO
20	10	Mean value	9.3627e−002	1.3901e−001	8.1573e−002	8.2364e−002
		St. Dev.	7.9415e−002	4.8233e−002	3.5967e−002	5.5734e−002
		Min	2.0163e−002	3.0994e−002	1.3029e−002	9.8573e−003
	20	Mean value	5.8212e−002	5.6745e−002	4.3195e−002	3.1448e−002
		St. Dev.	6.4934e−002	4.2267e−002	2.7830e−002	2.9442e−002
		Min	2.0407e−005	00000e+000	00000e+000	0.0000e+000
	30	Mean value	1.9235e−001	4.1733e−000	1.7908e−001	6.6079e−002
		St. Dev.	2.7506e−001	3.7786e−000	2.6839e−001	1.2421e−001
		Min	2.9618e−009	1.3545e−001	1.1424e−005	3.4376e−010

(*continued*)

Table 4. (*continued*)

40	10	Mean value	7.6583e−002	7.2025e−002	6.8570e−002	6.9333e−002
		St. Dev.	5.9771e−002	4.3162e−002	1.6405e−002	3.7984e−002
		Min	6.7088e−003	5.9476e−002	1.8311e−002	2.7037e−002
	20	Mean value	5.4734e−002	4.3979e−002	4.0720e−002	2.0134e−002
		St. Dev.	5.1682e−002	3.1848e−002	2.8183e−002	1.7913e−002
		Min	8.6707e−004	00000e+000	0.0000e+000	6.5454e−003
	30	Mean value	3.4395e−002	3.2599e−002	2.4759e−002	1.8896e−002
		St. Dev.	4.1504e−002	3.8676e−002	3.5383e−002	1.5956e−002
		Min	6.2816e−003	7.9302e−003	3.3418e−013	2.1289e−005
80	10	Mean value	7.2003e−002	6.5499e−002	5.8677e−002	6.2356e−002
		St. Dev.	4.1688e−002	5.4212e−002	3.2431e−002	2.2412e−002
		Min	5.4053e−003	3.3827e−002	8.5972e−004	2.4603e−002
	20	Mean value	2.8475e−002	2.5688e−002	2.0084e−002	1.0083e−002
		St. Dev.	1.7142e−002	2.0213e−002	1.5755e−002	1.5652e−002
		Min	0.0000e+000	00000e+000	0.0000e+000	0.0000e+000
	30	Mean value	2.1659e−002	1.7706e−002	9.5268e−003	8.8063e−003
		St. Dev.	1.8374e−002	1.8195e−002	1.5169e−002	1.3302e−002
		Min	0.0000e+000	00000e+000	0.0000e+000	0.0000e+000

The average best fitness in the following iterations for both DWPSO, EWPSO, RNW-PSO algorithms and LDW-PSO model for 40 particles (swarm size) and 30 dimensions is illustrated in Figs. 1, 2, 3 and 4. The vertical coordinates indicate the average best fitness in a logarithmic scale.

The results of simulations show that the proposed DWPSO method is more effective than the other methods investigated in this study. The dynamic strategy for inertia weight introduced for DWPSO facilitates the algorithm to maintain a diversity of the individuals in the search space and helps overcome the problem premature convergence.

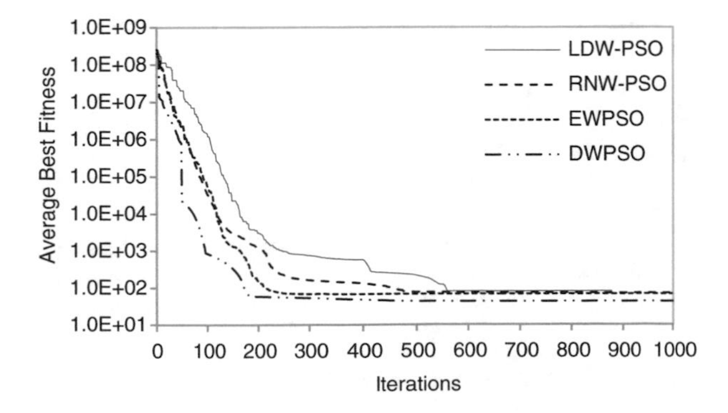

Fig. 1. The average best fitness for Rosenbrock30 and the population of 40 particles

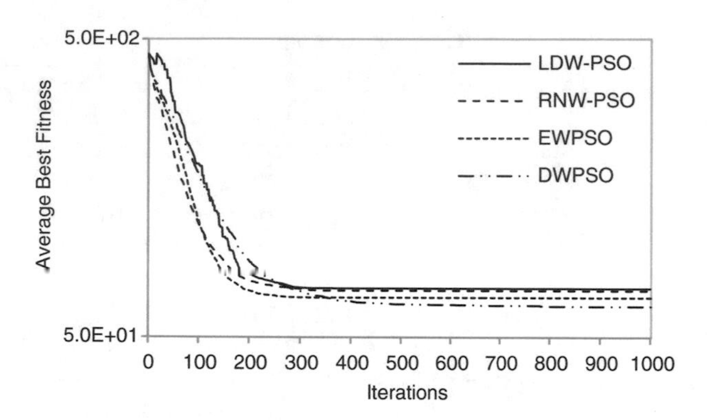

Fig. 2. The average best fitness for Rastrigin30 and the population of 40 particles

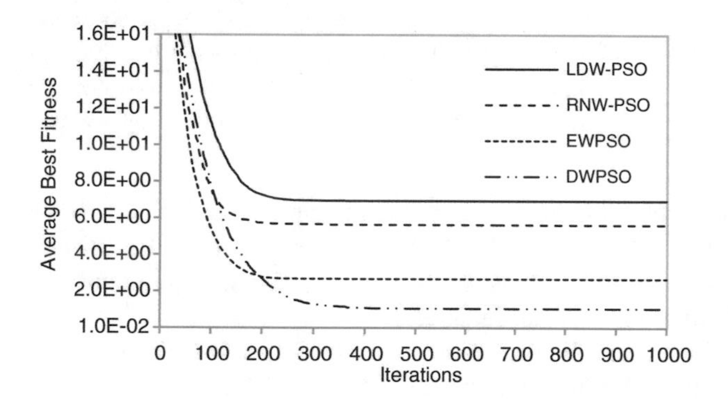

Fig. 3. The average best fitness for Ackley30 and the population of 40 particles

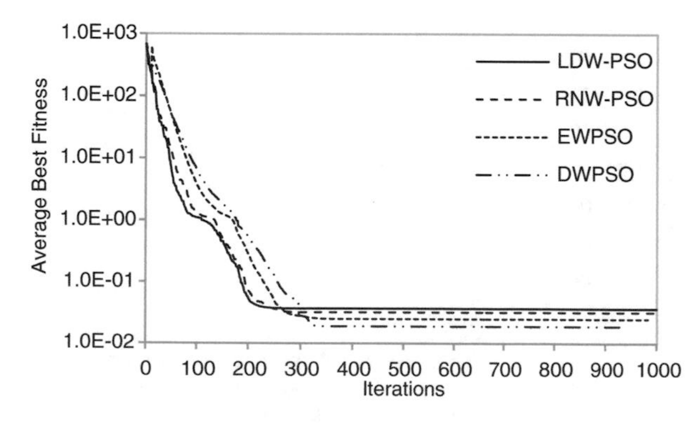

Fig. 4. The average best fitness for Griewank30 and the population of 40 particles

In almost all cases (except D = 10), the average function values found by DWPSO were lower than the results achieved by LDW-PSO and RNW-PSO, and lower or in rare cases comparable to those obtained by EWPSO. Moreover, the lowest standard deviation for DWPSO reported in most cases indicates its better stability compared to remaining investigated method. Furthermore, the minimum value was also lower in case of the DWPSO algorithm.. Additionally, in most simulations, DWPSO converged faster than LDW-PSO, RNW-PSO and EWPSO (Figs. 1, 2, 3 and 4) and only at the beginning, in the first two hundred iterations, the algorithm converged a bit slower than EWPSO or RNW-PSO (After first two hundred iterations DWPSO was the fastest). The RNW-PSO algorithm converged slower than EWPSO but still better than LDW-PSO.

Different performance of the algorithms have been noticed only for Griewank and Rastrigin functions and small dimensions. In case of Griewank function with D = 10 dimension size, the DWPSO algorithm performed a bit worse than EWPSO but better than RNW-PSO and LDW-PSO. For Rastrigin function with D = 10 for 20 and 40 particles, DWPSO achieve worse results than the remaining algorithms even when the minimum value was lower.

5 Summary

In this study, a modified particle swarm optimization algorithm named DWPSO with a novel strategy for inertia weight has been proposed. In the considered approach, instead of a commonly used constant or linear decreasing inertia weight, a dynamically changing weight was adopted. Values of the inertia weight coefficient depend on fitness of the individuals of the population. The effectiveness of the proposed strategy was tested on a set of benchmark test functions. The results of the simulations were compared with those obtained through the nonlinear EWPSO method, the RNW-PSO method with a random inertia weight and the LDW-PSO method with a linear decreasing inertia weight.

The usc of the proposed strategy helps maintain diversity of individuals of the population and performs better than the other investigated methods. Furthermore, the DWPSO algorithm is also faster and more efficient in avoiding the premature convergence, compared to the other methods.

References

1. Robinson, J., Rahmat-Samii, Y.: Particle swarm optimization in electromagnetics. IEEE Trans. Antennas Propag. **52**, 397–407 (2004)
2. Kennedy, J., Eberhart, R.C.: Particle swarm optimization. In: IEEE International Conference on Neural Networks, Perth, Australia, pp. 1942–1948 (1995)
3. Kennedy, J., Eberhart, R.C., Shi, Y.: Swarm Intelligence. Morgan Kaufmann Publishers, San Francisco (2001)
4. Guedria, N.B.: Improved accelerated PSO algorithm for mechanical engineering optimization problems. Appl. Soft Comput. **40**, 455–467 (2016)

5. Dolatshahi-Zand, A., Khalili-Damghani, K.: Design of SCADA water resource management control center by a bi-objective redundancy allocation problem and particle swarm optimization. Reliab. Eng. Syst. Saf. **133**, 11–21 (2015)
6. Mazhoud, I., Hadj-Hamou, K., Bigeon, J., Joyeux, P.: Particle swarm optimization for solving engineering problems: a new constraint-handling mechanism. Eng. Appl. Artif. Intell. **26**, 1263–1273 (2013)
7. Yildiz, A.R., Solanki, K.N.: Multi-objective optimization of vehicle crashworthiness using a new particle swarm based approach. Int. J. Adv. Manuf. Technol. **59**, 367–376 (2012)
8. Hajforoosh, S., Masoum, M.A.S., Islam, S.M.: Real-time charging coordination of plug-in electric vehicles based on hybrid fuzzy discrete particle swarm optimization. Electr. Power Syst. Res. **128**, 19–29 (2015)
9. Yadav, R.D.S., Gupta, H.P.: Optimization studies of fuel loading pattern for a typical pressurized water reactor (PWR) using particle swarm method. Ann. Nucl. Energ. **38**, 2086–2095 (2011)
10. Shi, Y., Eberhart, R.: A modified particle swarm optimizer. In: IEEE International Conference on Evolutionary Computation, pp. 69–73 (1998)
11. Shi, Y., Eberhart, R.C.: Parameter selection in particle swarm optimization. In: Proceedings of the Seventh Annual Conference on Evolutionary Programming, New York, pp. 591–600 (1998)
12. Clerc, M.: The swarm and the queen: towards a deterministic and adaptive particle swarm optimization. In: Proceedings of ICEC, Washington, D.C., pp. 1951–1957 (1999)
13. Trelea, I.C.: The particle swarm optimization algorithm: convergence analysis and parameter selection. Inf. Process. Lett. **85**, 317–325 (2003)
14. Shi, Y., Eberhart, R.C.: Empirical study of particle swarm optimization. In: Proceedings of the Congress on Evolutionary Computation, vol. 3, pp. 1945–1950 (1999)
15. Han, Y., Tang, J., Kaku, I., Mu, L.: Solving uncapacitated multilevel lot-sizing problems using a particle swarm optimization with flexible inertial weight. Comput. Math Appl. **57**, 1748–1755 (2009)
16. Zhang, L., Yu, H., Hu, S.: A new approach to improve particle swarm optimization. In: Proceedings of the International Conference on Genetic and Evolutionary Computation, pp. 134–139. Springer, Berlin (2003)
17. Niu, B., Zhu, Y.L., He, X., Wu, H.: A multi-swarm cooperative particle swarm optimizer. Appl. Math. Comput. **185**, 1050–1062 (2007)
18. Eberhart, R.C., Shi, Y.: Tracking and optimizing dynamic systems with particle swarms. In: IEEE Congress on Evolutionary Computation, Seoul, Korea, pp. 94–97 (2001)
19. Arumugan, M.S., Rao, M.V.C.: On the performance of the particle swarm optimization algorithm with various inertia weight variants for computing optimal control of a class of hybrid systems. Discrete Dyn. Nat. Soc. **2006**, 1–17 (2006)
20. Arumugan, M.S., Rao, M.V.C., Chandramohan, A.: A new and improved version of particle swarm optimization algorithm with global-local best parameters. Knowl. Inf. Syst. **16**, 331–357 (2008)
21. Umapathy, P., Venkataseshaiah, C., Arumugam, M.S.: Particle swarm optimization with various inertia weight variants for optimal power flow solution. Discrete Dyn. Nat. Soc. **2010**, 1–15 (2010)
22. Yang, X., Yuan, J., Yuan, J., Mao, H.: A modified particle swarm optimizer with dynamic adaptation. Appl. Math. Comput. **189**, 1205–1213 (2007)
23. Miao, A., Shi, X., Zhang, J., Wang, E., Peng, S.: A modified Particle Swarm Optimizer with Dynamical Inertia Weight, pp. 767–776. Springer, Berlin (2009)
24. Chauhan, P., Deep, K., Pant, M.: Novel inertia weight strategies for particle swarm optimization. Memetic Comput. **5**, 229–251 (2013)

25. Ememipour, J., Nejad, M.M.S., Rezanejad, M.M.J.: Introduce a new inertia weight for particle swarm optimization. In: Proceedings of Fourth International Conference on Computer Sciences and Convergence Information Technology, pp. 1650–1653 (2009)
26. Borowska, B.: Exponential inertia weight in particle swarm optimization. In: Advances in Intelligent Systems and Computing, vol. 524, pp. 265–275. Springer International Publishing, Heidelberg (2017)
27. Ghali, I., El-Dessouki, N., Mervat, A.N., Bakrawi, L.: Exponential particle swarm optimization approach for improving data clustering. Int. J. Electr. Electron. Eng. **3–4**, 208–212 (2009)
28. Shi, Y., Eberhart, R.C.: Fuzzy adaptive particle swarm optimization. In: Proceedings of the Congress on Evolutionary Computation, vol. 1, pp. 101–106 (2001)
29. Tian, D., Li, N.: Fuzzy particle swarm optimization algorithm. In: International Joint Conference on Artificial Intelligence, pp. 263–267 (2009)
30. Neshat, M.: FAIPSO: Fuzzy Adaptive Informed Particle Swarm Optimization. Neural Comput. Appl. **23**, 95–116 (2013)
31. Chen, T., Shen, Q., Su, P., Shang, C.: Fuzzy rule weight modification with particle swarm optimization. In: Soft Computing, pp. 1–15 (2015)
32. Mohiuddin, M.A., Khan, S.A., Engelbrecht, A.P.: Fuzzy particle swarm optimization algorithms for the open shortest path first weight setting problem. Appl. Intell. 1–24 (2016)
33. Chaturvedi, D.K., Kumar, S.: Solution to electric power dispatch problem using fuzzy particle swarm optimization algorithm. J. Inst. Eng. India **96**, 101–106 (2015)

Mathematical Method of Translation into Ukrainian Sign Language Based on Ontologies

Maksym Davydov(✉) and Olga Lozynska(✉)

Information Systems and Networks Department, Lviv Polytechnic National University, Lviv, Ukraine
{Maksym.V.Davydov,Olha.V.Lozynska}@lpnu.ua

Abstract. This paper introduces the mathematical method for translation into sign language based on ontologies. The modification of affix context-free grammar (AGFL) that adds semantical attribute and a new form of production called the "template production" is discussed. This new form helps to represent ontology-based productions in a short and computationally inexpensive way. The mathematical method utilizes dictionaries, ontology database, weighted affix context-free grammar (WACFG) parser, algorithm for transformation of constituency tree into dependency tree, and an algorithm for synthesis of Ukrainian sign language glosses. The algorithm for selection and convertion of grammatically augmented ontology (GAO) expressions into the set of WACFG productions is suggested. The major increase in percentage of correctly parsed sentences was achieved for Ukrainian sign language (UKL) and Ukrainian spoken language (USpL). All algorithms are components of the translation system for Ukrainian sign language. Simple video sequencing is utilized for sign language synthesis, however any other sign animation tool can be used. Tasks that require further research are defined.

Keywords: Sentence parser · Ontology · Machine translation · Sign language translation

1 Introduction

Development and study of methods for automatic translation into Ukrainian Sign Language is an urgent task today. Due to results of a survey conducted by Lviv Polytechnic National University, more than 90% of people who communicate using sign language would benefit from devices for automatic translation into sign language and more than 60% of them prefer using smartphone for performing this task. Nevertheless there are many known scientific articles on this issue, the problem of translation into Ukrainian Sign Language there are no viable solution for mobile devices yet.

Successful solution of Sign Language translation should be computationally effective, use limited storage, and require minimal bandwidth for communication with a server.

In the proposed solution translation from Ukrainian spoken language into annotated Ukrainian sign language and vice versa is divided into several steps: parsing sentences

N. Shakhovska and V. Stepashko (eds.), *Advances in Intelligent Systems and Computing II*, Advances in Intelligent Systems and Computing 689,
https://doi.org/10.1007/978-3-319-70581-1_7

using weighted affix context free grammar, the transformation of constituency trees into dependency trees using tree transformation algorithm, transformation of the dependency trees into the sequence of UKL glosses using translation rules for sign language translation, animation of glosses.

The use of ontologies and common sense databases play valuable role in parsing and translation of such languages as Ukrainian spoken language that have flexible word order. Grammatically augmented ontology is an ontology extension that links phrases to their meaning. The link is established via special expressions that connect phrase meaning to grammatical and semantical attributes of words that constitute it. The paper discusses an approach to sentence parsing that is based on integration of ontology relations into productions of weighted affix context-free grammar.

2 Related Work

A lot of linguistic problems such as machine translation, text recognition, information retrieval and extraction require the automatic analysis of sentences.

There are two main approaches for machine translation: the rule-based approach and the statistical approach. In the first approach, human experts specify a set of rules to describe the translation process. In another approach, large parallel corpora are used as source of knowledge. Each of these approaches has its own advantages and challenges. One of the steps of rule-based machine translation is the parsing input sentences.

The problem of sentence parsing has been already studied for a long time. There are a lot of approaches for sentence parsing: syntactic sentence parsing based on generative grammars [1], extended affix grammar (EAG) [2], and stochastic context-free grammar [3]; semantic sentence parsing based on predicate logic [4] and sub-domain driven parsing [5]; syntactic sentence parsing based on semantic relations using statistics [6], ontologies [7], tensor factorization [8] or mixed methods [9]; sentence parsing based on ontologies.

The idea of ontology integration to the process sentence parsing is not new. In [10] the generation of productions from ontologies for LTAG grammar parser was studied. In the article by Faten Kharbat [11] the WordNet ontology [12] is utilized to be the syntactic guide along with the Transition Network Grammar that helps to get better translation from English to Arabic language. The approach based on rich ontologies was used by Murat Temizsoy and Ilyas Cicekli [13] for Turkish language parsing. The approach uses ontologies to improve text meaning representation model for parsed sentences.

The work [14] describes a semi-automatic method for associating a Japanese lexicon with a semantic concept taxonomy called an ontology, using a Japanese-English bilingual dictionary. They developed three algorithms to associate a Japanese lexicon with the concepts of the ontology automatically: the equivalent-word match, the argument match, and the example match. These algorithms were tested on a dataset of 980 nouns, 860 verbs and 520 adjectives and can be effective for more than 80% of the words.

The new architecture for the translation (Italian – Italian Sign Language) that performs syntactic analysis, semantic interpretation and generation are proposed in [15].

They present some general issues of the ontological semantic interpretation that is based on a syntactic analysis that is a dependency tree.

However, the problem of integrating hypernymy/hyponymy relations into the process of sentence parsing was not previously studied. This article introduces a new method that extends the system of productions using ontology relations. These relations are expressions of GAO and hypernymy/hyponymy relations.

3 Main Part

3.1 Architecture and Components of the Developed System

The developed system consists of dictionaries, ontology database, weighted affix context-free grammar parser, algorithm for transformation of constituency tree into dependency tree, and an algorithm for synthesis of Ukrainian sign language glosses.

There are two kinds of used dictionaries: Ukrainian morphology dictionary and Ukrainian Sign Language dictionary.

There are two open-source dictionaries that are suitable for parsing Ukrainian morphology: Spell-uk dictionary maintained by Andriy Rysin and a dictionary supported by Mariana Romanyshyn and others from Grammarly. The first dictionary is widely known as it is used in OpenOffice spell checker. It is very memory effective and requires only 3 MB uncompressed and 500 KB when compressed. Unfortunately it has no direct method to obtain word tags that are required for further parsing of sentences. The second dictionary has more words and contains all necessary tags, but its size is 150 MB uncompressed and 17 MB in compressed state. Thus, the first dictionary is preferred for usage in mobile devices due to its compactness. In the developed system both dictionaries were used: the Spell-uk dictionary was extended with special word tagging rules, and words that were missing or were incorrectly tagged by spell-uk are used from the second dictionary.

The result of tagging a word is a set of hypotheses for its base form and possible grammatical attributes. Some mutually exclusive attributes can be included into the same hypothesis in order to decrease possible search space. This attributes can be refined later in the syntax parser.

For example parsing of Ukrainian word "мати" (mother) leads to the following productions that where added to the set of WACFG productions:

noun[gf c1 n1] → <мати>[gf c1 n1] → мати[r],
noun[gm c1 c4 n*] → <мат>[gm c1 c4 n*] → мати[r],
verb[i n1 m-] → <мати>[i n1 m-] → <мати>[r],

where tags "gm" and "gf" mean male and female gender, "n1" and "n*" mean singular and multiple number, "c1" and "c4" mean nominative and accusative cases respectively, tag "i" means infinitive, and tag "r" means word as it was written in the sentence (i.e. terminal symbol of the grammar). All of these productions have the same structure:

part_of_speech[TAGS] → <base_form>[TAGS] → <word>[r].

Base form of the word can be used later when ontology hyponymy and hypernymy relations are used.

The sentence parsing is done be means of WACFG parser that turned out to be very effective. Grammar productions are written in compact "template form" that assures low memory usage and computationally effective parsing. The system utilizes 230 productions for parsing of Ukrainian language. Two examples of these rules with small description are given below.

For example, production

$$\mathrm{NG}[=] \rightarrow \mathrm{adj}[=]?\ \mathrm{AN}(*)[=]\ \mathrm{NG}[c2]?$$

is used to describe a noun group that can be a noun with optional general adjective and adjective of place. One of possible phrases that can be handled by this production is "розумний студент Львівської політехніки" (smart student of Lviv Polytechnic University). In the given production NG stands for Noun Group, AN stands for Annotated Noun, "c2" means genitive case, symbol "?" means optional symbol, "=" means all standard attributes of this part of speech, and "*" denotes head word of the phrase that is used later to obtain dependency tree.

Production

$$\mathrm{NG}\left[\mathrm{C\,n} * \mathrm{p3\ gm\,gn\,gf}\right] \rightarrow \mathrm{NG}(*)[\mathrm{C}]\mathrm{conj\,NG}[\mathrm{C}]$$

is used to describe pair of nouns, for example "мило і рушник" (a soap and a towel). Here "C" means that all words should have the same case and the result phrase should have the same case, "p3" means the third person, "conj" means conjunction.

Besides regular productions that are used to describe the grammar of language being parsed, grammatically augmented ontology productions are used. These productions help to identify language constructions specific for particular subject area usage. For example, in phrases "PLAY THE PIANO" and "PLAY FOOTBAL" the word PLAY means completely different actions that are translated into sign language differently. Although such a colocation can be effectively solved using statistical approach it can't be applied right now due to the lack of large parallel corpuses for sign language. Instead of that the approach based on grammatically augmented ontologies is adopted. This approach requires creation of ontology dictionaries for specific target areas that can be effectively achieved by means of the developed GAODL language described earlier in [16].

The ontology is used to incorporate word abstraction rules into the grammar. For example, productions that were generated for subject area education had the following form:

<людина> → <школяр> (human → pupil)
<може-вчитись> → <людина> (can-learn → human)
<наука> → <математика> (science → math)
<містить-знання> → <наука> (contains-knowledge → science)
<вчити-навчати> → <вчити> (teach-proc → teach)
<вчити-навчати> → <вчити-навчати>(*) <може-вчитись>[c4] (teach-smb → teach-proc can-learn)

<вчити-навчати> → <вчити-навчати> <містить-знання>[c3] (teach-smth → teach-proc contains-knowledge)

These rules can be effectively used for semantical parsing of phrases like "вчити школярів математики" (teach math to pupils). More information about use of grammatically augmented ontologies for sentence parsing can be found in [17].

3.2 The Algorithm for Parsing Sentences

The problem of sentence parsing is formulated as a problem of finding a sequence of productions that has the maximum weight and can be applied sequentially to some starting attributed symbol (S, A_s) to produce a given sequence of terminals $t_1 t_2 \ldots t_n$. The weight of the sequence is calculated as a multiplication of weights of all contained productions.

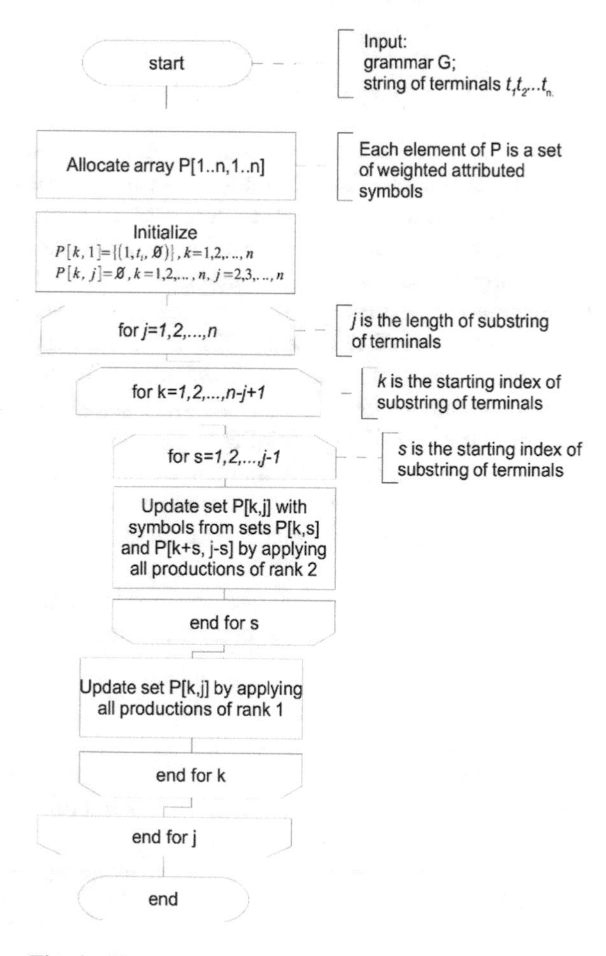

Fig. 1. The block scheme of sentence parsing algorithm.

The block scheme of parsing algorithm is shown on Fig. 1.

The algorithm provided above uses internal procedure for updating set of weighted attributes symbols Q with set of possible left symbols L and set of possible right symbols R by applying of rank 2. The block scheme of this procedure is depicted on Fig. 2.

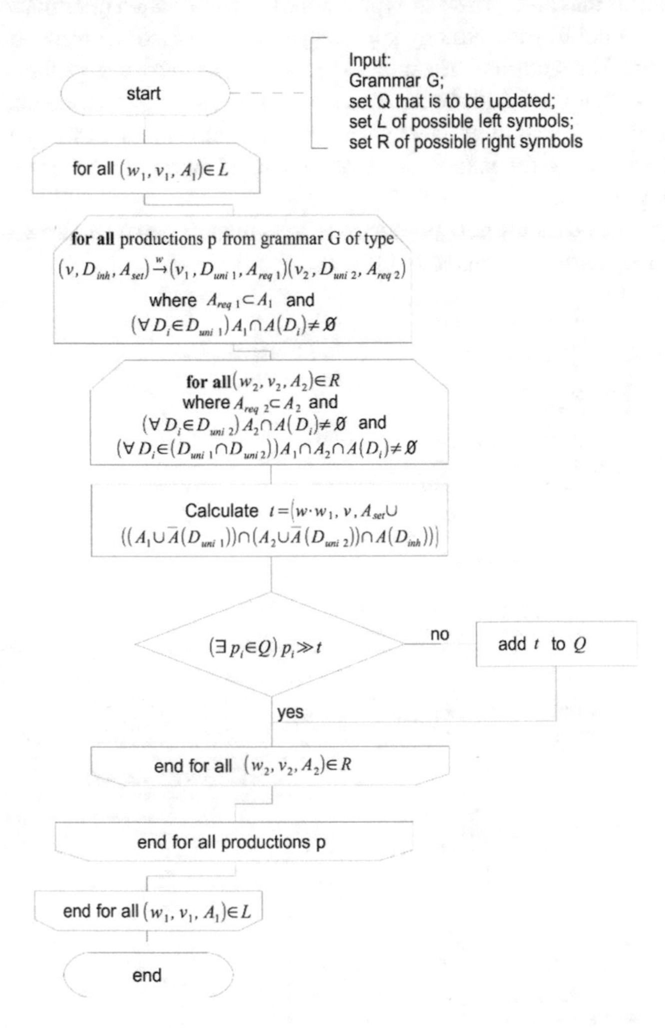

Fig. 2. The block scheme of procedure for updating set of weighted attributes symbols Q.

If the right part of a production contains only one symbol, the weight of the production should not exceed 1 in order to avoid cyclic productions that increase weight of non-terminal symbols during the bottom-up parsing procedure.

3.3 Extending the Set of Productions with Ontology Relations

The grammar augmented ontology was introduced in [18]. Along with relations that are common to ontology databases (hyponymy, hypernymy, meronymy, holonymy) GAO contains relations that link synsets to expressions with associated grammatical attributes.

In order to benefit from ontology knowledge new productions were added to generative grammar. The addition of ontology productions into the generative grammar extends the set of semantic attributes. Each synset of ontology was treated as semantic attribute. For the purpose of efficiency semantic attributes and corresponding productions were added only for hierarchies that contained words that were present in the sentence being parsed.

The algorithm that adds new productions to syntactic parser is shown on Fig. 3. A more detailed algorithm presented in [19].

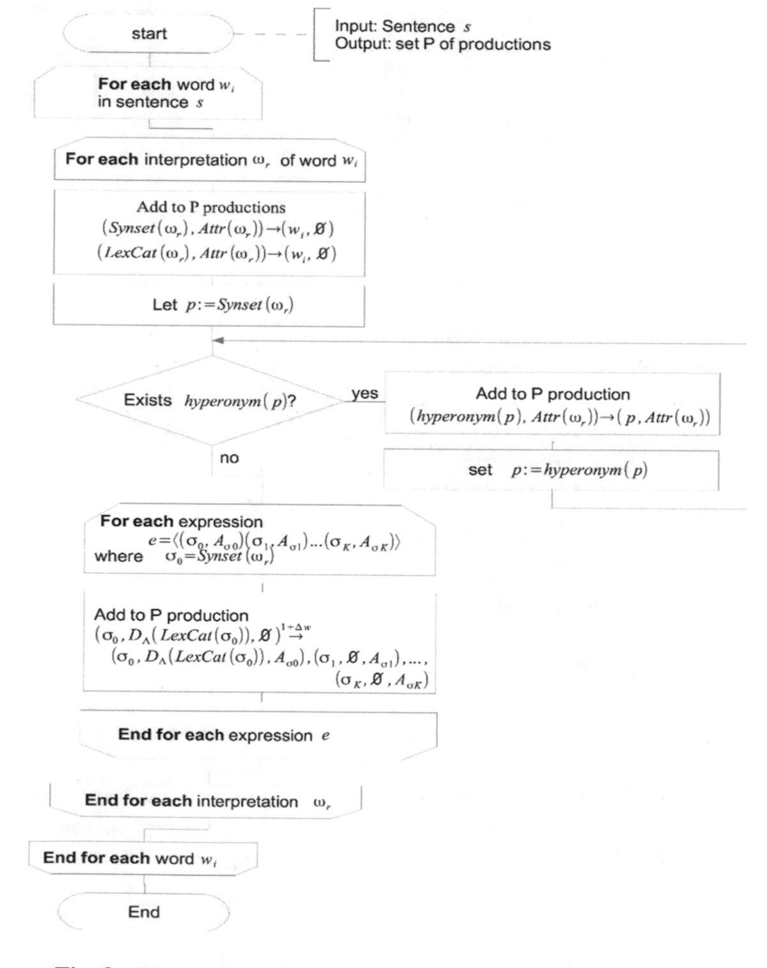

Fig. 3. The algorithm that adds new productions to syntactic parser.

Each word w in the sentence can have several interpretations $\omega_1, \omega_2, \ldots, \omega_r \in \Omega$, where Ω is a dictionary of all interpretations. Each interpretation ω uniquely defines its semantic attribute *SemAttr*(ω), lexical category *LexCat*(ω) and the set of grammatical attributes *GrAttr*(ω). A tuple $e = \langle (\sigma_1, A_{\sigma 1})(\sigma_2, A_{\sigma 2}) \ldots (\sigma^*, A_\sigma^*) \ldots (\sigma_K, A_{\sigma K}) \rangle$ – each expression in GAO, where σ_i is a synset that narrows the set of words that can appear in the given position of the expression e and $A_{\sigma i}$ is a set of grammar attributes the word is required to possess; σ^* is a head word of the expression. Let $\Lambda = \{noun, verb, adjective, adverb, \ldots\}$ be a set of all lexical categories and $D_\Lambda : \Lambda \to 2^D$ be a mapping from lexical category to the set of its attribute domains.

Productions that are generated from ontology expression have larger weight than simple syntactic productions in order to dominate over them. Additional weight Δw in expression is devised from the admissibility of the expression in the given context or text topic.

The result of paring the sentences “My father bought several candies at the table” and “My father bought several candies at the shop” using ontologies is depicted on Fig. 4.

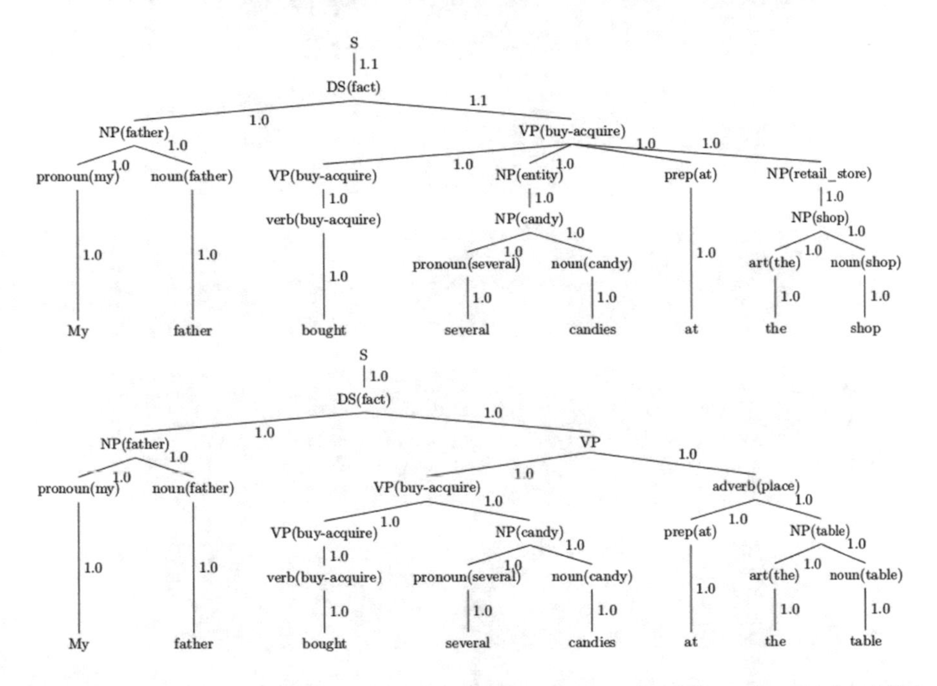

Fig. 4. The result of parsing sentences “My father bought several candies at the table” and “My father bought several candies at the shop”. The weight of the first parse is higher because one ontology expression was used while the second parsing tree is based only on syntactic productions.

The next step after parsing a sentence is to convert constituency tree to dependency tree. The algorithm of conversion takes constituency tree as input. The head of each phrase is used to determine main word of the sentence and main word of each sub-phrase. The core of the algorithm is to identify the head of each phrase in the constituency tree and establish its relation with the head of its parent node.

The transformation algorithm comprises of the following steps (Fig. 5):

1. Find word that is a head of the entire tree and mark all vertices in the tree that have the same head.
2. Join vertices obtained in step 1 into a single vertex of the dependency tree.
3. Apply steps 1–2 recursively for every sub-tree of marked vertices.
4. Verify dependencies in the obtained dependency tree and mark edges where words could not be in parent-child relationship as improper.
5. For all improper relations find better correspondence by width-first search in the dependency tree for other possible parent word.

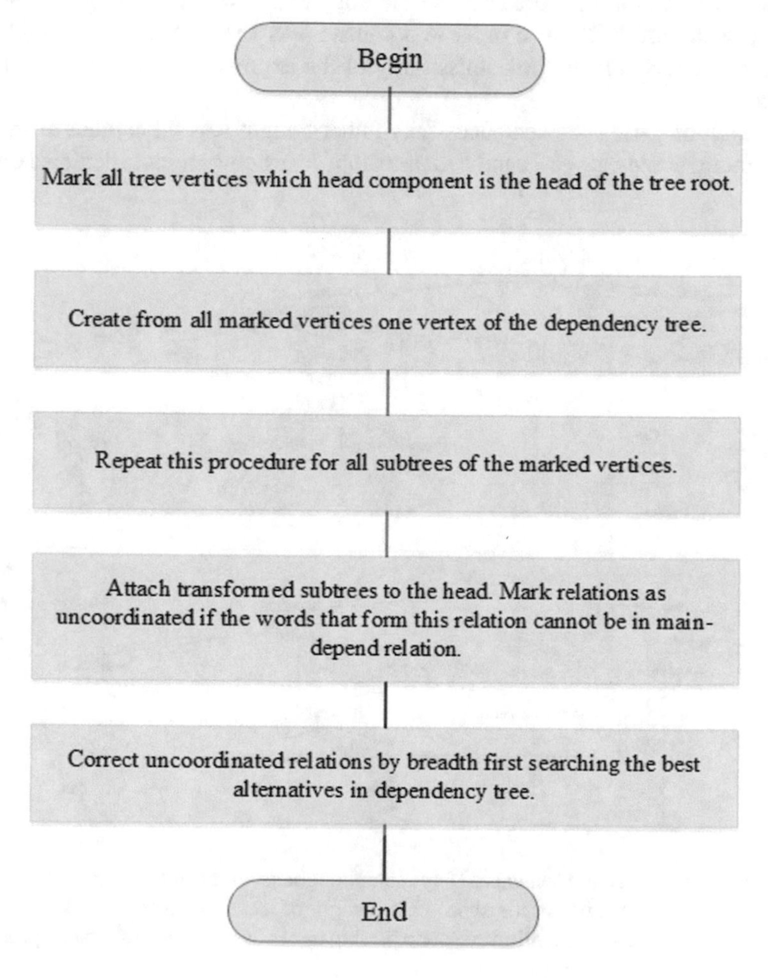

Fig. 5. Algorithm for transformation of constituency tree to dependency tree.

4 Results

The result of the transformation step is a dependency tree that can be used to produce translation. An example of a tree for sentence "My father bought several candies at the shop" is shown in Fig. 6.

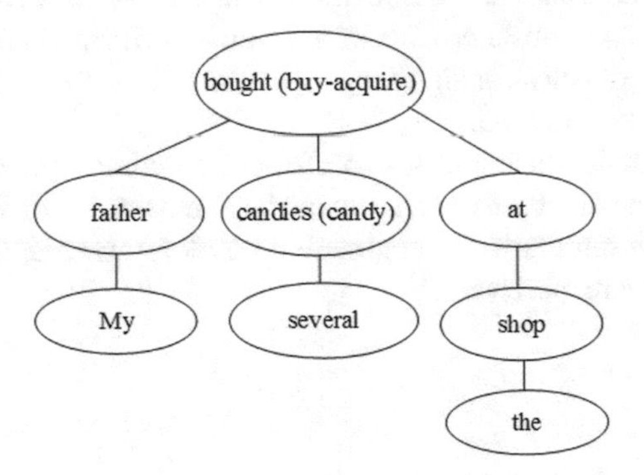

Fig. 6. Dependency tree obtained for sentence "My father bought several candies at the shop".

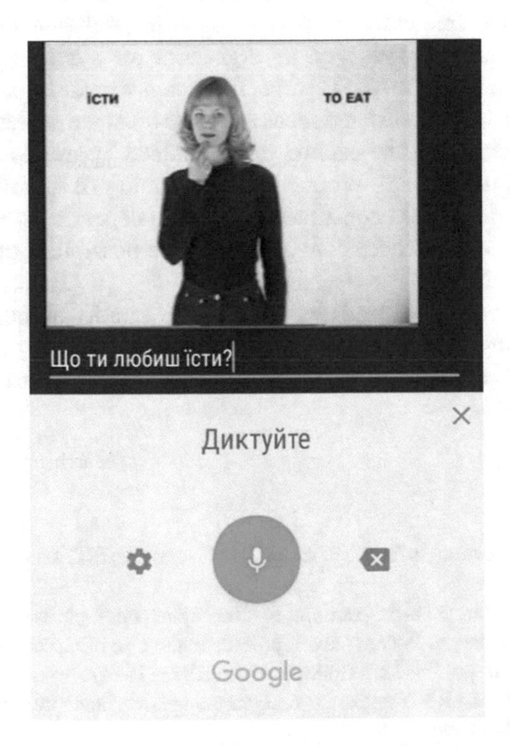

Fig. 7. Mobile app interface for entering text and translating into sign language.

The translation is produced using transformation rules described in [20].

The information system of translation takes an input text of Ukrainian spoken language. The input can be obtained by typing word with a screen keyboard or by means of speech recognition (Fig. 7).

After entering the text, it is split into sentences and their topics are determined. Productions from grammatically augmented ontology are added when possible and sentences are parsed and translated. The translation is performed involving transfer rules base, which consists of rules for sign language translation and reordering rules that are used to generate the text in UKL.

Evaluation of translation results was performed by comparing sentences with translations available in the database of test sentences. The database of WACFG productions and the database of grammatically augmented ontology were updated by adding new rules and new synsets respectively.

5 Conclusion

The mathematical method for translation into sign language based on ontologies are described in the paper.

The developed information system consists of dictionaries, ontology database, weighted affix context-free grammar parser, algorithm for transformation of constituency tree into dependency tree, and an algorithm for synthesis of Ukrainian sign language glosses. The use of WACFG parser for sentence parsing has allowed to increase the percentage of correctly translated sentences. The obtained sentence parsing trees are more semantically rich than the parsing trees obtained by means of regular syntactic parser. The system utilizes 230 productions for parsing of Ukrainian language. The transformation algorithm from constituency tree into dependency tree has shown high efficiency (89% correct sentences converted) and the possibility of its use in machine translation systems.

Further research can be focused on improving the quality of the information system for translation, adding new rules into WACFG parser, extending the set of synsets in grammatically augmented ontology and devising new rules for the transformation algorithm.

References

1. Chomsky, N.: Three models for the description of language. IRE Trans. Inf. Theor. **2**(3), 113–124 (1956)
2. Oostdijk, N.: An extended affix grammar for the english noun phrase. In: Aarts, J., Meijs, W. (eds.) Corpus Linguistics. Recent Developments in the Use of Computer Corpora in English Language Research, pp. 95–122. Rodopi, Amsterdam (1984)
3. Eddy, S.R., Durbin, R.: RNA sequence analysis using covariance models. Nucleic Acids Res. **22**(11), 2079–2088 (1994)
4. Blackburn, P., Bos, J.: Representation and Inference for Natural Language: A First Course in Computational Semantics. CSLI Publications, Stanford (2005)

5. Plank, B., Sima'an, K.: Subdomain sensitive statistical parsing using raw corpora. In: Proceedings Sixth International Conference on Language Resources and Evaluation, pp. 465–469. European Language Resources Association, Marrakech (2008)
6. Jiang, J.J., Conrath, D.W.: Semantic similarity based on corpus statistics and lexical taxonomy. In: Proceedings of the International Conference on Research in Computational Linguistics, Taiwan, pp. 19–33 (1997)
7. Rhee, S.K., Lee, J., Park, M.-W.: Ontology-based semantic relevance measure. In: Proceedings of the First International Workshop on Semantic Web and Web 2.0 in Architectural, Product and Engineering Design, Busan, Korea, pp. 63–68 (2007)
8. Anisimov, A., Marchenko, O., Vozniuk, T.: Determining semantic valences of ontology concepts by means of nonnegative factorization of tensors of large text corpora. Cybern. Syst. Anal. **50**(3), 327–337 (2014)
9. Nagarajan, M., Sheth, A.P., Aguilera, M., Keeton, K., Merchant, A., Uysal, M.: Altering document term vectors for classification - ontologies as expectations of co-occurrence. In: Proceedings of the 16th International Conference on World Wide Web, Banff, Alberta, Canada, pp. 1225–1226 (2007)
10. Unger, C., Hieber, F., Cimiano, P.: Generating LTAG grammars from a lexicon-ontology interface. In: Proceedings of the 10th International Workshop on Tree Adjoining Grammars and Related Formalisms (TAG+10), Yale University, pp. 61–68 (2010)
11. Kharbat, F.: A new architecure for translation engine using ontology: one step ahead. In: Proceedings of The International Arab Conference on Information Technology (ACIT 2011), Saudi Arabia, pp. 169–173 (2011)
12. Miller, G.A.: WordNet: a lexical database for English. Commun. ACM **38**(11), 39–41 (1995)
13. Temizsoy, M., Cicekli, I.: An ontology based approach to parsing Turkish sentences. In: Proceedings of AMTA 1998-Conference, pp. 124–135. Springer, Langhorne (1998)
14. Okumura, A., Hovy, E.H.: Building Japanese-English dictionary based on ontology for machine translation. In: HLT 1994 Proceedings of the Workshop on Human Language Technology, Plainsboro, pp. 141–146 (1994)
15. Lesmo, L., Mazzei, R., Radicioni, D.P.: An ontology based architecture for translation. In: Proceedings of the Ninth International Conference on Computational Semantics, Oxford, UK, pp. 345–349 (2011)
16. Lozynska, O.V., Davydov, M.V.: Domain-specific language for describing grammatically augmented ontology. Control Syst. Mach. **4**, 31–40 (2015)
17. Lozynska, O., Davydov, M.: Information technology for Ukrainian Sign Language translation based on ontologies. Econtechmod Int. Q. J. **4**(2), 13–18 (2015)
18. Davydov, M., Lozynska, O.: Spoken and sign language processing using grammatically augmented ontology. Appl. Comput. Sci. **11**(2), 29–42 (2015)
19. Davydov, M., Lozynska, O., Pasichnyk, V.: Partial semantic parsing of sentences by means of grammatically augmented ontology and weighted affix context-free grammar. Econtechmod Int. Q. J. **6**(2), 27–32 (2017)
20. Lozynska, O.V., Davydov, M.V., Pasichnyk, V.V.: Rule-based machine translation into Ukrainian Sign Language. Inf. Technol. Comput. Eng. Sci. J. VNTU **1**(29), 11–17 (2014)

Method of Parametric Identification for Interval Discrete Dynamic Models and the Computational Scheme of Its Implementation

Mykola Dyvak[1], Natalia Porplytsya[1], Yurii Maslyak[1](✉), and Mykola Shynkaryk[2]

[1] Department of Computer Science, Ternopil National Economic University, Ternopil, Ukraine
mdy@tneu.edu.ua, ocheretnyuk.n@gmail.com, yuramasua@gmail.com
[2] Department of Economical and Mathematical Methods, Ternopil National Economic University, Ternopil, Ukraine
shmi@tneu.edu.ua

Abstract. A method of parametric identification of interval discrete dynamic models is considered. In the case of an interval data set, finding estimations for parameters of such models requires solving an interval system of nonlinear algebraic equations for some known vector of basic functions. The solution of these equations forms a non-convex area in the parameter space which can consist of several unconnected subareas. For solving this parametric identification problem, methods of random search are widely used including that based on the procedure of the Rastrigin's director cone having high time complexity. Therefore, the detailed analysis of this parametric identification method was carried out in this work to reduce the time complexity. A new improved scheme of computational implementation of the method is proposed which takes into account areas of permissible values of the modeled characteristic. Results of the comparative efficiency analysis of implementation scheme of the proposed method and the known one are presented demonstrating that the time complexity of the improved scheme of the method is at least twice less compared to the known implementation scheme.

Keywords: Parametric identification · Random search procedures
Interval data analysis · Discrete dynamic model

1 Introduction

There are many real-world tasks solution of which requires building mathematical models in the form of discrete dynamic models (DDM). In particular, these are the prediction task of the humidity distribution on the surface of the drywall sheet during its manufacturing, the prediction task of the distribution of information signal maximal amplitude of the surgical wound surface during thyroid surgery and prediction task of dispersion of harmful vehicle emissions in the surface layer of the atmosphere [1–3].

N. Shakhovska and V. Stepashko (eds.), *Advances in Intelligent Systems and Computing II*, Advances in Intelligent Systems and Computing 689,
https://doi.org/10.1007/978-3-319-70581-1_8

When solving such kind of problems, the task of parametric identification of the DDM arises being, as it is well known, an NP-complexity task [4, 5]. Solving it requires using an exhaustive search of all possible solutions until fulfilling some specific criteria [6]. It is known that in case when a data set is of interval form, this means finding estimations for DDM parameters by solving an interval system of nonlinear algebraic equations (ISNAE) for some known vector of basic functions [7].

Note, that the solution of ISNAE is an area in the parameter space that is not convex and can consist of several unconnected subareas. When solving the task of estimation the solutions area of ISNAE, in case when structure of the model is presented in form of DDM, some difficulties arise, in particular, related to the recurrent scheme of the calculations [8, 9]. That is why for solving the problem of parametric identification, methods based on random search procedures are widely used, for example, based on the Rastrigin's director cone [10, 11]. Using this method assumes, instead of searching the whole area of ISNAE solutions, searching at least one of its solutions, that is only one point from this area [12].

However, the practice shows that even if this method gives an opportunity to build an adequate model in the form of DDM, it has still high time complexity. The solution time can be very varied even in one series of experiments due to the «random character» of the used procedures. Therefore, the detailed analysis of this parametric identification method was carried out in this work to reduce the time complexity of its using.

2 Statement of the Problem

Let us consider the problem of parametric identification of the DDM in such general form [2]:

$$\begin{gathered}\left[\hat{v}_{i,j,h,k}\right] = \left[\hat{v}^{-}_{i,j,h,k}; \hat{v}^{+}_{i,j,h,k}\right] = \vec{f}^{T}\left(\left[\hat{v}_{0,0,0,0}\right], \ldots, \left[\hat{v}_{0,0,h-1,0}\right], \ldots,\right. \\ \left.\left[\hat{v}_{i-1,j-1,h-1,k-1}\right], \vec{u}_{i,j,h,0}, \ldots, \vec{u}_{i,j,h,k}\right) \cdot \hat{\vec{g}}, \\ i = 1, \ldots, I, j = 1, \ldots, J, h = 1, \ldots, H, k = 1, \ldots, K,\end{gathered} \tag{1}$$

where $\vec{f}^{T}(\bullet)$ is vector of known basis functions defining the DDM structure; $v_{i,j,h,k}$ is the modeled characteristic at the point with discrete spatial coordinates i, j, h in discrete time k; $\vec{u}_{i,j,h,0}, \ldots, \vec{u}_{i,j,h,k}$ are vectors of input variables; $\vec{g}$ is a vector of unknown parameters of DDM. Below, the model (1) will be called as interval discrete dynamic model (IDDM).

Based on the requirements of ensuring accuracy of the model within the interval of the experiment accuracy, the identification of IDDM (1) will be realized using such criterion [12]:

$$\begin{gathered}\left[\hat{v}^{-}_{i,j,h,k}; \hat{v}^{+}_{i,j,h,k}\right] \subset \left[z^{-}_{i,j,h,k}; z^{+}_{i,j,h,k}\right] \\ \forall i = 1, \ldots, I, \forall j = 1, \ldots, J, \forall h = 1, \ldots, H, \forall k = 1, \ldots, K\end{gathered} \tag{2}$$

where $\left[\hat{v}^{-}_{i,j,h,k}; \hat{v}^{+}_{i,j,h,k}\right]$ is an interval estimation of the modeled characteristic; $\left[z^{-}_{i,j,h,k}; z^{+}_{i,j,h,k}\right]$ is an interval of possible measured values of the characteristic at the point with discrete coordinates i, j, h in the time moment k.

By substituting recurrent expression (1) into (2), instead of the interval estimates $\left[\hat{v}^{-}_{i,j,h,k}; \hat{v}^{+}_{i,j,h,k}\right]$ together with the given initial interval values of the set elements.

$$\left[\hat{v}_{0,0,0,0}\right] \subseteq \left[z_{0,0,0,0}\right], \ldots, \left[\hat{v}_{0,0,h-1,0}\right] \subseteq \left[z_{0,0,h-1,0}\right], \ldots, \\ \left[\hat{v}_{i-1,j-1,h-1,k-1}\right] \subseteq \left[z_{i-1,j-1,h-1,k-1}\right], \tag{3}$$

and given vectors of input variables $\vec{u}_{i,j,h,0}, \ldots, \vec{u}_{i,j,h,k}$, we obtain the ISNAE:

$$\begin{cases} \left[\hat{v}^{-}_{0,0,0,0}; \hat{v}^{+}_{0,0,0,0}\right] \subseteq \left[z^{-}_{0,0,0,0}; z^{+}_{0,0,0,0}\right], \\ \vdots \\ \left[\hat{v}^{-}_{i-2,j-2,h-2,k-2}; \hat{v}^{+}_{i-2,j-2,h-2,k-2}\right] \subseteq \left[z^{-}_{i-2,j-2,h-2,k-2}; z^{+}_{i-2,j-2,h-2,k-2}\right]; \\ \left[\hat{v}_{i-1,j-1,h-1,k-1}\right] = \vec{f}^{T}\left(\left[\hat{v}_{0,0,0,0}\right], \ldots, \left[\hat{v}_{i-2,j-2,h-2,k-2}\right], \vec{u}_{0}, \ldots, \vec{u}_{k-1}\right) \cdot \hat{\vec{g}}; \\ z^{-}_{i,j,h,k} \leq \vec{f}^{T}\left(\left[\hat{v}_{0,0,0,0,}\right], \ldots, \left[\hat{v}_{i-1,j-1,h-1,k-1}\right], \vec{u}_{i,j,0}, \ldots, \vec{u}_{i,j,k}\right) \cdot \hat{\vec{g}} \leq z^{+}_{i,j,h,k}; \\ z^{-}_{i+1,j,h,k} \leq \vec{f}^{T}\left(f^{T}\left(\left[\hat{v}_{0,0,0,0,}\right], \ldots, \left[\hat{v}_{i-1,j-1,h-1,k-1}\right], \vec{u}_{i,j,h,0}, \ldots, \vec{u}_{i,j,h,k}\right) \cdot \hat{\vec{g}}\right) \leq z^{+}_{i+1,j,h,k}; \\ i = 2, \ldots I, j = 2, \ldots, J, h = 2, \ldots, H, k = 2, \ldots, K. \end{cases} \tag{4}$$

Then, the ISNAE (4) can be transformed into an optimization problem of searching the minimum of such objective function:

$$\delta(\hat{\vec{g}}) \xrightarrow{\hat{\vec{g}}} 0. \tag{5}$$

Figure 1 illustrates a fragment of solutions area of IDDM parametric identification problem. In the space of parameters, each point from definition domain of objective function has coordinates $\left(\hat{g}_1; \hat{g}_2\right)$ that are known values of IDDM parameter vector, and $\delta\left(\hat{\vec{g}}\right)$ is a value of objective function calculated based on the IDDM (1), that determines the quality of current IDDM parameter vector.

As it is shown in the figure, the objective function contains large number of local and global extremums with $\delta\left(\hat{\vec{g}}\right) = 0$ which increases the probability of finding problem solution. But such large number of local extremum points substantially complicates the problem of IDDM parametric identification [2, 4].

Note that the points in the parameter space for which the condition $\delta\left(\hat{\vec{g}}\right) = 0$ is fulfilled provide the possibility to build an adequate model, namely such model that satisfies condition (2).

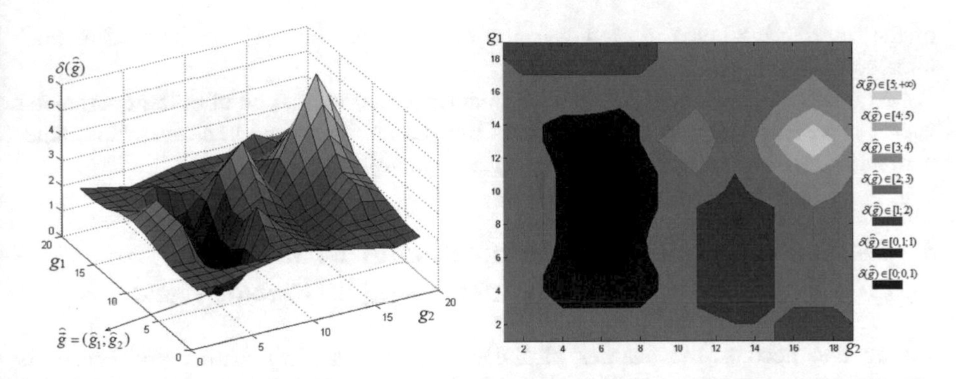

Fig. 1. Fragment of the two-dimensional space of solutions of the parametric identification problem.

Besides that, the objective function in the parameter space can be "torn", namely it can contain few unconnected subareas that confirms the complexity of problem of IDDM parameters identification [12].

To calculate values of the objective function $\delta\left(\hat{\vec{g}}\right)$, the following expressions are used:

$$\delta\left(\hat{\vec{g}}\right) = \max_{i=1,\ldots,I,j=1,\ldots,J,h=1,\ldots,H,k=1,\ldots,K}\left\{\left|mid\left(\vec{f}^{Ts}\left(\left[\hat{v}_{0,0,0,0}\right],\ldots,\right.\right.\right.\right.$$
$$\left.\left.\left.\left.\left[\hat{v}_{i-1,j-1,h-1,k-1}\right],\vec{u}_{i,j,h,0},\ldots,\vec{u}_{i,j,h,k}\right)\cdot\hat{\vec{g}}^{s}\right) - mid\left(\left[z_{i,j,h,k}\right]\right)\right|\right\}$$
$$if\left[\hat{v}_{i,j,h,k}\right]\cap\left[z_{i,j,h,k}\right] = \emptyset, \exists i = 1,\ldots,I, \forall j = 1,\ldots,J, \exists h = 1,\ldots,H, \exists k = 1,\ldots,K; \tag{6}$$

$$\delta\left(\hat{\vec{g}}\right) = \max_{i=1,\ldots,I,j=1,\ldots,J,h=1,\ldots,H,k=1,\ldots,K}\left\{wid\left(\vec{f}^{Ts}\left(\left[\hat{v}_{0,0,0,0}\right],\ldots,\right.\right.\right.$$
$$\left.\left.\left[\hat{v}_{1,j-1,0,0}\right],\ldots,\left[\hat{v}_{i-1,j-1,h-1,k-1}\right],\vec{u}_{i,j,h,0},\ldots,\vec{u}_{i,j,h,k}\right)\cdot\hat{\vec{g}}^{s}\right)$$
$$- wid\left(\left(\vec{f}^{Ts}\left(\left[\hat{v}_{0,0,0,0}\right],\ldots,\left[\hat{v}_{0,0,h-1,0}\right],\left[\hat{v}_{i-1,0,0,0}\right],\ldots,\right.\right.\right.$$
$$\left.\left.\left.\left.\left[\hat{v}_{i-1,j-1,h-1,k-1}\right],\vec{u}_{i,j,h,0},\ldots,\vec{u}_{i,j,h,k})\cdot\hat{\vec{g}}^{s}\right)\right)\cap\left[z_{i,j,h,k}\right]\right)\right\}$$
$$if\left[\hat{v}_{i,j,h,k}\right]\cap\left[z_{i,j,h,k}\right] \neq \emptyset, \forall i = 1,\ldots,I, \forall j = 1,\ldots,J, \forall h = 1,\ldots,H, \forall k = 1,\ldots,K. \tag{7}$$

The functions $mid(\bullet)$, $wid(\bullet)$ are the operations of determining the centers and the widths of the intervals, respectively.

Expression (6) describes the approximation of the current vector $\hat{\vec{g}}$ to satisfactory one as the difference between the most remote centers of predicted and experimental intervals in the case when they do not intersect, and expression (7) is the smallest width of the intersection between predicted and experimental intervals in the case of their intersection at all discrete points.

Expressions (6) and (7) describe the quantified approximation of current parameter vector to satisfactory one which provides the possibility to build an adequate mathematical model in terms of fulfilling the condition (2).

3 Method of Parametric Identification of IDDM Based on a Random Search Procedure Using a Director Cone

Taking into account the features of IDDM parametric identification problem, it is expedient to use random search methods for its solving. In [12], the method of IDDM parametric identification is built based on procedure of random search of minimum of the objective function $\delta\left(\widehat{\vec{g}}\right)$ using Rastrigin's director cone [10, 11]. Let us consider it in more details.

At the initial iteration of the method ($l = 0$), the initial approximation of IDDM parameter vector $\widehat{\vec{g}}_0$ is randomly set. Then, P random points are generated in the neighborhood of this approximation on the surface of imaginary hypersphere with radius r, scilicet, within distance r from point $\widehat{\vec{g}}_0$ in the parameter space based on the uniform distribution law:

$$\widehat{\vec{g}}_p = \widehat{\vec{g}}_0 + r \cdot \vec{\xi}_p, p = 1, \ldots, P. \tag{8}$$

Among all generated points, a point is selected that provides the minimal value of the objective function:

$$\widehat{\vec{g}}_1 = \underset{p=1,\ldots,P}{\arg\min}\left(\delta\left(\widehat{\vec{g}}_0 + r \cdot \vec{\xi}_p\right)\right). \tag{9}$$

This obtained estimation of IDDM parameter vector is an approximation for the next iteration. Additionally, a memory vector is calculated that determines successful direction of search, in this procedure:

$$\vec{w} = \left(\widehat{\vec{g}}_1 - \widehat{\vec{g}}_0\right)/r. \tag{10}$$

At the next iteration, an imaginary hypercone is built in the parameter space with top $\widehat{\vec{g}}_l$ which is current estimation of IDDM parameter vector with opening angle ψ and axis $\vec{w}_l$. This hypercone "cuts" some surface from hypersphere with center in the point $\widehat{\vec{g}}_l$ and radius r. Again, P points are randomly generated on the obtained surface in the parameter space based on uniform distribution law using expression (8). Note, in expression (8) vector $\vec{\xi}_p$ is calculated based on cone parameters limitations. Then again, the point is selected that provides the minimum value of the objective function among all generated points:

$$\vec{\hat{g}}_{l+1} = \arg\min_{p=1,\ldots,P}(\delta(\vec{\hat{g}}_l + r \cdot \vec{\xi}_p)). \tag{11}$$

The obtained estimation of the IDDM parameter vector is an approximation for the next iteration $l + 1$ of the search procedure. Moreover, an additional memory vector is determined in this procedure:

$$\vec{w}_{l+1} = \alpha \cdot \vec{w}_l + \beta \cdot \frac{\vec{\hat{g}}_{l+1} - \vec{\hat{g}}_l}{r}, \tag{12}$$

where α, $0 \leq \alpha \leq 1$, is a forgetting coefficient, β is a coefficient of intensity of taking into account of new information.

The search continues until the objective function decreases. If the value of the function does not decrease at a particular iteration, we need to use a hypersphere instead of using a cone as it is in the initial iteration for a given vector of parameter estimations. If it is impossible to find a point among all generated ones which would decrease the objective function, one should usually reduce the radius r.

4 Scheme of Computational Implementation of the Method of Parametric Identification of IDDM

For implementation of the parametric identification method described above, the following scheme of its computational implementation is proposed based on the dividing of experimental sample described in [12]. Namely, under condition of large quantity of interval data in the whole set, a procedure of dividing the entire set of interval data into parts is used with further operating on the set parts. However, conditions of providing of the given accuracy (2) for model in IDDM form should be fulfilled for the entire set of the interval data.

Using the principle of data sample division being traditional for GMDH, the interval data set is divided into two parts: main and testing.

If the quantity of interval data points in the main set is $N_0 = I_0 \times J_0 \times H_0 \times K_0$, then the quantity of points in the testing set remains:

$$N_P = I \times J \times H \times K - I_0 \times J_0 \times H_0 \times K_0. \tag{13}$$

Under these conditions, the problem of IDDM parametric identification is solved in two stages. At the first stage, we solve the problem (14) for the main part of the interval data set:

$$\delta(\vec{\hat{g}}_l) \xrightarrow{\vec{\hat{g}}_l} \min, \tag{14}$$

where objective function $\delta(\widehat{\vec{g}}_l)$ now is determined by expressions (6) and (7) at the set of discrete values: $i = 1, \ldots, I_0, j = 1, \ldots, J_0, h = 1, \ldots, H_0, k = 1, \ldots, K_0$.

When such estimation $\widehat{\vec{g}}_{l=L} \in \Omega_0$ (which provides $\delta(\widehat{\vec{g}}_{l=L}) = 0$) will be found at a current iteration of the method for main part of interval data set, then we move to the second stage of searching of IDDM parameters. This stage implies analysis of the obtained current estimation of parameters for the testing part of the interval data set.

If the condition of ensuring of IDDM given accuracy for testing part of interval data set for $\widehat{\vec{g}}_{l=L} \in \Omega_0$ is fulfilled, then the found estimation of IDDM parameter vector provides the possibility to build an adequate model. Otherwise, it is necessary to continue searching for other optimal estimation of the IDDM parameters in the main sample, scilicet, move to the first stage of solving the IDDM parametric identification problem.

Results of analysis of the presented scheme of computational implementation for the method of IDDM parametric identification has shown that for all points generated at a current iteration in the parameter space it is necessary to calculate the value of the objective function. It is well known that this procedure is the most complex in the method of IDDM parametric identification [12]. In addition, as it was noted earlier, to calculate the value of the objective function $\delta(\widehat{\vec{g}})$ using expressions (6) and (7), firstly it is necessary to predict the modeled characteristic values using the IDDM (1).

Therefore, we propose in this paper to consider an area of permissible values of the modeled characteristic when implementing the parametric identification method. This will allow reducing the number of calculations of the objective function values $\delta(\widehat{\vec{g}})$ which, in turn, will reduce the time complexity of implementation of the IDDM parametric identification method.

To do that, before using the method, researcher should specify an area of permissible values of the modeled characteristic as follows:

$$[\vec{v}_{i,j,h,k}^{\min}, \vec{v}_{i,j,h,k}^{\max}],$$
$$\forall i = d,\ldots,I, \forall j = d,\ldots,J, \forall h = d,\ldots,H, \forall k = d,\ldots,K; \quad (15)$$

where $\vec{v}_{i,j,h,k}^{\min}, \vec{v}_{i,j,h,k}^{\max}$ are vectors of minimal and maximal permissible values of the modeled characteristic, respectively; d is the IDDM order.

Note that the value of the interval (15) should be set by researcher empirically based on the analysis of physical characteristics of the modeled process. Then the following step of checking condition should be added to the computational scheme of implementation of parametric identification method:

$$[\hat{v}_{i,j,h,k}] \subset [v_{i,j,h,k}^{\min}, v_{i,j,h,k}^{\max}],$$
$$\forall i = d, \ldots, \mathrm{I}, \forall j = d, \ldots, J, \forall h = d, \ldots, H, \forall k = d, \ldots, K. \quad (16)$$

If condition (16) is fulfilled, then we calculate the objective function $\delta(\widehat{\vec{g}})$ value for the current estimation of the IDDM parameter vector, otherwise this point of the space of solutions will not be taken into account in further computations and the objective function $\delta(\widehat{\vec{g}})$ will not be calculated for it.

Thus, the following scheme will be used for each generated point in the parameter space $\widehat{\vec{g}}_p$ (p = 1,…, P). The below system (17) is formed that includes initial conditions, an interval equation for the first discrete step and limitation for permissible area of the modeled characteristic in this discrete step:

$$\begin{cases} [\widehat{v}_{0,0,0,0}^{-};\widehat{v}_{0,0,0,0}^{+}] \subseteq [z_{0,0,0,0}^{-};z_{0,0,0,0}^{+}] \\ \vdots \\ [\widehat{v}_{d,d,d,k}^{-};\widehat{v}_{d,d,d,k}^{+}] \subseteq [z_{d,d,d,k}^{-};z_{d,d,d,k}^{+}] \\ [\widehat{v}_{d+1,d,d,k}] = \vec{f}^{T}([\widehat{v}_{0,0,0,0}],\ldots,[\widehat{v}_{d,d,d,k}],\vec{u}_0,\ldots,\vec{u}_k)\cdot\widehat{\vec{g}} \\ v_{d+1,d,d,k}^{\min} \le \vec{f}^{T}([\widehat{v}_{0,0,0,0,}],\ldots,[\widehat{v}_{d,d,d,k}],\vec{u}_{i,j,0},\ldots,\vec{u}_{i,j,k})\cdot\widehat{\vec{g}} \le v_{d+1,d,d,k}^{\max} \end{cases} \tag{17}$$

Then the system (18) is formed which includes initial conditions, the result of solving of the previous system (17), the interval equation for the next discrete step and limitation for permissible area of the modeled characteristic in this step:

$$\begin{cases} [\widehat{v}_{0,0,0,0}^{-};\widehat{v}_{0,0,0,0}^{+}] \subseteq [z_{0,0,0,0}^{-};z_{0,0,0,0}^{+}] \\ \vdots \\ [\widehat{v}_{d,d,d,k}^{-};\widehat{v}_{d,d,d,k}^{+}] \subseteq [z_{d,d,d,k}^{-};z_{d,d,d,k}^{+}] \\ [\widehat{v}_{d+1,d,d,k}^{-};\widehat{v}_{d+1,d,d,k}^{+}] \subseteq [v_{d+1,d,d,k}^{\min};v_{d+1,d,d,k}^{\max}] \\ [\widehat{v}d+1,d+1,d,k] = \vec{f}^{T}([\widehat{v}_{0,0,0,0}],\ldots,[\widehat{v}_{d+1,d,d,k}],\vec{u}_0,\ldots,\vec{u}_k)\cdot\widehat{\vec{g}} \\ v_{d+1,d+1,d,k}^{\min} \le \vec{f}^{T}([\widehat{v}_{0,0,0,0}],\ldots,[\widehat{v}_{d+1,d,d,k}],\vec{u}_{i,j,0},\ldots,\vec{u}_{i,j,k})\cdot\widehat{\vec{g}} \le v_{d+1,d+1,d,k}^{\max} \end{cases} \tag{18}$$

Generating such ISNAEs is continued for $\forall i = d,\ldots,I$, $\forall j = d,\ldots,J$, $\forall h = d,\ldots,H$, $\forall k = d,\ldots,K$ under condition that each previously generated system is consistent. In such case, the value of the objective function $\delta(\widehat{\vec{g}})$ is calculated for current parameter vector. After that, one should repeat the procedure for the next parameter vector $\widehat{\vec{g}}_p$, p = 1, …, P.

In the case if any of generated systems is inconsistent, the process of generating of such ISNAEs is stopped and the value of objective function is not calculated for the current parameter vector.

Overall, consistency of all generated systems ensures fulfilling condition (16) on a set of all discrete steps. In turn, the implementation of condition (16) means that the current vector of parameters $\widehat{\vec{g}}$ provides possibility to build such mathematical model for which the predicted values of the modeled characteristic be matched with physical features of this characteristic.

Overall, taking into account permissible values area when solving problem of IDDM parametric identification at all iterations of the method implementation will provide reducing the time complexity of its using. Figure 2 shows the flowchart of

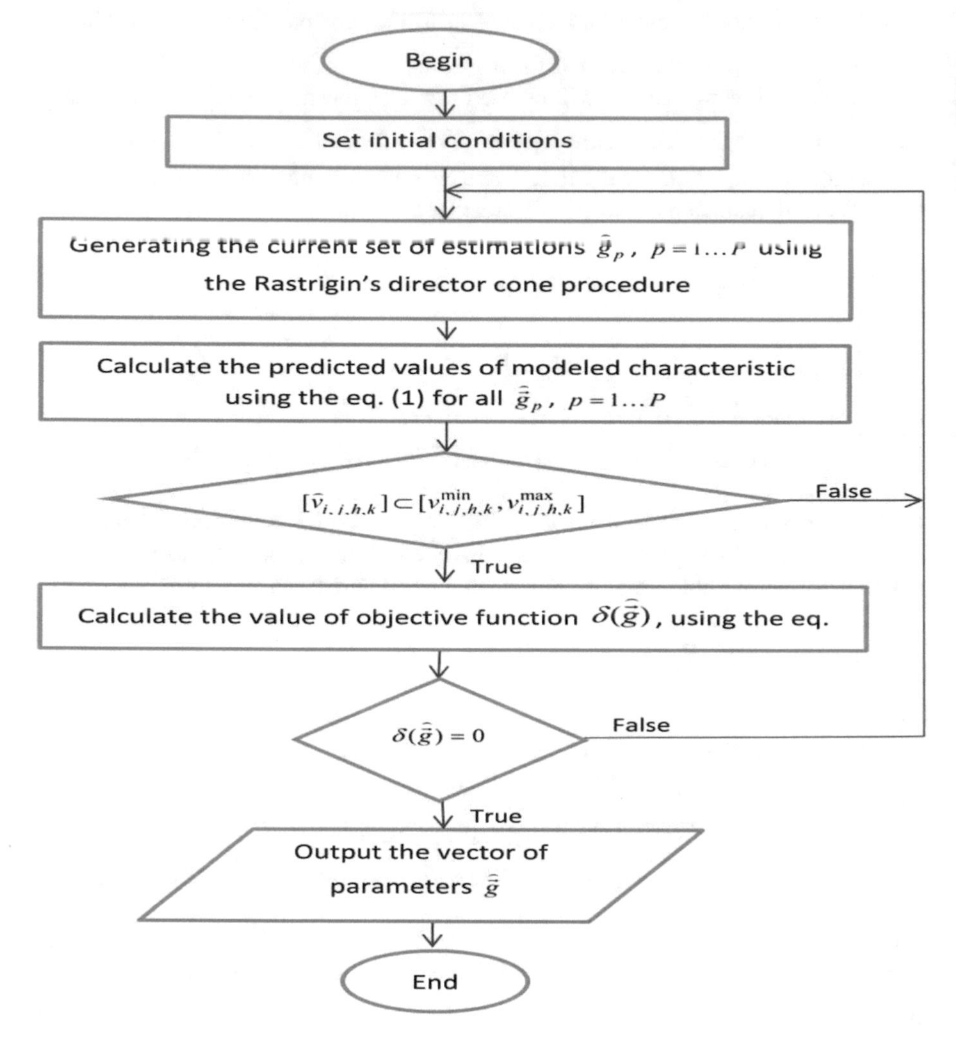

Fig. 2. Flowchart of implementation of the improved computational scheme of the IDDM parametric identification method

computational implementation scheme of parametric identification of the IDDM method with improved procedure for calculating values of the objective function $\delta(\hat{\vec{g}})$.

5 Experimental Research

To conduct the experiments, a software system was developed for IDDM parametric identification. For development of the software system, the object-oriented approach, C# programming language and .NET technology were used. For comparison of efficiency of the known and improved computational implementation scheme of the parametric identification method, five computational experiments were carried out. The experiments were conducted on the example of solving the prediction task of the humidity distribution on the surface of the drywall sheet during its manufacturing [3]. The values of relative humidity on the drywall sheet surface for ensuring manufacturing of high quality product must be in the range from 0.6% to 0.9%.

During each experiment, 40 researches were conducted, 10 of which using known implementation scheme of the IDDM parametric identification method and another 30 using the described improved scheme with different intervals of permissible values of humidity on the surface of the drywall sheet.

When conducting researches using the improved scheme of the method implementation, the following variants of permissible values of the relative humidity were used: $[v_k^{\min} = 0.001, v_k^{\max} = 1]$, $[v_k^{\min} = 0.001, v_k^{\max} = 2.5]$, $[v_k^{\min} = 0.001, v_k^{\max} = 5]$.

Note that based on physical characteristics of the modeled process, the permissible values of the relative humidity were set constant at all grid nodes.

The following structure of the IDDM was used during the research:

$$\begin{aligned}[\hat{v}^-_{i,j,k}; \hat{v}^+_{i,j,k}] = {} & g_1 + g_2 \cdot (u_{1,0} \cdot u_{2,k}/u_{2,0} \cdot u_{1,k}) \cdot [\hat{v}^-_{i-1,j-1,k}; \hat{v}^+_{i-1,j-1,k}] \\ & + g_3 \cdot [\hat{v}^-_{i-1,j-2,k}; \hat{v}^+_{i-1,j-2,k}] + g_4 \cdot [\hat{v}^-_{i,j-1,k}; \hat{v}^+_{i,j-1,k}] \\ & + g_5 \cdot (u_{1,0} \cdot u_{2,k}/u_{2,0} \cdot u_{1,k}) \cdot [\hat{v}^-_{i-1,j,k}; \hat{v}^+_{i-1,j,k}],\end{aligned} \tag{19}$$

where $\hat{v}_{i,j,k}$ is the humidity level at the point with coordinates i, j on the surface of the drywall sheet in the time moment k; $u_{1,0}$, $u_{1,k}$ are temperatures in the drying oven at the moment k; $u_{2,0}$, $u_{2,k}$ are predicted moving speeds of drywall sheet in a drying oven for the moment k.

The following initial values of the temperature and moving speed in a drying oven were used for calculation to predict the humidity level: $u_{1,0} = 120$ °C, $u_{2,0} = 0.25$ m/min, $u_{1,k} = 125$ °C, $u_{2,k} = 0.28$ m/min.

The described software system was used for conducting of the experiments. Figure 3 illustrates a screenshot of the software window with results of a research.

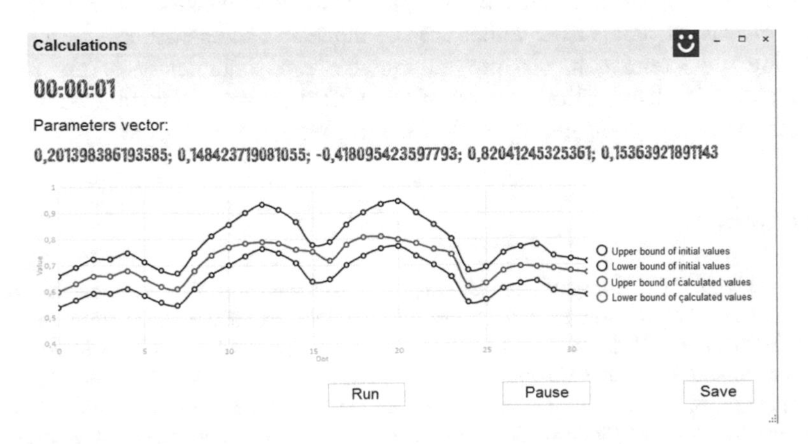

Fig. 3. Window of the software system with results of one conducted research

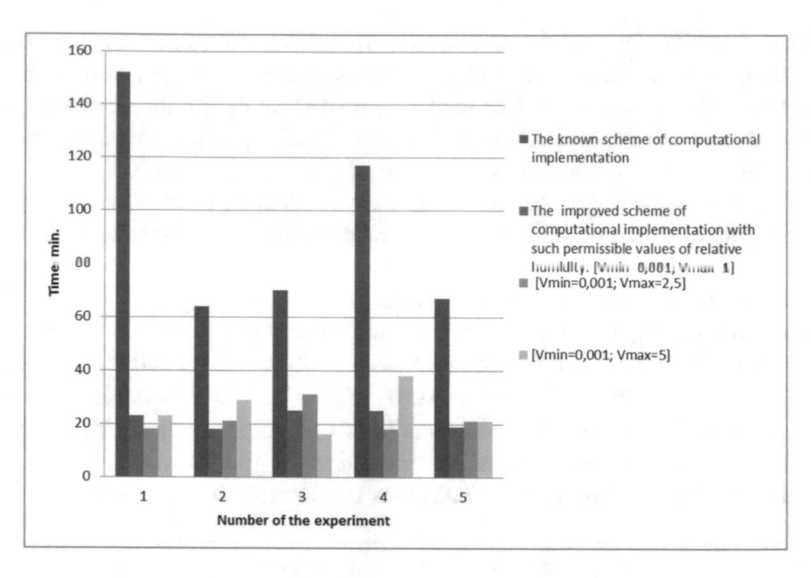

Fig. 4. Comparison of time complexity

Figure 4 demonstrates comparison of efficiency of the known implementation scheme of the parametric identification method and improved one with different intervals of permissible values of humidity on the drywall sheet surface. The highest values of time complexity indicator (in minutes) among ten researches within each experiment are shown in the diagram.

6 Conclusion

The method of parametric identification of interval discrete dynamic models is considered in the paper. An improved scheme of its computational implementation is proposed. The improved scheme of the parametric identification method takes into account an area of permissible values for the modeled characteristic.

During the research, it was proved that the time complexity of the improved computational implementation scheme of the IDDM parametric identification method is at least twice less compared to the known implementation scheme.

References

1. Ocheretnyuk, N., Voytyuk, I., Dyvak, M., Martsenyuk, Y.: Features of structure identification the macromodels for nonstationary fields of air pollutions from vehicles. In: Proceedings of XIth International Conference on Modern Problems of Radio Engineering, Telecommunications and Computer Science (TCSET 2012), p. 444. Lviv-Slavske (2012)
2. Porplytsya, N., Dyvak, M.: Interval difference operator for the task of identification recurrent laryngeal nerve. In: Proceedings of the 16th International Conference on Computational Problems of Electrical Engineering (CPEE 2015), pp. 156–158 (2015)

3. Porplytsya, N., Dyvak, M., Dyvak, T., Voytyuk, I.: Structure identification of interval difference operator for control the production process of drywall. In: Proceedings of 12th International Conference on the Experience of Designing and Application of CAD Systems in Microelectronics, CADSM 2013, pp. 262–264 (2013)
4. Fliess, M., Sira-Ramirez, H.: Closed-loop parametric identification for continuous-time linear systems via new algebraic techniques. In: Garnier, H., Wang, L. (eds.) Identification of Continuous-time Models from sampled Data, pp. 362–391. Springer (2008)
5. Graupe, D.: Identification of systems. Technol. Eng. (1976). Journal no. 12205
6. Sean, L.: Essentials of Metaheuristics, 2nd edn. Lulu, Raleigh (2013)
7. Bowden, R.: The theory of parametric identification. Econometrica **41**, 1069–1074 (1973)
8. Shary, S.P.: Algebraic approach to the interval linear static identification, tolerance, and control problems, or one more application of Kaucher arithmetic. Reliable Comput. **2**(1), 3–33 (1996)
9. Walter, E., Pronzato, L.: Identification of Parametric Model From Experimental Data. Springer, Heidelberg (1997)
10. Rastrigin, L.A.: A Random Search. Znanie, Moscow (1979). (in Russian)
11. Rastrigin, L.A.: Adaptation of Complex Systems. Zinatne, Riga (1981). (in Russian)
12. Dyvak, T.: Method of parametric identification of macro model in the form of interval difference operator with dividing of data sample. Inductive Model. Complex Syst. **3**, 49–60 (2011). (in Ukrainian)

Mechanisms to Enhance the Efficiency of Maritime Container Traffic Through "Odessa" and "Chernomorsk" Ports in the Balancing of Portfolios

Anatoly J. Gaida, Elena A. Zarichuk, and Konstantin V. Koshkin(✉)

National Shipbuilding University, Geroiv Ukrainy Ave., 9, Mykolaiv, Ukraine
{cetus,kkoshkin}@ukr.net, milena012@mail.ru

Abstract. Approaches to development classification mechanisms in condition of projects of marine container transportations by means of artificial neural networks are investigated. It is established that growth rates of transportation volumes not linearly depend on a ratio of volumes of import and export cargoes with obviously expressed maximum. Its offered mechanism of efficiency increase of the companies by a choice of a rational ratio volumes of import and export cargoes on the basis of a neural network assessment of the previous activity results of the company and its competitors which, unlike known mechanisms, provides possibility of an effective assessment of the situation which developed in the market of freight transportation.

Keywords: Management of projects · Sea transport · Classification
Neural network

1 Introduction

Maritime transport provides significant amounts of internal and external freight over long distances, it is an important component of the economy of coastal and maritime countries. In a competitive transport companies are faced with the problem of choosing an effective business acquisition strategy that provides adequate load vessels at admissible idle times.

The project management knows the position of the connection between resource efficiency and uniformity of loading. The high cost of ships and their operation generates interest maritime transport companies to the fullest use of available resources through the formation of portfolios of orders for shipping, which will provide a possible continuous and uniform loading of vessels [4, 5]. In addition in order to increase profits, such as loading of vessels can significantly reduce the additional costs associated with the peculiarities of navigation and safety of maritime transport, for example, caused by the need to idle, as the vessel is anchored or admission to under loading ship ballast water.

N. Shakhovska and V. Stepashko (eds.), *Advances in Intelligent Systems and Computing II*, Advances in Intelligent Systems and Computing 689,
https://doi.org/10.1007/978-3-319-70581-1_9

1.1 Relation of Highlighted Issue to Scientific and Practical Tasks

In the simplest case, the condition of continuous and uniform loading of vessels is achieved in a situation where in each port of loading and unloading cargo discharged is replaced by an equivalent volume, weight and requirements for transporting cargo for delivery to the ports on the route of the vessel. Unfortunately, such a balancing circuit is often unachievable due to the impact of a number of factors, such as competition between carriers, seasonal traffic fluctuations in the imports and exports volume of the countries, the orientation of certain ports for certain types of goods, and etc. As a consequence, marine transport companies have to contend with incomplete congestion of vessels, lost profits and the risk of possible loss of the transport market [7, 9].

The aim of the article: research of load balancing indicators for resources transportation companies and the development of mechanisms to improve project management effectiveness of marine transport by improving the balance of the load on the resources in accordance with the state of each company and the transport market as a whole.

1.2 Analysis of Recent Researches and Publications

The methodology of the research. To achieve this goal will perform data collection based on the results of marine transport companies work in "Odessa" and "Chernomorsk" ports, to analyze the data to identify the relationships between levels of balanced portfolio of orders and the growth in maritime freight transport methods of comparative analysis, time series analysis, correlation and regression analysis, the method of neural models.

Statement of the main research material. Delivery of goods through "Odessa" and "Chernomorsk" ports provided by shipping lines: CMA, ACOL, ANL, APL, ARKAS, BUL, COSCO, CSCL, ECM, EMES, EVERGR, HJS, HLC, HMM, KLINE, MAE, MISC, MOL, MSC, NOR, NYK, OOCL, PIL, UASC, WHL, YMG, ZIM and others [8–12]. Due to the significant volume of initial data on volumes of freight traffic through "Odessa" and "Chernomorsk" ports on mentioned shipping lines such data are represented in graphs (Fig. 1 – cargo turnover data; Fig. 2 – data on the ratio of import and export cargo volume).

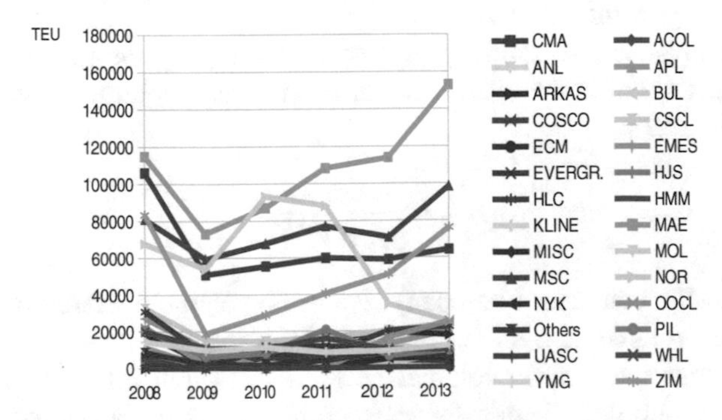

Fig. 1. The volume of freight traffic (in TEU) through "Odessa" and "Chernomorsk" ports along the lines of (2008–2013)

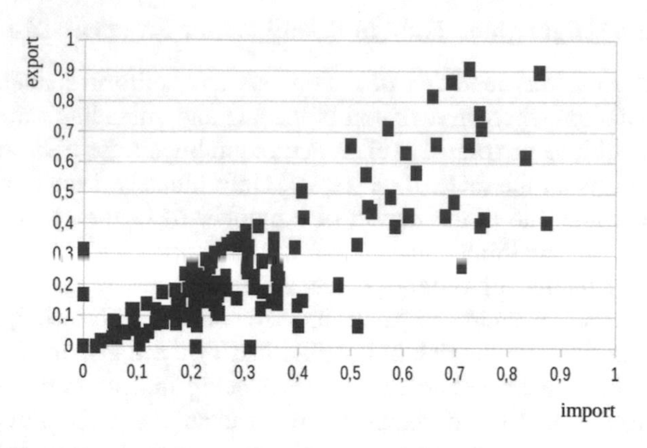

Fig. 2. Allocation of import and export cargoes (TEU) through "Odessa" and "Chernomorsk" ports (2008–2013)

As is evident from Fig. 1, the volume of container shipping for the past few years in general increased slightly – despite the considerable drop of transportation in 2009 caused by global crisis 2008–2009. After a recession in 2009 traffic volumes generally rose slightly, and according to the data for 2013 almost reached pre-crisis levels.

2 The Main Objectives of the Research and Their Meaning

Taking into account available data project management efficiency of each company can be assessed through growth indicator volume transportation. Total volume of traffic of each company is:

$$V_i = \sum_{j=s}^{f} v_{i,j}, \tag{1}$$

where: *f*, *s* – year end and beginning of the period, respectively; $v_{i,j}$ – the volume of traffic "*i*" company in the "*j*" year.

Then, the average annual traffic growth of each company for a certain period can be calculated as the ratio of in the relative increment in traffic volume during that period to a long period in years:

$$k_i = \frac{v_{\mathrm{i,f}} - v_{\mathrm{i,s}}}{(f - s)V_i}, \tag{2}$$

where: k_i – growth coefficient i company; $v_{i,f}$ – the volume of traffic at the end of the period; $v_{i,s}$ – the volume of traffic at the end of the period.

Under the impact on traffic volumes of random factors (as is the case here), the average annual traffic growth rate for each company for the analyzed period can be more accurately expressed by the coefficient a of the linear regression $y = ax + b$:

$$a_i = \frac{\sum_{j=1}^{m} v_{i,j} * \sum_{j=1}^{m} j - m * \sum_{j=1}^{m} j * v_{i,j}}{\left(\sum_{j=1}^{m} v_{i,j}\right)^2 - m * \sum_{j=1}^{m} v_{i,j}^2}, \quad (3)$$

where: a_i – the growth factor of traffic with linear regression, j – the position of the indicator of traffic volumes $v_{i,j}$ in the time series, m – the number of elements of the time series.

The calculated gain coefficients traffic volumes for expression (2) are shown in Table 1, for expression (3) – in Table 2. As can be seen from Fig. 1 and Tables 1, 2, despite the unfavorable conjuncture in 2009, some companies were able to significantly increase the volume of traffic. Among the leaders highlighted Maersk company (shipping line MAE), which is not only the largest carrier, but also has a high average annual growth of 8.8% (Table 2) and absolute growth of 33.2% (Fig. 1). Impressive results have also demonstrated the company Mediterranean Shipping Company (MSC line, respectively, 5.0% and 21.6%) and the company ZIM (eponymous line). The latter, although it showed a slight decline in general, showed the best among large companies after the collapse of the growth rate of 2009.

Table 1. The coefficients of the average annual traffic growth for expression (2)

№	Line	Grow rate	№	Line	Grow rate
1	ACOL	−0,106	15	KLINE	−0,129
2	ANL	0,089	16	MAE	0,059
3	APL	0,163	17	MISC	0,123
4	ARKAS	0,189	18	MOL	−0,046
5	BUL	0,265	19	MSC	0,039
6	CMA	−0,990	20	NOR	−0,118
7	COSCO	−0,103	21	NYK	−0,426
8	CSCL	−0,073	22	OOCL	0,042
9	ECM	−0,719	23	PIL	−0,164
10	EMES	−0,719	24	UASC	−0,751
11	EVERGR	0,224	25	WHL	−0,132
12	HJS	−0,842	26	YMG	−0,040
13	HLC	−0,092	27	ZIM	−0,024
14	HMM	0,149	28	Others	−0,081

In order to identify efficiency of resource management consider the connection between the import and export cargo volume (Fig. 2). In the most simple case, this link can be expressed in terms of linear correlation coefficient [1–3] calculated between the amount of import and export goods for every single company. Company's results of correlation calculations are shown in Table 2 (in the calculation of available authors full data have been used for months).

Table 2. The coefficients of the average annual traffic growth for expression (3)

№	Line	Grow rate	№	Line	Grow rate
1	ACOL	0,099	15	KLINE	−0,113
2	ANL	0,306	16	MAE	0,088
3	APL	0,300	17	MISC	0,191
4	ARKAS	0,386	18	MOL	−0,019
5	BUL	−0,854	19	MSC	0,050
6	CMA	−0,078	20	NOR	−0,131
7	COSCO	−0,055	21	NYK	−0,331
8	CSCL	−0,039	22	OOCL	0,150
9	ECM	−0,761	23	PIL	−0,109
10	EMES	−0,761	24	UASC	−0,772
11	EVERGR	0,312	25	WHL	−0,109
12	HJS	−0,803	26	YMG	−0,047
13	HLC	−0,008	27	ZIM	−0,024
14	HMM	0,217	28	Others	−0,041

As can be seen from Tables 1 and 2 growth data for individual companies have significant discrepancies. Further we will use the data from Table 2, which, unlike Table 1, take into account the traffic volumes for all the years of the period, and therefore more accurately reflect the growth trends in the volume of traffic companies.

As can be seen from Fig. 1 and Table 2, between the volume of import and export cargo takes place significant correlation due to aspiration provide company managers possible more complete and balanced load of vessels. Obviously, this regularity is achieved by the replacement of discharged goods at ports (most of which – import) taking on board cargo (export).

Nevertheless, individual companies, this correlation is far enough from 1.0. For example, for a large and rapidly increase volumes of traffic, Maersk correlation coefficient of 0.959, this corresponds to the level of the eighth balanced load on resources of the positions among more than 27 companies that provide transportation through "Odessa" and "Chernomorsk" ports. Similar deviations are reviewed on several other companies, which in this case indicates a mixed growth dependence on the level of traffic load balanced resource.

Table 2 shows that Maersk Company (MAE line) in terms of growth takes only eighth position, which indicates the existence of reserves for future growth. The position of Maersk becomes even more significant because companies with smaller volumes of freight traffic and, consequently, larger relative overhead costs, able to get ahead of Maersk in terms of growth.

In order to determine the relationship between the values of uniformity level of loading resources and value of the relative growth of the company's share in cargo transportation, consider the relationship between the growth in traffic volume and the companies defined above correlation coefficients reflecting the regularity loading of resources (Table 3, Fig. 3). Figure 3 also shows the linear regression (dashed line). As can be seen from the figure, "a" coefficient of linear regression equation $y = ax + b$ is

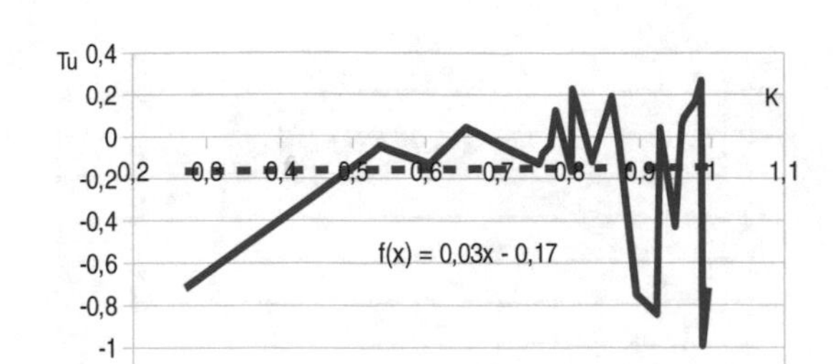

Fig. 3. Dependence of the rate of the relative growth companies *Tu* from the uniform level of loading resources *K* (dotted trend line set)

Table 3. The coefficients of the linear correlation between the volume of import and export cargo

№	Line	Correl.	№	Line	Correl.
1	ACOL	0,962	15	KLINE	0,757
2	ANL	0,977	16	MAE	0,959
3	APL	0,861	17	MISC	0,781
4	ARKAS	0,986	18	MOL	0,538
5	BUL	0,987	19	MSC	0,929
6	CMA	0,760	20	NOR	0,832
7	COSCO	0,740	21	NYK	0,949
8	CSCL	0,529	22	OOCL	0,656
9	ECM	0,271	23	PIL	0,803
10	EMES	0,995	24	UASC	0,895
11	EVERGR	0,805	25	WHL	0,605
12	HJS	0,923	26	YMG	0,773
13	HLC	0,830	27	ZIM	0,871
14	HMM	0,811	28	Others	0,762

0.03, which generally indicates a slight increase with the growth of freight traffic load balanced.

In order to detect changes in traffic volume trend performed polynomial smoothing of rates depending on the company's growth from a uniform level of resource utilization. Smoothing results are presented in Fig. 4. A polynomial curve flattening is designed for smoothing exponent 3. Here, the trend line indicates that increasing resource utilization level of uniformity results in a corresponding significant change in the rate of growth of the company's operations. In the low level of resources balanced load values (here – up to the value of the correlation coefficient of 0.7) achieved significant growth, and above – a significant drop of transportation volumes. Polynomial fourth-order gives close to the Fig. 4, curve smoothing, which confirms the convex nature of the trend distribution with one peak.

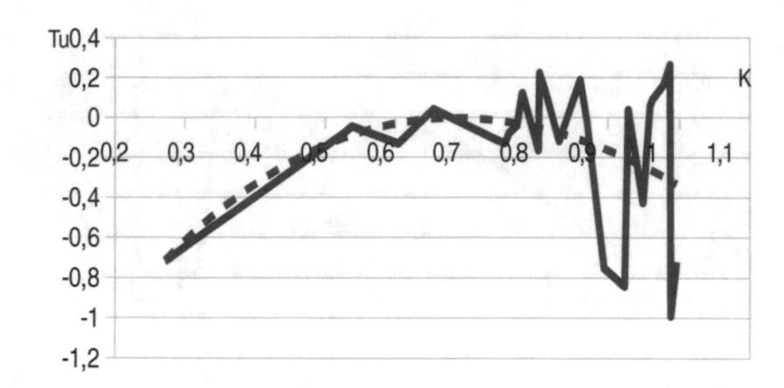

Fig. 4. The results of the smoothing depending on the relative rates of growth of the company *Tu* from the uniform level of loading resources *K* (third-degree polynomial trend line set a dotted line)

Considering the significant fluctuations in traffic volumes and, above all, their import component, in order to represent the company's previous history in the traffic market, it is important to clear the performance of companies from changes caused by fluctuations in total traffic volumes. This can be achieved most simply by calculating the correlation coefficient between the total volumes of traffic and the company's share (including import and export shipments). Such data are presented below: for import and export operations (Fig. 4), separately for import (Fig. 5) and separately for export (Fig. 6) operations.

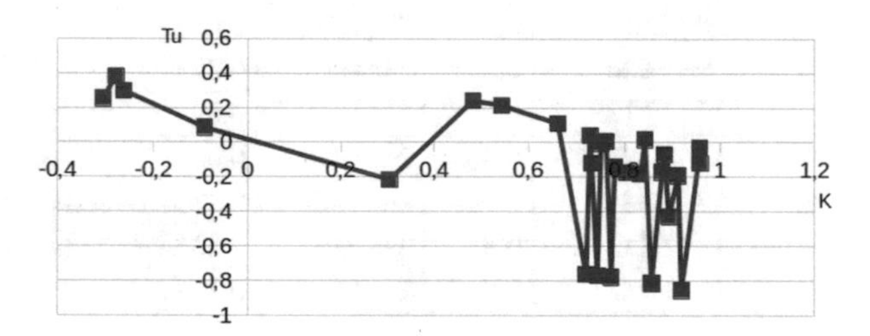

Fig. 5. Dependence of the rates of relative growth of companies on the consistency of the volumes of import traffic

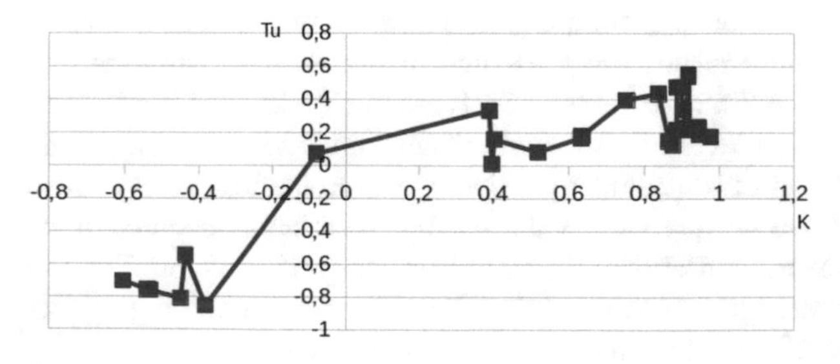

Fig. 6. Dependence of the rates of relative growth of companies on the consistency of the volumes of export traffic

As can be seen from Fig. 5, with a general drop in the volume of traffic by import, the growth rates of companies are also decreasing, thus the growth of companies is in a negative correlation with the company following the current market trends. Similarly, from Fig. 6, with the overall growth in the volume of traffic by export, the growth rates of companies also increase, so the growth of companies is in a positive correlation with the company following the current market trends. At the same time, the correlation between traffic volumes and growth rates is insignificant and is 0.24. Obviously, differently directed dependencies are a consequence of a significant excess of imports over exports to these ports.

The obtained results allow us to consider the problem of finding the optimum ratio of volumes received/discharged as the problem of maximizing on a convex curve, and the task of improve management efficiency – as the task of identifying dependencies, ensuring high growth rates in the most successful from this point of view of companies.

Mathematical formulation of this problem can be worded as follows: it is necessary to find a value for the level of load evenly on resources which in the current market position and market conditions will maximize the value of the index rate of growth of the company. This problem can be solved as a problem of clustering and classification [13, 14].

Based on the available data for each prediction cargo turnover in some of the following moment in time $n + 1$ can be performed. This forecast can be viewed as a function of the time series of indicators of its previous cargo turnover viewed through the ports, many companies state, working in these ports, and the ratio of values of import and export goods (see Fig. 4).

$$v_{i,n+1} = G\left(V_{i,n}, S_n, x_{n+1}\right), \tag{4}$$

where: x_{n+1} – expected ratios of values of import and export cargo at time $n + 1$.

Imagine the state of the "i" – company at the time "n" as a function of time series the values of achieved cargo turnover:

$$s_{i,n} = g\left(v_{i,s}, v_{i,s+1}, \ldots, v_{i,s+n}\right), \tag{5}$$

where: g – calculation function state – cargo turnover (see (1)).

The states of all companies can be represented as a matrix:

$$S_n = \left\{s_{1,n}, s_{2,n}, \ldots, s_{m,n}\right\}, \tag{6}$$

The time series of cargo turnover – as a matrix:

$$V_{i,n} = \left\{v_{i,s}, v_{i,s+1} \ldots, v_{i,s+n}\right\}, \tag{7}$$

To substitute (5) and (6) in (4), we find the expected cargo turnover of each company at some assignment of ratios of import and export cargo.

The problem of calculating the forecast values can be solved by means of an artificial neural network with back-propagation errors. The number of network inputs should be chosen according to the number of time series values of n ($n - 1$ value of a time series and one coefficient value of import and export cargo relations), the last value of the time series in the training must be seen as a prediction result. With a significant number of time-series elements can be introduced horizon cyclically shifted training the time series. Solution of the problem is a value ratio of received and discharged cargoes, which ceteris paribus will provide the greatest increase in cargo turnover.

Popular neural networks are networks with back-propagation errors. In the simplest case, a single-layer network its output "*s*" can be represented by a vector of input signals and "*x*" coefficient vector "*k*" as:

$$s_c = f\left(\sum_{i=0}^{n} x_i k_{i,c} + b_{i,c}\right), \quad (8)$$

where: s_c – the output status for the class "*c*"; x_i – *i* value of the parameter at the input; $k_{i,c}$ – transfer coefficient of "*i*" entry on "*c*"-output; "*f*" – the activation function; $b_{i,c}$ – displacement.

As can be seen from (4), a single-layer network is able to reproduce the linear dependence of the output values from the input and their subsequent transformation activation function. In the case of more complex dependencies apply multilayer network. For a two-layer network output value s:

$$s_c = f\left(\sum_{j=0}^{n}\left(k_{j,c} f\left(\sum_{i=0}^{m} x_i k_{\mathrm{j,i}} + b_{\mathrm{j,i}}\right)\right)\right), \quad (9)$$

where: $k_{\mathrm{j,c}}$ – transmission coefficient intermediate "*j*" element of the intermediate layer on "*c*" element output; x_i – "*i*" value of the parameter at the input; $k_{j,i}$ – transmission ratio "*i*" input of the input layer to the intermediate "*j*"-input.

Test ANNs trained on known data sets (Figs. 1 and 2, Table 2) has been implemented with the aim of testing the quality of cargo turnover of volumes forecasting growth.

Justification of results. The results predict cargo turnover of individual companies received for fixed values of the correlation of import and export of goods are shown in Fig. 7. In Fig. 6 shows the results of simulation of the relative of volumes cargo turnover in some companies in 2013 with different ratios of volumes import and export cargo.

From the modeling results can be seen that generally increase in the level of resources provides balanced load cargo handling increase growth (Fig. 7).

At the same time, for ACOL lines, MAE, MSC growth reserves are almost exhausted (Figs. 8 and 9), which clearly indicates the high quality of management in these companies. For other companies, the growth of reserves are significant.

The selection task factor of a parity of import and export goods, providing the greatest increase in competition between companies can be resolved by known methods of research.

In general, testing results have shown the ability to improve maritime transport companies' management.

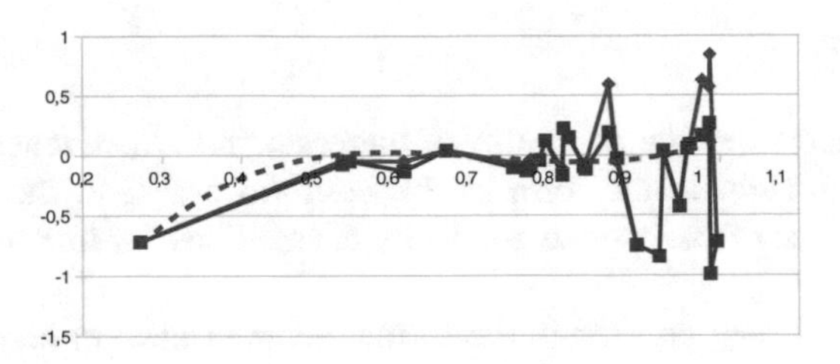

Fig. 7. The results of forecasting for companies that constantly were present on shipping market via the "Odessa" and "Chernomorsk" ports (trend line shown in phantom)

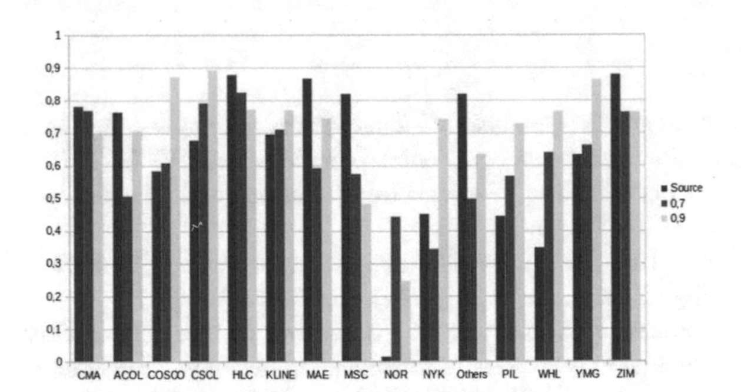

Fig. 8. Forecast of the relative traffic volumes for 2013 through the "Odessa" and "Chernomorsk" for some companies at different values of the correlation of "imports" and "exports" amounts (for testing)

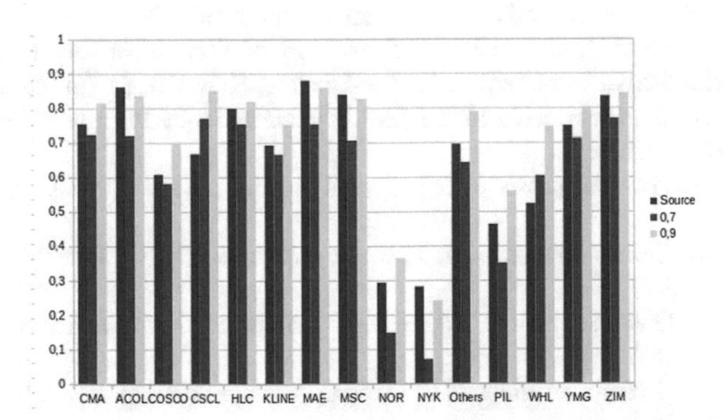

Fig. 9. Forecast of the relative traffic volumes for 2014 through the "Odessa" and "Chernomorsk" for some companies at different values of the correlation of "imports" and "exports" amounts

3 Conclusion

The obtained results show the possibility of increasing the effectiveness of management of marine transportation in the form of the portfolio orders to the best, in terms of growth in the volume of transported goods, by the relations "of import" and "export" of goods.

Obtained results may be used in the formation and transport companies balancing portfolios of projects, as well as in dealing with other project management tasks for which the available data on the levels of uniformity of loading of resources and associated performance of the company.

References

1. Boks, D., Dzhenkins, G.: Analysis of Time Series. Prognosis and Management [Analiz vremennyih ryadov: prognoz i upravlenie]. Mir, Moscow (1974). (in Russian)
2. Byvshev, V.A.: Econometrics [Ekonometrika]. Finance and Statistics, Moscow (2008). (in Russian)
3. Venikov, V.A.: The Theory of Similarity and Modeling [Teoriya podobiya i modelirovaniya]. High School, Moscow (1976). (in Russian)
4. Gaida, A.J., Gordeev, B.N., Koshkin, K.V.: Features of Management Knowledge-Intensive Industries in Shipbuilding Projects [Osobennosti upravleniya resursami proektov naukoemkih proizvodstv v sudostroenii]. NUS, Nikolaev (2011). (in Russian)
5. Koshkin, K.V., Gaida, A.J.: Recourse Management Portfolio Knowledge-Intensive Production Projects in the System with Predictive Model [Upravlenie resursami portfelya proektov naukoemkih proizvodstv v sisteme s prognoziruyuschey modelyu]. Publishing House "Science Book", Voronezh (2013). (in Russian)
6. Korn, G., Korn, E.: Mathematical Handbook for Scientists and Engineers [Spravochnik po matematike dlya nauchnyih rabotnikov i inzhenerov]. Nauka, Moscow (1984). (in Russian)
7. Grebenik, E.: Two of Brooklyn: The Success Story of Yuri Gubankova [Dvoe iz Bruklina: istoriya uspeha Yuriya Gubankova]. Forbes-Ukraine, Kiev (2015). (in Russian)
8. BlackSeaLines: Golden container-2008 [Zolotoy konteyner-2008]. Black Sea Lines, Odessa (2009). (in Russian)
9. BlackSeaLines: Peak fall passed [Pik padeniya proyden]. Black Sea Lines, Odessa (2010). (in Russian)
10. BlackSeaLines: Golden container 2010 [Zolotoy konteyner 2010]. Black Sea Lines, Odessa (2011). (in Russian)
11. BlackSeaLines: Golden container 2011 [Zolotoy konteyner 2011]. Black Sea Lines, Odessa (2012). (in Russian)
12. BlackSeaLines: Golden Container 2012 [Zolotoy konteyner 2012]. Black Sea Lines, Odessa (2013). (in Russian)
13. Karpov, L.E., Yudin, L.E.: Adaptive management on precedents, based on the classification of the states of managed objects [Adaptivnoe upravlenie po pretsedentam, osnovannoe na klassifikatsii sostoyaniy upravlyaemyih ob'ektov]. In: Proceedings of the Institute for System Programming of the Russian Academy of Sciences, Moscow (2007). (in Russian)
14. Zipkin, J.Z.: Fundamentals of the Theory of Learning Systems [Osnovyi teorii obuchayuschihsya sistem]. Nauka, Moscow (1970). (in Russian)

Porting a Real-Time Objected Oriented Dependable Operating System (RODOS) on a Customizable System-on-Chip

Muhammad Faisal(✉) and Sergio Montenegro

Institute of Aerospace Information Technology, Julius-Maximilians-University Wuerzburg, Würzburg, Germany
{muhammad.faisal,sergio.montenegro}@uni-wuerzburg.de

Abstract. Modern semiconductor chips offer a FPGA, A Hard Microcontroller and a programmable Analog circuitry all integrated on a single chip, which gives the system designer a full featured, easy-to-use design and development platform where all the units are programmable and under full control of the designer, Combining this state of the art silicon chip with a Real Time Operating Systems (RTOS) gives an Engineer full power and all degree of freedoms to design end-use applications with an unparalleled performance characteristic as far as speed, Security, Simplicity, Flexibility and reliability is concerned. In this paper we describe.

Keywords: FPGA · Real-Time-Operating-System · Porting

1 Introduction

Last decade has witnessed cutting edge inventions in the field of Systems-on-Chip which has provided engineers an out-of-the-box solution for high configurability, Reliability, Accuracy, Area and power efficiency in an embedded systems. Over the years engineers have relied on the traditional microcontrollers to design their systems but the micro-controllers have a bottleneck for the speed as the clock rate cannot go very high because of large amount of heat generation so to overcome this limitation FPGAs had come in this domain but the complex parallel design flow of the FPGAs slows the development time and dealing with the analog signals with external ADC and DAC is always challenging but the fusion of the microcontroller, FPGA and programmable analog in a single chip is an integrated and comprehensive approach to design an ideal platform for control and processing (Fig. 1).

This fused single chip as shown in Fig. 2 has all the intelligence of a microcontroller, Speed through FPGA and programmable Analog circuitry to deal with the real world applications. Microcontroller is relieved from many initializations of the analog components as the analog controller takes this responsibility on his own shoulders. All the programmable units on the single chip make the system highly flexible and reduce the footprint of the hardware. All the data is locally transferred between these programmable

N. Shakhovska and V. Stepashko (eds.), *Advances in Intelligent Systems and Computing II*, Advances in Intelligent Systems and Computing 689,
https://doi.org/10.1007/978-3-319-70581-1_10

Fig. 1. Block diagram of mixed-signal integrated circuit

computational units with no overhead as described in Fig. 2. This fused integrated chip provides engineer great power but great power comes with great responsibility and to fulfill this responsibility a RTOS is considered to be the best resource -manager of an embedded systems. RODOS is a real-time-operating-system for embedded systems and it was designed for application domains demanding high dependability [2]. RODOS is jointly developed by the Central Core Avionics department of German Aerospace Center and Chair of Aerospace Information Technology University of Wuerzburg Germany. RODOS is specifically developed for aerospace applications as it has minimal footprint but it is also very well suited to all applications that demand high dependability [3]. It is used for the current micro satellite programs of German Aerospace Center (DLR). RODOS runs on the operational satellite TET-1(Technology Experiment Carrier-1) [4] and it will be used for the currently developed satellite BIROS (Bi-spectral Infrared Optical System) [5]. RODOS is further enhanced and extended at the German Aerospace Center as well as the department for aerospace information technology at the University of Wuerzburg.

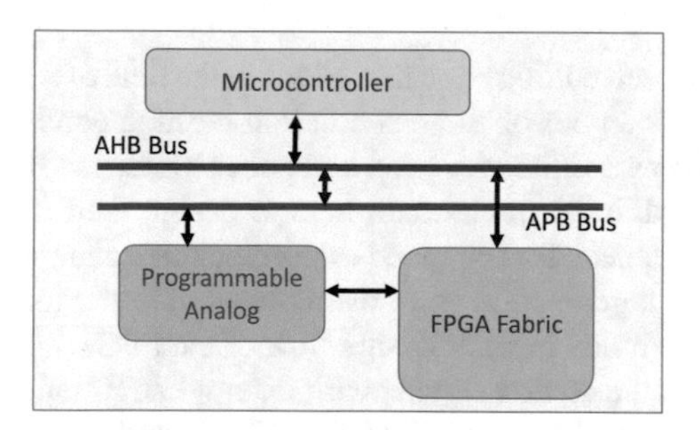

Fig. 2. Inter-system communication in mixed signal IC

This project is performed as a part for the Moon Landing mission of the Part-Time-Scientist team of Germany which are building an Autonomous Landing and Navigation Module (ALINA) as the first private mission to the Moon [6]. Being the project partner the Aerospace Information Technology department in University of Wuerzburg is developing the software for the navigation and control, It programs both the lander as well as rover on the moon, The department will provide interfaces to monitor the flight to the moon, beginning at low earth orbit (LEO) until the landing. The Part-Time-Scientist has chosen the ACTEL Smartfusion chip [7] as their on board computer. Aerospace Information

Technology department in University of Wuerzburg is using RODOS in all the recently launched satellites which it plans to use again for ALINA mission, That's why porting RODOS on Smart-Fusion platform was a big requirement of ALINA mission. Figure 3 shows the conceptual design of the on-board-processor for the ALINA project with RODOS on Smart-Fusion. Basic strategy to port the RODOS on Smart-Fusion platform is described in Fig. 4. First of all the hardware platform must be configured and in this case the micro-controller will be configured by programming the FPGA. The tool-chain has to be set for cross-compiling and flashing the binaries on the target. The other important step is getting the CPU and other peripheral register definitions in the relevant header files, Setting the memory-map and also mentioning the starting address and sizes of volatile and non-volatile memory, Providing the mechanism for Timing, Context-Switching, Debugging and in the end performing the test cases to verify the correct operation of RODOS on the Smart-Fusion platform. Libero and Soft-Console were used as the software component from the manufacturer of the chip in the design flow of the whole porting process. The experienced which is gained in this porting process will set the guide line of the future porting of RODOS on the new hardware architecture.

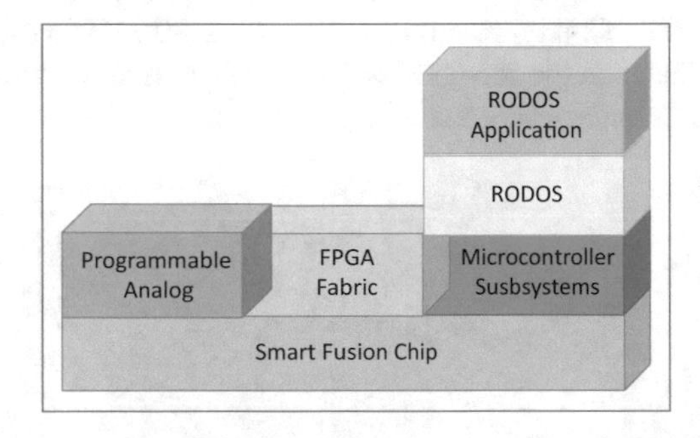

Fig. 3. RODOS application running on a mixed signal IC

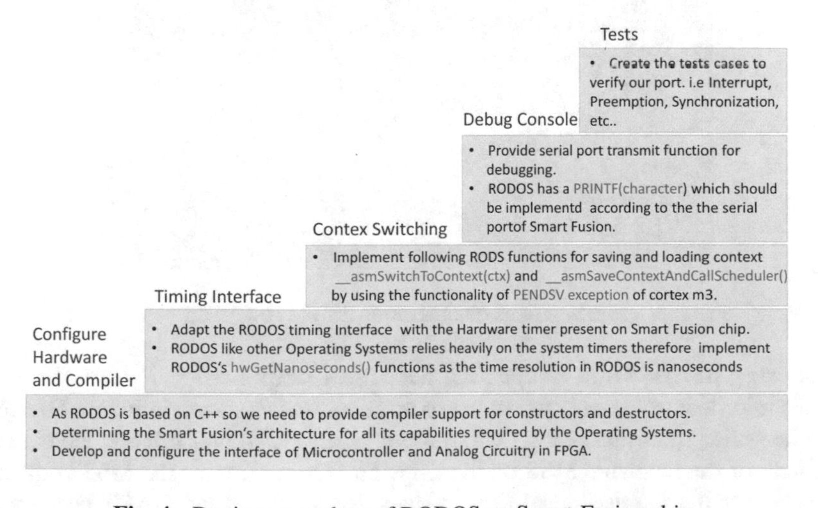

Fig. 4. Porting procedure of RODOS on Smart-Fusion chip.

2 Designing a Programmable Hardware

Before porting the RTOS on the customizable chips we must configure the peripherals on Smart Fusion chip. These peripherals includes clock, SRAM, ENVRAM, AC (Analog Controler) and I/O interfaces such as UART, SPI, I2c etc. Besides these peripherals configuration, three most important steps has to be performed as shown in the list below:

- Setting Clock Speed.
- Configuring the microcontroller-sub-systems (MSS) and Generate the MSS Component.
- Generating the Binary Files to Program the FPGA.

2.1 Setting the Clock Speed

Clock Management must be performed to configure the On-Chip oscillator. The input clock frequency for the PLL has to be set as shown in the Fig. 5, CLKA the clock to the PLL is set as 100 MHz, Configure the PLL to get the 80 MHz MSS clock frequency. These settings are taken as the recommendations of the manufacturer's startup guide, Later it can be modified according to the requirement.

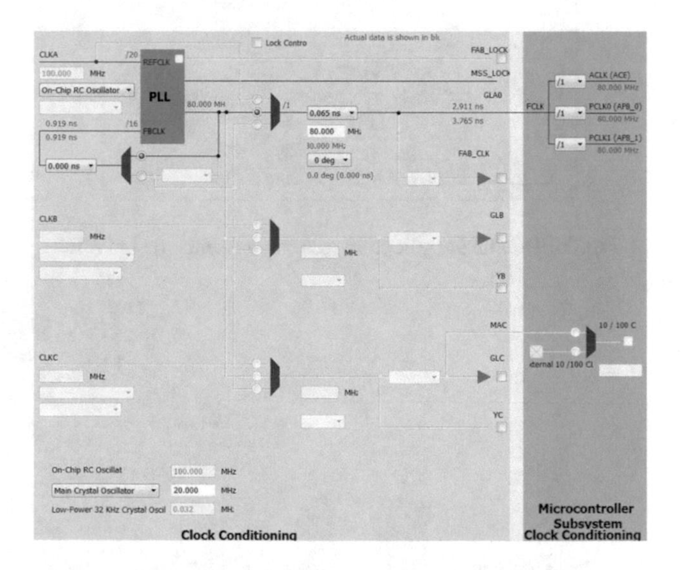

Fig. 5. Clock configuration for the controller.

Configuring the MSS and Generating the MSS Component

First the microcontroller-subsystem has to be generated as shown in Fig. 6. This microcontroller-subsystem must be set with proper clock speed and it has to configure the peripheral of the available MSS on the chip. The Generation of the MSS components means generating the peripheral drivers specially for timer and UART because these drivers are required for the scheduling and debugging respectively as the result of this step the linker script and the startup files are also generated which are required for initialization and flashing the high level code (C or C++) on the chip (Fig. 7).

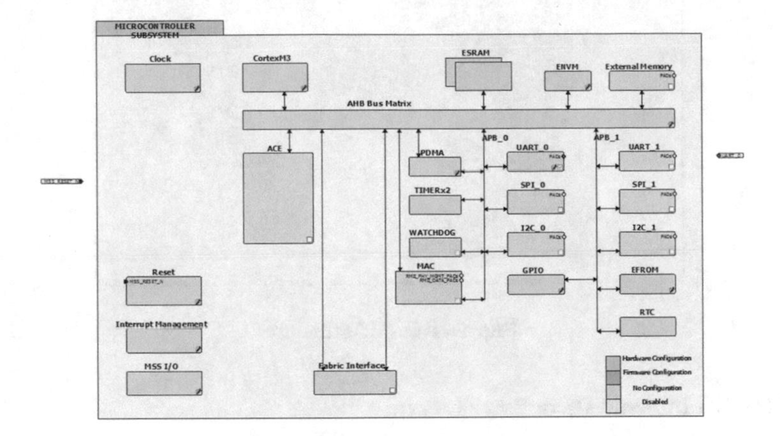

Fig. 6. Micro-controller-subsystems

Fig. 7. MSS component

2.2 Programming the FPGA

Depending on the selection and the configuration of the user, Hardware must be synthesized and routed and later the binary file should be generated to program the FPGA. After this step all the communication path is established between the FPGA fabric and the Microcontrollers as the FPGA get able to access the external memory controller (EMC) of the microcontroller to communicate with external memory devices [8].

3 RODOS

RODOS is a real time operating systems which was purposely designed for Aerospace Applications, It is already used in a real time satellite TET of German Aerospace Center, which is already in the orbit. It is hoped that this thesis work will be a significant step to manage more sophisticated tasks in RODOS related to the modern high speed space instruments and ever increasing complexities with the help of customizable-high-performance-architecture in the future space missions (Fig. 8).

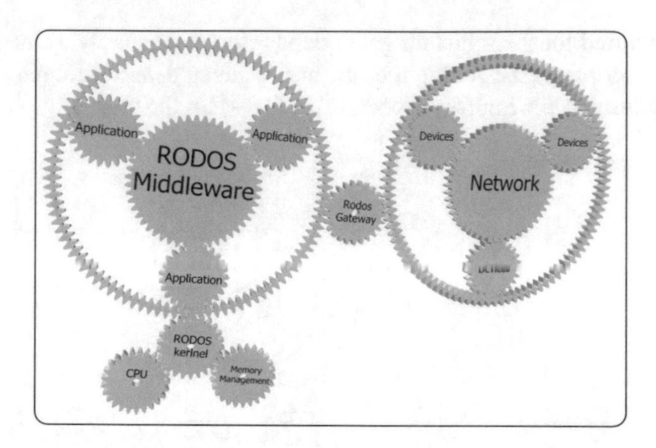

Fig. 8. RODOS structure

3.1 RODOS Internal Directory Structure

RODOS kernel was developed using C++ but it is also compatible with C files without any changes. RODOS directory structure is shown in the Fig. 9.

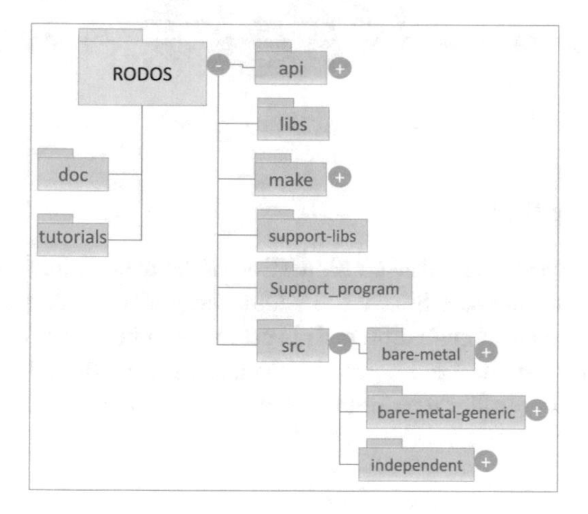

Fig. 9. RODOS main directory structure

The folder “api” contains the header files of RODOS kernel. This folder contains all the files related to the declaration of all the kernel components like thread, FIFO, Semaphore, Timers, Gateways etc. The “doc” directory contains the documentation about the RODOS structure. “Make” folder contains all the shell scripts of RODOS compilation on the different platforms. The folder “support_libs” contains the libraries which are supported by RODOS. Although there are numerous folders in RODOS but here only those folders are described which are directly related to the porting process.

Figure 10 shows the "src" directory which contains the hardware related implementation of RODOS porting. It contains the Smart-Fusion directory which is related to the hardware structure, All the porting steps are related to the files in this folder, It contains the peripherals drivers, definition of debug board and speed, Stack definition and its size, The hal-rodos folder contains all the files related to the implementation of the communication protocol like UART, SPI, CAN, UDP etc. These protocols are provided with the abstraction level i.e. the functions prototype, parameters, their names are same but the user has to implement the functionality depending on the architecture of the processor.

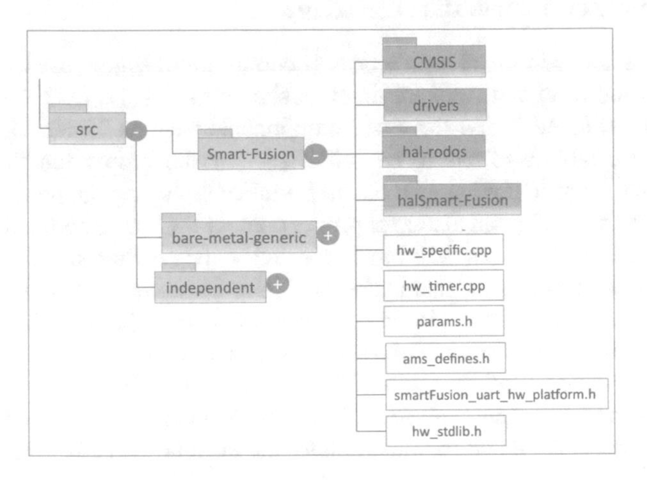

Fig. 10. RODOS sub-directories

4 Porting RODOS on Smart-Fusion

RODOS is already ported on a wide range of processor architecture, like PowerPC, Leon, ARM, Blackfin, AVR etc. In this work we the cortex-M controller is being used which is present on the Smart-Fusion chip. In this section the approach to port RODOS on Smart-Fusion SOC is described in detail.

4.1 C++ Support of Compiler

As RODOS is based on C++ that's why compiler support has to be provided to call the global constructors and destructors. This call will initialize all the data members of a C ++ class before any objects of the class are created. The constructors are invoked before calling the "main" function and destructors must be called after the "return" function [9], Which is mainly invoking global Constructors in the _start function before executing the main(). _start symbol is responsible to manage all the initialization tasks related to the booting. This task is achieved through five Object files: crt0.o, crti.o, crtbegin.o, crtend.o and crtn.o

- crtbegin.o specifies the starting point of the constructors in the memory.
- crtend.o represents the starting point of the destructors in the memory.

crtbegin.o and crtend.o are part of compiler. These files are also referred as auxiliary startup files [10]. These constructors and destructors are called in the order. These static constructors and destructors are addressed through the linker script. In RODOS compilation it is explicitly configured not to use these functions provided by the compiler as RODOS uses the functions from the manufacturer who provides this in the startup files.

4.2 Hard and Soft Initialization at Startup

Before calling the main function in RODOS certain initialization has to be performed both at hardware level and software level as shown in Fig. 11. On hardware-side the clock must be enabled for all the basic components such as Timer, UART etc. The Interrupt vector table also has to be initialized for the proper functionality of the controller. One more important step has to be taken regarding the pendSV exception, its priority must be configured to low for proper context-switching during the scheduling. On RODOS's side developer will have to provide the mechanism to call the global constructors before calling the main. The stack size for each thread must be defined in RODOS's params.h file, Here it is also required to identify the UART port to be used, In the Smart-Fusion chip there are two serial ports and in this work UART0 is used for printing the debugging messages. In RODOS's hw_specific.h file the baud-rate also to be defined for the serial-communication. From the point of view of thread-initialization the thread name, Its ID, Its priority and delay time must be set before it is called by the scheduler.

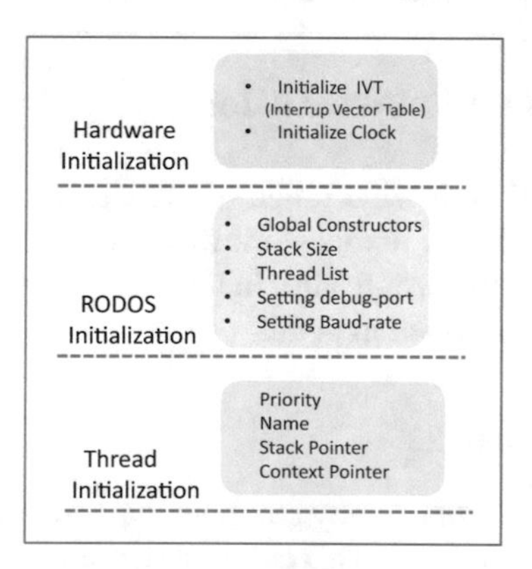

Fig. 11. Iitialization before executing tread

4.3 RODOS Timing Interface

Like any other operating system, RODOS is required to use a timer mechanism for Scheduling and timing.

There are two 32 bit-timers and one watchdog-timer available in SmartFusion which can be used as one-shot and one periodic timer [13]. In this porting procedure Timer0 is used which is a periodic timer, The driver for this timer is produced as the result of MSS component generation. The thread timing in RODOS is managed by following three functions:

- void TIMx_init()
- void TIMx_IRQHandler()
- unsigned long long hwGetNanoseconds()

For porting RODOS on any architecture the developer must provide the implementation of above three functions as far as timing operations are concerned: RODOS gives time in nanoseconds related to the startup time. These function are defined in hw-timer.cpp file. The implementations of these functions is depicted in Figs. 12, 13, 14 and 15.

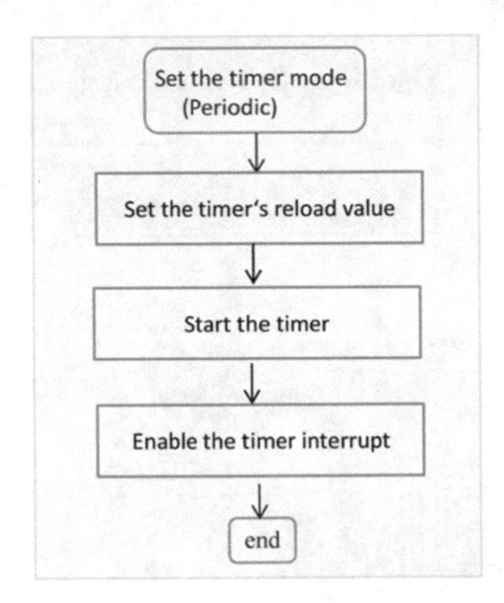

Fig. 12. Timer initialization function (TIMx_init())

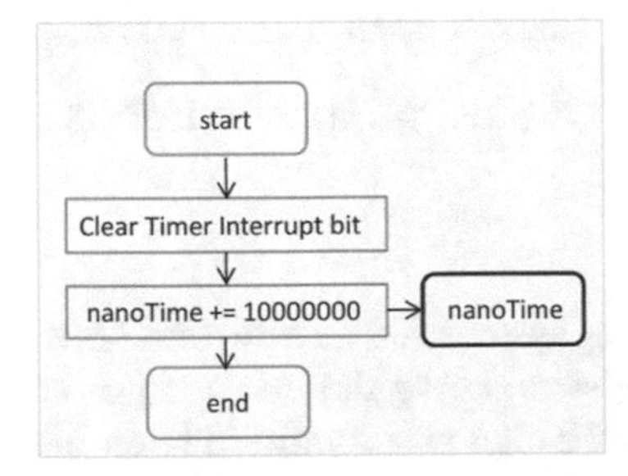

Fig. 13. Timer interrupt handler (TIMx_IRQHandler())

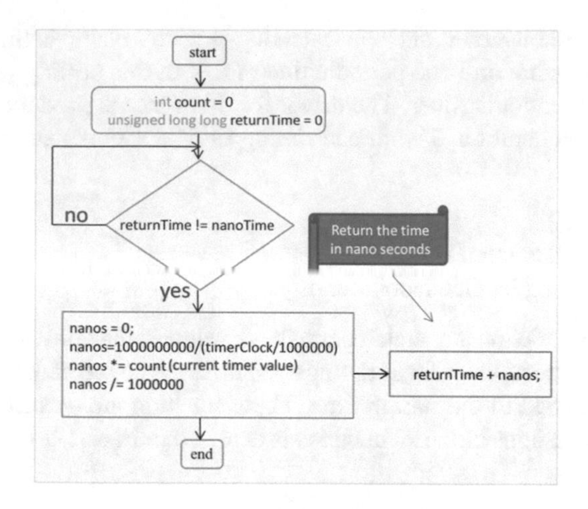

Fig. 14. hwGetNanoseconds()

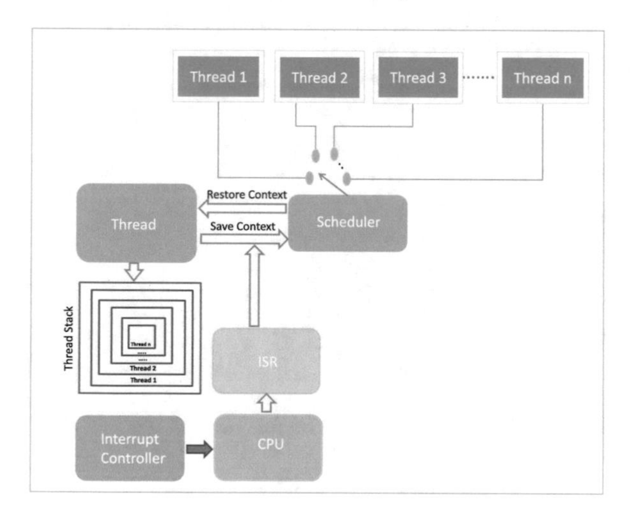

Fig. 15. Scheduling in RODOS

4.4 Context Switching

Every thread in any operating systems has its own stack i.e. the state of the CPU registers, This state of the CPU registers during the execution of any thread is referred as the context of the thread. When a high priority thread interrupts a low priority thread then the first job of CPU is to save the context of interrupted thread before serving the interrupts and when the interrupt service routine (ISR) is serviced then the CPU first reload the context (i.e. the values of core registers) of the blocked thread and then restore the execution of the blocked thread as show in Fig. 16. In order to handle the context-switching two functions must be implemented in RODOS:

- void __asmSaveContextAndCallScheduler()
- void __asmSwitchToContext()

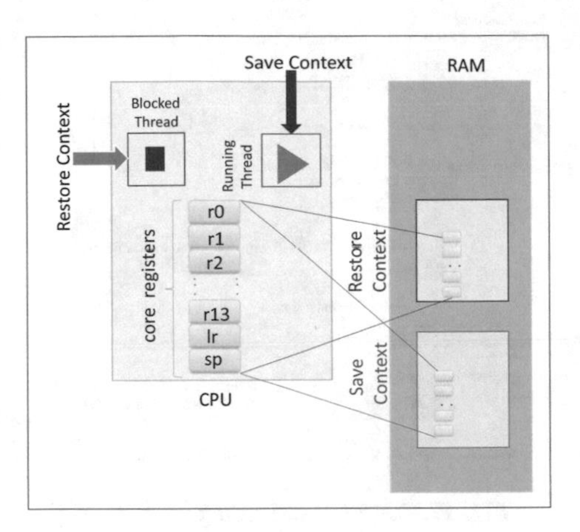

Fig. 16. Context-switching in ARM

RODOS uses the PendSV Exception for context switching. These functions for context switching are implemented in hw-specific.cpp file. As the CPU in Smart-Fusion is based on ARM cortex-m3 and the context switch is based on the handling of ARM-core registers that's why all the context switching functionality was taken from previously ported RODOS versions [14]. The pendsv software exception form ARM was used to avoid the use of systick exception.

The systick exception was not used to avoid the danger of delay that might be introduced by the using systick timer because CPU can not be halted too long. PendSV is a software interrupt mechanism provided by ARM to reduce the contex switching time [15], PendSV exception provides a brute-force approach for context-switching.

First RODOS is in idle thread mode and when its time slot is over the timer timeout occur then pendsv exception occurs by the yield function of RODOS, In this pendsv handler all the context is saved. The context switching is managed by the pendSV exception to avoid processor's usage fault exception. Then the scheduler find the next thread to run, It loads the thread's context and run the first thread as shown in the Fig. 16 and when its time slot is over the scheduler again searches next thread to run and in this case threads2 starts execution and if during the execution of thread2 if any ISR comes then the control is transferred to pendsv which saves the context and after servicing the interrupt service routine the context of interrupted thread is loaded back and thread2 resumes and If there is no other thread in the list then after thread2 completes the Idle threads is called as shown in Fig. 17.

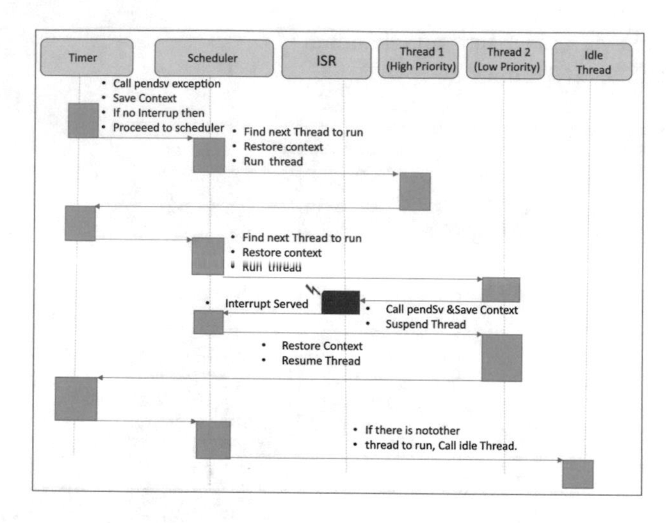

Fig. 17. Context switching in RODOS

4.5 Serial Console (for Debugging)

To provide the runtime debugging for fault detection and correction a debug mechanism is required. Normally there is a "printf" function on host computers but to get the interaction with the embedded board, The developer has to implement the RODOS's own "PRINTF" function which is a serial output of the strings. There are two UART port present on the Smart-Fusion and for the serial console but here in this porting procedure UART0 is used. First the Initialization of Serial port for Debugging for printing Messages is required as shown in Fig. 17 then the developer has to implement the UART Transmit function of the chip in the RODOS PutcharNoWait function as shown in Fig. 19. These debugging functions has to be implemented in RODOS's hal_uart.cpp (Fig. 18).

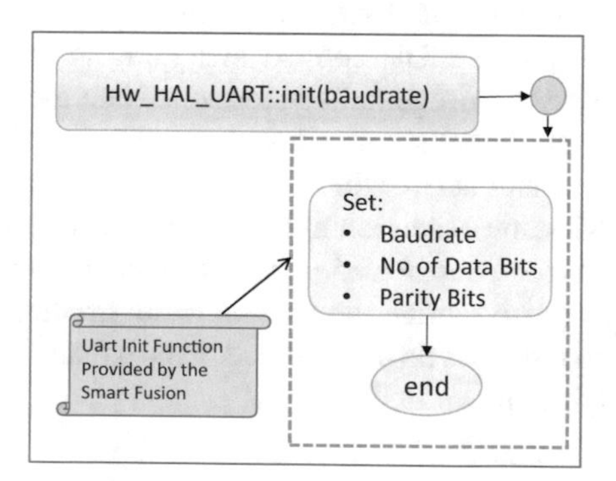

Fig. 18. Initialization of UART for Debugging in RODOS

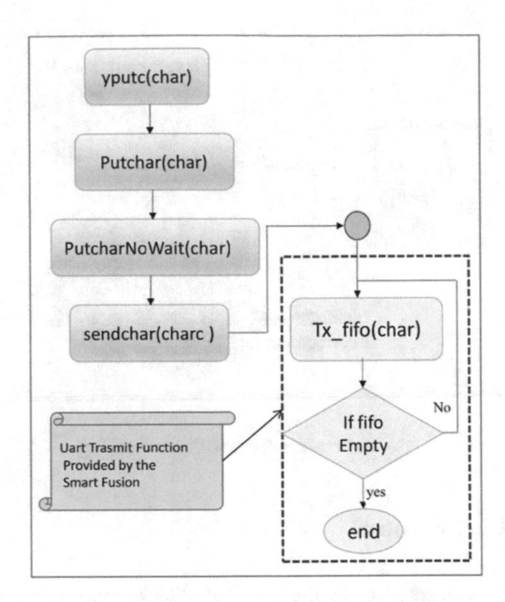

Fig. 19. RODOS debug function (PRINTF)

5 Hardware Setup

To perform the porting procedure the Smart-Fusion hardware development kit from Microsemi was used in combination with the GNU-ARM-Toolchain, the details of the hardware and software components is given in the following list:

- GNU arm-toolchain
- A 8086 AMD 64-bit Host Computer
- Smart-Fusion AF200 Development Kit
- Lauterbach Power-Debug Module

The Hard and soft components were used as shown in the Fig. 20. ARM GNU tool-chain was used for compilation, assembling and linking, A low cast Smart-Fusion development kit was used as prototype to run the applications. The porting was tested both on Windows and Linux, On Windows Smart-Fusion can be programmed and debugged through in-system USB based programming but on Linux platform an external debugger [16] was used as it was not possible to program it with USB.

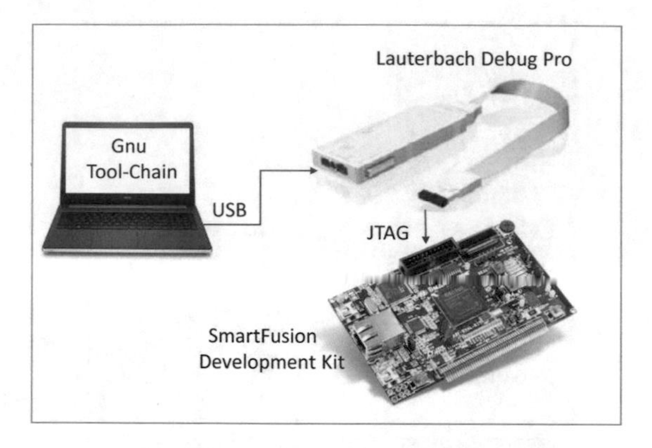

Fig. 20. Hardware-setup for porting

6 Execution of Test Cases

RODOS has some test cases to verify the porting. These tests are related to the Thread's timing, Preemption mechanism, Local Communication, Middleware, Gateway and link-interface of RODOS. Here the results of the basics tests are described which verifies the timing, priority and memory sharing mechanism of RODOS.

6.1 Threads and Time

This is a thread which waits for a time point. It loops and waits 2 s in each loop in order to check if the time difference of each execution is more than 2 s, because of the time consumed in the thread execution. The code and result to verify the timing is given in Figs. 21 and 22 respectively.

```
1  #include "rodos.h"
2  static Application module01("TestTime");
3
4  class TestTime : public Thread {
5  public:
6    TestTime() : Thread("waitfor") { }
7    void run(){
8      int cnt=0;
9      while(1){
10       cnt++;
11       suspendCallerUntil(NOW() + 2*SECONDS);
12       PRINTF("After 2 Seconds  : %3.9f %d\n", SECONDS_NOW(), cnt);
13     }
14   }
15   void init() { PRINTF("Waiting 2 seconds");}
16 };
17
18 static TestTime testTime;
```

Fig. 21. Timer testing thread

	SECONDS_NOW()	cnt
After 2 Seconds :	2.039963311	1
After 2 Seconds :	6.039969550	3
After 2 Seconds :	10.039958586	5
After 2 Seconds :	14.039966350	7
After 2 Seconds :	18.039968224	9
After 2 Seconds :	22.039959837	11
After 2 Seconds :	26.039961512	13
After 2 Seconds :	28.039965450	14
After 2 Seconds :	32.039964675	16

Fig. 22. Result of timing test

```
#include "rodos.h"

static Application module01("PreemptiveTest", 2000);
class HighPriorityThread: public Thread{
public:
  HighPriorityThread() : Thread("HiPriority", 25) {
  }

  void init() {
    xprintf(" hipri = '*'");
  }

  void run() {
    while(1) {
      xprintf("*");
      FFLUSH();
      suspendCallerUntil(NOW() + 500*MILLISECONDS);
    }
  }
};

class LowPriorityThread: public Thread {
public:
  LowPriorityThread() : Thread("LowPriority", 10) {
  }

  void init() {
    xprintf(" lopri = '.'");
  }

  void run() {
    int64_t cnt = 0;
    int32_t intervalToPrint = getSpeedKiloLoopsPerSecond() * 10;
    while(1) {
      cnt++;
      if (cnt % intervalToPrint == 0) {
        xprintf(".");
  FFLUSH();
      }
    }
  }
};

/*****************/
HighPriorityThread highPriorityThread;
LowPriorityThread  lowPriorityThread;
/*****************/
```

Fig. 23. Preemption code

6.2 Preemption

To perform the priority testing, two threads were created as given in Fig. 23, One with a high priority which is executed very shortly every 3 s and one with a low priority which is executed constantly. The high priority thread (printing "*") shall assume the CPU when it needs it even if the low priority thread (printing ".") does not suspend or yield as depicted in Fig. 24.

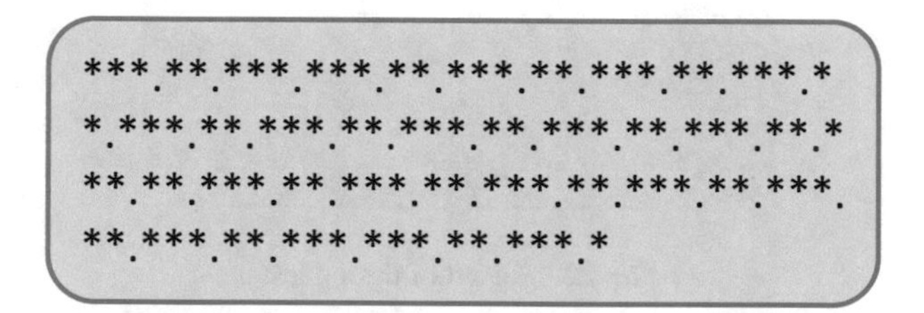

Fig. 24. Preemption result

6.3 Local Communication

Communication between the different threads can be established by using shared memory, RODOS provides two mechanism to implement the data sharing among threads, these mechanism includes, FIFO and Semaphore. Here simple FIFO test were performed as shown in Fig. 25. For Synchronous communication from one single writer to one single reader. Neither will be suspended. Writing to a full FIFO has no effect and returns 0. Reading from an empty FIFO returns 0. Figure 26 refers the outcome of the local communication test.

```
 1 #include "rodos.h"
 2
 3 static Application applic("FifoTest");
 4
 5 Fifo<int, 10> fifo;
 6
 7 class Sender : public Thread {
 8   void run () {
 9     int  cnt = 0;
10     xprintf("sender\n");
11     while(1) {
12       cnt++;
13       bool ok = fifo.put(cnt);
14       if (ok) {
15         PRINTF("Sending %d\n", cnt);
16       } else {
17         PRINTF("Fifo full\n");
18       }
19       if ((cnt % 15) == 0) {
20         PRINTF("Wainting 3 seconds\n");
21   suspendCallerUntil(NOW() + 3*SECONDS);
22       }
23     }
24   }
25 };
26
27 class Receiver : public Thread {
28   void run () {
29     int cnt;
30     xprintf("receiver\n");
31
32     while(1) {
33       bool ok = fifo.get(cnt);
34       if (ok) {
35         PRINTF("reading %d\n", cnt);
36       } else {
37         suspendCallerUntil(NOW() + 1*SECONDS);
38       }
39     }
40   }
41 };
42
43 /******************************/
44 Sender   sender;
45 Receiver receiver;
```

Fig. 25. Local communication thread

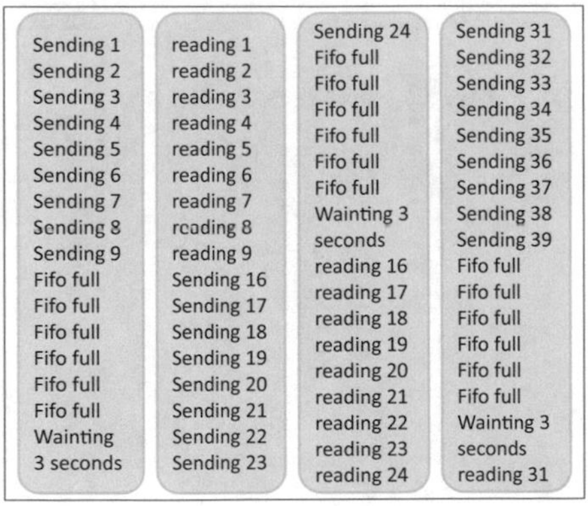

```
Sending 1       reading 1       Sending 24      Sending 31
Sending 2       reading 2       Fifo full       Sending 32
Sending 3       reading 3       Fifo full       Sending 33
Sending 4       reading 4       Fifo full       Sending 34
Sending 5       reading 5       Fifo full       Sending 35
Sending 6       reading 6       Fifo full       Sending 36
Sending 7       reading 7       Fifo full       Sending 37
Sending 8       reading 8       Wainting 3      Sending 38
Sending 9       reading 9       seconds         Sending 39
Fifo full       Sending 16      reading 16      Fifo full
Fifo full       Sending 17      reading 17      Fifo full
Fifo full       Sending 18      reading 18      Fifo full
Fifo full       Sending 19      reading 19      Fifo full
Fifo full       Sending 20      reading 20      Fifo full
Fifo full       Sending 21      reading 21      Fifo full
Wainting        Sending 22      reading 22      Wainting 3
3 seconds       Sending 23      reading 23      seconds
                                reading 24      reading 31
```

Fig. 26. Local communication result

6.4 Stack Usage

The stack utilization was also examined on the Smart-Fusion board with RODOS's threads. Every RODOS's thread has its own stack which might be set in the parameter file (params.h).

Stack overflow might be lethal for any application, In this demo five threads were created in RODOS to analyze the stack usage as shown in Figs. 27 and 28, One of thread (Stack Consumer) was designed to consume all the stack memory and the results are shown in Fig. 29.

```
#include "rodos.h"

#define LOW_STACK_LIMIT 000

/** Checks percent usage of each stack **/

class ThreadChecker : public Thread {
public:
    ThreadChecker() : Thread("ThreadChecker") { }

    void init() { }

    void run() {

        long minStack = DEFAULT_STACKSIZE;
        Thread* dangerousThread = 0;
        TIME_LOOP(0, 3*SECONDS) {
            PRINTF("TST: Threads and stacks:\n");
            ITERATE_LIST(Thread, Thread::threadList) {
                long index = 0;
                char* stk = iter->stackBegin;
                for(index = 16; index < iter->stackSize; index++)
                if(stk[index]!= 0) break;
                if(index < minStack) minStack = index;
                if(index < LOW_STACK_LIMIT) dangerousThread = iter;
PRINTF("%sPrio= %7ld Stack= %6ld StackFree= %6ld,(min = %6ld)\n",
iter->getName(),iter->priority,iter->stackSize,index,minStack);
            }
            if(minStack < LOW_STACK_LIMIT) {
                PRINTF("TST: Dager! Stack to low\n");

            }
        }
    }
} threadChecker;
```

Fig. 27. Stack-usage test part 1

```
/**************************************************/
class Innocent1 : public Thread {
public:
    Innocent1() : Thread("Innocent1") { }
    void run() { }
} innocent1;
class Innocent2 : public Thread {
public:
    Innocent2() : Thread("Innocent2") { }
    void run() { }
} innocent2;
class Innocent3 : public Thread {
public:
    Innocent3() : Thread("Innocent3") { }
    void run() { }
} innocent3;



/************************/


/********** Otehre Threads to test stack occupation ***/

void stackUser(int len) {
    char variableOnStack[len];
    for(int i = 0; i < len; i++) variableOnStack[i] = 0x5a;
}


/** Consumes more an more stack until it crases *****/

class StackConsumer : public Thread {
public:
    StackConsumer() : Thread("StackConsumer", 400, 2000) { }
    void run() {
        long consumer = 1000;
        TIME_LOOP(6*SECONDS, 4*SECONDS) {
            stackUser(consumer);
            consumer+=100;
        }
    }
} stackConsumer;
```

Fig. 28. Stack-usage test part 2

name	low	high	sp	%	lowest	spare	max	0 10 20 30 40 50 60 70 80 90 100
IdleThread	2000272C	20002B10	20002AB4	9%	20002A48	0000031C	20%	
StackConsumer	20001F5C	20002728	2000267C	8%	20001F5C	00000000	100%	
Innocent3	20001B74	20001F58	20001F0C	7%	20001EC0	0000034C	15%	
Innocent2	2000178C	20001B70	20001B24	7%	20001AD8	0000034C	15%	
Innocent1	200013A4	20001788	2000173C	7%	200016F0	0000034C	15%	
ThreadChecker	20000FBC	200013A0	200012C4	22%	20001238	0000027C	36%	

Fig. 29. Result for stack consumption test 1

```
1
2  #include "rodos.h"
3
4  static class TestTime2 : public Thread {
5  public:
6    TestTime2() : Thread("waitAT") { }
7
8    void init(){this->priority=50;}
9    void run(){
10
11     PRINTF("Hello WOrld\n");
12
13     TIME_LOOP(NOW(),0.0 SECONDS) {
14         PRINTF("\n I am Thread_B and I am ALive \n");
15     }
16
17 }
18
19 }a;
20
21
22 static class TestTime3 : public Thread {
23 public:
24   TestTime3() : Thread("waitAT_3") { }
25
26   void init(){this->priority=80;}
27   void run(){
28
29  PRINTF("Hello WOrld\n");
30   TIME_LOOP(NOW(),0.1*SECONDS) {
31       PRINTF("\n I am Thread_C and I am ALive \n");
32     }
33 }
34
35 }b;
```

Fig. 30. Code for CPU-time consumption

Analysis of CPU consumption

The CPU-time consumption or CPU-usage was also analyzed with RODOS threads as shown in Fig. 29 where two threads were created, One of the thread TestTime2 is printing a message every 300 ms and the thread TestTime3 is printing the message every 100 ms. The result for this test is shown in Fig. 31.

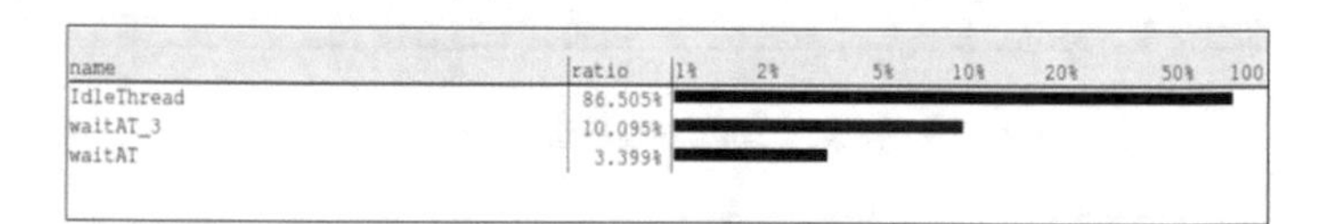

Fig. 31. Result for CPU-time consumption test

The thread named "waitAT_3" is printing more often as compared to thread named "waitAt" that's why its CPU consumption was larger whereas thread named "IdleThread" shows the time when the CPU was not busy or in IDLE state (Fig. 30).

7 Conclusion

This paper provides the details of porting a real time operating system RODOS on a fully programmable mixed signal system-on-chip. RODOS was already ported on Cortex-M0 based microcontroller so many of the routines such as context-switching were reusable. Here the test code for Thread Timing, Preemption and local communication and result of these test cases were also presented which verifies proper functionality. The highly configurable nature of Smart-Fusion chip allows porting of a real time operating system with best run-time characteristics with extremely optimized hardware resources. As RODOS is designed specifically for Aerospace Applications that's why it is highly optimized in terms of performance and size which is the prime requirement in aerospace application so RODOS running on this highly reconfigurable SOC delivers immense capabilities to an On-board computer. In this work the FPGA was programmed with minimal resources i.e. Clock, External SRAM, External Non-volatile memory, UART and Timer the porting results of RODOS on this chip shows correct execution of RODOS's thread, Later for other functionalities FPGA program can be easily extended to include I2C, SPI, Analog controller etc., The FPGA design flow environment will generate the relevant driver files which can be easily integrated into RODOS.

Acknowledgment. The authors would like to thank Mr. Erik Dilger who shared his understanding about the internal structure of RODOS and how its threads work and also many thanks to the support team in Lauterbach [16] who have provided significant tips and help for debugging the Smart-Fusion kit.

References

1. Rodos (operating system). https://en.wikipedia.org/wiki/Rodos
2. RODOS, Real time kernel design for dependability. http://www.dlr.de/irs/en/Portaldata/46/Resources/dokumente/produkte/rodos.pdf
3. Dannemann, F., Montenegro, S.: Embedded Logging Frame-Work For SpaceCrafts DASIA (Data Systems In Aerospace), Porto, Portugal (2013)
4. TET-1 (Technology Experiment Carrier-1). https://directory.eoportal.org/web/eoportal/satellite-missions/t/tet-1
5. BIROS (Bi-spectral InfraRed Optical System). https://directory.eoportal.org/web/eoportal/satellite-missions/b/biros
6. ALINA, The Autonomous Landing and Navigation Module, First Private Mission to the Moon. http://ptscientists.com/
7. SmartFusion SoC FPGAs. https://www.microsemi.com/products/fpga-soc/soc-fpga/smartfusion
8. Microsemi Application Note AC346 SmartFusion cSoC: Loading and Booting from External Memories
9. Dassen, J.H.M.: Constructors and Destructors in C++. http://l4u-00.jinr.ru/usoft/WWW/wwwdebian.org/Documentation/elf/node4.html
10. Rob Williams Section C++ global constructors and destructors in ELF in book Real-Time Systems Development (2006)
11. Master thesis: RODOS in a Multicore environment by Tiago Lus Gonalves Duarte, University of Minho in coordination with Julius–Maximilians University Wrzburg (2013)
12. Master thesis: Design and Implementation of Multi-core Support for an Embedded Real-time Operating System for Space Applications by Wei Zhang, KTH Royal Institute of Technology, May 2015
13. Actel SmartFusion MSS Timer Driver User's Guide Version 2.0

14. Contribution of Miroslaw Sulejczak and by Michael Ruffer for the context switching for ARMcore in RODOS
15. Choi, H., Park, S.: Evaluations of hardware and software-based context switching methods in Cortex-M3 for embedded applications. Int. J. Smart Home **9**(2), 111–122 (2015)
16. Lauterbach Development Tools. http://www.lauterbach.com/frames.html?home.htm

A General Game-Theoretic Approach to Harmonization the Values of Project Stakeholders

Tigran G. Grigorian(✉), Sergey D. Titov, Anatoliy Y. Gayda, and Vladimir K. Koshkin

National Shipbuilding University, Geroev Ukrainy Ave., 9, Mykolaiv, Ukraine
grigorian.tigran@gmail.com, {ssl-ssl0,cetus}@ukr.net, koshkin-vladimir@mail.ru

Abstract. The problem of project stakeholders values harmonization as a solution of non-cooperative game between two players is stated. The concept of the value balancing operation and the value harmonization process are presented. The alternative strategies of project team and stakeholders as players, allowing to typify situations in real projects and reduce the variety of possible behavior into fairly small amount of combinations are presented. The developed model allows to obtain the recommendations for the use of pure and mixed strategies aimed to maximization players' values under different circumstancies on the basis of the models of zero sum and bimatrix games. The models and method presented allow to ensure the sustainability of project execution and finalization. Further research tasks should be aimed at developing the means for the increase the effectiveness of the payoff matrices building.

Keywords: Project management · Value · Stakeholders' values
Value harmonization · Matrix games · Zero-Sum game · Bimatrix games

1 Introduction

The main project manager task is to ensure the implementation of the project and its completion within the triangle of basic restrictions, taking into account the characteristics of the environment. However, the concept of value-driven management is gaining popularity, according to which the main task of the manager is to ensure the creation of value in the form of project output and its delivery to stakeholders [1, 2]. This idea is accentuated in IT-projects managed in accordance with the Agile and Lean methodologies [3–5]. Today the value becomes a key driver for project initiation, implementation and finalization. One of the most important tasks in ensuring of value creation and delivering is its harmonization.

The reason to solve the value harmonization problem is a conflict of interests of a systemic nature, which is almost inevitable in the process of project management, and is due to the difference in stakeholder values and different perceptions of the situation around the project and its output. Effectiveness and efficiency of project management are directly related to the possibilities of forecasting such value conflicts, developing

N. Shakhovska and V. Stepashko (eds.), *Advances in Intelligent Systems and Computing II*, Advances in Intelligent Systems and Computing 689,
https://doi.org/10.1007/978-3-319-70581-1_11

scenarios for minimizing their negative impact on project implementation. The solution of the problem is exacerbated by the unique character of the project and the turbulence of its environment, which certainly affects the stakeholders. First of all harmonization should be aimed at eliminating these conflicts, caused by the discrepancy of value expectations and the perception of the product by stakeholders. *The goal of stakeholders' value harmonization* is to ensure the support and participation of stakeholders during project implementation and, ultimately, the adoption of a product, aimed at creating and delivering value. Thus, there is a need for models that will help us to predict the influence of value conflicts in projects and develop decision support when choosing the most effective models of manager behavior in certain situations.

2 Literature Review

A considerable amount of scientific work has been devoted to the study of value. It must be noted that it is possible to allocate conditionally researches of values of conceptual and descriptive types. The first include the fundamental works of Rokich [6], Schwartz [7], Hofstede [8]. Special works in the field of project management are devoted to studying values of the second type. The issues of value management in project management are generalized and consistently set out in the international standard P2M [9]. Besides considerable attention to the value is paid in the basic principles and practical recommendations of Agile [3, 4]. Kerzner and Saladis also point to the need to eliminate the value conflict, noting that balancing the needs of stakeholders becomes especially difficult due to the internationalization of projects – it is necessary to take cultural, ethical, religious and other factors into consideration [1]. And this also confirms the advisability of applying "soft" management techniques in solving the problem of harmonization. They identify 6 types of conflicts by pair combinations of 4 groups of stakeholders, and indicate that balancing aimed at eliminating conflicts is an extremely difficult task. However, they do not provide any specific recommendations for its solving.

In [2] the model of value harmonization, which allows assessing the stability of the organization and facilitates the analysis of alternatives when choosing strategies for the development of financial institutions is presented. However, this model is used to harmonize the values of the development of organizations not projects themselves, and, moreover, does not sufficiently take into account the dynamic nature of the values themselves. In addition, the application of various approaches aimed at harmonizing the values proposed and used in general management cannot be applied to project management because of uniqueness [1].

In general, two problems of value harmonization are declared and being solved in project management [1, 10, 11]: (a) harmonization in accordance with the enterprise development strategy, which can be reduced to ranking projects (subprojects) and giving preferences to those that are more in line with the strategic values of the organization; (b) harmonization between project stakeholder values, which is the sort of balancing operation. The first problem is successfully solved with the help of verbal decision analysis methods [12]. This method and corresponding models show good results in various applied fields, including management of project portfolios in the nuclear power industry and municipal administration, decision-making in the management of outsourcing project teams, scope and schedule management in Agile projects, etc. [13–15]. In contrast, in the field of solving the problem of harmonization of stakeholder values, there is a clear lack of research due to the complexity of this problem, related to the following features: the uniqueness of the project, which significantly complicates the development of models, the logic of which is based on precedents; the turbulence of project environment and, as a result, the high dynamics of stakeholders' evaluations; the subjective nature of value, complicating the processes of its identification, systematization and evaluation for making project decisions; the insufficient level of development of the methodological basis for value forecasting, due to the lack of research aimed at developing models and methods for of value management. Another feature that significantly complicates the solution of value harmonization is the presence of a "soft" component due to the need to work with the stakeholders of the project. In the works listed above and other studies, there are not enough solutions aimed at balancing the value to maximize the effectiveness of projects. Thus, models and tools that will allow to harmonize project works in accordance with stakeholder values and make decisions aimed at maximizing the satisfaction of expectations through creating value in the project and managing its delivery to stakeholders. The elements of the game-theoretic approach to harmonization the values of project stakeholders on the ground of bimatrix games are considered in [16].

3 The Purpose of the Study

The aim of the article is to develop the models of decision support in the planning and management of project works aimed at harmonization in accordance with the values of stakeholders to improve the efficiency and effectiveness of project execution. To achieve this goal, it is necessary to solve a number of tasks, the most important of which are: the disclosure of the essence of harmonization of stakeholder values, the choice of methods and the development of a harmonization model, and the definition of logic and scenarios for its application.

4 The Research Method

The fundamental research of the nature of value is based on philosophical concepts and psychological methods and models owing to the features of value. From the viewpoint of socio-technical systems to which project management belongs, value management models based on social methods for interviewing, data structuring and analysis, statistical models etc.

In general, the concept of harmonization is related to the notion of balance (from fr. *balance* – balance), defined through an equilibrium, by which is meant the stable state in which the body is located under the influence of static or dynamic forces or actions [17]. In the context of this study, under the harmonization of stakeholders' values, we mean a set of actions that lead to an increase in one or more values for one or more project stakeholders at a given time. Since stakeholders' attitude can be either positive or negative, on the one hand, we have a certain unity of opinions and value relations to the project and its output (cooperativeness), and on the other hand – the difference of stakeholders' views caused by their different value orientations (antagonism).

There are a significant number of approaches, methods and models of the analysis and management behavior in conflict situations. One of the approaches that show successful results in solving the problem of conflict harmonization is game theory. Its foundations are made in the fundamental work of Von Neumann and Morgenstern [18]. The use of game theory makes it possible to predict and choose the most effective strategies in conflict situations. There are hundreds of different examples of conflict situations, effectively modeled and managed on the basis of game theory. The choice of a particular model is of fundamental importance for achieving the stated goal – ensuring the efficiency and effectiveness of project management. And the concept of Nash equilibrium plays a great role in solving this problem.

5 Main Research Material

A general expression for estimating the product value changes by a set of balancing operations. In the issues of value harmonization in projects, it is necessary to take into account that, as it usually applied to the value, its reduction is inappropriate. This reduces the attractiveness of the project output for stakeholders and adversely affects their participation in the project and, as a result, on its progress and compliance with the requirements of the project management triangle [19]. Therefore, we assume that any set of k value-balancing operations should ultimately increase the total amount of value B_k of project output:

$$B_k = \sum_{k=1}^{K} z_j \cdot v_{ij} \cdot \Delta b_{ij} > 0 \quad (1)$$

where Δb_{ij} – the increment of i-th value in the evaluation made by j-th stakeholder, v_{ij} – the evaluation of i-th value made by j-th stakeholder, z_j – power ("weight") of j-th stakeholder, $i = 1 \ldots I$, I – the number of values chosen, $j = 1 \ldots J$, J – number of stakeholders, K – total number of value-balancing operations.

In general, any elementary operation of value balancing can be aimed at increasing the value of an output through a change in it or in a stakeholder relation, the evaluations of which are analyzed. When changing an output, it is important to understand that the task of increasing value at the initial stages leads to changing the product model from the point of stakeholders views, and after delivering the minimum viable product to further modification of a product itself [5, 20].

The concept of project stakeholders' value harmonization. The uniqueness of the project management is determined by its nature: the features of a project, the characteristics of its team, the turbulence of the environment and other factors. This, in turn, exerts in the uniqueness of each project situation, which influences making appropriate decisions.

The development of models for stakeholders' values harmonization should take into account key features of the value and processes of its management, which influence the approach to balancing it. Various project stakeholders, driven by their values, relate differently to project and its output. And for the convenience of obtaining and evaluating information on project stakeholders' preferences they need to be grouped on different grounds during project analysis. According to the recommendations of PMBoK, most often they use the attitude towards the project and its output. This attitude can be positive or negative; however, it is present in any case. It is by the presence of different attitude that we include the person in some group of project stakeholders.

Rational structuring of project situations allows to develop typical form of analyzed value conflict situations and to reduce the influence of project uniqueness in decision-making on the basis of game theory. As a result, the application of this method makes it possible to reduce the variety of possible situations in value-driven balancing in project management into fairly small amount of combinations and thus to level out the impact of the uniqueness of situations in project management.

Taking into account the project management features (uniqueness, turbulence, subjectivity), it is most expedient to use harmonization of project works in accordance with stakeholders' values on the basis of non-cooperative matrix games theory. It gives an opportunity to treat the absence of strict opposition in stakeholder interests while balancing value in projects (i.e. the conflict is not strictly antagonistic).

The preservation of strategies leading to a state of equilibrium provides an opportunity to balance the interests of stakeholders and thus ensure the continuity of project management aimed at the project implementation and completion, with the subsequent product delivery to the customer. The strength of this approach is the ability to obtain on its basis recommendations about the use of particular strategies by conflict participants. Thus, the harmonization of stakeholder' values can be reduced to a final non-cooperative matrix game of two players.

The model of the game problem solution fot the project stakeholders' values harmonization in general form. Players A and B have m and n strategies respectively: A_1, A_2, ..., A_m and B_1, B_2, ..., B_n. In general case payoffs of players A and B are given by matrices $\mathbf{A}$ and $\mathbf{B}$ consequently:

$$\mathbf{A} = \begin{pmatrix} a_{11} & a_{12} & \dots & a_{1n} \\ a_{21} & a_{22} & \dots & a_{2n} \\ \dots & \dots & \dots & \dots \\ a_{m1} & a_{m2} & \dots & a_{mn} \end{pmatrix}, \mathbf{B} = \begin{pmatrix} b_{11} & b_{12} & \dots & b_{1n} \\ b_{21} & b_{22} & \dots & b_{2n} \\ \dots & \dots & \dots & \dots \\ b_{m1} & b_{m2} & \dots & b_{mn} \end{pmatrix} \quad (2)$$

If player A applies his strategy a_i and player B uses strategy b_j, then the players' payoffs will be in the appropriate intersection of row i and column j in the payoff matrices: a_{ij} for the A and b_{ij} – for B. From the viewpoint of balancing the stakeholders' values, the player's payoff is nothing more than an increment in value for the stakeholder group defined in (1).

In accordance with the theorem of Nash, any matrix game has at least one equilibrium in pure or mixed strategy [21, 22]. The identification of an equilibrium is important, since it allows to determine whether there exists such a combination of strategies of players, the deviation from which does not lead to an payoff increase, provided that the second player retains his choice [21–23]. Therefore, the main result of solving the matrix game model shows pairs of player' pure ($i^{\#}, j; i, j^{\#}$) and mixed ($\mathbf{x}, \mathbf{y}$) strategies, which leads to a stable state of the project management system and guarantees the highest game values v_A and v_B for both players A and B respectively. Therefore, in general the problem of project stakeholders' values harmonization can be reduced to the linear programming problem on the polyhedral set.

$$W \rightarrow opt, \{\mathbf{x}, \mathbf{y} \in \Omega\}. \tag{3}$$

The conceptual model of project stakeholders' values harmonization on the basis of matrix games models is presented in Fig. 1.

There are five logical levels in the model (project, gamers, strategies, combinations of gamers' strategies and recommended strategies) that describe the concept and are combined by the single hierarchy. This general approach gives an opportunity to represent, interpret and solve various tasks of value harmonization for any project. In every particular case this generalized model can be clarified taking into account the features of a project and harmonization purpose so as it will be discussed below.

The structure of project stakeholders' values. Based on the generalization of value analysis research, the grounds for determining the structure of the values balanced in projects are identified. The basis of this set is basic human *values*, developed by Schwartz and Hofstede's cultural dimensions. Their combination allows us to comprehensively determine the project values for each individual. These values overlap with the value structures proposed in P2M, PMBoK, Agile principles. However, the latter complement their specific values associated with the project activities. To solve the task of balancing the values of project stakeholders, two players are singled out (conflict parties – a project manager with his team and a sponsor with customers of a product), whose interaction determines the efficiency and dynamics of the project implementation. Despite the fact that in general, the values and interests of a project manager and team members do not coincide, in this solution they are grouped, as there is a proximity of interests. In addition, a project manager has a certain amount of power, allowing him to persuade team members to his side. Based on these reasons, the final structure of values determined for each of the players is developed and presented in Tables 1 and 2.

The developed internal structure of value allows to make decisions aimed at prioritizing and balancing project works according to evaluation of project stakeholders' values.

The general strategies of projectstakeholders' value balancing participant. To select the type of game model, we will consider the features of the value conflict in project management by the example of constructing a bimatrix game model for two players corresponding to the project manager and his team (player A) and stakeholders groups (player B). Then, in accordance with the general logic of behavior in a conflict and taking into account the features of project management, for player A, such behavior

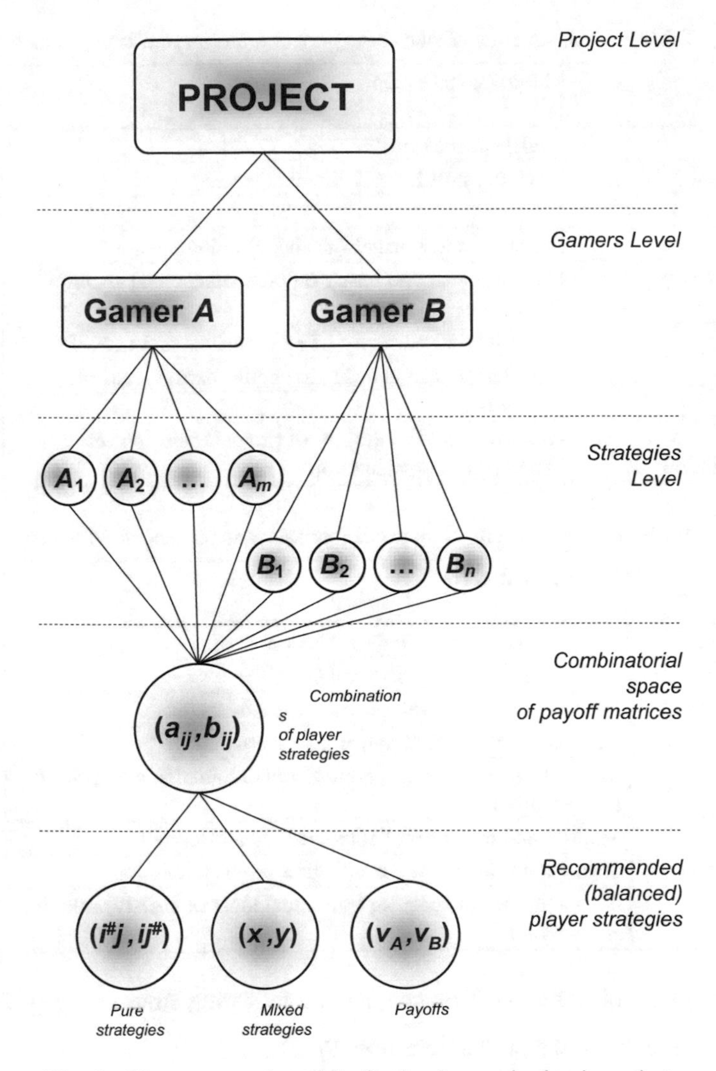

Fig. 1. The conceptual model of value harmonization in project.

strategies can change the levels of value estimations from stakeholders' viewpoint (Fig. 2a):

- change product A_1;
- change the attitude (perception) of stakeholders A_2;
- change manager and team attitude A_3;
- change product and the stakeholders' perception (the combination of A_1 and A_2);
- change product and manager and team attitude (A_1 and A_3);
- change stakeholders' perception and manager and team attitude (A_2 and A_3).

Table 1. The structure of player A (project manager and team) values

Values	Motivational goals	Value type
Autonomy	Independence of thinking and action	Social
Engagement	Passion, novelty and life challenges	Social
Agility		Personal
Comfort	Pleasure and sensual (aesthetic) enjoyment	Personal
Realization of personal strategy	Personal success, based on competence and social standards	Personal
Safety	Security and harmony of relationships and myself	Personal
Ethics	Self-restraint in actions that could harm or violate social norms	Social
Standards implementation	Improvement the quality of project management standards implementation	Personal

Table 2. The structure of player B (sponsor and stakeholders) values

Values	Motivational goals	Value type
Competence	Independence of thinking and action	Personal
Engagement	Passion, novelty and life challenges	Personal
Achievements	Personal success, based on social standards	Personal
Wealth	Abundance of material and intangible assets	Personal
Power	Social status and prestige, control and domination over people and resources	Social
Safety	Security and harmony of relationships and myself	Personal
Communication	Interaction oriented to the effective problem solution	Social
Sustainability	Acceptance and respect for traditional ideas of society, culture and religion	Social

Accordingly, project stakeholders can use the following strategies (Fig. 2b):

- active promotion of the product creation B_1;
- positive attitude to the project and product B_2;
- neutral attitude B_3;
- negative attitude to the project and product B_4;
- active resistance to the project and product B_5.

The combination of these strategies makes it possible to simulate any project situation and gives grounds for the use of game models because of characterization and repeatability. The variants of strategy combinations for the behavior of players are determined by conditions of particular project environment: stakeholders' relations, experience and knowledge of a project manager, the standards of a company or a team, the requirements declared in the project charter, etc. At the same time, the stakeholders are grouped in accordance with their attitude to the project and its product, taking into account the level of power and influence of each side on the recommendations of PMBoK as it is shown in (1).

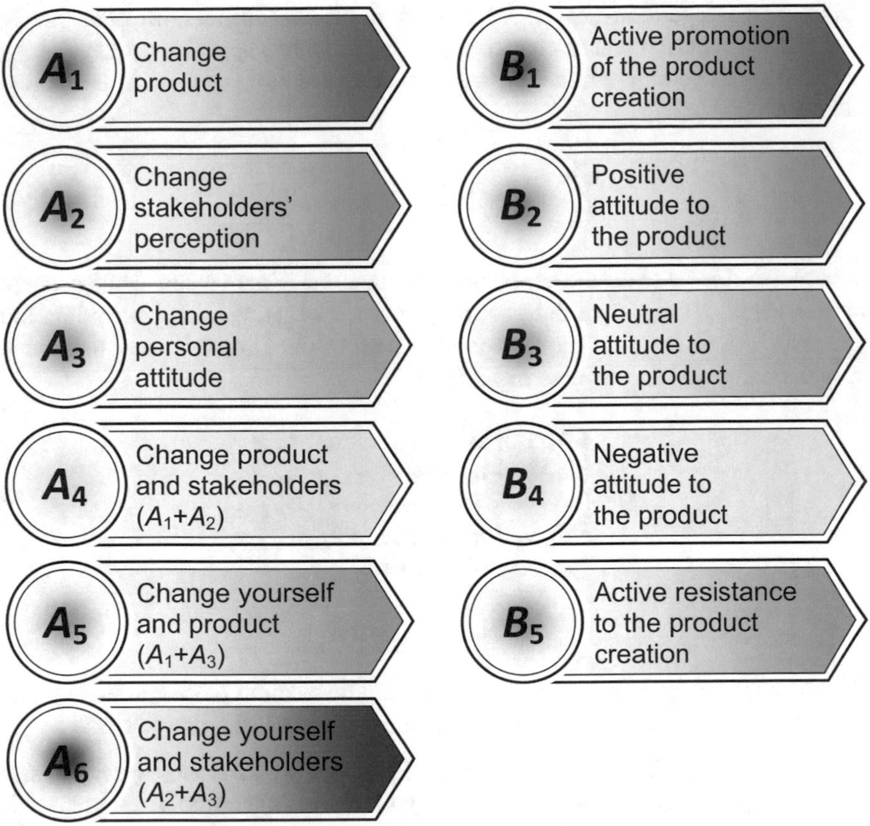

A_1 Change product

A_2 Change stakeholders' perception

A_3 Change personal attitude

A_4 Change product and stakeholders (A_1+A_2)

A_5 Change yourself and product (A_1+A_3)

A_6 Change yourself and stakeholders (A_2+A_3)

a) The strategies of player *A* (manager and project team)

B_1 Active promotion of the product creation

B_2 Positive attitude to the product

B_3 Neutral attitude to the product

B_4 Negative attitude to the product

B_5 Active resistance to the product creation

b) The strategies of player *B* (sponsor and stakeholders)

Fig. 2. The strategies of value balancing game players *A* and *B*.

The model of project stakeholders' value harmonization as a solution of the zero-sum game. The game theory is closely related to linear optimization. It is known, that each zero-sum game of two players *A* and *B* (winnings of player *A* are equal to losses of player *B* and conversely) can be represented in the form of a linear optimization task.

The mathematical model for solving the primal problem of harmonizing project stakeholders' values a as zero-sum game. Let $\mathbf{A} = [a_{ij}]_{m \times n}$ denote the payoff matrix (2) of the first player *A*. According to the general problem (3) the goal of the first player or objective function W_I is to maximize his winning – the game value *v*:

$$W_I = v \rightarrow \max \quad (4)$$

In the game with payoff matrix $\mathbf{A} = [a_{ij}]_{m \times n}$ the strategies of *A* player are given by an ordered set of **x** probabilities (frequencies) or *m*-dimensional vector. In case of multiple game repetition player A have to implement its set of pure strategies $[A_1, A_2, \ldots A_m]$ with the values corresponding to $\mathbf{x}$. Such a behavior of player A guarantees that game value v will not be reduced. A vector $\mathbf{x} = [x_1, x_2, \ldots x_m]$ satisfies the conditions:

$$\sum_{i=1}^{m} x_i = 1,\ x_i \geq 0,\ i = 1, \ldots, m. \tag{5}$$

The player A have to ensure payoffs not less than the game value v while using pure strategies $[A_1, A_2, \ldots A_m]$ with different frequencies $\mathbf{x} = [x_1, x_2, \ldots x_m]$ according to the payoff matrix $\mathbf{A} = [a_{ij}]_{m \times n}$. The constraint system of the classical linear optimization task (polyhedron Ω_I) in this case has a form:

$$\Omega_I : \begin{cases} a_{11}x_1 + a_{21}x_2 + \ldots + a_{m1}x_m \geq v, \\ a_{12}x_1 + a_{22}x_2 + \ldots + a_{m2}x_m \geq v, \\ \cdots\cdots\cdots\cdots\cdots\cdots\cdots\cdots \\ a_{1n}x_1 + a_{2n}x_2 + \ldots + a_{mn}x_m \geq v, \\ x_1 + x_2 + \ldots x_m = 1, \end{cases} \tag{6}$$

$$x_i \geq 0,\ i = 1, \ldots, m.$$

This task is usually called the primal problem. The solution of this problem gives us the frequencies $\mathbf{x} = [x_1, x_2, \ldots x_m]$ and the game value v for the first player A.

The mathematical model for solving the dual problem of harmonizing project stakeholders' values a as zero-sum game. The payoff matrix (2) for the second player B is also $\mathbf{A} = [a_{ij}]_{m \times n}$. The goal of the second player B or objective function W_{II} is to minimize his loss that is equal to the game value v:

$$W_{II} = v \rightarrow \min$$

The strategies of the player B are given by an ordered set $\mathbf{y}$ of probabilities (frequencies) or n-dimensional vector. In case of multiple game repetition player B have to implement its set of pure strategies $[B_1, B_2, \ldots B_n]$ with the values corresponding to $\mathbf{y}$. Such a behavior of player B guarantees that his loss v will not be increased. The vector $\mathbf{y} = [y_1, y_2, \ldots y_n]$ satisfies the conditions:

$$\sum_{j=1}^{n} y_j = 1,\ y_j \geq 0, j = 1, \ldots, n.$$

The player B have to ensure losses not more than the game value v while using pure strategies $[B_1, B_2, \ldots B_m]$ with different frequencies $\mathbf{y} = [y_1, y_2, \ldots y_n]$ according to the payoff matrix $\mathbf{A}$. The constraint system of the classical linear optimization task (polyhedron Ω_I) in this case has a form:

$$\Omega_{II}: \begin{cases} a_{11}y_1 + a_{21}y_2 + \ldots + a_{1n}x_n \leq v, \\ a_{21}y_1 + a_{22}y_2 + \ldots + a_{2n}x_n \leq v, \\ \cdots\cdots\cdots\cdots\cdots\cdots\cdots\cdots \\ a_{n1}y_1 + a_{n2}y_2 + \ldots + a_{mn}y_n \leq v, \\ y_1 + y_2 + \ldots y_n = 1, \end{cases}$$

$$y_j \geq 0, j = 1, \ldots, n.$$

The task presented is usually called the dual problem. The solution of this problem gives us the frequencies $\mathbf{y} = [y_1, y_2, \ldots y_n]$ and the game value v for the second player B.

The features of the choice and application of strategies in gaming models in the project management tasks. It is common to distinguish between solutions in pure or mixed strategies for the zero-sum matrix game. The solution in pure strategies is always associated with at least one saddle point – the equilibrium point. The saddle point is defined after calculating upper and lower values of a game. The lower game value α is lowest guaranteed winning of the player A:

$$\alpha = \max_i \min_j (a_{ij})$$

You should select the smallest values for each payoff matrix row (the guaranteed winning of the first player A) $\min(a_{ij}) = [\alpha_1, \alpha_2, \ldots \alpha_m]$ for calculating this expression. The second step is to choose the largest one among selected guaranteed winnings $\max_i \min_j (a_{ij})$. The upper game price β is the lowest guaranteed loss of the second player B:

$$\beta = \min_j \max_i (a_{ij})$$

You should select the largest values for each payoff matrix column $\max(a_{ij}) = [\beta_1, \beta_2, \ldots \beta_n]$ for calculating this expression. The next step is to choose the smallest loss among the selected losses $\min_j \max_i (a_{ij})$.

In case of the coincidence of numerical values $\alpha = \beta$ we have a solution in pure strategies. The corresponding strategies A_k, B_l are called optimal and are recommended for the prior use by gamers

$$A_k = A^{opt}, \; B_l = B^{opt}.$$

In this case the game value is $v = a_{kl}$.

The very point with the position (k, l) in payoff matrix is the equilibrium point. Note that the zero-sum game could have several equilibrium points, but they all have the same numerical value.

It is known that $\alpha \leq \beta$. In case of obtaining different values for α and β, we have a solution in so-called mixed strategies $\mathbf{x} = [x_1, x_2, \ldots x_m]$ of player A and $\mathbf{y} = [y_1, y_2, \ldots y_n]$ of player B. The situations $(\mathbf{x}, \mathbf{y})$ allows us to calculate the game value on the basis of multiple game repetitions as a mathematical expectation $M(\mathbf{x},\mathbf{y})$ of a random process $(\mathbf{x},\mathbf{y})$ with discrete random variables in a form of payoff matrix $\mathbf{A}$ that is calculated by formula

$$M(\mathbf{x},\mathbf{y}) = \sum_{i=1}^{m}\sum_{j=1}^{n} a_{ij}x_i y_j = v.$$

This number is the mean expected value that is equal to the game value v. It must be pointed out that

$$\alpha \leq = \sum_{i=1}^{m}\sum_{j=1}^{n} a_{ij}x_i y_j \leq \beta.$$

The presence of at least one zero value among mixed strategies known as the degenerate solution. It is well known that the largest number of non-zero values in a situation $(\mathbf{x},\mathbf{y})$ is equal to the payoff matrix $\mathbf{A}$ rank (the number of linearly independent rows or columns) [24, 25]. Since

$$\text{rang}\,(A_{m\times n}) \leq \min(m,n),$$

this maximum number will be equal to the maximum of the smallest measurability of payoff matrix. With this in mind, the question arises as to the usefulness of further research of non-square payoff matrices from the viewpoint of mixed strategies existence. However, from a practical point of view, we can make a conclusion that it is more profitable for a player to have more strategies since it gives additional agility to a project manager.

The application of the model for project stakeholders' value balancing while solving the primal problem in a zero-sum game. Let's suppose that for a project the payoff matrix describing the strategies presented in Fig. 3 and value estimations of the stakeholders has the following form:

$$\mathbf{A} = \begin{pmatrix} 1 & 7 & 9 & 6 & 3 \\ 9 & 1 & 1 & 2 & 1 \\ 6 & 5 & 2 & 7 & 5 \\ 8 & 7 & 1 & 4 & 1 \\ 5 & 1 & 8 & 5 & 5 \\ 5 & 5 & 4 & 1 & 8 \end{pmatrix}$$

The general primal optimization problem for player A stated in (4–6) is as follows:

$$W_I = v \rightarrow \max$$

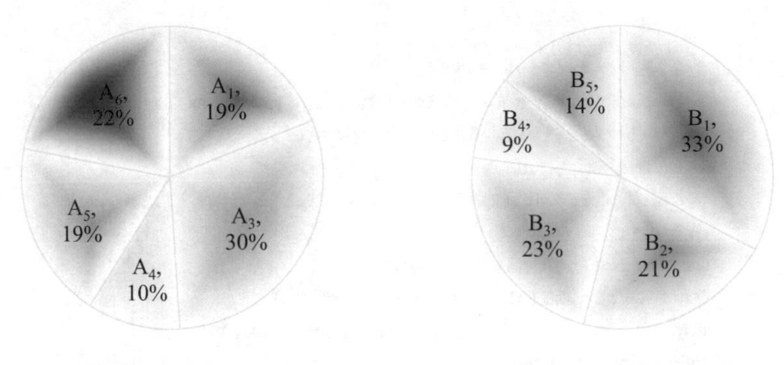

Fig. 3. The recommended frequencies of the mixed strategies use in zero-sum game W_I.

$$\begin{cases} x_1 + 9x_2 + 6x_3 + 8x_4 + 5x_5 + 5x_6 \geq v; \\ 7x_1 + x_2 + 5x_3 + 7x_4 + x_5 + 5x_6 \geq v; \\ 9x_1 + x_2 + 2x_3 + x_4 + 8x_5 + 4x_6 \geq v; \\ 6x_1 + 2x_2 + 7x_3 + 4x_4 + 5x_5 + x_6 \geq v; \\ 3x_1 + x_2 + 5x_3 + x_4 + 5x_5 + 8x_6 \geq v; \\ x_1 + x_2 + x_3 + x_4 + x_5 + x_6 = 1; \end{cases}$$

$$x_i \geq 0, \ \ i = 1, \ \ldots, \ 6.$$

The lower and upper game values are $\alpha = 2$, $\beta = 7$. The problem has a solution in mixed strategies:

$$\mathbf{x} = (0.19, \ 0, \ 0.30, \ 0.10, \ 0.19, \ 0.22), \ \mathbf{y} = (0.33, \ 0.21, \ 0.23, \ 0.09, \ 0.14), \ v = 4.8.$$

The decision warranties the following distribution (frequencies) of application of mixed strategies in order to ensure stakeholders' values harmonization (Fig. 3).

It is necessary to pay attention that the strategy A_2 is degenerate and therefore does not fall into the list of strategies recommended for player A.

The presented model is implemented in the Maple® – mathematical software environment. The source code fragment associated with the primal problem solving is presented below.

```
#######################
# Determination of player A winnings (by lines)
zfA:=add(xz[ip1], ip1=1..im):
for jp1 to jn do
  eqA[jp1]:=add(xz[ip1]*c(ip1,jp1), ip1=1..im)>=1
end do:
ura:={seq(eqA[j],j=1..jn)}:
wra:=minimize(zfA,ura,NONNEGATIVE):
nup:=1/(add(rhs(wra[ip1]), ip1=1..im)):
for i to im do
  xp[i]:=rhs(wra[i])*nup:
end do:
X=[seq(xp[i],i=1..im)], nu=nup, {evalf[2](nup)};
X=evalf[2] (seq((xp[i],i=1..im)));
# Determination of player B losses (by columns)
zfB:=add(yz[jp3], jp3=1..jn):
for ip4 to im do
  eqB[ip4]:=add(yz[jp4]*c(ip4,jp4), jp4=1..jn)<=1
end do:
urb:={seq(eqB[i],i=1..im)}:
wrb:=maximize(zfB,urb,NONNEGATIVE):
nup:=1/(add(rhs(wrb[jp5]), jp5=1..jn)):
for j to jn do
  yp[j]:=rhs(wrb[j])*nup:
end do:
Y=[seq(yp[j],j=1..jn)], nu=nup,{evalf[2](nup)};
Y=evalf[2] (seq((yp[j],j=1..jn)));
#######################
```

The given solution of the primal problem of harmonizing project stakeholders' values a as zero-sum game of two players have a dual solution the analysis of which lies outside of this study.

The model of project stakeholders' value harmonization as a solution of the bimatrix game. Let's consider at a game in which two players *A* and *B* have their own non-antagonistic interests. Each player chooses an independent strategy of behavior. Player *A* has its own behavior strategies $[A_1, A_2, \ldots A_m]$ and player *B* $[B_1, B_2, \ldots B_n]$ as well.

The fundamental issue of bimatrix game theory is that in the case of choosing *i*-strategy by the player *A* and *j*-strategy by the player *B* both players get different values a_{ij} and b_{ij}. That is, unlike the matrix game, where $b_{ij} = -a_{ij}$ each bimatrix game player obtains personal subjective value. The brute-force search of combinatorial space in a bimatrix game is given by two payoff matrices **A** and **B** (2).

For bimatrix games there can be solutions in both pure and mixed strategies. The solution in pure strategies is related with the Nash equilibrium point respective to the payoff matrix element $(i^{\#}, j^{\#})$ when the following conditions are met:

$$a_{ij^{\#}} \leq a_{i^{\#}j^{\#}},\ i = 1, \ldots, m,\ b_{i^{\#}j} \leq b_{i^{\#}j^{\#}},\ j = 1, \ldots, n. \tag{7}$$

The optimization approach can be used for calculating A player mixed strategies $\mathbf{x} = [x_1, x_2, \ldots x_m]$. The optimization task in this case has a form:

$$W_I^A = v_A \rightarrow \max \tag{8}$$

$$\Omega_I^A : \begin{cases} a_{11}x_1 + a_{21}x_2 + \ldots + a_{m1}x_m \geq v, \\ a_{12}x_1 + a_{22}x_2 + \ldots + a_{2m}x_m \geq v, \\ \cdots\cdots\cdots\cdots\cdots\cdots\cdots\cdots \\ a_{m1}x_1 + a_{m2}x_2 + \ldots + a_{mn}x_m \geq v, \\ x_1 + x_2 + \ldots x_m = 1, \end{cases} \tag{9}$$

$$x_i \geq 0,\ i = 1, \ldots, m. \tag{10}$$

The solution of this task gives us the frequencies $\mathbf{x} = [x_1, x_2, \ldots x_m]$ and game value v_A for the first player A. We also can use an optimization approach for definition the mixed strategies $\mathbf{y} = [y_1, y_2, \ldots y_n]$ of player B. The optimization task has the form:

$$W_I^B = v_B \rightarrow \max, \tag{11}$$

$$\Omega_I^B : \begin{cases} b_{11}y_1 + b_{21}y_2 + \ldots + b_{m1}y_n \geq v, \\ b_{12}y_1 + b_{22}y_2 + \ldots + b_{m2}y_n \geq v, \\ \cdots\cdots\cdots\cdots\cdots\cdots\cdots\cdots \\ b_{m1}y_1 + b_{m2}y_2 + \ldots + b_{mn}y_n \geq v, \\ y_1 + y_2 + \ldots y_n = 1, \end{cases} \tag{12}$$

$$y_j \geq 0,\ j = 1, \ldots, n. \tag{13}$$

The solution of the task gives us the frequencies $\mathbf{y} = [y_1, y_2, \ldots y_n]$ and game value v_B for the second player B.

Combining calculations for pure and mixed strategies provides a comprehensive vision of gamers' optimal behavior.

The application of the model for project stakeholders' value balancing while solving the problem in a bimatrix game. Let's suppose that for a project the payoff matrices describing the strategies presented in Fig. 2 and value estimations of two players A (project manager and his team) and B (project stakeholders) have the following forms:

$$\mathbf{A} = \begin{pmatrix} 4 & 6 & 6 & 1 & 7 \\ 8 & 2 & 6 & 5 & 6 \\ 4 & 6 & 3 & 9 & 7 \\ 3 & 4 & 9 & 9 & 5 \\ 3 & 9 & 5 & 6 & 1 \end{pmatrix}, \mathbf{B} = \begin{pmatrix} 2 & 8 & 2 & 1 & 5 \\ 2 & 9 & 1 & 6 & 3 \\ 2 & 5 & 5 & 2 & 6 \\ 5 & 4 & 1 & 8 & 5 \\ 8 & 2 & 8 & 6 & 5 \end{pmatrix},$$

The general optimization task for player A in accordance with (8–10) is as follows:

$$W_I^A = v_A \to \max,$$

$$\Omega_I^A : \begin{cases} 4x_1 + 8x_2 + 4x_3 + 3x_4 + 3x_5 \geq v_A; \\ 6x_1 + 2x_2 + 6x_3 + 4x_4 + 9x_5 \geq v_A; \\ 6x_1 + 6x_2 + 3x_3 + 9x_4 + 5x_5 \geq v_A; \\ x_1 + 5x_2 + 9x_3 + 9x_4 + 6x_5 \geq v_A; \\ 7x_1 + 6x_2 + 7x_3 + 5x_4 + x_5 \geq v_A; \\ x_1 + x_2 + x_3 + x_4 + x_5 = 1; \end{cases}$$

$$x_i \geq 0, \ i = 1, \ldots, 5.$$

The lower and upper game values for player A are $\alpha_A = 3$, $\beta_A = 7$, and for player B $\alpha_B = 2$, $\beta_B = 6$. The task has a solution in mixed strategies: $\mathbf{x}_\mathrm{A} = (0.20, 0.37, 0.19, 0.01, 0.23)$, $\mathbf{y}_\mathrm{B} = (0.08, 0.27, 0.08, 0.14, 0.43)$, and game values $v_A = 5.2$, $v_B = 4.8$.

The solution obtained warranties the presence of equilibrium for the payoff matrices at which the balance of the value interests of players A and B subject provided by applying these strategies with given probability distributions (Fig. 4).

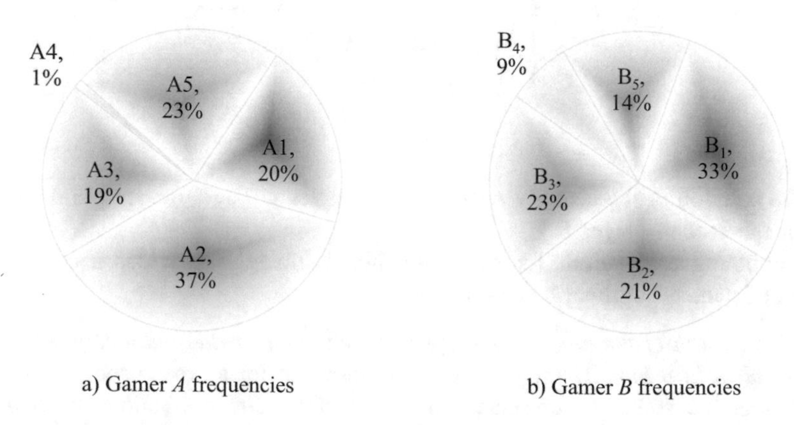

Fig. 4. The recommended frequencies of the mixed strategies use in bimatrix game W_I^A.

The general optimization task for player B in accordance with (11–13) is as follows:

$$W_I^B = v_B \to \max,$$

$$\Omega_I^B : \begin{cases} 2y_1 + 8y_2 + 2y_3 + y_4 + 5y_5 \geq v_B; \\ 2y_1 + 9y_2 + y_3 + 6y_4 + 3y_5 \geq v_B; \\ 2y_1 + 5y_2 + 5y_3 + 2y_4 + 6y_5 \geq v_B; \\ 5y_1 + 4y_2 + y_3 + 8y_4 + 5y_5 \geq v_B; \\ 8y_1 + 2y_2 + 8y_3 + 6y_4 + 5y_5 \geq v_B; \\ y_1 + y_2 + y_3 + y_4 + y_5 = 1; \end{cases}$$

$$y_j \geq 0, \ j = 1, \ldots, 5.$$

The task has a solution in mixed strategies: $\mathbf{x}_B = (0.14,\ 0.19,\ 0.17,\ 0.08,\ 0.42)$, $\mathbf{y}_A = (0.28,\ 0.31,\ 0.20,\ 0.07,\ 0.14)$, and game values $v_A = 4.8$, $v_B = 5.2$. The frequencies of the gamers' mixed strategies use for balancing value of the players A and B subject is shown on Fig. 5.

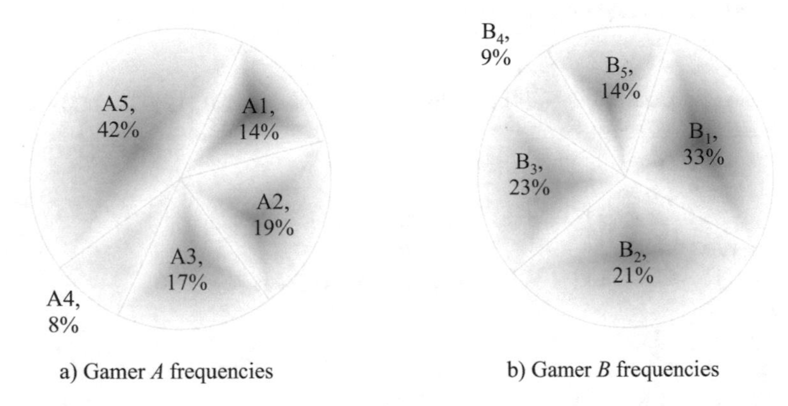

Fig. 5. The recommended frequencies of the mixed strategies use in bimatrix game W_I^B.

The equilibrium in pure strategies is determined according to (7) for these payoff matrices by two situations:

(1) for $i^{\#} = 3$ and $j^{\#} = 5$ – $\mathbf{x} = (0, 0, 1,\ 0,\ 0)$, $\mathbf{y} = (0, 0, 0, 0,\ 1)$, $v_A = 7$, $v_B = 6$.
(2) for $i^{\#} = 4$ and $j^{\#} = 4$ – $\mathbf{x} = (0, 0, 0, 1,\ 0)$, $\mathbf{y} = (0, 0, 0, 1, 0)$, $v_A = 9$, $v_B = 8$.

Each equilibrium situation is characterized by its payoffs. In the 2 solution $(i^{\#}, j^{\#}) = (4, 4)$ the equilibrium is characterized by a greater benefit for project team (player A). It is advisable to adhere the second pair of strategies during periods of stable operation and while project is being implemented within the triple constraints. The 1solution should be used at the project milestones or during the period of deviations from the project base plan or in the circumstances of stakeholder dissatisfaction. This is one of the possibilities of project "soft" management based on the bimatrix game applications.

The presented model is implemented in the Maple® environment. The source code fragment associated with the search for mixed strategies is presented below.

```
#######################
# Determination probabilities of mixed strategies use
xa:=Matrix(1,im,[[seq(p[i],i=1..im)]]):
yb:=Matrix(1,jn,[[seq(q[j],j=1..jn)]]):
#X[A]=xa, Y[B]=yb;
# Expected values
mx:=0:
for i to im do
  for j to jn do
    mx:=mx+ca[i,j]*xa[1,i]*yb[1,j]
  end do:
end do:
mx: mxe:=expand(mx):  Mx[A](seq(p[i],i=1..im),seq(q[j],j=1..jn))=mxe:
my:=0:
for i to im do
  for j to jn do
    my:=my+cb[i,j]*xa[1,i]*yb[1,j]
  end do:
end do:
my: mye:=expand(my): My[B](seq(p[i],i=1..im),seq(q[j],j=1..jn))=mye:
# Nash equilibrium equation
for i to im do
  for j to jn do
    if i=j then pp[j]:=1 else pp[j]:=0 fi:
  end do:
  eqq[i]:=-mxe+subs( {seq(p[j]=pp[j],j=1..jn)}, mxe )=0:
end do:
for j to jn do
  for i to im do
    if j=i then qq[i]:=1 else qq[i]:=0 fi:
  end do:
  eqq[im+j]:=-mye+subs( {seq(q[i]=qq[i],i=1..im)}, mye )=0:
end do:
seq(p[i]>=0,i=1..im):
seq(p[i]<=1,i=1..im):
seq(q[j]>=0,j=1..jn):
seq(q[j]<=1,j=1..jn):
add( p[i], i=1..im )=1;
add( q[j], j=1..jn )=1:
# Searching for Nash equilibrium mixed strategies
zqq:=solve( {seq(eqq[i],i=1..im),   seq(eqq[im+j],j=1..jn),
  add( p[i], i=1..im )=1,add( q[j], j=1..jn )=1,
  seq(p[i]>=0,i=1..im),  seq(q[j]>=0,j=1..jn)},
  {seq(p[i],i=1..im),seq(q[j],j=1..jn)} );
#######################
```

The joint application of the equilibrium conditions obtained through the use of the model proposed and software developed makes it possible to solve various problems of value-driven project management. The pure strategies allow to choose the logic of behavior that ensures the most stable implementation of the project. And mixed strategies help to control the balance of forces and contributions of stakeholder groups to different attitudes to a project and its product.

6 Conclusions

The proposed game-theoretic approach to modeling the processes of value-driven conflict management in projects, based on the use of zero-sum and bimatrix games, allows to balance the execution of project works taking into account the values of project stakeholders. The use of the apparatus of matrix games while solving the problem of project's work harmonization in accordance with the values of stakeholders allows to "softly" manage project implementation processes at different stages. The proposed method allows to form a general model of a typical situation of value conflict that can be adapted to conditions of any particular project and used to the analysis and adoption of appropriate management decisions. Further research should to be directed to the analysis of the features of decision making and to increase the effectiveness of determining the stakeholders' values for the payoff matrices building.

References

1. Kerzner, H., Saladis, F.P.: Value-Driven Project Management, p. 281. Wiley, New-York (2009)
2. Bushuev, S.D., Bushueva, N.S., Yaroshenko, R.F.: Model garmonizatsii tsennosteў programm razvitiya organizatsiĭ v usloviyakh turbulentnosti okruzheniya [The Model of Value Harmonization for Program of Organization Development in Turbulent Environment]. Upravlinnia rozvytkom skladnykh system. [The Management of Complex Systems Development]. Kyiv, vol. 10, pp. 9–13. KNUBA Publisher (2012)
3. Principles behind the Agile Manifesto. Manifesto for Agile Software Development (2001). http://agilemanifesto.org/principles.html. Accessed 10 Dec 2016
4. Turner, M.: Microsoft Solutions Framework Essentials: Building Successful Technology Solutions. Microsoft Press (2006)
5. Rice, E.: The Lean Startup: How Today's Entrepreneurs Use Continuous Innovation to Create Radically Successful Businesses, p. 336. Crown Business, New York (2011)
6. Rokeach, M.: The Nature of Human Values. Free Press, New York (1973)
7. Schwartz, S.: Basic Human Values: An Overview. The Hebrew University of Jerusalem (2004)
8. Hofstede, G.: Dimensionalizing Cultures: The Hofstede Model in Context. Universities of Maastricht and Tilburg (2011)
9. A Guidebook of Project & Program Management for Enterprise Innovation. PMJA (2005). http://www.pmaj.or.jp/ENG/P2M_Download/P2MGuidebookVolume1_060112.pdf. Accessed 10 Dec 2016

10. Rach, V.A.: Tsinnist' yak bazova katehoriya suchasnoyi metodolohiyi upravlinnya proektamy [The value of a basic category of modern project management methodology]. In: Proceedings of VII international conference "Upravlinnya proektamy v rozvytku suspil'stva" [Project management in social development], Kyiv, KNUBA Publisher, pp. 167–168 (2010)
11. Anshin, V.M.: Issledovanie metodologii i faktorov tsennostno-orientirovannogo upravleniya proektami v rossiyskih kompaniyah [The Research of methodologies and factors of value-driven project management in Russian companies] Upravlenie proektami i programmami [Project and Program Management]. Izd. Dom "Grebennikov", vol. 2(38), pp. 104–110 (2014)
12. Larichev, O.I.: Verbalnyiy analiz resheniy [Verbal Decision Analysis]. In-t sistemnogo analiza RAN [System Analysis Institute of Russian Academy of Sciences]. Nauka, Moscow, p. 181 (2006)
13. Grigorian, T.G., Koshkin, K.V.: Value-driven decision-making while choosing outsourcers in the projects of municipal water supply systems reconstruction. In: Proceedings of the 2015 IEEE 8th International Conference on Intelligent Data Acquisition and Advanced Computing Systems: Technology and Applications, IDAACS 2015, pp. 527–530
14. Grigorian, T.G., Koshkin, K.V.: Improved models of value-oriented managing portfolios of projects for reconstruction of water supply. Eastern Eur. J. Enterp. Technol. **2/3**(74), 43–49 (2015)
15. Grigorian, T.G., Shatkovskiy, L.Y.: Modely protsessov prynyatyya reshenyy pry tsennostno-oryentyrovannom upravlenyy trebovanyyamy v IT-proektakh [The models of decision-making processes with value-driven requirement management in IT-projects]. Upravlinnya proektamy ta rozvytok vyrobnytstva [Project management and development of production]. Lugansk, vol. 2(58), pp. 81–98 (2016)
16. Grigorian, T.G., Titov, S.D., Gayda, A.Y., Koshkin, V.K.: A game-theoretic approach to harmonization the values of project stakeholders. In: Proceedings of the XII-th International Scientific and Technical Conference: Computer Sciences and Information Technologies, CSIT 2017, pp. 527–530 (2017)
17. Merriam-Webster Dictionary. https://www.merriam-webster.com. Accessed 1 Mar 2017
18. Von Neumann, J., Morgenstern, O.: Theory of Games and Economic Behavior. Princeton University Press (1953)
19. A Guide to the Project Management Body of Knowledge (PMBOK® Guide), Fifth Edition, PMI, p. 590 (2013)
20. Denne, M., Cleland-Huang, J.: Software by Numbers, Low-Risk, High-Return Development. Prentice-Hall (2003)
21. Nash, J.: Non-cooperative Games. Ann. Math. **54**(2), 286–295 (1951)
22. Vorobyev, N.N.: Situatsii ravnovesiya v bimatrichnyh igrah [The situation of equilibrium in bimatrix games]. Teoriya veroyatnostey i ee primeneniya [Probability Theory and its Applications], vol. 3, pp. 318–331. Nauka Publisher, Moscow (1958)
23. Vorobyev, N.N.: Beskoalitsionnyie igryi [Noncooperative games], p. 495. Nauka Publisher, Moscow (1984)
24. Lau, D.: Algebra und Diskrete Mathematik 1. Grundbegriffe der Mathematik, Algebraische Strukturen 1, Lineare Algebra und Analytische Geometrie, Numerische Algebra. Zweite, korrigierte und erweiterte Auflage, p. 485. Springer, Berlin (2007)
25. Lax, P.D.: Linear Algebra and Application, 2nd edn, p. 377. Wiley, New York (2007)

Information Technology for Assurance of Veracity of Quality Information in the Software Requirements Specification

Tetiana Hovorushchenko(✉)

Khmelnitsky National University, Institutska Street, 11, Khmelnitsky 29016, Ukraine
tat_yana@ukr.net

Abstract. The aim of this study is the development of information technology of evaluating the sufficiency of quality information in the software requirements specification (SRS) for assurance of veracity of quality information in the SRS. This study is also devoted to design and research of the subsystem of evaluating the sufficiency of the SRS information for software quality assessment based on the comparative analysis of ontologies. The developed information technology and subsystem provide: evaluating the sufficiency of the SRS information for software quality assessment by the standard ISO 25010:2011 and based on the metric analysis; identifying the missing (in the SRS) measures and (or) indicators (if the SRS information is insufficient); prioritization of the addition of the missing measures and (or) indicators in the SRS; quantify evaluating the veracity of the available in the SRS information for software quality assessment; increasing the veracity of the quality information in the SRS; increasing the software quality assessment at the early lifecycle stages.

Keywords: Software · Software Requirements Specification (SRS) · Software quality · Software quality information · Sufficiency of quality information · Ontology

1 Introduction

Today almost all spheres of human activity are connected with information systems, the basis of which is software. A key factor in ensuring the effective using of software products and one of the main user requirements to modern software is to achieve high values of its quality. The need to ensure the quality of software follows from the fact that software bugs and failures threaten by catastrophes resulting in human casualties, environmental disasters, significant time and financial losses.

As statistics [1–5] show, there are currently problems in the field of software quality assurance – the large projects are still performed with the lag of schedule or cost overruns, the developed software products often lack the necessary functionality, their performance is low, and the quality doesn't suit consumers.

A large number of software bugs occurs at the stage of requirements formation and formulation – these errors constitute 10–23% of all bugs, and the greater the size of

N. Shakhovska and V. Stepashko (eds.), *Advances in Intelligent Systems and Computing II*, Advances in Intelligent Systems and Computing 689,
https://doi.org/10.1007/978-3-319-70581-1_12

software, the more errors are made in the stage of requirements formation and formulation [2, 6]. The vast majority of software-related crashes occurred due to false requirements, not because of coding bugs [2, 6]. The earlier the defect (bug, trouble, drawback, malfunction) will be revealed, the cheaper it will cost its correction – the cost of correcting the incorrect requirements of the specification, discovered after the release of the product, is almost 100 times the cost of correcting the defects of the specification, assumed in the process of formation and formulation requirements [4].

In the process of formating and formulating the requirements there are the information losses due to incomplete and different understanding of the needs and context of information – especially these losses are significant for software projects that are developed at the junction of subject domains (for example, software for medicine), when it is necessary to consider as standards for development of software, and the standards of the subject domain, for which software is being developed. It's difficult to implement such standards, and it is even more difficult to verify the degree of consideration of the recommendations of these standards.

Software projects with incomplete requirements and specifications cannot be successful [2]. Under such circumstances, the analysis of the SRS, the ability to "cut off" the software projects with the incomplete (with insufficient information) specification is the actual and very important task. Sufficiency of information is one of the most important aspects of software quality assessment. The quality and success of the software project implementation significantly depend on the SRS, and on the sufficiency of the SRS information (the presence of all the information elements, which are necessary to the software quality assessment). The insufficiency, inaccuracy and distortion of the SRS information lead, respectively, to a decrease in the veracity of software quality assessments, as well as to increase the gap of knowledge about software that results in unpredictable emergent properties of software systems.

Currently, the software quality evaluation by standard ISO 25010:2011 [7] is as follows (Fig. 1) – the software quality is evaluated on the basis of the characteristics, the characteristics are evaluated on the basis of subcharacteristics, the subcharacteristics are evaluated on the basis of measures, that are described in ISO 25023:2016 [8]. Evaluation of software quality and complexity based on the metric analysis is as follows (Fig. 2) – the software quality and complexity are calculated on the basis of the metrics, and the metrics are calculated on the basis of the indicators. For metric analysis, 14 quality metrics and 10 complexity metrics with exact or predicted values at the design stage have been selected [9]. The software quality measures, the software quality and complexity indicators, which are defined in the SRS, constitute the quality information of the SRS.

So, *the sufficiency of quality information in the SRS* is the presence in the specification of all information elements (measures and indicators), which are necessary to the software quality assessment.

Today the evaluation of measures for the software quality subcharacteristics and characteristics, indicators for the software quality and complexity metrics is conducted only at the stage of the quality evaluation for the ready source code [5]. But the software requirements determine the required characteristics of the software quality, and also affect the methods of quantitative evaluation of software quality [5]. So, the SRS have

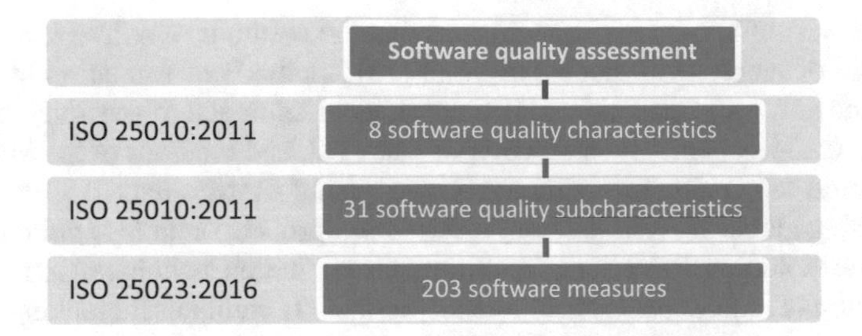

Fig. 1. The modern concept of software quality assessment by ISO 25010:2011

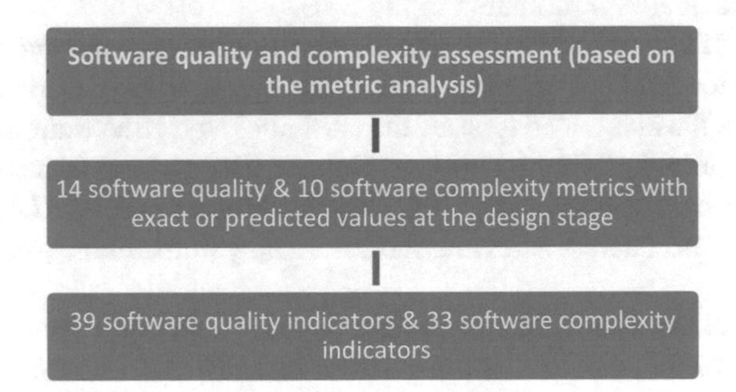

Fig. 2. The modern concept of assessment of software quality and complexity based on the metric analysis

all measures and indicators, which are needed to the subcharacteristics and metrics calculation [5]. So the information sufficiency for future software quality assessment can be evaluated on the basis of the SRS. And if some measures or indicators are absent, then the SRS has insufficient information for software quality assessment and the developers have to make the necessary adjustments in the SRS.

The conducted analysis of standards [7, 8] showed that they are presented in natural language in the textual form, so there is no mechanism for verification of the results of the implementation of these standards in the software development process. It has been established that quality information is conveniently presented as ontologies, which provide the reflection of cause-effect relationships between concepts.

The analysis of known ontological models in the field of software engineering has shown that, at present, the ontological models of profile for software certification [10], ontological models of intelligent decision support systems [11], ontological models for a single coherent underpinning for all ISO/IEC JTC1's SC7 standards [12, 13], the model of domain ontology for ISO/IEC 24744 [14] and the model of domain ontology in the software analysis and reengineering tools [15] have been developed. But nowadays there aren't ontological models of software quality based on ISO 25010:2011, ontological models of software quality and complexity based on metric analysis, and ontological models of the SRS in terms of the availability of information for the software quality assessment.

The analysis of known methods showed that the methods of software development on the basis of ontological models of tasks [16], methods of formation of normative profile in the software certification [10], ontological approach to specification of properties of software systems and their components [17], and methods of the SRS analysis (Using natural language processing technique, Using CASE analysis method, QAW-method, Using global analysis method, O'Brien's approach, Method to discover missing requirement elicitation, Selection of requirements elicitation technique, Comparison and categorization of requirements elicitation techniques, Techniques for ranking and prioritization of software requirements) [18–20] have been developed. But these methods are devoted to monitor the implementation of requirements rather than on evaluating the sufficiency of the quality information in the SRS.

The analysis of known tools has shown that the number of tools have been developed, in particular, the tools for constructing the software systems based on ontological models of tasks [16], and the automated tools of the SRS analysis (IBM Rational RequisitePro, IBM Rational/Telelogic DOORS, Borland Caliber RM, Sybase PowerDesigner, Open Source Requirements Management Tool, Sigma Software, DEVPROM) [18–20]. But these tools are not oriented to assessing the sufficiency of the quality information in the SRS.

Consequently, the known models, methods and tools don't solve the problem of evaluating the sufficiency of quality information in the SRS. In addition, they all belong to different methodological approaches and don't integrate among ourselves, that is, nowadays the information technology of evaluating the sufficiency of quality information in the SRS is absent.

The lack of the information technology of evaluating the sufficiency of quality information in the SRS creates the *actual scientific problem*, one of the ways of solving which is the development of the models, methods and tools of analyzing the sufficiency of quality information in the SRS. Therefore, the *aim of this study* is the development of the information technology (models, methods and tools) of evaluating the sufficiency of quality information in the SRS.

2 Information Technology of Evaluating the Sufficiency of Quality Information in Software Requirements Specification

The structure of the information technology (models, methods and tools) of evaluating the sufficiency of quality information in the SRS can be represented as follows – Fig. 3.

Figure 3 shows that the developed information technology consists of: (1) mathematical and ontological models of the software quality by the standard ISO 25010:2011 (were developed and represented in [21]); (2) mathematical and ontological models of the software complexity and quality based on the metric analysis (were developed and represented in [22]); (3) mathematical and ontological models of the SRS (were developed and represented in [21, 22]); (4) methods of evaluating the sufficiency of the SRS information for software quality assessment (by the standard ISO 25010:2011) based on the ontologies; (5) methods of evaluating the sufficiency of the SRS information for software complexity and quality assessment (on the basis of the metric analysis results)

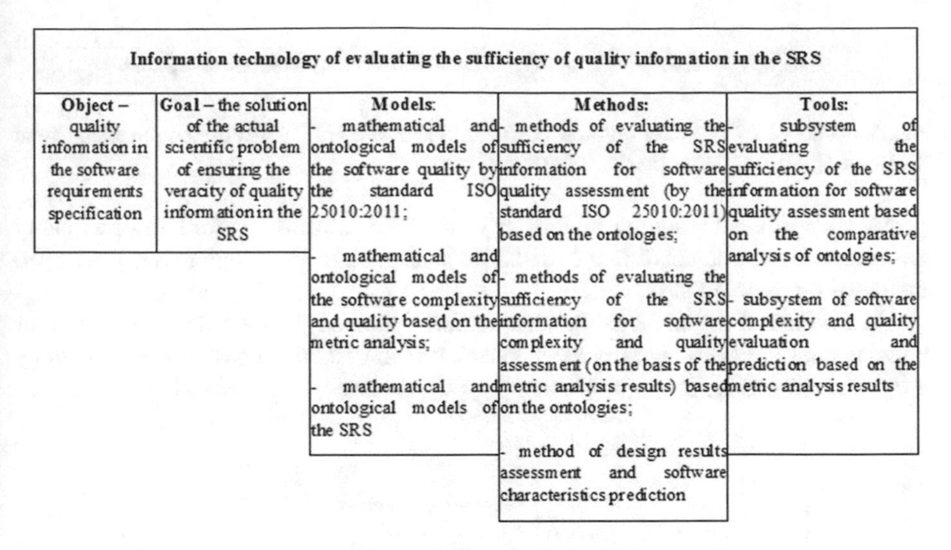

Fig. 3. The structure of information technology of evaluating the sufficiency of quality information in the SRS

based on the ontologies; (6) method of design results assessment and software characteristics prediction (was developed and represented in [23]); (7) subsystem of evaluating the sufficiency of the SRS information for software quality assessment based on the comparative analysis of ontologies; (8) subsystem of software complexity and quality evaluation and prediction based on the metric analysis results (was developed and represented in [23]).

The practical implementation of the developed base (universal) ontological model of the subject domain "Software Engineering" (part "Software Quality") is the base ontology of the subject domain "Software Engineering" ("Software Quality"), the concept of which is represented on Fig. 4. The practical implementation of the developed base (universal) ontological models of the subject domain "Software Engineering" (part "Sofware Quality and Complexity") is the base ontology of the subject domain "Software Engineering" (part "Software Quality and Complexity"), the concept of which is represented on Fig. 5. The components of the base ontology of the subject domain "Software Engineering" ("Software Quality") are represented in [21], and the components of the base ontology of the subject domain "Software Engineering" ("Software Quality and Complexity") are represented in [22].

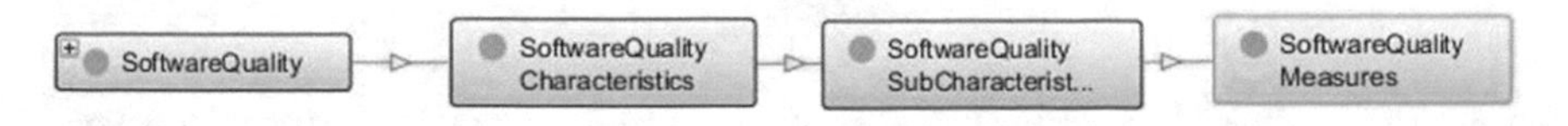

Fig. 4. Concept of the base ontology of the subject domain "Software Engineering" ("Software Quality")

Fig. 5. Concept of the base ontology of the subject domain "Software Engineering" (part "Software Quality and Complexity. Metric analysis")

Methods of evaluating the sufficiency of the SRS information for software quality assessment (by the standard ISO 25010:2011) based on the ontologies were developed and detail represented in [21, 24].

The scheme of the method of evaluating the sufficiency of the SRS information for software quality assessment (by the standard ISO 25010:2011) based on the ontology is represented on Fig. 6.

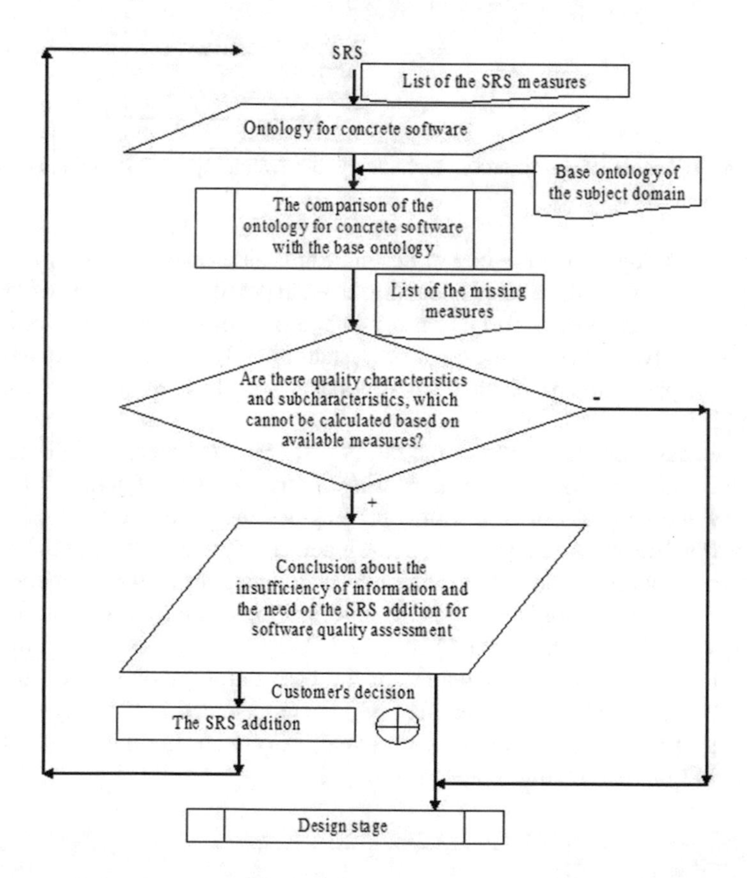

Fig. 6. The scheme of the method of evaluating the sufficiency of the SRS information for software quality assessment (by the standard ISO 25010:2011) based on the ontology

For the eliminate of the subjective evaluation and formal satisfaction of the software quality, it's necessity to consider the degree of severity of quality characteristics and subcharacteristics, and their significance. One of the problems of the known quality models is the calculation of the significance of the quality measures and characteristics. Quality characteristics and subcharacteristics correlate with each other by the measures. It was proven during the above software quality modeling. The existence of such correlations between subcharacteristics increases the significance and weight of software quality measures. Scheme of the method of evaluating the weights of software quality measures is represented on Fig. 7.

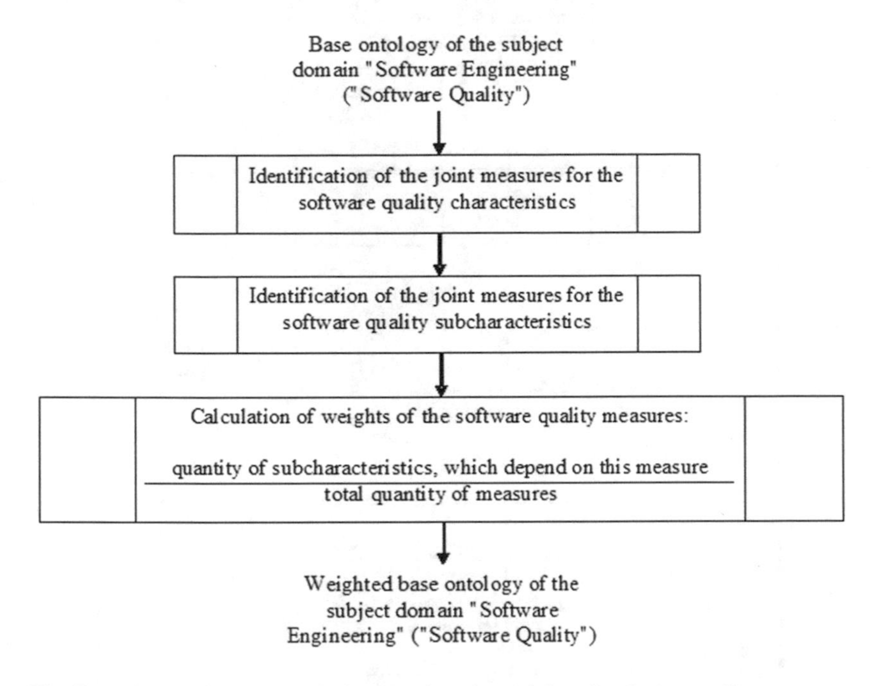

Fig. 7. Scheme of the method of evaluating the weights of software quality measures

The weights of the software quality measures were estimated by the method of evaluating the weights of software quality measures [21]. During the software quality assessment by ISO 25010:2011, it's important to satisfy the availability of measures with larger weights in the SRS for ensuring the appropriate level of veracity of infromation. Weighted ontology of the subject domain "Software Engineering" (part "Software quality") is the ontology, in which the software quality measures have weights with purpose of the recommendation of further satisfaction of these measures in the SRS. The weighted base ontology of the subject domain "Software Engineering" (part "Software quality") was developed on the basis on the base ontology of the subject domain "Software Engineering" (part "Software quality") with addition of information about the weights of software quality measures [21].

The scheme of the method of evaluating the sufficiency of the SRS information for software quality assessment (by the standard ISO 25010:2011) based on the weighted ontology is represented on Fig. 8.

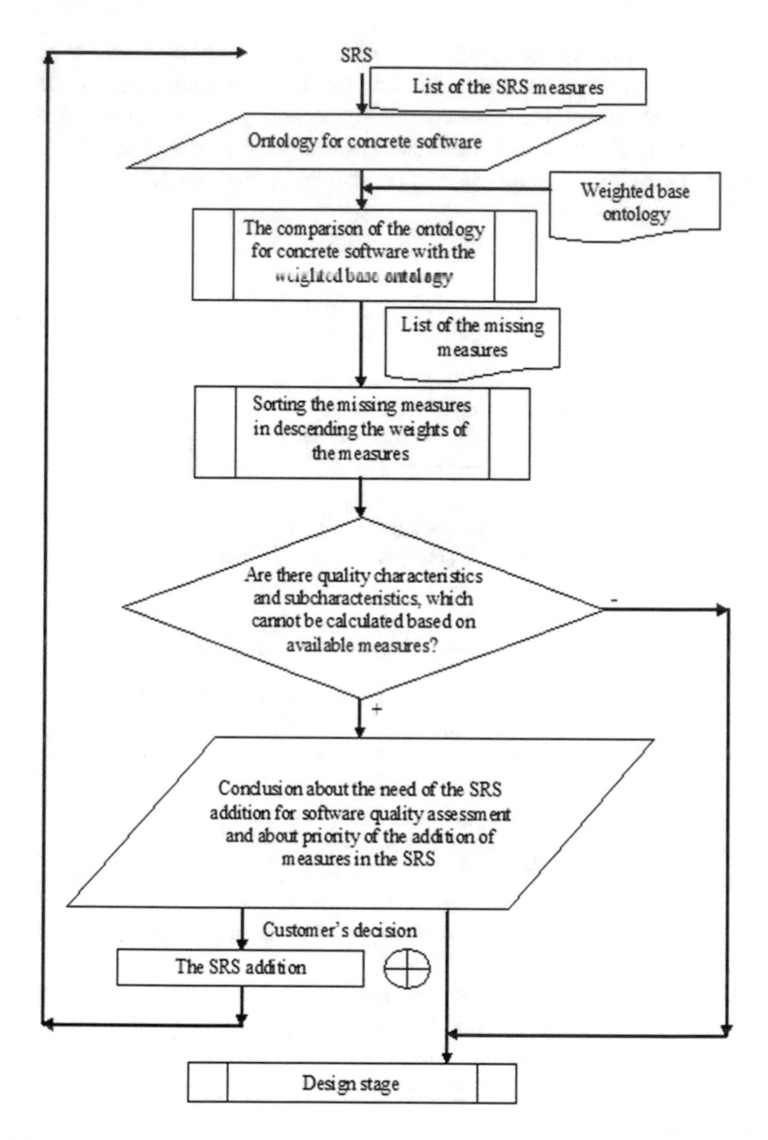

Fig. 8. Scheme of the method of evaluating the sufficiency of the SRS information for software quality assessment (by the standard ISO 25010:2011) based on the weighted ontology

For forming the logical conclusion about sufficiency of the SRS information for software quality assessment by ISO 25010:2011 the production rules were formed on the basis of the developed base and the weighted base ontologies for subject domain "Software engineering" (part "Software quality"). 138 production rules (for the each of measures) have the form "if-then" and were constructed as follows: if measure is missing in the concrete SRS, then: the counters of missing measures for appropriate subcharacteristics are increased by 1 and the counters of missing measures for appropriate characteristics are increased by the quantity of subcharacteristics of this characteristic, for calculation of which the SRS information is insufficient; the weight of the focused measure is assigned to the element of array of the missing measures weights (index of which is the focused measure).

The rule No. 139 has the form: if counters of missing measures for the all 31 software quality subcharacteristics are simultaneously equal to 0, then the SRS information is sufficient for calculation of all software quality subcharacteristics, else: the SRS information is insufficient for the calculation of some software quality subcharacteristics (with indicating the subcharacteristics, for calculation of which the SRS measures are insufficient). The rule No. 140 has the form: if counters of missing measures for the all 8 software quality characteristics are simultaneously equal to 0, then the SRS information is sufficient for calculation of all software quality characteristics, else: the SRS information is insufficient for the calculation of some software quality characteristics (with indicating the characteristics, for calculation of which the SRS measures are insufficient); array of the missing measures weights should be sorted in descending the values of elements (weights of missing measures); indices of those elements of the sorted array of the missing measures weights, which aren't equal 0, should be displayed – as the recommended priority of addition of the missing measures in SRS [24].

The scheme of method of forming the logical conclusion about sufficiency of the SRS information for software quality assessment by ISO 25010:2011 is represented on Fig. 9.

The methods of evaluating the sufficiency of the SRS information for software complexity and quality assessment (on the basis of the metric analysis results) based on the ontologies were developed and detail represented in [22, 25]. These methods are similar to the above methods of evaluating the sufficiency of the SRS information for software quality assessment (by the standard ISO 25010:2011) based on the ontologies (Figs. 6 and 8). The weights of software complexity and quality indicators were calculated in [22] by the method of evaluating the weights of software quality measures (Fig. 7).

For forming the logical conclusion about sufficiency of the SRS information for software quality and complexity assessment by the metric analysis results the production rules were formed on the basis of the developed base and the weighted base ontologies for subject domain "Software engineering" (part "Software quality and complexity. Metric analysis"). 42 production rules (for the each of indciators) have the form "if-then" and were constructed as follows: if indicator is missing in the concrete SRS, then: the counters of missing indicators for appropriate metrics are increased by 1; the weight of the focused indicator is assigned to the element of array of the missing indicators weights (index of which is the focused indicator). The rule No. 43 has the form: if counters of missing indicators for the all 24 software quality and complexity metrics are simultaneously equal to 0, then the SRS information is sufficient for calculation of all software metrics, else: the SRS information is insufficient for the calculation of some software metrics (with indicating the metrics, for calculation of which the SRS indicators are insufficient); array of the missing indicators weights should be sorted in descending the values of elements (weights of missing indicators); indices of those elements of the sorted array of the missing indicators weights, which aren't equal 0, should be displayed – as the recommended priority of addition of the missing indicators in the SRS [25]. The

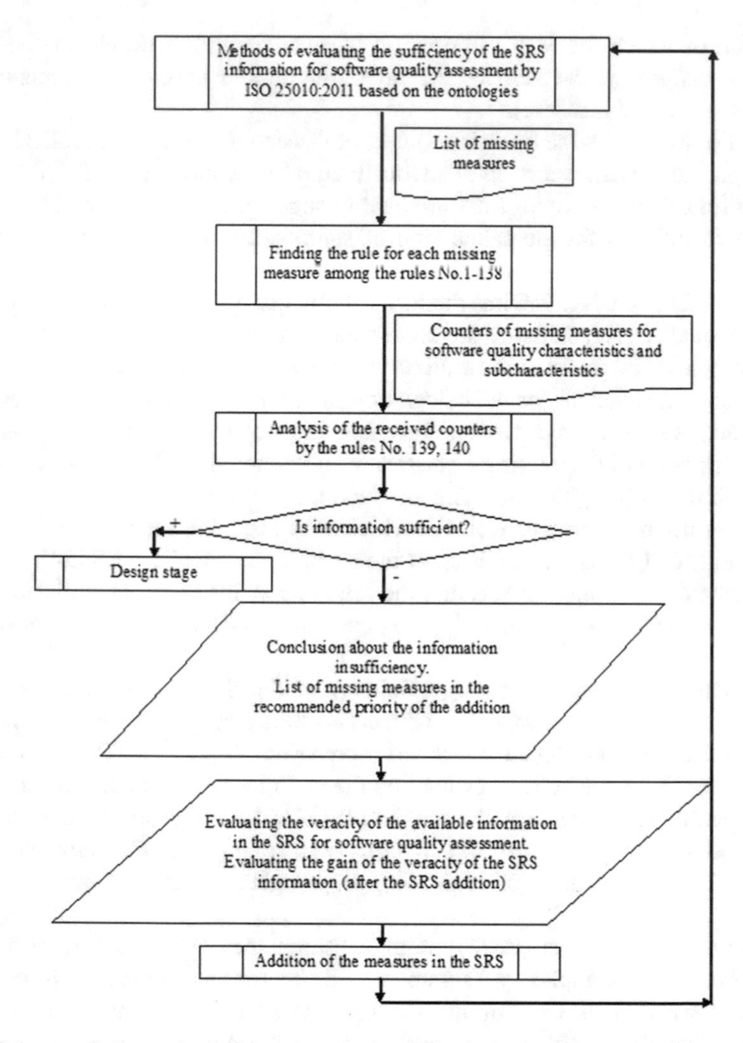

Fig. 9. Scheme of the method of forming the logical conclusion about sufficiency of the SRS information for software quality assessment by ISO 25010:2011

method of forming the logical conclusion about sufficiency of the SRS information for software complexity and quality assessment by the metric analysis results is similar to the above method of forming the logical conclusion about sufficiency of the SRS information for software quality assessment by ISO 25010:2011 (Fig. 9).

Concept of method and subsystem of design results assessment and software characteristics prediction is represented on Fig. 10.

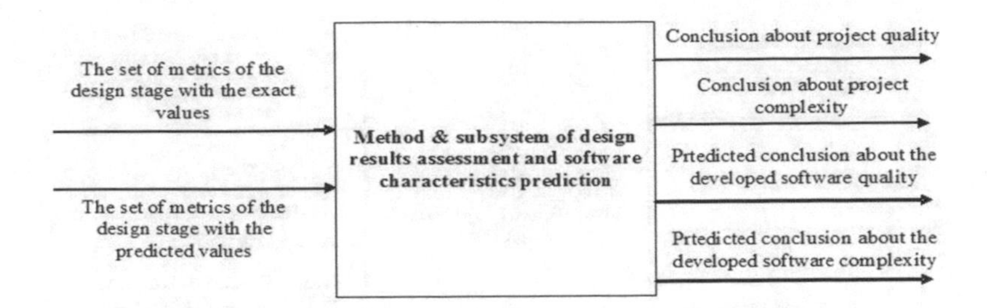

Fig. 10. The concept of method and subsystem of design results assessment and software characteristics prediction

For the completion of the proposed information technology of evaluating the sufficiency of quality information in the SRS, it's necessary to develop (to design and realize) the subsystem of evaluating the sufficiency of the SRS information for software quality assessment based on the comparative analysis of ontologies.

3 Subsystem of Evaluating the Sufficiency of Software Requirements Specification Information for Software Quality Assessment Based on the Comparative Analysis of Ontologies

The inputs of the subsystem of evaluating the sufficiency of the SRS information for software quality assessment based on the comparative analysis of ontologies are the sets: (1) $\{qms_1, \ldots, qms_{nm}\}$ ($nm \leq 138$) available in the SRS software quality measures (according to standards [7, 8], the software quality subcharacteristcs depend on 203 measures, but only on 138 different measures); (2) $\{sqcxi_1, \ldots, sqcxi_{ni}\}$ ($ni \leq 42$) available in the SRS software quality and complexity indicators (the selected in [9, 23] software metrics depend on 72 indicators, but only on 42 different indicators).

The results of the developed subsystem are: (1) conclusion about the sufficiency of the SRS information for software quality assessment by the standard ISO 25010:2011; (2) recommendations about necessity and priority of the addition of the measures in the SRS for software quality assessment by ISO 25010:2011; (3) evaluation of the veracity of the available in the SRS information for software quality assessment by ISO 25010:2011; (4) conclusion about the sufficiency of the SRS information for software complexity and quality assessment by the metric analysis results; (5) recommendations about necessity and priority of the addition of the indicators in the SRS for determining the software complexity and quality by the metric analysis results; (6) evaluation of the veracity of the available in the SRS information for software quality assessment by the metric analysis results.

The concept of subsystem of evaluating the sufficiency of the SRS information for software quality assessment based on the comparative analysis of ontologies is represented on Fig. 11. The structure of subsystem of evaluating the sufficiency of the SRS information for software quality assessment based on the comparative analysis of ontologies is represented on Fig. 12.

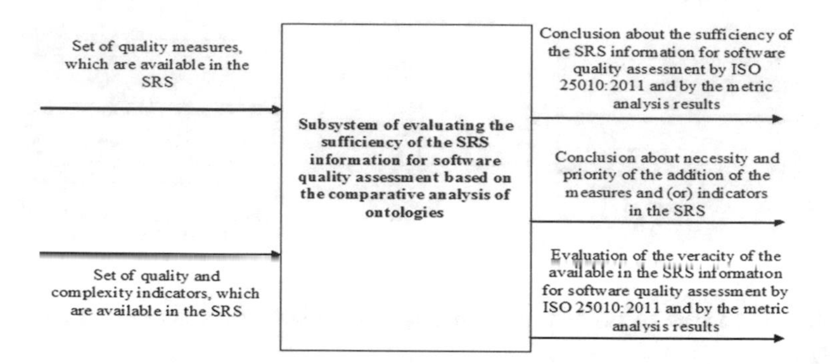

Fig. 11. The concept of subsystem of evaluating the sufficiency of the SRS information for software quality assessment based on the comparative analysis of ontologies

The developed subsystem consists of the next components: (1) *module of introduction of the SRS measures* – collects the user information about the available values of measures $\{qms_1, \ldots, qms_{nm}\}$ $(nm \leq 138)$ in the SRS for concrete software; (2) *module of introduction of the SRS indicators* – collects the user information about the available values of indicators $\{sqcxi_1, \ldots, sqcxi_{ni}\}$ $(ni \leq 42)$ in the SRS; (3) *module of the user support* – provides to the user the information about the structure of the SRS; about the SRS measures, which are necessary for software quality assessment by ISO 25010; about the SRS indicators, which are necessary for determining the software complexity and quality based on the metric analysis results; about the process of the forming the results of the described subsystem; (4) *module of evaluating the sufficiency of the SRS information for software quality assessment by the standard ISO 25010:2011* – works according to methods of evaluating the sufficiency of the SRS information for software quality assessment (by ISO 25010:2011) based on the ontologies. The generation and filling of the ontology template for assessing the quality of the concrete software are performed, considering introduced the available measures $\{qms_1, \ldots, qms_{nm}\}$ $(nm \leq 138)$. The comparative analysis of the ontology for the concrete software with the developed base ontology for subject domain "Software engineering" (part "Software quality") is performed. The result of this comparative analysis is the list of missing measures (in the concrete SRS). If during the comparative analysis of ontologies the differences were not identified, then information of the SRS is sufficient for software quality assessment by ISO 25010. If during the comparative analysis of ontologies the differences were identified, then the available in the SRS measures are insufficient for some subcharacteristics and characteristics calculation, then the comparative analysis of the ontology for the concrete software with the developed weighted base ontology for subject domain "Software engineering" (part "Software quality") is performed, and sorting of all missing (in the SRS) measures in descending the values of weights is conducted, i.e. priority of the addition of these measures in the SRS is established. The quantitative evaluation of the veracity of the available in the SRS information for the software quality assessment is calculated; (5) *module of evaluating the sufficiency of the SRS information for determining the software complexity and quality based on the metric analysis results* – works according to methods of evaluating the sufficiency of the SRS

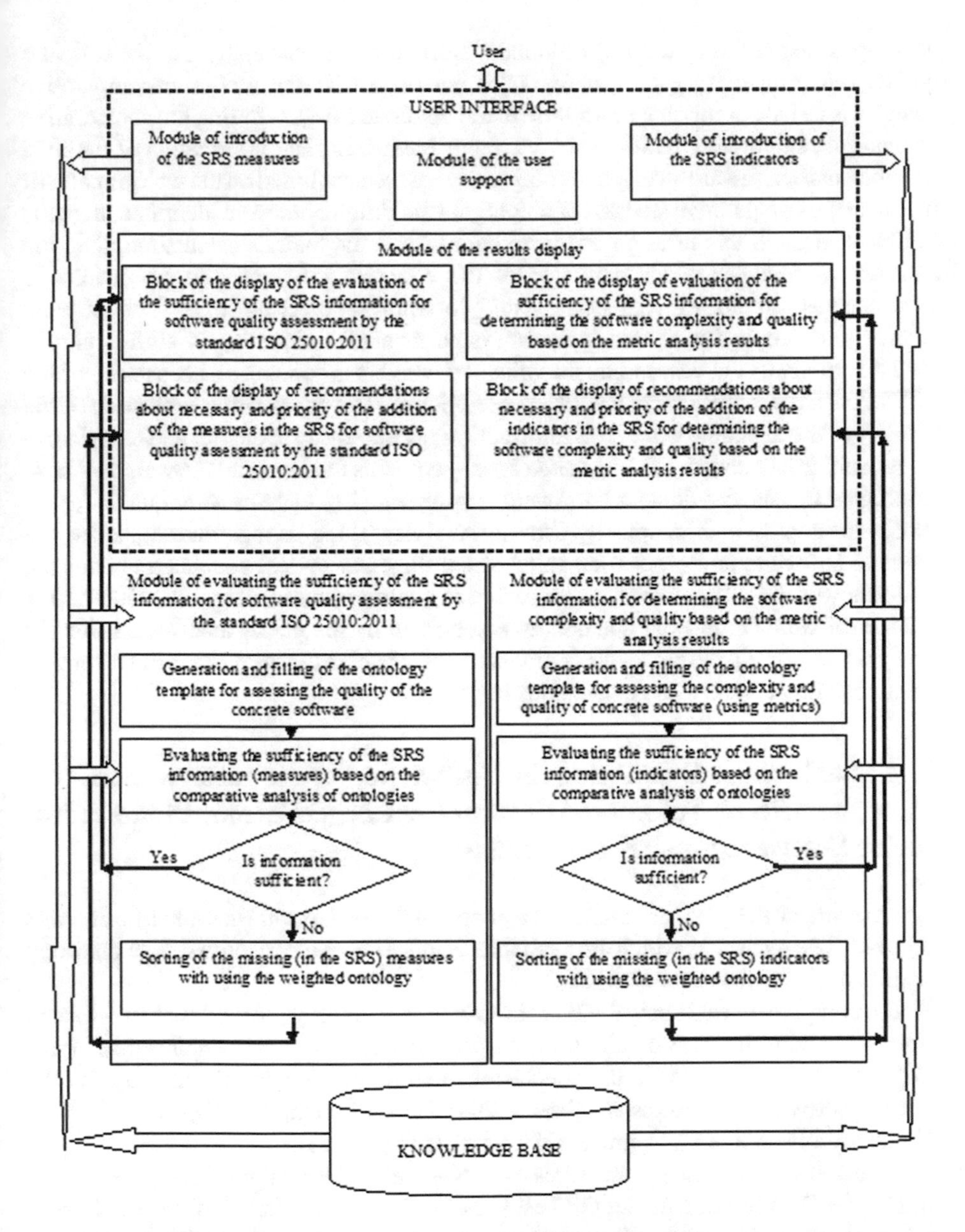

Fig. 12. The structure of subsystem of evaluating the sufficiency of the SRS information for software quality assessment based on the comparative analysis of ontologies

information for software complexity and quality assessment (on the basis of the metric analysis) results based on the ontologies. The generation and filling of the ontology template for assessing the quality and complexity of the concrete software are performed, considering introduced the available indicators $\{sqcxi_1, \ldots, sqcxi_{ni}\}$ ($ni \leq 42$). The comparative analysis of the ontology for the concrete software with the developed base ontology for subject domain "Software engineering" (part "The software quality and complexity. Metric analysis") is performed. The result of this comparative analysis is the list of missing indicators (in the concrete SRS). If during the comparative analysis of ontologies the differences were not identified, then information of the SRS is sufficient for software quality and complexity assessment based on the metric analysis. If during the comparative analysis of ontologies the differences were identified, then the available in the SRS indicators are insufficient for some metrics calculation, then the comparative analysis of the ontology for the concrete software with the developed weighted base ontology for subject domain "Software engineering" (part "The software quality and complexity. Metric analysis") is performed, and sorting of all missing (in the SRS) indicators in descending the values of weights is conducted, i.e. priority of the addition of these indicators in the SRS is established. The quantitative evaluation of the veracity of the available in the SRS information for the software quality and complexity assessment is calculated; (6) *knowledge base* – contains the base and the weighted base ontologies for subject domain "Software engineering" (part "Software quality", part "Software quality and complexity. Metric Analysis"), the formed ontologies for the concrete software, and production rules of forming the logical conclusion about the sufficiency of the SRS information for software quality assessment by ISO 25010:2011 and for software complexity and quality assessment by the metric analysis results; (7) *module of the results display* – the components of this block display the formed conclusions, recommendations and evaluations to user.

4 Experiments: Evaluating the Sufficiency of the Information of the SRS of Automated System for Large-Format Photo Print for Software Quality Assessment

For experiment the SRS of automated system (AS) for large-format photo print was analyzed. The measures, which are available in this SRS, were identified. The ontology for this software was developed [21].

The comparison (in Protégé 4.2) of the developed ontology for AS for large-format photo print with the base ontology for subject domain "Software engineering" (part "Software quality") provides the conclusion, that in the developed ontology for the concrete software 4 measures are absent: "Number Of Functions", "Operation Time", "Number Of Data Items", "Number Of Test Cases" (Fig. 13).

Then the set of missing measures is: {Number Of Functions, Operation Time, Number Of Data Items, Number OF Test Cases}. The finding the rule for each element of this set among 138 rules for the measures is performed. According to these rules, the counters of missing measures are counted.

According to the rule No. 139, the fact was established, that the available measures in the SRS of AS for large-format photo print are insufficient for calculation of following subcharacteristics: Functional Completeness, Functional Correctness, Functional Appropriateness, Maturity, Availability, Fault Tolerance, Recoverability, Time Behaviour, Resource Utilization, Capacity, Appropriateness Recognisability, Learnability,

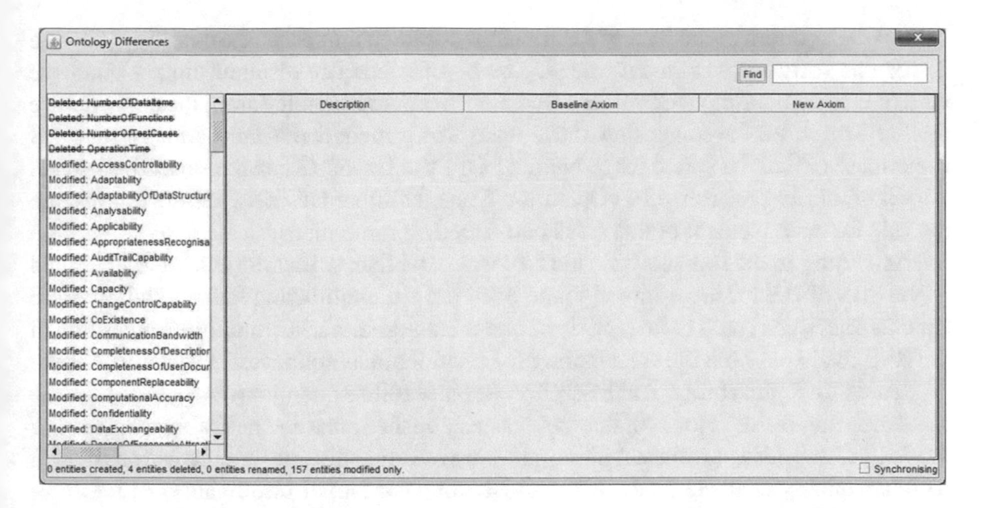

Fig. 13. Comparison of ontology for concrete software (AS for large-format photo print) with the base ontology of the subject domain "Software Engineering" (part "Software Quality")

Operability, Modularity, Analysability, Modifiability, Testability, Confidentiality, Integrity, CoExistence, Interoperability, Adaptability, Replaceability.

According to the rule No. 140, the fact was established, that the available measures in the SRS AS for large-format photo print are insufficient for calculation of all 8 software quality characteristics. Thus, the lack of 4 measures in the SRS led: to the impossibility of calculating the 23 (from 31) subcharactersitics, to the impossibility of calculating all 8 software quality characteristics with high veracity and, respectively, to the impossibility of software quality assessment with high veracity. After establishing the fact of insufficiency of information of the SRS of AS for large-format photo print: sorting the array of the missing measures weights in descending the values of elements was conducted; displaying the indices of those elements of the sorted array of the missing measures weights, which aren't equal 0. Sorted list of missing in the SRS measures in descending the weights: (1) Operation Time (17/138); (2) Number of Functions (11/138); (3) Number of Data Items (8/138); (4) Number of Test Cases (5/138). This list represents the recommended priority of the addition of missing measures in the SRS of AS for large-format photo print.

Next, the evaluation of the veracity of the available in the SRS information for software quality assessment is done (according to the method of forming the logical conclusion about sufficiency of the SRS information for software quality assessment by ISO 25010:2011 [21, 24]). So, for the analyzed SRS of AS for large-format photo print the conclusion about insufficient data for software quality assessment was formed by the developed subsystem, and the veracity of the available in the SRS information for the software quality assessment by ISO 25010:2011 is 76%.

Because the proposed methods of evaluating the sufficiency of the SRS information for software quality assessment (by ISO 25010:2011) based on the ontology are iterative, and there are subcharacteristics and characteristics, for calculation of which the measures of SRS are insufficient, then the addition of the necessary measures in the SRS was held. After addition of the SRS of AS for large-format photo print, the ontology (version 2) for this software was re-developed. The comparison of the re-developed ontology with the base ontology for subject domain “Software engineering” (part “Software quality”) provides the conclusion, that 2 measures were added in the SRS: “Number Of Functions” (2nd in the sorted list), “Number Of Data Items” (3rd in the sorted list). Then the set of missing measures is: {Operation Time, Number Of Test Cases}. The finding the rule for each element of this set is performed.

According to the rule No. 139, the fact was established, that the available measures in the SRS of AS for large-format photo print are still insufficient for calculation of 18 subcharacteristics (with indicating these subcharacteristics), but addition 2 measures in the SRS made possible the calculation of: Functional Completeness, Capacity, Appropriateness Recognisability, Analyzability, Replaceability.

According to rule No. 140, the fact was established, that the available measures in the SRS of AS for large-format photo print are still insufficient for calculation of all 8 software quality characteristics. After establishing the fact of insufficiency of information of the SRS: sorting the array of the missing measures weights in descending the values of elements was conducted; displaying the indices of those elements of the sorted array of the missing measures weights, which aren’t equal 0. Sorted list of missing (after addition) in the SRS measures in descending the weights: (1) Operation Time; (2) Number of Test Cases.

Next, the evaluation of the veracity of the available (after addition) in the SRS information for software quality assessment is done (according to the method of forming the logical conclusion about sufficiency of the SRS information for software quality assessment by ISO 25010:2011 [21, 24]). So, for the analyzed SRS of AS for large-format photo print the conclusion about still insufficient information for software quality assessment was formed by the developed subsystem, and the veracity of the available (after addition) in the SRS information for the software quality assessment by ISO 25010:2011 is 88%.

The process of addition the necessary measures in the SRS is iterative. It can be continued until all quality characteristics and subcharacteristics will be possible to calculate or until the conclusion will be formed, that the SRS information are insufficient for software quality assessment. The customer of developed AS for large-format photo print has decided that further complement of SRS is economically inexpedient.

The gain of the veracity of the SRS information for software quality assessment by ISO 25010:2011 after addition of necessary measures in the SRS is 12% (according to the method of forming the logical conclusion about sufficiency of the SRS information for software quality assessment by ISO 25010:2011 [21, 24]). So, *the developed information technology and subsystem of evaluating the sufficiency of the quality information in the SRS provides the increase of the veracity of the SRS information for the software quality assessment by ISO 25010:2011 by 12% for AS for large-format photo print.*

Let’s consider the functioning of the developed information technology and subsystem for evaluating the sufficiency of the quality information in the SRS for metric analysis. For experiment the SRS of AS for large-format photo print was analyzed. The indicators, which are available in this SRS, were identified. The ontology for this software metric analysis was developed [22].

The comparison (in Protégé 4.2) of the developed ontology for AS for large-format photo print with the base ontology for the subject domain “Software Engineering” (part “Software complexity and quality. Metric analysis”) provides the conclusion, that in the developed ontology for the concrete software metric analysis 9 indicators are absent: “Control Variables”, “Cost Of One Line”, “Project Duration”, “Project Type”, “Quantity Of Code Lines”, “Quantity Of Links Of Each Module”, “Quantity Of Modules”, “Share Of Design Stage In Lifecycle”, “Total Quantity Of Operators”. Then the set of missing indicators is: {Control Variables, Cost Of One Line, Project Duration, Project Type, Quantity Of Code Lines, Quantity Of Links Of Each Module, Quantity Of Modules, Share Of Design Stage In Lifecycle, Total Quantity Of Operators}. The finding the rule for each element of this set among 42 rules for the indicators is performed. According to these rules, the counters of missing indicators are counted.

According to the rule No. 43, the fact was established, that the available indicators in the SRS of AS for large-format photo print are insufficient for calculation of the 20 (from 24) metrics with high veracity and, respectively, and for metric analysis with high veracity. After establishing the fact of insufficiency of information of the SRS of AS for large-format photo print: sorting the array of the missing indicators weights in descending the values of elements was conducted; displaying the indices of those elements of the sorted array of the missing indicators weights, which aren’t equal 0. So, for increasing the veracity of the SRS information for software metric analysis the next indicators should be added in the SRS in this consistency: (1) Quantity Of Code Lines, (2) Quantity Of Modules, (3) Project Duration, (4) Total Quantity Of Operators, (5) Cost Of One Line, (6) Project Type, (7) Share Of Design Stage In Lifecycle, (8) Control Variables, (9) Quantity Of Links Of Each Module.

Next, the evaluation of the veracity of the available in the SRS information for metric analysis is done (according to the method of forming the logical conclusion about sufficiency of the SRS information for software quality assessment by metric analysis results [22, 25]). So, for the analyzed SRS of AS for large-format photo print the conclusion about insufficient data for metric analysis was formed by the developed subsystem, and the veracity of the available in the SRS information for the metric analysis is 42%.

The addition of the necessary indicators in the SRS was held. After addition of the SRS of AS for large-format photo print, the ontology (version 2) for this software was re-developed. The comparison of the re-developed ontology with the base ontology provides the conclusion, that 2 indicators were added in the SRS: “Quantity Of Modules” (2nd in the sorted list), “Total Quantity Of Operators” (4th in the sorted list). So, for increasing the veracity of the SRS information for the metric analysis the next indicators should be added in the SRS in this consistency: (1) Quantity Of Code Lines, (2) Project Duration, (3) Cost Of One Line, (4) Project Type, (5) Share Of Design Stage In Lifecycle, (6) Control Variables, (7) Quantity Of Links Of Each Module.

Next, the evaluation of the veracity of the available (after addition) in the SRS information for metric analysis is done (according to the method of forming the logical conclusion about sufficiency of the SRS information for metric analysis [22, 25]). So, for the analyzed SRS of AS for large-format photo print the conclusion about still insufficient information for metric analysis was formed by the developed subsystem, and the veracity of the available (after addition) in the SRS information for the metric analysis is 56%.

The customer of developed AS for large-format photo print has decided that further complement of the SRS is economically inexpedient.

The gain of the veracity of the SRS information for metric analysis after addition of necessary indicators in the SRS is 14% (according to the method of forming the logical conclusion about sufficiency of the SRS information for metric analysis [22, 25]). So, *the developed information technology and subsystem of evaluating the sufficiency of the quality information in the SRS provides the increase of the veracity of the SRS information for the metric analysis by 14% for AS for large-format photo print.*

5 Conclusions

The information technology of evaluating the sufficiency of quality information in the SRS are first time proposed in this paper. It designed to the support of the software quality assessment at the early lifecycle stages. They provides: the conclusion about the sufficiency of the SRS information for software quality assessment by ISO 25010:2011 and based on the metric analysis results; the prioritization of the additions of the SRS by the measures and (or) by the indicators (if the SRS information is insufficient); the quantitative evaluations of the veracity of the available in the SRS information for software quality assessment by ISO 25010 and by the metric analysis; the increasing the software quality at the early lifecycle stages.

The subsystem of evaluating the sufficiency of the SRS information for the software quality assessment on the basis of the comparative analysis of the ontologies is first time proposed in this paper. It is the decision support system that provides the decision about: the sufficiency of the SRS information for the software quality assessment, the necessity of the addition(s) of the measures and(or) the indicators in the SRS, the veracity of the available in the SRS information for the software quality assessment by ISO 25010:2011 and by the metric analysis.

The experiments proved, that the use of the developed information technology of evaluating the sufficiency of quality information in the SRS provides the increase of the veracity of the SRS information for software quality assessment based on ISO 25010:2011 by 12%, and based on the metric analysis results by 14% even after one addition of the SRS for AS for large-format photo print.

The proposed information technology provides the increasing the veracity of the quality information in the SRS, and improving the software quality at the early stages of the lifecycle.

References

1. Jones, C., Bonsignour, O.: The Economics of Software Quality. Pearson Education, Boston (2012)
2. Ishimatsu, T., Levenson, N., Thomas, J., Fleming, C., Katahira, M., Miyamoto, Y., Ujiie, R.: Hazard analysis of complex spacecraft using systems-theoretic process analysis. J. Spacecraft Rockets **51**, 509–522 (2014)
3. Yourdon, E.:: Death March: The Complete Software Developer's Guide to Surviving "Mission Impossible" Projects, 2nd edn. Prentice Hall (2003)
4. Shamieh, C.: Systems Engineering for Dummies. Wiley Publishing (2011)
5. Maevskiy, D., Kozina, Y.: Where and when is formed of software quality? Electr. Comput. Syst. **18**, 55–59 (2016). (in Russian)
6. McConnell, S.: Code Complete. Microsoft Press (2013)
7. ISO/IEC 25010:2011: Systems and software engineering. Systems and software Quality Requirements and Evaluation (SQuaRE). System and software quality models (2011)
8. ISO 25023:2016: Systems and software engineering. Systems and software Quality Requirements and Evaluation (SQuaRE). Measurement of system and software product quality (2016)
9. Fenton, N.: Software Metrics: A Rigorous Approach, 3rd edn. CRC Press (2014)
10. Shostak, I., Butenko, I.: Ontology approach to realization of information technology for normative profile forming at critical software certification. Herald of the Military Institute of Kiev National University named after Taras Shevchenko, vol. 38, pp. 250–253 (2012)
11. Lytvyn, V.: Modeling of intelligent decision support systems using the ontological approach. Radio Electr. Comput. Sci. Manage. **2**, 93–101 (2011). (in Ukrainian)
12. Henderson-Sellers, B., Gonzalez-Perez, C., McBride, T., Low, G.: An ontology for ISO software engineering standards: 1) creating the infrastructure. Comput. Stan. Interfaces. **36**(3), 563–576 (2014)
13. Ruy, F.B., Falbo, R.A., Barcellos, M.P., Guizzardi, G., Quirino, G.K.S.: An ISO-based software process ontology pattern language and its application for harmonizing standards. ACM SIGAPP Appl. Comput. Rev. **15**(2), 27–40 (2015)
14. Hamri, M.M., Benslimane, S.M.: Building an ontology for the metamodel ISO/IEC24744 using MDA process. Int. J. Mod. Educ. Comput. Sci. **8**, 48–60 (2015)
15. Jin, D., Cordy, J.R.: Ontology-based software analysis and reengineering tool integration: the OASIS service-sharing methodology. In: IEEE International Conference on Software Maintenance Proceedings, Budapest, pp. 613–616. IEEE (2005)
16. Burov, E.: Complex ontology management using task models. Int. J. Knowl. Based Intell. Eng. Syst. **18**(2), 111–120 (2014)
17. Babenko, L.: Ontological approach to specification of software systems features and components. Cybern. Syst. Anal. **1**, 180–187 (2009). (in Russian)
18. Chen, A., Beatty, J.: Visual Models for Software Requirements. MS Press, Washington (2012)
19. Fatwanto, A.: Software requirements specification analysis using natural language processing technique. In: IEEE International Conference on Quality in Research Proceedings, Yogyakarta, pp. 105–110. IEEE (2013)
20. Rehman, T., Khan, M.N.A., Riaz, N.: Analysis of requirement engineering processes, tools/techniques and methodologies. Int. J. Inf. Technol. Comput. Sci. **5**(3), 40–48 (2013)
21. Hovorushchenko, T.: Methodology of evaluating the sufficiency of information for software quality assessment according to ISO 25010. J. Inf. Organ. Sci. (2017, to appear)

22. Hovorushchenko, T.: Models and methods of evaluation of information sufficiency for determining the software complexity and quality based on the metric analysis results. Centr. Eur. Res. J. **2**, 42–53 (2016)
23. Pomorova, O., Hovorushchenko, T.: Artificial neural network for software quality evaluation based on the metric analysis. In: IEEE East-West Design and Test Symposium Proceedings, Kharkiv, pp. 200–203. IEEE (2013)
24. Hovorushchenko, T.: Forming the logical conclusion about sufficiency of information of software requirements specification for software quality assessment by ISO 25010:2011. In: IEEE First Ukraine Conference on Electrical and Computer Engineering Proceedings, Kyiv, pp. 789–794. IEEE (2017)
25. Hovorushchenko, T.: The rules and method of forming the logical conclusion about sufficiency of information for software metric analysis. In: IEEE International Scientific and Technical Conference on Computer Science and Information Technologies, Lviv. IEEE (2017, to appear)

A Multidimensional Adaptive Growing Neuro-Fuzzy System and Its Online Learning Procedure

Zhengbing Hu[1], Yevgeniy V. Bodyanskiy[2], and Oleksii K. Tyshchenko[2](✉)

[1] School of Educational Information Technology, Central China Normal University, 152 Louyu Road, 430079 Wuhan, China
hzb@mail.ccnu.edu.cn
[2] Control Systems Research Laboratory, Kharkiv National University of Radio Electronics, 14 Nauky Ave., 61166 Kharkiv, Ukraine
yevgeniy.bodyanskiy@nure.ua, lehatish@gmail.com

Abstract. The paper presents learning algorithms for a multidimensional adaptive growing neuro-fuzzy system with optimization of a neuron ensemble in every cascade. A building block for this architecture is a multidimensional neo-fuzzy neuron. The demonstrated system is distinguished from the well-recognized cascade systems in its ability to handle multidimensional data sequences in an online fashion, which makes it possible to treat non-stationary stochastic and chaotic data with the demanded accuracy. The most important privilege of the considered hybrid neuro-fuzzy system is its trait to accomplish a procedure of parallel computation for a data stream based on peculiar elements with upgraded approximating properties. The developed system turns out to be rather easy from the effectuation standpoint; it holds a high processing speed and approximating features. Compared to acclaimed countertypes, the developed system guarantees computational simpleness and owns both filtering and tracking aptitudes. The proposed system, which is ultimately a growing (evolving) system of computational intelligence, assures processing the incoming data in an online fashion just unlike the rest of conventional systems.

Keywords: Learning method · Cascade system · Ensemble of neurons
Multidimensional neo-fuzzy neuron · Computational intelligence
Adaptive neuro-fuzzy system

1 Introduction

A great combination of different neuro-fuzzy systems is of considerable use nowadays for a large variety of data processing problems. This fact should be highlighted by a number of preferences that neuro-fuzzy systems hold over other existing methods, and that comes from their abilities to get trained as well as their universal approximating capacities.

A degree of the training procedure may be refined by adapting both a network's set of synaptic weights and its topology [1–8]. This notion is the ground rules for evolving

N. Shakhovska and V. Stepashko (eds.), *Advances in Intelligent Systems and Computing II*, Advances in Intelligent Systems and Computing 689,
https://doi.org/10.1007/978-3-319-70581-1_13

(growing) systems of computational intelligence [9–11]. It stands to mention that probably one of the most prosperous actualizations of this attitude is cascade-correlation neural networks [12–14] by reason of their high level of efficacy and learning simplicity for both a network scheme and for synaptic weights. In general terms, such sort of a network gets underway with a rather simple architecture containing an ensemble of neurons to be trained irrespectively (a case of the first cascade). Every neuron in an ensemble can possess various activation functions as well as learning procedures. Nodes (neurons) in the ensemble do not intercommunicate while they are being learnt.

Eventually, when all the elements in the ensemble of the first cascade have had their weights adapted, the best neuron in relation to a learning criterion builds up the first cascade, and its synaptic weights are not able of being configured any longer. In the next place, the second cascade is commonly formed by means of akin neurons in the training ensemble. The sole difference is that neurons to be learnt in the ensemble of the second cascade own an additional input (and consequently an additional synaptic weight) which proves to be an output of the first cascade. In similar fashion to the first cascade, the second one withdraws all elements except a single one, which gives the best performance. Its synaptic weights should be fixed afterwards. Nodes in the third cascade hold two additional inputs, namely the outputs of the first and second cascades. The growing network keeps on adding new cascades to its topology until it gains the required quality of the results received over the given training set.

By way of evading multi-epoch learning [15–23], various kinds of neurons (preferably their outputs should depend in a linear manner on synaptic weights) may be utilized as the network's elements. This could give the opportunity to exploit some optimal in speed learning algorithms and handle data as it arrives to the network. In the meantime, if the system is being trained in an online manner, it looks impossible to detect the best neuron in the ensemble. While handling non-stationary data objects, one node in a training ensemble may be confirmed to be the best element for one part of the training data sample (but it cannot be selected as the best one for the other parts). It may be recommended that all the units should be abandoned in the training ensemble, and some specific optimization method (selected in agreement with a general quality criterion for the network) is meant to be used for estimation of an output of the cascade.

It will be observed that the widely recognized cascade neural networks bring into action a non-linear mapping $R^n \to R^1$, which means that a common cascade neural network is a system with a single output. By contrast, many problems solved by means of neuro-fuzzy systems demand a multidimensional mapping $R^n \to R^g$ to be executed, that finally accounts for the fact that a number of elements to be trained in every cascade is g times more by contrast to a common neural network, which makes this sort of a system too ponderous. Hence, it seems relevant to operate a specific multidimensional neuron's topology as the cascade network's unit with multiple outputs instead of traditional.

The described growing cascade neuro-fuzzy system of computational intelligence is actually an effort to develop a system for handling a data stream that is fed to the system in an online way and that is in possession of a far smaller amount of parameters to be set as opposed to other widely recognized analogues.

2 An Architecture of the Hybrid Growing System

A scheme of the introduced hybrid system is represented in Fig. 1. In fact, it coincides with architecture of the hybrid evolving neural network with an optimized ensemble in every cascade group of elements to have been developed in [24–29]. A basic dissimilarity lies in a type of elements utilized and learning procedures respectively.

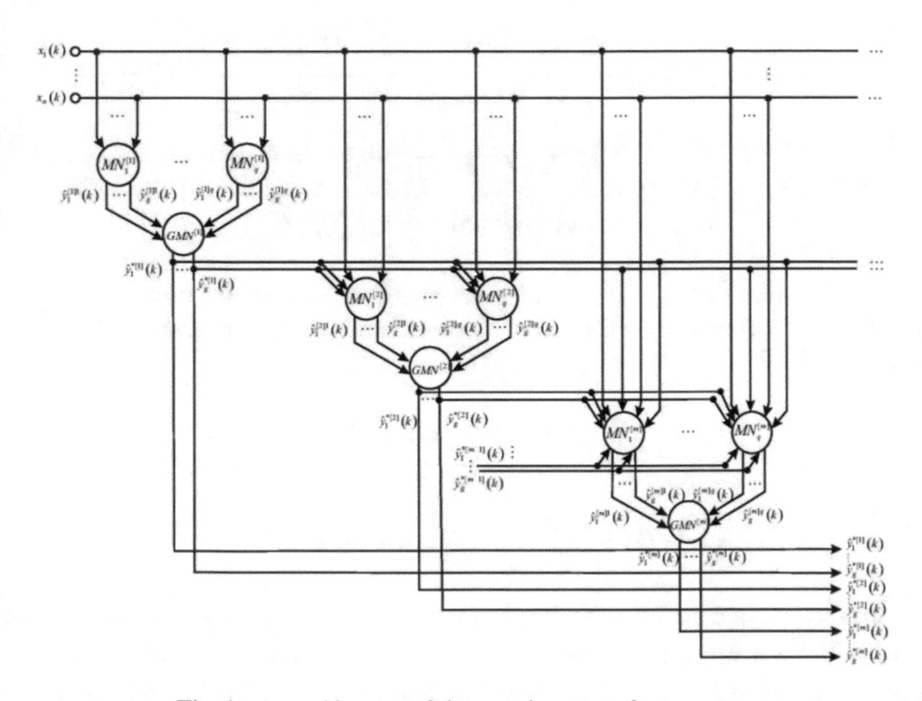

Fig. 1. An architecture of the growing neuro-fuzzy system.

A network's input can be described by a vector signal $x(k) = (x_1(k), x_2(k), \ldots, x_n(k))^T$, where $k = 1, 2, \ldots$ stands for either a plurality of observations in the "object-property" table or an index of the current discrete time. These signals are moved to inputs of each neuron $MN_j^{[m]}$ in the system ($j = 1, 2, \ldots, q$ denotes a number of neurons in a training ensemble, $m = 1, 2, \ldots$ specifies a cascade's number). A vector output $\hat{y}^{[m]j}(k) = \left(\hat{y}_1^{[m]j}(k), \hat{y}_2^{[m]j}(k), \ldots, \hat{y}_d^{[m]j}(k), \ldots, \hat{y}_g^{[m]j}(k)\right)^T$ is eventually produced, $d = 1, 2, \ldots, g$. These outputs are in the next place fed to a generalizing neuron $GMN^{[m]}$ to reproduce an optimized vector output $\hat{y}^{*[m]}(k)$ for the cascade m. Just as the input of the nodes in the first cascade is $x(k)$, elements in the second cascade take g additional arriving signals for the obtained signal $\hat{y}^{*[1]}(k)$, neurons in the third cascade have $2g$ additional inputs $\hat{y}^{*[1]}(k), \hat{y}^{*[2]}(k)$, whilst neurons in the m-th cascade own $(m-1)g$ additional incoming signals $\hat{y}^{*[1]}(k), \hat{y}^{*[2]}(k), \ldots, \hat{y}^{*[m-1]}(k)$. New cascades are becoming a part of the hybrid system within the learning procedure just as it turns out to be clear that an architecture with a current amount of cascades does not provide the required accuracy.

Since a system signal in a conventional neo-fuzzy neuron [30–32] is governed by the synaptic weights in a linear manner, any adaptive identification algorithm [33–35] may actually be applied to learning the network's neo-fuzzy neurons (like either the exponentially-weighted least-squares method in a recurrent form

$$\begin{cases} w_d^{[m]j}(k+1) = w_d^{[m]j}(k) + \dfrac{P_d^{[m]j}(k)\left(y^d(k+1) - \left(w_d^{[m]j}(k)\right)^T \mu_d^{[m]j}(k+1)\right)}{\alpha + \left(\mu_d^{[m]j}(k+1)\right)^T P_d^{[m]j}(k)\mu_d^{[m]j}(k+1)} \mu_d^{[m]j}(k+1), \\ P_d^{[m]j}(k+1) = \dfrac{1}{\alpha}\left(P_d^{[m]j}(k) - \dfrac{P_d^{[m]j}(k)\mu_d^{[m]j}(k+1)\left(\mu_d^{[m]j}(k+1)\right)^T P_d^{[m]j}(k)}{\alpha + \left(\mu_d^{[m]j}(k+1)\right)^T P_d^{[m]j}(k)\mu_d^{[m]j}(k+1)}\right) \end{cases} \quad (1)$$

(here $y^d(k+1)$, $d = 1, 2, \ldots, g$ specifies an external learning signal, $0 < \alpha \leq 1$ marks a forgetting factor) or the gradient learning algorithm with both tracking and filtering properties [35])

$$\begin{cases} w_d^{[m]j}(k+1) = w_d^{[m]j}(k) + \dfrac{y^d(k+1) - \left(w_d^{[m]j}(k)\right)^T \mu_d^{[m]j}(k+1)}{r_d^{[m]j}(k+1)} \mu_d^{[m]j}(k+1), \\ r_d^{[m]j}(k+1) = \alpha r_d^{[m]j}(k) + \left\|\mu_d^{[m]j}(k+1)\right\|^2, \; 0 \leq \alpha \leq 1. \end{cases} \quad (2)$$

An architecture of a typical neo-fuzzy neuron (Fig. 2) as part of the multidimensional neuron $MN_g^{[1]}$ in the cascade system is abundant, since a vector of input signals $x(k)$ (the first cascade) is sent to same-type non-linear synapses $NS_{di}^{[1]j}$ of the neo-fuzzy neurons, where each neuron obtains a signal $\hat{y}_d^{[1]j}(k)$, $d = 1, 2, \ldots, g$ at its output. As a result, components of the output vector $\hat{y}^{[1]j}(k) = \left(\hat{y}_1^{[1]j}(k), \hat{y}_2^{[1]j}(k), \ldots, \hat{y}_g^{[1]j}(k)\right)^T$ are computed irrespectively.

This fact can be missed by introducing a multidimensional neo-fuzzy neuron [36], whose architecture is shown in Fig. 3 and is a modification of the system proposed in [37]. Its structural units are composite non-linear synapses $MNS_i^{[1]j}$, where each synapse contains h membership functions $\mu_{li}^{[1]j}$ and gh tunable synaptic weights $w_{dli}^{[1]j}$. In this way, the multidimensional neo-fuzzy neuron in the first cascade contains ghn synaptic weights, but only hn membership functions. That's g times smaller in comparison with a situation if the cascade is formed of common neo-fuzzy neurons.

Assuming a $(hn \times 1)$ – vector of membership functions

$$\mu^{[1]j}(k) = \left(\mu_{11}^{[1]j}(x_1(k)), \mu_{21}^{[1]j}(x_1(k)), \ldots, \mu_{h1}^{[1]j}(x_1(k)), \ldots, \mu_{li}^{[1]j}(x_i(k)), \ldots, \mu_{hn}^{[1]j}(x_n(k))\right)^T$$

and a $(g \times hn)$ – matrix of synaptic weights

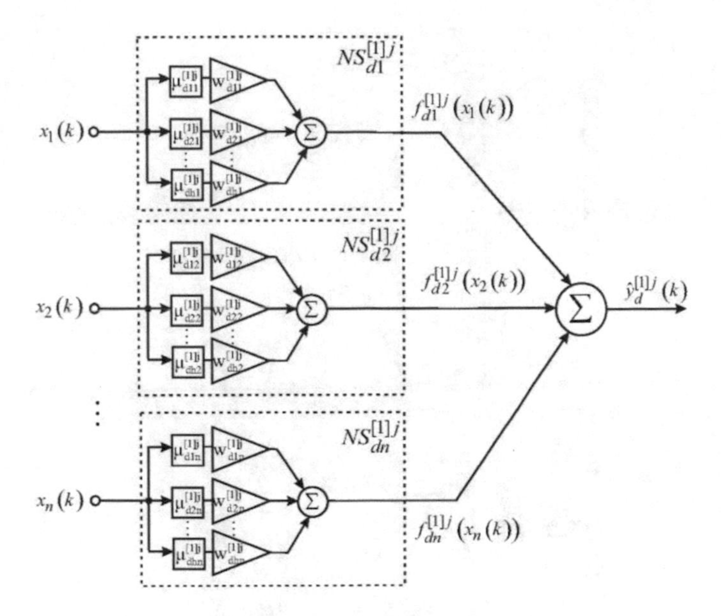

Fig. 2. An architecture of the traditional neo-fuzzy neuron.

$$W^{[1]j} = \begin{pmatrix} w_{111}^{[1]j} & w_{112}^{[1]j} & \cdots & w_{1li}^{[1]j} & \cdots & w_{1hn}^{[1]j} \\ w_{211}^{[1]j} & w_{212}^{[1]j} & \cdots & w_{2li}^{[1]j} & \cdots & w_{2hn}^{[1]j} \\ \vdots & \vdots & \vdots & \vdots & \vdots & \vdots \\ w_{g11}^{[1]j} & w_{g12}^{[1]j} & \cdots & w_{gli}^{[1]j} & \cdots & w_{ghn}^{[1]j} \end{pmatrix},$$

the output signal $MN_j^{[1]}$ can be written down at the k – th time moment in the form of

$$\hat{y}^{[1]j}(k) = W^{[1]j}\mu^{[1]j}(k). \tag{3}$$

Learning the multidimensional neo-fuzzy neuron may be carried out applying either a matrix modification of the exponentially-weighted recurrent least squares method (1) in the form of

$$\begin{cases} W^{[1]j}(k+1) = W^{[1]j}(k) + \dfrac{\left(y(k+1) - W^{[1]j}(k)\mu^{[1]j}(k+1)\right)\left(\mu^{[1]j}(k+1)\right)^T P^{[1]j}(k)}{\alpha + \left(\mu^{[1]j}(k+1)\right)^T P^{[1]j}(k)\mu^{[1]j}(k+1)}, \\ P^{[1]j}(k+1) = \dfrac{1}{\alpha}\left(P^{[1]j}(k) - \dfrac{P^{[1]j}(k)\mu^{[1]j}(k+1)\left(\mu^{[1]j}(k+1)\right)^T P^{[1]j}(k)}{\alpha + \left(\mu^{[1]j}(k+1)\right)^T P^{[1]j}(k)\mu^{[1]j}(k+1)}\right), 0 < \alpha \le 1 \end{cases} \tag{4}$$

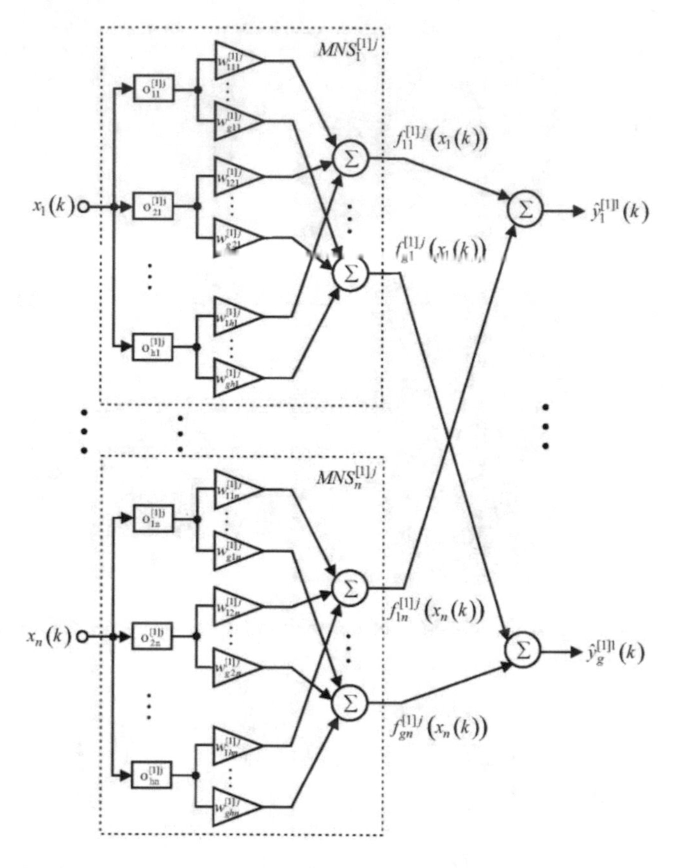

Fig. 3. An architecture of the multidimensional neo-fuzzy neuron.

or a multidimensional version of the algorithm (2) [38]:

$$\begin{cases} W^{[1]j}(k+1) = W^{[1]j}(k) + \dfrac{y(k+1) - W^{[1]j}(k)\mu^{[1]j}(k+1)}{r^{[1]j}(k+1)} \left(\mu^{[1]j}(k+1)\right)^T, \\ r^{[1]j}(k+1) = \alpha r^{[1]j}(k) + \left\| \mu^{[1]j}(k+1) \right\|^2, \ 0 \leq \alpha \leq 1, \end{cases} \tag{5}$$

here $y(k+1) = (y^1(k+1), y^2(k+1), \ldots, y^g(k+1))^T$.

The rest of cascades are trained in a similar fashion, while a vector of membership functions $\mu^{[m]j}(k+1)$ in the m-th cascade enlarges its dimensionality by $(m-1)g$ elements which are guided by the preceding cascades' outputs.

3 Output Signals' Optimization of the Multidimensional Neo-fuzzy Neuron Ensemble

Outputs generated by the neurons in each ensemble are combined by the corresponding neuron $GN^{[m]}$, whose output accuracy $\hat{y}^{*[m]}(k)$ must be higher than the accuracy of any output $\hat{y}_j^{[m]}(k)$. This task can be solved through the use of the neural networks' ensembles approach. Although the well-recognized algorithms are not designated for operating in an online fashion, in this case one could use the adaptive generalizing forecasting [39, 40].

Let's introduce a vector of ensemble inputs for the m-th cascade

$$\hat{y}^{[m]}(k) = \left(\hat{y}_1^{[m]}(k), \hat{y}_2^{[m]}(k), \ldots, \hat{y}_q^{[m]}(k)\right)^T;$$

then an optimal output of the neuron $GN^{[m]}$, which is intrinsically an adaptive linear associator [1–8], can be defined as

$$\hat{y}^{*[m]}(k) = \sum_{j=1}^{q} c_j^{[m]} \hat{y}_j^{[m]}(k) = c^{[m]T} \hat{y}^{[m]}(k)$$

or with additional constraints on unbiasedness

$$\sum_{j=1}^{q} c_n^{[m]} = E^T c^{[m]} = 1 \tag{6}$$

where $c^{[m]} = \left(c_1^{[m]}, c_2^{[m]}, \ldots, c_q^{[m]}\right)^T$ and $E = (1, 1, \ldots, 1)^T$ are $(q \times 1)$ – vectors.

The vector of generalization coefficients $c^{[m]}$ can be found with the help of the Lagrange undetermined multipliers' method. For this reason, we'll introduce a $(k \times g)$ – matrix of reference signals and a $(k \times gq)$ – matrix of ensemble's output signals

$$Y(k) = \begin{pmatrix} y^T(1) \\ y^T(2) \\ \vdots \\ y^T(k) \end{pmatrix}, \hat{Y}^{[m]}(k) = \begin{pmatrix} \hat{y}_1^{[m]T}(1) & \hat{y}_2^{[m]T}(1) & \cdots & \hat{y}_q^{[m]T}(1) \\ \hat{y}_1^{[m]T}(2) & \hat{y}_2^{[m]T}(2) & \cdots & \hat{y}_q^{[m]T}(2) \\ \vdots & \vdots & \vdots & \vdots \\ \hat{y}_1^{[m]T}(k) & \hat{y}_2^{[m]T}(k) & \cdots & \hat{y}_q^{[m]T}(k) \end{pmatrix},$$

a $(k \times g)$– matrix of innovations

$$V^{[m]}(k) = Y(k) - \hat{Y}^{[m]}(k) I \otimes c^{[m]}$$

and the Lagrange function

$$
\begin{aligned}
L^{[m]}(k) &= \frac{1}{2}Tr\left(V^{[m]T}(k)V^{[m]}(k)\right) + \lambda\left(E^T c^{[m]} - 1\right) \\
&= \frac{1}{2}Tr\left(Y(k) - \hat{Y}^{[m]}(k)I \otimes c^{[m]}\right)^T \left(Y(k) - \hat{Y}^{[m]}(k)I \otimes c^{[m]}\right) + \lambda\left(E^T c^{[m]} - 1\right) \\
&= \frac{1}{2}\sum_{\tau=1}^{k} \|y(\tau) - \hat{y}^{[m]}(\tau)c^{[m]}\|^2 + \lambda\left(E^T c^{[m]} - 1\right)
\end{aligned}
\tag{7}
$$

Here I is a $(g \times g)-$ identity matrix, $\otimes$ is the tensor product symbol, λ stands for an undetermined Lagrange multiplier.

Solving the Karush-Kuhn-Tucker system of equations

$$
\begin{cases}
\nabla_{c^{[m]}} L^{[m]}(k) = \sum_{\tau=1}^{k} \left(-\hat{y}^{[m]T}(\tau)y(\tau) + \hat{y}^{[m]T}(\tau)\hat{y}^{[m]}(\tau)c^{[m]}\right) + \lambda E = \vec{0}, \\
\dfrac{\partial L^{[m]}(k)}{\partial \lambda} = E^T c^{[m]} - 1 = 0
\end{cases}
$$

allows obtaining the desired vector of generalization coefficients as follows

$$
c^{[m]}(k) = c^{*[m]}(k) + P^{[m]}(k)\frac{1 - E^T c^{*[m]}(k)}{E^T P^{[m]}(k)E}E
\tag{8}
$$

where

$$
\begin{cases}
P^{[m]}(k) = \left(\sum_{\tau=1}^{k} \hat{y}^{[m]T}(\tau)\hat{y}^{[m]}(\tau)\right)^{-1}, \\
c^{*[m]}(k) = P^{[m]}(k)\sum_{\tau=1}^{k} \hat{y}^{[m]T}(\tau)y(\tau) = P^{[m]}(k)p^{[m]}(k),
\end{cases}
$$

$c^{*[m]}(k)$ is an estimate of the traditional least squares method obtained by the previous k observations.

In order to research vector properties of the obtained generalization coefficients, we should make some obvious transformations. Considering that a vector of learning errors for the neuron $GMN^{[m]}$ can be written down in the form

$$
\begin{aligned}
e^{[m]}(k) &= y(k) - \hat{y}^{*[m]}(k) = y(k) - \hat{y}^{[m]}(k)c^{[m]} = e^{[m]}(k) \\
&= y(k)E^T c^{[m]} - \hat{y}^{[m]}(k)c^{[m]} = \\
&= \left(y(k)E^T - \hat{y}^{[m]}(k)\right)c^{[m]} = \upsilon^{[m]}(k)c^{[m]},
\end{aligned}
$$

the Lagrange function (7) can be also put down in the form

$$L^{[m]}(k) = \frac{1}{2}\sum_{\tau=1}^{k} c^{[m]T} \upsilon^{[m]}(\tau)\upsilon^{[m]T}(\tau)c^{[m]} + \lambda\left(E^T c^{[m]} - 1\right)$$
$$= \frac{1}{2} c^{[m]T} R^{[m]}(k) c^{[m]} + \lambda\left(E^T c^{[m]} - 1\right)$$

and then solving a system of equations

$$\begin{cases} \nabla_{c^{[m]}} L^{[m]}(k) = R^{[m]}(k)c^{[m]} + \lambda E = \vec{0}, \\ \dfrac{\partial L^{[m]}}{\partial \lambda} = E^T c^{[m]} - 1 = 0, \end{cases}$$

we receive

$$\begin{cases} c^{[m]}(k) = \left(R^{[m]}(k)\right)^{-1} E\left(E^T\left(R^{[m]}(k)\right)^{-1} E\right)^{-1}, \\ \lambda = -2E^T\left(R^{[m]}(k)\right)^{-1} E \end{cases}$$

where $R^{[m]}(k) = \sum_{\tau=1}^{k} \upsilon^{[m]}(\tau)\upsilon^{[m]T}(\tau) = V^{[m]T}(k)V^{[m]}(k)$.

The Lagrange function's value can be easily written down at a saddle point

$$L^*(k) = \left(E^T\left(R^{[m]}(k)\right)^{-1} E\right)^{-1},$$

analyzing which by the Cauchy-Schwarz inequality, it can be shown that the generalized output signal $\hat{y}^{*[m]}(k)$ is not inferior to accuracy of the best neuron $\hat{y}^{[m]j}(k)$, $j = 1, 2, \ldots, q$ in an ensemble of output signals.

In order to provide information processing in an online manner, the expression (8) should be performed in a recurrent form which acquires the view of (by using the Sherman-Morrison-Woodbery formula)

$$\begin{cases} P^{[m]}(k+1) = P^{[m]}(k) - P^{[m]}(k)\hat{y}^{[m]T}(k+1)\left(I + \hat{y}^{[m]}(k+1)P^{[m]}(k)\hat{y}^{[m]T}(k+1)\right)^{-1} \\ \cdot\, \hat{y}^{[m]}(k+1)P^{[m]}(k) = \left(I - P^{[m]}(k)\hat{y}^{[m]T}(k+1)\hat{y}^{[m]}(k+1)\right)^{-1} P^{[m]}(k), \\ p^{[m]}(k+1) = p^{[m]}(k) + \hat{y}^{[m]T}(k+1)y(k+1), \\ c^{*[m]}(k+1) = P^{[m]}(k+1)p^{[m]}(k+1), \\ c^{[m]}(k+1) = c^{*[m]}(k+1) + P^{[m]}(k+1)\left(E^T P^{[m]}(k+1)E\right)^{-1}\left(1 - E^T c^{*[m]}(k+1)\right)E. \end{cases} \tag{9}$$

Unwieldiness of the algorithm (9), that is in fact the Gauss-Newton optimization procedure, has to do with inversion of $(g \times g)$ – matrices at every time moment k. And when this value g is large enough, it is much easier to use gradient learning algorithms to tune the weight vector $c^{[m]}(k)$. The learning algorithm can be obtained easily enough if the Arrow-Hurwitz gradient algorithm is used for a search of the Lagrange function's saddle point which takes on the form in this case

$$\begin{cases} c^{[m]}(k+1) = c^{[m]}(k) - \eta_c(k+1)\nabla_{c^{[m]}} L^{[m]}(k), \\ \lambda(k+1) = \lambda(k) + \eta_\lambda(k+1)\dfrac{\partial L^{[m]}(k)}{\partial \lambda} \end{cases} \tag{10}$$

or specifically for (10)

$$\begin{cases} c^{[m]}(k+1) = c^{[m]}(k) + \eta_c(k+1)\left(\hat{y}^{[m]T}(k)e^{[m]}(k) - \lambda(k)E\right), \\ \lambda(k+1) = \lambda(k) + \eta_\lambda(k+1)\left(E^T c^{[m]}(k+1) - 1\right) \end{cases} \tag{11}$$

where $\eta_c(k+1)$, $\eta_\lambda(k+1)$ are some learning rate parameters.

The Arrow-Hurwitz procedure converges to a saddle point of the Lagrange function when a range of learning rate parameters $\eta_c(k+1)$ and $\eta_\lambda(k+1)$ is sufficiently wide. However, one could try to optimize these parameters to reduce training time. For this purpose, we should write down the expression (10) in the form

$$\begin{cases} \hat{y}^{[m]}(k)c^{[m]}(k+1) = \hat{y}^{[m]}(k)c^{[m]}(k) - \eta_c(k+1)\hat{y}^{[m]}(k)\nabla_{c^{[m]}} L^{[m]}(k), \\ y(k) - \hat{y}^{[m]}(k)c^{[m]}(k+1) = y(k) - \hat{y}^{[m]}(k)c^{[m]}(k) + \eta_c(k+1)\hat{y}^{[m]}(k)\nabla_{c^{[m]}} L^{[m]}(k). \end{cases} \tag{12}$$

A left side of the expression (12) describes an a posteriori error $\tilde{e}^{[m]}(k)$, which is obtained after one cycle of parameters' tuning, i.e.

$$\tilde{e}^{[m]}(k) = e^{[m]}(k) + \eta_c(k+1)\hat{y}^{[m]}(k)\nabla_{c^{[m]}} L^{[m]}(k).$$

Introducing the squared norm of this error

$$\begin{aligned} \left\|\tilde{e}^{[m]}(k)\right\|^2 = {} & \left\|e^{[m]}(k)\right\|^2 + 2\eta_c(k+1)e^{[m]T}(k)\hat{y}^{[m]}(k)\nabla_{c^{[m]}} L^{[m]}(k) \\ & + \eta_c^2(k+1)\left\|\hat{y}^{[m]}(k)\nabla_{c^{[m]}} L^{[m]}(k)\right\|^2 \end{aligned}$$

and minimizing it in $\eta_c(k+1)$, i.e. solving a differential equation

$$\frac{\partial \left\|\tilde{e}^{[m]}(k)\right\|^2}{\partial \eta_c} = 0,$$

we come to an optimal value for a learning rate parameter

$$\eta_c(k+1) = -\frac{e^{[m]T}(k)\hat{y}^{[m]}(k)\nabla_{c^{[m]}}L^{[m]}(k)}{\|\hat{y}^{[m]}(k)\nabla_{c^{[m]}}L^{[m]}(k)\|^2}.$$

Then the algorithms (10) and (11) can be finally put down as follows

$$\begin{cases} \nabla_{c^{[m]}}L(k) = -\left(\hat{y}^{[m]T}(k)e^{[m]}(k) - \lambda(k)E\right), \\ c^{[m]}(k+1) = c^{[m]}(k) + \dfrac{e^{[m]T}(k)\hat{y}^{[m]}(k)\nabla_{c^{[m]}}L^{[m]}(k)}{\|\hat{y}^{[m]}(k)\nabla_{c^{[m]}}L^{[m]}(k)\|^2}\nabla_{c^{[m]}}L(k), \\ \lambda(k+1) = \lambda(k) + \eta_\lambda(k+1)\left(E^T c^{[m]}(k+1) - 1\right). \end{cases} \quad (13)$$

The procedure (13) is computationally much easier than (9), and if there are no constraints (6) it turns into a multidimensional modification of the Kaczmarz-Widrow-Hoff algorithm which is widely spread in the problems of ANNs learning.

Elements of a generalization coefficients' vector can be interpreted as membership levels, if a constraint on synaptic weights' non-negativity for the generalizing neuron $GMN^{[m]}$ is introduced into the Lagrange function to be optimized, i.e.

$$\sum_{j=1}^{q} \tilde{c}_j^{[m]} = E^T\tilde{c}^{[m]} = 1, \quad 0 \le \tilde{c}_j^{[m]} \le 1 \quad \forall j = 1, 2, \ldots, q. \quad (14)$$

Introducing the Lagrange function with additional constraints-inequalities

$$\begin{aligned} \tilde{L}^{[m]}(k) &= \frac{1}{2}Tr\left(V^{[m]T}(k)V^{[m]}(k)\right) + \lambda\left(E^T\tilde{c}^{[m]} - 1\right) - \rho^T\tilde{c}^{[m]} \\ &= \frac{1}{2}Tr\left(Y(k) - \hat{Y}^{[m]}(k)I \otimes \tilde{c}^{[m]}\right)^T\left(Y(k) - \hat{Y}^{[m]}(k)I \otimes \tilde{c}^{[m]}\right) + \lambda\left(E^T\tilde{c}^{[m]} - 1\right) - \rho^T\tilde{c}^{[m]} \\ &= \frac{1}{2}\sum_{\tau=1}^{k}\|y(\tau) - \hat{y}^{[m]}(\tau)\tilde{c}^{[m]}\|^2 + \lambda\left(E^T\tilde{c}^{[m]} - 1\right) - \rho^T\tilde{c}^{[m]} \end{aligned}$$

(here ρ is a $(q \times 1)$ – vector of non-negative undetermined Lagrange multipliers) and solving the Karush-Kuhn-Tucker system of equations

$$\begin{cases} \nabla_{\tilde{c}^{[m]}}\tilde{L}^{[m]}(k) = \vec{0}, \\ \dfrac{\partial \tilde{L}^{[m]}(k)}{\partial \lambda} = 0, \\ \rho_j \ge 0 \quad \forall j = 1, 2, \ldots, q, \end{cases}$$

an analytical solution takes on the form

$$\begin{cases} \tilde{c}^{[m]}(k) = P^{[m]}(k)\Big(p^{[m]}(k) - \lambda E + \rho\Big), \\ \lambda = \dfrac{E^T P^{[m]}(k) p^{[m]}(k) - 1 + E^T P^{[m]}(k)\rho}{E^T P^{[m]}(k) E} \end{cases}$$

and having used the Arrow-Hurwicz-Uzawa procedure, we obtain a learning algorithm of the neuron $GMN^{[m]}$ in the view of

$$\begin{cases} \tilde{c}^{[m]}(k+1) = c^{*[m]}(k+1) - P^{[m]}(k+1)\dfrac{E^T c^{*[m]}(k+1) - 1 + E^T P^{[m]}(k+1)\rho(k)}{E^T P^{[m]}(k+1)E} E \\ + P^{[m]}(k+1)\rho(k), \\ \\ \rho(k+1) = \Pr_+\Big(\rho(k) - \eta_\rho(k+1)\tilde{c}^{[m]}(k+1)\Big). \end{cases} \tag{15}$$

The first ratio (15) can be transformed into the form of

$$\begin{aligned} \tilde{c}^{[m]}(k+1) &= c^{[m]}(k+1) - P^{[m]}(k+1)\frac{E^T P^{[m]}(k+1)\rho(k)}{E^T P^{[m]}(k+1)E} E + P^{[m]}(k+1)\rho(k) \\ &= c^{[m]}(k+1) + \left(I - \frac{P^{[m]}(k+1)EE^T}{E^T P^{[m]}(k+1)E}\right) P^{[m]}(k+1)\rho(k) \end{aligned} \tag{16}$$

where $c^{[m]}(k+1)$ is defined by the ratio (8), $\left(I - P^{[m]}(k+1)EE^T\left(E^T P^{[m]}(k+1)E\right)^{-1}\right)$ is a projector to the hyperplane $\tilde{c}^{[m]T}(k+1)E = 1$. It can be easily noticed that the vectors E and $\left(I - P^{[m]}(k+1)EE^T\left(E^T P^{[m]}(k+1)E\right)^{-1}\right)P^{[m]}(k+1)\rho(k)$ are orthogonal, so we can write down the ratios (14) and (15) in a simpler form

$$\begin{cases} \tilde{c}^{[m]}(k+1) = c^{[m]}(k+1) + \Pr_{c^{[m]T}E=1}\Big(P^{[m]}(k+1)\rho(k)\Big), \\ \rho(k+1) = \Pr_+\Big(\rho(k) - \eta_\rho(k+1)\tilde{c}^{[m]}(k+1)\Big). \end{cases}$$

Then the learning algorithm of the generalizing neuron with the constraints (14) finally takes on the form

$$\begin{cases} P^{[m]}(k+1) = P^{[m]}(k) - P^{[m]}(k)\hat{y}^{[m]T}(k+1)\Big(I + \hat{y}^{[m]}(k+1)P^{[m]}(k)\hat{y}^{[m]T}(k+1)\Big)^{-1} \\ = \Big(I - P^{[m]}(k)\hat{y}^{[m]T}(k+1)\hat{y}^{[m]}(k+1)\Big)^{-1} P^{[m]}(k), \\ p^{[m]}(k+1) = p^{[m]}(k) + \hat{y}^{[m]T}(k+1)y(k+1), \\ c^{*[m]}(k+1) = P^{[m]}(k+1)p^{[m]}(k+1), \\ c^{[m]}(k+1) = c^{*[m]}(k+1)P^{[m]}(k+1)\Big(E^T P^{[m]}(k+1)E\Big)^{-1}\Big(1 - E^T c^{*[m]}(k+1)\Big)E, \\ \tilde{c}^{[m]}(k+1) = c^{[m]}(k+1) - P^{[m]}(k+1)\dfrac{E^T P^{[m]}(k+1)\rho(k)}{E^T P^{[m]}(k+1)E}E + P^{[m]}(k+1)\rho(k), \\ \rho(k+1) = \mathrm{Pr}_+\Big(\rho(k) - \eta_\rho(k+1)\tilde{c}^{[m]}(k+1)\Big). \end{cases} \tag{16}$$

The learning procedure (16) can be considerably simplified similar to the previous one with the help of the gradient algorithm

$$\begin{cases} \tilde{c}^{[m]}(k+1) = \tilde{c}^{[m]}(k) - \eta_c(k+1)\nabla_{\tilde{c}^{[m]}}\tilde{L}^{[m]}(k), \\ \lambda(k+1) = \lambda(k) + \eta_\lambda(k+1)\Big(E^T\tilde{c}^{[m]}(k+1) - 1\Big), \\ \rho(k+1) = \mathrm{Pr}_+\Big(\rho(k) - \eta_\rho(k+1)\tilde{c}^{[m]}(k+1)\Big). \end{cases}$$

Carrying out transformations similar to the abovementioned ones, we finally obtain

$$\begin{cases} \nabla_{\tilde{c}^{[m]}}\tilde{L}(k) = -\Big(\hat{y}^{[m]T}(k)e^{[m]}(k) - \lambda(k)E + \rho(k)\Big), \\ \tilde{c}^{[m]}(k+1) = \tilde{c}^{[m]}(k) + \dfrac{e^{[m]T}(k)\hat{y}^{[m]}(k)\nabla_{\tilde{c}^{[m]}}\tilde{L}^{[m]}(k)}{\left\|\hat{y}^{[m]}(k)\nabla_{\tilde{c}^{[m]}}\tilde{L}^{[m]}(k)\right\|^2}\nabla_{\tilde{c}^{[m]}}\tilde{L}^{[m]}(k), \\ \lambda(k+1) = \lambda(k) + \eta_\lambda(k+1)\Big(E^T\tilde{c}^{[m]}(k+1) - 1\Big), \\ \rho(k+1) = \mathrm{Pr}_+\Big(\rho(k) - \eta_\rho(k+1)\tilde{c}^{[m]}(k+1)\Big). \end{cases} \tag{17}$$

The algorithm (17) comprises the procedure (13) as a particular case.

4 Experimental Results

To illustrate the effectiveness of the suggested adaptive neuro-fuzzy system and its learning procedures, we have actualized an experimental test by means of handling the chaotic Lorenz attractor identification. The Lorenz attractor is a fractal structure which matches the Lorenz oscillator's behavior. The Lorenz oscillator is a three-dimensional dynamical system that puts forward a chaotic flow that is also renowned for its lemniscate shape. As a matter of fact, a state of the dynamical system (three variables of the three-dimensional system) is evolving with the course of time in a complex non-repeating pattern.

The Lorenz attractor may be exemplified by a differential equation in the form of

$$\begin{cases} \dot{x} = \sigma(y - x), \\ \dot{y} = x(r - z) - y, \\ \dot{z} = xy - bz. \end{cases} \tag{18}$$

This system of Eq. (18) can be also put down in the recurrent form

$$\begin{cases} x(i+1) = x(i) + \sigma(y(i) - x(i))dt, \\ y(i+1) = y(i) + (rx(i) - x(i)z(i) - y(i))dt, \\ z(i+1) = z(i) + (x(i)y(i) - bz(i))dt \end{cases} \tag{19}$$

where parameter values are: $\sigma = 10$, $r = 28$, $b = 2.66$, $dt = 0.001$.

A data set was acquired with the benefit of (19) which comprises 10000 samples, where 7000 points establish a training set, and 3000 samples make up a validation set.

In our system, we had 2 cascades containing 2 multidimensional neurons each and a generalized neuron in each cascade. The first neuron in each cascade involves 2 membership functions. The graphical results are represented in Figs. 4, 5 and 6. One can basically see the forecasting results for the last cascade in Table 1.

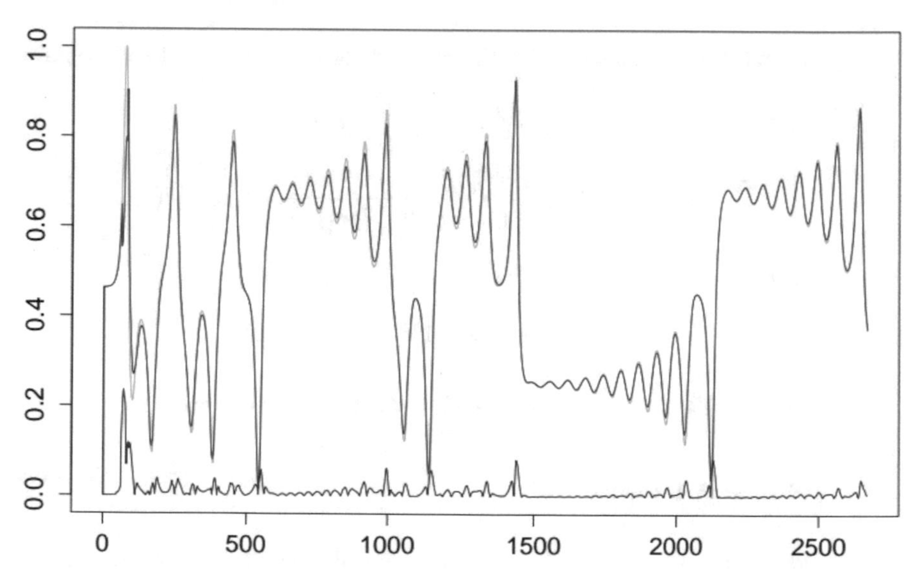

Fig. 4. Identification by means of the Lorenz attractor. The X-component results.

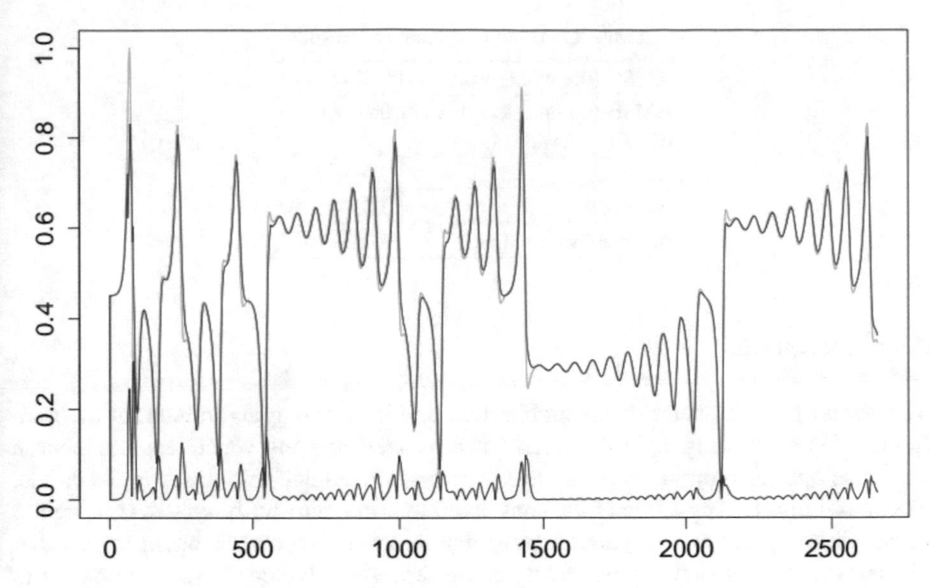

Fig. 5. Identification by means of the Lorenz attractor. The Y-component results.

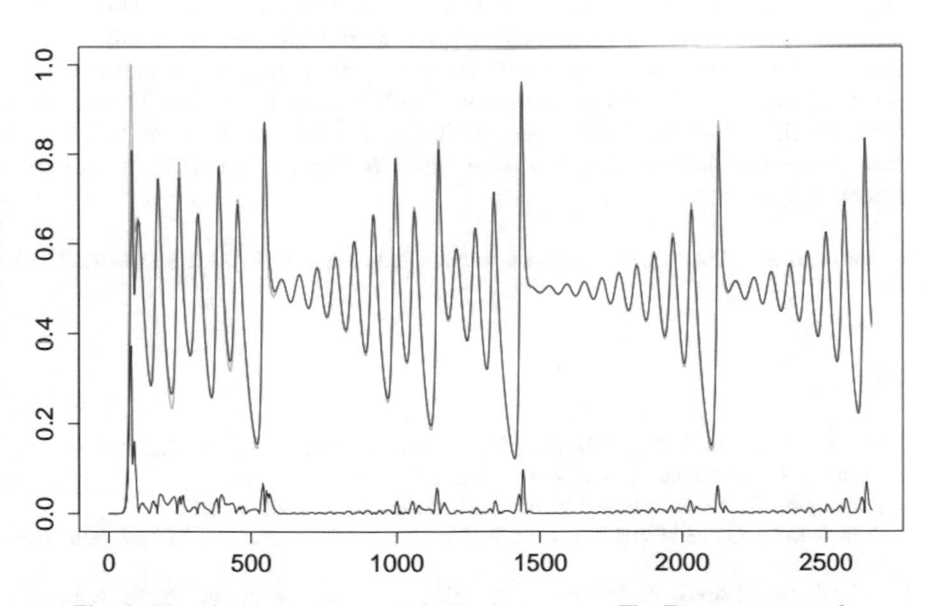

Fig. 6. Identification by means of the Lorenz attractor. The Z-component results.

Table 1. Table of forecasting results

RMSE (the whole data set)	0.03453924
RMSE (on an interval when the last cascade was added)	
Neuron1	0.01954961
Neuron2	0.01975597
A generalized output	0.0191146

5 Conclusion

The hybrid growing neuro-fuzzy architecture and its learning algorithms for the multidimensional growing hybrid cascade neuro-fuzzy system which enables neuron ensemble optimization in every cascade were considered and introduced in the article. The most important privilege of the considered hybrid neuro-fuzzy system is its trait to accomplish a procedure of parallel computation for a data stream based on peculiar elements with upgraded approximating properties. The developed system turns out to be rather easy from the effectuation standpoint; it holds a high processing speed and approximating features. It can be described by a rather high training speed which makes it possible to process online sequential data. The distinctive feature of the introduced system is the fact that every cascade is put together by an ensemble of neurons, and their outputs are joined with the optimization procedure of a specific sort. Thus, every cascade produces an output signal of the optimal accuracy. The proposed system, which is ultimately a growing (evolving) system of computational intelligence, assures processing the incoming data in an online fashion just unlike the rest of conventional systems.

Acknowledgment. This research project is partially subvented by RAMECS and CCNU16A 02015.

References

1. da Silva, I.N., Spatti, D.H., Flauzino, R.A., Bartocci Liboni, L.H., dos Reis Alves, S.F.: Artificial Neural Networks: A Practical Course. Springer, Cham (2017)
2. Cartwright, H.: Artificial Neural Networks. Springer, New York (2015)
3. Suzuki, K.: Artificial Neural Networks: Architectures and Applications. InTech, New York (2013)
4. Koprinkova-Hristova, P., Mladenov, V., Kasabov, N.K.: Artificial Neural Networks: Methods and Applications in Bio-/Neuroinformatics. Springer, Cham (2015)
5. Borowik, G., Klempous, R., Nikodem, J., Jacak, W., Chaczko, Z.: Advanced Methods and Applications in Computational Intelligence. Springer, Cham (2014)
6. Graupe, D.: Principles of Artificial Neural Networks. Advanced Series in Circuits and Systems. World Scientific Publishing Co. Pte. Ltd., Singapore (2007)
7. Haykin, S.: Neural Networks and Learning Machines, 3rd edn. Prentice Hall, New Jersey (2009)

8. Hanrahan, G.: Artificial Neural Networks in Biological and Environmental Analysis. CRC Press, Boca Raton (2011)
9. Lughofer, E.: Evolving Fuzzy Systems and Methodologies: Advanced Concepts and Applications. Springer, Heidelberg (2011)
10. Angelov, P., Filev, D., Kasabov, N.: Evolving Intelligent Systems: Methodology and Applications. Willey, Hoboken (2010)
11. Kasabov, N.: Evolving Connectionist Systems. Springer, London (2003)
12. Fahlman, S., Lebiere, C.: The cascade-correlation learning architecture. Adv. Neural. Inf. Process. Syst. **2**, 524–532 (1990)
13. Avedjan, E.D., Barkan, G.V., Levin, I.K.: Cascade neural networks. J. Avtomatika i Telemekhanika **3**, 38–55 (1999)
14. Prechelt, L.: Investigation of the CasCor family of learning algorithms. Neural Netw. **10**, 885–896 (1997)
15. Bodyanskiy, Y., Dolotov, A., Pliss, I., Viktorov, Y.: The cascaded orthogonal neural network. Int. J. Inf. Sci. Comput. **2**, 13–20 (2008)
16. Bodyanskiy, Y., Viktorov, Y.: The cascaded neo-fuzzy architecture using cubic-spline activation functions. Int. J. Inf. Theor. Appl. **16**(3), 245–259 (2009)
17. Bodyanskiy, Y., Viktorov, Y.: The cascaded neo-fuzzy architecture and its on-line learning algorithm. Int. J. Intel. Process. **9**, 110–116 (2009)
18. Bodyanskiy, Y., Viktorov, Y., Pliss, I.: The cascade growing neural network using quadratic neurons and its learning algorithms for on-line information processing. Int. J. Intell. Inf. Eng. Syst. **13**, 27–34 (2009)
19. Kolodyazhniy, V., Bodyanskiy, Y.: Cascaded multi-resolution spline-based fuzzy neural network. In: Angelov, P., Filev, D., Kasabov, N. (eds.) Proceedings of the International Symposium on Evolving Intelligent Systems, pp. 26–29. De Montfort University, Leicester (2010)
20. Bodyanskiy, Y., Kharchenko, O., Vynokurova, O.: Hybrid cascaded neural network based on wavelet-neuron. Int. J. Inf. Theor. Appl. **18**(4), 335–343 (2011)
21. Bodyanskiy, Y., Grimm, P., Teslenko, N.: Evolving cascaded neural network based on multidimensional Epanechnikov's kernels and its learning algorithm. Int. J. Inf. Technol. Knowl. **5**(1), 25–30 (2011)
22. Bodyanskiy, Y., Vynokurova, O., Teslenko, N.: Cascaded GMDH-wavelet-neuro-fuzzy network. In: Proceedings of the 4th International Workshop on Inductive Modelling, Kyiv, pp. 22–30 (2011)
23. Bodyanskiy, Y., Vynokurova, O., Dolotov, A., Kharchenko, O.: Wavelet-neuro-fuzzy-network structure optimization using GMDH for the solving forecasting tasks. In: Proceedings of the International Workshop Inductive Modelling, Kyiv, pp. 61–67 (2013)
24. Bodyanskiy, Y., Tyshchenko, O., Kopaliani, D.: A hybrid cascade neural network with an optimized pool in each cascade. Soft. Comput. **19**(12), 3445–3454 (2015)
25. Bodyanskiy, Y., Tyshchenko, O., Kopaliani, D.: An evolving connectionist system for data stream fuzzy clustering and its online learning. Neurocomputing (2017). http://www.sciencedirect.com/science/article/pii/S0925231217309785
26. Bodyanskiy, Y., Tyshchenko, O., Kopaliani, D.: Adaptive learning of an evolving cascade neo-fuzzy system in data stream mining tasks. Evolving Syst. **7**(2), 107–116 (2016)
27. Bodyanskiy, Y., Tyshchenko, O., Deineko, A.: An evolving radial basis neural network with adaptive learning of its parameters and architecture. Autom. Control Comput. Sci. **49**(5), 255–260 (2015)
28. Hu, Z., Bodyanskiy, Y.V., Tyshchenko, O.K., Boiko, O.O.: An evolving cascade system based on a set of neo-fuzzy nodes. Int. J. Intell. Syst. Appl. (IJISA) **8**(9), 1–7 (2016)

29. Bodyanskiy, Y., Tyshchenko, O., Kopaliani, D.: A multidimensional cascade neuro-fuzzy system with neuron pool optimization in each cascade. Int. J. Inf. Technol. Comput. Sci. (IJITCS) **6**(8), 11–17 (2014)
30. Miki, T., Yamakawa, T.: Analog implementation of neo-fuzzy neuron and its on-board learning. In: Mastorakis, N.E. (ed.) Computational Intelligence and Applications, pp. 144–149. WSES Press, Piraeus (1999)
31. Uchino, E., Yamakawa, T.: Soft computing based signal prediction, restoration and filtering. In: Ruan, D. (ed.) Intelligent Hybrid Systems: Fuzzy Logic, Neural Networks, and Genetic Algorithms, pp. 331–349. Kluwer Academic Publishers, Boston (1997)
32. Yamakawa, T., Uchino, E., Miki, T., Kusanagi, H.: A neo fuzzy neuron and its applications to system identification and prediction of the system behavior. In: Proceedings of the 2nd International Conference on Fuzzy Logic and Neural Networks, pp. 477–483 (1992)
33. Hoff, M., Widrow, B.: Adaptive switching circuits. In: Neurocomputing: Foundations of Research, pp. 123–134 (1988)
34. Kaczmarz, S.: Approximate solution of systems of linear equations. Int. J. Control **53**, 1269–1271 (1993)
35. Ljung, L.: System Identification: Theory for the User. Prentice Hall, Upper Saddle River (1999)
36. Bodyanskiy, Y., Tyshchenko, O., Wojcik, W.: A multivariate non-stationary time series predictor based on an adaptive neuro-fuzzy approach. Elektronika - konstrukcje, technologie, zastosowania **8**, 10–13 (2013)
37. Caminhas, W.M., Silva, S.R., Rodrigues, B., Landim, R.P.: A neo-fuzzy-neuron with real time training applied to flux observer for an induction motor. In: Proceedings 5th Brazilian Symposium on Neural Networks, pp. 67–72 (1998)
38. Bodyanskiy, Y.V., Pliss, I.P., Solovyova, T.V.: Multistep optimal predictors of multidimensional non-stationary stochastic processes. Doklady AN USSR **A**(12), 47–49 (1986)
39. Bodyanskiy, Y.V., Pliss, I.P., Solovyova, T.V.: Adaptive generalized forecasting of multidimensional stochastic sequences. Doklady AN USSR **A**(9), 73–75 (1989)
40. Bodyanskiy, Y., Pliss, I.: Adaptive generalized forecasting of multivariate stochastic signals. In: Proceedings of the International Conference on Latvian Signal Processing, vol. 2, pp. 80–83 (1990)

Method of Integration and Content Management of the Information Resources Network

Olga Kanishcheva[1(✉)], Victoria Vysotska[2], Lyubomyr Chyrun[2], and Aleksandr Gozhyj[3]

[1] National Technical University "KhPI", Kharkiv, Ukraine
kanichshevaolga@gmail.com
[2] Lviv Polytechnic National University, Lviv, Ukraine
{Victoria.A.Vysotska,Lyubomyr.V.Lyubomyr}@lpnu.ua
[3] Black Sea National University named after Petro Mohyla, Nikolaev, Ukraine
alex.gozhyj@gmail.com

Abstract. The paper describes the integrated processing method of heterogeneous information resources of web systems, which allows for further integration and management. Our method involves the decomposition of the overall process on subprocesses of value integration, data syntax, semantics, and structure. The advantage of this approach to integration processes is the ability to execute them at the metadata level. It allows reducing the number of hits to the data of web systems, the volumes of which can be significant. The paper presents a model of the content life cycle in web systems. The model describes the processes of processing information resources in web systems and simplifies the technology of integrated automation and content management. In this work, the main problems of functional services of integration and content management were analyzed. The proposed method gives the opportunity to create tools for working out of information resources in web systems and implement the subsystem of integration and content management.

Keywords: Content · Content analysis · Content monitoring · Content retrieval · Web system · Web resource · Data integration · Distributed data systems · Heterogeneous data · Business process · Content management system · Content lifecycle

1 Introduction

One of the main features of our time is a constant growth rate of production information [1–3]. This process is objective and usually definitely positive [4–8]. However, today humanity faced with paradoxical situation. Progress in the information field leads to a decrease in the general level of awareness [9–11]. Besides the problem of information increasing, there were a number of specific problems related with the rapid development of information technology [12, 13]. It caused the such problems as: disproportionate increase in information noise due to lack of structure of the information, appearance of parasitic information (received as attachments), lack of relevance of formally relevant

N. Shakhovska and V. Stepashko (eds.), *Advances in Intelligent Systems and Computing II*, Advances in Intelligent Systems and Computing 689,
https://doi.org/10.1007/978-3-319-70581-1_14

(thematically appropriate) information to the actual needs of consumers, multiple duplication of information (a typical example is the publication of the same message in multiple publications) [14–18]. Because of these reasons, the traditional information retrieval system gradually began to lose its relevance. The reason is not in physical volumes of information flow, but in their dynamics and a constant systematic recovery of information. This is not always apparent regularity. Media generates large dynamic information flows and requires new approaches for information processing [19–24]. The solution can only be found in automation of identifying the most important components in content streams [26, 27]. That is why resource-monitoring systems closely related to content analysis. They have used over the recent years. This promising trend is obtained through content monitoring. Its appearance was caused primarily by challenges in the systematic tracking of trends and processes in an updated environment. Content monitoring often realizes a meaningful analysis of information flows in order to obtain the necessary qualitative and quantitative sections. However, the quality is not constant over a predefined period [4–8]. The most important methodological component content monitoring is content analysis [9–13].

In our work we propose the architecture of intellectual system of content formation and additional modules of content processing. It allows automating of identifying the most important components in content streams.

The paper is organized as follows: Sect. 2 discusses related work and summarizes different approaches for creation of intellectual information systems of content formation. In Sects. 3 and 4 we describe our method for content creation, general scheme of content management module and our experiments analysis. Finally, in Sect. 5 we present the conclusion and future work.

2 Related Works

Information system – a set of organizational and technical means for storing and processing content in order to ensure information needs of users. On the Internet, the information system or web-based system acts as a collection of websites that are part of the same network for a particular factor [28–36]. A website is a collection of web pages available on the Internet which are combined both in content and navigation. Physically, a website can be hosted on one or several servers. The network of websites on the theme of "urban sites" should cover the cities of Ukraine. There are many information systems which specialize in submitting pages related to cities. In the Internet environment, there are three types of traffic to the site. The first is from the search engines, the second is the site navigation, the third is the navigation from the bookmarks and direct hits to the site through the URL of the browser's tape. We will focus on the first type of traffic for this Internet project. For this, customers need to update the website in time and add only the current content; a moderator cannot do it manually because of information changes in the Internet very fast space.

The work [37] introduce the current state of Internet provider development of tourist services. Also, the factors which are an influence on the future of Internet tourism: the institutional tension; competition for end users and investors; end-user expectations. It's important for further developing the provision of tourist services in Europe, where the Internet plays a significant role. This situation is facilitated by the rapid processing of Internet information, which needs to be integrated in a timely manner from reliable sources – information resources. The paper [38] defines a number of key changes in information and communication technologies that gradually change the tourism industry. Electronic tourism and the Internet, in particular, support interaction between tourism companies and consumers, and as a result, they form a new approach to the entire process of developing, managing and marketing tourist products and destinations [39]. Therefore, all interested parties related to tourism and hospitality, gradually see how their role changes, new opportunities and challenges emerge. Authors in work [40] demonstrate the future of Internet tourism and show that the Internet tourism is centered on consumer-centric technology to provide the services of new experienced and prompt consumers. Therefore, traffic strategies are needed both at strategic and tactical levels of management to develop an "information structure" for tourism organizations to manage their internal functions, their relationships with partners, and their interaction with all stakeholders, including consumers. Only those organizations that appreciate the opportunities that modern ITs successfully bring and manage their resources in Internet tourism can enhance their innovation and competitiveness in the future [41]. The e-tourism industry is an experienced provider, and more of this experience needs to be customized. Authors [42] analyze the using of Data Mining in Internet marketing for the field of e-tourism and e-management of customer relationships. In particular, profiling clients, query routing, email filtering, online auctions, and updates to electronic directories are explained. But there are a number of problems related to the implementation of the integration method and content management in the field of Internet tourism. These problems are *(i)* unstructured and the lack of one template sources of information; *(ii)* the lack of research on demand for categories of operational tourist information among consumers; *(iii)* the lack of requirements for the overall architecture of the integration system, management, processing and *(iv)* providing the end user with timely and reliable tourist information.

3 Our Method of Integration and Content Management

Web-system of the network consists of subsites, connected in a network. It should be built in a modern style with the necessary functionality, optimized for search engines and services, distributed by servers. The model of such a Web-system presents as

$$W = < X, C, T, H, S, U, Q, Y, \alpha, \beta >, \tag{1}$$

where $X = \{x_1, x_2, \ldots, x_n\}$ – incoming integrated data from various information resources; $C = \{c_1, c_2, \ldots, c_m\}$ – the generated content (information content, news, articles, restaurants, etc.); $T = \{t_1, t_2, \ldots, t_l\}$ – time indicator of the relevance of the generated content to maintain its relevance; $H = \{h_1, h_2, \ldots, h_p\}$ – a list of classified URLs of information resources; $S = \{s_1, s_2, \ldots, s_k\}$ – social content (results of conducting groups in social networks, repositions of records, etc.); $U = \{u_1, u_2, \ldots, u_r\}$ – UGC or results of the content generated by users (the ability to add content to the site through the promotion, as well as comments without promotion); $Q = \{q_1, q_2, \ldots, q_w\}$ – user requests of the necessary Web-system content; $Y = \{y_1, y_2, \ldots, y_f\}$ – relevant content as a result of system user queries; α – process of necessary content integration from different information sources; β – a process of content management to generate lists of relevant content according to user requests of the web system.

The formation of a list of relevant content directly proportional to the competently rules for integrating data from different sources $Y = \beta \circ \alpha$ (filtering, spamming and duplication, classification, etc.), i.e. Y

$$Y = \beta(Q, C, S, U, \alpha(X, H, T)). \tag{2}$$

The process of integrating data from different sources provide a superposition of functions

$$C = \alpha(X, H, T) = \alpha_7 \circ \alpha_6 \circ \alpha_5 \circ \alpha_4 \circ \alpha_3 \circ \alpha_2 \circ \alpha_1, \tag{3}$$

where α_1 – collection of data from various sources, predetermined by the system moderator; α_2 – filtering the collected data (dividing the content into two arrays of recognized/unrecognized, or identified/not identified, with the former working on the system, with the other – the moderator); α_3 – spamming and elimination (check with dictionaries of filters for blocked linguistic variables); α_4 – duplication selection and its elimination (checking with existing content in the database); α_5 – data formatting in accordance with the requirements laid down in the particular web system (most often in the XML format); α_6 – a general classification (for example, content of a type of restaurant, accommodation, location, tourism, help, etc.), if necessary, a detailed classification (for example, for the content of the type of restaurant – cafes, bars, pubs, restaurants, pizzerias, bistros, clubs, coworking, time-clubs) and the storage of relevant data; α_7 – display of new data on the information resource of the web system.

The content management process of the web system gives a superposition

$$Y = \beta_6 \circ \beta_5 \circ \beta_4 \circ \beta_3 \circ \beta_2 \circ \beta_1, \tag{4}$$

where β_1 – the function of processing the user's request (identification, that is, spam/non-spam, with no spam save (for further analysis of statistics hits and searches), spam save the request for further analysis by the moderator, the allocation of word markers (classification linguistic variables) and the classification, that is, to which column of the content belongs (for example, the search for the restaurant), β_2 – the selection of data from the database of the web system according to the words specified in the request of the markers and the formation relevant content; β_3 – sorting relevant content before displaying certain criteria (frequency, reading time, for registered users, taking into account their preferences, etc.); β_4 – displaying sorted relevant content; β_5 – collecting and storing statistics about the user's activity with the displayed relevant user; β_6 – a recount of the content history, according to the latest statistics of user manipulation.

The Fig. 1 shows the conceptual model of city network. The network is divided on sites by the specific context.

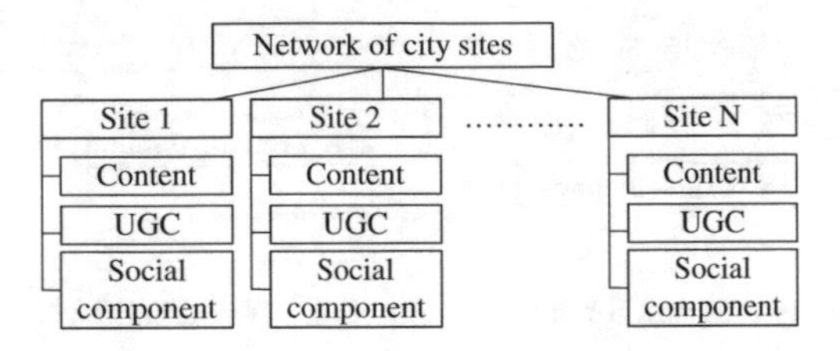

Fig. 1. The conceptual model of information system "Networks of city sites"

The Fig. 2 presents the usage diagram, where we present all actors for our information system and all action variants.

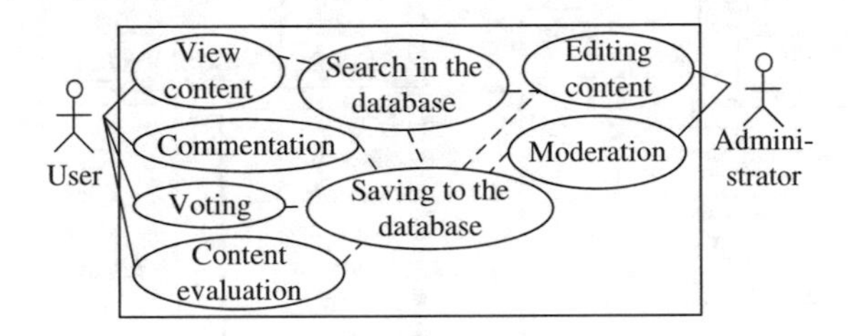

Fig. 2. The usage diagram of information system "Networks of city sites"

The Fig. 3 shows classes and attributes of information system "Networks of city sites". The network consists of sites, where each site consists of certain classes: content, UGC, social component etc. The diagrams (Figs. 4 and 5) present the structure of one site throughout the network, it includes the "Functional Site" package, "Restaurants", "Locations", "Accommodation", "Tourism" etc.

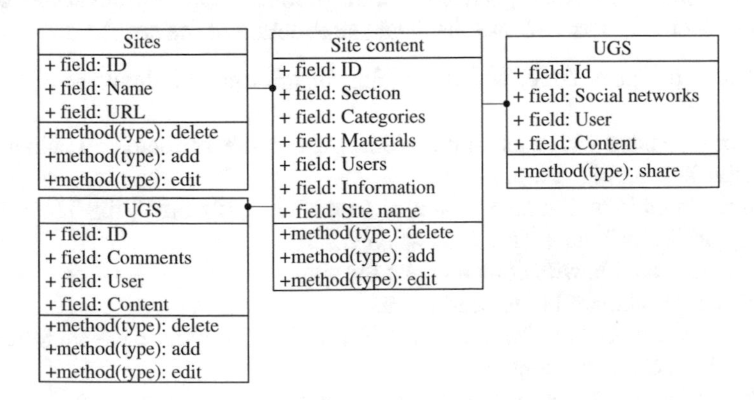

Fig. 3. The class diagram of information system "Networks of city sites"

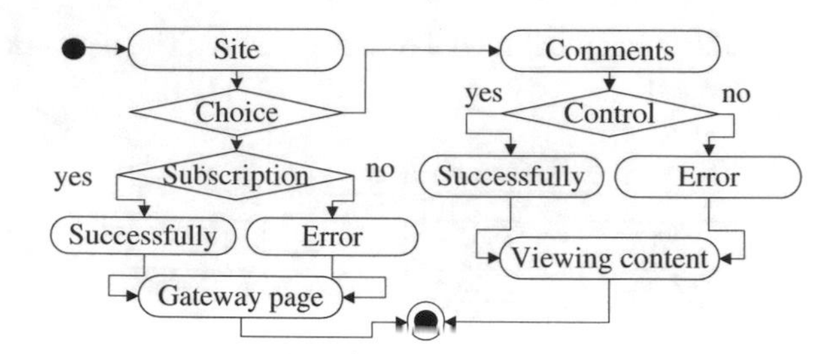

Fig. 4. The activity diagram for customer functions of system "Networks of city sites"

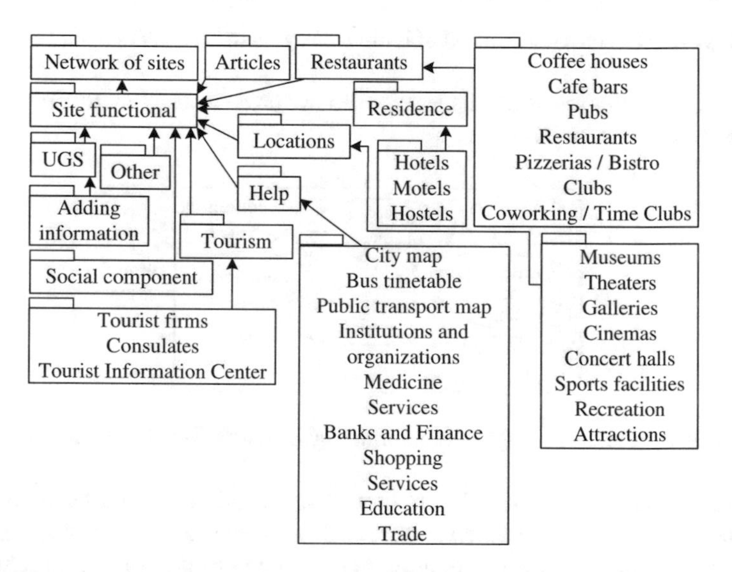

Fig. 5. The package diagram of information system "Networks of city sites"

Information System is designed to solve the problem of finding the necessary information in the city. In Fig. 6 shows the database structure of the site for one city.

1. Content – the main table, which contains all information about the city from all Sections.
2. Category – a table with a list of partitions, the cat_sub attribute will help to make a recursive subcategory output.
3. Ratings – a table with material ratings. In this case, it's the "I like" button. In the rating_points attribute +1 is added when voting.
4. Comments – a table with comments on the site.
5. User – a table about all registered users.
6. Pages – a separate table from the general functional for static pages, for example, such as bus schedules, contacts, etc.
7. Polls – a table about voting and voting lists in the widget on the site.

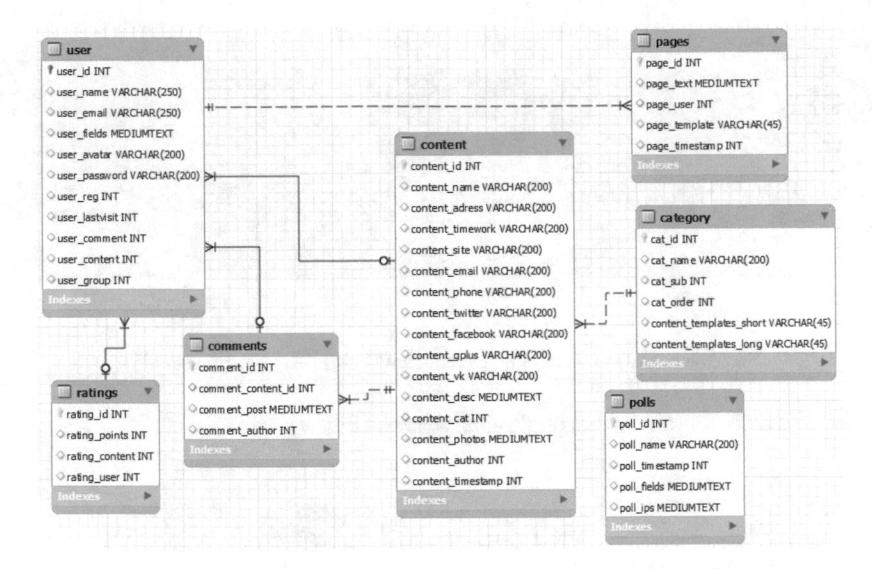

Fig. 6. The example of database structure

4 Experiments and Results

The Fig. 7 shows the prototype of structure and site makeup for all networks.

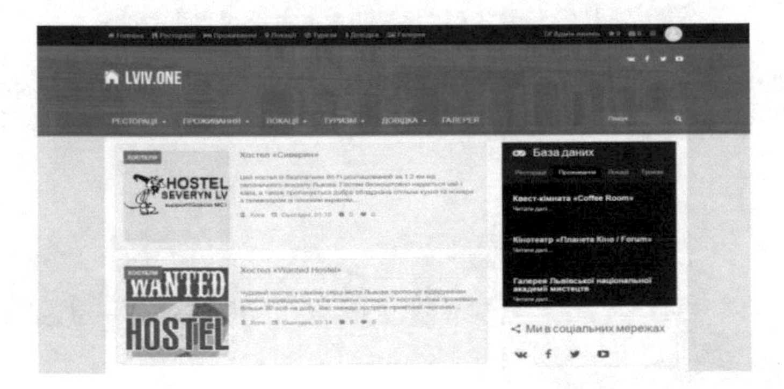

Fig. 7. The prototype of site makeup for the first site

The design is divided into two columns. The main content and side bar. The sidebar has navigation and widgets of different types, and the main content. Each site has its own structure for each city. And it is divided into the following main sections: Restaurants, Accommodation, Locations, Tourism, Help, Gallery and subsidiaries. The example of the site section "Restaurants" is shown in the Fig. 8a and the example of the site section interface "Accommodation" is shown in the Fig. 8b.

Ресторації
Категорії:
Проживання
Категорії:
Готелі
Мотелі
Хостели
Апартаменти
Хостел «Сиверин»
HOSTEL
SEVERYN LV

a) b)

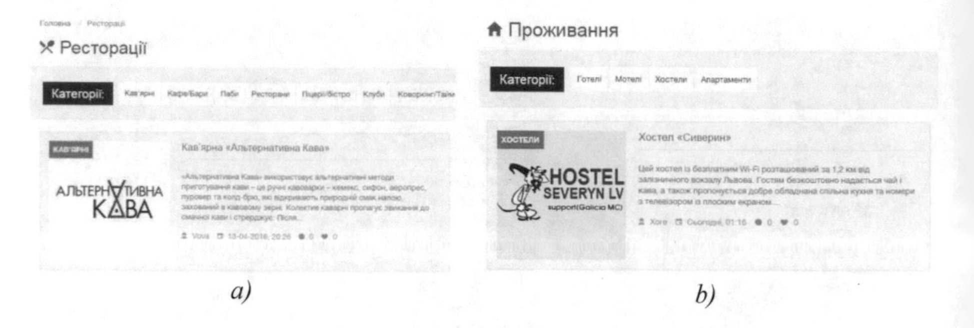

Fig. 8. The site sections "Restaurants" and "Accommodation"

The site also contains a morphological search (Fig. 9).

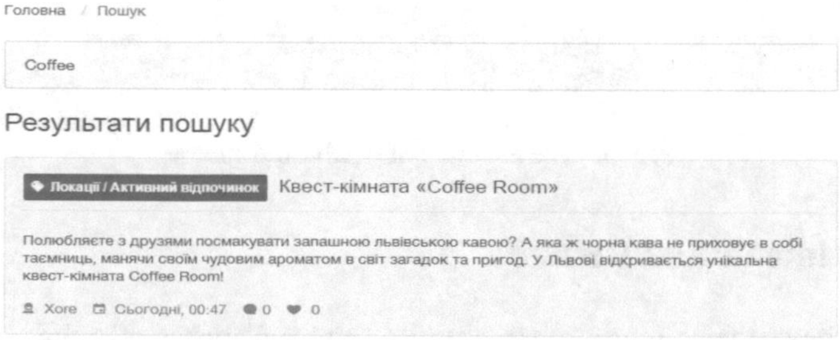

Головна / Пошук
Coffee
Результати пошуку
Локації / Активний відпочинок
Квест-кімната «Coffee Room»
Полюбляєте з друзями посмакувати запашною львівською кавою? А яка ж чорна кава не приховує в собі таємниць, манячи своїм чудовим ароматом в світ загадок та пригод. У Львові відкривається унікальна квест-кімната Coffee Room!
Xore
Сьогодні, 00:47
0
0

Fig. 9. The example of site with search form and results.

Our site has the polling module, social functions such as authorization, commentary, subscription to materials, evaluation of materials and the photo gallery (Fig. 10).

Місто Львів
Галерея

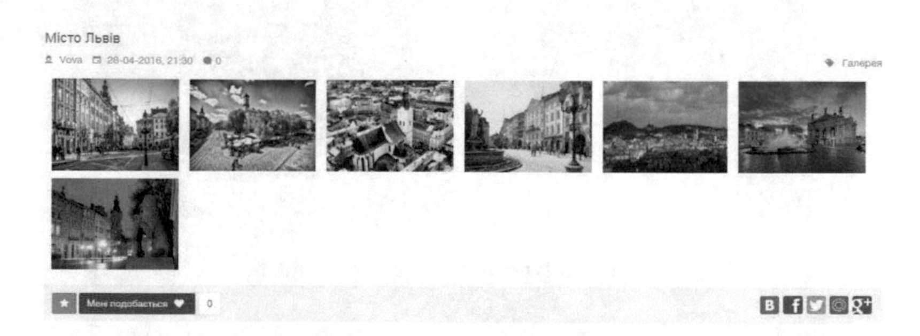

Fig. 10. The photo gallery of our site.

Our site also has information widgets with the Private API and the Yandex API (weather). Data from the API is collected into the database, and then outputs with using the user's geolocation, and upload on the server is reduced. Each site has registration, profile and a form for adding materials. The main feature is function search of similar materials; it is generated by the collected data from the user. On the base of these data, we can receive similar materials (Fig. 11).

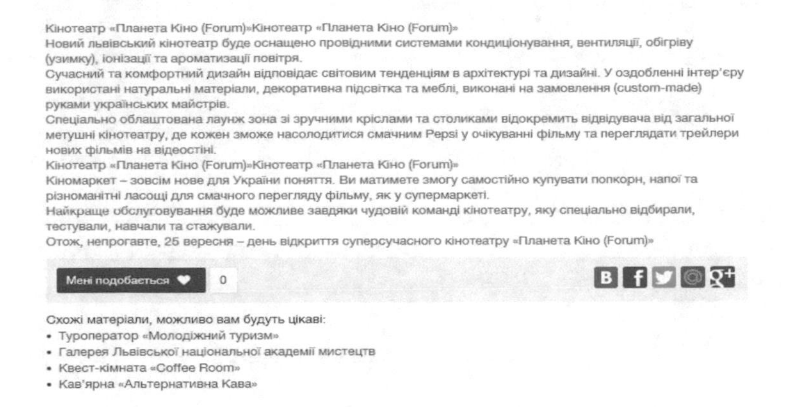

Fig. 11. The window with similar materials.

We used the loadimpact.com service for load testing of our site. The results are shown in Figs. 12 and 13. The server operates in regular mode with 25 simultaneous connections.

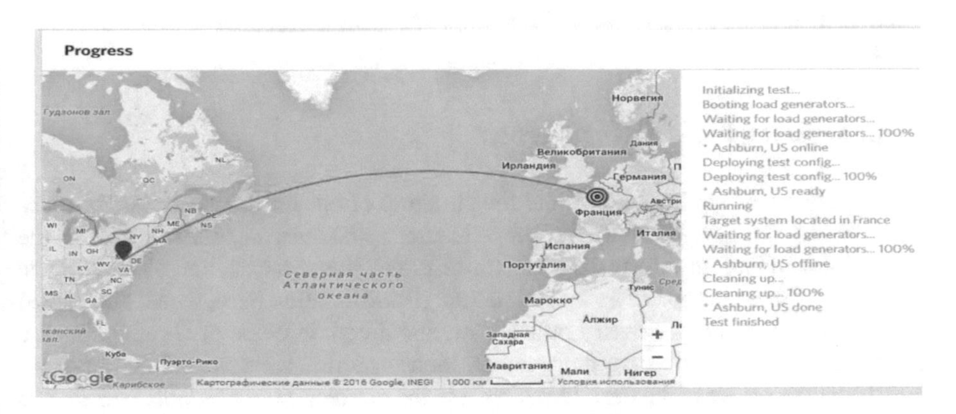

Fig. 12. Results of load testing, part 1.

The next step in testing is the validation of the web portal for validity, using the free service onlinewebcheck.com. We received 0 errors, but 22 attention messages.

The possibility of this information system is scalability. It can be easily scaled for any site and under any subject. Everything is done for maximum speed and work with large data sets. One of the best features of the information system in comparison with analogs are:

- speed of page generation;
- an SSL certificate and TLS encryption;
- better content (by editors), not just a parser;

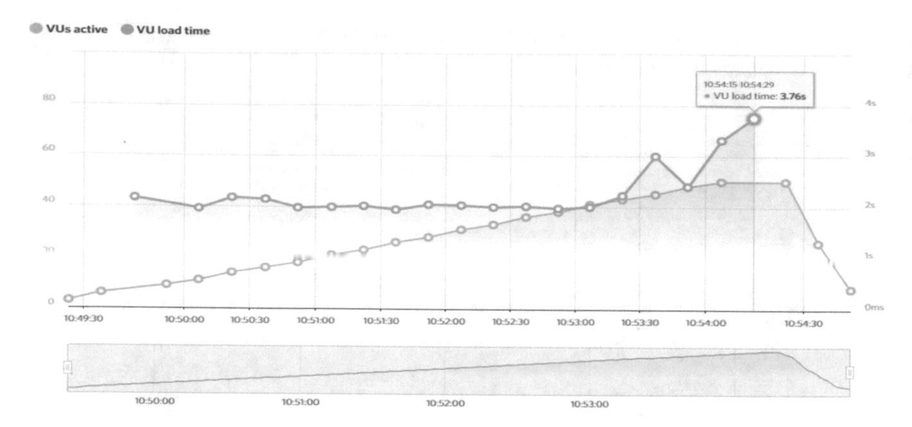

Fig. 13. Results of load testing, part 2.

- lack of news, all the news flowed into groups in social networks;
- there are no such inactive sections as a bulletin board, a directory with vacancies, real estate, etc. Since these sections are basically a parser from sites that are monopolies (olx.ua, work.ua, lun.ua etc.);
- mobile site layout without dubbing content on a subdomain;
- RSS feed for each section;
- the main focus will be on social networks and visitors (mobile users) from these networks.

5 Conclusions and Future Work

This work describes information about the structure of our information system, UML diagrams and the subject area. We show methods and instruments for solving the problem of building an information system of urban site network, and also show the advantages of these tools. We have created a single information base, which allows you to find and provide the necessary information about certain cities of Ukraine. Such a system allows users to observe the weather, look for a timetable for public transport, view the movie poster in theaters, news, browse the directories of available phones, the catalog of restaurants, all about tourism, city photos, monuments, addresses of institutions etc. In this case, the information system is an interactive catalog with dynamic operational information, designed as a reference for each site.

References

1. Mobasher, B.: Data mining for web personalization. In: The Adaptive Web, pp. 90–135. Springer, Heidelberg (2007)
2. Dinucă, C.E., Ciobanu, D.: Web Content Mining, vol. 85. University of Petroşani, Economics (2012)
3. Xu, G., Zhang, Y., Li, L.: Web content mining. In: Web Mining and Social Networking, pp. 71–87. Springer, Boston (2011)

4. Lande, D., Furashev, V., Braychevskiy, S., Grigoriyev, O.: Osnovy modelirovaniya i otsenki yelektronnykh informatsionnykh potokov. Inzhiniring, Kyiv (2006)
5. Lande, D.: Nekotoriye metody analiza novostnykh informatsionnykh potokov. IKVT-2005, vol. 93, pp. 277–287. DonNTU, Donetsk (2005)
6. Lande, D.: Skaner sistemy kontent-monitoringa, Otkrytyye informatsionnyye i kompyuterny integrirovannyye tekhnologii, vol. 28, pp. 53–58. NAKU «KHAI», Kharkov (2005)
7. Bolshakova, Y., Klyshinskiy, E., Lande, D., Noskov, A., Peskova, O., Yagunova, Y.: Avtomaticheskaya obrabotka tekstov na yestestvennom yazyke i kompyuternaya lingvistika. MIEM, Moskva (2011)
8. Grigoriyev, A., Lande, D.: Adaptivnyy interfeys utochneniya zaprosov k sisteme kontent-monitoringa. In: KHÍ mezhdunarodnaya nauchno-prakticheskaya konferentsiya, pp. 17–20. UkrINTEI, Kyiv (2005)
9. Vysotska, V., Chyrun, L., Chyrun, L.: Information technology of processing information resources in electronic content commerce systems. In: Computer Science and Information Technologies, CSIT 2016, pp. 212–222 (2016)
10. Vysotska, V., Chyrun, L., Chyrun, L.: The commercial content digest formation and distributional process. In: Proceedings of the XI-th International Conference on Computer Science and Information Technologies, CSIT 2016, pp. 186–189 (2016)
11. Vysotska, V., Chyrun, L.: Analysis features of information resources processing. In: Proceedings of the X-th International Conference on Computer Science and Information Technologies, CSIT 2015, pp. 124–128 (2015)
12. Vysotska, V., Chyrun, L., Lytvyn, V., Dosyn, D.: Methods Based on Ontologies for Information Resources Processing. LAP Lambert Academic Publishing, Germany (2016)
13. Lytvyn, V., Pukach, P., Bobyk, I., Vysotska, V.: The method of formation of the status of personality understanding based on the content analysis. East. Eur. J. Enterp. Technol. **5/2**(83), 4–12 (2016)
14. McGovern, G., Norton, R.: Content Criticsl. FT Press, Upper Saddle River (2001)
15. McKeever, S.: Understanding Web content management systems: evolution, life cycle and market. Ind. Manage. Data Syst. **103**(9), 686–692 (2003)
16. Rockley, A.: Managing Enterprise Content: A Unified Content Strategy. New Riders Press, Reading (2002)
17. Khomytska, I., Teslyuk, V.: The Method of statistical analysis of the scientific, colloquial, belles-lettres and newspaper styles on the phonological level. Adv. Intell. Syst. Comput. **512**, 149–163 (2017)
18. Khomytska, I., Teslyuk, V.: Specifics of phonostatistical structure of the scientific style in English style system. In: Proceedings of the XI-th International Conference on Computer Science and Information Technologies, CSIT 2016, pp. 129–131 (2016)
19. Lytvyn, V., Vysotska, V., Veres, O., Rishnyak, I., Rishnyak, H.: Classification methods of text documents using ontology based approach. In: Advances in Intelligent Systems and Computing, vol. 512, pp. 229–240. Springer, Cham (2017)
20. Vysotska, V.: Linguistic analysis of textual commercial content for information resources processing. In: Modern Problems of Radio Engineering, Telecommunications and Computer Science, TCSET 2016, pp. 709–713 (2016)
21. Lytvyn, V., Vysotska, V., Pukach, P., Bobyk, I., Pakholok, B.: A method for constructing recruitment rules based on the analysis of a specialist's competences. East. Eur. J. Enterp. Technol. **6**(84), 4–14 (2016)
22. Lytvyn, V., Vysotska, V., Pukach, P., Brodyak, O., Ugryn, D.: Development of a method for determining the keywords in the slavic language texts based on the technology of web mining. East. Eur. J. Enterp. Technol. **2/2**(86), 4–12 (2017)

23. Lytvyn, V., Vysotska, V, Veres, O., Rishnyak, I., Rishnyak, H.: Content linguistic analysis methods for textual documents classification. In: Proceedings of the XI-th International Conference on Computer Science and Information Technologies, CSIT 2016, pp. 190–192 (2016)
24. Lytvyn, V., Vysotska, V.: Designing architecture of electronic content commerce system. In: Proceedings of the X-th International Conference on Computer Science and Information Technologies, CSIT 2015, pp. 115–119 (2015)
25. Basyuk, T.: The main reasons of attendance falling of internet resource. In: Proceedings of the X-th International Conference on Computer Science and Information Technologies, CSIT 2015, pp. 91–93 (2015)
26. Burov, E.: Complex ontology management using task models. Int. J. Knowl. Based Intell. Eng. Syst. **18/2**, 111–120 (2014). IOS Press, Amsterdam
27. Lytvyn, V., Uhryn, D., Fityo, A.: Modeling of territorial community formation as a graph partitioning problem. East. Eur. J. Enterp. Technol. **1/4**(79), 47–52 (2016)
28. Kravets, P., Kyrkalo, R.: Fuzzy logic controller for embedded systems. In: Proceedings of the 5th International Conference on Perspective Technologies and Methods in MEMS Design, MEMSTECH (2009)
29. Mykich, K., Burov, Y.: Algebraic framework for knowledge processing in systems. In: Advances in Intelligent Systems and Computing, pp. 217–228. Springer, Cham (2017)
30. Shakhovska, N., Vysotska, V., Chyrun, L.: Features of e-Learning realization using virtual research laboratory. In: Proceedings of the XI-th International Conference on Computer Science and Information Technologies, CSIT 2016, pp. 143–148 (2016)
31. Shakhovska, N., Vysotska V., Chyrun, L.: Intelligent systems design of distance learning realization for modern youth promotion and involvement in independent scientific researches. In: Advances in Intelligent Systems and Computing, vol. 512, pp. 175–198. Springer, Cham (2017)
32. Lytvyn, V., Vysotska, V., Chyrun, L., Chyrun, L.: Distance learning method for modern youth promotion and involvement in independent scientific researches. In: Proceedings of the IEEE First International Conference on Data Stream Mining and Processing (DSMP), pp. 269–274 (2016)
33. Lytvyn, V., Tsmots, O.: The process of managerial decision making support within the early warning system. Actual Probl. Econ. **11**(149), 222–229 (2013)
34. Chen, J., Dosyn, D., Lytvyn, V., Sachenko, A.: Smart data integration by goal driven ontology learning. In: Advances in Big Data. Advances in Intelligent Systems and Computing, pp. 283–292. Springer, Cham (2016)
35. Lytvyn, V., Dosyn, D., Smolarz, A.: An ontology based intelligent diagnostic systems of steel corrosion protection. Elektronika **8**, 22–24 (2013)
36. Pasichnyk, V., Shestakevych, T.: The model of data analysis of the psychophysiological survey results. In: Advances in Intelligent Systems and Computing, vol. 512, pp. 271–282. Springer, Cham (2016)
37. Rayman-Bacchus, L., Molina, A.: Internet-based tourism services: business issues and trends. Futures **33**(7), 589–605 (2001)
38. Buhalis, D., O'Connor, P.: Information communication technology revolutionizing tourism. Tourism Recreation Res. **30**(3), 7–16 (2005)
39. Buhalis, D., Law, R.: Progress in information technology and tourism management: 20 years on and 10 years after the Internet - the state of eTourism research. Tourism Manage. **29**(4), 609–623 (2008)
40. Werthner, H., Ricci, F.: E-commerce and tourism. Commun. ACM **47**(12), 101–105 (2004)

41. Cardoso, J.: E-tourism: Creating dynamic packages using semantic web processes. In: W3C Workshop on Frameworks for Semantics in Web Services (2005)
42. Olmeda, I., Sheldon, P.J.: Data mining techniques and applications for tourism internet marketing. J. Travel Tourism Mark. **11**(2-3), 1–20 (2002)

Geoinformation Technology for Analysis and Visualisation of High Spatial Resolution Greenhouse Gas Emissions Data Using a Cloud Platform

Vitalii Kinakh[1(✉)], Rostyslav Bun[1,2], and Olha Danylo[3]

[1] Lviv Polytechnic National University (LPNU), Lviv, Ukraine
kinakh.vitalii@gmail.com

[2] University of Dąbrowa Górnicza, Dąbrowa Górnicza, Poland

[3] International Institute for Applied Systems Analysis, Laxenburg, Austria

Abstract. The geoinformation technology for spatial analysis and visualisation of greenhouse gas (GHG) emissions is proposed using Google Earth Engine cloud technology as a key component of interaction with high-resolution spatial data. This technology includes a website for spatial analysis and visualisation of vector data, as well as an interactive site for deeper analysis of raster data on GHG emissions. We use high-resolution vector data of emissions at the level of point, line and areal emission sources, which are converted into raster emission data. Emissions can be analysed within user-created polygons including calculation of the total, specific, maximum or average emission magnitudes. There is also the possibility to fix and select pixels containing a certain interval of emission magnitudes. Using Python's Google Earth Engine module, we have created a website where users can clip raster data from hand-drawn polygons that can be saved on Google Drive. We have also used Python modules (Matplotlib, Pandas, Numpy) for statistical analysis of raster data and histogram construction. Geoinformation technology includes many sectors and categories of human activity included in national inventory reports on GHG emissions, such as those regarding the burning of fossil fuels for power and heat production, within the industrial, agricultural, construction, residential, institutional and waste sectors, as well reports addressing emissions caused by chemical processes. Implementation of the proposed technology is presented using high spatial resolution greenhouse gas emissions data from Poland.

Keywords: Geoinformation technology · Greenhouse gas · Emission data · High resolution · Google Earth Engine · Spatial analysis · Raster map · Histogram · Pandas · Matplotlib

1 Introduction

In recent decades, scientists have identified global warming as a main cause of climate change and many disastrous changes in different regions of our planet. The temperature increase is caused by anthropogenic greenhouse gas effects, especially those due to the

N. Shakhovska and V. Stepashko (eds.), *Advances in Intelligent Systems and Computing II*, Advances in Intelligent Systems and Computing 689,
https://doi.org/10.1007/978-3-319-70581-1_15

emissions of carbon dioxide, methane, nitrous oxide and other greenhouse gases (GHG), which are produced as a result of intense human activity. Scientists are now trying to find ways to stabilise and decrease GHG emissions into the atmosphere and increase absorption of these gases.

For scientists, as well as practical specialists, effective methods and tools for the estimation of GHG emissions (also known as GHG inventory) are required to better understand emission processes and structure in order to provide effective control mechanisms necessary to meet emission obligations, emission trade quotes, etc. [1]. Such an inventory will also help to estimate the effectiveness of low carbon technologies, renewable energy sources, etc.

Emission inventories are usually performed at the country or regional level. However, spatial estimates are highly needed for the analysis of geospatial patterns of emissions [2]. Such inventories may be used in atmospheric flow and climate models, and in tools to support administrative decision making at the regional or local level [3, 4]. The main benefit of such an approach is a spatial representation of emissions in the territories where they actually occur. In the most cases, spatial analysis of emissions is carried out using a grid with a determined size of cells. Such analyses include the calculation of emission magnitudes, as well as uncertainties [5], and the results are represented as raster maps. A huge effort has been made to increase the spatial resolution of GHG estimates since a higher resolution better reflects the specifics of territorial emission and absorption processes. Accordingly, the grid cell sizes have decreased from 1° latitude and longitude for global fossil fuel CO_2 emissions [6] to 1 km for a global fossil fuel CO_2 emissions inventory derived using a point source database and satellite observations of night time lights as proxy data [2].

Nevertheless, other approaches also exist which present GHG emissions at the level of emission sources and are based on vector maps of emissions [7] instead of a grid. Such approaches provide increased resolution of inventory and decreased depth of disaggregation of activity and proxy data, and make possible a spatial inventory for separate categories of human activity, individual GHG, etc. [8].

An increasing number of people are now interested in the reduction of GHG emissions. As a result, there is demand for creating simple and handy solutions for the analysis of emission processes and visualisation of results. The cloud-based platform Google Earth Engine [9] provides many new opportunities for analysis and visualisation of geospatial data, including GHG emissions [10]. At the same time, Python modules (Pandas, Numpy and Matplotlib) provide statistical analysis for creating simple histograms for data distribution, etc.

The aim of this study is to present how cloud platform Google Earth Engine and Python modules can be used to create simple and handy geoinformation technology (GIS) for rapid analysis and visualisation of high spatial resolution GHG emissions data in an easily understandable form.

2 Specifics of Spatial Data Used

Spatial vector data on GHG emissions in Poland [11] was used as input data for the geoinformation system developed in this study. These data were calculated for different types of emission sources depending on the specifics of each category of human activity. Objects with significant emissions (e.g. power plants, cement plants, refineries, etc.) are considered as point-type emission sources and compared with the surrounding area [12, 13]. Lengthy objects with small widths that cause GHG emissions (e.g. road segments) are considered as line-type emission sources [14]. Objects that emit GHG or absorb the carbon dioxide from some area (e.g. settlements in the residential sector [15], arable lands in the agricultural sector [16], etc.) are considered as area-type source/sink emission sources.

The above mentioned vector data include emissions of carbon dioxide, methane, nitrous oxide and other GHG, as well as the total emission of all gases in CO_2-equivalent. The spatial resolution of the emissions data from point- and line-type sources is quite high (i.e. several metres) as all point sources were inspected visually using Google Earth, and line-type sources were built on the basis of Open Street Maps [17]. In contrast, the spatial resolution of area-type sources is lower because they were created on the basis of the Corine Land Cover Maps [18] with 100 m spatial resolution. Therefore, the spatial resolution of these data does not generally exceed 100 m.

3 GIS Technology

3.1 Structure of Technology

The main components of the proposed GIS technology for spatial analysis and visualisation of high-resolution GHG emissions data are:

- operations with vector data on GHG emissions including website for spatial analysis and visualisation of vector data on GHG emissions at the emission source level;
- conversion of vector data to raster including grid creation and calculation of gridded emissions;
- operations with raster data on GHG emissions including: spatial analysis and visualisation of high-resolution raster data on GHG emissions at grid level using the interactive website; band choice for deep analysis of GHG emissions; polygon creation for deeper analysis of GHG emissions; calculation of the main parameters of emission; fixing interval (i.e. min, max) and selecting pixel values within this interval.

The main structure of this technology is shown in Fig. 1. We can see that all of the above mentioned components are connected with the Google Earth Engine cloud platform as a core of the developed geoinformation system.

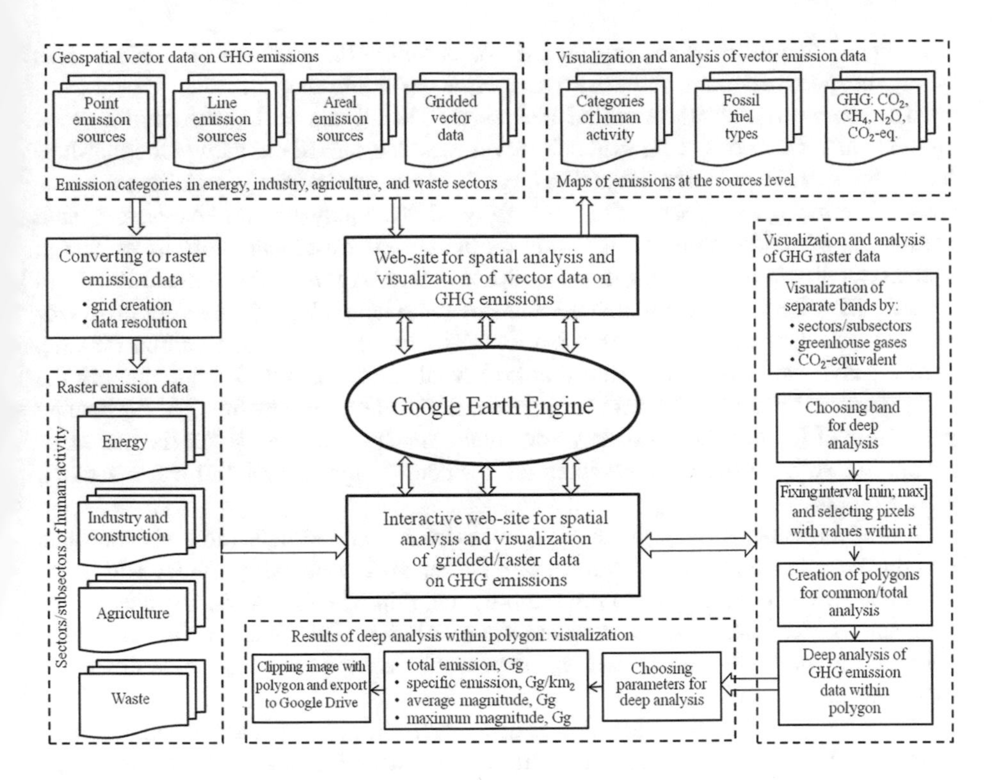

Fig. 1. Structure of the GIS technology for spatial analysis and visualisation of high-resolution GHG emissions data

3.2 Main Components

The website for spatial analysis and visualisation of vector data on GHG emissions at the emission source level was carried out using Google Maps JavaScript API. The vector spatial data were uploaded onto Google Fusion Tables and made accessible from Google Earth Engine. These data are based on a high-resolution spatial inventory [11–16] and include emission magnitudes of separate GHG from point-, line- and area-type emission sources, as well as total emission of all gases in CO_2-equivalent. Additionally, all spatial data on emissions were aggregated into cells of a fixed size. A 2×2 km grid was used for the emissions data from Poland. Cells are further divided into smaller vector objects if they are crossed by administrative boundaries of a province, district or municipality, which helps to maintain administrative assignment of each elementary object and provides the opportunity to sum GHG emissions within the bounds of each administrative object without loss of precision.

The main raster parameters, which are related to spatial resolution and map extent, are set during the conversion of vector data (usually in tab MapInfo or ESRI shape file formats) into raster/gridded data (in tiff file format). High spatial resolution GHG emissions data from different sectors and categories of human activity are organised as bands of raster data. The bands contain information regarding the emission of different GHG (carbon dioxide, methane and nitrous oxide), as well as the total emission of all gases taking the global warming potentials of separate gases into account.

The interactive website for spatial analysis and visualisation of high-resolution raster data on GHG emissions at the grid cells level was also carried out using Google Maps JavaScript API. The additional functionality of this site provides the possibility to choose the corresponding raster data band, create any arbitrary polygon and analyse the main parameters of emission processes within its bounds. The user can chose an operation such as calculation of the total, specific, maximum or average emission magnitude. Users can also fix minimum and maximum values of some interval and then select pixels for which the magnitude is within this interval. The raster data include carbon dioxide, methane and nitrous oxide emissions, as well total emissions in CO_2-equivalent in the energy sector, separately for electricity generation and heat production, extraction and processing of fossil fuels, petroleum refinement, transport, as well as for the manufacturing, industry, construction, commercial, residential, agricultural and waste sectors, among others.

The second interactive website for spatial analysis and visualisation of raster data was developed using Google Earth Engine Python API. This tool provides additional functionality such that users can draw a polygon, clip all cells within the polygon as a separate raster image and export it to Google Drive. This functionality is a great opportunity for users who do not have powerful computers as all operations are processed in the cloud.

There are several positive features of using such a tool for spatial analysis of GHG emissions. First, as the tool is based on Google Earth Engine, there are no strong requirements of the user's computer for interacting with it, aside from Internet access and the availability of a browser with JavaScript support. Second, all data are stored on the server such that users do not need to download and pre-process the data. Finally, the specified emissions in a selected sector can be calculated for the area of interest without handling the entire dataset.

In order to calculate total emissions, the analytical results at the level of separate emission sources are aggregated to vector data in a regular 2×2 km grid. As a result of this operation, emissions from point-, line- and area-type sources that are completely or partially within these cells are summed [11]. The easy operation of these data and quick implementation of analysis of the main parameters of the emission processes within the user-determined territories demonstrates the need to create corresponding geoinformation technology using Google Earth Engine.

4 Practical Implementation of Gridded Emissions

Practical implementation of the geoinformation technology for spatial analysis and visualisation of high-resolution GHG emissions data is presented in Fig. 2 using the example of total GHG emissions in the carbon dioxide equivalent from the residential sector in Poland. The red-yellow colour scheme was selected for better visualisation of the results.

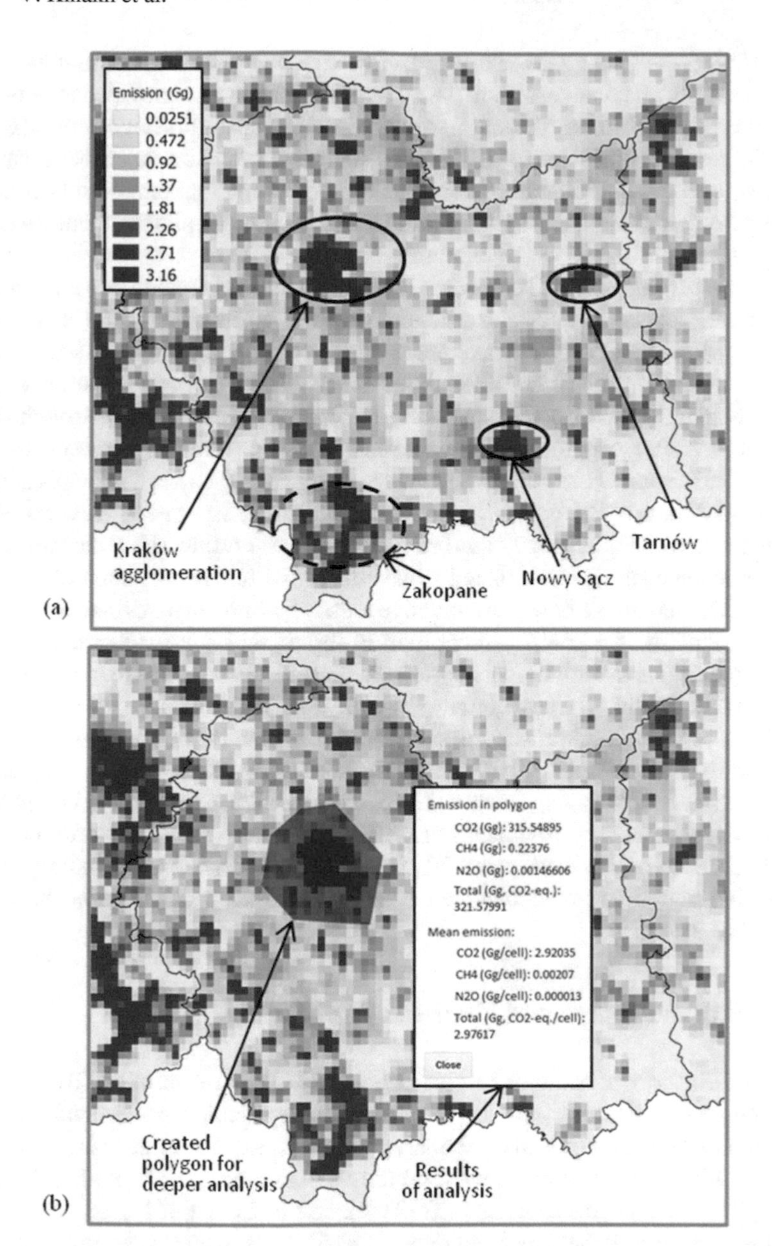

Fig. 2. Example of geospatial analysis of total GHG emissions in the residential sector in the Lesser Poland province: (a) initial raster data and (b) polygon created with visual results of a deep analysis (2010, 2 × 2 km grid, CO_2-equivalent, Gg)

These GHG emissions data were obtained at the level of settlements as area-type emission sources. For this purpose, the statistical data on fuel consumption at the municipal level were disaggregated into the level of settlements using proxy data including: maps of heating degree days and population density in Poland; data on access to energy sources; living area data; percentage of living area equipped with central heating and hot water supply and data regarding the amount of heat energy provided to households. A full description of this approach for high-resolution spatial inventory of GHG in the residential sector is presented in [15]. Following this approach, data on the consumption of coal, natural gas and biomass, as well as the corresponding emission factors, emissions of carbon dioxide, methane, and nitrous oxide at the level of settlements and total CO_2-equivalent emissions were computed [19].

These vector data on GHG emissions in Poland were aggregated into 2 × 2 grid cells as vector map elements and then converted into a similar raster grid in tiff format. Implementation of these spatial data into the developed geoinformation system is illustrated in Fig. 2 using the Lesser Poland province as an example. The upper part of Fig. 2 demonstrates gridded emissions in the residential sector. One can see the highest emissions in Kraków agglomeration and emissions of the largest towns of this region including Tarnów and Nowy Sącz. In the bottom of Fig. 2, one can see emissions from the neighbourhood of Zakopane. Although this territory is a resort area, it is densely populated and wood is primarily used as fuel within the residential sector, which has a higher emission coefficient compared with other fossil fuel energy sources.

In Fig. 2b, one can see how the interactive website works. In particular, a user-formed polygon around the Kraków agglomeration is shown with a box of results from additional deep analysis of emission processes. Generalised parameters (e.g. mean and total emission values within the polygon) are indicated within this box.

In order to create the website for analysis and extraction of images/maps, we used Google Earth Engine Python's API, Jinja2 HTML template (i.e. website templates) and Google Drive API for saving images. Users can download images clipped within the hand-drawn polygon for further analysis. This functionality is very helpful for inexperienced users who do not have powerful computers as such operations using a desktop GIS involve significant computational costs and all operations on our website are processed in the cloud.

5 Statistical Analysis of Gridded Emissions

The geoinformation technology for spatial analysis of high-resolution GHG emissions data developed in this study provides the possibility to calculate some additional parameters of the raster data. For statistical analysis of emission magnitudes in pixels and the creation of histograms and plots, we used the Python modules Matplotlib, Pandas and Numpy [20]. We used raster maps in ESRI shape file format of high-resolution GHG emissions data from Poland as input data [12–16]. Then we selected pixels with non-zero emission magnitudes using the Pandas module. Due to the specifics of emission processes in some categories of human activity (e.g. electricity generation, petroleum refinement, metallurgy, etc.), there are a significant number of cells with a zero emission magnitude. Using this dataset and the Matplotlib module, we created histograms of emission magnitudes in CO_2-equivalent for all sectors/subsectors covered by the high-resolution spatial inventory of GHG. Plots and maps can be manipulated in two different ways using Matplotlib. The first is an interactive way where one can zoom into a plot, view a chart from different angles and view specific emission values or regions. The second way is to save and work with the plot as a picture.

We show histogram examples in Fig. 3 for GHG emission magnitudes in Poland per each 2 × 2 km cell created for:

- fossil fuels (i.e. liquid, solid and gaseous fuels) in the residential sector;
- wood/biomass in the residential sector;
- other categories including agriculture, forestry, fishery etc.

We show similar histograms in Fig. 4 for GHG emission magnitudes per cell from the use of fossil fuels in the:

- manufacturing industry and construction sectors;
- transport sector;
- energy sector (total emissions).

Both figures include calculated histograms for total emissions of all GHG (carbon dioxide, methane and nitrous oxide) in CO_2-equivalent terms. For better visualisation of the results along the ordinate axis, the occupied area is displayed instead of the number of pixels.

The histogram for each sector consists of two parts. The upper part demonstrates the distribution of data from minimum to mean value. This is the main part of the GHG emissions data and reflects emissions from the majority of cells. The lower part of each histogram illustrates the full data distribution (i.e. from minimum to maximum). The majority of the data is clearly within the first part. This feature is characteristic for most sectors including residential, agriculture/forestry/fishery, manufacturing industry, construction and transportation.

Emissions within the pixels from the residential sector are relatively small, both from the burning of fossil fuels and from the use of wood/biomass (Fig. 3). For fossil fuels, a large area (124,000 km^2) is occupied by pixels with emissions less than 150 $Mg_{CO2\text{-}eq}$. Similarly, for the use of wood, an area of 127,000 km^2 is occupied by pixels with emissions less than 80 $Mg_{CO2\text{-}eq}$. Fossil fuel emissions within the agriculture/forestry/fishery sector are insignificant such that more than 89,000 km^2 occupy pixels with emissions less than 20 $Mg_{CO2\text{-}eq}$.

For non-zero pixels of the manufacturing industry and construction sector in Poland, the mean magnitude is higher (4.34 $Gg_{CO2\text{-}eq.}$) (Fig. 4), which is influenced by high emissions point sources (e.g. metallurgical and chemical plants, etc.) even though the total area occupied by these pixels is relatively small. The most 'uniform' emissions distribution is in the transport sector, for which the mean value for non-zero pixels is 654 $Mg_{CO2\text{-}eq}$.

As can be seen in Fig. 4c, the distribution of GHG total emissions in the energy sector in Poland is close to lognormal. The magnitudes of most pixels with a total area larger than 25,000 km are within the range of 279–419 $Mg_{CO2\text{-}eq}$. The tail of this distribution, however, reaches a maximum value of 26,000 $Gg_{CO2\text{-}eq}$, as can been seen in bottom part of the histogram in Fig. 4c. The highest emissions are caused by power plants as emission point sources although there are only a few such pixels, whereas there are 80 power plants in Poland that use fossil fuels with power greater than 20 MW.

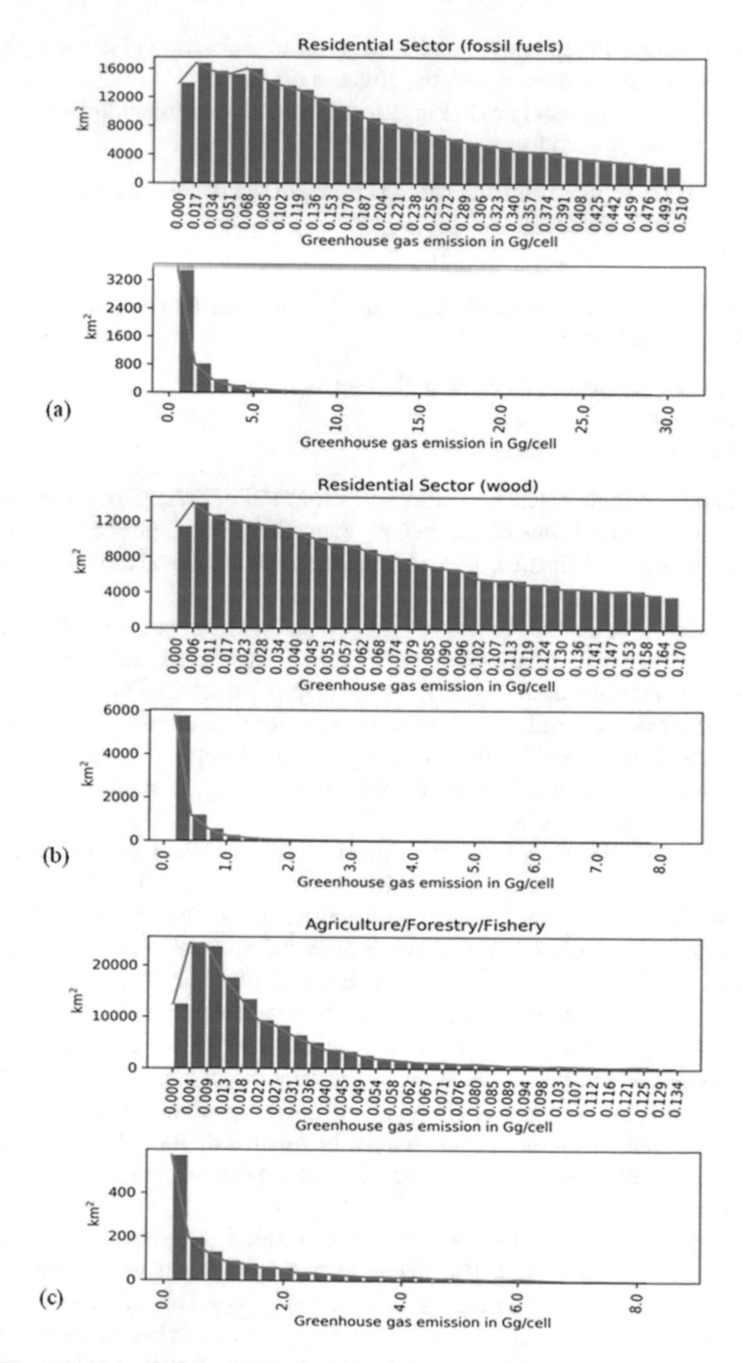

Fig. 3. Histograms of emission magnitudes per cell in the residential sector due to (a) fossil fuels, (b) wood and (c) in other categories such as agriculture, forestry etc. (Poland, 2010, 2 × 2 km grid, CO_2-equivalent, Gg/cell)

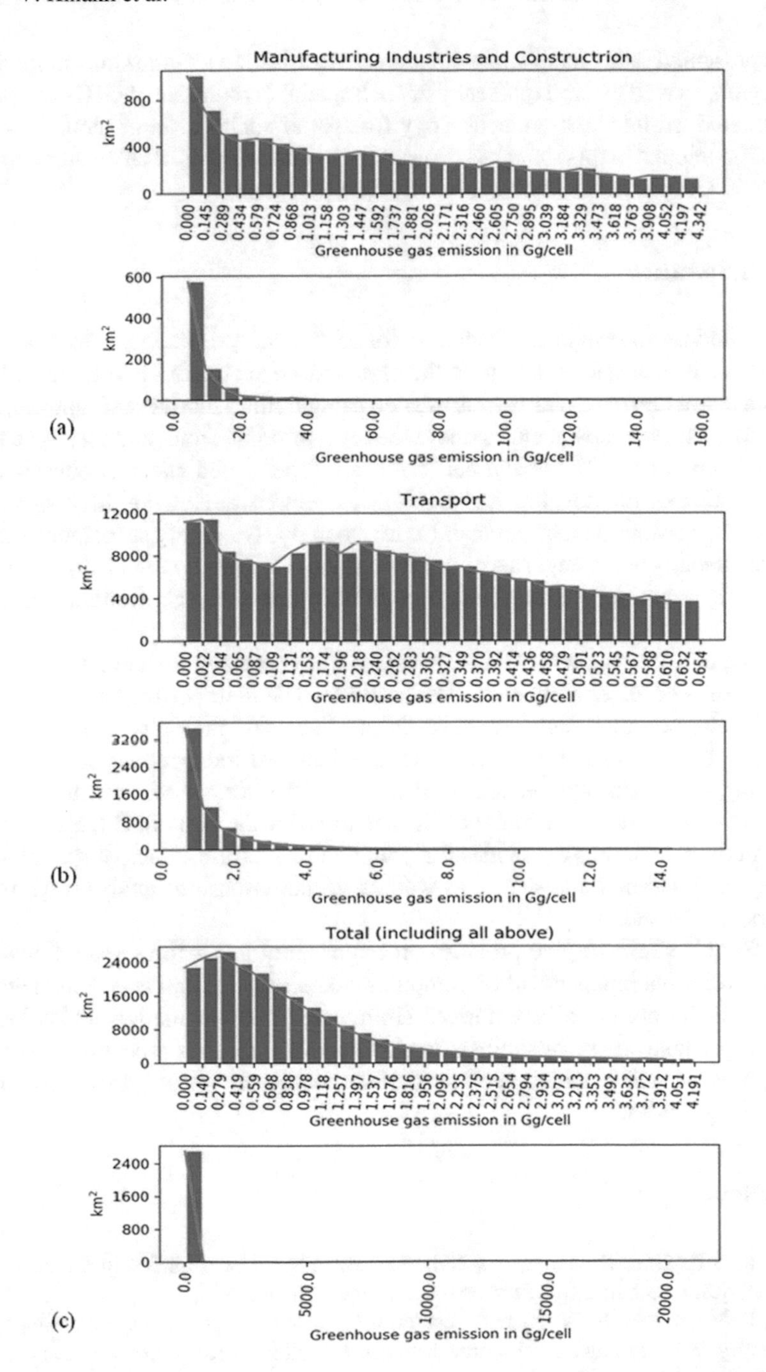

Fig. 4. Histograms of emission magnitudes per cell in the (a) manufacturing industry and construction sector, (b) transport sector and (c) energy sector (total emissions) including all subsectors (Poland, 2010, 2 × 2 km grid, CO_2-equivalent, Gg/cell)

The presented data characterise the uneven distribution of emissions in the investigated region, as well as the expediency of such spatial inventories of GHG and potential of the created geoinformation technology for spatial analysis. In general, all websites and tools developed in this study can be used for spatial analysis of any raster map/image available in Google Earth Engine, not only for GHG data.

6 Conclusions

The presented geoinformation technology for spatial analysis and visualisation of high-resolution GHG emissions data uses the cloud-based platform Google Earth Engine. Presented histograms and charts were created using Python modules Matplotlib, Pandas and Numpy. Vector maps of emissions at the level of point-, line- and area-type sources are used as input data. The analytical results are represented as raster emission maps. Implementation of the developed technology is demonstrated using high spatial resolution GHG emissions data from Poland as a case study. This analysis includes emissions in all main sectors and categories of human activity covered by the National Inventory Reports [21] submitted to the United Nations Framework Convention on Climate Change.

The use of vector data on GHG emissions in the created technology yields better results in terms of uncertainty as well as usability. The methodology allows utilisation of high-resolution input vector maps of point-, line-, and area-type emission sources, and maintains high resolution in the results, which are independent of the grid size, overlapping grids, etc. Consequently, it excludes the source location uncertainty as a component of the total uncertainty. The use of raster data on GHG emissions in this geoinformation technology provides the potential to perform effective spatial analysis at the regional and national scales, as well as deeper emissions analyses within user-created polygons, etc.

The Google Earth Engine provides an opportunity to perform spatial analysis of emission processes independent of computer characteristics and user experience with regards to GHG emission inventories. The created geoinformation technology also allows deeper analysis of generalised parameters of emission processes within user-formed polygons with calculation of the total or specific, as well as maximum or average emission magnitudes.

References

1. Ometto, J.P., Bun, R., Jonas, M., Nahorski, Z. (eds.): Uncertainties in Greenhouse Gas Inventories: Expanding Our Perspective. Springer, Cham (2015)
2. Oda, T., Maksyutov, S.: A very high-resolution (1 km × 1 km) global fossil fuel CO_2 emission inventory derived using a point source database and satellite observations of nighttime lights. Chem. Phys. **11**, 543–556 (2011)
3. Déqué, M., Somot, S., Sanchez-Gomez, E., Goodess, C.M., Jacob, D., Lenderink, G., Christensen, O.B.: The spread amongst ENSEMBLES regional scenarios: regional climate models, driving general circulation models and interannual variability. Clim. Dyn. **38**(5), 951–964 (2012)

4. Neale, R.B., Richter, J., Park, S., Lauritzen, P.H., Vavrus, S.J., Rasch, P.J., Minghua, Z.: The mean climate of the community atmosphere model (CAM4) in forced SST and fully coupled experiments. J. Clim. **26**, 5150–5168 (2013)
5. Hogue, S., Marland, E., Andres, R.J., Marland, G., Woodard, D.: Uncertainty in gridded CO2 emissions estimates. Earth's Future **4**, 225–239 (2016)
6. Andres, R.J., Marland, G., Fung, I., Matthews, E.: A 1° × 1° distribution of carbon dioxide emissions from fossil fuel consumption and cement manufacture, 1950–1990. Global Biogeochem. Cycles **10**(3), 419–429 (1996)
7. Bun, R., Gusti, M., Kujii, L., Tokar, O., Tsybrivskyy, Y., Bun, A.: Spatial GHG inventory: analysis of uncertainty sources. A case study for Ukraine. Water Air Soil Pollut. Focus **7**(4–5), 483–494 (2010)
8. Boychuk, K., Bun, R.: Regional spatial inventories (cadastres) of GHG emissions in the energy sector: accounting for uncertainty. Clim. Change **124**, 561–574 (2014)
9. Google Earth Engine: A planetary-scale platform for Earth science data and analysis. https://earthengine.google.com. Accessed 5 July 2017
10. Lemoine, G., Léo, O.: Crop mapping applications at scale: using Google Earth Engine to enable global crop area and status monitoring using free and open data sources. In: Remote Sensing: Understanding the Earth for a Safer World, IGARSS 2015, Milan, pp. 1496–1499 (2015)
11. Bun, R., Nahorski, Z., Horabik-Pyzel, J., Danylo, O., Charkovska, N., Topylko, P., Halushchak, M., Lesiv, M., Striamets, O.: High resolution spatial inventory of GHG emissions from stationary and mobile sources in Poland: summarized results and uncertainty analysis. In: Proceedings of the 4th International Workshop on Uncertainty in Atmospheric Emissions, Kraków, Poland, 7–9 October 2015, pp. 41–48. SRI PAS, Warsaw (2015)
12. Topylko, P., Halushchak, M., Bun, R., Oda, O., Lesiv, M., Danylo, O.: Spatial greenhouse gas (GHG) inventory and uncertainty analysis: a case study for electricity generation in Poland and Ukraine. In: Proceedings of the 4th International Workshop on Uncertainty in Atmospheric Emissions, Kraków, Poland, 7–9 October 2015, pp. 49–56. SRI PAS, Warsaw (2015)
13. Charkovska, N., Halushchak, M., Bun, R., Jonas, M.: Uncertainty analysis of GHG spatial inventory from the industrial activity: a case study for Poland. In: Proceedings of the 4th International Workshop on Uncertainty in Atmospheric Emissions, Kraków, Poland, 7–9 October 2015, pp. 57–63. SRI PAS, Warsaw (2015)
14. Boychuk, P., Nahorski, Z., Boychuk, K., Horabik, J.: Spatial analysis of greenhouse gas emissions in road transport of Poland. Econtechmod **1**(4), 9–15 (2012)
15. Danylo, O., Bun, R., See, L., Topylko, P., Xu, X., Charkovska, N., Tymków, P.: Accounting uncertainty for spatial modeling of greenhouse gas emissions in the residential sector: fuel combustion and heat production. In: Proceedings of the 4th International Workshop on Uncertainty in Atmospheric Emissions, Kraków, Poland, 7–9 October 2015, pp. 193–200. SRI PAS, Warsaw (2015)
16. Charkovska, N., Bun, R., Danylo, O., Horabik-Pyzel, J., Jonas, M.: Spatial GHG inventory in the Agriculture sector and uncertainty analysis: a case study for Poland. In: Proceedings of the 4th International Workshop on Uncertainty in Atmospheric Emissions, Kraków, Poland, 7–9 October 2015, pp. 16–24. SRI PAS, Warsaw (2015)
17. Arsanjani, J., Zipf, A., Mooney, P., Helbich, M. (eds.): OpenStreetMap in GIScience - Experiences, Research, and Applications. Springer, Cham (2015)
18. Corine Land Cover data (2016). http://www.eea.europa.eu/. Accessed 20 June 2017

19. Eggleston, H.S., Buendia, L., Miwa, K., Ngara, T., Tanabe, K. (eds.): IPCC Guidelines for National Greenhouse Gas Inventories, Prepared by the National Greenhouse Gas Inventories Programme, IPCC (2006)
20. McKinney, W.: Python for Data Analysis, pp. 120–256. O'Reilly Media, Sebastopol (2012)
21. Poland's National Inventory Report 2012, KOBIZE, Warsaw (2012). http://unfccc.int/national_reports. Accessed 1 July 2017

Comparative Analysis of Conversion Series Forecasting in E-commerce Tasks

Lyudmyla Kirichenko(✉), Tamara Radivilova, and Illya Zinkevich

Kharkiv University of Radioelectronics, Kharkiv, 61166, Ukraine
Lyudmyla.kirichenko@nuze.ua, tamara.radivilova@gmail.com

Abstract. The characteristic features of time series conversion, which arise in the tasks of e-commerce are described. It is shown that these series are weakly correlated, which does not allow to use traditional methods for their prediction. Forecasting of the series is performed by methods of exponential smoothing, neural network and decision tree using data from an online store. A comparative analysis of the results is carried out. The advantages and disadvantages of each method are considered.

Keywords: Time series · Conversion rate · Machine learning · Forecasting · Exponential smoothing · Decision tree · Long-term memory neural network

1 Introduction

Time series describe a wide range of phenomena, for example, they are the stock prices, solar activity, the overall incidence rate and much more. Economic indicators can also be considered as time series and you can try to find not visible at first glance laws, hidden periodicity, to predict the moments when peaks appear, etc. At the moment is urgent time-series analysis in the field of e-commerce. E-commerce is in process of development, which is facilitated by new technologies, services and tactical tools [1]. For successful sale in online stores, web analytics is used, which allows to work on optimization, increase conversion and attendance of the electronic store.

To "survive" and stand out among the many online stores, it is important to understand the user's behavior from the moment of the first arrival on the site: to track his movements, to know what products he looked at, put in the basket, where he clicked, what saw, the time he left, how and when he returned. Web-analytics will help in this, which involves ongoing collection, analysis and interpretation of data about visitors, work with basic metrics. Careful analysis of the online store and user behavior is a necessary stage of business development.

Quality web analytics of online store always begins with the visitor's way that he passed before making a purchase. The order processing consists of the following steps: (1) product search; (2) add item to shopping cart; (3) go to the checkout page; (4) fill out and submitting the form; (5) go to the page of the order, payment. The main task of analytics is to periodically find and fix the weak points in this chain. At each stage, the

N. Shakhovska and V. Stepashko (eds.), *Advances in Intelligent Systems and Computing II*, Advances in Intelligent Systems and Computing 689,
https://doi.org/10.1007/978-3-319-70581-1_16

user can stop without having made a purchase. Each of these stages is represented in the form of a time series.

It is impossible to work on optimization, increase of conversion and site attendance without web analytics. By using key performance indicators, it can be significantly improved profit site. These indicators show how quickly and efficiently the business grows. One of the advantages of running an online store is the transparency of key performance indicators tracking and the ability to optimize processes for business growth [2]. The competition in the field of e-commerce is so great that an online store that does not use analytics and metrics will not last long. Here are the main success indicators that any online store should measure:

1. Attendance of the site - the number of users visiting the site, which are measured in terms of daily audience, weekly and monthly. This will allow to evaluate the incidence and bursts of site traffic and identify their causes.
2. Views commodity page – which pages visitors view on the site often and which less. Analyzing the attendance of commodity pages, it is possible to understand the customers' shopping preferences and the way of their interacting with the site.
3. The average time on the site and the average number of pages viewed – if these indicators are low, it is worth assessing the quality of traffic on site.
4. Exit pages - by analyzing the exit points of the site visitors (registration, shopping cart, ordering), can better understand the reasons for the low conversion and optimize the site so that users stay on the site and complete purchases.
5. Channels to attract visitors - you need to track not just a source of attraction of visitors, and their impact.
6. The overall conversion rate of an online store is the number of visitors who made a purchase.
7. Indicator of return of visitors – the number of not only new, but also returned visitors to the online store are analyzed. These will allow evaluate how the site is interesting for the target audience.
8. Profit from the buyer – profit minus costs. This indicator provides an understanding of how successful online store.
9. The failure rate is the number of orders that have been started, but not completed.
10. Number of products in the order – the number of products per order.
11. Average order value - total sales/orders quantity.

A store that is already selling on the Internet, often to increase profits, it is sufficient to pay attention to only a few metrics, one of which is the conversion. Conversion is the most important parameter that characterizes the effectiveness of the website promotion process. Conversion is the ratio of the number of users who made purchase of a product or service on your site to the number of users who came to your site for an advertising link, ad, or banner. For example, if a site was visited by 100 people, but only 2 people bought the product, then the conversion rate is equal to 2%. The increase of conversion percent depends on many factors: from the design of the page to its functionality. Monitoring conversion allows to understand in time that the e-shop needs to be improved. Conversion is the main metric in the web analytics of all commercial sites [3].

Analysis and forecasting of time series of daily value conversion percent plays crucial value for optimizing the efficiency of the online business [4]. However, it should be noted that almost all of the classical methods of time series forecasting based on the calculation of the correlation between the time series values [5]. In the case of weakly correlated time series, and also in the case when the time series has sparse zero values structure, which is typical for many electronic sales sites, these methods do not fit or have a large error.

Neural network approach has been widely used to solve forecasting problems. Neural networks allow you to model complex relationships between data as result of learning by examples. However, the prediction of time series using neural networks has its drawbacks. First, for training the majority of neural networks, time series of a large length are required. Secondly, the result essentially depends on the choice of the architecture of the network, as well as the input and output data. Third, neural networks require preliminary data preparation, or preprocessing. Preprocessing is one of the key elements of forecasting: the quality of the forecast of a neural network can depend crucially on the form in which information is presented for its learning. The general goal of preprocessing is to increase the information content of inputs and outputs. An overview of the methods for selecting input variables and preprocessing is contained in [6].

Recently, for the analysis of the regularities of the time series, the methods of machine learning [7, 8] have been increasingly used to detect various patterns in the time series. In this case, logical methods are of particular value in the detection of such patterns. These methods allow us to find logical if-then rules. They are suitable for analyzing and predicting both numerical and symbolic sequences, and their results have a transparent interpretation.

The goal of the presented work is to carry out a comparative analysis of weakly correlated time series forecasting, based on classical prediction methods and machine learning ones such as neural networks and decision trees, using the data of a real online store.

2 Input Data

Input data in the work were daily data from the online sales site, which included the number of clicks on the site from social networks, the number of sales and the corresponding conversion rate. In addition, there was information about which language the customer used, from which country the order was and other data [9].

Figure 1(at the top) shows typical time series of conversion rate. Series of conversion rate are characterized by zero values, which significantly complicates forecasting the next day. The correlation function of the rate series is shown on Fig. 1(at the bottom). Obviously, there is no correlation between the time series values.

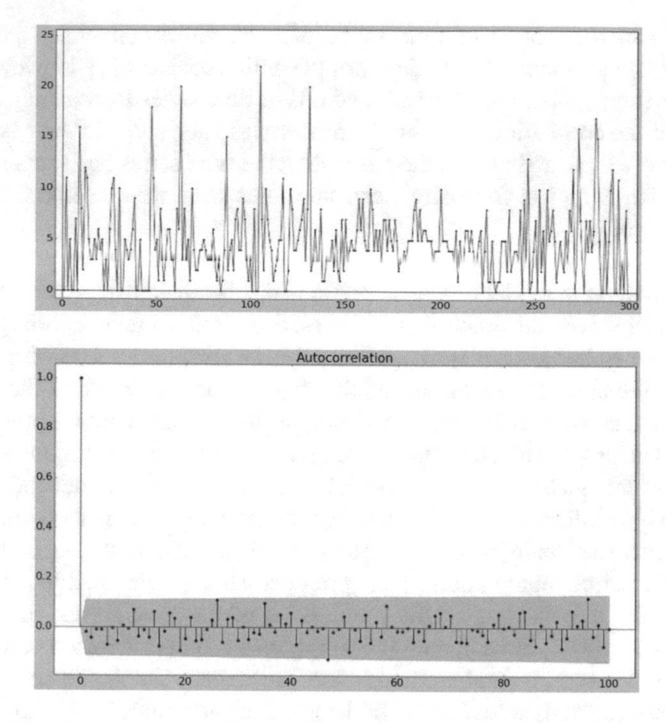

Fig. 1. Time series of conversion rate and the correlation function

Figure 2 shows the histogram of the distribution density of typical conversion rate series. It is easy to see that the percent from 0 to 2 is the highest, then there is a more even distribution, but for each series of conversion rate there are bursts that are most difficult to forecast.

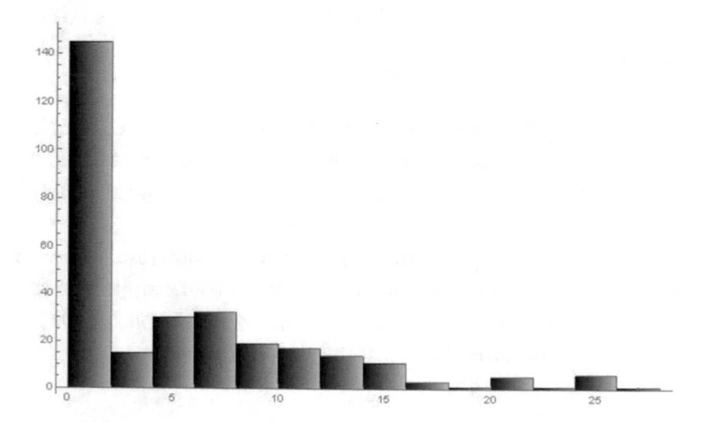

Fig. 2. Distribution density of typical conversion rate series

3 Forecasting Methods

E-commerce is constantly evolving, it is facilitated by new technologies, services and tactical tools. Suppliers, range of buyers, assortment of goods change regularly, that leads to a rapid obsolescence of information. Therefore, forecasting methods that require time series of great length, such as, for example, autoregressive and moving average models, work poorly [10].

Methods of Exponential Smoothing. The basis for exponential smoothing is the idea of a constant revision of the forecast values as the actual ones arrive. The model of exponential smoothing assigns exponentially decreasing weights to observations as they become outdated [5, 11]. Thus, the latest available observations have greater influence on the forecast value than older observations.

The model of exponential smoothing has the form:

$$Z(t) = S(t) + \varepsilon_t,\ S(t) = \alpha \cdot Z(t-1) + (1-\alpha) \cdot S(t-1), \quad (1)$$

where α is smoothing factor; $0 < \alpha < 1$; Z(t) is projected time series; S(t) is smoothed time series; initial conditions are defined as $S(1) = Z(0)$. In this model, each subsequent smoothed value S(t) is the weighted average between the previous value of the time series $Z(t-1)$ and the previous smoothed value $S(t-1)$.

The value α is determined by how much the current series value should affect the next value. The closer α is to unity, the stronger the forecast takes into account the value in the previous step. To find the optimal value α, it is required to minimize the mean error of the forecast. When the value α is automatically selected, all forecasts are calculated for α that change with a given step, the mean error is calculated, and the value α at which the error has the smallest value is selected.

3.1 Decision Tree Method

Methods of machine learning are an extremely broad and dynamically developing field of research using a lot of theoretical and practical methods. One of these methods is the decision tree method [7, 12, 13]. The decision tree is a decision support tool used in statistics and data analysis for predictive models.

In intellectual data analysis, decision trees can be used as mathematical and computational methods to help describe, classify and summarize a set of data that can be written as follows: $(x, Y) = (x_1, x_2, x_3, \ldots, x_k, Y)$. The dependent variable Y is the objective variable that needs to be analyzed, forecast and generalized. A vector x consists of input variables x_1, x_2, x_3, etc., which are used to perform the task.

The decision tree method for classification or prediction task is the process of dividing the original data into groups until homogeneous (pure) subsets are obtained. The set of rules, due to which there is such division, allows to make a forecast obtained as a result of evaluating some input features x_1, x_2, x_3 for new data.

The decision tree is a model that represents a set of rules for decision-making. Graphically, it can be represented in tree structure form, where decision-making moments correspond to the decision nodes. The data to be classified are at the root of tree. In nodes, depending on decision made, a branching process occurs. Terminal nodes are called leaf nodes. Each leaf is final result of consistent decision-making and represents value of objective variable, which was modified during movement from root to leaf. Each internal node corresponds to one of input variables. Depending on decision made at nodes, the process eventually stops in one of leaves, where a variable of response is assigned a particular value.

The algorithm of learning (forming the tree) operates according to the principle of recursive partitioning. The partitioning of data set (i.e., splitting into disjoint subsets) is performed on the basis of the most suitable for this feature. A corresponding decision node is created in tree, and process continues recursively to the stopping criterion.

There are various numerical algorithms for constructing decision trees. One of the most famous is the algorithm called C5.0, developed by the programmer J.R. Quinlan. In fact, the C5.0 is the standard for construction of decision trees. This program is implemented on a commercial basis, but version built into the Python (and some other packages) is available for free.

The algorithm implements the principle of recursive partitioning. The algorithm starts with empty tree and complete data set. In nodes, starting from the root node, feature is selected whose value is used to divide all data into two classes. After the first iteration, the tree appears with one node dividing the data set into two subsets. After that, this process can be performed repeatedly, with respect to each of subsets for creating subtrees. To separate data, we use conditions of form: $\{x < a\}$, $\{x > a\}$, where x is feature, and a is some fixed number. Such partitions are called "axis-parallel splits". Essentially, with each condition check, the data samples are sorted in such way that each data element is determined to correspond only to one branch. Decision criteria divide the original data set into disjoint subsets. The recursion terminates if subset in node has same values of objective variable, so it does not add values to forecasts.

To create a decision tree, its needed to determine the features by which the partition will be performed. In the case of the classification of data samples, these values may be the sampling values. From the set of attributes for the partition, it is required to choose those that would allow to obtain as homogeneous (pure) sets as possible. Algorithm C5.0 uses as impurity measure of entropy concept, which is a measure of data disorder.

Using the entropy as measure of impurity sets that are result of partitioning, algorithm can select feature on which partitioning will give the purest set (i.e., the set with lowest entropy). These calculations are called "information gain". The feature is determined by the search method. For each feature, the value of information gain is calculated as difference in entropy of sets before and after partitioning.

The higher information gain for selected feature, better this feature is suitable for partitioning, since such partition will ensure that the most pure set is obtained. If, for the selected characteristic, the value of information gain is close to zero, it means that the partition by this feature is unpromising, since it does not lead to entropy decrease. On the other hand, the maximum possible value of information gain is equal to the value of entropy before the partition. This means that the entropy after partition will be zero, i.e. resulting sets will be completely pure.

The main advantages of the C5.0 algorithm for forecasting tasks: it is universal, it solves well the problems of classification and forecasting from different areas; to construct a decision tree, it selects from set of features only those that strongly influence the result; it requires a relatively small amount of training sample. One of the significant advantages of the C5.0 algorithm is its ability to post-pruning of the built decision tree, that is, cutting off those nodes and branches that have little impact on the forecast results. The disadvantages are that the algorithm "gravitates" to split based on a large number of levels; inaccuracies in classification can arise from the fact that only "parallel-axis axes" splits are used; decision trees sometimes turn out to be very large.

Neural Networks. It can be said that any neural network (NN) acts as follows: iteration after iteration, it deforms the vector of input data in this way that as a result of deformation, the input data fall into the zones where we expect to see them at the output. In ordinary neural networks, each individual sample is processed without taking into account the influence of past information on current result. To solve this problem, recurrent neural networks (RNN) were developed in the 1980s. These are networks that contain feedback and allow to take into account the previous iterations. A recurrent network can be viewed as several copies of the same network, each of which transmits information of a subsequent copy. RNN resembles a chain, and its architecture is well suited for working with sequences, lists and time series.

The scheme of the RNN operation looks like this: there is an input layer of neurons that is projected onto a hidden layer (one or several), the outputs of the hidden layer are transferred to output layer, and also copied to context layer, which at the next iteration is perceived together with input layer and connected to hidden layer. Accordingly, cycle is produced: a hidden layer - a context layer - a hidden layer. In progress new samples will come in RNN, they will change the context and context circulating within the network, will retain this information, which will affect the current classification. Over the past few years, RNN has successfully applied to a lot of tasks: speech recognition, language modeling, translation, image recognition, etc. [14]. But the main disadvantage of classic RNN is to reduce the effect of samples with increasing time delay. As a rule, the maximum impact on response of RNN samples that were at previous iteration, two iterations back, etc. have and the further, the less this effect decreases. While quite often there are situations when information important for correct forecasting is not on the nearest samples, but on 10-20-30 iterations back.

Recently, the architecture of RNN which are called long short-term memory neural networks (LSTM) has become popular. This is a special kind of recurrent neural networks, which are capable of learning long-term dependencies. LSTM are specifically designed to avoid the problem of long-term dependencies. Remembering information for a long period of time is practically their default behavior [6, 15, 16].

A method that some elements in the context in the previous iterations provide a greater influence on the result, while other elements have a smaller effect was suggested. In LSTM, it is proposed to extend the classical RNN schema with notion gate, which is a memory gate and forget gate, and which determines how likely the given sample should be forgotten or remembered for next iteration. The previous samples affect saving or deleting of sample. If they indicate that this information is important for future classification, more importance will be given. If the current information plays a weak role for forecasting at subsequent iterations, impact will decrease.

Currently LSTM work incredibly well on a wide variety of tasks and are widely used. Many impressive results of work of RNN were achieved precisely on the basis of LSTM architecture [16].

The Forecast Errors. To obtain quantitative characteristics of the comparative analysis of the models, the following characteristics of forecast errors were chosen [5, 11, 17]. The Mean Absolute Deviation (MAD) measures the accuracy of the forecast by averaging the values of the forecast errors. Using MAD is most useful when the analyst needs to measure the forecast error in the same units as the original series. This error is calculated as follows:

$$MAD = \frac{1}{n}\sum_{t=1}^{n}\left|X(t) - \hat{X}(t)\right|. \tag{2}$$

The average deviation (Mean Deviation, MD) allows to see how the forecast value is overvalued or undervalued on average:

$$MD = \frac{1}{n}\sum_{t=1}^{n} X(t) - \hat{X}(t). \tag{3}$$

Mean squared error (MSE) is another way of estimating the forecasting method. Since each deviation value is squared, this method emphasizes large forecast errors. The MSE error is calculated as follows:

$$MSE = \frac{1}{n}\sum_{t=1}^{n} (X(t) - \hat{X}(t))^2. \tag{4}$$

The Mean Absolute Percentage Error (MAPE) is calculated by finding the absolute error at each time and dividing it by the actual observed value, with subsequent averaging of the obtained absolute percent errors. This error is calculated as follows:

$$MAPE = \frac{1}{n}\sum_{t=1}^{n} \frac{\left|X(t) - \hat{X}(t)\right|}{X(t)}. \tag{5}$$

This approach is useful when the size or value of the predicted value is important for estimating the accuracy of the forecast. MAPE emphasizes how large the forecast errors are in comparison with the actual values of the series. This approach is useful when the size or value of the predicted value is important for estimating the accuracy of the forecast. MAPE emphasizes how large the forecast errors are in comparison with the actual values of the series.

4 Software Implementation of Machine Learning Methods

The methods and algorithms of Data Mining and machine learning must be implemented in a certain programming language, in a certain environment, calculated on a certain type of computer elements, etc. There are many tools available today to implement Data Mining and Machine Learning algorithms [18].

One of the most widely used programming languages for solving application problems is the Python language. Python is a general purpose programming language, which means that people have built modules to create websites, interact with a variety of databases, and manage users. Python uses a large number of people and organizations around the world, so it develops and is well documented; it is cross-platform and you can use it free [12].

This language has several advantages. It is quite easy to learn and, as a rule, is a language that needs a low entry level. A person who has basic knowledge in the theory of algorithms and mathematics can simply master the basic functionality, methods and syntax to solve applied problems. Python has implemented a large number of libraries, which provide most of the available algorithms in a convenient way.

In general, for machine learning there are several basic Python libraries, which have quite a big advantage compared to libraries of other programming languages. The main one is: very detailed and qualitative documentation. Most of these libraries use the NumPy library [19]. This is a library that allows you to quickly and effectively work with numeric data matrices, tables of numbers in different formats, to carry out a large number of typical operations that are required in the process of solving applied machine learning tasks.

One of the great documentation libraries that implements most of the typical machine learning methods is scikit-learn [20]. Dozens of algorithms are implemented in this library for clustering, regression, classification, reference vector method, linear and logistic regression, and dozens of other algorithms. Each of the available algorithms has a large number of parameters that can be customized to your task.

One very convenient language libraries in Python, which help to work with lots of tabular data (often training and testing sample look like .csv-table with hundreds of thousands and millions of rows and columns parameters) is a Pandas library [21]. It allows to download data very quickly, preprocess it (to prepare it in a suitable format), to send in a convenient form for processing by our algorithm, which we, for example, have chosen from the Scikit-learn library.

Currently, due to the popularity of neural networks in Python, there are many libraries at quite different levels of abstraction (from low to high-abstraction architecture descriptions) allow construction of various neural networks.

One of the most commonly used libraries for low-level operations for the implementation of neural network algorithms is the Theano Library [22]. It implements complex matrix cartoons, rapid methods of convolution with multiplication, sampling, regression methods, and all backend-logic of neural networks.

One key bonus of libraries is that in addition to CPU-realization (i.e., implementation of algorithm's operation on processor), Theano or TensorFlow libraries (by Google) are open source (you can see and add modules that you need).

It should be noted that between the graphics card, the Python language and the library written in this language there is one more layer - CUDA - a set of libraries, implemented by NVidia, which allow effectively and quickly perform calculations on its graphics cards.

To work with web resources, the Scrapy library is used [23]. Scrapy is an application framework for crawling web sites and extracting structured data which can be used for a wide range of useful applications, like data mining, information processing or historical archival. Scrapy provides a lot of powerful features for making scraping easy and efficient, such as: built-in support for selecting and extracting data from HTML/XML sources using extended CSS selectors and XPath expressions, with helper methods to extract using regular expressions; for generating feed exports in multiple formats (JSON, CSV, XML) and storing them in multiple backends (FTP, S3, local filesystem); an interactive shell console (IPython aware) for trying out the CSS and XPath expressions to scrape data, very useful when writing or debugging your spiders.

5 Research Results

The analysis of the conversion series for compliance with ARMA models was performed. Figure 3 shows the typical values of the Akaike information criterion (AIC), which is used exclusively for the selection of several statistical models for one set of data [5, 9]. The values of the criterion indicate that the use of ARMA models in this case is not appropriate.

	Candidate	AIC
1	**MAProcess[0]**	**1411.31**
2	MAProcess[1]	1412.81
3	ARProcess[1]	1412.85
4	ARMAProcess[1, 1]	1414.7

Fig. 3. Values of the Akaike information criterion

To carry out the forecasting, time series were divided into two parts, where the first one was used to train the model, and the second one was applied to assess its plausibility. The models were trained on the S last values of time series.

Checking the models for forecasting m values was carried out in the following way: take the window of last S values from the first part series and will do the forecast one value ahead; then will move window one value forward, including the forecast of new value in the window, and will again do the forecast, and so m times.

Figures 4, 5 and 6 presents the results of the forecasts of each model for 7 values ahead or $S = 20$ and $m = 1$. The solid line shows the actual values. Figure 4 shows the values obtained by the method of exponential smoothing. On Fig. 5 presents ones based on decision tree. On Fig. 6 values obtained with the help of the LSTM neural network are shown.

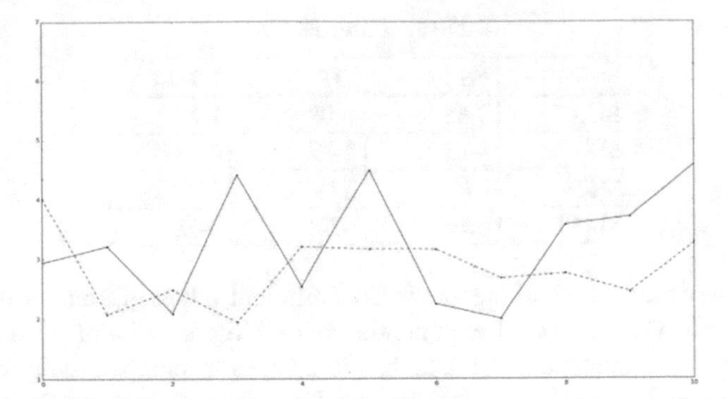

Fig. 4. Forecasted values obtained by the method of exponential smoothing

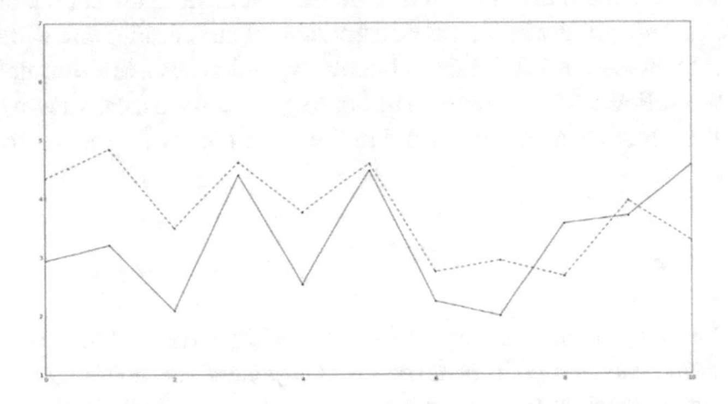

Fig. 5. Forecasted values based on decision tree

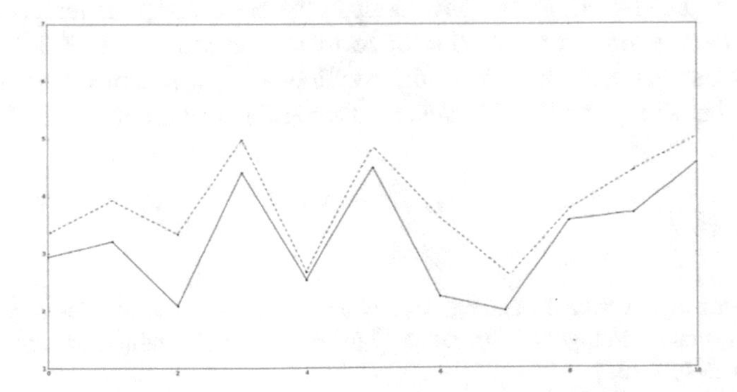

Fig. 6. Forecasted values obtained with the help of the LSTM

The predicted values for S = 20 and m = 1 (this choice of parameters is determined by the requirements of the online store) were computed for 100 values of the daily data conversion rate and corresponding values of clicks and sales number and other data. The results of calculations typical for most series are given in Table 1.

Table 1. Forecast errors

	ES	DT	LSTM
MAD	7.65	6.08	1.2
MA	−1.99	1.32	0.91
MSE	64	53	22
MAPE	0.49	0.38	0.07

As a result of the analysis of the forecasts of different values of S and m, the following was established. The method of exponential smoothing, in spite of its simplicity and non-exactingness in the amount of data, has in most cases comparatively small prediction errors. But at the same time, with the use of this method, some predicted values are significantly removed from real ones.

The decision tree method has proved to be inconvenient in the choice of parameters and has errors comparable with errors of exponential smoothing, but without strongly remote forecast values. The LSTM neural network, which has a more complex structure and needs to be preliminarily trained on a rather large time series, has shown good results, as well as in the overall forecast error, and in the remoteness of forecasts from real time series values.

6 Conclusion

The results of a study of methods for predicting weakly correlated time series typical of e-commerce conversion series have shown that exponential smoothing is the simplest, fastest and most convenient to set up predictive method, but in the cases of complex or long-term dependencies, it does not apply. The decision tree method is fast in learning, not difficult to understand, but inconvenient in the choice of parameters and does not work well when learning on data that have many characteristics. The LSTM neural network is a cumbersome, long learning, requires a lot of parameters that need to be selected, but has a very good performance in forecasting and order of magnitude smaller errors.

References

1. LPgenerator - Professional Landing Page is a platform to increase sales of your business. http://lpgenerator.ru/blog/2015/07/02/kakoj-dolzhna-byt-veb-analitika-internet-magazina. Accessed 10 July 2017
2. Stillwagon, A.: 14 Key Performance Indicators (KPIs) to Measure Customer Service. https://smallbiztrends.com/2015/03/how-to-measure-customer-service.html. Accessed 10 July 2017
3. Conversion Probability Forecast. https://www.searchengines.ru/prognoz_veroyat.html. Accessed 10 July 2017
4. Wei, D., Geng, P., Ying, L., Shuaipeng, L.: A prediction study on e-commerce sales based on structure time series model and web search data. In: 26th Chinese Control and Decision Conference, Changsha, China, pp. 1–4. IEEE (2014)

5. Hanke, J.E., Wichern, D.: Business Forecasting, 9th edn. Pearson Prentice Hall, Upper Saddle River (2008)
6. Guyon, I.J., Elisseeff, A.: An introduction to variable and feature selection. J. Mach. Learn. Res. **3**, 1157–1182 (2003)
7. Ian, H.W., Frank, E., Hall, M.A., Pal, C.J.: Data Mining: Practical Machine Learning Tools and Techniques, 4th edn. Morgan Kaufmann, Elsevier (2011)
8. Ismail, M., Mansur Ibrahim, M., Mahmoud Sanusi, Z., Nat, M.: Data Mining in electronic commerce: benefits and challenges. Int. J. Commun. Netw. Syst. Sci. **8**, 501–509 (2015)
9. Kirichenko, L., Radivilova, T., Zinkevich, I.: Forecasting weakly correlated time series in tasks of electronic commerce. In: 12th International Conference Computer Sciences and Information Technologies, Lviv, Ukraine. IEEE (2017)
10. Qiang, X., Rui-Chun, H., Hui, L.: Data mining research on time series of e-commerce transaction. Int. J. u- and e- Serv. Sci. Technol. **7**(1), 9–18 (2014)
11. Hyndman, R., Koehler, A.B., Keith Ord, J., Snyder, R.D.: Forecasting with Exponential Smoothing. Springer, Heidelberg (2008)
12. Cielen, D., Meysman, A., Ali, M.: Introducing data science: Big Data, machine learning, and more, using Python tools. Manning Publications, Greenwich (2016)
13. Lantz, B.: Machine Learning with R, 2nd edn. Packt Publishing, Birmingham (2015)
14. The Unreasonable Effectiveness of Recurrent Neural Networks. http://karpathy.github.io/2015/05/21/rnn-effectiveness/. Accessed 10 July 2017
15. Understanding LSTM Networks. http://colah.github.io/posts/2015-08-Understanding-LSTMs. Accessed 10 July 2017
16. LSTM - network of long short-term memory. https://habrahabr.ru/company/wunderfund/blog/331310/. Accessed 10 July 2017
17. White, D.S., Ariguzo, G.: A time-series analysis of U.S. e-commerce sales. Rev. Bus. Res. **11**(4), 134–140 (2011)
18. Hongjiu, G.: Data mining in the application of e-commerce website. In: Du, Z. (ed.) Intelligence Computation and Evolutionary Computation. Advances in Intelligent Systems and Computing, vol. 180, pp. 493–497. Springer, Heidelberg (2013)
19. Numpy and Scipy Documentation. https://docs.scipy.org/doc/. Accessed 10 July 2017
20. Documentation of scikit-learn 0.18. http://scikit-learn.org/stable/documentation.html. Accessed 10 July 2017
21. pandas: powerful Python data analysis toolkit. http://pandas.pydata.org/pandas-docs/stable/. Accessed 10 July 2017
22. Theano Python library. http://deeplearning.net/software/theano/. Accessed 10 July 2017
23. Scrapy 1.4 documentation. https://docs.scrapy.org/en/latest/. Accessed 10 July 2017

Performance Analysis of Open Source Machine Learning Frameworks for Various Parameters in Single-Threaded and Multi-threaded Modes

Yuriy Kochura(✉), Sergii Stirenko, Oleg Alienin, Michail Novotarskiy, and Yuri Gordienko

National Technical University of Ukraine "Igor Sikorsky Kyiv Polytechnic Institute", Kiev, Ukraine
iuriy.kochura@gmail.com

Abstract. The basic features of some of the most versatile and popular open source frameworks for machine learning (TensorFlow, Deep Learning4j, and H2O) are considered and compared. Their comparative analysis was performed and conclusions were made as to the advantages and disadvantages of these platforms. The performance tests for the de facto standard MNIST data set were carried out on H2O framework for deep learning algorithms designed for CPU and GPU platforms for single-threaded and multithreaded modes of operation Also, we present the results of testing neural networks architectures on H2O platform for various activation functions, stopping metrics, and other parameters of machine learning algorithm. It was demonstrated for the use case of MNIST database of handwritten digits in single-threaded mode that blind selection of these parameters can hugely increase (by 2–3 orders) the runtime without the significant increase of precision. This result can have crucial influence for optimization of available and new machine learning methods, especially for image recognition problems.

Keywords: Machine learning · Deep learning · TensorFlow · Deep learning4j · H2O · MNIST · Multicore CPU · GPU · Neural network · Classification · Single-threaded mode

1 Introduction

Machine learning (ML) is a subfield of Artificial Intelligence (AI) discipline. This branch of AI involves the computer applications and/or systems design that based on the simple concept: get data inputs, try some outputs, build a prediction. Nowadays, machine learning (ML) has advanced many fields like pedestrian detection, object recognition, visual-semantic embedding, language identification, acoustic modeling in speech recognition, video classification, fatigue estimation [1], generation of alphabet of symbols for multimodal human-computer interfaces [2], etc. This success is related to the invention and application of more sophisticated machine learning models and the development of software platforms that enable the easy use of large amounts of computational resources

N. Shakhovska and V. Stepashko (eds.), *Advances in Intelligent Systems and Computing II*, Advances in Intelligent Systems and Computing 689,
https://doi.org/10.1007/978-3-319-70581-1_17

for training such models [3]. The main aims of this paper are to review some available open source frameworks for machine learning, analyze their advantages and disadvantages, and test one of them in various computing environments including CPU and GPU-based platforms.

Also, we have tested the H2O system by using the publicly available MNIST dataset of handwritten digits. This dataset contains 60,000 training images and 10,000 test images of the digits 0 to 9. The images have grayscale values in the range 0:255. Figure 3 gives an example images of handwritten digits that were used in testing. We have trained the net by using the host with Intel Core i7-2700K insight. The computing power of this CPU approximately is 29.92 GFLOPs.

In this paper, we also present testing results of various net architectures by using H2O platform for single-threaded mode. Our experiments show that net architecture based on cross entropy loss function, tanh activation function, logloss and MSE stopping metrics demonstrates better efficiency by recognition handwritten digits than other available architectures for the classification problem.

This paper is structured as follows. The Sect. 2 State of the Art contains the short characterization of some of the most popular and versatile available open source frameworks (TensorFlow, Deep Learning4j, and H2O) for machine learning and motivation for selection of one of them for the performance tests. The Sect. 3 Performance Tests includes description of the testing methodology, data set used, and results of these tests. The Sect. 4 Discussion dedicated to discussion of the results obtained and lessons learned. Also we present here our experimental results where we apply different acti vation functions and stopping metrics to the classification problem with use case in single-threaded mode. Section 5 contains the conclusions of the work.

2 State of the Art

During the last decade numerous frameworks for machine learning appeared, but their open source implementations are seeming to be most promising due to several reasons: available source codes, big community of developers and end users, and, consequently, numerous applications, which demonstrate and validate the maturity of these frameworks. Below the short characterization of the most versatile open source frameworks (Deep Learning4j, TensorFlow, and H2O) for machine learning is presented along with their comparative analysis.

2.1 Deep Learning4j

Deep Learning4j (DL4J) is positioned as the open-source distributed deep-learning library written for Java and Scala that can be integrated with Hadoop and Spark [4]. It is designed to be used on distributed GPUs and CPUs platforms, and provides the ability to work with arbitrary n-dimensional arrays (also called tensors), and usage of CPU and GPU resources. Unlike many other frameworks, DL4J splits the optimization algorithm from the updater algorithm. This allows to be flexible while trying to find a combination that works best for data and problem.

2.2 TensorFlow

TensorFlow is an open source software library for numerical computation was originally developed by researchers and engineers working on the Google Brain Team within Google's Machine Intelligence research organization [5] for the purposes of conducting machine learning and deep neural networks research. This software is the successor to DistBelief, which is the distributed system for training neural networks that Google has used since 2011. TensorFlow operates at large scale and in heterogeneous environments. This system uses dataflow graphs to represent computation, shared state, and the operations that mutate that state. It maps the nodes of a dataflow graph across many machines in a cluster, and within a machine across multiple computational devices, including multicore CPUs, general purpose GPUs, and custom-designed ASICs known as Tensor Processing Units (TPUs). Such architecture gives flexibility to the application developer: whereas in previous "parameter server" designs the management of shared state is built into the system, TensorFlow enables developers to experiment with novel optimizations and training algorithms.

2.3 H2O

H2O software is built on Java, Python, and R with a purpose to optimize machine learning for Big Data [6]. It is offered as an open source platform with the following distinctive features. Big Data Friendly means that one can use all of their data in real-time for better predictions with H2O's fast in-memory distributed parallel processing capabilities. For production deployment a developer need not worry about the variation in the development platform and production environment. H2O models once created can be utilized and deployed like any Standard Java Object. H2O models are compiled into POJO (Plain Old Java Files) or a MOJO (Model Object Optimized) format which can easily embed in any Java environment. The beauty of H2O is that its algorithms can be utilized by various categories of end users from business analysts and statisticians (who are not familiar with programming languages using its Flow web-based GUI) to developers who know any of the widely used programming languages (e.g. Java, R, Python, Spark). Using in-memory compression techniques, H2O can handle billions of data rows in-memory, even with a fairly small cluster. H2O implements almost all common machine learning algorithms, such as generalized linear modeling (linear regression, logistic regression, etc.), Naive Bayes, principal components analysis, time series, k-means clustering, Random Forest, Gradient Boosting, and Deep Learning.

2.4 Parameters of Machine Learning

The Activation Functions. Activation functions also known as transfer functions are used to map input nodes to output nodes in certain fashion [7] (see the conceptual scheme of an activation function in Fig. 1).

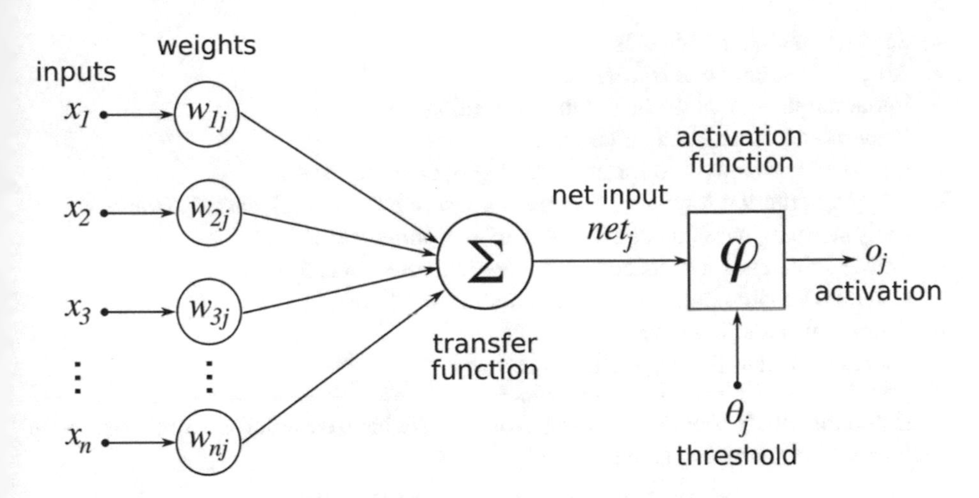

Fig. 1. The role of activation function in the process of learning neural net.

Functions with dropout are used for reducing overfitting by preventing complex co-adaptations on training data. This technique is known as regularization. Figure 2 demonstrate the difference between standard neural net and neural net after applying dropout [8].

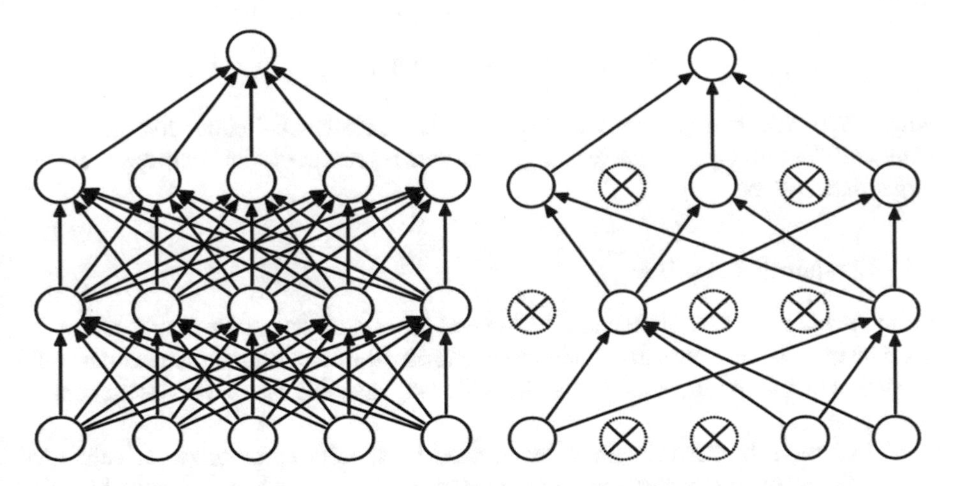

Fig. 2. An example of standard neural net on the left and neural net with dropout on the right with 2 hidden layers.

Constant Parameters of the Training Model. We have used the network model with such constant parameters, namely:

- Response variable column is *C785*
- Hidden layer size is *[50, 50]*
- Epochs are *500*

- Seed for random numbers is *2*
- Adaptive learning rate is *false*
- Initial momentum at the beginning of training is *0.9*
- Final momentum after the ramp is *0.99*
- Input layer dropout ratio for improving generalization is *0.2*
- Stopping criterion for classification error fraction on training data is *disable*
- Early stopping based on convergence of stopping metric is *3*
- Relative tolerance for metric-based stopping criterion is *0.01*
- Compute variable impotence for input features is *true*
- Sparse data handling is *true*
- Force reproducibility on small data is *true*

Variable Parameters of the Training Model. We have used the network model with such variable parameters, namely:

- Activation function: *Tanh, TanhWithDropout, Maxout, MaxoutWithDropout, Rectifier, RectifierWithDropout*
- Metric to use for early stopping: *logloss, misclassification, MAE, MSE, RMSE* and *RMSLE*
- Loss function: *Cross Entropy*

Loss function is a function that used to measure the degree of fit. The cross entropy loss function for the distributions *p* and *q* over a given set is defined as follows:

$$\mathrm{H(p, q)} = \mathrm{H}(P) + \mathrm{D}_{KL}(\mathrm{p} \parallel \mathrm{q}) \tag{1}$$

where *H(p)* is the entropy of *p*, and $\mathrm{D}_{KL}(\mathrm{p} \parallel \mathrm{q})$ is the Kullback–Leibler divergence of *q* from *p* (also known as the relative entropy of *p* with respect to *q)*. Cross entropy is always larger than entropy.

2.5 Comparative Analysis

From the point of view of an end user, several aspects of these frameworks are of the main interest. Except for performance and maturity, the open source frameworks could be attractive and useful, if they have the wide language and operating system support (see Table 1).

All of these frameworks are characterized by a quite wide ranges of supported languages and operating systems. But nowadays it is not enough in the view of the fast development of parallel and distributed computing like cluster and, especially, GPGPU computing. In this connection, TensorFlow has clear notification as to the pre-requisites for NVIDIA GPGPU cards, that should have CUDA Compute Capability (CC) 3.0 or higher. As to DL4J this is not clear because the developers stated just general support of NVIDIA GPGPU cards from GeForce GTX to Titan and Tesla that have various CC from 2.0 to 3.5. For H2O types of supported NVIDIA cards and CC are not specified, but proposed in the branching sub-framework Deep Water. The additional important aspects are the low entrance barrier and fast learning curve. They usually are based on the convenient graphical user interface, workflow management, and visualization tools. Now these features become "de facto standard" tools for integration of end users, workflows, and resources. The examples of their implementations (like WS-PGRADE/gUSE [9], KNIME [10], etc.) and applications in physics [11], chemistry [12], astronomy [13], brain-computing [14], eHealth [15] can be found elsewhere. In this context TensorFlow and H2O propose web-based graphic user interfaces TensorBoard and Flow, respectively, which are actually workflow management and visualization tools. In contrast to other frameworks H2O proposes the much shorter learning curve due to Flow, the web-based and self-explanatory user interface. In general, Flow allows end users without experience in software programming even to import remote data, create model, train it, validate it, and then save the whole workflow. In addition, the machine learning model developed in Flow can be compiled into Plain Old Java Files (POJO) format, which can be easily embedded in any Java environment. Due to these advantages, now more than 5000 organizations currently use H2O, and many well-known companies (like Cisco, eBay, PayPal, etc.) are using it for big data processing. This data set contains 785 columns. The final column is the correct answer, 0 to 9. The first 784 are the 28 × 28 grid of grayscale pixels, and each is 0 (for white) through to 255 (for black).

Table 1. Comparison of machine learning frameworks.

System (initial release)	GPU support	GUI	Operating system	Language support
TensorFlow (2015)	NVIDIA GPUs (CC 3.0 or higher)	TensorBoard (workflow, visualization)	Linux, macOS, Windows, Android, iOS	Python, C++
DL4J (2013)	NVIDIA GPUs (Tesla, Titan)	–	Linux, macOS, Windows, Android	Java, Scala, CUDA, C, C++, Python
H2O (2011)	Deep Water, NVIDIA GPUs (CC not stated)	Flow (workflow, visualization, POJO)	Linux, macOS, Windows	Java, Python, R

3 Performance Tests

The performance of the mentioned frameworks was a topic of many investigations performed by developers of these frameworks and independent end users [16]. But performance of H2O was not investigated thoroughly except for its developers for unknown CPU and GPU platforms [17]. That is why H2O was selected for performance tests in this paper.

The data set used in this work, called the "MNIST data," was proposed in 1998 to identify handwritten numbers. We have tested the H2O system by recognizing the handwritten digits (Fig. 3) from the publicly available MNIST data set for machine learning methods [18]. Now it is well-known "de facto standard" data set for a typical "easy-for-humans-but-hard-for-machine" problem. The used MNIST database of handwritten digits has a training set of 60,000 examples, and a test set of 10,000 examples. Each digit is represented by $28 \times 28 = 784$ gray-scale pixel values (features).

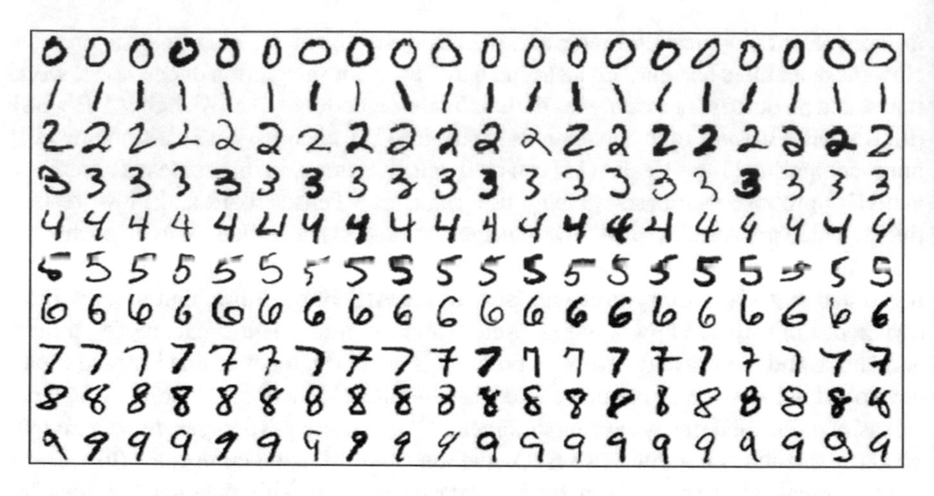

Fig. 3. The examples of the handwritten digits from MNIST data set.

The tests were performed on different platforms including Intel Core i5-7200U with 4 cores (CPU1), Intel Core i7-2700K with 8 cores (CPU2), NVIDIA Tesla K40 GPU accelerator using single-threaded and multi-threaded modes of operation. The parameters of neural network were the same for the Deep Learning (CPU only) and Deep Water (CPU+GPU) algorithms. The details of these platforms and modes of operation are given above in Tables 2 and 3.

Table 2. Multi-threaded operation on CPUs.

Parameter's name	Intel Core i5-7200U (CPU1)	Intel Core i7-2700K (CPU2)
GFLOPs	13.85	29.92
Duration	2 min 18 s	2 min 32 s
Training speed, obs/sec	23746	78972
Epochs	48.5953	108.3821
Iterations	103	65
Training logloss	0.0407	0.0297
Validation logloss	0.1584	0.1616

Table 3. Single-threaded operation on CPUs.

Parameter's name	Intel Core i5-7200U (CPU1)	Intel Core i7-2700K (CPU2)
GFLOPs	13.85	29.92
Runtime	2 min 15 s	2 min 5 s
Training speed, obs/sec	13820	15174
Epochs	26	26
Iterations	26	26
Training logloss	0.0577	0.0577
Validation logloss	0.1664	0.1664

The performance tests were carried out with Rectifier activation function for two algorithms Deep Learning (CPU only) and Deep Water (CPU+GPU). The stopping criterion was based on convergence of stopping_metric (equal to misclassification). The stop event occurs, if simple moving average of length k of the stopping_metric does not improve for k: = stopping_rounds (equal to 3) scoring events. The relative tolerance for metric-based stopping criterion was equal to 0.01. The typical convergence of training (lower) and validation (upper) logloss values with epochs is shown on Fig. 4. The results of these performance tests using H2O system are presented above in Tables 2, 3 and 4. It should be noted that the results of learning neural network to recognize the handwritten digits on CPUs and GPU by using multi-threaded mode of operation are inherently not reproducible due to randomization. To estimate data scattering in multi-threaded modes of operation the runs were repeated for 12 times with determination of mean and standard deviation (Table 4).

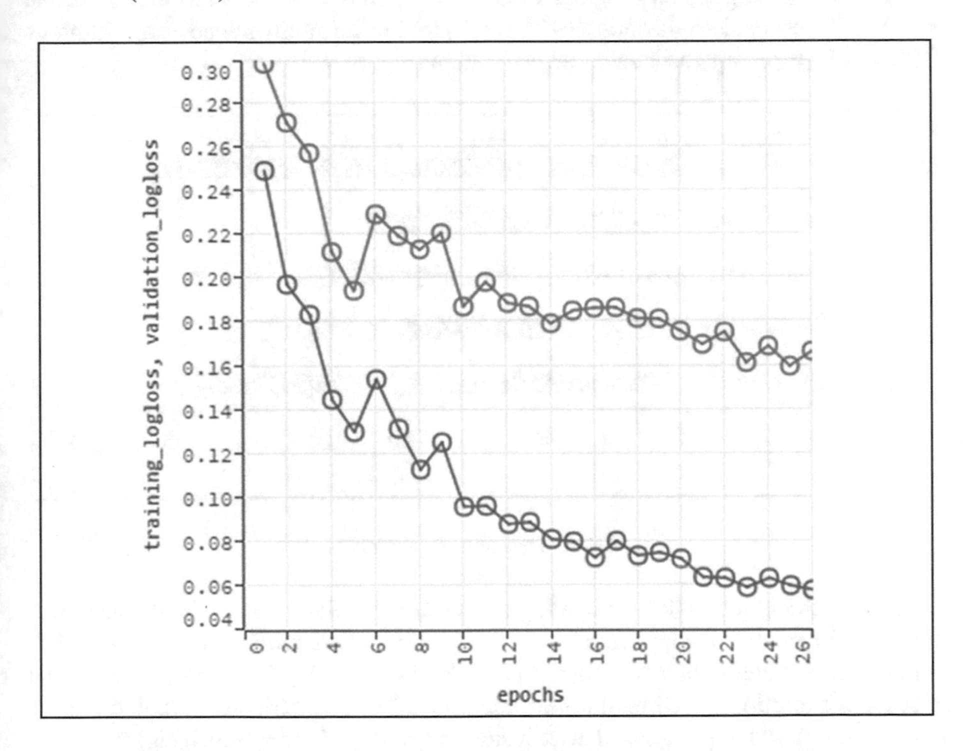

Fig. 4. Evolution of training (lower) and validation (upper) logloss values.

Table 4. Multi-threaded operation on GPU (1.43 TFLOPs)

Parameter's name	Mean value	Standard deviation
Runtime	2 min 29 s	17.2 s
Training speed, obs/sec	18707	520
Epochs	42.24	5.66
Iterations	2475	332
Training logloss	0.285	0.0192
Validation logloss	0.437	0.0236

4 Discussion

The time of convergence for logloss values with epochs was not very different for all regimes, if the standard deviation (~17 s) of duration for multi-threaded operation on GPU will be taken into account as an estimation (Fig. 5).

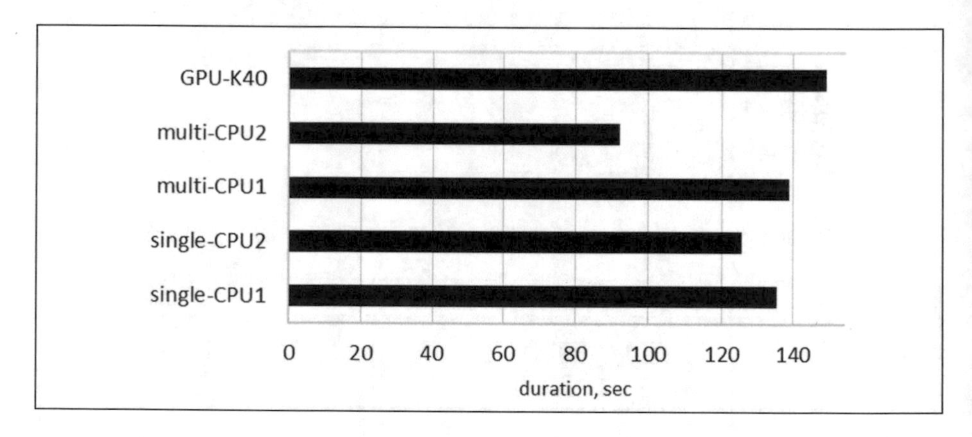

Fig. 5. Duration of training.

Despite the much higher computing power of GPU the better training speed was observed for multi-threaded regime for CPU2 with 8 cores with speedup up to 5.2 in comparison to single-threaded regime (Fig. 6). For CPU1 with 4 cores the similar speedup for multi-threaded regime was equal to 1.7 in comparison to single-threaded regime. As to GPU training speed these results can be explained by much bigger number (by ~100 times) of performed iterations.

As it is well-known the logloss values are very sensitive to outliers and this tendency is very pronounced in the case of GPU, where the much bigger iterations were used and higher training logloss values were found (Fig. 7).

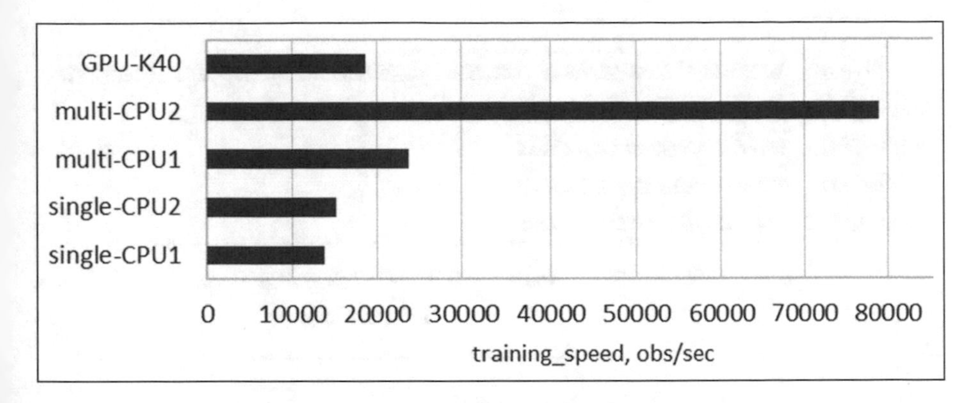

Fig. 6. Training speed.

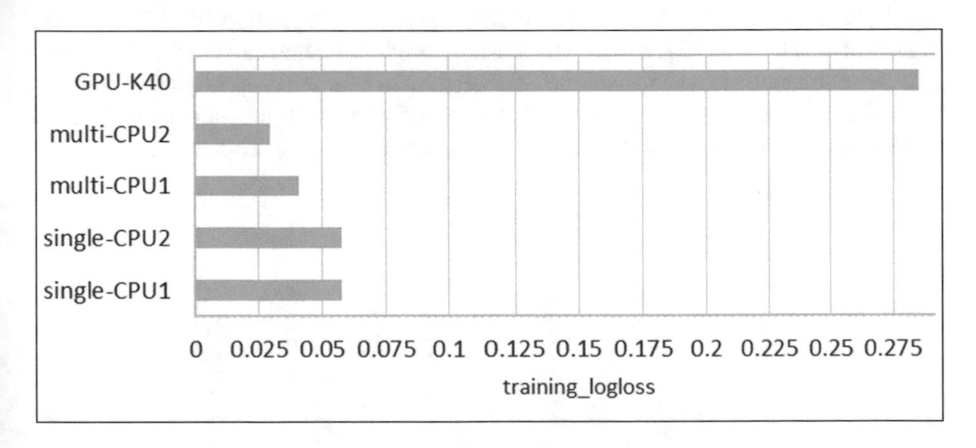

Fig. 7. Training logloss values.

The ratio of validation logloss (Fig. 8) to training logloss is equal to 1.53 for Deep Water case, which is much lower in comparison to the same ratio 2.88 for Deep Learning single-threaded case, and 3.89 and 5.44 even for Deep Learning multi-threaded case CPU1 and CPU2, respectively. This allows to make assumption that the more iterations in GPU mode give the more realistic model with the lower risk of overfitting.

Finally, in this paper we described the basic features of some open source frameworks for machine learning, namely TensorFlow, Deep Learning4j, and H2O. For usability and performance tests H2O framework was selected. It was tested on several platforms like Intel Core i5-7200U (4 cores), Intel Core i7-2700K (8 cores), Tesla K40 GPU with the goal to evaluate their performance in the context of recognizing handwritten digits from MNIST data set. To reach this goal the same parameters of the neural network were used for Deep Learning and Deep Water algorithms. The influence of many other aspects like nature of data (for example, sparsity level and sparsity pattern), number of hidden layers and their sizes should be taken into account for the better comparative analysis, but these aspects were out of scope of the current work and will be published in separate paper elsewhere [19].

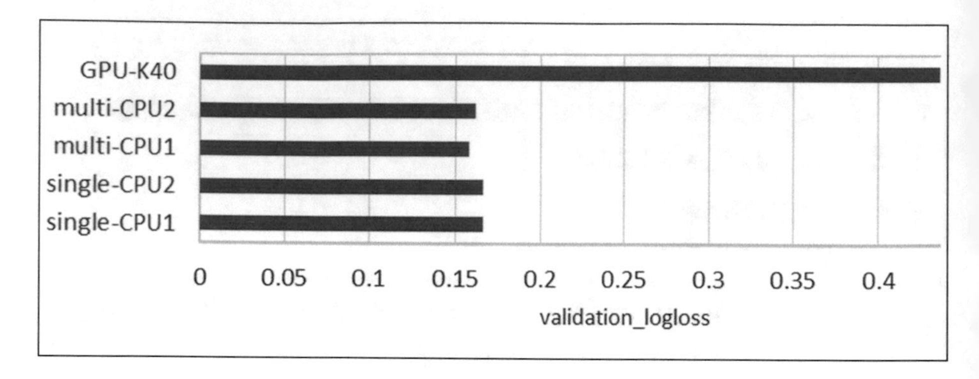

Fig. 8. Validation logloss values.

We trained neural networks for classification problems on publicly available MNIST dataset of handwritten digits with use case in single-threaded mode. We found that generalization performance has very strong dependence on activation function and very slight dependence on stopping metric. Figure 9 shows the runtime values on the logarithm scale obtained for these different architectures as training progresses.

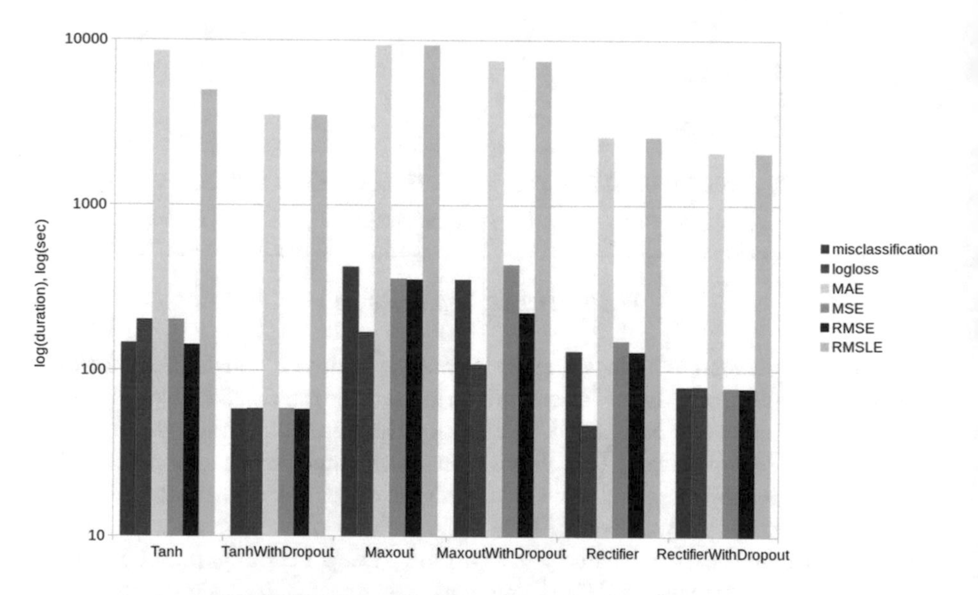

Fig. 9. Runtime of learning nets different architectures.

Figure 10 demonstrates the effectiveness of using tanh activation function for all stopping metrics that considered in this paper. In the case of the learning net based on the tanh activation function, MAE and RMSLE stopping metric has achieved the logloss value of 0.0104. These architectures demonstrate better training prediction ability than others but take much time for building model.

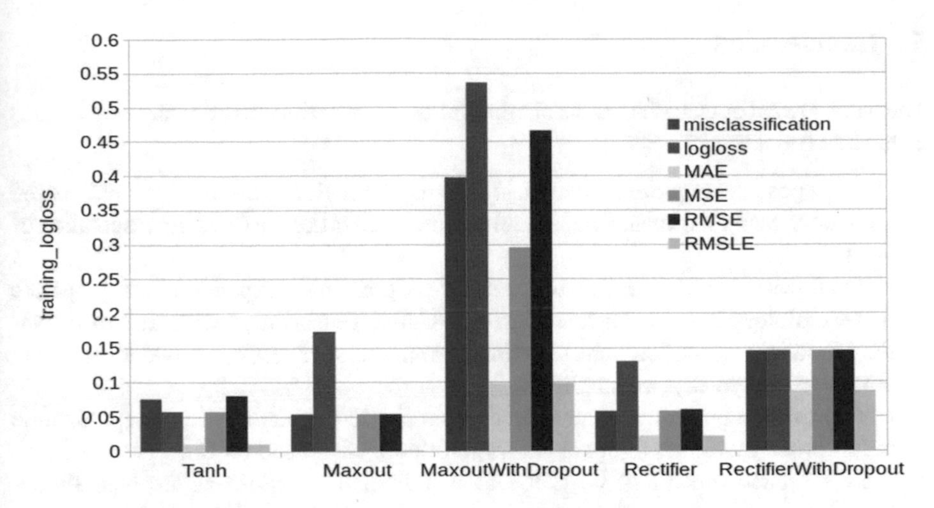

Fig. 10. Training logloss of learning nets different architectures.

In order to find the best neural net architecture for digits recognition just needs to look at the behavior of models on unknown data should be checked. Figure 11 shows the validation error rates for different architectures that are considered here. We see, the best digit's recognition results were achieved in the case of tanh activation function. The type of stopping metric is very slightly effects on the values of the validation error but it does very much on the runtime of building model.

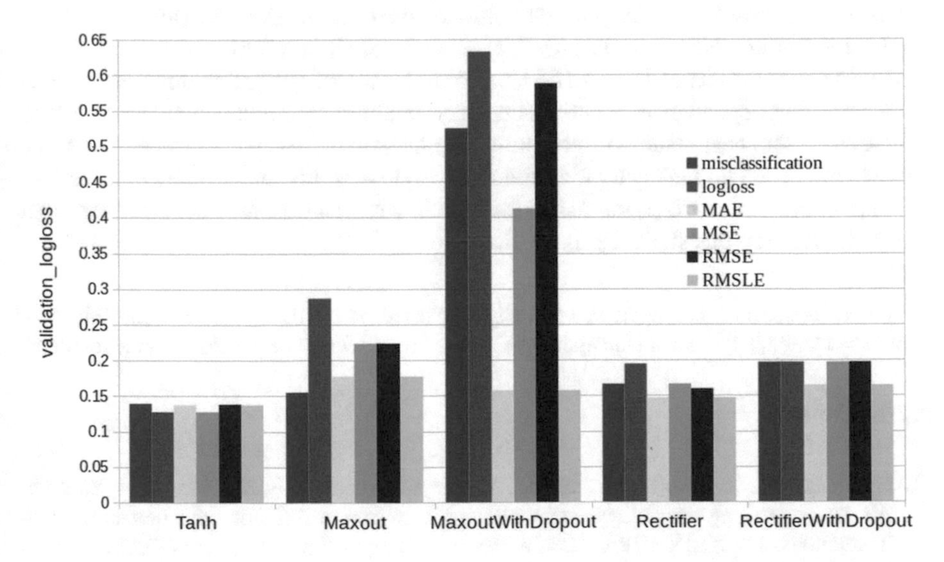

Fig. 11. Validation logloss of learning nets different architectures.

5 Conclusions

The work carried out and the results obtained allow us to make the following conclusions as to H2O framework:

- H2O propose the unprecedentedly fast learning curve due to the available web-based GUI, easy workflow management tools, and visualization tools for representation of data.
- H2O allows the data scientists without any programming experience easily operate by several deep learning backends (mxnet, Caffe, TensorFlow) with various activation functions (rectifier, tahn), various parameters of neural network, stopping criteria, and convergence conditions.
- H2O propose opportunities for reproducible single-threaded and non-reproducible multi-threading modes of operation for multicore CPUs and GPUs.
- Multi-threaded operations on CPUs give the smaller logloss values than single-threaded operations, but the ratio of validation logloss to training logloss is much lower in comparison to multi-threaded operations on GPU, which gives the more realistic model with the lower risk of overfitting.

In this paper, we present the results of testing neural networks architectures on H2O platform for various activation functions, stopping metrics, and other parameters of machine learning algorithm. It was demonstrated for the use case of MNIST database of handwritten digits in single-threaded mode that blind selection of these parameters can hugely increase (by 2–3 orders) the runtime without the significant increase of precision. This result can have crucial influence for optimization of available and new machine learning methods, especially for image recognition problems.

During the process of testing H2O, we found out that generalization performance has very strong dependence on activation function and very slight dependence on stopping metric. The best results of recognition digits were achieved in case of using nets architecture based on tanh activation function, logloss and MSE stopping metrics.

This paper summarizes the activities which were started recently and described shortly in the previous student paper [20].

Acknowledgements. The work was partially supported by NVIDIA Research and Education Centers in National Technical University of Ukraine "Igor Sikorsky Kyiv Polytechnic Institute".

References

1. Gordienko, N., Stirenko, S., Kochura, Y., Alienin, O., Novotarskiy, M., Gordienko, Y., Rojbi, A.: Deep learning for fatigue estimation on the basis of multimodal human-machine interactions. In: XXIX IUPAP Conference on Computational Physics, CCP2017, Paris, France (2017)
2. Hamotskyi, S., Rojbi, A., Stirenko, S., Gordienko, Y.: Automatized generation of alphabets of symbols for multimodal human computer interfaces. In: Proceedings of Federated Conference on Computer Science and Information Systems, FedCSIS-2017, Prague, Czech Republic (2017)

3. Witten, I.H., Frank, E., Hall, M.A., Pal, C.J.: Data Mining: Practical Machine Learning Tools and Techniques. Morgan Kaufmann (2016)
4. Team, D.J.D.: Deep Learning4j: open-source distributed deep learning for the JVM. Apache Software Foundation License
5. Abadi, M., et al.: TensorFlow: a system for large-scale machine learning. In: 12th USENIX Symposium on Operating Systems Design and Implementation, OSDI 2016, Savannah, GA, USA, pp. 265–283 (2016)
6. Candel, A., Parmar, V., LeDell, E., Arora, A.: Deep Learning with H2O. AI Inc. (2016)
7. Activation Functions. https://medium.com/towards-datascience/activation-functions-in-neural-networks-58115cda9c96
8. Srivastava, N., et al.: Dropout: a simple way to prevent neural networks from overfitting. J. Mach. Learn. Res. **15**(1), 1929–1958 (2014)
9. Kozlovszky, M., et al.: DCI bridge: executing WS-PGRADE workflows in distributed computing infrastructures. In: Science Gateways for Distributed Computing Infrastructures, pp. 51–67. Springer, Cham (2014)
10. O'Hagan, S., Kell, D.B.: Software review: the KNIME workflow environment and its applications in genetic programming and machine learning. Genet. Program. Evolvable Mach. **16**(3), 387–391 (2015)
11. Gordienko, Y., et al.: IMP science gateway: from the portal to the hub of virtual experimental labs in e-science and multiscale courses in e-learning. Concurrency Comput. Pract. Experience **27**(16), 4451–4464 (2015)
12. Herres-Pawlis, S., et al.: Quantum chemical meta-workflows in MoSGrid. Concurrency Comput. Pract. Experience **27**(2), 344–357 (2015)
13. Castelli, G., et al.: VO-compliant workflows and science gateways. Astron. Comput. **11**, 102–108 (2015)
14. Stirenko, S., et al.: User-driven intelligent interface on the basis of multimodal augmented reality and brain-computer interaction for people with functional disabilities. arXiv: 1704.05915 (2017)
15. Gordienko, Y., et al.: Augmented coaching ecosystem for non-obtrusive adaptive personalized elderly care on the basis of Cloud-Fog-Dew computing paradigm. In: 40th International Convention on Information and Communication. Technology, Electronics and Microelectronics (MIPRO) (2017)
16. Fox, J., Zou, Y., Qiu, J.: Software Frameworks for Deep Learning at Scale, Internal Indiana University Technical Report (2016)
17. H2O Deep Learning Benchmark. https://github.com/h2oai/h2o-3/tree/master/h2o-algos/src/main/java/hex/DeepLearning
18. LeCun, Y., Cortes, C., Burges, C.J.: The MNIST database of handwritten digits (1998). http://yann.lecun.com/exdb/mnist
19. Kochura, Y., Stirenko, S., Rojbi, A., Alienin, O., Novotarskiy, M., Gordienko, Y.: Comparative analysis of open source frameworks for machine learning with use case in single-threaded and multi-threaded modes. In: IEEE XII International Scientific and Technical Conference on Computer Sciences and Information Technologies, CSIT 2017, Lviv, Ukraine (2017)
20. Kochura, Y., Stirenko, S., Gordienko, Y.: Comparative performance analysis of neural networks architectures on H2O platform for various activation functions. In: YSF-2017, Lviv, Ukraine

On Scalability of Predictive Ensembles and Tradeoff Between Their Training Time and Accuracy

Pavel Kordík[(✉)] and Tomáš Frýda

Department of Computer Science, Faculty of Information Technology, Czech Technical University in Prague, Prague, Czech Republic
kordikp@fit.cvut.cz

Abstract. Scalability of predictive models is often realized by data subsampling. The generalization performance of models is not the only criterion one should take into account in the algorithm selection stage. For many real world applications, predictive models have to be scalable and their training time should be in balance with their performance. For many tasks it is reasonable to save computational resources and select an algorithm with slightly lower performance and significantly lower training time. In this contribution we made extensive benchmarks of predictive algorithms scalability and examined how they are capable to trade accuracy for lower training time. We demonstrate how one particular template (simple ensemble of fast sigmoidal regression models) outperforms state-of-the-art approaches on the Airline data set.

Keywords: Combining classifiers · Regression models
Model blending · Scalability · Map reduce

1 Introduction

There are not many predictive modeling algorithms that can be considered scalable. Often, data subsampling in both dimensionality (number of attributes) and numerosity (number of instances) is required in order to train model in reasonable time. Data reduction is however not always good strategy, especially when redundancies, correlations and noise cannot be further eliminated.

In this case, it is beneficial to use a scalable predictive algorithm. Scalability can be realized by clever training and base algorithms that can be scaled to millions of instances [1]. Other approach is to use some form of famous map-reduce strategy [2]. More general data manipulation strategies can be found in ensemble methods. The popularity of ensemble methods in machine learning is rising steadily and very complex forms of ensembles, including hierarchical ensembles are described in the literature.

N. Shakhovska and V. Stepashko (eds.), *Advances in Intelligent Systems and Computing II*, Advances in Intelligent Systems and Computing 689,
https://doi.org/10.1007/978-3-319-70581-1_18

Moreover, it is necessary to take into account the computational complexity of the algorithm selection process. It is not feasible to run all candidate algorithms on a new data set to select the best performing one, simply because there are infinite number of available algorithms.

One approach how to select a training algorithm for a new data set in a reasonable time is to use a meta-data collected during training on similar data sets. Meta-learning approaches [3] utilizing meta-data have been studied intensively in past few decades. They can predict performance of algorithms on new data sets allowing to select good performing algorithm among multiple candidates.

Majority of meta-learning approaches [3–6] simply select from a set of few predefined fully specified data mining algorithms the one, producing models with the best generalization performance for the given data.

More advanced meta-learning approaches combine algorithm selection and hyperparameter optimisation such as CASH [7] elaborated within the *INFER project*. In [8–10] data mining workflows are optimized including data cleaning and preprocessing steps together with selected hyperparameters of modeling methods.

We focus on modeling stage only and take one step further. In our approach we optimize structure of algorithmic ensembles together with their hyperparameters as explained in Sect. 4. In this way, we can discover new algorithmic building blocks.

The hierarchical structure of algorithmic ensembles is represented by *Meta-learning algorithm templates*. Our templates indicate which learning algorithms are used and how their outputs are fed into other learning algorithms.

We use the genetic programming [11] to evolve structure of templates and a parameter optimization to adjust parameters of algorithms for specific data sets. Our approach allocates exponentially more time for evolution of above-average templates similarly to the Hyperband approach [12].

Furthermore, templates evolved (discovered) on small data sets can be reused as building blocks for large data sets.

Building predictive models on large data samples is a challenging task. Learning time of algorithms often grows fast with number of training data samples and dimensionality of a data set. Hyper-parameter optimization can help us to generate more precise models for given task, but it adds significant computational complexity to the training process. We show that templates evolved on small data subsamples can outperform state of the art algorithms including complex ensembles in terms of performance and scalability.

First, we briefly describe building blocks of predictive ensembles.

2 Base Algorithms and Ensembling Strategies

We build ensembles from fast weak learners [13]. Many of our base models resemble neurons with different activation functions. We can use base models to construct both classification and regression ensembles. In this contribution, we focus on classification tasks only, however regression models can also be present in classification ensembles.

The classification task itself can be decomposed into regression subproblems by separation of single classes from the others. These binary class separation problems can be solved by regression models – by estimating continuous class probabilities. The maximum probability class is then considered as output value. The classifier consisting of regression models is further referred to as **Classifier-Model**.

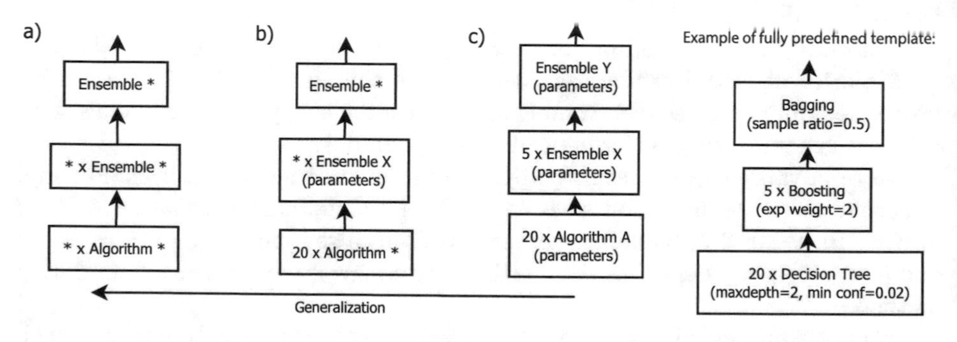

Fig. 1. Nested ensembles can be represented by a template. Using wildcards, specific (or predefined in other words) template can be generalized to represent set of templates.

2.1 Base Algorithms

Training regression models is fast and straightforward. We use several activation functions, namely **Sigmoid, SigmoidNorm, Sine, Polynomial, Gaussian, Exponential** and **Linear**.

To train coefficients of linear or polynomial models, the General Least Squares method [14] is applied. For models non-linear in their coefficients, an iterative optimization process is needed. We compute analytic gradients of error for all fast regression models and employ quasi-Newton method [15] to optimize their parameters.

As an example, we show how SigmoidNorm models are trained. The output of the model y_j for jth instance with target variable d_j and input vector $\boldsymbol{x}_j$ can be computed as

$$y_j = \left[1 + \exp\underbrace{\left(\sum_{i=0}^{n-1} a_i x_{ij} + a_n\right)}_{\rho_{ij}}\right]^{-1}$$

where coefficients $\boldsymbol{a}$ should be optimized to fit a training data and reduce the error $E = \sum_{j=1}^{m} (y_j - d_j)^2$ of the model.

The gradient of error can be computed as $\nabla \boldsymbol{E} = \left(\frac{\partial \boldsymbol{E}}{\partial a_0}, \frac{\partial \boldsymbol{E}}{\partial a_1}, \cdots, \frac{\partial \boldsymbol{E}}{\partial a_n}\right)$, where $\frac{\partial \boldsymbol{E}}{\partial a_i} = \sum_{j=1}^{m} \frac{\partial \boldsymbol{E}}{\partial y_j} \cdot \frac{\partial y_j}{\partial a_i}$ and $\frac{\partial \boldsymbol{E}}{\partial y_j} = 2\sum_{j=1}^{m} (y_j - d_j)$. The last partial derivative $\frac{\partial y_j}{\partial a_i}$ has to be computed for each coefficient a_i.

In case of the SigmoidNorm model from Eq. 2.1, we can compute partial derivatives as follows $\frac{\partial \boldsymbol{E}}{\partial a_i} = \sum_{j=1}^{m} \frac{\partial \boldsymbol{E}}{\partial y_j} \cdot \frac{\partial y_j}{\partial \rho_j} \cdot \frac{\partial \rho_j}{\partial a_i}$. Then the components of the gradient are

$$\frac{\partial \boldsymbol{E}}{\partial a_n} = -2 \sum_{j=1}^{m} (1 + e^{\rho_{ij}})^{-2} \cdot e^{\rho_{ij}},$$

$$\frac{\partial \boldsymbol{E}}{\partial a_i} = -2 \sum_{j=1}^{m} \left[(y_j - d_j) \cdot (1 + e^{\rho_{ij}})^{-2} \cdot e^{\rho_{ij}} \cdot a_i \right].$$

Accordingly, we derived analytic gradients of error on training data for Exponential, Gaussian, Sigmoid and Sine models. Gradients are then supplied together with errors to the quasi-Newton optimization method [16] during the training to speed up the convergence. More details can be found in [17] and the source code is also available [18].

The **LocalPolynomial** base model as well as **Neural Network (NN), Support Vector Machine (SVM), Naive Bayes classifier (NB), Decision Tree (DT), K-Nearest Neighbor (KNN)** were adopted from the Rapidminer environment [19].

2.2 Ensembling Algorithms

The performance of models can often be further increased by combining or ensembling [20–25] base algorithms, particularly in cases where base algorithms produce models of insufficient plasticity or models overfitted to training data [26].

A detailed description of the large variety of ensemble algorithms can be found in [27]. We briefly describe the ensembling algorithms that are used in our experiments. **Bagging** [28] is the simplest one; it selects instances for base models randomly with repetition and combines models with simple average. **Boosting** [23] specializes models on instances incorrectly handled by previous models and combines them with weighted average. **Stacking** [22] uses a meta model, which is learned from the outputs of all base models, to combine them. Another ensemble utilizing meta models is the **Cascade Generalization** [29], where every model except the first one uses a data set extended by the output of all preceding models. **Delegating** [30] and **Cascading** [31,32] both use a similar principle: they operate with certainty of model output. The latter model is specialized not only in instances that are classified incorrectly by previous models, but also in instances that are classified correctly, but previous models are not certain in terms of their output. Cascading only modifies the probability of selecting given instances for the learning set of the next model. **Arbitrating** [33] uses a meta-model called referee for each model. The purpose of this meta-model is to predict the probability of correct output. All methods used in this study were implemented within the FAKE GAME open source project [18].

3 Meta-learning Templates

The meta-learning template [3] is a prescription how to build hierarchical supervised models. In the most complex case, it can be a collection of ensembling algorithms and base algorithms combined in a hierarchical manner, where base algorithms are leaf nodes connected by ensembling nodes. Regression models or classifiers deeper in the hierarchy can be more specialized to a particular subset of data samples or attributes. This scheme decomposes the prediction problem into subproblems and combines the final solution (model) from subsolutions. The procedure of problem decomposition depends on ensembling methods. Typically, it distributes data to member models and when all outputs are available, they are combined to the ensemble output.

Similarly to the Holland's schema theorem [34], we can define fitness of a template as average/maximum fitness of individual algorithms represented by this particular template. Wildcards here are used just as placeholders for random decisions on type of ensembles or base algorithms and their parameters. On the contrary, in rooted tree schema theory [35] wildcards represent sub-trees.

4 Discovering Templates

The meta-learning template can be designed manually using an expert knowledge (for example, bagging boosted decision trees showed good results on several problems) so it is likely to perform well on a new data set. This is however not guaranteed.

In our approach we optimize templates on data sub-samples using a genetic programming [11]. In this way, we can search the space of possible architectures of hierarchical ensembles and optimize their parameters simultaneously.

5 Templates at Scale

Recent rise of big data modeling challenges scalability of predictive modeling algorithms and tools. One obvious approach is to reduce dimensionality and numerosity of data [36]. This approach works in most of the cases because big data are often redundant. However for some data sets, the performance of predictors increase significantly with growing number of instances used for training. For such data, scalable algorithms [37] and tools [38,39] have been developed.

Most of these approaches are based on a map-reduce technique [40].

In this section, we show, that meta-learning can be also used at scale. Our approach is inspired by [41], where classifier selected on sub-samples work reasonably well on larger data sets. We evolve templates on a subset of 3000 randomly selected instances. Then, evolved template can be executed on full data. When we do not have enough time for the meta-learning template evolution, it is also possible to generate the subset just for computing meta-features. Then we can use a best performing template for the data set with most similar meta-features.

For the template execution we split large data into multiple disjoint subsets and then use the map-reduce paradigm to train multiple instances of the template. Prediction is made by reducing (majority voting) of models generated from templates.

This approach is very similar to bagging except that we do not use the bootstrap sampling.

5.1 Experiments

We have conducted experiments to get insight into the scalability of several machine learning algorithms from H2O as well as our parallel training of templates. Our motivation is to show that proper algorithm selection is important especially for large data sets and can be often done using a fraction of the data set.

We have chosen two public data sets—Airline Delays which is available through H2O [42] and HIGGS [43]. Those data sets are used for binomial classification of selected output attributes.

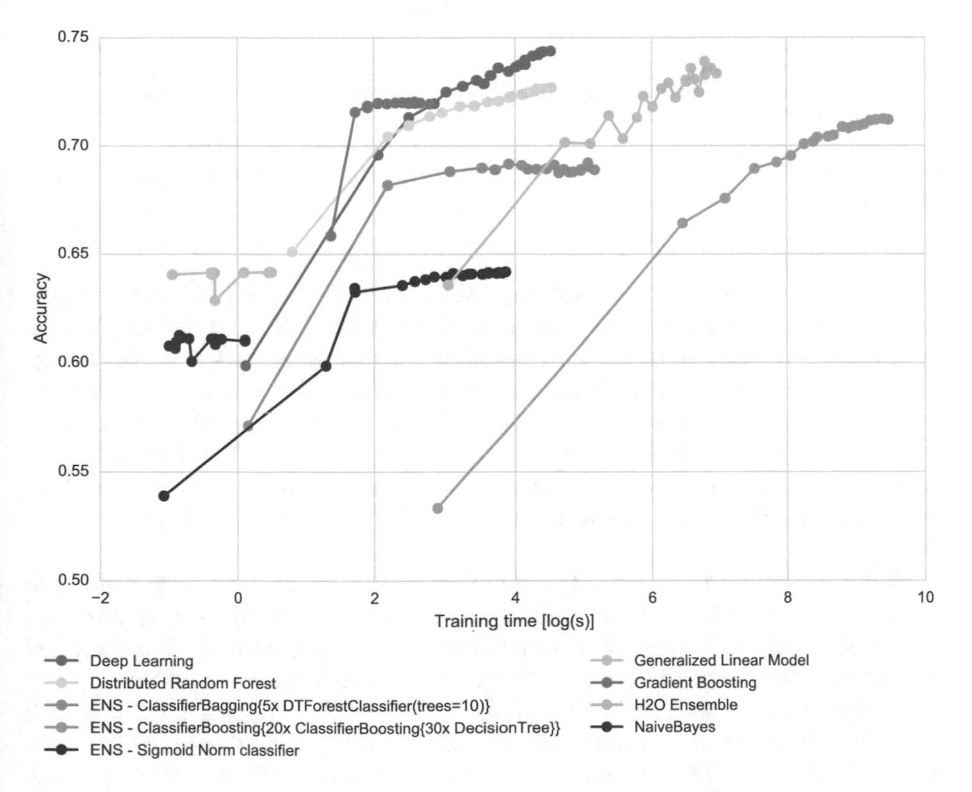

Fig. 2. Comparison of several machine learning algorithms in H2O.ai trained on samples with various sizes from Higgs [43] data set

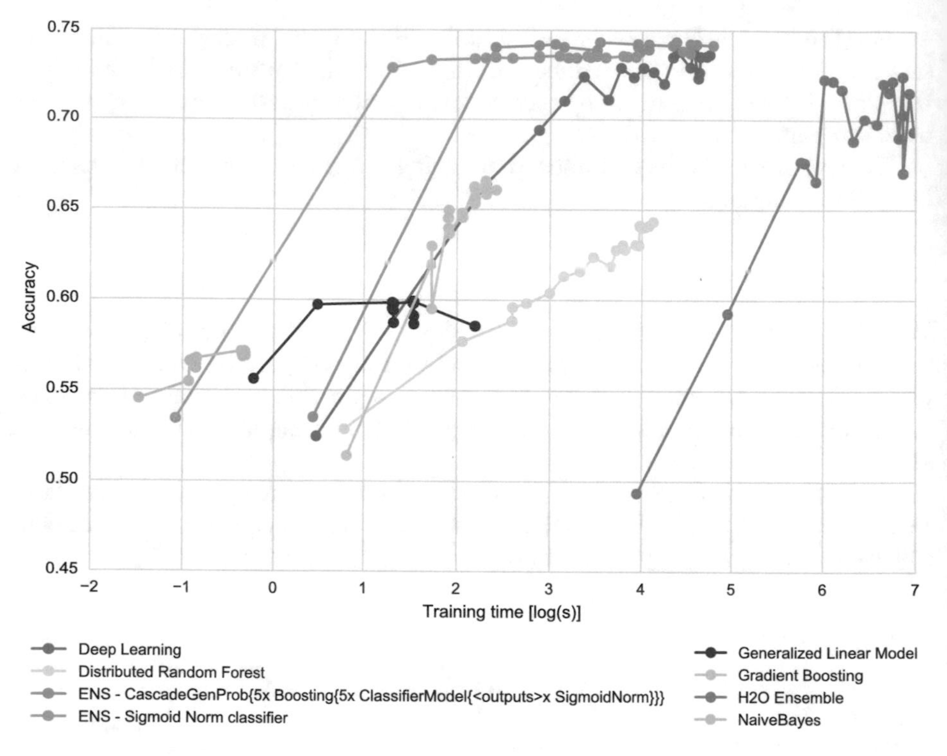

Fig. 3. Predicting IsArrDelayed on Airline data set: comparison of algorithms in H2O.ai trained on subsamples of increasing size.

We benchmark our paralelized templates to models available in H2O.ai implemented using the map reduce approach. **Generalized Linear Model** [44] is using logistic regression to deal with classification problems. **Naive Bayes** classifier assumes Independence of input attributes and classifies based on conditional probabilities obtained from training data. **Deep learning** [38] is a feedforward neural network with various activation functions in neurons. **Distributed Random Forest** and **Gradient Boosted Machine** [45] are ensembles based on decision trees. **H2O Ensemble** is an ensemble classifier called Super Learner by [46].

Following experiments use 1 000 000 randomly selected rows from each data set. Then 50% rows is randomly selected as test set and the rest is then sampled to subsets of growing size to examine scalability of algorithms. This sampled data are randomly split to training set (80%) and validation set (20%).

At first, we examined scalability of algorithms on the Higgs data set. Figure 2 shows learning time and performance of individual algorithms executed on subsets of growing size. The best performance was achieved by Deep Learning which was also reasonably fast. Gradient Boosting is faster, but it does not have capacity to improve with bigger data subsets. Distributed Random Forest is also reasonably accurate and fast, but it is dominated by Deep Learning on Higgs.

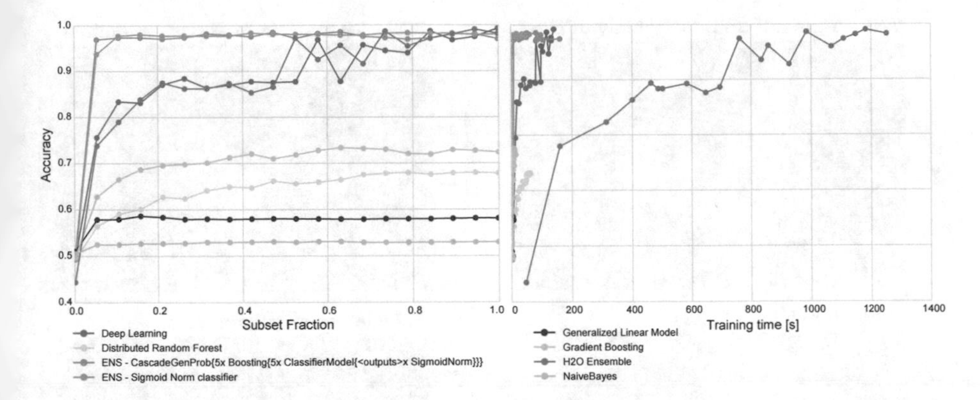

Fig. 4. Predicting IsDepDelayed on Airline data set: comparison of algorithms in H2O.ai trained on subsamples of increasing size.

Ensembles produced from templates are not very competitive on this data set. Only complex hierarchical ensemble of decision trees is approaching the performance of Distributed Random Forest, but it is much slower. Our implementation is not optimized for H2O.ai.

Looking at the Fig. 3, where arrival delay is predicted on the Airlines data set, results are completely different. Our ensembles are both more accurate and faster. The difference is so big, that we decided to analyze these results further.

We even simplified the prediction task by predicting the departure time without removing the DepTime attribute.

The prediction problem becomes then quite trivial, because you can obtain the target (is departure delayed?) by comparing DepTime and CRSDepTime attribute. It is quite surprising that most of the classifiers are mislead by other

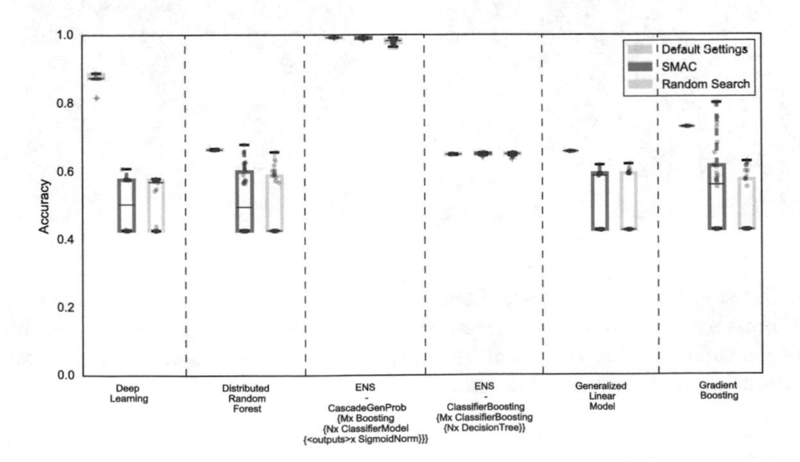

Fig. 5. Optimization of several machine learning algorithms using Random Search and SMAC

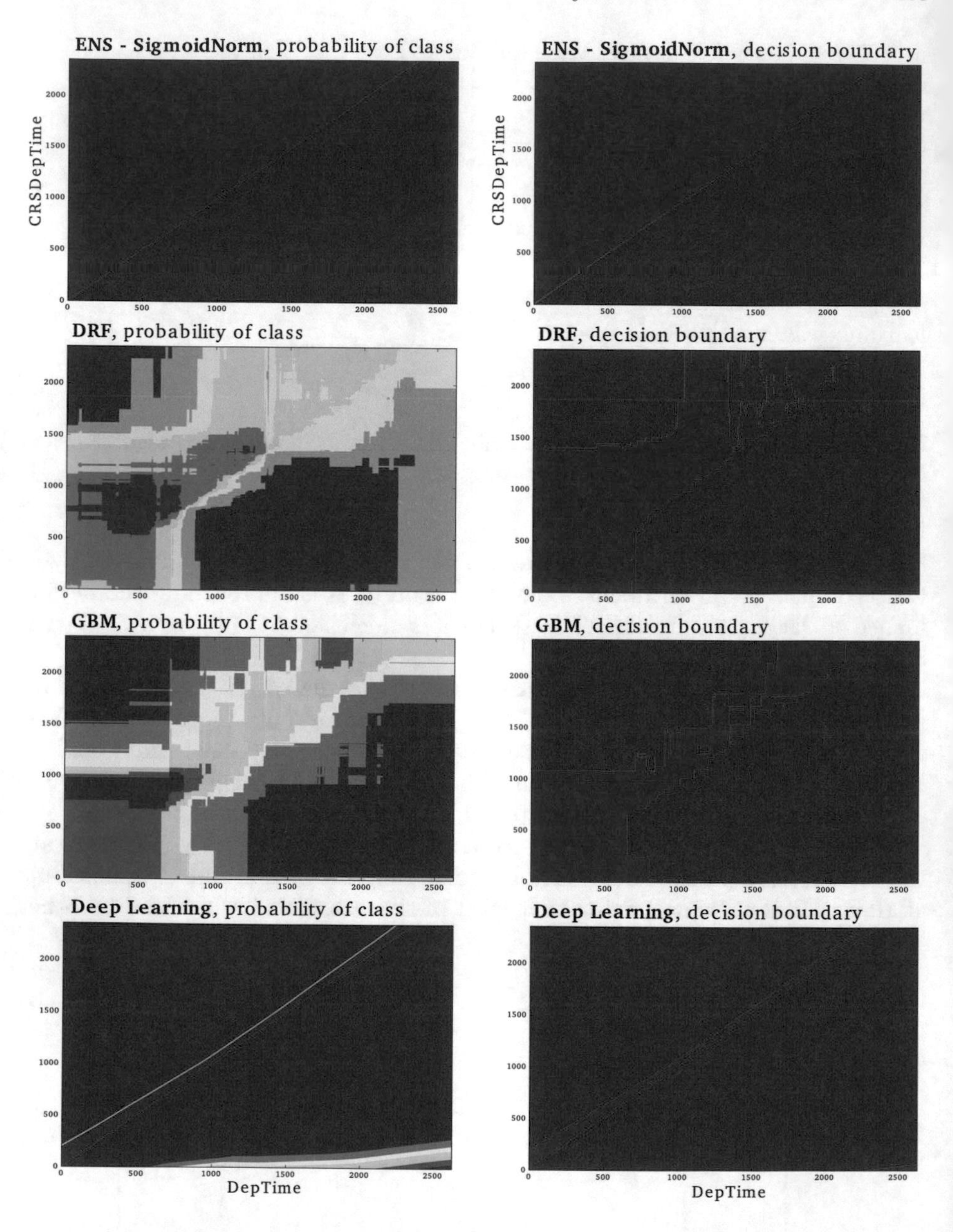

Fig. 6. Decision boundaries of algorithms on problem of predicting aircraft departure delay. Simple ensemble of sigmoid classifiers was able to generalize the relationship well, whereas decision tree based ensembles overfitted the data. Deep Learning discovered the relationship only on large data samples.

attributes and fail to discover this simple relationship. Figure 4 shows that again our simple ensembles based on Sigmoidal model are able to learn fast and solve the problem even on small subsets. H2O Ensemble and Deep Learning discovered the relationship on 500 thousand instances and their learning time was significantly higher.

To ensure that the problem is not caused by improper parameter settings, we run optimization of parameters on a subset of 100 thousand instances. The list of parameters and their ranges are available [47]. Figure 5 shows that most of the H2O algorithms are very sensitive to improper parameter settings. Deep learning was able to converge in default parameter setting only, our assumption is that parameters are controlled adaptively by default. Similarly, negative impact was observed for Generalized Linear Model. For Gradient Boosting and Distributed Random Forest, optimization discovered better performing configuration, however the difference was not significant. We also optimized number of models in our hierarchical ensembles but apparently it had almost no effect on performance. Decision Tree based ensemble was unable to solve the task in any configuration which is consistent with poor performance of DT based ensembles from H2O. On the other hand the Sigmoid based ensemble was able to discover the relationship even with minimal number of models in the ensemble which is consistent with previous experiments. From boxplots and distribution of individual results (red dots) the Bayesian Optimization (SMAC) method outperformed the Random search.

Plots of class probabilities and decision boundaries helped us to reveal the reason of poor performance of decision tree based ensembles. Figure 6 shows that successful classifiers (ensemble of sigmoid models, Deep Learning) were able to identify simple relation of two input attributes to departure delay prediction. The relationship (decision boundary) is hard for decision trees to model with their orthogonal decisions. It is also impossible to solve for Naive Bayes classifier assuming independence of input attributes.

Apparently, we were able to discover very efficient template for this trivial problem. We believe that our approach can contribute to evolve (discover) templates for diverse data sets and predictive tasks. Building library of algorithmic templates can improve capacity of predictive modeling systems to solve diverse tasks efficiently.

6 Conclusions

We show how ensembles of predictive models (meta-learning templates) can be scaled up for large data sets modeling using the map-reduce approach. Benchmarks revealed that our approach is able to produce algorithms competitive with state of the art approaches for large scale predictive modeling. Ensembles of simple regression models can outperform popular algorithms in both generalization ability and scalability as demonstrated on the Airlines data set.

Acknowledgments. This research was partially supported by the Data mining from unstructured data ($SGS16/119/OHK3/1T/18$) grant of the Czech Technical University in Prague.

References

1. Segata, N., Blanzieri, E.: Fast and scalable local kernel machines. J. Mach. Learn. Res. **11**(June), 1883–1926 (2010)
2. Dean, J., Ghemawat, S.: MapReduce: simplified data processing on large clusters. Commun. ACM **51**(1), 107–113 (2008)
3. Kordík, P., Černý, J.: Self-organization of supervised models. In: Jankowski, N., Duch, W., Graczewski, K. (eds.) Meta-learning in Computational Intelligence. Studies in Computational Intelligence, vol. 358, pp. 179–223. Springer, Heidelberg (2011)
4. Sutherland, A., Henery, R., Molina, R., Taylor, C.C., King, R.: StatLog: Comparison of Classification Algorithms on Large Real-World Problems. Springer, Heidelberg (1993)
5. Bensusan, H., Kalousis, A.: Estimating the predictive accuracy of a classifier. In: Proceedings of the 12th European Conference on Machine Learning. Springer (2001)
6. Botia, J.A., Gomez-Skarmeta, A.F., Valdes, M., Padilla, A.: METALA: a meta-learning architecture. In: Proceedings of the International Conference, Seventh Fuzzy Days on Computational Intelligence, Theory and Applications (2001)
7. Thornton, C., Hutter, F., Hoos, H.H., Leyton-Brown, K.: Auto-WEKA: combined selection and hyperparameter optimization of classification algorithms. In: Proceedings of the 19th ACM SIGKDD International Conference on Knowledge Discovery and Data Mining, pp. 847–855 (2013)
8. Salvador, M.M., Budka, M., Gabrys, B.: Automatic composition and optimisation of multicomponent predictive systems. arXiv preprint arXiv:1612.08789 (2016)
9. Salvador, M.M., Budka, M., Gabrys, B.: Towards automatic composition of multicomponent predictive systems. In: International Conference on Hybrid Artificial Intelligence Systems, pp. 27–39. Springer (2016)
10. Salvador, M.M., Budka, M., Gabrys, B.: Adapting multicomponent predictive systems using hybrid adaptation strategies with auto-WEKA in process industry. In: International Conference on Machine Learning. AutoML Workshop (2016)
11. Koza, J.R.: Genetic programming. IEEE Intell. Syst. **14**(4), 135–84 (2000)
12. Li, L., Jamieson, K., DeSalvo, G., Rostamizadeh, A., Talwalkar, A.: Efficient Hyperparameter Optimization and Infinitely Many Armed Bandits. arXiv preprint (2016)
13. Duffy, N., Helmbold, D.: A geometric approach to leveraging weak learners. In: European Conference on Computational Learning Theory, pp. 18–33. Springer (1999)
14. Marquardt, D.W.: An algorithm for least-squares estimation of nonlinear parameters. J. Soc. Ind. Appl. Math. **11**(2), 431–441 (1963)
15. Shanno, D.F.: Conditioning of Quasi-Newton methods for function minimization. Math. Comput. **24**(111), 647–656 (1970)
16. Bičík, V.: Continuous optimization algorithms. Master's thesis, CTU in Prague (2010)

17. Kordík, P., Koutník, J., Drchal, J., Kovářík, O., Čepek, M., Šnorek, M.: Metalearning approach to neural network optimization. Neural Netw. **23**(4), 568–582 (2010). 2010 special issue
18. The fake game environment for the automatic knowledge extraction, February 2011. http://www.sourceforge.net/projects/fakegame
19. Software: Rapid miner, data mining. http://rapid-i.com/
20. Brazdil, P., Giraud-Carrier, C., Soares, C., Vilalta, R.: Metalearning: Applications to Data Mining. Cognitive Technologies. Springer, Heidelberg (2009)
21. Kuncheva, L.: Combining Pattern Classifiers: Methods and Algorithms. John Wiley and Sons, New York (2004)
22. Wolpert, D.H.: Stacked generalization. Neural Netw. **5**, 241–259 (1992)
23. Schapire, R.E.: The strength of weak learnability. Mach. Learn. **5**(2), 197–227 (1990)
24. Woods, K., Kegelmeyer, W., Bowyer, K.: Combination of multiple classifiers using local accuracy estimates. IEEE Trans. Pattern Anal. Mach. Intell. **19**, 405–410 (1997)
25. Holeňa, M., Linke, D., Steinfeldt, N.: Boosted neural networks in evolutionary computation. In: Neural Information Processing. LNCS, vol. 5864, pp. 131–140. Springer, Heidelberg (2009)
26. Brown, G., Wyatt, J., Tino, P.: Managing diversity in regression ensembles. J. Mach. Learn. Res. **6**, 1621–1650 (2006)
27. Brazdil, P., Giraud-Carrier, C., Soares, C., Vilalta, R.: Metalearning, Applications to Data Mining. Cognitive Technologies. Springer, Heidelberg (2009)
28. Breiman, L.: Bagging predictors. Mach. Learn. **24**(2), 123–140 (1996)
29. Gama, J., Brazdil, P.: Cascade generalization. Mach. Learn. **41**(3), 315–343 (2000)
30. Ferri, C., Flach, P., Hernández-Orallo, J.: Delegating classifiers. In: Proceedings of the Twenty-First International Conference on Machine Learning, ICML 2004, p. 37. ACM, New York (2004)
31. Alpaydin, E., Kaynak, C.: Cascading classifiers. Kybernetika **34**, 369–374 (1998)
32. Kaynak, C., Alpaydin, E.: Multistage cascading of multiple classifiers: one man's noise is another man's data. In: Proceedings of the Seventeenth International Conference on Machine Learning, ICML 2000, pp. 455–462. Morgan Kaufmann Publishers Inc., San Francisco (2000)
33. Ortega, J., Koppel, M., Argamon, S.: Arbitrating among competing classifiers using learned referees. Knowl. Inf. Syst. **3**(4), 470–490 (2001)
34. Holland, J.H.: Adaptation in Natural and Artificial Systems: An Introductory Analysis with Applications to Biology, Control, and Artificial Intelligence. U Michigan Press, Ann Arbor (1975)
35. Rosca, J.P.: Analysis of complexity drift in genetic programming. In: Genetic Programming, pp. 286–294 (1997)
36. Borovicka, T., Jirina Jr., M., Kordik, P., Jirina, M.: Selecting representative data sets. In: Advances in Data Mining Knowledge Discovery and Applications. Intech (2012)
37. Basilico, J.D., Munson, M.A., Kolda, T.G., Dixon, K.R., Kegelmeyer, W.P.: Comet: a recipe for learning and using large ensembles on massive data. In: 2011 IEEE 11th International Conference on Data Mining, pp. 41–50. IEEE (2011)
38. Arora, A., Candel, A., Lanford, J., LeDell, E., Parmar, V.: Deep Learning with H2O. H2O.ai, Mountain View (2015)
39. Meng, X., Bradley, J., Yuvaz, B., Sparks, E., Venkataraman, S., Liu, D., Freeman, J., Tsai, D., Amde, M., Owen, S., et al.: MLlib: machine learning in apache spark. JMLR **17**(34), 1–7 (2016)

40. Chu, C., Kim, S.K., Lin, Y.A., Yu, Y., Bradski, G., Ng, A.Y., Olukotun, K.: Map-Reduce for machine learning on multicore. Adv. Neural Inf. Process. Syst. **19**, 281 (2007)
41. van Rijn, J.N., Abdulrahman, S.M., Brazdil, P., Vanschoren, J.: Fast algorithm selection using learning curves. In: International Symposium on Intelligent Data Analysis, pp. 298–309. Springer (2015)
42. H2O.ai: H2O: Scalable Machine Learning (2015)
43. Baldi, P., Sadowski, P., Whiteson, D.: Searching for exotic particles in high-energy physics with deep learning. Nat. Commun. **5** (2014). Article no. 4308
44. Hussami, N., Kraljevic, T., Lanford, J., Nykodym, T., Rao, A., Wang, A.: Generalized linear modeling with H2O (2015)
45. Click, C., Malohlava, M., Candel, A., Roark, H., Parmar, V.: Gradient boosting machine with H2O (2016)
46. LeDell, E.: Scalable super learning. In: Handbook of Big Data, p. 339 (2016)
47. Software: Algorithmic templates for H2O.ai. https://github.com/kordikp

Agent DEVS Simulation of the Evacuation Process from a Commercial Building During a Fire

Andrzej Kułakowski(✉) and Bartosz Rogala

Faculty of Electrical Engineering, Automation Control and Computer Science, Kielce University of Technology, Kielce, Poland
a.kulakowski@tu.kielce.pl

Abstract. The paper describes a simulation application for the process of evacuating people from a building in the event of a fire. This simulation was developed as an agent system using the DEVS formalism. The following sections show the design and implementation of such a system using the Adevs library. A series of simulation tests were then performed for different simulation scenarios.

Keywords: Computer simulation · Agent simulation · Building evacuation · DEVS

1 Introduction

This Simulating the evacuation process of crowded rooms enables to improve human safety when designing new buildings. The ability to study the impact of changes significantly improves the safety of newly designed objects. Simulation of a dangerous situation, i.e. a fire, requiring the evacuation of people from the building allows anticipating its effects and introducing changes to the project at an early stage. In recent years many solutions have been developed that allow performing very advanced and meticulous simulations. In the case of behavioral simulations, the integration with the multi-agent system (M.A.S.) is a good solution. This type of simulation is gaining flexibility because each agent has its own perception, autoname, and the ability to communicate [1]. Each agent has its own state, a set of behaviours and rules that determine the status of the agent in the next step. The concept of agents represents the idea of an autonomous system that perceives the environment and acts in it. The agent has an internal state representing their knowledge and purpose, usually maximizing its usability [2–4]. Thanks to the independence of individuals, multi-agent systems are most commonly used in social and behavioral sciences, but also in such disciplines as epidemiology and ecology [2].

Examples of existing evacuation simulation solutions are described in [5–7].

2 Specifying the Simulation Environment

Creating each simulation requires specifying the requirements and assumptions of the project. In the case of multi-agent simulations, apart from specifying the simulation stage

N. Shakhovska and V. Stepashko (eds.), *Advances in Intelligent Systems and Computing II*, Advances in Intelligent Systems and Computing 689,
https://doi.org/10.1007/978-3-319-70581-1_19

itself, i.e. the corresponding model, tools and parameters, consideration should also be given to creating the environment itself.

Designing an evacuation system application requires the use of map representation methods and path search algorithms. Mapping motion of an individual in a computer program requires the implementation of one of the map representation methods. The most commonly used solution is to create a navigation mesh that maps the area where one can move. To designate a path on such a map, it is necessary to create a hybrid forming a grid of triangles on the basis of its own shape [8]. To create a grid of triangles, representing the map discussed, Delaunay triangulation has been used [9].

Agents of multi-agent systems need to adapt to the changing environment they are in. In the case of human motion, this includes, eg. the observation of other agents, avoiding collision and searching destination. Representations of the map serve as a basis for finding a path while selecting an algorithm.

The main direct search strategies are: uniform search cost and heuristic search. The A* algorithm combines both approaches. This allows A* to find the optimal path while being much more efficient than other algorithms [8].

3 Tools Used

Creating any software development project requires the use of appropriate tools. For the agent simulation of the evacuation process, the Adevs library was selected as the basis for the simulation [12]. This library allows programming the behavior and relationships between agents as well as the entire simulation. Because Adevs is written in C++, the whole project has also been developed in this programming language. Creating a simulation environment and its visualization requires the use of tools for programming 2D graphics and graphical user interfaces. In this case, the Qt library was selected to implement all required elements.

Delaunay triangulation was used to obtain the Triangle [10] library. It allows creating several types of high quality mesh using different Delaunay and Voronoi triangulation algorithms. The library uses its own files describing the polygons and the result files with node points generated by the grid and the triangle neighbours. An important feature of the program is the ability to handle "holes" in the polygons because they represent obstacles on the map. In addition, the increase in the number of generated triangles is transferred to the quality of the path being created [10].

The final tool needed to create a simulation environment is the simulation library itself. In this case, the Adevs library was selected. Adevs (A Discrete EVent System Simulator) is a C++ library for modelling and simulation of DEVS models [11–13]. The library supports the basic model DEVS, Parallel DEVS and Dynamic Structure DEVS (DSDEVS). Adevs is a free library described by [11] as the fastest among the available ones. This is very important for large multi-model projects. Developing discrete simulations in Adevs involves combining atomic models into more complex structures [12].

The purpose of the formalism was to enable the construction of discrete models controlled by events in a hierarchical, modular system. DEVS This is a formal specification for a system scheme that enables modelling and analysis of discrete systems that can be described as transient status tables [13].

4 Program Algorithm

Creating the application allows carrying out the entire simulation process, from creating the whole environment, through simulation to visualizing the results. The application design assumes that the whole process can be carried out in a few steps: creating a simulation environment, drawing navigation nets for each floor of a building: determining emergency exits and obstacles, connecting emergency exits between floors, creating an area of fire, then adding agents to the floor. The next step is setting and performing the simulation parameters, then analyzing the results of the simulation and its automatic visualization or manual visualization of each step.

4.1 Decision Agent Model

The agent model analyzes the environment in which it is located and communicates with other agents to make a decision to change their own status. The current agent state is determined by the decision tree. Agents have three basic parameters: knowledge of the environment, degree of courage and speed with which they move. The courage and knowledge parameters are used in taking the decision, additionally the environment and other agents are taken into account. When the decision is made, the agent performs the selected motion model [14, 15].

The agent determines the motion vector to the next point on the found route or to the agent they follow. The next steps of the agent along the route are calculated by adding motion vectors to the current position. If the next step is not possible, the agent checks whether the points shifted by 45- and 90-degrees on both sides are possible to occupy, if not, they remain stationary. Otherwise, the agent occupies the desired place and updates its route.

The simulation uses the A* algorithm. The triangular grid found in this way will not be optimal and does not correspond to the actual path travelled by the person possibly moving in straight line. In addition, smoothening of the found route needs to be done, including collisions with fire and obstacles.

Simulation of real fire is one of the most complex algorithmic problems. Since the paper focuses on the evacuation process in multi-agent simulation, a simpler solution is used. Fire is represented by an ellipse that grows with time. The simulation foresees the fire expansion, blocking subsequent triangles of the grid, thereby eliminating them from locations available to agents. In addition, on the way agents check whether the next steps do not lead them towards the fire, if so, they update the route.

4.2 Simulation Process

The simulation algorithm is based on the diagram model available as part of the Adevs library [12]. The simulation assumption is independent mapping of the building floors

(Fig. 1). Every floor has a list of agents who are currently on it. The agent can move from one floor to another using the additional Atomic teleport class. The next steps in the simulation are managed by the clock updating time, synchronizing the simulation on the floors, and sending a command to calculate the next step in the simulation.

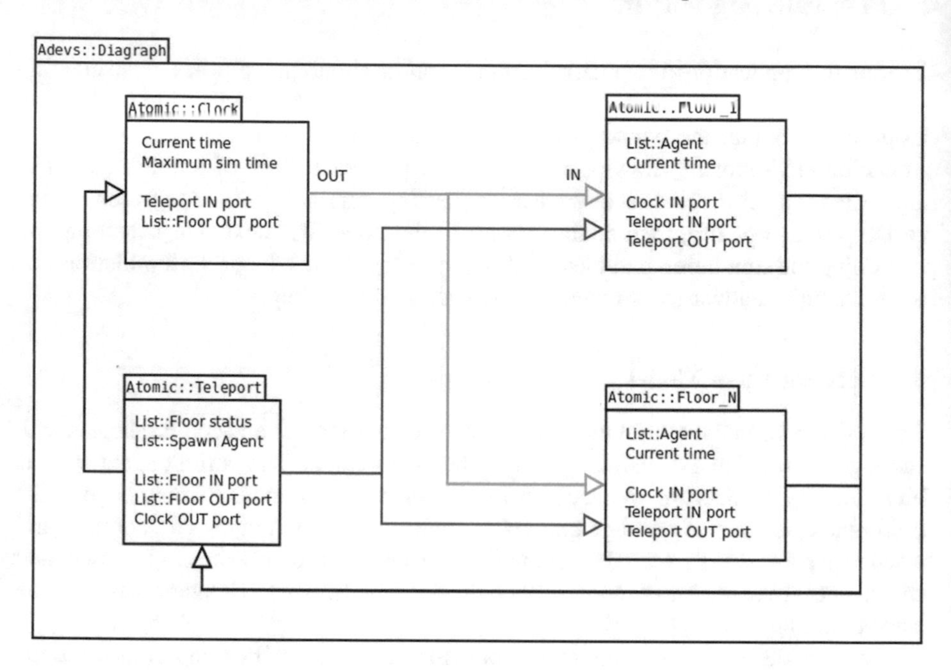

Fig. 1. Diagram of simulation model created with the use of DEVS formalism in the Adevs library [14, 15].

The next simulation steps are initiated by the clock, sending the current time to all floors. The floors perform the movements of all agents, checking whether the agent does not leave the floor. In this case, the agent is sent by teleport to another floor if there is enough space on the map. On the last stage of the step, the teleport checks the status of each floor and then sends the information to the clock, continuing the loop or ending the whole process.

5 Application Implementation

5.1 Creating the Environment Project

Each floor of the project is based on a plan loaded from the graphic file. Ready-made design of the floor taking into account people, obstacles, emergency exits and fires (Fig. 2).

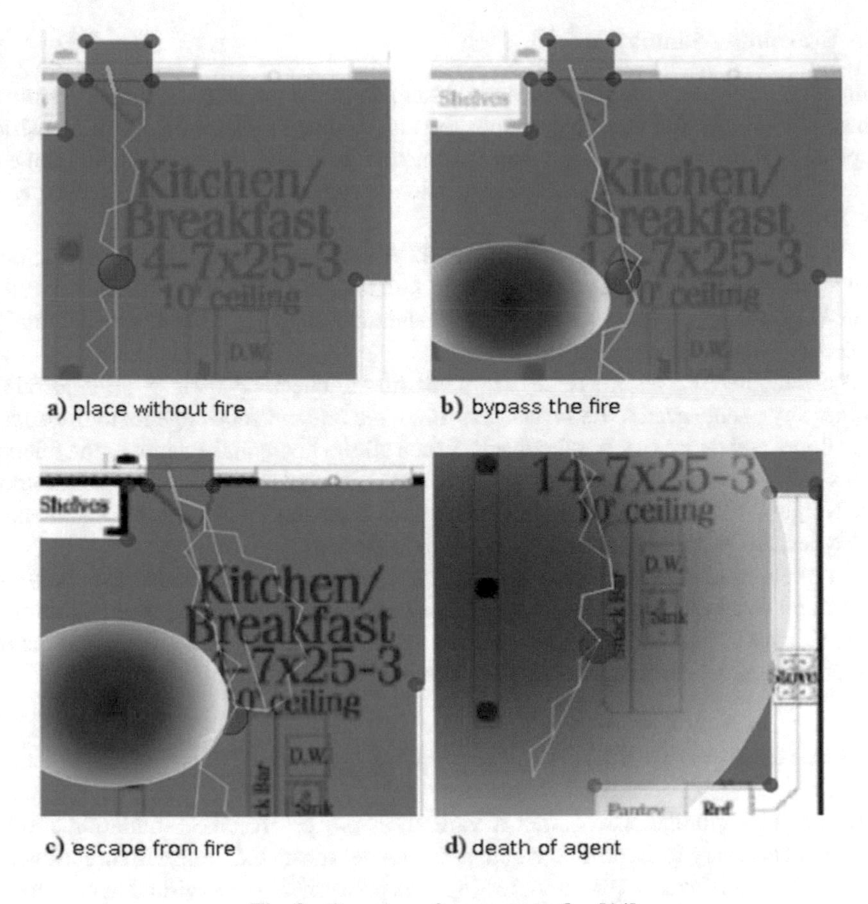

Fig. 2. Reaction of an agent to fire [14].

The graphic plan serves as a guide for creating the navigation grid. It consists of three types of polygons that define the surface type: blue for floor, red for emergency exits, and gray for obstacles. Polygons can be freely joined and modified by adding or subtracting vertices. The program will automatically create the missing polygons of the obstacle and will generate a new grid.

Then, on the ready map there are placed agents with the specified parameters and an ellipse representing the fire. Once all the floors are complete, the appropriate passages must be connected.

The last step is generating a triangle grid and checking the connections. The mesh compression can be controlled by setting the maximum area of the triangles. Triangulation is performed by Triangle external program and its results are loaded into the program.

5.2 Evacuation Simulation

Simulation takes two parameters: the maximum time and the time defining the simulation step. The simulation step determines the unit of time by which the agent's position is updated. The simulation ends when the maximum time is reached or when none of the agents is able to move. The implementation in Adevs maps the model presented in Fig. 1.

The simulation starts by sending an UP_TIME message containing the current time to all floors by the clock. Each of them triggers all their agents to complete the next step, defined by their status. The floor examines the number of agents moving and the number of the ones wanting to move to another floor.

The teleport receives messages from all floors, checking their type, determines whether any agents need to be forwarded. If so, the teleport sends agents to the appropriate floors and then expects a feedback. When all the floors make a move, the teleport checks their status and sends the information about the next step to the clock. If all floors have NO_MOVE responses, the simulation ends prematurely because all agents have already left the building or are no longer able to move.

Each subsequent simulation step calculates the point at which the agent should be located and writes them to the list. These values are used to recreate the positions of agents at a given time and visualize their movements. The program allows playing the simulation automatically or manually by moving the time slider.

6 Testing the Simulation

Each computer simulation requires a validation process for the simulation model. Testing is necessary to determine whether the model meets the design assumptions. In the case of the analyzed application, testing was performed on individual agent motion systems. Three simulation scenarios are then presented to validate the appropriacy and suitability of the applications in real-life situations [14].

6.1 Motion Systems

The agents' motion required the implementation of collision and analysis of the surrounding systems to represent reality most accurately. The first is to bypass the fire while mapping out the agent's route, escaping from the fire when it is close, and dying in the event of a rapidly expanding flame (Fig. 2).

The other systems are the search for the leader (a trained person), the closest visible exit, and the collision check. The program checks collisions with walls and other obstacles to determine what the agent can see. When an agent does not know the distribution of rooms, but can see another agent possessing the knowledge, he follows him. If an agent detects a collision possibility during a movement, it bypasses the obstacle. Of course, it avoids colliding with another agent as well (Fig. 3).

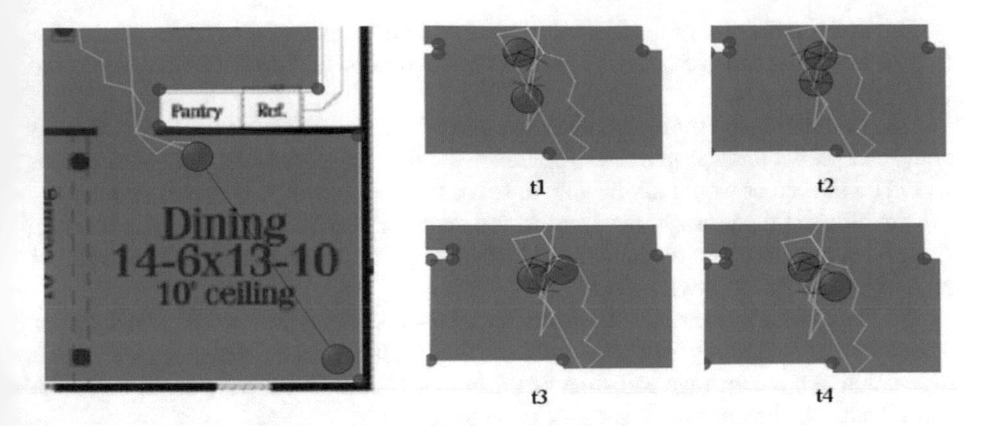

Fig. 3. Agent mobility: following the leader (on the left) and avoiding collisions between agents (on the right) [14].

6.2 Simulation Scenarios

The first scenario examines the effect of the exit system in a commercial building during the spread of a fire (Fig. 4). Three simulations have been carried out with the following assumptions: a two-speed fire spreading in a restaurant building is being tested; the setting and all agent parameters are the same for each simulation.

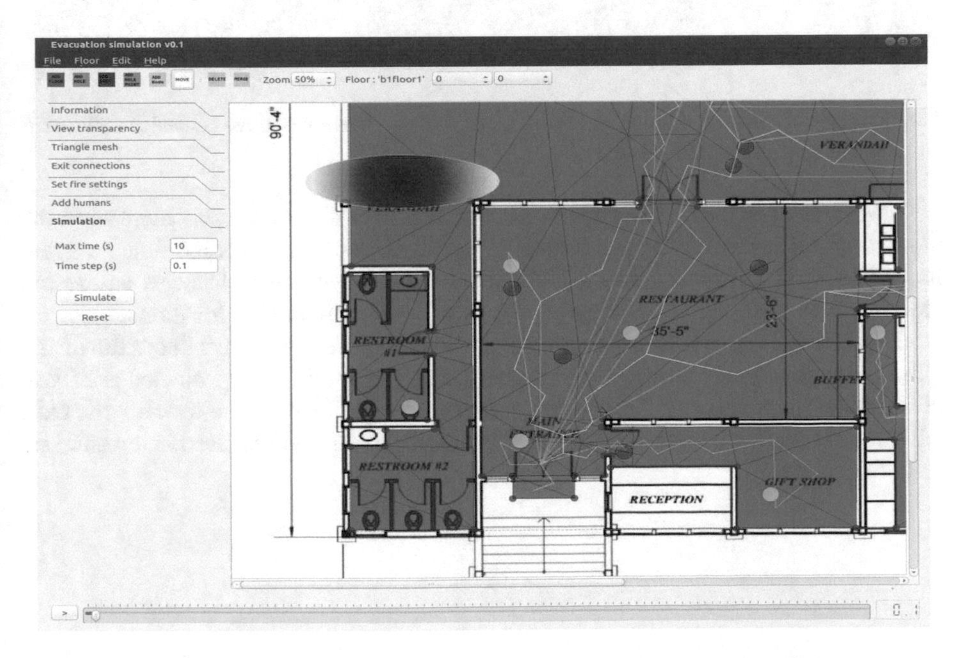

Fig. 4. Ready-made application during the evacuation simulation of a commercial building [14].

Runs of simulations: fire spreads at a speed of 1 cm/ s, the building has two exits; the fire spreads at a speed of 5 cm/s, the building has two exits; the fire spreads at 5 cm/s, the building has four exits.

The first simulation showed that when the fire develops slowly, potentially everyone could escape, but three people had to approach the fire very closely. In fact, these people would be injured or would not be able to leave the building. In the second simulation, the fire blocked the passage, resulting in the death of four people in the 11th second. Additional door in the third simulation significantly shortened the evacuation time and provided a safe route for all agents.

The second scenario analyzes the simulations of restaurant evacuation with different quantities and locations of obstacles (tables) (Fig. 5). It checks the time needed for total evacuation and the average exit time of thirty one agents. There were six simulations carried out. The fire spreading speed was set to 0.1 cm/s.

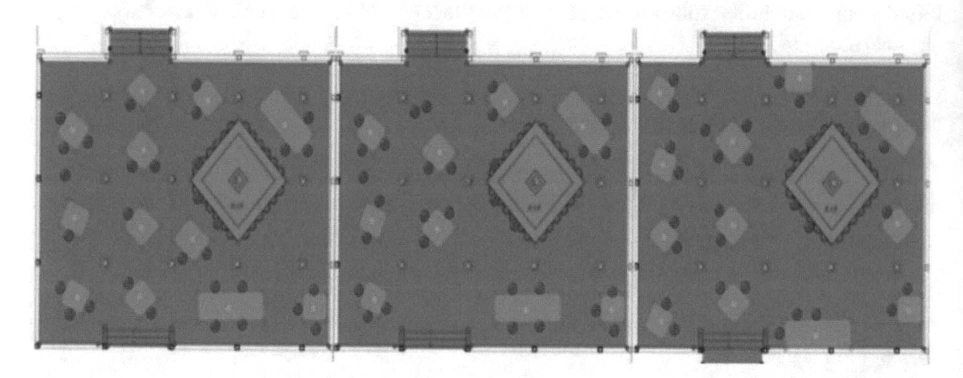

Fig. 5. Layout of the simulation room, left: initial setting, deletion of three obstacles at the exits and modification of the setting [14].

The greatest impact on the speed of evacuation of all agents was increasing the number of exits. By rearranging obstacles, you can reduce the average time an agent needed to leave the room. Another solution is to remove some obstacles e.g. in the vicinity of the exits, thus providing greater flow capacity in critical locations.

In public buildings like banks or offices, the clients do not always know the distribution of rooms. The last simulation checks the influence of agents' knowledge in case of changing the number of "employees" whom the agent can follow towards the exit. All other simulation parameters are controlled. Office plans will be used in the analysis (Fig. 6).

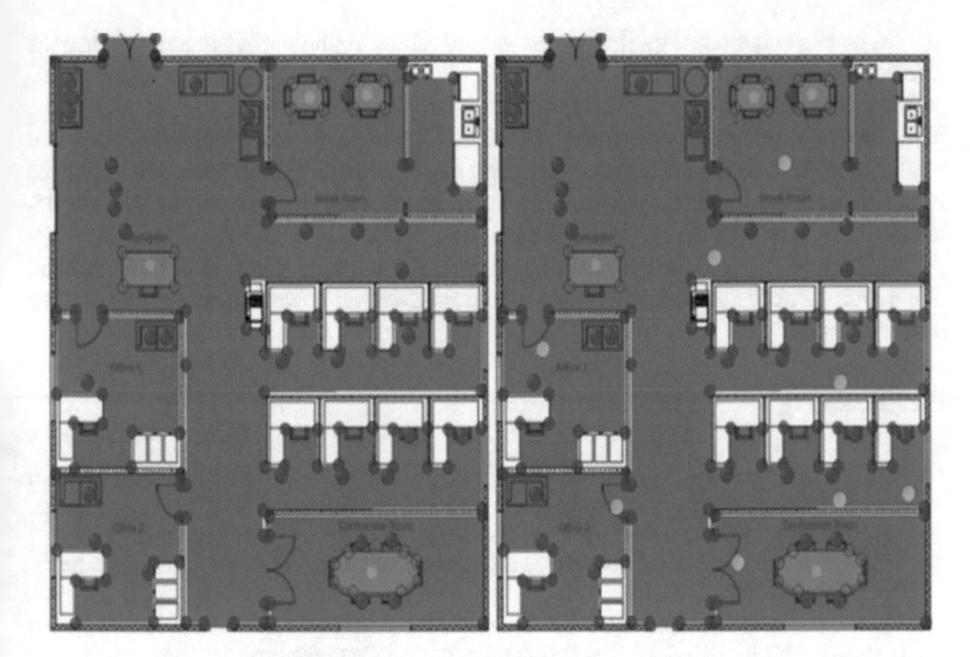

Fig. 6. Room plan with clients only (pink) and "officials" (blue ones) [14].

The simulation shows that the agents who are unfamiliar with the layout of the premises are not able to find the way out themselves. Some panic or wander aimlessly looking for the way. Increasing the number of agents who know the way and whom the rest can follow, affects the number of people rescued and the time of simulation. In the third simulation, it can be observed that the time of the last evacuation and average time are higher than before. This is due to the following and later loss of visual contact with the "employee". This results in an autonomous search for the way, which takes more time. With the large number of agents who know the way and their proper location, the clients follow them, leaving the building much faster.

The above simulations show how important building design is and how computer simulation is helpful in this process. Simulation results analysis allows you to design buildings in a faster, cheaper, and most importantly more secure way. In addition, the simulation data helps in the event of an actual evacuation, allowing to predict the potential number of people injured.

7 Summary

The paper presents selected theoretical aspects of computer simulation applied to human evacuation. An algorithm and outline of application implementation is then presented to create and simulate the evacuation process in the event of fire. The tools used allow you to create an entire simulation environment and modify it while conducting further tests. Meeting the evacuation system requirements and testing it, shows the suitability of such solutions in the real world.

Several evacuation scenarios are presented, showing how environmental changes affect the evacuation process. It has been shown that while designing a building anticipating potential fire sources and appropriate matching number of exits and evacuation paths have a large impact on security. Such modifications not only secured the survival of all agents, but also reduced the time needed to leave the building. Even minor changes to the obstacle setting or the addition of trained agents (fire wardens) allow for a significant improvement in the speed of evacuation.

The present form of the application does not allow for the creation of fully realistic simulations, but the presented solution of the multi-agent system allows for its further development.

References

1. Klügl, F., Klubertanz, G., Rindsfüser, G.: Agent-based pedestrian simulation of train evacuation integrating environmental data. In: KI 2009: Advances in Artificial Intelligenc, Lecture Notes in Computer Science, vol. 5803, pp. 631–638. Springer, Heidelberg (2009)
2. Uhrmacher, A.M., Weyns, D.: Multi-Agent Systems: Simulation & Applications. CRC Press, Boca Raton (2009)
3. Alonso, E., Karcanias, N., Hessami, A.G.: Multi-agent systems: a new paradigm for systems of systems. In: ICONS 2013: The Eighth International Conference on Systems
4. Weiss, G.: Multiagent Systems: A Modern Approach to Distributed Artificial Intelligence. The MIT Press, Cambridge (2000)
5. Van Schyndel, M.: Evacuation for a building with multiple levels. In: SYSC 5104, Carleton University (2011)
6. Smith, J.L.: Agent-based simulation of human movements during emergency evacuations of facilities, 30 September 2015. https://www.wbdg.org/pdfs/agent_based_sim_paper.pdf
7. Pathfinder Thunderhead Technical Reference, 30 Septemeber 2015 (2015). http://www.thunderheadeng.com/downloads/pathfinder/tech_ref.pdf
8. Cui, X., Shi, H.: Direction oriented pathfinding in video games. Int. J. Artif. Intell. Appl. (IJAIA) **2**, 1 (2011)
9. Weatherill, N.P.: Delaunay triangulation in computational fluid dynamics. Comput. Math Appl. **24**, 129–150 (1992)
10. Triangle, 10 June 2015 (2015). https://www.cs.cmu.edu/~quake/triangle.html
11. Van Tendeloo, Y., Vangheluwe, H.: The modular architecture of the Python(P)DEVS simulation kernel work in progress paper. In: Wainer, G.A. (ed.) DEVS: Proceedings of the Symposium on Theory of M&S, SCS International, pp. 387–392 (2013
12. Nutaro, J.: Adevs Manual, A Discrete EVent system Simulator, 8 June 2015 (2014). http://web.ornl.gov/~1qn/adevs/index.html
13. Song, H.: Infrastructure for DEVS modelling and experimentation. Master thesis draft. McGill'a Univesity, Montréal, Canada (2006). Supervisor: prof. Hans Vangheluwe
14. Rogala, B.: Project and application for building evacuation agent simulation process during fire. Master thesis, Kielce, Kielce University of Technology, Poland (2015). Supervisor: dr.eng. A. Kułakowski
15. Kułakowski, A., Rogala, B.: Agent simulation of the evacuation process from a building during a fire. In: CSIT 2017, Lviv, Ukraine (2017)

The Alpha-Procedure as an Inductive Approach to Pattern Recognition and Its Connection with Lorentz Transformation

Tatjana Lange(✉)

University of Applied Sciences, 06217 Merseburg, Germany
tanja.28.lange@gmail.com

Abstract. The paper deals with problems which appear when solving the task of pattern recognition in a feature space that is not identical with the space of the unknown decisive key features. In a common sense it deals with the correctness of solutions which are found in different coordinate systems. Even more, the opportunity of constructing a feature space for the final separation of classes by selecting the features in pairs, how it is done by the Alpha-procedure, will be investigated. Thereby, the problem of stability during the solving of pattern recognition tasks will be considered from the point of view of transformation groups. The possibility of avoiding the necessity of regularization by using the geometric equiaffine Lorentz transformation will be shown, exploiting for that aim as example the alpha-procedure.

Keywords: Pattern recognition · Alpha-procedure · Stability of reverse tasks Kernel transform · Lorentz transformation

1 Introduction

The pattern recognition and classification methods using the supervisor approach developed strongly at the beginning of the sixties [1–10] and they have got a second push with the development of computer technology [11–19]. The automatic methods of pattern recognition and classification are also often called Machine Learning.

Nowadays, pattern recognition methods are used in nearly all areas of social and industrial life. They are not only applied to automatic reading and picture recognition, voice recognition and translation, e.g. in automotive applications. We can find them in banking (e.g. fraud recognition), in agriculture and ecology (e.g. influence of different factors on growth), in medicine (e.g. diagnostics or survival analyses), in molecular biology (e.g. prediction of cellular location sites of protein), in pharmacology (e.g. efficiency of medicine including placebo) or in manufacturing (e.g. material synthesis or technology identification). An exceptional topic is there current and future use in criminology and early identification of potential terrorists (e.g. with the help of the so-called dragnet investigation).

The task of classification and machine learning can be formulated in a most common way as follows: In a training phase, a limited number of objects and their affiliation to a certain pattern or class (e.g. out of 2 classes) is given. An amount of

N. Shakhovska and V. Stepashko (eds.), *Advances in Intelligent Systems and Computing II*, Advances in Intelligent Systems and Computing 689,
https://doi.org/10.1007/978-3-319-70581-1_20

features is proposed which can be owned by **any of the objects of the two classes**. The task is to find such a "decision rule", that allows a routine classification without repeated thinking of completely new objects, which did not belong to the initial "training sample" that was used during the training (or learning) phase.

For automatic computer-based recognition it is natural and comfortable to represent such a task in a geometric manner where the objects together with the values of their features are vectors in an n-dimensional vector space of the n proposed features.

Together with it there are two approaches known the classical deductive approach [2, 8, 10] and the more common inductive approach [3–5, 11].

In case of the *deductive approach* the separating hyperplane is searched in the **complete** feature space or even in an extended space of all original features and additional extended features (combinations of the original "elementary" features). Such an extended space is called rectifying feature space. Together with it the dimension of the space remains *unchanged during the whole search process*.

In contrast to the deductive approach the *inductive methods* search an appropriate **subspace** of key features which separate the classes in an optimal way. These key features are selected either from the original feature space or from the rectifying space.

The inductive methods of recognition as well as of identification and modelling [24–30], which are mathematically very similar sciences, are sometimes also called selective methods depending on the kind of bottom-up or top-down selection. Both kinds search the solution in reduced spaces or rather subspaces.

The inductive recognition method that is called Alpha-procedure [11] executes a bottom-up selection of the feature space.

But there is one problem that needs to be considered: When we select the feature space the solution of the task may be instable in the sense of Hadamard [34]. For pattern recognition instability means that the rule separating the classes faultlessly during the training phase can make a lot of mistakes for new objects with new data. Thus, for the reduction of the feature space we need a scheme for the correct comparison of features by their separating (or recognizing) ability. Such a comparison is normally performed stepwise by pairs or a by a small number of features. Consequently we have to deal with **different** feature subspaces on the different selection stages, both in sense of the dimensions and of the semantics of the features.

Therefore, if using an inductive selection of features serious problems with respect to convergence/ asymptotic can appear.

But concerning the Alpha-procedure as well as the inductive methods at all the question appears: Why do these methods work stable and well during the exploitation phase, i.e. during the **phase of classifying new** objects, despite of the open issue concerning the justifiability of the comparison of solutions in **sub**spaces of the original features during the training phase?

The aim of the current paper is the finding of an answer to the question above showing the connection between the inductive construction of the separating plane and the special Lorentz transformation.

Thereby, the paper is structured as follows:

- Section 2 describes the classification task in a common form and focuses thereby on the semantics of the notion key feature as an important element of the

Alpha-procedure which differentiates the Alpha-procedure from other methods and which gives a geometric description of the solution independently of the search method.

- Section 3 briefly considers the stability of the solution of a mathematical task from the point of view of the theory of group transformations and their invariants.
- Section 4 describes the algorithm of the Alpha-procedure from a geometric point of view. It also gives all what is needed for the understanding of the nature of the Lorentz transformation und for its appropriate comparison with the algorithm of the Alpha-procedure, again in a geometric manner.
- Section 5 pointers to sources where numeric examples are published which were executed by a team around the author of this paper. It also provides a link to a public R-package every one can use to test the Alpha-procedure with own data.
- Finally, Sect. 6 consists of some conclusions and a short outlook.

2 Common Characterization of the Pattern Recognition Task

The task of pattern recognition with the help of measured or observed data can be described as follows:

Normally, we start with given (measured) data for different objects and their features that are assigned to different classes by a "trainer" or "supervisor". We call this set of classified data the "training set". The Table 1 gives shows an example consisting of measured data x_{ij} of the features $\mathbf{y}_j$ characterizing the objects $\mathbf{x}_i$, and the supervisor's assignments of the objects to the classes A or B.

Table 1. Example of a "training set".

Objects $\mathbf{x}_i$	Features $\mathbf{y}_j$					Assignments by supervisor
	$\mathbf{y}_1$	$\mathbf{y}_2$	...	...	$\mathbf{y}_n$	
$\mathbf{x}_1$	x_{11}	x_{12}	...	...	x_{1n}	A
$\mathbf{x}_2$	x_{21}	x_{22}	...	...	x_{21}	B
....	...	...	...	...	...	...
....	...	...	...	...	...	...
$\mathbf{x}_{p-1}$	...	...	...	...	...	B
$\mathbf{x}_p$	x_{p1}	...	...	...	x_{pn}	A

Note: Each row i corresponding to the object No. i can be considered as vector $\mathbf{x}_i$ in the n-dimensional feature space. The x_{ij} are the elements of this vector.

The aim of the pattern recognition is to find such a stable rule in form of a hyperplane (Fig. 1) for the separation of objects using as basis just these training data and the statements of the supervisor concerning the affiliation of the objects to the classes A or B. This should be done in a way that in case of appearing new objects with

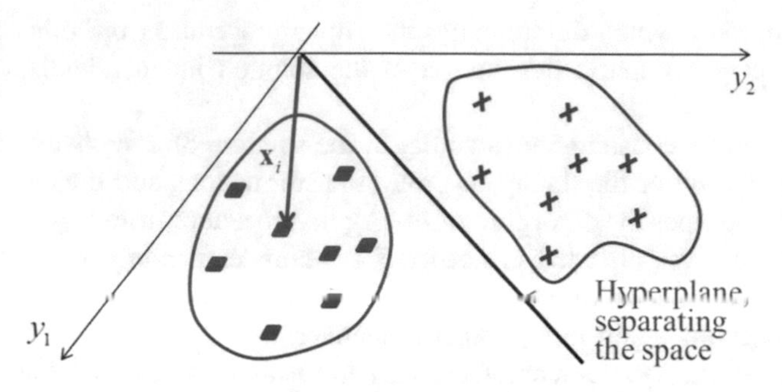

Fig. 1. Object separation.

their own measured values of the features one can automatically define with the help of that rule to which class the new objects belong.

But the objects $X(x_1, x_2, \ldots, x_p)$ can be located in the feature space in a way that they can be separated in the **given** space of features $Y(y_1, y_2, \ldots, y_n)$ only with the help of a complicated curve (Fig. 2) or of a fractured line.

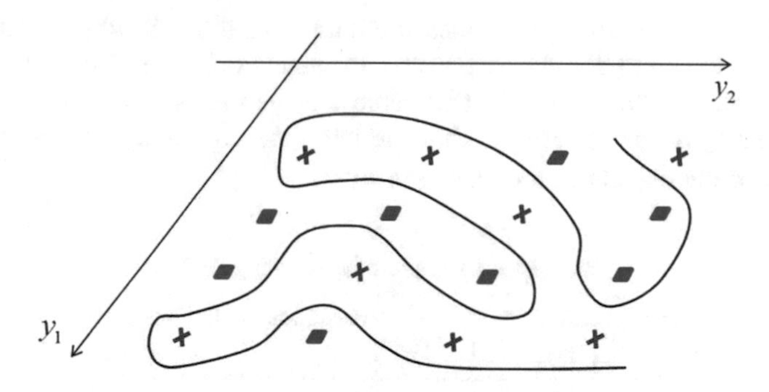

Fig. 2. Complicated separation of objects.

The reason for that problem can be, for example, one of the following:

1. A feature that is very important for the separation and which we will call key feature has not been recorded with the original data. But there may be some other features which are somehow mathematically connected with the missing key feature.
2. All key features are available in the original data table but together with that there are other features recorded, which are not only useless but they are even more disturbing the recognition procedure. In some applications, for example in case of criminological dragnet investigations such harmful features may be even invaded intentionally.
3. There also may be a combination of the first two reasons.

Although the term key feature is self explanatory, the understanding of this term should be described in a bit more detail. In Fig. 2 a separation of classes is shown that was performed without the participation of the key feature y_3. In Fig. 3 the same classes are shown now clearly separated with the help of the y_3. In Fig. 3 the key feature y_3 is demonstrated in a "clean" way as a feature that is independent of the features y_1 and y_2 and the separating plane is parallel to the plane built by y_1 and y_2. The separating rule corresponding to Fig. 2 has a random character which reflects the randomness of the choice of the training data set but not the dependence on the interrelation law of the classes.

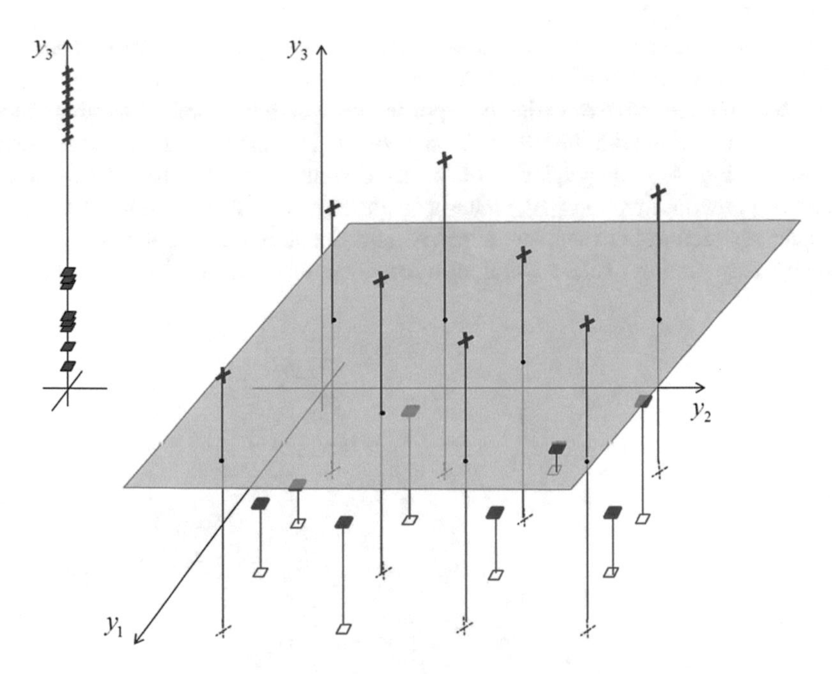

Fig. 3. Separation of classes by key feature.

In case of Alpha-procedure it is assumed that the interrelation law of classes can only be represented by a modest ensemble of key features. Thereby, the Alpha-procedure automatically searches for this ensemble of key features.

For the dealings with the first reason a so-called *rectifying space with rectifying features* is used. These rectifying features are combinations of the original features in form of simple functions, for example $\mathbf{y}_k = \mathbf{x}_1 \cdot \mathbf{x}_5^2$. Together with the original features they will be expressed in a new extended data table where they appear as new columns together with the original ones.

The use of the rectifying space can be interpreted as recognition by indirect measurements.

In the second case only a reduction of the original feature space, best to a space of only key features, makes sense. Only if there is an absolute need new rectifying features, which are created from some remaining features, are added to the reduced space.

The problem is that we do not know in advance with which of the cases we have to deal.

But normally the kind of application helps us a little with the indication. For example, in case of automatic reading the second reason described above will hardly appear. Thus, the majority of pattern recognition methods uses the rectifying space. With it, the search of the separating hyperplane is performed in a constant feature space that has been defined only once before constructing the separating hyperplane. That means the number and the semantics of the features during the optimization remain unchanged (e.g. [10]).

Now, let us consider in a more detailed way the geometric appearance of the solution of pattern recognition tasks in a common form.

For both the inductive or deductive approaches which are used for solving the tasks of pattern recognition with the help of a "supervisor", the final result can be represented as shown in Fig. 4 – independently of the used solution method (e.g. SVM [10, 12], method of potential functions [8], alpha-procedure [8, 11–16, 31], DD-alpha [18] etc.). The main distinction between the inductive and deductive approaches is lying in the different interpretation of the separating hyperplane $D(y)$ and its directing vector Ψ.

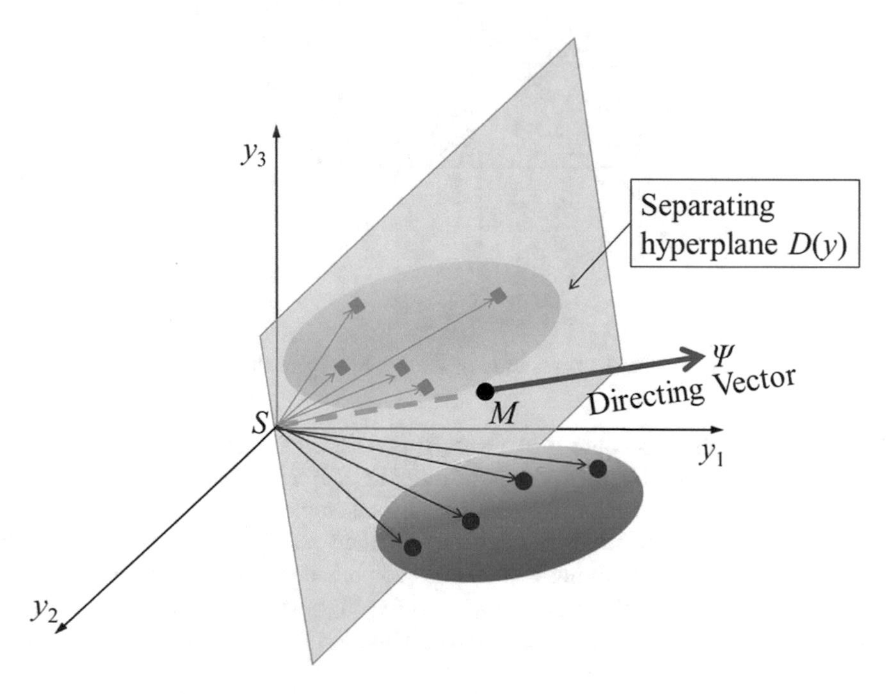

Fig. 4. Geometric depiction of the separation of classes.

In case of deductive approach the feature space corresponds to the complete set of measured object features.

In the contrast, a reduced set of features is characteristic for the inductive approach. This reduced feature set is selected in a certain way from the original (complete) amount of features offered by the "customer" in form of measured data.

This reflects the differences between the possible mathematical approaches for solving the tasks of pattern recognition as well as the different practical application fields and it is just the subject of the current paper.

Ψ (shown in Fig. 4) is an n-dimensional vector within the pattern space that is called directing vector and which is perpendicular to the separating hyperplane that separates the class A from class B.

Class A and class B represent different objects (points) within the feature space: any of the p objects (1, 2, …, k, …, p) is defined by n coordinates of the Euclidean space, i.e. the measured values of the corresponding features (1, 2, …, i, …, n).

With other words, the objects are points, i.e. vectors $x_1, \ldots, x_k, \ldots, x_p$ within the space Y, where p equals the number of objects of the training set.

During an initial training phase a "supervisor" defines to which of the classes each of the objects belongs.

The abscissae of the space Y, $y_1, y_2, \ldots, y_i, \ldots, y_n$, represent the quantities which characterize the objects to be recognized. They are called features or features. They can be colors, weight etc. That means they can be expressed quantitatively. But they can also be function of the characteristic quantities $\varphi_{j(y_i)}$.

That means, each point (object) x_k is represented by the projections y_{ki} of an n-dimensional vector in the pattern space.

After the separation hyperplane $D(y)$ and the directing vector Ψ are found, the points $x_1, \ldots, x_k, \ldots, x_p$ can be projected onto the straight line Ψ (see Fig. 5). This can be interpreted as transformation of the points from the form of expression in Fig. 4 into the form, shown in Fig. 5. We will name this kind of transformation R-transformation (or Recognition transformation).

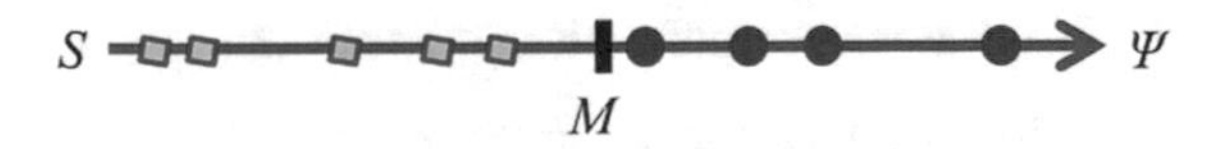

Fig. 5. Projection of objects onto the directing vector Ψ.

Applying one of the recognition methods and getting the separating hyperplane that is represented by its directing vector Ψ we will afterwards use this hyperplane for the separation of new objects, but now without the "supervisor".

The task of pattern recognition can be mathematically reworded in the way that the hyperplane $D(y)$ can be interpreted as divisor that divides the feature space into two parts. This space divisor will be then represented by the following equation:

$$F(y, a) = \sum_{i=1}^{n} a_i \varphi_i(y) = 0 \tag{1}$$

This can correspond, for example, to an n-dimensional regression plane in the n-dimensional rectifying space.

The search of such a function $F(y, a)$ is often called *reconstruction of generating function* or *reconstruction of regularity*.

Let us stick to the example of regression plane. Both tasks, the **pattern recognition** and the **reconstruction of generating function**, can be solved either in a "deductive" or in an inductive way, depending on whether the morphology of the solution feature space **is known** (case 1) or is **not known** (case 2).

In case 1 the reconstruction of generation function leads to the finding of coefficients from Eq. (1) and the constructed hyperplane. With it, the hyperplane is searched in a space of **constant dimension** and features, i.e. in a "deductive" way. A large number of applications work just in this way, e.g. automated reading, recognition of acoustic or optic patterns.

But there is also a big number of tasks where the morphology of the feature space is not known exactly. Such situations we meet in medical diagnostics, in criminal search, in financials etc., which means wherever the so-called "subjective features" play their role.

3 Short Survey of Problems and Different Ideas in the Development of the Theory of Reconstructing Generative Functions or Rather Regularities

Naturally it is expected that the generating function or rather the regularity should be invariant, i.e. the law and the bonds should be unchanged if some of the surrounding conditions change.

It is desirable to know *what* will remain invariant (unchanged) by *which* changes and under *which* conditions. The answers to this question lifted mathematics, physics and technology of twenties century to a new level of development.

The following notions are connected with these answers:

- Transformation, Groups of Transformation, Symmetric/ Reverse Group Elements, Invariant of Group etc.,
- Kinds of space geometry where the transformation is performed (Euclidean, Hilbert, Lorentz, Minkowski, Lobachevsky spaces, complex space etc.),
- Integral transform or rather kernel transform and the convergence of model function to its generating function (singular integral),
- Kernel transform and minimization of the mean risk for reconstructing of generating functions,
- Methods of searching the separating hyperplane for recognition tasks using the kernel transform, such as
  - Neuronal Networks of Novikoff,
  - Methods of potential functions (Rozonoer),
  - SVM Support Vector Machine (Vapnik),
- Stability of solution, regularization, metric.

The following questions are also tied to this issue:

- What is transformation, what is group of transformations and its invariant?
- Why did the notion "groups of transformations" push the modern sciences so far?

Several mathematicians (Galois, Abel, Poincaré, Cartan, Lie, Pontryagin, Kolmogorov, Heaviside, Dirac) worked on the question of a stable representation of laws from a common mathematical point of view when they developed the transformation theory, the transformation groups, the invariants of transformation groups, long time before the task of pattern recognition was formulated mathematically.

These theories help us to look at many different and well-known for a long time mathematical instruments (e.g. algebraic, differential and integral equations) and ideas in geometry from a very general point what allows us to use the rich experience of geometric representations including the projective geometry.

The borders of the adequate work of the law, which is represented by the transformation, are defined by the so-called invariant of the transformation (or group of transformations): The invariant describes what will remain unchanged when the transformation (or group of transformations) is applied to a mathematical or geometric object.

For example, in case of using a differential transformation (differential equation) the invariants of the transformation are the eigenfunctions of the differential equation.

More concrete, in automatic control the invariant is given with the independence of the eigenfunctions from the initial state and from the kind of the action at the input of the object of the automatic control, which is described by a differential equation. In case of operator equation the invariant is the eigenvector of the fundamental function.

As another example, the **homothetic** transformation (or resemblance transformation) of leafs shown in Fig. 6 belongs to the **group of conformal transformations**.

In case of the group of homothetic transformations the invariant, i.e. the feature of the figures that does not change while performing the transformation, is the angle between two tangents to any two intersecting curves of these figures.

Let us highlight another important invariant which geometric interpretation has served the well-known idea of the Lorentz transformation that is described in more detail in Sect. 4 of this paper. This is the integral invariant of dynamic systems

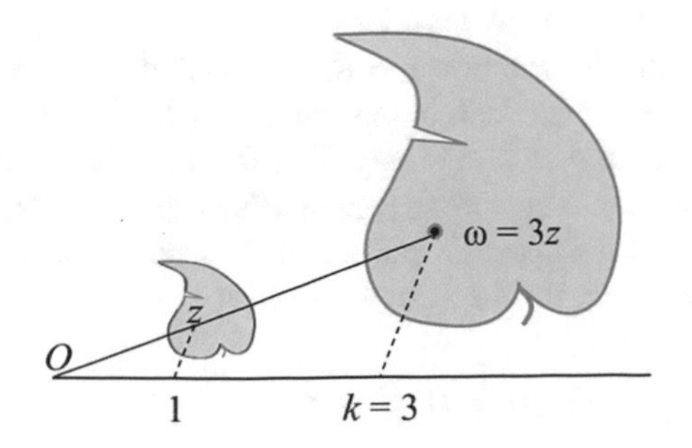

Fig. 6. Homothetic transformation.

(Poincaré [31], Cartan [33]) and also the concept of the phase space (Poincaré [32], Lie, Kolmogorov, Pontryagin [32]), which is based on the integral invariant, are the basis for the advanced theory of optimal control.

Further more, the integral invariant is also the basis for the equiaffine Lorentz transformation conserving the volume as invariant.

In the further course the Sect. 4 of this paper tries to justify the correctness of the reduction or rather of the selection of the feature space by the Alpha-procedure with the help of the Lorentz transformation that can be considered as theoretical basis of this inductive pattern recognition method.

The thing is that for thousands of years a lot of single experiences of work with natural simple spatial (geometric) representations have been accumulated in mathematics and technology. But only with the appearance of the notion **group of transformations** (about the turn of $18^{th}/19^{th}$ centuries) introduced by Gaspard Monge, Évariste Galois and Niels Abel, it became possible to unite algebraic representations with geometric representations in a very general way. The inventor of descriptive geometry, Monge, had his origin in geometrics, while Abel and Galois came from algebra. But only for the last 30 years, with the development of computers, it became possible to represent the Galois field (as a special group of transformation) *in a geometric manner* within the complex space, e.g. for the investigation of the stability of automated control systems.

Let us consider in a bit more detailed way the notions kernel transform, stability of solution and metric.

The kernel transform was investigated as singular integral by Lebesgue in 1909. One of the means of representing functions is given with the **singular integral**:

$$f_n(x) = \int_a^b K_n(x,t) \cdot f(t) \cdot dt \tag{2}$$

that converges to its generating function $f(x)$ if $n \to \infty$ (with these or those limits of function *f*). The function $K_n(x,t)$ is called kernel.

More or less at the same time J. Hadamard [34] invented the notion "stability of the solution of a reverse task by measured data":

Let us assume that the notion "solution" is defined and only one solution $z = R(u)$ out of the space F corresponds to any element $u \in U$.

The task of defining the solution $z = R(u)$ out of the space F using the original (measured) data will be a stable task in the spaces (*F*, *U*), if for any number $\varepsilon > 0$ one can find such a number $\delta(\varepsilon) > 0$ in a way, that from the inequality $\rho_U(u_1, u_2) \leq \delta(\varepsilon)$ follows $\rho_F(z_1, z_2) \leq \varepsilon$, where $z_1 = R(u_1)$, $z_2 = R(u_2)$, $u_1, u_2 \in U$, $z_1, z_2 \in F$.

ρ_U is the distance in the space U and ρ_F is the distance in the space *F*.

Solving the reverse task for the kernel equation

$$\int_a^b K(x,t) \cdot z(t) \cdot dt = u(x), \qquad c \leq x \leq d \tag{3}$$

means to search for the solution $z(t)$ using the measured data $u(t)$.

The tasks of reconstructing a generating function or a model of a dynamic system, the tasks of finding a separating hyperplane in pattern recognition, the tasks of finding a probability density function etc. – all these tasks are reverse tasks.

In the sixties Tikhonov [35] proposed a regularization method for the stabilization of an instable (according to Hadamard) solution of kernel equations using measured data. Then, Tikhonov generalized that method for other reverse tasks as a search of the solution z out of $A \cdot z = u$ by measured u, where A is any linear operator.

Here, z is the argument of the optimization functional

$$\min\left[\rho^2(Az, u) + \alpha \cdot \Omega(z)\right] \tag{4}$$

where $\Omega(z)$ is a convex regularizing ("penalty") functional.

The parameter α must be found with the help of additional information what appears to be a quite difficult task.

Concurrently with Tikhonv's regularization another method of regularization was invented – the **validation** (Jackknife, Bootstrap etc.).

4 The Recognizing Algorithm of Alpha-Procedure and the Lorentz Transformation

All recognition methods mentioned above either use a stabilization/ regularization of solution or they use special information that exclude instability. For example, in case of Novikoff's Neuronal Networks the stabilization is performed by a given (in advance) point of rotation of the separating hyperplane. The method of potential functions uses for the same purpose the predetermined width of the potential function. Besides, these methods are developed and well applied for the search of the separating hyperplane in the cases when the space is of constant dimension.

On the other hand in recognition cases of the type of "criminal search" with so-called "subjective features" an automatic reduction of the space of the original features is usually necessary because some of the features may be only disturbing.

But we know from the group theory of topologic transformations that the main invariant of continuous group guarantying the convergence is the dimension of the space. That means that the methods permitting a reduction of the space tend to have instable solutions from a mathematical point of view. Therefore, even the use of validation-type regularization may be problematic.

As known from the physics (R. Penrose, S.W. Hawking) and from Tikhonov's theory the choice of the right geometry of the space can solve the stability problem.

Now, let as show how the **Alpha-procedure** [11–16, 18] uses the equiaffine Lorentz transformation with an invariant that fits the given recognition task.

For that purpose we will, first, shortly describe the Alpha-procedure. But before starting an important note must be placed here: *During the following description of the Alpha-procedure where a new (reduced) feature space is created we will use the term "**property**" for the original features and the term "feature" only for the selected and for the so-called artificial features.*

The Alpha-procedure is an advanced method for classifying objects belonging to two different classes. The recognition algorithm is defined during a training phase using a training data set and a "supervisor" showing us which object belongs to which class. All objects are characterised by properties or rather measured data.

The Alpha-procedure is based on a step-by-step selection of the best properties and uses consistently only a subset of these properties. That means we work normally with a reduced property or feature space.

The idea can be roughly described as follows:

First, we select the property with the best discrimination power, i.e. the property that gives the best separation, **as a basic features** f_0 and represent it together with its values for the objects as an axis (Fig. 7).

Then we add a second property p_k to the coordinate system and define the positions of the objects in the plane that is built by the axes f_0 and p_k (Fig. 8).

After that we create **a new axis** and turn it around the origin of the coordinate system **by the angle** α up to the moment when the **projections** of the objects onto this new axis give us the best separation of the objects.

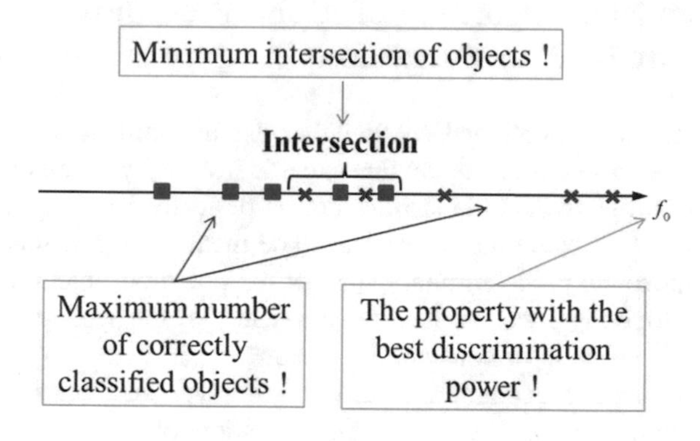

Fig. 7. Alpha-procedure (1)

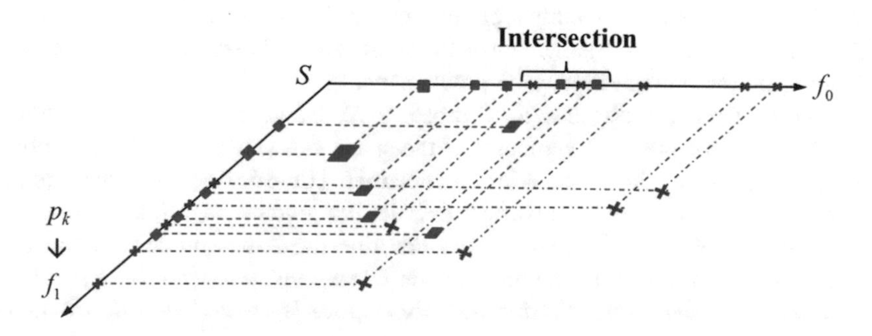

Fig. 8. Alpha-procedure (2)

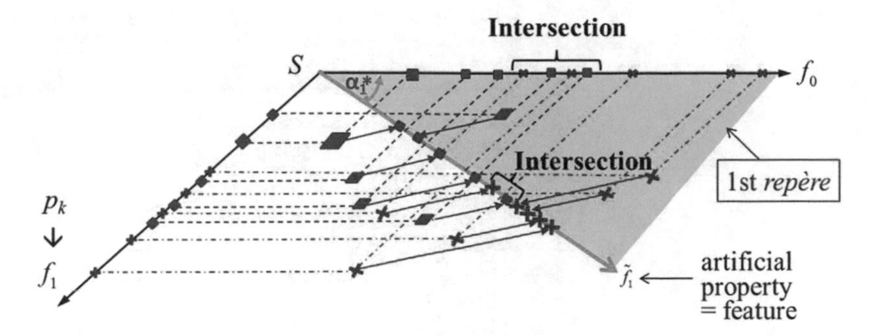

Fig. 9. Alpha-procedure (3)

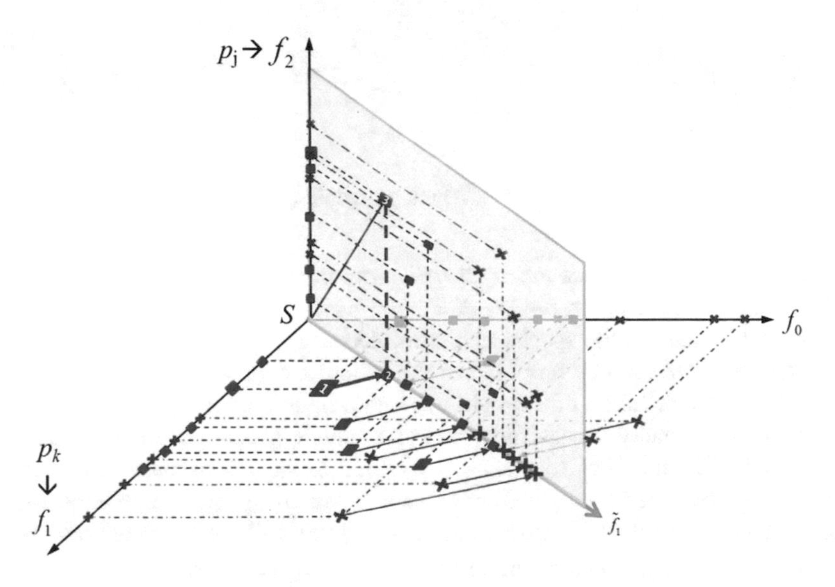

Fig. 10. Alpha-procedure (4)

We repeat this procedure for all properties and select the property that gives the best separation on the new (artificial) axis. $\tilde{f}_1$ This way we construct the 1st *repère* (Fig. 9).

In the third stage we add another property p_j as a third axis and define the position of the objects in a new plane that is created by the axes $\tilde{f}_1$ and p_j (Fig. 10).

Again we create a **new axis** on this new plane and turn it around the origin of the coordinate system **by the angle α** up to the moment when the **projections** of the objects onto this new axis provide the best separation. We repeat this procedure for all remaining properties and select the best one (Fig. 11).

Using each time the last created artificial feature $\tilde{f}_r$ together with the next property we create step by step 2-dimensional planes where we define the next artificial feature $\tilde{f}_{r+1}$. The procedure stops when a faultless or at least the best separation is reached.

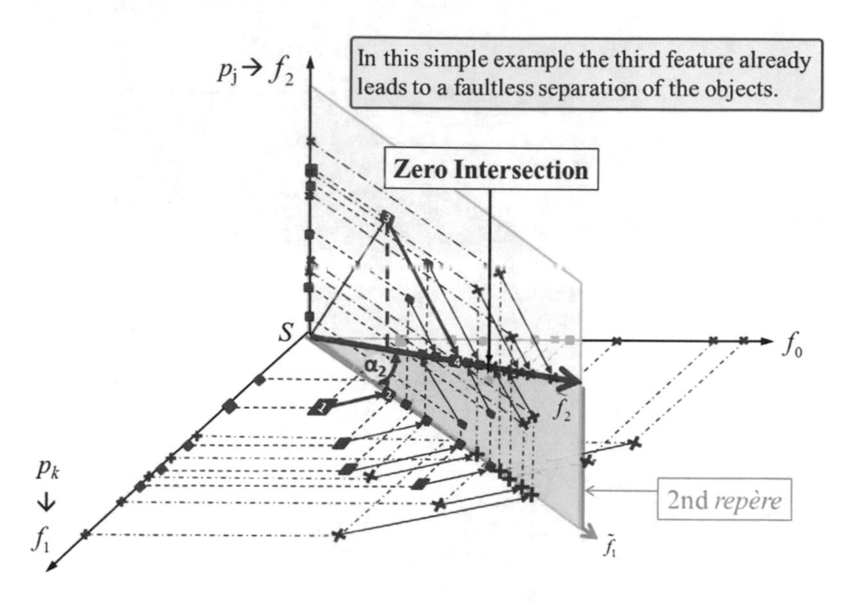

Fig. 11. Alpha-procedure (5)

Note: *As one can see from the described above the step-by-step bottom-up selection is performed without leaving the bundle S.*

Now, let us look at the special equiaffine Lorentz transformations. What are they and which invariant do they propose for the pattern recognition?

For the characterization of a certain geometric figure and its location with the help of numbers it is usually necessary to invent an ancillary reference system or a coordinate system. The numbers $x_1, x_2, ..., x_n$ we get this way do not only characterize the geometric figure (to be investigated) but also its relationship with the reference system. In case of changing the reference system the figure will be characterized by other numbers $x_1', x_2', ..., x_n'$. When a certain expression $f(x_1, x_2, ..., x_n)$ describes the figure in itself then it must not depend on the reference system, i.e. the following relation should be valid:

$$f(x_1, x_2, ..., x_n) = f\left(x_1', x_2', ..., x_n'\right) \tag{5}$$

All expressions which satisfy the relation above are called invariant.

There are three special Lorentz transformations known which are equiaffine, i.e. these affine transformations do not change the objects:

1. Any simple rotation of the space like a solid around the axis of a cone (see Fig. 12) by a certain angle ω is an equiaffine transformation of the space transforming the cone into itself. We will call it ω-transformation.
2. The reflection of the space on any plane π, intersecting the axis of the cone is also a Lorentz transformation. We will call it π-transformation.

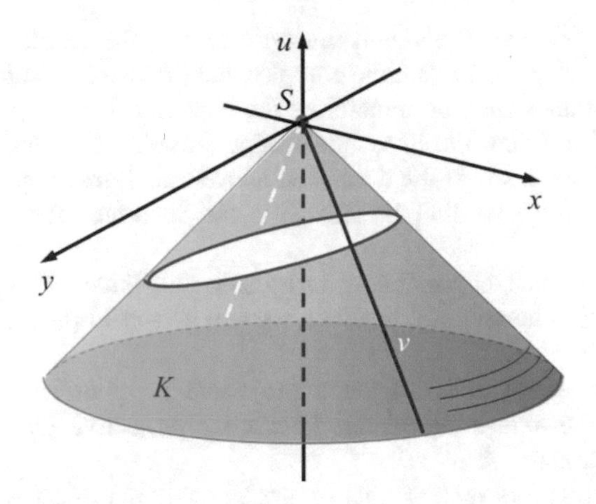

Fig. 12. Illustration to ω- and π-transformation

3. And, finally, let us consider the L-transformation (from Lorentz). Figure 13 illustrates the following: The compression (or rather the stretching) of the space towards the plane P and the stretching (or rather the compression) away from the plane do not change the volumes, because for any cut of the cone by the plane R that is parallel to the plane Sxu, the equation $p \cdot q = const.$ is valid. Here p and q are the distances of any point of the hyperbola to its asymptotes (i.e. to the planes P and Q) and P and Q there selves are the tangents of the cone connecting its generating line.

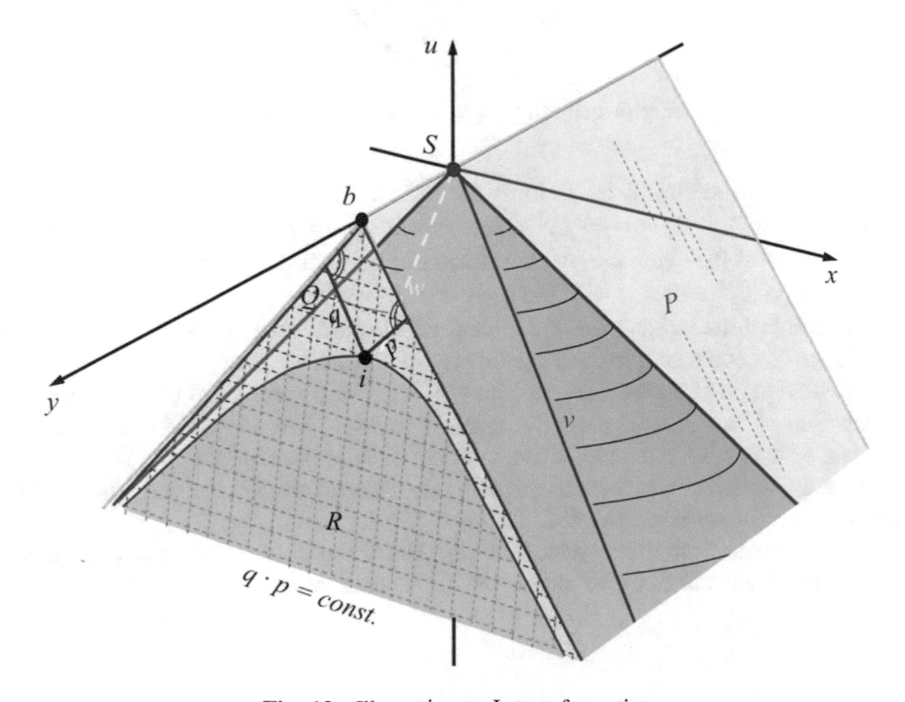

Fig. 13. Illustration to L-transformation

Any point of the hyperbola remains on one and the same hyperbola, i.e. the hyperbola goes over into itself. But the complete cone consists of such hyperbolas what means that the L-transformation transforms the cone into itself.

The line bundle S (i.e. the lines intersecting in point S) maps onto itself unambiguously in all three cases of the described above transformations.

With other words, the line bundle S is the invariant for **each** of the three transformations.

But the proof of that is possible only by using all three of them. Therefore, not stopping on the well-known proof in detail, we will here only show the train of thought of that proof.

In the Lorentz space one can get a projective Π^*-plane to any cut Π that is perpendicular to the axis of the cone. To such a projective plane corresponds the projective transformation Λ.

The basis for the proof is the ability to break down the projective transformation Λ into a series of Lorentz transformations which correspond to the converting of any repère M on the circular into any other repère M' on the same circular (Fig. 14).

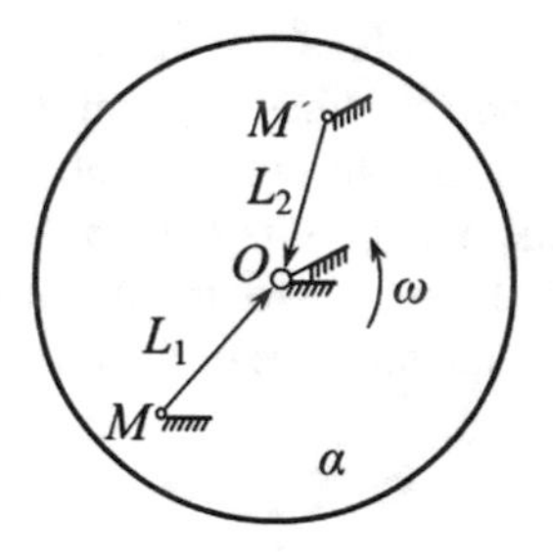

Fig. 14. Illustration to repère transformation

For this it is sufficient to perform the transformations $\Lambda = L_1 \cdot \omega \cdot L_2^{-1}$ (or the transformations $\Lambda = L_1 \cdot \omega \cdot \pi \cdot L_2^{-1}$). The transformation L_1 moves the first repère M to the centre O of the circular α, the transformation ω turns it as needed and, finally, the transformation L_2^{-1} unites it with the second repère M′.

The proof of the uniqueness of the identical transformation is completely based on the rules of the common projective geometry onto a plane.

It is obvious that the Alpha-procedure uses at every step the ω-transformation of Lorentz and at the turns in pairs the π-transformation. The vectors of the measured objects are the line bundle S that is the invariant of these transformations. This way it is guaranteed that two single features are stepwise comparable, i.e. the correctness of the stepwise comparison is guaranteed.

The statements of the “supervisor” during the training process play not only the role of automatic feedback but give also a visual feedback on the plane.

5 On Numerical Examples

A lot of numerical examples executed by team of Ukrainian postgraduates at University of Cologne under the lead of K. Mosler, R. Dyckerhoff and the author of this article are published in [18, 20–23] and others.

Thereby the paper [20] focuses on the original Alpha-procedure briefly described above and in [18, 21, 22] on an advanced version that is known as DD-Alpha (combining the Alpha-procedure with the Data Depth approach). Most of the examples also compare the effectiveness of the Alpha- and DD-Alpha-procedures with other known classification methods.

The source [23] refers to an R-package including its description. This package is free accessible and can be used by everyone for testing the DD-Alpha-Procedure with his/her own data.

6 Summary and Further Investigations

It is known that the construction of a solution of the task of pattern recognition in a subspace or, in a common sense, in a "non-adequate" space (e.g. in an overfitting space) requires measures for ensuring the convergence and the stability of the solution. It is shown above that such measures can be

- a "post-factum" regularization of Tikhonov type [2, 35] what means a two-criterial optimization in fixed **sub**spaces or spaces [12],
- a "preventative" or "post-factum" validation or
- the use of geometric transformations together with an invariant that is appropriate for the task (and this was the main topic of this paper).

Concerning the further investigations we can say the following:

- The principle of the geometric representation of the differential transformation can be used for the construction of the separating plane.
- Pontryagin's Maximum Principle can be probably used for the classification of functions which are represented in form of measured points.
- It is desirable to investigate the concept of Data Depth as transformation together with it invariants with respect to its use for pattern recognition.

References

1. Rosenblatt, F.: Principles of Neurodynamics. Spartan Books, New York (1962)
2. Novikoff, A.: On convergence proofs for perceptrons. In: Proceedings of Symposium on Mathematical Theory of Automata (XII). Polytechnic Institute of Brooklyn (1963)
3. Ivakhnenko, A.G.: Recognizing system "Alpha" as predicting learning filter and extremum controller without searching oscillations. Automatics **4**, 17–24 (1966)
4. Vassilev, V.I., Ivakhnenko, A.G., Reyzkii, V.E.: Algorithm of a recognizing system of perceptron type with a correlator at the input. Automatics **1**, 20–35 (1966). (in Russian)

5. Vassilev, V.I., Ivakhnenko, A.G., Reyzkii, V.E.: Recognition of moving bodies. Automatics **6**, 47–52 (1967). (in Russian)
6. Ivakhnenko, A.G.: Self-learning systems of recognition and of automatic control. Tekhnika, Kyiv (1969). (in Russian)
7. Vassilev, V.I.: Recognizing systems. Naukova Dumka, Kyiv (1969). (in Russian)
8. Aizerman, M.A., Braverman, E.M., Rozonoer, L.I.: The method of potential functions in the theory of machine learning. Nauka, Moscow (1970). (in Russian)
9. Bock, H.H.: Automatische Klassifikation. Vandenhoeck & Ruprecht, Göttingen (1974)
10. Vapnik, V., Chervonenkis, A.Y.: The Theory of Pattern Recognition. Nauka, Moscow (1974)
11. Vassilev, V.I.: The reduction principle in pattern recognition learning (PRL) problem. Pattern Recogn. Image Anal. **1**(1), 23–32 (1991)
12. Cortes, C., Vapnik, V.: Support-vector networks. Mach. Learn. **20**, 273–297 (1995)
13. Vassilev, V.I., Lange, T.: The principle of duality within the training problem during pattern recognition (in Russian). Cybern. Comput. Eng. **121**, 7–16 (1998)
14. Vassilev, V.I., Lange, T., Baranoff, A.E.: Interpretation of diffuse terms. In: Proceedings of the VIII, International Conference KDS 1999, pp. 183–187, Kaziveli, Krimea, Ukraine (1999). (in Russian)
15. Vassilev, V.I., Lange, T.: Reduction theory for identification tasks. In: Proceedings of International Conference on Control, Automatics-2000, pp. 49–53, Lviv (2000). (in Russian)
16. Vassilev, V.I., Lange, T.: The mutual supplementability of the Group Method of Data Handling (GMDH) and the Method of Extreme Simplifications (MES). In: Proceedings of International Conference on Inductive Modelling ICIM-2002, Lviv, vol. 1, pp. 68–71 (2002). (in Russian)
17. Steinwart, I., Christmann, A.: Support Vector Machines. Springer, New York (2008)
18. Lange, T., Mosler, K., Mozharovskyi, P.: Fast nonparametric classification based on data depth. Stat. Papers **55**, 49–69 (2014). Springer
19. Hennig, C.: How many bee species? A case study in determining the number of clusters. In: Spiliopoulou, M., Schmidt-Thieme, L., Janning, R. (eds.) Data Analysis, Machine Learning and Knowledge Discovery, pp. 41–49. Springer, Heidelberg (2014)
20. Lange, T., Mozharovskyi, P.: The alpha-procedure: a nonparametric invariant method for automatic classification of multi-dimensional objects. In: Spiliopoulou, M., Schmidt-Thieme, L., Janning, R. (eds.) Data Analysis, Machine Learning and Knowledge Discovery, pp. 79–86. Springer, Heidelberg (2014)
21. Lange, T., Mosler, K., Mozharovskyi, P.: DDα-classification of asymmetric and fat-tailed data. In: Spiliopoulou, M., Schmidt-Thieme, L., Janning, R. (eds.) Data Analysis, Machine Learning and Knowledge Discovery, pp. 71–78. Springer, Heidelberg (2014)
22. Mozharovskyi, P., Mosler, K., Lange, T.: Classifying real-world data with the DDα-procedure. Adv. Data Anal. Classif. **9**(3), 287–314 (2015)
23. Pokotylo, O., Mozharovskyi, P., Dyckerhoff, R.: Depth-Based Classification and Calculation of Data Depth. R package version 1.2.1, ddalpha. https://cran.r-project.org/web/packages/ddalpha/index.html. Accessed 05 June 2017
24. Ivakhnenko, A.G.: Long-Term Prediction and Control of Complex Systems. Tekhnika, Kyiv (1975). (in Russian)
25. Ivakhnenko, A.G.: An Inductive Method of Self-Organization of Models of Complex Systems. Naukova Dumka, Kyiv (1982). (in Russian)
26. Farlow, S.J. (ed.): Self-Organizing Methods in Modelling. GMDH type Algorithms. Marcel Dekker Inc., New York and Basel (1984)

27. Ivakhnenko, A.G., Stepashko, V.S.: Noise-Immunity of Modelling. Naukova Dumka, Kyiv (1985). (in Russian)
28. Akaike, H.: Experiences on development of time series models. In: Bozdogan, H. (ed.) Proceedings of the First US/Japan Conference on the Frontiers of Statistical Modeling: An Information Approach, vol. 1, pp. 33–42. Kluwer Academic Publishers, Dordrecht (1994)
29. Lange, T.: New structure criteria in GMDH. In: Bozdogan, H. (ed.) Proceedings of the First US/Japan Conference on the Frontiers of Statistical Modeling: An Information Approach, vol. 3, pp. 249–266. Kluwer Academic Publishers, Dordrecht (1994)
30. Stepashko, V.S.: Method of critical variances as analytical tool of theory of inductive modeling. J. Autom. Inf. Sci. **40**(3), 4–22 (2008)
31. Poincaré, H.: Les methods nouvelles de la mecanique celeste. Gauthier-Villars, Paris (1899)
32. Pontryagin, L.S., Boltyansky, V.G., Gamkrelidze, R.V., Mishchenko, E.F.: The Mathematical Theory of Optimal Processes. Nauka, Moscow (1969). (in Russian)
33. Cartan, E.: Leçons sur les invariants intégraux. Editions Hermann, Paris (1971)
34. Hadamard, J.: Sur les problèmes aux dérivées partielles et leur signification physique. Bull. Univ. Princeton **13**, 49–52 (1902). Princeton
35. Tikhonov, A.N., Arsenin, V.Y.: Methods of Solving Incorrect Tasks. Nauka, Moscow (1974). (in Russian)

Analyzing Project Team Members' Expectations

Vira Liubchenko(✉)

Odessa National Polytechnic University,
1 Shevchenko av., Odessa 65044, Ukraine
lvv@opu.ua

Abstract. This paper describes an approach for analyzing the project team members' expectations to achieve the personal goals as well as the project objectives. There are described four types of expectations and suggested the expectation map as an analytical tool. The paper introduces the important antipatterns and the process of expectation map analysis.

Keywords: Project manager · Team member · Project objective Personal expectation · Expectation map

1 Introduction

The key to a successful team is the alignment of objectives within the team. The challenge of the project manager is setting a common goal the entire team is willing to pursue. If the case of lack of a common goal, team members who disagree with the objective in hand will feel reluctant to utilize their full effort, leading to failure to reach the goal.

Software project teams coalesce and become more productive when they are coordinated [1]. It takes time for teams to progress through the Tuckman stages of forming, storming, norming, and finally to performing to optimize team output [2]. Developing the project team improves the people skills, technical competencies, and overall team environment and project performance. The project managers should identify, build, maintain, motivate, lead, and inspire project teams to achieve high team performance and to meet the project's objectives.

The team can work towards attaining the goals only if they exactly know what management expects from them and what role they hold in the project. The project manager needs to provide a structure for the project team and set expectations and priorities as well as assign roles carefully. The expectations, as well as the overall goal, should not be fuzzy.

The problem is the team members are not the predictable systems, and the project manager is not able to use optimization techniques to achieve the goal. Therefore, the project manager needs the tool supported the process of coordination the personal objectives with the project goal; he/she has to be careful about the persons' objectives.

For establishing persons' objectives, the understanding of persons' expectations that will impact the effectiveness and motivation of team members is a constructive way.

N. Shakhovska and V. Stepashko (eds.), *Advances in Intelligent Systems and Computing II*, Advances in Intelligent Systems and Computing 689,
https://doi.org/10.1007/978-3-319-70581-1_21

People are motivated if they feel they are valued in the project team and this value is demonstrated by the attention to their expectation.

In the paper, we propose a simple tool for coordination of project objectives with personal expectations and describe how to use it for analysis of current situation.

2 Team Common Goal

Team common goal is what separates a high performing team from a bad project experience. Common goals are important because they bring people together and encourage them to communicate problems and results. They allow for a much earlier and faster recognition of problems in the project development.

To fully complete individual's task roles, one needs to have clear expectations about his subgoals, the paths to accomplish these subgoals, and the link between his work and the work of others [3]. Because individuals' roles are embedded in the larger context of teams, the clarity of team goals and individual members' roles in working toward meeting the goals has a powerful impact on team effectiveness.

In [4] there was demonstrated that team goal setting was an effective team-building tool for influencing cohesiveness in the teams. Cohesion had been defined as "a dynamic process that is reflected in the tendency for a group to stick together and remain united in the pursuit of its instrumental objectives and/or for the satisfaction of member affective needs" [5]. Cohesiveness is the extent to which team members stick together and remain united in the pursuit of a common goal. A team is said to be in a state of cohesion when its members possess bonds linking them to one another and to the team as a whole.

When agreeing and prioritizing workloads with members of the team, the project manager needs to ensure that individual members of the team are happy with their workloads, and not working under undue stress. It is crucial both to the effectiveness and the quality of the working atmosphere of the project team. The project manager should ensure that team members are well prepared, that they understand what their current objectives mean, how they will be measured and what manager's expectations are.

Involving individuals in the process of their objectives setting will improve understanding of manager's expectations and the expectations of the organization. The project manager has to motivate to obtain the objectives as well as clearly demonstrate the coherence between individual objectives and project goal. In other cases, team members might divert themselves to other tasks due to a lack of belief or interest in the goal.

The biggest and most important common goal for a team is to finish the project successfully. However, this should not be the only goal. Many things can bring people together and focus them on results and not on personal comfort. To negotiate the individual objectives effectively, the project manager needs to know the expectations of team members.

3 Expectation Mapping

Academics and practitioners agree that expectations play a central role in project organizations. For the latter, these expectations are typically set at the start of initiatives and comprise iron triangle measurements, as well as more refined metrics for expected benefits or user satisfaction. Usually, it pointed at the expectation of stakeholders and senior management [6]. In addition, the researchers and practitioners pay attention to customer expectation management [7]. Let us focus on the expectations of team members, because of them define the personal objectives and cause the behavior during the project implementation.

Usually, there are recognized four expectation types: must, will, should, and could [8]. This classification explains how expectations affect relationships and define potential gain or damage due to gaps between expectation and performance. Different expectation types have different impacts on interpersonal relations and relate to various areas of the project implementation.

Also, by analogy with [9], we can distinguish four types of expectations:

- ideal expectations are visions, aspirations, needs, hopes and desires, related to the participation in the project;
- normative expectations are expectations about what should or ought to happen, mostly derived from what colleagues are told, or led to believe;
- predicted expectations are beliefs about what will happen and are likely to result from individual experiences;
- unformulated expectations are not articulated expectations.

The most valuable for project manager are normative and predicted expectation because they affect the personal preferences and can be articulated.

The best practice to discover personal expectations is one-to-one meetings. During the series of such meetings, the project manager (or team leader) is gathering the unstructured set of notes (mental or hand-written) about personal expectations of team members. After finishing, he should arrange the notes into a useful model to help understand the expectations of team members, identify holes and omissions in relations between project objectives and personal expectations, and successfully plan the motivation strategy for each team member.

As the model of team members' expectations, we propose the expectation map. The expectation map tool is a diagram representing how the personal expectations correspond to the project objectives (see Fig. 1).

On the expectation map a set $\{v_1, \ldots, v_M\}$ is a set of project objectives, a set $\{s_{j1}, \ldots, s_{jp}\}$ is a set of expectations of *j*th team member, an arrow between v_i and s_{jk} represents correspondence between *i*th objective and *k*th expectation of *j*th team member.

For example, the *i*th objective is formulated as "To improve customer satisfaction rates by 50 percent by September 06 through improving user interface usability." Let us suppose *k*th expectation of *j*th team member is "To climb on career ladder as UI/UX designer." To visualize the existed correspondence the nodes v_i and s_{jk} should be connected by the arrow from s_{jk} to v_i.

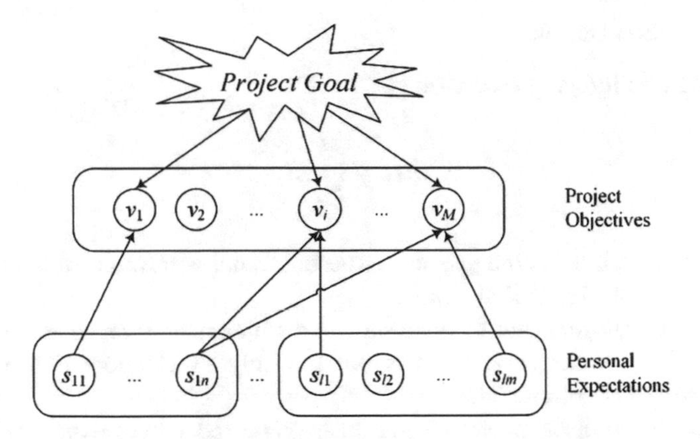

Fig. 1. The structure of expectation map.

Expectation maps clearly lay out the mental perceptions of team members so that project manager may identify disconnects in the personal expectations and project objectives.

4 Formal Analysis of Expectation Map

Expectation map is not only the visualization tool but also analysis support tool. Let us introduce the mapping M of expectations set onto objectives set such as $M(s_{jk}, v_i)$ takes place when kth expectation of jth team member corresponds to ith objective. Define a set S_{v_i} as a set of personal expectations corresponding to the objective v_i: $S_{v_i} = \{s_{jk} : M(s_{jk}, v_i)\}$.

The ideal expectation map is formally described as

$$\forall v_i | S_{v_i} \neq \emptyset \ \ \& \ \ \forall s_{jk} \in \bigcup_{i=1}^{M} S_{v_i} \tag{1}$$

In other words, for each project objective there was discovered the set of corresponded personal expectations. The case looks like the most comfortable for a project manager because he can find motivated team members for all objectives. However, sometimes, the ideal map contains the set of conflicting expectations (the antipattern Sect. 4.3 described below).

The project manager can prioritize the project objectives accordingly with cardinal numbers of the sets S_{v_i}. Therefore, the highest priority defines the objective interested for the majority of team members. In other words, the highest prioritized objectives are the most valuable ones from the team point of view.

Below we describe the important antipatterns for the expectation map analysis.

4.1 Unmet Expectation

The antipattern is formally described as

$$\exists s_{jk} \notin \bigcup_{i=1}^{M} S_{v_i} \tag{2}$$

In other words, there is the personal expectation not corresponded to any objective (for example, s_{l2} at Fig. 1) at the map.

The project manager needs to make sure no team member is an island, or the project might fail. If the unmet expectation is not the only expectation of particular team member, the case is not crucial.

However, the project manager should negotiate the unmet expectation with the team member as well as the vision of its realization. In the best case, the unmet expectation is achievable during the project implementation, although it is not directly related to the project objectives. Otherwise, the project manager should suggest the ways realize the unmet expectation in the future.

Additionally, the project manager should analyses whether the unmet expectation is an unreasonable one. Unreasonable expectations are those, which are impossible or highly unlikely for any individual to meet. The project manager should understand possible reasons behind unreasonable expectations, as well as their impact on the motivation of the team member. Also, he has to keep in mind discovered expectation gap during all project period.

4.2 Missed Objective

The antipattern is formally described as

$$\exists v_i | \, S_{v_i} = \emptyset. \tag{3}$$

In other words, there is the project objective not corresponded to any personal expectation (for example, v_2 at Fig. 1) at the map.

The case is essential from project manager's point of view. The lack of personal interest in particular objective usually causes its ignoring. If it is not possible to drop the missed objective, project manager should find the way to make the objective attractive for the team members.

In this situation, we can offer two solutions.

- The best solution. The project manager negotiates with the team member the reasons for his expectations and shows that some his/her reasons are related to the achievement of the missed project objective.
- The deferred solution. The project manager explains to the team member that his/her knowledge and experience are necessary to achieve the missed project objective. At the same time, both participants are looking for a compromise solution that could potentially lead to a negative result in the future.

4.3 Conflicting Expectations

The antipattern should be taken into account when the cardinality of set S_{v_i} is greater than one possibility,$|S_{v_i}| > 1$. In this case, the relations between personal expectations S_{v_i} and project objective v_i can be different: positive in the case when the objective achieving satisfy the expectations, negative in the case when failure to meet objective satisfy the expectations.

The project manager has to understand that this antipattern indicates a precondition for conflict in the team. Therefore, the right solution is the involvement of conflict management techniques. It is the very first stage of the conflict, so the probability of finding win/win solution is high.

5 The Process of the Expectation Map Analysis

Finally, we describe the process of the expectation map analysis (see Fig. 2).

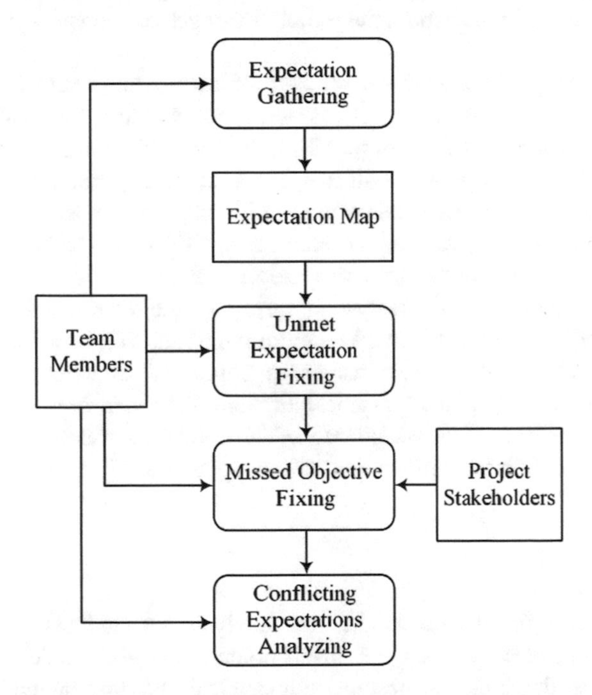

Fig. 2. DFD-diagram – expectation map analysis

Firstly, the project manager has to find out the personal expectations of the team members and to visualize their relationship with project objectives. As mentioned above, with this purpose, the project manager should use the one-to-one meeting; the best condition is an informal atmosphere. The project manager has to pay attention to the cognitive biases, which can lead to deviation from real expectations (for example, attentional bias, bandwagon effect, choice-supportive bias, etc.). The cognitive biases can influence not only on the team members when they formulate their expectations, but also on project manager when he/she is building the expectation map.

Then the project manager should proceed to antipatterns analysis. It requires additional one-to-one meetings with some team members as well as with project stakeholders.

As a first step of the analysis, Unmet Expectations antipattern is examined. The project manager should negotiate each unmet expectation with its bearer. If the expectation is unreasonable, it is necessary to explain the situation to team member and control his behavior during project implementation period. Otherwise, the project manager should negotiate with team member his expectations and try to find a connection between them and project objectives. Sometimes it is impossible; it is not supposed to happen in the frame of the project. Such case should be negotiated carefully to avoid falling short of expectation.

After this step, the expectation map usually is modified; we should point to the particular situation. It appears when there are unmet expectations impossible for satisfaction in the frame of the project. If the corresponding team member agrees with the case for ineligibility of expectation, the project manager can consider it as conditionally met.

As a second step of the analysis, Missed Objective antipattern is examined. It is necessary to negotiate missed objectives with stakeholders to decide whether this objective can be dropped. If it is not possible, the stakeholder should consider whether the objective could be formulated with respect to some expectations.

If the ideal or conditionally ideal (with conditionally met expectations) map was not gotten after previous steps, the project manager should return to the first step. Otherwise, he continues with Conflicting Expectation analysis.

The project manager should analyze subgroup of the team members corresponded to the set of conflicting expectations. The team members with positive "expectations – objective" correlation are possible resources appointed to work for corresponding project objective. If such solution is not realizable, then it is necessary to understand why team members have different interests. These reasons should be negotiated with team members to define the actions that are acceptable to all parties.

6 Case Study

Let us see an example of expectation map analysis for students' expectation management in the project on degree work development. The problem decided in the frame of the project is the analysis of the geographical distribution and evaluation of employment successfulness of universities graduates. It is an important issue not only for particular University but also for whole regions. But there has not still developed the automated tool, so that the solutions are usually time and cost ineffective because of need in the organization of primary research with survey distribution, collecting and analyzing covered a large number of people in different cities and countries. To reduce the investigation efforts, the analyzing software based on data from social networks should have been developed.

Respectively the software development project had four objectives:

- v_1 – to avoid the need for the primary research for gathering the source data for analysis;
- v_2 – to realize the automated tool for collecting and analyzing data;
- v_3 – to support the big data processing and renewal;
- v_4 – to provide the different kinds of information representation.

There were five students involved in the project implementation. At the early stage, their expectations were quite different; mined expectations are listed below:

- for student S_1:
  - s_{11} – to get the experience of teamwork;
  - s_{12} – to improve the software design and development skills;
  - s_{13} – to study new technologies;
- for student S_2:
  - s_{21} – to understand is it realistic to implement the software architecture;
  - s_{22} – to get the experience in multiservice application development as a team member;
  - s_{23} – to work with big data;
- for student S_3:
  - s_{31} – to get expertise in the project implementation in Java from scratch to the end;
  - s_{32} – to enhance the teamwork skills;
  - s_{33} – to develop the communication skills;
  - s_{34} – to expand the professional horizons.
- for student S_4:
  - s_{41} – to learn how to use the API of social network fully;
  - s_{42} – to discover the teamwork on joint project;
  - s_{43} – to strengthen the knowledge in software architecture design;
  - s_{44} – to examine some design patterns implemented them in code;
  - s_{45} – to get the comments and pieces of advice from experienced mentors;
- for student S_3:
  - s_{51} – to get the experience of teamwork on large system development;
  - s_{52} – to gain the skill of development the big data analysis systems;
  - s_{53} – to learn how to use the modern development frameworks;
  - s_{54} – to improve the knowledge of technologies for web application development;
  - s_{55} – to enhance the skills in group working with version control systems;

The initial expectation map is shown in Fig. 3.

As we see the initial expectation map is not ideal because the condition (1) does not hold. Therefore, we should have started with unmet expectation fixing.

Nine unmet expectations could have been divided into three semantic groups – teamwork, technology and specific groups. The teamwork group is the biggest one, it contents $\{s_{11}, s_{22}, s_{32}, s_{33}, s_{42}, s_{51}, s_{55}\}$ and cannot be ignored. The development of the analytical system was too great for individual graduate work and was planned for group work. Therefore, the graduate work supervisor (as project manager) articulated it as v_5 – to organize teamwork on the software project. The technology group is represented by s_{13}, s_{34} (it was cleared up that it concerns new technologies and design patterns) and

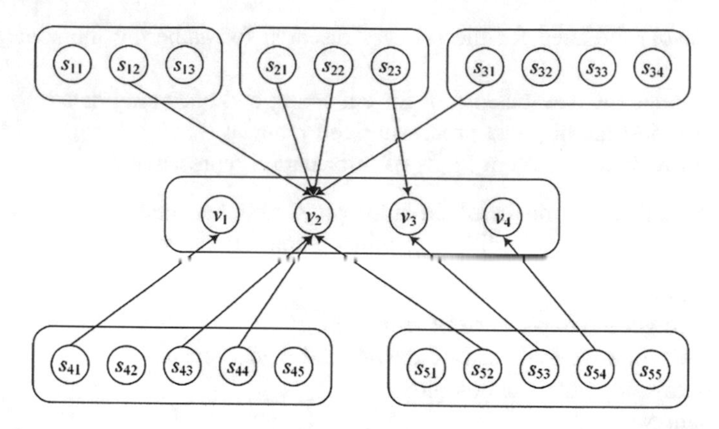

Fig. 3. Initial expectation map for students group

partially s_{55}. During the one-to-one meeting, there were negotiated the requirement to the developed system and found that the most required technologies are new for students. The specific expectation s_{45} was not directly related to the project objectives, and student understood the fact. Anyway, s_{45} was taken into account, and expertises from industry gave mentor session for students' team four times at the design stage of project implementation. However, during the expectation map analysis, s_{45} was left as conditionally met expectation.

As result of unmet expectation fixing, we got the modified expectation map (see Fig. 4).

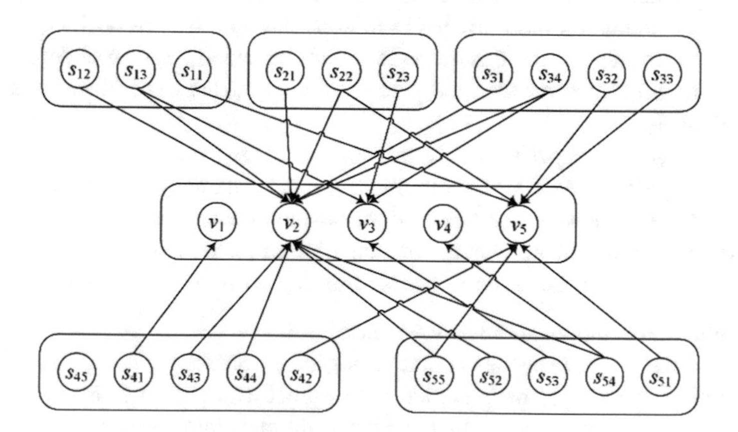

Fig. 4. Modified expectation map for students group

As we see, the modified map does not include the missed objective. We described above the specific of s_{45}, so the modified map is conditionally ideal. It is possible to prioritize the project objectives in order v_2, v_5, v_3, v_1 and v_4. The conflicting expectation analysis did not find the conflicting expectations in sets S_{v_2}, S_{v_5}, S_{v_3}. Therefore, the process of expectation map analysis was finished after the first iteration.

7 Conclusion

This paper introduced an empirically grounded approach to analyzing team members' expectations with the aim of coordinate them with project objectives. It focused on two types of expectations: normative and predictive expectations. It also suggested the expectations map as a model of team members' expectation. The important antipatterns for expectation map analysis were formulated and discussed. The process of expectation map analysis was finally described.

In the process of expectation map analysis, there is three complicated moments. The first moment appears during the initial one-to-one meeting when project manager eliminates expectations of team members. Sometimes team members are not able to formulate their expectations, sometimes they strive to be approved and express not relevant expectations, and sometimes project manager does not understand right team member and fixes irrelevant expectations. The project manager has to use some checking procedures because the principle "garbage in – garbage out" works perfectly also for expectations map analysis. The second moment appears when project manager tries to fight with Unmet Expectation antipattern. He/she have to negotiate unmet expectations with corresponded team members, and it is possible to arrive at the problems presented above. The third moment appears when project manager works with stakeholders to fight with Missed Objective antipattern. The project manager has to persuade stakeholders to reconsider (in some way) the project objectives, to understand the points of view of all stakeholders and generalize them. Therefore, we should pay attention at the process subjectivity and dependence on communication skills of the project manager.

The pilot exploitation of expectation map analysis was realized for teams of software development projects and study projects. Nevertheless, there are no limitations to make use of it in other kinds of projects.

References

1. Software Extension to the PMBOK® Guide, 5th edn. Project Management Institute, Inc. (2013)
2. Tuckman, B.: Developmental sequences in small groups. Psychol. Bullet. **63**, 384–399 (1965)
3. Hu, J., Liden, R.C.: Antecedents of team potency and team effectiveness: an examination of goal and process clarity and servant leadership. J. Appl. Psychol. **96**(4), 851–862 (2011)
4. Senécal, J., Loughead, T.M., Bloom, G.A.: A season-long team-building intervention: examining the effect of team goal setting on cohesion. J. Sport Exerc. Psychol. **30**, 186–199 (2008)
5. Carron, A.V., Brawley, L.R., Widmeyer, W.N.: Measurement of cohesion in sport and exercise. In: Duda, J.L. (ed.) Advances in Sport and Exercise Psychology Measurement, pp. 213–226. Fitness Information Technology, Morgantown (1998)

6. Schiff, J.L.: 11 project management tips for setting and managing expectations. http://www.cio.com/article/2378680/project-management/11-project-management-tips-for-setting-and-managing-expectations.html. Accessed 24 July 2017
7. Ojasalo, J.: Managing customer expectations in professional services. Manag. Serv. Qual. Int. J. **11**(3), 200–212 (2011)
8. Olkkonen, L., Luoma-aho, V.: Public relations as expectation management? J. Commun. Manag. **18**(3), 222–239 (2014)
9. Geurts, J.W., Willems, P.C., Lockwood, C., van Kleef, M., Kleijnen, J., Dirksen, C.: Patient expectations for management of chronic non-cancer pain: a systematic review. Health Expect. 1–17 (2016)

The Contextual Search Method Based on Domain Thesaurus

Vasyl Lytvyn(✉), Victoria Vysotska(✉), Yevhen Burov(✉), Oleh Veres, and Ihor Rishnyak

Lviv Polytechnic National University, Lviv, Ukraine
{Vasyl.V.Lytvyn, Victoria.A.Vysotska, Yevhen.V.Burov, Oleh.M.Veres, Ihor.V.Rishnyak}@lpnu.ua

Abstract. The growth of volume of text resources on Internet and natural limitations of human cognition justifies the development of search enhancement systems. This paper proposes a contextual search method based on domain thesaurus. The method uses semantic metrics defined on thesaurus and weighted conceptual graph model for search enhancement. As an example a task of searching for partners working in similar research area is analyzed. The effectiveness of proposed method when evaluated using precision metric was better compared to Jaccard's and WUP methods.

Keywords: Thesaurus · Semantic metric · Search method · Linguistics Graph model

1 Introduction

The number of Internet users is estimated to have reached several billion and continue to grow [1]. World Wide Web contains billions of documents, and size of textual information in them amounts to hundreds of terabytes [2]. However, our ability to understand and process information is rather limited and remains the same [3]. Therefore, the demand for contextual search enhancing systems (CSES) continues to grow [4]. It is important to develop automated systems, which are able to analyze and represent the informational needs of their users with more precision [5]. For example, in the area of scientific research it is important to find partners for grant application with specified competence or find a reviewer for an article, having similar scientific results [6–8]. In order to resolve this problem the usage of domain area thesaurus was proposed [9, 10]. The system functions as follows. The user types a set of keywords from thesaurus, which, according to his opinion, fully describes the specifics of research area where partner's search is performed [11–13]. CSES, based on those keywords, thesaurus and some semantic metric, finds a set of relevant textual documents, rated according to this metric [14–16]. The authors of those documents will form a pool of potential partners [17–20]. The central part of proposed system is occupied by domain thesaurus acting as a kernel for system's knowledge base [21–24]. Another important module will calculate the semantic distance between user's request

N. Shakhovska and V. Stepashko (eds.), *Advances in Intelligent Systems and Computing II*, Advances in Intelligent Systems and Computing 689,
https://doi.org/10.1007/978-3-319-70581-1_22

and textual documents. The architecture of proposed system is shown on Fig. 1. The module, which calculates the semantic distance, uses some semantic distance metric. Let us review some semantic distance metrics that are used when resolving similar problems.

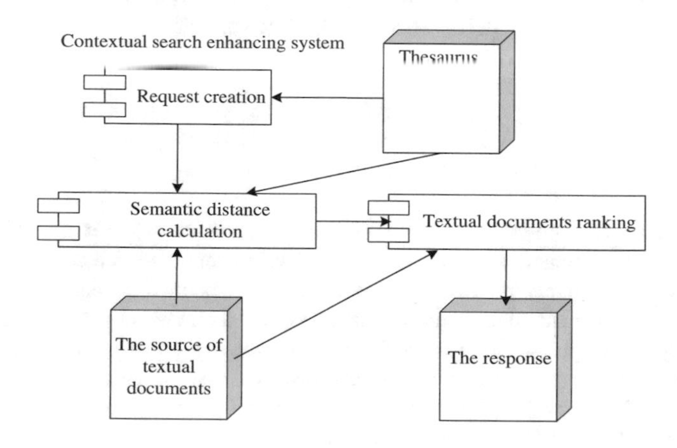

Fig. 1. The architecture of CSES

2 Previous Research

There are several definitions of semantic metrics. In Table 1 are presented the methods for evaluating text documents similarity that are based on:

- Frequencies of words usage in text document;
- The distances between words in a taxonomy;
- Frequencies and distances simultaneously.

Google distance is a degree of semantic coherence, which is calculated based on the number of pages obtained by pursuing Google search for a given set of keywords. The Table 1 shows the formula for calculation of the normalized Google distance (NGD) for two terms: x i y, where M is the total number of web-pages indexed by Google; $f(x)$ and $f(y)$ – number of pages containing keywords x i y, respectively $f(x, y)$ – number of pages containing both x, and y. If x and y are found on all pages together, then we consider NGD = 0, if they occur only separately, then we consider NGD = ∞. Let us select a class of metrics that compute similarity based on taxonomy data. These metrics are used to compute the similarity of concepts in WordNet [27], GermaNet, Wikipedia [25]. In [28] a formula is proposed that takes into account both the depth in the hierarchy of concepts, and the depth of the *lcs* (least common subsume function):

$$wup(C_1, C_2) = \frac{lcs(C_1, C_2)}{depth(C_1) + depth(C_2)} \quad (1)$$

Table 1. Semantic metrics classification

Formula/description of the algorithm	Title
1. Word frequency in text document	
$NGD(x,y) = \frac{\max(\log f(x),\log f(y)) - \log f(x,y)}{\log M - \min(\log f(x),\log f(y))}$	Normalized distance Google (NGD)
$jaccard(x,y) = \frac{Hits(x \wedge y)}{Hits(x) + Hits(y) - Hits(x \wedge y)}$	Jaccard [25]
2. Distances in the taxonomy of terms	
Distance corresponds to the number of edges in the shortest path between concepts	Metrics was used for the concepts of Roget's thesaurus [26]
$lch(C_1, C_2) = -\log \frac{length(C_1,C_2)}{2D}$	Leacock and Chodorov 1997, [27] pp. 265–283
$wup(C_1, C_2) = \frac{lcs(C_1,C_2)}{depth(C_1) + depth(C_2)}$	Wu and Palmer [28]
$res_{hypo}(C_1, C_2) = 1 - \frac{\log(hypo(lcs(C_1,C_2)) + 1)}{\log(C)}$	Metrics *res* [29], adapted to the taxonomy of the Wikipedia categories
3. Frequency words and distances in the taxonomy	
$res(C_1, C_2) = \max_{C \in S(C_1,C_2)} [-\log(P(C))]$	Distance *res* [30]
$lin(C_1, C_2) = \frac{2 \cdot \log(P(C_0))}{\log(P(C_1)) + \log(P(C_2))}$	Distance *lin* [31]
4. Text overlap	
Text overlap (based on WordNet)	Lesk [32]
Extended gloss overlap – text overlap using the neighbouring concepts from WordNet	Banerjee and Pedersen 2003 [33]
$relate_{gloss/text}(T_1, T_2) = \tanh \frac{overlap(T_1,T_2)}{length(T_1) + length(T_2)}$	Distance *relate* [25]

Ryeznyk [29] proposed to assume that two words are more similar if concept which relate these two words is more informative, that is placed lower in the taxonomy (synset in WordNet). When constructing probabilistic function $P(C)$, it is assumed that the concept's probability should not be changed while moving up the hierarchy: $res(C_1, C_2) = \max_{C \in S(C_1,C_2)} [-\log(P(C))]$. Thus, abstract concepts are less informative. Ryeznyk proposed to estimate the probability using the frequency of concept's synonyms in a text document (TD), so:

$$P(C) = \frac{freq(C)}{N} \cdot freq(C) = \sum_{n \in words(C)} count(n), \tag{2}$$

where *words* (C) are nouns with value C; N is total number of nouns in text document.

In the paper [30] Ryeznik's metric has been applied to Wikipedia and informative category was calculated as a function of the hyponyms number (categories in Wikipedia), but not statistically:

$$res_{hypo}(C_1, C_2) = 1 - \frac{\log(hypo(lcs(C_1, C_2)) + 1)}{\log(C)}, \tag{3}$$

where *lcs* is the nearest common subsumer of concepts C_1 and C_2, hypo – number of hyponyms of this subsumer, and C – total number of concepts in the hierarchy.

In [31] *lin* defines the similarity of objects *A* and *B* as the ratio of the amount of information required to describe the similarity of *A* and *B*, to the amount of information that fully describes *A* and *B*. To measure the similarity between words *lin* takes into account the frequency distribution of words in text (similar to the Reznik's measure):

$$lin(C_1, C_2) = \frac{2 \cdot \log(P(C_0))}{\log(P(C_1)) + \log(P(C_2))}, \tag{4}$$

where C_0 – nearest common super class in the concept hierarchy for both concepts C_1 and C_2, P – probability of concept, calculated on the basis of its frequency in the text document. It differs from the formula *res* by normalization method, correct computation *lin*(*x*, *x*) (independent of the concept's position in the hierarchy), takes into account existence of common and distinctive properties in objects.

In the paper [25] similarity of the two texts T_1 and T_2 is calculated from the double normalization (the length of the text and using hyperbolic tangent) as:

$$relate_{gloss/text}(T_1, T_2) = \tanh\frac{overlap(T_1, T_2)}{length(T_1) + length(T_2)}, \tag{5}$$

$$overlap(T_1, T_2) = \sum_n m^2, \tag{6}$$

where *n* phrases *m* words overlap.

Thus, the analysis of existing metrics has shown that no semantic metric is based on thesauri, and only a few of them take into account the taxonomy of concepts.

This article aims to develop and evaluate semantic metric based on domain thesaurus and method of contextual search enhancement using this metric.

3 Thesaurus-Based Search Enhancement Method

Thesaurus is a list of logical-semantic relations between linguistic terms. The thesaurus embraces not only the set of terms provided in the form of an alphabetical list of their definitions, but also contains the models, which represent relationships between terms. Based on the achievements of modern linguistics it gives in a compact and accessible form the interpretation of terminological units from terminological dictionaries and encyclopedias. The thesaurus contains terms in main research areas of theoretical and applied linguistics: grammar, word formation, lexicology, semantics, lingvosemiotic, computational linguistics, lexicography etc. For the purpose of this article, the terms were selected from the abstracts of papers, published in the Ukrainian linguistic periodicals in the 2009–2011. Building a thesaurus provides for the elucidation of the main types of relations between concepts, the main ones are correlation, synonymy, hyponymy-hypernymy, holonymy-meronymy.

The title of relation is a double predicate $R(A, B)$, which binds headword from article (A) and predicate term (B) [34]. The set of relations R is divided into types (correlation, hyperonymy - hyponymy, synonymy, holonymy-meronymy) - $R = \{R_1, R_2, \ldots, R_k\}$. n_i is the number of relations of type R_i in the thesaurus. So, the total number of relations is $N = \sum_{i=1}^{k} n_i$. Let us assume that the weight of the relation is more, when this type of relation is more frequent in the thesaurus. This weight of the ratio is defined as $L_i = \frac{n_i}{N}$. Let us weigh our semantic network that defines the thesaurus. For this purpose we define the weight of the relation between thesaurus terms. The smaller the weight, the terms are more similar. Therefore, the weight of the arcs in semantic network is defined as inversely proportional to the weight of such relation that defines this arc: $l_i = \frac{K}{L_i} = \frac{K \cdot N}{n_i}$, where K is some constant that specifies the unit of weight measurement for arcs in semantic network [35–37].

The weighted semantic network is used to find potential partners who are engaged in similar research issues in the domain for which the thesaurus was built.

To do this, a set of key terms $C = \{C_1, C_2, \ldots, C_n\}$ from the thesaurus should be created, which is defining the specific research issues. Search engine finds a set of documents, which contain terms from the thesaurus. For each such document T_s a set with capacity m is built containing the terms from the thesaurus that are frequently used in T_s: $\hat{C}^s = \left\{\hat{C}_1^s, \hat{C}_2^s, \ldots, \hat{C}_m^s\right\}$. Using the Floyd-Warshall or Deikstra method [38] $n \times m$ of the shortest distance $d_{ij}^s = d\left(C_i, C_j^s\right)$ between terms from sets C and $\hat{C}^s$ are obtained. Then the distance to the document found T_s according to the formula: $d^s = \sum_{i=1}^{n} \sum_{j=1}^{m} d_{ij}^s$ is calculated. The found documents are ranked according to increasing values d^s. The authors of the documents with the higher rank may be our potential partners. [39–46]. Let us consider an example of thesaurus for text's linguistics domain. Corresponding semantic network is shown on Fig. 2. In vertexes of the network are placed terms. The links between terms are denoted by "S" – synonyms, "CR" – correlation, "Hol" – holonym, "M" – meronym, "Hyp" – hypernym.

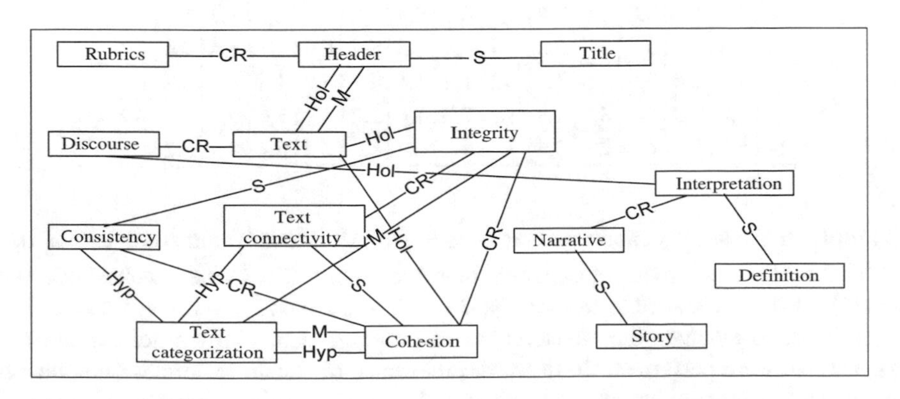

Fig. 2. Semantic network for thesaurus of text linguistics

After the application of the formulae (1) we obtain the values for networks edges: for synonimical link – 1; correlation link – 1,2; holonym and meronim links – 1,4; hypernym – 1,5. The conceptual graph with correspondingly weighted vertexes is shown on Fig. 3. Let us find the shortest paths between terms of this graph using Floyd-Warshall algorithm [38]. The resulting matrix of distances is shown in Table 2.

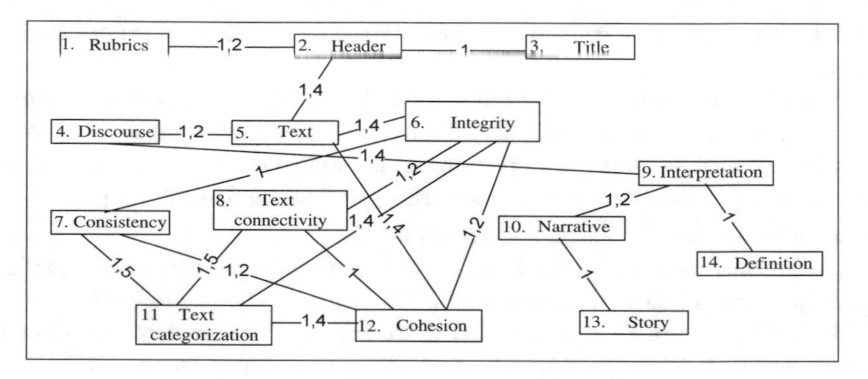

Fig. 3. The weighted conceptual graph

Table 2. The distances between terms in conceptual graph.

	1	2	3	4	5	6	7	8	9	10	11	12	13	14
1		1,2	2,2	3,8	2,6	4	5	5,2	5,2	6,4	5,4	4	7,4	6,2
2	1,2		1	2,6	1,4	2,8	3,8	4	4	5,2	4,2	2,8	6,2	5
3	2,2	1		3,6	2,4	3,8	4,8	5	5	6,2	5,2	3,8	7,2	6
4	3,8	2,6	3,6		1,2	2,6	3,6	3,8	1,4	2,6	4	2,6	3,6	2,4
5	2,6	1,4	2,4	1,2		1,4	2,4	2,6	2,6	3,8	2,8	1,4	4,8	3,6
6	4	2,8	3,8	2,6	1,4		1	1,2	4	5,2	1,4	1,2	6,2	5
7	5	3,8	4,8	3,6	2,4	1		2,2	5	6,2	1,5	1,2	7,2	6
8	5,2	4	5	3,8	2,6	1,2	2,2		5,2	6,4	1,5	1,4	7,4	6,2
9	5,2	4	5	1,4	2,6	4	5	5,2		1,2	5,4	4	2,2	1
10	6,4	5,2	6,2	2,6	3,8	5,2	6,2	6,4	1,2		6,6	5,2	1	2,2
11	5,4	4,2	5,2	4	2,8	1,4	1,5	1,5	5,4	6,6		1,4	7,6	6,4
12	4	2,8	3,8	2,6	1,4	1,2	1,2	1,4	4	5,2	1,4		6,2	5
13	7,4	6,2	7,2	3,6	4,8	6,2	7,2	7,4	2,2	1	7,6	6,2		3,2
14	6,2	5	6	2,4	3,6	5	6	6,2	1	2,2	6,4	5	3,2	

Example. If the set C contain two terms C = {'rubric', 'text'} and some textual document T is described by a set $\hat{C}$, which has three terms $\hat{C}$ = {'title', 'cohesion', 'narrative'}, then the distance to this text is: $d = 2{,}2 + 6{,}4 + 4{+}2{,}4 + 3{,}8 + 1{,}4 = 20{,}2$.

In order to evaluate the effectiveness of proposed semantic metric, the series of experiments were performed. In them, the relevance of scientific articles (annotations, documents), found in Internet was evaluated related to query, containing keywords from developed thesaurus. For comparison, the relevance of found texts to keywords was also calculated, using other knows metrics – the metrics of Jaccard and WUP.

Ten experiments were executed for each method. The results were additionally evaluated by an expert. From the set of found relevant documents he selected ones which were truly relevant according to his opinion and search criteria with a purpose to analyze the precision of compared methods.

$$precision = \frac{number\_found\_relevant}{number\_all\_found} \quad (7)$$

So, the effectiveness of semantic metrics were evaluated using precision parameter τ: $\tau = r_e/r_m \cdot 100$, where $r_e = |R_e|$ – the cardinality of set R_e of relevant documents, found with corresponding method (according to opinion of domain expert); $r_m = |R_m|$ – the cardinality of set R_m of all documents, found with corresponding method, $R_e \subseteq R_m$. The results of all ten experiments are shown on Fig. 4. The precision of relevant documents search performed with developed system is higher (dark columns) than precision when using Jaccard and WUP methods.

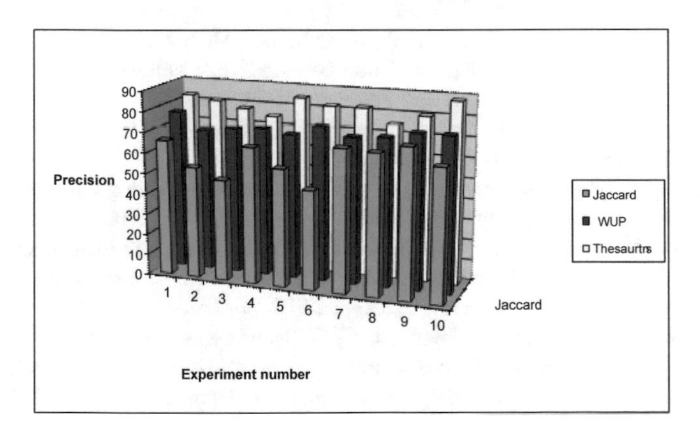

Fig. 4. The results of experiments aiming to determine the precision of search for relevant documents with different methods

Overall, the effectiveness of approach using thesaurus-based semantic metric defined by parameter τ, in average is in the range from 10 to 20% better when compared with other metrics. Higher efficiency is obtained due to usage of weighted conceptual graph representing thesaurus.

4 Conclusion

The continuous growth of number and size of textual informational resources in Internet requires development of machine-assisted search methods allowing to enhance the relevance of search results with domain knowledge. An approach for building a contextual search using semantic metric based on domain thesaurus was proposed. This thesaurus is represented as semantic network graph. The vertexes of the graph are weighted reverse proportionally to the number of relations of specified type. Semantic metric is built using this semantic network graph. The application of proposed method to the task of searching research partners is demonstrated.

Experiments have shown that the search using Jaccard's metric only in 40% of cases found the documents having the largest number of words in common with a search query. The search using WUP metric based on number of common connections also did not provide satisfactory results. On the other hand, the usage of domain information from thesaurus and weighting of vertexes in conceptual graph allowed to find the documents which are more relevant to keywords provided by end-user.

References

1. The Internet Economy in the G-20 – BCG. https://www.bcg.com/documents/file100409.pdf
2. Vysotska, V., Chyrun, L., Chyrun, L.: Information technology of processing information resources in electronic content commerce systems. In: Computer Science and Information Technologies, CSIT 2016, pp. 212–222 (2016)
3. Vysotska, V., Chyrun, L., Lytvyn, V., Dosyn, D.: Methods Based on Ontologies for Information Resources Processing. LAP Lambert Academic Publishing, Saarbrücken (2016)
4. Vysotska, V., Chyrun, L.: Analysis features of information resources processing. In: Proceedings of Xth International Conference on Computer Science and Information Technologies, CSIT 2015, pp. 124–128 (2015)
5. Gladun, A.: Building thesaurus of subject area as a tool for modeling information needs of a user for Internet search. Comput. Inf. Technol. Rev. **1**, 26–33 (2007)
6. Shakhovska, N., Vysotska, V., Chyrun, L.: Features of E-learning realization using virtual research laboratory. In: Proceedings of the XIth International Conference on Computer Science and Information Technologies, CSIT 2016, pp. 143–148 (2016)
7. Shakhovska, N., Vysotska V., Chyrun, L.: Intelligent systems design of distance learning realization for modern youth promotion and involvement in independent scientific researches. In: Advances in Intelligent Systems and Computing, pp. 175–198. Springer (2017)
8. Lytvyn, V., Vysotska, V., Chyrun, L., Chyrun, L.: Distance learning method for modern youth promotion and involvement in independent scientific researches. In: Proceedings of the IEEE First International Conference on Data Stream Mining and Processing (DSMP), pp. 269–274 (2016)
9. Gruber, T.: A translation approach to portable ontologies. Knowl. Acquisition **5**(2), 199–220 (1993)
10. Gavrilova, T.: Knowledge Bases in Intelligent Systems. Piter, Saint Petersburg (2001). 384 pages
11. Lytvyn, V., Pukach, P., Bobyk, I., Vysotska, V.: The method of formation of the status of personality understanding based on the content analysis. Eastern Eur. J. Enterp. Technol. **5/2** (83), 4–12 (2016)
12. Lytvyn, V., Vysotska, V., Veres, O., Rishnyak, I., Rishnyak, H.: Classification methods of text documents using ontology based approach. In: Advances in Intelligent Systems and Computing, vol. 512, pp. 229–240. Springer International Publishing AG (2017)

13. Lytvyn, V., Vysotska, V., Pukach, P., Bobyk, I., Pakholok, B.: A method for constructing recruitment rules based on the analysis of a specialist's competences. Eastern Eur. J. Enterp. Technol. **6/2**(84), 4–14 (2016)
14. Lytvyn, V., Vysotska, V., Pukach, P., Brodyak, O., Ugryn, D.: Development of a method for determining the keywords in the Slavic language texts based on the technology of web mining. Eastern Eur. J. Enterp. Technol. **2/2**(86), 4–12 (2017)
15. Lytvyn, V., Vysotska, V., Veres, O., Rishnyak, I., Rishnyak, H.: Content linguistic analysis methods for textual documents classification. In: Proceedings of the XIth International Conference on Computer Science and Information Technologies, CSIT 2016, pp. 190–192 (2016)
16. Lytvyn, V., Vysotska, V.: Designing architecture of electronic content commerce system. In: Computer Science and Information Technologies, CSIT 2015, pp. 115–119 (2015)
17. Khomytska, I., Teslyuk, V.: The method of statistical analysis of the scientific, colloquial, Belles-Lettres and newspaper styles on the phonological level. In: Advances in Intelligent Systems and Computing, vol. 512, pp. 149–163 (2017)
18. Khomytska, I., Teslyuk, V.: Specifics of phonostatistical structure of the scientific style in english style system. In: Proceedings of the XIth International Conference on Computer Science and Information Technologies, CSIT 2016, pp. 129–131 (2016)
19. Vysotska, V.: Linguistic analysis of textual commercial content for information resources processing. In: Modern Problems of Radio Engineering, Telecommunications and Computer Science, TCSET 2016, pp. 709–713 (2016)
20. Vysotska, V., Chyrun, L., Chyrun, L.: The commercial content digest formation and distributional process. In: Proceedings of the XIth International Conference on Computer Science and Information Technologies, CSIT 2016, pp. 186–189 (2016)
21. Lytvyn, V., Uhryn, D., Fityo, A.: Modeling of territorial community formation as a graph partitioning problem. Eastern Eur. J. Enterp. Technol. **1/4**(79), 47–52 (2016)
22. Lytvyn, V., Tsmots, O.: The process of managerial decision making support within the early warning system. Actual Prob. Econ. **11**(149), 222–229 (2013)
23. Chen, J., Dosyn, D., Lytvyn, V., Sachenko, A.: Smart data integration by goal driven ontology learning. In: Advances in Big Data. Advances in Intelligent Systems and Computing, pp. 283–292. Springer (2016)
24. Lytvyn, V., Dosyn, D., Smolarz, A.: An ontology based intelligent diagnostic systems of steel corrosion protection. Elektronika **8**, 22–24 (2013)
25. Strube, M.: WikiRelate! Computing semantic relatedness using Wikipedia. In: Proceedings of the 21st National Conference on Artificial Intelligence. http://www.eml-research.de/english/research/nlp/public
26. Jarmasz, M.: Roget's thesaurus and semantic similarity. In: Proceedings of Conference on Recent Advances in Natural Language Processing, Borovets, Bulgaria, pp. 212–219 (2003)
27. Fellbaum, C.: WordNet: an Electronic Lexical Database. MIT Press, Cambridge (1998). 423 pages
28. Wu, Z.: Verb semantics and lexical selection. In: Proceedings of ACL-94, pp. 133–138 (1994)
29. Resnik, P.: Disambiguating noun groupings with respect to WordNet senses. http://xxx.lanl.gov/abs/cmp-lg/9511006
30. Resnik, P.: Semantic similarity in a taxonomy: an information-based measure and its application to problems of ambiguity in natural language. J. Artif. Intell. Res. (JAIR) **11**, 95–130 (1999)
31. Lin, D.: An information-theoretic definition of similarity. In: Proceedings of International Conference on Machine Learning (1998). http://www.cs.ualberta.ca/~lindek/papers.htm

32. Smirnov, A.: Ontologies in artificial intelligence systems. In: Artificial Intelligence News, vol. 2, pp. 3–9. (in Russian)
33. Sovpel, I.: The system for automatic extraction of knowledge from text and it's applications. Artif. Intell. **3**, 668–677 (2004). (in Russian)
34. Nikitina, S.E.: Thesaurus on theoretical and applied linguistics. Science, 14 (1978)
35. Lytvyn, V., Shakhovska, N., Pasichnyk, V., Dosyn, D.: Searching the relevant precedents in dataspaces based on adaptive ontology. Comput. Prob. Electric. Eng. **2**(1), 75–81 (2012)
36. Lytvyn, V., Dosyn, D.: Planning of intelligent diagnostics systems based domain ontology. In: VIIIth International Conference on Perspective Technologies and Methods in MEMS Design, vol. 103 (2012)
37. Lytvyn, V., Dosyn, D., Medykovskyj, M., Shakhovska, N.: Intelligent agent on the basis of adaptive ontologies construction. In: Signal Modelling Control (2011)
38. Svami, M.: Graphs, Networks and Algorithms (1984). 256 pages
39. Montes-y-Gómez, M., Gelbukh, A., López-López, A.: Comparison of Conceptual Graphs. Lecture Notes in Artificial Intelligence, vol. 1793 (2000)
40. Knappe, R., Bulskov, H., Andreasen, T.: Perspectives on ontology-based querying. Int. J. Intell. Syst. (2004). http://akira.ruc.dk/~knappe/publications/ijis2004.pdf
41. Basyuk, T.: The main reasons of attendance falling of internet resource. In: Proceedings of the Xth International Conference on Computer Science and Information Technologies, CSIT 2015, pp. 91–93 (2015)
42. Kravets, P., Kyrkalo, R.: Fuzzy logic controller for embedded systems. In: Proceedings of 5th International Conference on Perspective Technologies and Methods in MEMS Design, MEMSTECH (2009)
43. Mykich, K., Burov, Y.: Algebraic framework for knowledge processing in systems. In: Advances in Intelligent Systems and Computing, pp. 217–228. Springer (2017)
44. Pasichnyk, V., Shestakevych, T.: The model of data analysis of the psychophysiological survey results. In: Advances in Intelligent Systems and Computing, vol. 512, pp. 271–282. Springer International Publishing AG (2016)
45. Lytvyn, V.: Design of intelligent decision support systems using ontological approach. Int. Q. J. Econ. Technol. New Technol. Model. Process. **II**(1), 31–38 (2013)
46. Burov, E.: Complex ontology management using task models. Int. J. Knowl. Based Intell. Eng. Syst. Amsterdam **18**(2), 111–120 (2014)

Optimizing Wind Farm Structure Control

Vitalii Kravchyshyn[1(✉)], Mykola Medykovskyy[1], Roman Melnyk[1], and Marianna Dilai[2]

[1] Department of Automated Control Systems, Lviv Polytechnic National University, 12, S. Bandera St., Lviv 79013, Ukraine
{vitalikl99lua,medykmo,hjvfyfyfy}@gmail.com

[2] Department of Applied Linguistics, Lviv Polytechnic National University, 12, S. Bandera St., Lviv 79013, Ukraine
mariannadilai@gmail.com

Abstract. The paper presents a solution to the wind farm structure optimization problem (determination of the optimal set of active wind turbines at a given time). The significance of wind parameters prediction for determining the wind farm active set is justified. The theoretical solution to this problem enables refining control algorithms for the power-dynamic modes of wind farms in order to increase their efficiency by reducing operational losses and electrodynamic overload when switching wind turbines. It can be achieved by estimating the future wind power capacity based on the short-term prediction of wind speed values in combination with other mode parameters. The findings are particularly relevant when electric power storage element is used in a wind farm structure. In this case, the key variables that determine the mode parameters are current and future wind speed, consumer load, current energy capacity of a storage battery, effectiveness and availability of wind turbines (WT). As a result, we receive the optimal number of active (operating at a given time) wind turbines not only for the current, but also the next parameters of power-dynamic modes minimizing switching operations. The effectiveness of the available methods of time series prediction depends on the knowledge and skills of experts, and subjectivity in the development of mathematical models. Hence, the feasibility of artificial neural networks application for prediction is surveyed. The research shows the results of the implementation of such networks as a classical back propagation network, the Elman network and the cascade neural network. The impact of the energy storage element capacity on the number of switching operations in the wind farm structure is investigated. We offer the solution to the problem of optimal choice of the energy storage element capacity which will balance meeting maximum load requirements and the cost of its installation and operation. As a practical result, the algorithm (rules of inference) for minimizing the number of switching operations of the active set of WF is developed taking into account such parameters as current wind speed, archival wind speed, predicted value of wind speed, current load value, next load value, nominal output of a wind farm, types of wind turbines, the number of wind turbines of each type,

The original version of this chapter has been revised: The reference number [9] and its citation name have been amended. A correction to this chapter can be found at https://doi.org/10.1007/978-3-319-70581-1_46

N. Shakhovska and V. Stepashko (eds.), *Advances in Intelligent Systems and Computing II*, Advances in Intelligent Systems and Computing 689,
https://doi.org/10.1007/978-3-319-70581-1_23

technical condition of each wind turbine, power capacity of a storage battery, etc. The results of the experimental studies of the developed algorithm efficiency are presented.

Keywords: Wind farm · Neural networks · Wind speed prediction
Wind turbine

1 Introduction

The estimated capacity of the modern wind farms (WF) is about tens or even hundreds of Megawatts. A wind farm consists of a set of automated wind turbines (WT), energy storage elements, dispatch control systems for power-dynamic modes of parallel operation with a distribution network [1, 2]. At the same time, increasingly frequently, in addition to generating active electric power, wind farms perform functions of controlling distribution networks modes in order to provide the required energy quality and reduce its process loss [3]. The possibility of performing these functions is predetermined by irregularity of load curve (there are idle capacities in the periods of daily, weekly and annual minimum loads) and wind power capacity (excess of power generation capacity). Moreover, a long-term operation of wind turbines requires scheduled outage and maintenance work which makes it impossible to use them for production functions for some period of time. Such restrictions may also be imposed during emergencies or in pre-emergency situations. Consequently, at a given time, there is a need and it is possible to use a certain set of wind turbines to generate active electric power, hereinafter they will be referred to as an active set of a wind farm.

As for the remaining set of wind turbines, there is either no need or no possibility to use them for the stated purpose. They currently perform other functions or are not available for use. Taking into account the fact that each wind turbine may have different nominal parameters (power capacity, operating wind speed range) and performance criteria (service life, active mode time, objective parameters of mechanical and electromagnetic systems), we can state that the formation of the active set of a wind farm is a topical problem of multi-parameter optimization researched in a number of studies [4, 5].

The overview of well-known works shows that today insufficient attention is devoted to issues concerning the efficient use of fixed assets, that is, the reduction of operating costs for energy production by ensuring uniform wear of system elements [6, 7]. Particularly acute this issue is for WF operation due to the stochastic nature of the wind and load curve which requires frequent switching of units. Switching operations of powerful electromechanical systems trigger transition processes which as a result speed up wearing out of elements. Consequently, uniform wear and minimization of WF elements' switching are important parameters for optimizing the modes and overall performance of WF.

This paper presents the research findings obtained with regards to the stochastic nature of the wind given a power storage element and the active power load curve.

The research goal is to optimize the control of power-dynamic modes of modern wind farms with the energy storage element by minimizing the number of switching operations based on the wind speed prediction, taking into account WT technical condition and load curve requirements.

2 Description of the Data

Recently, the problems of real-time data prediction have become particularly relevant in the field of optimization of various processes due to the rapid development of algorithms for finding an optimal and effective solution. The main idea of using prediction methods for determining the active set of a wind farm is to take into account the future values of the wind speed looking for an efficient solution, in order to minimize the number of switching operations for the WF active set, and to enable more flexible control of the individual components of the system in order to improve their performance.

To date there are a large number of solutions that allow us to make predictions employing probabilistic methods and subjective knowledge of experts [8–10]. The main drawbacks of such methods are as follows: insufficient accuracy of prediction, dependence of the results on the knowledge and qualifications of the expert in a particular subject area, subjectivity in the development of the mathematical model and structure, and, accordingly, the difficulties with the implementation of such models, etc.

That is why, in order to obtain an adequate solution for time series prediction, it is appropriate to use models with artificial neural networks [11, 12]. The construction of a neural network consists of two stages: the choice of the network type (architecture) and the selection of weights, that is, the neural network training.

A specific type of transformation in the neural network is determined not only by the characteristics of neural-like elements (activation functions, etc.), but also directly depends on its architecture.

At the first stage, it is necessary to select neurons that will be used in the network (taking into account the number of inputs, activation functions), how they will be interconnected, and which will be the input and output of the neural network. Based on the functions performed by neurons in the network, they can be divided into three types: input, output and intermediate [13]. In most networks, the type of neuron is associated with its position in the network.

At the second stage, it is necessary to "train" the chosen network, that is, to select the weight values, so that the network operates properly.

Three types of artificial neural networks are used to solve the problem of dynamic wind speed prediction, namely the classical unidirectional artificial back propagation network, the Elman neural network and the cascade neural network.

Two types of computer experiments were carried out to predict wind speed in different time intervals. The data were obtained from [14], measurements were taken daily at 2:00, 5:00, 8:00, 11:00, 14:00, 17:00, 20:00, 23:00 at the height of 10–12 m:

1. For the first experiment, a test sample of wind speed statistics was created for the city of Askania-Nova for the period from January 1, 2014 to December 31, 2014 inclusive.
2. For the second experiment, four test samples of wind speed statistics were created for the city of Askania-Nova for the periods from March 1 to March 31, 2014, from June 1 to June 30, 2014, from September 1 to September 31, 2014, and from December 1 to December 31, 2014.

The output reference data for both experiments were represented by the samples of the wind speed statistics for Askania-Nova for similar periods of 2015, namely, from March 1 to March 31, 2015, from June 1 to June 30, 2015, from September 1 to September 30, 2015 and from December 1 to December 31, 2015.

As a result, the structure of the neural network which allows us to predict with a relatively constant accuracy the wind speed value for the studied geographical location was justified. The flowchart of unidirectional back propagation network is given in Fig. 1.

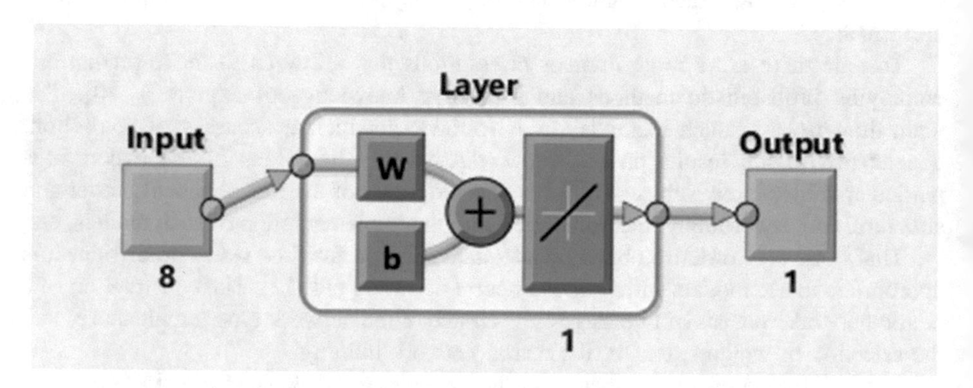

Fig. 1. The flowchart of the unidirectional back propagation neural network

The flowchart shows training neural network without hidden layers, 8 serial values of the wind speed are fed into the input layer, and 1 new output value is obtained. According to this scheme, the neuron uses a linear activation function to get the output value. To train the neural network, a function that modifies weights and biases according to the elastic back propagation algorithm is used.

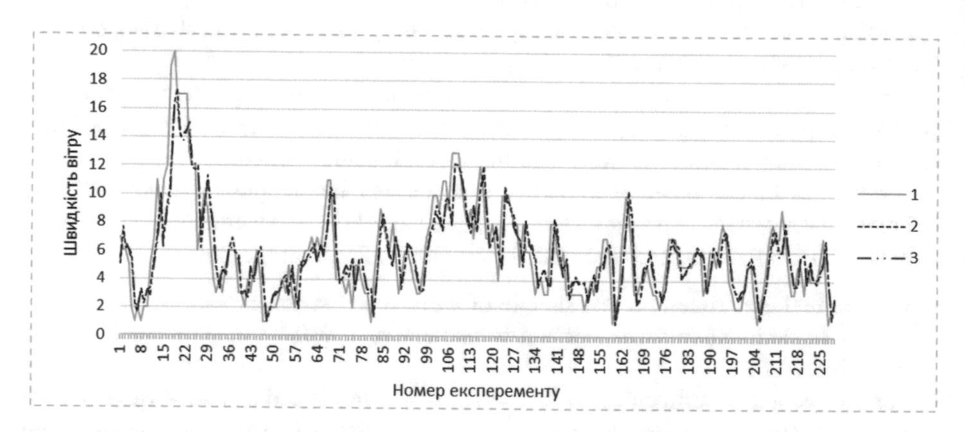

Fig. 2. The dynamics of wind speed variations on real-time and training samples for Askania-Nova for the period of March 1–31, 2015 (1 - the reference values, 2 - the prediction obtained as a result of the first type of experiments, 3 - the prediction obtained as a result of the second type of experiments)

The results of the conducted experiments are given in Figs. 2, 3, 4 and 5:

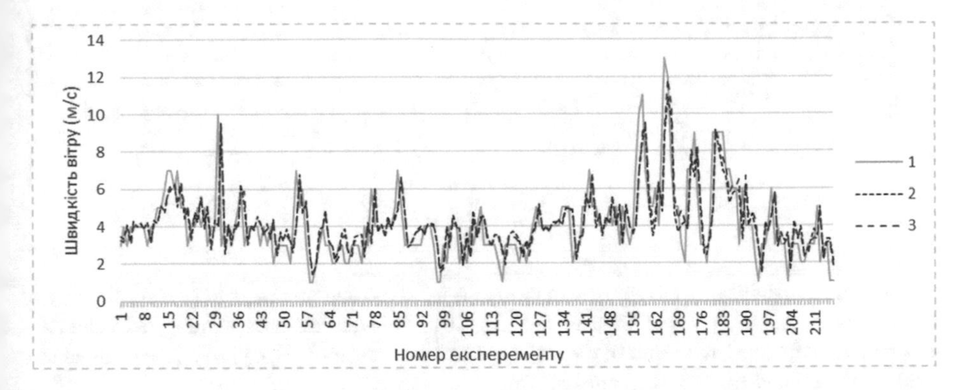

Fig. 3. The dynamics of wind speed variations on real-time and training samples for the city of Askania-Nova for the period of June 1–30, 2015 (1- the reference wind speed values, 2- the prediction obtained as a result of the first of type experiments, 3 - the prediction obtained as a result of the second type of experiments)

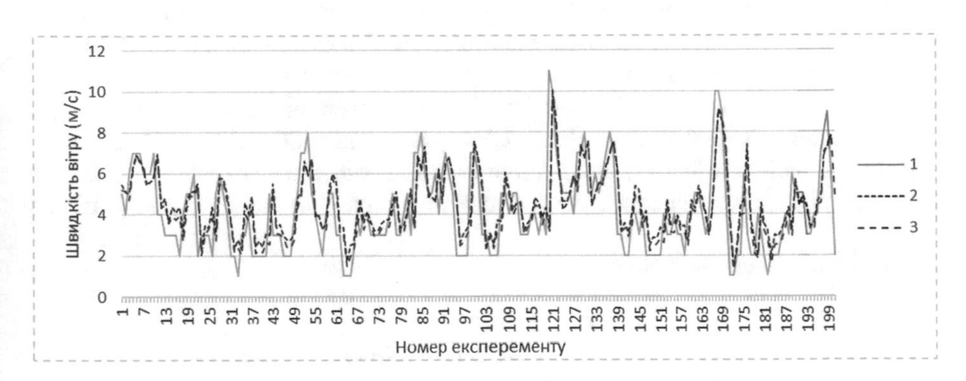

Fig. 4. The dynamics of wind speed variations for the city of Askania-Nova for the period of September 1–30, 2015 (1 – the reference wind speed values, 2 – the prediction obtained as a result of the first type of experiments, 3 – the prediction obtained as a result of the second type of experiments)

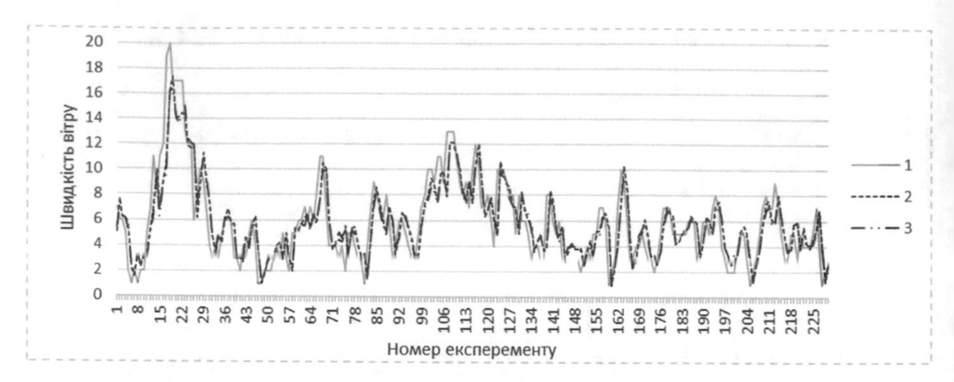

Fig. 5. The dynamics of wind speed variations on real-time and training samples for the city of Askania-Nova for the period of December 1–31, 2015 (1 – the reference wind speed value, 2 – the prediction obtained as a result of the first type of experiments, 3 – the prediction obtained as a result of the second type of experiments)

3 Data Processing

The objective prediction quality assessment is an important point in the process of prediction. Since prediction assessments are random variables, a set of statistical criteria is used to determine it. To assess the prediction, the mean square error (MSE) can be used. However, the MSE value is not sufficient to assess the prediction quality, since it depends on the scale of the data. An in-depth assessment of the prediction quality is achieved using criteria that give relative quality assessment (Theil's coefficient, or the forecast inequality coefficient) and relative assessment in percentage terms. The main advantage of such assessments is the fact that they do not depend on the scale of data [15].

Theil's coefficient is defined as:

$$U = \frac{\sqrt{\frac{1}{s}\sum_{k=1}^{s}\left[y(k+i) - \hat{y}(k+i)\right]^2}}{\sqrt{\frac{1}{s}\sum_{i=1}^{s} y^2(k+i)} + \sqrt{\frac{1}{s}\sum_{i=1}^{s} \hat{y}^2(k+i)}}, \quad (1)$$

where S is the prediction horizon, $y(k+i)$ is actual data values, $\hat{y}(k+i)$ is prediction assessments. Theil's coefficient is an important prediction model quality index. The values of this coefficient by definition are in the range from zero to one.

In this way prediction assessments approximate the actual values of the series and, accordingly, the model has a high degree of adequacy. Based on the assessment it is possible to determine the appropriateness of a model or a method for a time series prediction.

In this research, in order to assess the prediction quality, Theil's coefficient, the mean square error and the mean absolute deviation are calculated.

The prediction mean square error is calculated as follows:

$$\varpi = \sqrt{\frac{\sum_{i=1}^{n} (y_i - y_i')^2}{n}} \quad (2)$$

where y_i is the reference value of the i-th element and y_i' is the predicted value of the i-th element, n is the number of values in the sample.

The average relative prediction error is calculated as:

$$\varphi = \frac{\sum_{i=1}^{n} \frac{|y_i - y_i'|}{y_i}}{n} * 100 \quad (3)$$

The research findings are given in Table 1. The analysis of the results shows the possibility of employing the structure of the studied neural network to predict wind speed at any time of the year.

Table 1. The prediction results

	March		June		September		December	
	exp1	*exp2*	*exp1*	*exp2*	*exp1*	*exp2*	*exp1*	*exp2*
Theil's coefficient	0,2708	0,2708	0,3299	0,3224	0,3186	0,3115	0,3276	0,3246
The mean square error	3,3347	3,3344	2,2398	2,139	2,1549	2,0607	2,4429	2,3984
The average relative error, %	34,6102	34,6039	34,6007	33,1866	36,8742	32,6494	30,5681	29,1537

Based on the research, the appropriateness of using unidirectional network with 8 input elements and 1 output element for wind speed prediction in the areas with high wind power potential was justified.

To train the neural network we used the wind speed data for the study area measured over the same period of the previous year in the first experiment, and over the period of one year in the second experiment.

The line charts of the Elman neural networks, the cascade neural network and the unidirectional back propagation neural network for the short-term wind speed prediction in the city of Drohobych, Lviv region, over the period from January 1 to January 31, 2015 inclusive are given in Fig. 6. Sixty four neurons are fed into the system's input layer; the result is one output wind speed value. To solve the problem of wind speed prediction in the study area, three line charts of neural networks were designed. Each of the line charts shows a different type of neural networks, namely the Elman network, the unidirectional back propagation neural network and the cascade neural network.

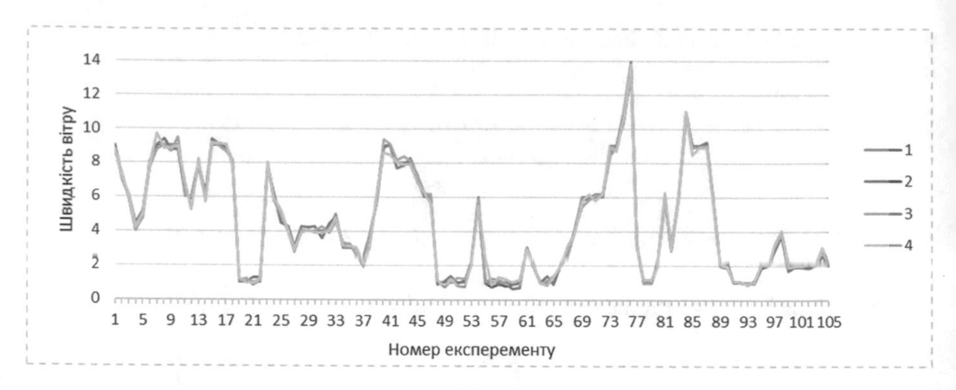

Fig. 6. The dynamics of wind speed on real-time and training samples for the city of Drohobych over the period of December 1–14, 2015 (1 – the reference values of wind speed, 2 – the data obtained by using the unidirectional back propagation neural network, 3 – the data obtained using the Elman network, 4 – the data obtained using the cascade neural network)

The training of the neural networks was carried out and the analysis of the obtained prediction data was conducted. The results of the wind speed prediction for the city of Drohobych during the period from December 1 to December 14, 2015 are shown on the charts.

The analysis of the results of wind speed prediction is presented in Table 2.

Table 2. The prediction results

Neural network structure	Back propagation	Elman	Cascade
Theil's coefficient	0,043	0,042	0,0474
The mean square error	0,0559	0,0535	0,068
The average relative error, %	17,6988	18,1302	17,1876

The analysis of the findings allows us to draw conclusions on the accuracy of prediction achieved by using each of the studied types of neural networks (NN). The discrepancy coefficient for each of NN types is less than 0.05%, and the average absolute error varies in the range of 17%–18%.

The previous section explores the neural network structure which can be used for the given wind power region to make the short-term wind speed prediction at any time. Of course, the universality of the neural network structure increases the prediction error, so in each particular case it may be necessary to improve the prediction accuracy by changing neural network structure. The prediction accuracy is improved by applying different types of neural networks, methods (types) of NN training, different number of input and output signals.

The analysis of the results shows that it is appropriate to use artificial neural networks for wind speed prediction. In addition, we have revealed the dependence of the deviation of wind speed predicted values on the selected artificial neural network structure, the size of training sample and the network learning function.

Wind parameters prediction is essential for determining the wind farm active set, as it allows us to revise solutions to the problems as follows:

- reducing the number of switching operations which is achieved based on the predicted values of wind speed in conjunction with other parameters such as current capacity of a storage battery, future load value and the number of wind turbines available in a wind farm. As a result, we obtain the optimal set of active wind turbines not only for the current but also the next parameters of the power-dynamic modes;
- improving wind farm overall operational efficiency by reducing operational losses.

As a result, we obtained advanced input data processing techniques, modification of dynamic programming method for determining wind farm active set, employing the method of wind speed prediction to enhance the performance of the set of active wind turbines, and improving basic energy parameters of a wind farm.

For the effective employment of the developed method, it is necessary to form inference rules which will ensure reduction in the number of the active wind turbines set determinations and optimize the use of power-generating equipment.

The key parameters that should be considered include current wind speed, archival wind speed, predicted value of wind speed, current load value, next load value, nominal output of a wind farm, types of wind turbines, the number of wind turbines of each type, technical condition of wind turbines, energy capacity of a storage battery, etc.

The algorithm for minimizing the number of switching operations of the active set of WF considering the above parameters is as follows:

1. The future value is predicted based on the wind speed statistics. The input parameters are as follows: P_1 is the current load value, P_2 is the next load value, υ_1 is the current wind speed, υ' is the next predicted wind speed value, Q - the energy capacity of a storage battery.
2. Given the load $P_1 \geq P_2$, and the wind speed $\upsilon_1 \leq \upsilon'$, the parameters P_1 and υ_1 are used as the input parameters for the method of dynamic determination of the active wind turbines.
3. Given the load $P_1 \geq P_2$, and the wind speed $\upsilon_1 > \upsilon'$, the parameters P_1 and υ' are used as the input parameters for the method of dynamic determination of the active wind turbines.
4. Given the load $P_1 < P_2$, and the wind speed $\upsilon_1 \leq \upsilon'$, the parameters P_2 and are used as the input parameters for the method of dynamic determination of the active wind turbines.
5. Given the load $P_1 < P_2$, and the wind speed $\upsilon_1 > \upsilon'$, the parameters P_2 and υ' are used as the input parameters for the method of dynamic determination of the active wind turbines.

The use of such inference rules will reduce the number of repackings as the active set of WF is determined taking into accounted current and next values of load and wind speed, and the wind farm power output is to fully meet the needs of consumers for a longer time interval. The obtained data are used as input parameters in determining the active set of wind farm employing the method of dynamic programming and its modification. Obtaining the next value of the wind speed υ_2 consistency check is carried out, the value is compared with υ_1, υ'. If the wind speed value is within the range of $[\upsilon_1, \upsilon']$ or higher than υ_1 and υ', any excess energy can be used to charge the storage element (in case of incomplete charge). In other cases, it is advisable to use the storage element to mitigate transients and reduce the number of switching operations of active wind turbines. Therefore, to reduce the number of switching operations of a wind turbine it is important to take into account the energy capacity of a storage battery.

The results of statistical analysis of wind speed, namely the probability of recurrence of a particular wind speed and wind power potential can be used for substantiating wind farm structure using energy storage element as a part of the input data in the analysis of environmental factors that affect the installation of wind farms, along with consumer electrical load which is planned to be transferred to the wind farm in the future. They also allow us to optimize the energy capacity of a storage element for the electrical load curve.

On the other hand, since at present the cost of power storage elements is high, it should be justified not only technically, but also economically. The problem of reasonable choice of an optimal efficient and cost-effective energy storage element which can ensure balance between meeting maximum load requirements and the cost of installation and operation is topical.

A computer simulation of wind farm control system was carried out using test data to determine the optimal, in terms of the number of switching operations, storage element size in percentage terms relative to the nominal output of the wind farm.

A random sample consists of 1000 load values in the interval [5000; 15000] kW. The operating modes of the wind farm which consists of 60 wind turbines were investigated. The key characteristics of the wind turbines are given in Table 3.

Table 3. The key characteristics of WF turbines

The type of WT	ENERCON E-48	ENERCON E-44	V52-850
The number of WT	20	20	20
Rotor power output	800 kW	900 kW	850 kW
Rotor diameter	48 m	44	52
Cut-in speed	3	3	3
Rated wind speed	17	14	14
Cut-out speed	25	25	25

Similarly, a random sample of 1000 wind speed values in the interval [5; 15] m/s was created.

For each energy capacity of a storage battery 1000 experiments were carried out.

The actual power output of each wind turbine depends on the current wind speed and operational characteristics of the wind turbine (the number of starts and stops, the amount of generated power, the operating range of wind speeds, turbine's technical condition, operating time).

The results of computer simulation of the wind farm operation were analyzed implementing the developed modification and the determined characteristics of the wind farm (the WT power output, the energy capacity of a storage element, the number of wind turbines of each type).

The results of computer simulation of the wind farm control system, using the energy storage element of different capacities is presented in Tables 4 and 5.

Table 4. The results of using storage batteries of different capacities in the structure of WF (classical method of dynamic programming)

The energy capacity of a storage battery (% of determined WF power output)	0	1	5	10	15	20	25	30
Maximum deviation %	−2,465	−2,405	−2,445	−2,445	−2,435	−2,435	−2,415	−2,42
The average percentage of underpacking/repacking (%)	1,29	1,17	0,92	0,58	0,44	0,52	0,41	0,38
The number of repackings	887	809	612	394	273	244	203	184
The number of experiments	1000							

Table 5. The results of using SB of different capacities in the structure of WF (modification of the method of dynamic programming)

The energy capacity of a storage battery (% of determined WF power output)	0	1	5	10	15	20	25	30
Maximum deviation %	−2,46	−2,45	−2,495	−2,44	−2,49	−2,125	−2,45	−2,355
The average percentage of underpacking/repacking (%)	0,68	0,58	0,46	0,32	0,21	0,27	0,28	0,28
The number of repackings	864	806	625	383	260	199	183	157
The number of experiments	1000							

Two types of experiments were carried out:

- employing the classical method of dynamic programming to solve the knapsack problem when determining the active set of the wind farm;
- employing the developed modification of the dynamic programming method for determining the active set of the wind farm.

The First Type of Experiments

Figure 7 shows the results of the simulation of wind farm control system operation with the energy storage element with different capacities (0%, 1%, 5%, 10%, 15%, 20%, 25%, 30%) employing the classical method of dynamic programming (CDP).

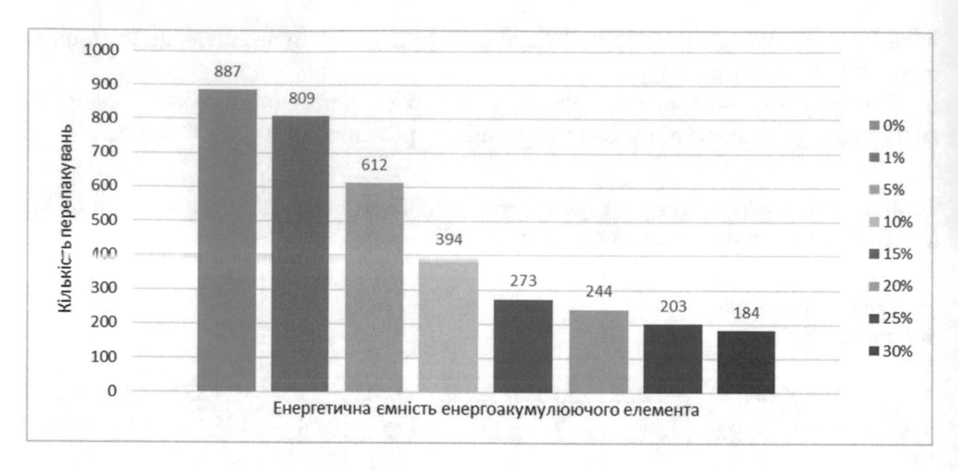

Fig. 7. The dependence of the number of switching operations of the WF active set on the energy capacity of the storage battery using the classical method of dynamic programming

Similarly, Fig. 8 shows the result of the simulation of wind farm control system operation with the energy storage element with different capacities (0%, 1%, 5%, 10%, 15%, 20%, 25%, 30%) implementing the modification of the method of dynamic programming (MDP).

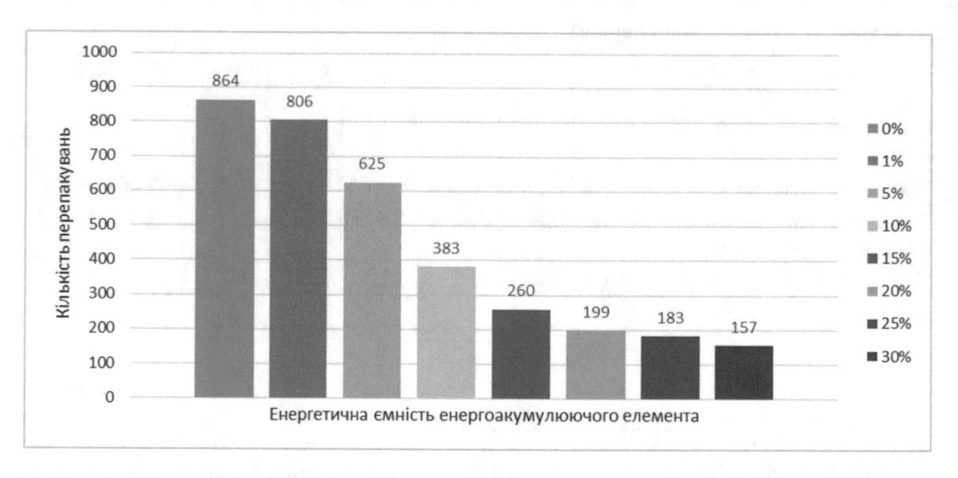

Fig. 8. The dependence of the number of switching operations of the WF active set on the energy capacity of the storage battery implementing the developed modification of the method of dynamic programming

The obtained results show the appropriateness of using energy storage element in the WF structure, as it ensures significant reduction in the number of redeterminations of the active set of the wind farm. Figures 7 and 8 show that the use of energy storage element (ESE) with the capacity of 10% of the nominal output of wind farm makes it possible to reduce the number of redeterminations 2.25 times from 887 to 394 by CDP and 2.25 times from 806 to 383 by MDP. The dependence of the number of redeterminations of WF on the energy capacity of the charging battery shows their continuous reduction with increasing SB energy capacity, however using the battery with the energy capacity of 10% of the nominal output of WF, the percentage of the reduction in the number of switching operations decreases gradually, increasing accordingly cost-effectiveness of the application.

4 Conclusion

The paper presents the solution to the problem of wind farm structure optimization (determination of the optimal set of active wind turbines at a given time). The importance of wind parameters prediction for determining the active set of a wind farm is justified. The theoretical solution to this problem enables developing algorithms for controlling power-dynamic modes of WF in order to increase their efficiency by reducing operational losses and electrodynamic overloads when switching wind turbines. The research presents the results of using such artificial neural networks as the classical unidirectional back propagation neural network, the Elman network and the cascade neural network.

The impact of the capacity of energy storage element on the number of switching operations in the wind farm structure is investigated. We offer the solution to the problem of the optimal choice of the energy storage element in terms of its efficiency and cost-effectiveness, ensuring the balance between meeting maximum load requirements and the cost of its installation and operation.

The algorithm (rules of logical output) for optimization of the number of switching operations of the WF active set was developed based on such parameters as current wind speed, archived wind speed, predicted wind speed value, current load value, next load value, nominal output of wind farm, types of wind turbines, the number of WT of each type, technical condition of each WT, energy capacity of a storage battery and others. The results of the experimental studies illustrate the efficiency of the developed algorithm.

References

1. Kravchyshyn, V.S., Medykovskyy, M.O., Melnyk, R.V., Shunevych, O.B.: Research of control modes of energy dynamical processes in power supply systems with a battery in structure. Sci. J. Natl. For. Univ. Ukraine **26**(7), 291–298 (2016). (in Ukrainian)
2. Kravchyshyn, V.S., Medykovskyy, M.O.: Control of wind power plant with battery. In: Intellectual Systems for Decision Making and Problems of Computational Intelligence ISDMCI 2016, Zaliznyy Port, Ukraine, 24–28 May 2016, pp. 83–85 (2016). (in Ukrainian)
3. Medykovskyy, M.O., Shunevych, O.B.: Modified Petri net for the structure analysis of wind power plant, Motrol – Commission of motorization and energetic in agriculture – Lublin, vol. 14. no. 4, pp. 178–184 (2012). (in Ukrainian)

4. Medykovskyy, M.O., Shunevych, O.B.: Multicriteria method for evaluating the effectiveness of wind energy installations. J. Eng. Acad. Ukraine Kyiv **3–4**, 240–245 (2010). (in Ukrainian)
5. Medykovskyy, M.O., Shunevych, O.B.: Investigation of the effectiveness of methods for determining weight coefficients importance. Journal of Khmelnytsky National University: Coll. scientifictechnical. papers – Khmelnytskyi, no. 5, pp. 176-182 (2011). (in Ukrainian)
6. Medykovskyy, M.O., Shunevych, O.B.: Using integer programming for determination of wind power. Model. Inf. Technol. Kyiv **57**, 230–233 (2010). (in Ukrainian)
7. Medykovskyy, M.O., Tesluk, V.M., Shunevych, O.B.: Using the dynamic programming for the task of equable using of wind turbines. J. "Tekhnichna Elektrodynamika" ("Technical Electrodynamics") **4**, 135–137 (2014). (in Ukrainian)
8. Baklan, I.V., Stepankova, H.A.: Probabilistic models for analyzing and predicting time series. In: Arts. Intelligence 2008, vol. 3, pp. 505–515 (2008). (in Ukrainian)
9. Kuchansky, O., & Biloshchytskyi, A.(2015). Selective pattern matching method for time-series forecasting. Eastern-European Journal of Enterprise Technologies, **6**(4(78), 13–18 https://doi.org/10.15587/1729-4061.2015.54812
10. Mazorchuk, M.S., Symonova, K.A., Hrekov, L.D.: Using of fuzzy logic methods and models for modeling economic processes. Inf. Process. Syst. **9**, 159–162 (2007). (in Russian)
11. Vernyhora, R.V., Yelnykova, L.O.: The possibility of using artificial neural networks when predicting train work train directions. In: Proceedings of the Dnipropetrovsk National University of Railway Transport named after Academician V. Lazaryan "The Problems of the Transport Economics", vol. 7, pp. 15–19 (2014). (in Ukrainian)
12. Kalinina, I.O.: Research of algorithms of training of neural networks in forecasting tasks, Scientific works of Petro Mohyla Black Sea State University, Sir: Computer Technology, vol. 104, pp. 160–171 (2009). (in Ukrainian)
13. Boychuk, V.O., Novakevych, V.Y.: Modern artificial neural networks and approaches to their modeling. Meas. Comput. Devices Technol. Process. **4**, 216–219 (2014). (in Ukrainian)
14. Online resource: Weather Forecast. http://rp5.ua
15. Vartanyan, V.M., Romanenkov, Y.A., Kashcheeva, V.Y.: Estimation of the frequency parameters of the Teyl-Weige model in short-range forecasting problems. Eastern Eur. J. Adv. Technol. **1**(5), 49–54 (2011). (in Russian)

Model of the System of Personalized Analysis of Financial Condition of the Enterprise

Melnykova Nataliia(✉)

Lviv Polytechnic National University, S. Bandery Street, 12, Lviv 79013, Ukraine
melnykovanatalia@gmail.com

Abstract. This article reflects the peculiarities of valuation the financial performance of the enterprise, which ensure the implementation of qualitative personalized analysis, through the application of economical-mathematical methods, methods of heuristic analysis and methods of artificial intelligence. This allowed to predict actions to optimize financial planning decisions. The model of a system of personalized analysis of financial state of the enterprise and the formation of objective decisions concerning the financial planning of its successful activities was proposed.

Keywords: Comprehensive approach · Personalized data · Analysis of financial status · Financial planning · Personalization of data

1 Introduction

In the field of financial and economic activity significantly increased the part of timely and qualitative analysis of the financial condition of the enterprises, the evaluation of their liquidity, solvency and financial durability, and finding the ways to consolidation of financial stability. Analysis of the enterprise financial condition characterizes degree of using the financial resources and capital, obligations before the Condition, the owners, staff and other entities. In addition, it is one of the methods of adaptation to market changeable conditions. Actuality of such investigation stipulated by high requirement of objective evaluation of the financial condition, which permit creation of personalized solution to increase potential planning of enterprise financial development.

There are number of approaches for evaluation the crisis conditions of enterprise, each of them has own disadvantages and advantages. The current condition of personalized evaluation of the enterprise financial condition depends on large number of parameters, which stipulate difficulties connected with detection the correlation structure of these parameters. In the terms, when decisions passed on the basis of stochastic incomplete information, the using of multidimensional statistical analysis and self-organizing Kohonen maps is not only justified, but necessary. The problem of condition analysis the and modelling of enterprise activity reflected in the works of many foreign and domestic scholars, including I.O. Blank, N.N. Bureeva, G. Debok, T. Kohonen S. Hayken [2, 3].

N. Shakhovska and V. Stepashko (eds.), *Advances in Intelligent Systems and Computing II*, Advances in Intelligent Systems and Computing 689,
https://doi.org/10.1007/978-3-319-70581-1_24

Looking the results of investigation to financial analysis of enterprise condition, remains unresolved question about methodology of analysis result influence to process of enterprise financial planning activity. At nowadays literature, scientists pay attention to the problem of achieving efficiency of financial planning mostly from positions of management and prediction of cash flows goes in according to existing theories, which in its turn are not capable of fully, quantitatively predict future cash inflows. Financial analysis and planning by specialists practically regarded as only the one effect-oriented planning system of enterprise [1, 4].

2 Formulation of the Problem

Today, any enterprise in one or another degree is constantly requires additional funding sources. You need to find them on the capital market, attracting potential investors and creditors by objective informing them of your financial and economic activity. To solve these and many other questions allows the methodology of financial analysis.

The process of enterprise financial analysis provides including: efficiency increasing of equity and profitability of equity; attracting debt capital; evaluation of earned revenue, working capital, financial stability, liquidity and solvency of the enterprise; possibilities of improving the efficiency of its functioning by using rational financial policy [5, 6].

This approach to evaluation the activity of the enterprise is one of the methods of observation and adaptation to market changeable conditions. Therefore, objective personalized evaluation of enterprise financial condition under the terms of different forms of property, acquires special significance, as no one of the owners should not ignore potential opportunities to increase the firm profits.

3 Comparative Analysis of Existing Methods of Personalized Approach

Theory and practice of financial condition analysis of the business object includes different types of economic-mathematical methods and models, which conditionally could be classified by groups: methods of correlation-regression analysis, methods of mathematical programming, matrix methods and models, nonlinear models, other economic-mathematical methods and models [6].

3.1 Methods of Correlation-Regression Analysis

Methods of correlation-regression analysis can be used for establishing the quantitative dependence of those or other objective and subjective parameters of investigated enterprise, where the nature of functional dependencies between them is uncertain. Pay attention that the most processes which take place in the economics are random, than the relationship between the factors that affect to resulting variable is random magnitude, at such case, the correlation expresses probabilistic dependence between variable parameters of communication algorithm. Frequently, the correlation-regression analysis used on the stage of forming of representative statistical sample. This allows to exclude the interdependent variables, there by decrease the dimension of table that contains the statistics, but not reducing its significance. As result, the researcher receives opportunity to apply to the investigated phenomenon the most adequate model that can effectively solve the original problem, without overloading of its incoming statistical data [5, 7] (Fig. 1).

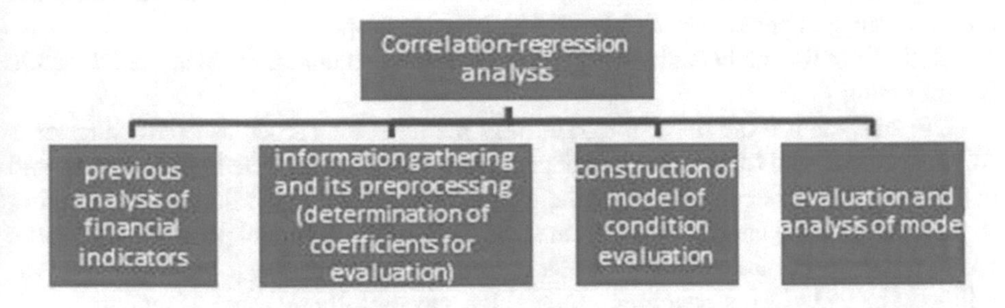

Fig. 1. Stages of correlation-regression analysis of enterprise financial condition

On the first stage of correlation-regression analysis, determine indicators-factors of correlation relationship. On the second, the communication density of effective indicators with indicators-factors, which identifies by appropriate coefficient values of pair or multiple correlation. The values of coefficients indicate the probability of changes of effective indicator when changing the value of input variable.

The dependence of indicator from one factor in the most simplified form could be expressed as linear dependence of type

$$P = k1p1 + k2, \tag{1}$$

Where P – evaluation indicator of the enterprise profitability, k1, k2 – some of the model parameters for determining the values of which is the used method of least squares. More complex dependencies can be presented as the form of multiple linear correlation

$$P = k0 + k1p1 + k2p2 + \ldots + knpn. \tag{2}$$

The main task put before the analyst - forming the task of analytical investigation and properly interpretation of its solution [8].

3.2 Methods of the Mathematical Programming

Methods of mathematical programming by its nature reduce to solving of conditional optimization tasks with multiple variables. Mostly, the methods of mathematical programming used during the solving of tasks. They are by planning range and assortment of products, determination of the optimal route, minimizing the production residues, adjusting the level of reserves in the calendar planning of production, etc. [2]. Thus, the methods of mathematical programming primarily designed for optimization of economic activity. It allows the analyst to evaluate the degree of achievement, to determine the limited resources, bottlenecks, the degree of competition and unprofitability.

As the mathematical methods cannot be applied directly to object under investigation, the necessary term is to build adequate to the objective the enterprise mathematical model. Under the mathematical model of object, phenomenon, system, means some artificial system, physical or abstract, which simply reflects the structure and basic development laws of real object by way that its investigation provides information about the condition and behaviour of object under investigation.

Methods of the mathematical programming include methods of linear and dynamic programming.

The analysts use the linear programming methods for numerous optimizing tasks solving, where apply the functional dependence of the investigated phenomenon and determined processes. Obtained results during the application of linear programming methods enable the analyst to determine and analyse the potential possibility of value changing of any of the parameters of object, and determine reserves of unrealized possibilities. The tasks of linear programming solve successfully by using of modern specialized software products. During the investigation of systems where the value of one or more of the variables changes randomly using the methods of stochastic programming.

Deterministic mathematical model - the analytic representation of patterns by which for given block of input data, on the output of the system could be obtained only one result. Such model can show as probabilistic system then it is its simplification and determined system [10].

3.3 Matrix Methods and Model of Economic Analysis

Matrix methods and model of economic analysis allow to approach to analysis the most systematically, by arranging as constituent elements of the system and the relationships between them. Ground of those methods based on the linear and vector-matrix algebra, which during the investigation of complex and large economic structures is often used.

The matrix models acquired widespread extension in the investigation field of intersector balance, matrixes of multicriterion optimization, portfolio analysis, etc. In the economic analysis model of intersector balance commonly used for calculation of technological standards, balance analysis of the production process, the management by resources and reserves of raw materials, etc. [6].

Matrix of multicriterion optimization used in economic analysis as method of comparative, the rating evaluation of possible changes of parameters of the economic system of enterprise activity to multicriterion basis. By form, it is singular matrix with separation of given number of comparison criteria.

3.4 Method with Using the Game Theory

As part of other economic-mathematical methods and models, could be separated game theory, the theory of mass service and the theory of fuzzy logic and neural networks.

Methods used during the solving game tasks widely used in the field of making management decisions on the stage of forming the alternatives and selection of the optimal strategy for enterprise functioning. Analytically, they could be presented as game of two, three or more players, each of whom wants to maximize its profits, minimize loss due to another, by choosing the optimal strategy [11].

3.5 Method of Operations Research

Method of operations research used in the analysis for obtaining of comparative evaluation of alternative solutions. The complexity of the method consist in that the researched operations are open systems, therefore related to other actions, which at some point of time are not interesting for researcher, however, affect the progress of the analysis. Running operations, analysis of factors related to specific tasks, comparison of the costs and results. This should give the basis to analyst for separation of the most important in the task and formulation conditions for the forming of decisions, after the selection of the required indicators [12].

3.6 Methods of Artificial Intelligence

The methods described above for a long time were the main tools of economic-mathematical modelling, which is actively used by researchers and analysts in the process of evaluating by them of various economic processes. However, most of them have a number of disadvantages that are primarily associated with lots of restrictions and conditions, leading to significant reduction of the effective solution of real tasks, which generally do not meet requirements for the analysis of enterprise financial condition. Another significant problem that occurs during using the classical methods of economic-mathematical modelling, linked to their failure of operating by quality indicators. [17] As result, analysts are forced to increasingly use methods of qualitative analysis of economic systems, which eliminates the quantitative methods of mathematical modelling of process of planning and optimization of their activity. Also the using of adequate mathematical approaches for the analysis and predicting of activity development of the economic systems allows increasing efficiency of their functioning and preventing additional economic effect.

So to change of the classic methods of economic-mathematical modelling become new methods, including methods of fuzzy logic and artificial intelligence. It is methodology and mathematical apparatus, which provides the ability to set and mathematically-grounded solve even the problem for which unavailable full statistics or among the most informative factors are only qualitative indicators, ensuring the possibility of adapting the economic and mathematical models to changeable conditions in the economics [5, 13].

The model is built on the foundation of artificial intelligence is quite well-established in solving complex tasks. Which consider the problems in the field of economic analysis and predicting changes in the stock indexes, during evaluation of reliability of borrower in financially-credit area, with defined probability of enterprise bankruptcy and in the researching of activity of the industrial and commercial enterprises by analysis of financial and insurance risks, etc. [9, 10].

4 Search for Solutions for Analysing the Enterprises Financial Condition

4.1 Evaluation of the Financial Success of the Enterprise

According to the results of the research of enterprise condition, determined that it would be advisable to use relative dynamic indicators of financial assets that calculated based on the information contained in report about its financial activity. The actual values of the pace of change of certain dynamic effectiveness indicator of enterprise activity will show the shortcomings in its activity and previously the problem areas -bottlenecks. Besides, proposed to analyze relative profitability indicators, which values are taken into account in coefficients or in percentages.

A quick and objective evaluation of enterprise financial success achieves by applying various optimizing methods of analysis process taking into account contradictory influence factors.

In conditions of unstable operating environment and high innovation of organizations development there are increasingly using methods of heuristic analysis based on professional judgment, experience and intuition of specialists. Using these methods allows effectively carry current and strategic analysis, to give balanced evaluation of property, financial status of the organization and substantiate the prospects of its development [11, 12].

During taking into account the criteria for evaluating the enterprise, there is problem in determining their amount for the analysis of financial condition and priority of their influence to process of decisions optimization, because increasing of their number simultaneously makes the analysis more accurate, but complicated. The possibility of using a multipurpose approach for solving planning tasks is to determine the optimal criterion, which is important stage for multicriteria optimization. Consequently, the final result of decision success depends on the correct criterion. During the choosing of criteria followed certain principles, such as independence, coordination, principle of completeness, etc. [8].

4.2 Characteristics of the Main Factors of the Analysis of Enterprise Success

To characterize the elements of the system sets of enterprise financial success, we take into account the relative and absolute condition indicators of enterprise activity condition. For this, we consider the characteristics of the system.

$$P = \langle p_1, \ldots, p_n \rangle, \tag{3}$$

Where p1, … , pn - are elements of sets of relative financial indicators of enterprise (P), n - the number of indicators-factors.

In the Table 1 given elements of the set P, which are often considered to be mostly used indicators-factors that reflect the results of processes: investment, liabilities, indebtedness, asset management, capital formation, cost planning, profit generation, etc.

$$K = \langle k_1, \ldots, k_m \rangle, \tag{4}$$

Table 1. Indicators-factors of enterprise valuation

Symbols	Indicators
p_1	Current financial investments
p_2	Short-term liabilities
p_3	Receivables
p_4	Current assets
p_5	Equity
p_6	Total amount of property of the enterprise
p_7	Attracted funds
p_8	Prepaid expenses
p_9	Net profit
p_{10}	Spending

Where $k_1, \ldots, k_m$ - elements of coefficients set of enterprise valuation (K), m - is the number of evaluation coefficients.

In the Table 2 given coefficients that reflects results of processes: investing, liabilities, debt, asset management, capital formation, cost planning, profit making, etc.

Table 2. The coefficients of enterprise valuation

Symbols	Coefficients of valuation
k_1	Absolute liquidity coefficient
k_2	Intermediate coefficient of coverage
k_3	Total coverage coefficient
k_4	Coefficient of ownership (autonomy)
k_5	Coefficient of borrowed funds
k_6	The coefficient of the correlation of borrowed and own funds
k_7	Coefficient of mobility (manoeuvrability) of own funds
k_8	Coefficient of collateral availability of reversible assets
k_9	Own reversible assets
k_{10}	Profit margin
k_{11}	Profitability indicator

$$S = \langle s_1, \ldots, s_l \rangle, \tag{5}$$

Where $s_1, \ldots, s_l$ are elements of set of strategic decisions for enterprise activity development (S), l - the number of solutions.

In the Table 3 given solutions which affect the enterprise activity and characterize strategic direction of changes for further development.

Table 3. Decisions of the development strategy

Symbols	Coefficients of valuation
s_1	Plan of the number and composition of employees
s_2	Wage fund planning
s_3	Use of the wage fund
s_4	Plan for calculating the cost of the capital
s_5	Plan of production and sales of products
s_6	Plan for increasing productivity of work
s_7	Plan for the use of material resources

H - characteristics of the enterprise determines the category of enterprise and its sphere of activity, the significance of which affect to determination of evaluation of enterprise activity condition.

$$A = \langle a_1, \ldots, a_i \rangle, \quad (6)$$

Where $a_1, \ldots, a_i$- elements of set of enterprise financial evaluation (A), i - the number of evaluations.

Table 4. The coefficients of enterprise performance evaluation

Symbols	Coefficients of valuation
a_1	Non-profit
a_2	Unprofitable
a_3	Profitable
a_4	Cost effective
a_5	Liquid
a_6	Illiquid
a_7	Critically-liquid
a_8	Not criticaly-liquid
a_9	Inefficient use of own funds
a_{10}	Efficient use of own funds
a_{11}	Creditworthiness
a_{12}	Non-creditworthy
a_{13}	A large share of capital
a_{14}	Insignificant share of capital
a_{15}	Investment-dependent
a_{16}	Independent of investments
a_{17}	Has low opportunities for financial maneuverability
a_{18}	Has high opportunities for financial maneuverability
a_{19}	Insolvency
a_{20}	Solvent

In the Table 4 given possible estimates of enterprise financial condition, which show the problem areas of enterprise, i.e. “bottlenecks”.

Based on description of characteristics of system analysis of financial success, namely: sets of system input data, sets of coefficients of enterprise valuation, sets of strategic decisions to enterprise financial success, sets of evaluations of enterprises financial success and enterprise characteristics can form set of product rules.

4.3 Financial Model of the Investigated Enterprise

For objective evaluation of enterprise it is necessary to construct its formal financial model, that allow to present it as some physical artificial system of enterprise success, which simplifies its financial structure and provides information about its condition and behaviour.

So the formal financial model of enterprise presented as model of product expert system, which is usually used to solve such class of tasks.

The knowledge base in accordance to structural scheme of system of personalized evaluation is to select certain set of rules R [14]:

$$R = \{R_1, \ldots, R_j\}, \quad (7)$$

Where products are

$$R_j = p_{j1} \bigwedge p_{j2} \bigwedge \cdots \bigwedge p_{jn} \to k_m \quad (8)$$

And the finite set of enterprise parameters P:

$$R_1 : \Psi \to K \quad (9)$$

Where

$$\Psi = \Psi(p_n), p_n \in P, \Psi = P \cap H, \quad (10)$$

Ψ- set of time-independent characteristics (H) and time-independent parameters of enterprise (P). An example of time-independent characteristic is: ownership, type of economic activity, enterprise category, etc.

An example of rules is searching for strategic decisions and financial policies of enterprise based on selected characteristics and parameters.

The system of enterprise’s success, which provides search for evaluation of enterprise financial condition and search for business optimization solutions, is presented as:

$$Fe = \langle P, K, H, A, S, G, R \rangle, \quad (11)$$

Where *Fe* - system of enterprise’s success, *P* - set of system input data that characterize relative financial indicators-factors of enterprise derived from dynamics of financial assets that calculated on the basis of the information contained in the report about its financial activities, *K* - set of enterprise evaluation coefficients, that shows to changes of effective indicator during changing the value of input variable, S - set of strategic decisions to enterprise financial success, R - product rules of decisions to strategy of enterprise development, A - set of evaluations of enterprises financial success, which depends on the evaluation coefficients, H - enterprise characteristics, G - indicator of enterprise financial policy [15].

At the same time, the relationship between the entities of system can be represented as the binary relation, the partial example, Table 5.

Table 5. The partial example of binary relation of determining the evaluation coefficients.

P	Indicators	K	Coefficients
p1	Current financial investments	k1	Absolute liquidity coefficient
p2	Short-term liabilities	k1	Absolute liquidity coefficient
p2	Short-term liabilities	k2	Intermediate coefficient of coverage
p2	Short-term liabilities	k3	Total coverage coefficient
p3	Receivables	k2	Intermediate coefficient of coverage
p4	Current assets	k3	Total coverage coefficient
p5	Equity	k4	Coefficient of ownership (autonomy)
p5	Equity	k6	The coefficient of the ratio of borrowed and own funds

The value of parameter-factors determined by coefficients of evaluation of financial success.

The values of the obtained coefficients determine the result of evaluation of enterprise financial condition.

$$\exists k K(k) \rightarrow \exists a A(a). \tag{12}$$

The results of the evaluation are determined by evaluation of condition, financial policy, enterprise characteristics and decisions to the actual directions of enterprise activity improvement.

They can be presented as form of product rule to development strategy, as n -relation:

$$R_2 = H \times A \times S \times G, \tag{13}$$

or in the form of four:

$$(h, a, s, g) \in R_2. \tag{14}$$

The received rules determine the decision to review and change the planning of the number and composition of employees, the wage fund, the calculation of capital cost, production and sales of products, increasing of work productivity, using of material resources and wage fund [16].

5 Conclusions

Consequently, during the process of financial analysis of enterprise condition used different techniques and methods of decisions optimization. We can assume that due to taking into account the individual performance indicators of the enterprise, its financial success and various factors of influence, the use of different optimization methods is foreseen for analyze financial success enterprise.

For differently tasks, appropriate to use:

- the multipurpose approach is expedient to use for complex analysis of enterprise condition;
- the heuristic methods are effectiveness in unstable environment of functioning and high innovation of development;
- new methods, such as methods of fuzzy logic and artificial intelligence become for determination of optimal solutions in conditions of unstable environment and solving complex problems in the field of economic analysis and forecasting modeling.

The proposed model of system of personalized analysis of enterprise success allows to provide processes for finding objective and qualitative strategic decisions regarding the determination of non-profitability, profitability, liquidity, efficiency of using own funds, credit potential, dependence on investments, solvency of the enterprise and finding opportunities for financial maneuvers.

Thus, such approach to objective personalized evaluation of enterprise financial condition, subject to various forms of ownership and activities, becomes of particular importance, since no owner abandons the potential opportunities to increase profits and develop his business.

References

1. Zavgorodnij, V.P.: Automatization of Business Accounting, Management, Analysis and Audit, 768 p. A.S.K., Kyiv (1998)
2. Computerization of enterprise main stages and leading of control-auditing work – pledge of its efficiency increasing. In: Financial Management, no. 1, pp. 53–56 (2002)
3. Yakovenko, S.I.: Re-engineering of business processes by way of management informatization at the enterprises of Ukraine. Actual Probl. Econ. **9**(39), 118–130 (2004)
4. Melnykova, N., Marikutsa, U., Slych, A.: The intelligent system architecture of personalized management. In: XXIV Ukrainian-Polish Conference on CAD in Machinery Design. Implementation and Educational Issues – CADMD 2016, Lviv, 21 October–22 October 2016, pp. 27–28 (2016)
5. Abdikeev, N.M., Tyhomyrova, N.P.: Designing of Intellectual Systems at Economics. Edition «Ekzamen», 528 p. (2004). Textbook/Under red
6. Mark, D.A.: Clement McGoen. Methodology of structure analysis and designing of SADT (Structured Analysis & Design Technique). http://www.interface.ru/fset.asp?Url=/case/sadt0.htm

7. Lytvyn, V., Shakhovska, N., Melnyk, A., Ryshkovets, Y.: Designing intelligent agents based on adaptive ontologies. In: Materials of VI International Scientific Conference Intelligent Decision Support Systems and Computational Intelligence Problems, Yevpatoria, 17–21 May 2010, vol. 2, pp. 401–404. KNTU, Kherson (2010)
8. Dunham, M.: Data Mining Introductory and Advanced Topics. Pearson Education Inc., Upper Saddle River (2003)
9. Shakhovska, N., Cherna, T.: The method of automatic summarization from different sources. Econtechmod. Int. Q. J. **5**(1), 103–109 (2016)
10. Shakhovska, N., Veres, O., Hirnyak, M.: Generalized formal model of big data. Econtechmod. Int. Q. J. **5**(2), 33–38 (2016)
11. Vinogradov, O.A.: Application information technology in providing marketing innovation. Actual Probl. Econ. **10**(52), 45–52 (2005)
12. Blundel, G.L.: Effective Business Communications: Principles and Practices in the Era of Informatization. Peter, St. Petersburg (2000)
13. Titorenko, G.A., Makarova, G.L., Daiitbegov, D.M., et al.: Information technologies in marketing, 335 p. (2000)
14. Kalyanov, G.N.: Case-Technology. Consulting for Automation of Business Processes, 15th edn., 320 p. Hot line - Telecom (2002)
15. Chernorutsky, I.G.: Methods of Decision Making, 416 p. BHV, St. Petersburg (2005)
16. Larichev, O.I.: Theory and Methods of Decision Making, and Chronicle of Events in Magical Countries, 296 p. Logos, Moscow (2000)

Hybrid Sorting-Out Algorithm COMBI-GA with Evolutionary Growth of Model Complexity

Olha Moroz(✉) and Volodymyr Stepashko

Department for Information Technologies of Inductive Modelling, International Research and Training Centre for Information Technologies and Systems of the NASU and MESU, Glushkov Avenue 40, Kyiv 03680, Ukraine
olhahryhmoroz@gmail.com, stepashko@irtc.org.ua

Abstract. Paper presents latest achievements in the development of the sorting-out hybrid COMBI-GA algorithm based on a new mechanism of evolutionary growth of models complexity to find the optimal model structure. The mechanism uses generation of model structures of GA initial population by binomial random number generator with low probability and specific mutation operator with adding some units in model structures. The effectiveness of this mechanism is compared with two another mechanisms of evolutionary simplification (using binomial random number generator with big probability to form the GA initial population and mutation operator with reducing amount of units in model structures) and conventional random changing the complexity (using binomial random number generator with average probability of forming the GA initial population and mutation operator where gene values in a chromosome are inverted according to a given probability). The presented experimental results demonstrate that this new algorithm with evolutionary complication of model structures performs quickly, accurately and reliably when solving both artificial and real inductive modelling tasks.

Keywords: GMDH · Sorting-out algorithm · Combinatorial algorithm COMBI
Genetic algorithm GA · COMBI-GA algorithm · Model complication
Model structure generator

1 Introduction

GMDH algorithms [1] belong to the most effective means of solving inductive modelling problems like structural and parametric identification, forecasting, building models of objects and processes from statistical or experimental data samples under uncertainty conditions. They are based on the principle of model's self-organization [1]: with a gradual increase of the model structure complexity, the value of an external criterion decreases, reaches a minimum, and then becomes unchanged or begins to increase. The smallest value of the criterion defines a unique model of optimal complexity. Here "model complexity" means the number of included variables in a linear model structure or the number of nonlinear model parameters.

N. Shakhovska and V. Stepashko (eds.), *Advances in Intelligent Systems and Computing II*, Advances in Intelligent Systems and Computing 689,
https://doi.org/10.1007/978-3-319-70581-1_25

GMDH algorithms may be divided into two main groups: sorting-out and iterative ones. The main difference between them consists in using, respectively, search-type and iteration-type generators of different model structures [2].

The very first sorting-out algorithm COMBI [2] is practically successful only in solving tasks with the number of input variables (arguments) less than 30 because of using exhaustive search of models of different structures in a given basic class.

Sorting-out GMDH algorithms with incomplete (directed) model search such as MULTI [3], BSS and FSS [4, 5], use specific ways of optimal structure search and demonstrate polynomial computational complexity. They are able to significantly reduce the time and amount of required resources to find the optimal model as compared to the conventional COMBI algorithm and to solve problems with hundreds of arguments. However, these algorithms have quite complex search schemes to find the optimal result and require significant efforts for their implementation.

The very first and the most known iterative-type GMDH algorithm of neural network type called now Multilayered Iterative Algorithm MIA GMDH was created in 1968 by Prof. A.G. Ivakhnenko [1]. But it has such main drawbacks as high probability of loss of relevant arguments and retention of irrelevant ones and it was the reason to construct new kinds of the iterative GMDH algorithms [1, 6]. For more than ten last years, several researchers actively develop hybrid structures of MIA and genetic algorithms to increase the productivity and overcome these shortcomings [7].

Genetic algorithms belong to evolutionary methods being a powerful tool to find global solution of optimization problems. The GAs have quite clear structure and fairly obvious realization. However, the use of GA in each task faces major difficulties: defining the method of coding solutions that form chromosome of an initial population, defining a size of the initial population and the number of selected individuals to apply genetic operators, selection of appropriate operators (crossover, mutation, etc.) and their probability, choosing stop criterion of the algorithm. These difficulties significantly affect the performance of any GA-based algorithm.

This paper considers only GMDH algorithms of sorting-out type which can find optimal model during small time. To enhance the productivity and accuracy of such algorithms we apply the idea of COMBI and GA hybridization. The first variant of a COMBI-GA algorithm was studied in [8] and demonstrated very promising results.

The aim of this paper is to consider the effectiveness of a new modification of the hybrid COMBI-GA algorithm based on the usage of binomial random number generator [9] with small, average and high probability to form GA initial population. Together with this, a new special mutation operator is examined allowing: (a) to evolutionary increase the model complexity; (b) to randomly change the complexity; and (c) to only evolutionary simplify the complexity.

The paper structure is as follows. Section 1 explains existing sorting-out algorithms and place of COMBI-GA among them. Section 2 describes the basic principles of GA and using genetic operators. Section 3 introduces a newly modified hybrid COMBI-GA algorithm. Section 4 explains approaches to the algorithm research. Sections 5 and 6 present the results of its investigation in artificial tasks and application to solving a real-world task respectively.

2 Sorting-Out GMDH Algorithms

The article [10] presents a survey of the available techniques of exhaustive and partial search of models in GMDH algorithms and their comparative analysis. It considers also the analysis and classification of models structure generators in most known sorting-out GMDH algorithms.

The greatest effect on the speed and amount of calculations in this kind of algorithms has the choice of the model structures generator, i.e. the mechanism of models sorting. The disadvantage of different types of searching algorithms in comparison with the combinatorial one is the risk of omitting the model of optimal complexity. But this is balanced by their significant advantages in speed and possibility to find a model close to the optimal one by the value of a given criterion.

Sorting-out GMDH algorithms enable to effectively solve actual problems of artificial intelligence, system identification and process prediction. They help to reduce a subjective influence of a user on the results of modeling. The main feature of sorting-out GMDH algorithms is that they search for the model of optimal structure minimizing error of forecasting in a finite set of models with different structures.

A classification of majority of known sorting-out GMDH algorithms can be presented by the diagram showed on Fig. 1 firstly published in [10].

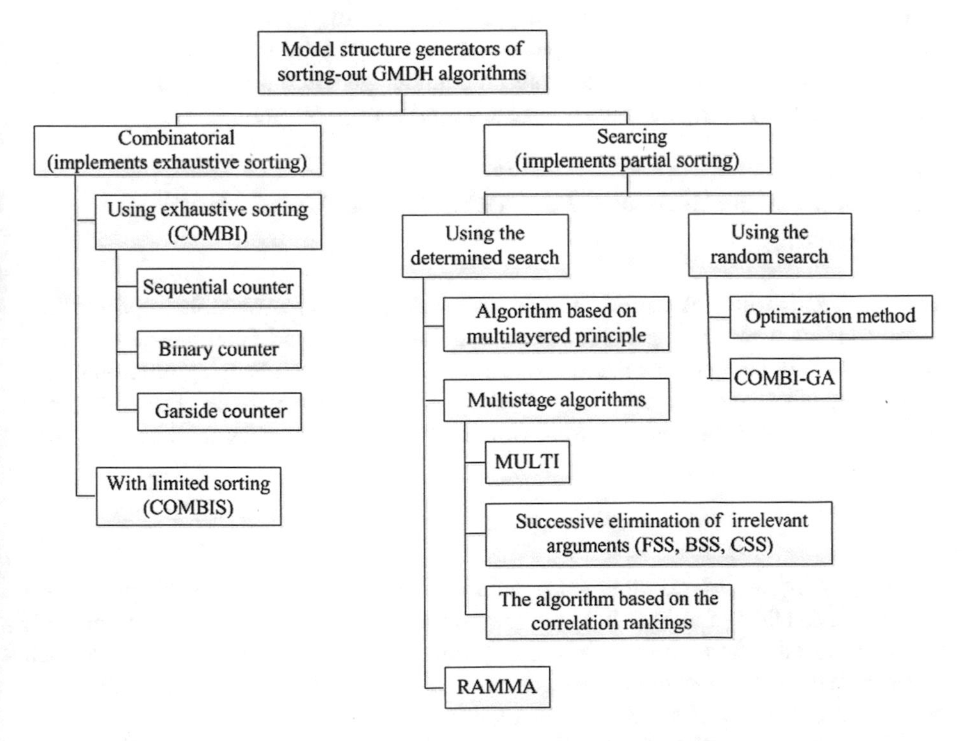

Fig. 1. Classification diagram of known GMDH algorithms of sorting-out type.

The diagram shows that most of structure generators in sorting-out algorithms are of searching type. Such sorting-out generators eliminate the main drawback of the combinatorial algorithm, namely exhaustive search of all possible models. Due to that they are able to solve high dimension problems, have much greater speed of calculations and require less computing resources.

The first algorithm based on the directed sequential search of the optimal model (the result of the exhaustive search by COMBI) was the «combinatorial-selective multistage algorithm» MULTI [3]. It implements a step-by-step procedure of forming model structures of incrementally increasing complexity of partial models and a recurrent method of parameter estimation. Generated models remain in the original basis and the number of their arguments corresponds to the current stage number.

Papers [4, 5] illustrate the effectiveness of sorting-out COMBI-based algorithms with successive selection of informative arguments and/or elimination of spurious ones in three cases: (1) passage from simple models to complex ones (Forward Successive Selection, FSS); (2) passage from complex to simple models (Backward Successive Selection, BSS); (3) combining of the two approaches (Combined Successive Selection, CSS). In the second case the result proved to be the best.

Another possible way to enhance the performance of sorting-out algorithms is constructing hybrid structures of COMBI with some computational intelligence algorithms. Our previous papers [8, 11, 12] showed that hybridization between COMBI and genetic algorithm (GA) is relevant and promising for solving various high-dimension modeling tasks with hundreds of input variables. The GA initial population was generated previously by the uniform random number generator. Here we show more efficient results by using the binomial random number generator.

3 Genetic Algorithm and Its Operators for Generating Offspring

The genetic algorithm is one of the meta-heuristic procedures of global optimization constructed as a result of generalization and simulation in artificial systems of such properties of living nature as natural selection, adaptability to changing environmental conditions, inheritance by offspring of vital properties from parents. The basic principles of GA were formulated by John Holland [13]. Formally, GA can be represented in such a way:

$$\mathrm{GA} = \{P_0, M, L, F, G, s\},$$

where $P_0 = (a_1^0, \ldots, a_M^0)$ is an initial population; a_i^0 is an individual of this population treated as a candidate for the solution of the optimization problem presented in the form of a chromosome; M is the population size (integer number); L is the length of each chromosome of the population (integer number); F is a fitness function (FF) of an individual; G is a set of genetic operators; s is the algorithm stopping rule.

Chromosomes consist of genes that are usually encoded as values from the binary alphabet $\{0, 1\}$, but other codes – alphabetic, decimal, etc., can be used too. In this paper, only binary coding of chromosomes is considered.

As input data for any GA initial population P_0, a finite set of chromosomes (elements, individuals, specimens, etc.) is used each of which represents a potential solution of the problem. Then the first population of offspring P_1 is formed from the parent chromosomes P_0 using some genetic operators, similarly the next population P_2 is formed from the population P_1 and so on. The process continues until the specified stopping rule of the algorithm will be satisfied. An important feature of the GA work is that with each step the mean FF value of the current population decreases and converges to the solution of the optimization problem.

The central role in the GA is played by two types of operators for creating new individuals (offspring) from the available ones in the current population, which can be divided into two main groups. Operators of the first type need two individuals (parents) to create two offspring and rules for the exchange of genetic material between them. Operators of the second type form one offspring from one individual, modifying it according to certain rules. Operators of the first type are for example crossover, translocation, segregation, etc.; the second type operators include mutation, inversion, transposition and others [13]. The most important among them are crossover (OC) and mutation (OM) operators implemented not in all cases of the offspring generation process but with a given probability which is close to 1 for crossover operators and to 0 for most mutation operators. In general, there are no strict limitations in the using of different models of offspring creation because in solving optimization problems it is not necessary to completely copy the laws of nature and to confine oneself only to them. It is more expedient to use the common sense and take into account the specificity of a problem as much as possible.

In what follows, there is a brief description of the operators that were investigated.

One-point crossover operator (opCO) [13]. Two chromosomes of the population are randomly selected, then a crossover point $l_k \in [1, L - 1]$ is randomly generated where L is the chromosome length. As a result of crossover of the pair of parental chromosomes relative to l_k we get a pair of offspring: the first one is a chromosome which in positions from 1 to l_k contains the genes of the first parent, and at the positions from $l_k + 1$ to L is genes of the second parent; the second one is a chromosome which, respectively, in positions from 1 to l_k contains the genes of the second parent, and in positions from $l_k + 1$ to L genes of the first parent.

Uniform crossover operator (ufCO) [13]. One of the existing descriptions of this kind of OC is the following. A mask of crossover is randomly formed as a string of zeros and units with the length equal to that of the chromosome. The mask indicates which genes should be inherited from the first parent and which from the second one, while the two new offspring are formed according to such rule: unit in i^{th} position of mask means that i^{th} element of the first parent should be located in i^{th} place of the first offspring, and zero means that i^{th} the element of the second parent should be in i^{th} place of the first offspring. Considering the first parent as the second, and the second as the first one, the second offspring is obtained according to this specified rule. The mask of crossover can be one and the same for all parents or a new for each pair.

Universal crossover operator (uvCO) [13]. Similarly to the uniform crossover, a binary mask is formed with the length equal to that of the given chromosomes. The first offspring is formed by a bitwise addition of the chromosome elements of the first parent and mask according to the following rules: 0 + 0 = 0, 0 + 1 = 1, 1 + 1 = 0. To obtain the second child, they act in the same way with the second parent.

Ring crossover operator (rgCO) [14]. In this crossover, the parent chromosomes are joined by the beginnings and ends forming a ring, and then a random point of its rupture is generated. The straight line that passes through this point and the center of the ring divides it into two parts. The first descendant is formed clockwise from the crossover point, and the second is against it.

Here we consider three gene mutation operators: M1, when according to a given probability, e.g. 0.2, only zero gene values in a chromosome are inverted [11]; M2 is standard gene mutation [13] of an individual with a given inversion probability of several bits of a chromosome according to; M3, when, on the contrary, only unit gene values in a chromosome are inverted with a given probability.

4 Hybrid COMBI-GA Algorithm

A general definition of the inductive modelling problem may be done as follows. Let us given: a data set of n observations after m inputs $x_1, x_2, \ldots, x_m$ and one output y variables. The GMDH task is to find a model $y = f(x_1, x_2, \ldots, x_m, \theta)$ with minimum value of a given model quality criterion $C(f)$, where θ is unknown vector of model parameters. The optimal model is defined as $f^* = argmin_{\Phi}C(f)$, where Φ is a set of models of various complexity, $f \in \Phi$. This problem consists of the discrete optimization task for finding the best model and a continuous optimization task for estimation of parameters.

The two tasks are generally solved by any of the sorting-out algorithms, including COMBI, in the following steps:

(1) transformation of the initial data according to the chosen system of base functions, typically polynomial;
(2) generation of the complete set of all possible structures of partial models or some subset of them in the chosen basis;
(3) use the least squares method (LSM) to estimate coefficients of a partial model;
(4) calculation of a given external selection criterion for each model;
(5) successive selection of best models by this criterion and choosing the model of optimal complexity with the minimum criterion value.

The hybrid COMBI-GA algorithm uses generally the same procedure, but steps 2 and 5 are modified by applying GA (see below in more details).

To avoid the exhaustive search typical for COMBI, the developed COMBI-GA algorithm forms gradually a set of the most promising structures of partial models and finds the optimal one using genetic operators of selection, crossover and mutation that establishes a specific sorting-out mechanism. Formally, the algorithm can be defined as follows:

$$\text{COMBI-GA} = \langle \mathbf{Z}, \mathbf{y}, \mathbf{f}, \mathbf{X}, D, CR, \mathbf{P}_0, H, M, \mathbf{G}, k, F \rangle,$$

where Z$[n \times r]$ is the measurement matrix of input variables of an object, r and n are numbers of inputs and measurements respectively; $\mathbf{y}[n \times 1]$ is vector of measurements of an output variable;

$\mathbf{f}[m \times 1]$ is vector of a given m base functions of input variables;
$\mathbf{X}[n \times m]$ is the measurement matrix of base set of arguments;
D is a given rule of dividing matrix $\boldsymbol{X}[n \times m]$ and vector $\boldsymbol{y}[n \times 1]$ to testing A and checking B parts;
CR is an external selection criterion (as fitness function) based on dividing the sample $(\boldsymbol{X}, \boldsymbol{y})$;
$\mathbf{P_0}$ is a set of model structures of GA initial population consisting of binary chromosomes (encoded structure of partial models);
H is size of initial population of models, $H < m$;
M is size of any next population, $M > H$;
$\mathbf{G}$ is set of genetic operators;
k is stopping rule of GA;
F is number of best partial models (freedom of choice) monitored during all iterations of the algorithm, $1 < F \leq H$.

This algorithm consists of the following steps:

Step 1. Calculating the matrix of the base set of arguments $\boldsymbol{X}[n \times m]$ using the input matrix $\boldsymbol{Z}$ and the vector of base functions $\boldsymbol{f}$ and dividing it and the output vector of measurements $\boldsymbol{y}[n \times 1]$ according to the rule D in testing $\boldsymbol{X_A}[n_A \times m]$ and checking $\boldsymbol{X_B}[n_B \times m]$ submatrices $(n_A + n_B = n)$. Obviously, in the case of linear polynomial, matrices $\boldsymbol{X}$ and $\boldsymbol{Z}$ are identical $(m = r)$.

Step 2. Randomly generating the initial population $\boldsymbol{P_0}$ of the genetic algorithm.

Step 3. Calculating the coefficients of each partial model by LSM or another method using the training matrix of base arguments $\boldsymbol{X_A}$ and output vector $\boldsymbol{y_A}$.

Step 4. Calculating the value of an external criterion CR (as the GA fitness function) for each partial model using the checking matrix $\boldsymbol{X_B}$ and output $\boldsymbol{y_B}$.

Step 5. Forming the current population of partial models (chromosomes) of the size H with better criterion values to form the next offspring. In addition, selection the best F partial models that are potential solutions of the task of model building.

Step 6. Forming new population of M individuals applying genetic operators of crossover and mutation to individuals of the current population.

Step 7. Checking a given GA stopping rule. If it is satisfied, then go to step 8, otherwise go to step 3.

Step 8. Choosing F best models from the current population of the size H.

Step 9. The end.

5 Main Ideas of Numerical Experiments

The purpose of these experiments is to determine the effectiveness of the hybrid COMBI-GA algorithm depending on the methods used to generate offspring of selected elite individuals by genetic operators.

In our numerical experiments, individuals (model structures) of GA initial population are encoded as binary structural vectors with elements 0 and 1 indicating absence or presence of a particular argument in the model respectively. As a fitness function of each individual we use standard GMDH external regularity criterion [1] based on dividing the data sample into two subsamples, training and checking.

The stop criterion of the hybrid algorithm is achieving given accuracy for the difference between minimum values of the fitness function of two adjacent iterations.

In this paper, we consider main obtained results for such two variants:

Variant 1. The use of four crossover operators (uvCO, opCO, rCO, ufCO) in the GA along with the 3-point mutation operator (3pMO) which performs most efficiently in the COMBI-GA algorithm [11].

The research was carried out in [8] for tasks with three types of test data without and with noise under such general conditions: the linear models class; the GMDH regularity criterion is the GA fitness function; the size of the initial population is 100 individuals; the probability of using a crossover operator is 0.9, the probability of 3pMO mutation is 0.2; the GA stopping rule is achieving the absolute value of the difference between minimum values of the fitness function of the previous and current populations less than 10^{-7}; the sample of the input data is divided into the training, checking and examination subsamples at the ratio 0.5: 0.3: 0.2.

Variant 2. According to the results of crossover comparison in [12], we use the best one-point crossover operator to form the offspring population from any pair of selected elite individuals. To produce a new offspring by the one-point crossover, a chromosome point (between neighboring bits in a row vector) is randomly selected, the fragment of binary string from the beginning of the first parent chromosome to the point is copied, and the rest (tale) is copied from the second parent.

Also we consider the following three types of mutation operators: M1 is the specific gene mutation [7] where, according to a given probability, only zero gene values in a chromosome are inverted; M2 is standard gene mutation [8] of an individual with the inversion of several bits of a chromosome according to a given probability, e.g. 0.2; M3, on the contrary to M1, only unit gene values in a chromosome are inverted according to a given probability.

This research variant was carried out on test tasks. Two data samples with 20 and 1000 input arguments were artificially generated and half of them were used in both cases as relevant ones. Both relevant and irrelevant arguments for the initial data sample are randomly generated. A test "true" linear model with random parameters is calculated using only relevant arguments.

6 Results of the Computational Experiments

Variant 1. *The case of non-noisy data.* Here we considered three problems with the number of input variables 20, 50, and 200 arguments.

In each case, approximately half of elements of the randomly generated input variable vector are true arguments. Consequently, they are taken into account when calculating the output variable values by the preassigned formula.

For example, for the data set with 20 input arguments, the model specified for the experiments contained 8 informative arguments in the linear dependence [8]:

$$y(x) = 0.5x_2 + 0.2x_4 + 2.4x_5 - 0.6x_9 + 0.3x_{13} + 2.1x_{15} + 2x_{17} + 0.8x_{19}.$$

The matrix of arguments X and the model parameters were obtained using a generator of uniformly distributed random numbers, the values of the arguments was in the range from 0 to 20, and the values of the parameters ranging from –1 to 6. The influence of genetic operators on the efficiency of the hybrid algorithm was investigated with taking into account the completeness of restoring the true (exact) structure of the model and the speed of the model constructing process.

The results of the research are given in Table 1 and show that the COMBI-GA works most effectively with opCO which gives a fairly rapid decrease in the minimum value of the fitness function of each GA population that indicates the direction of the search for the optimal model. The graphs of COMBI-GA convergence depending on some genetic operators are shown in Fig. 2.

Table 1. The COMBI-GA efficiency dependent on a used GA operator (not-noisy data).

Number of input variables	Time to find optimal model, c				
	COMBI	COMBI-GA			
		opCO	uvCO	rCO	ufCO
20	84.9	0.08	0.084	0.09	0.081
50	–	1.36	4.7	4.4	2.2
200	–	74.15	100.5	80.63	84.1

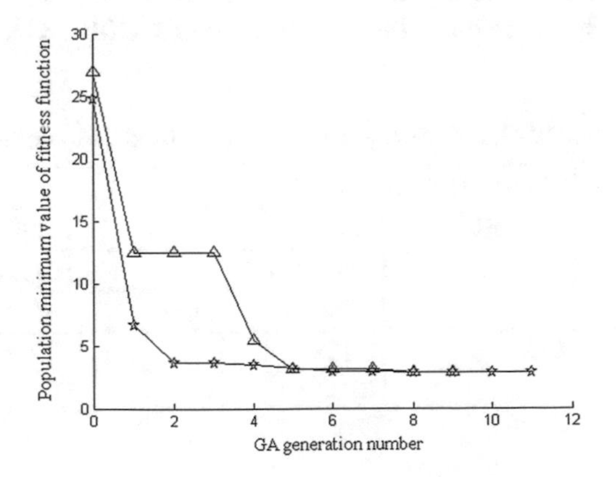

Fig. 2. Convergence of the COMBI-GA using operators (opCO + 3pMO) (☆) and (rgCO + 3pMO) (Δ).

The case of noisy data. For the case with 20 input arguments, the test model was the same as in the example without noise.

With 20% noise, the COMBI and COMBI-GA algorithms with ufCO found, respectively, the following optimal models of the same structure (except a few extra uninformative arguments with small coefficients):

$$y_1(x) = \mathbf{0.5}\boldsymbol{x_2} + \mathbf{0.26}\boldsymbol{x_4} + \mathbf{2.39}\boldsymbol{x_5} + 0.01x_7 - \mathbf{0.61}\boldsymbol{x_9} + 0.03x_{10} - 0.01x_{12} \\ + \mathbf{0.3}\boldsymbol{x_{13}} + \mathbf{2.08}\boldsymbol{x_{15}} + \mathbf{1.99}\boldsymbol{x_{17}} + \mathbf{0.86}\boldsymbol{x_{19}}$$

$$y_2(x) = -0.01x_1 + \mathbf{0.59}\boldsymbol{x_2} + \mathbf{0.16}\boldsymbol{x_4} + \mathbf{2.38}\boldsymbol{x_5} - 0.02x_7 - \mathbf{0.63}\boldsymbol{x_9} + 0.04x_{12} \\ + \mathbf{0.39}\boldsymbol{x_{13}} - 0.05x_{14} + \mathbf{2.14}\boldsymbol{x_{15}} + \mathbf{2}\boldsymbol{x_{17}} + \mathbf{0.74}\boldsymbol{x_{19}}$$

Using other genetic operators, the COMBI-GA algorithm worked less efficiently. In the case of 50% noisy data, COMBI built such optimal model:

$$y(x) = \mathbf{0.49}\boldsymbol{x_2} + \mathbf{0.35}\boldsymbol{x_4} + \mathbf{2.37}\boldsymbol{x_5} + 0.02x_7 - \mathbf{0.62}\boldsymbol{x_9} - 0.08x_{10} - 0.02x_{12} \\ + \mathbf{0.19}\boldsymbol{x_{13}} + 0.11x_{14} + \mathbf{2.06}\boldsymbol{x_{15}} + \mathbf{1.97}\boldsymbol{x_{17}} + 0.01x_{18} + \mathbf{0.94}\boldsymbol{x_{19}}$$

The COMBI-GA with opCO worked most efficiently and found the optimal model:

$$y(x) = \mathbf{0.48}\boldsymbol{x_2} + \mathbf{0.211}\boldsymbol{x_4} + \mathbf{2.312}\boldsymbol{x_5} \\ - \mathbf{0.63}\boldsymbol{x_9} + \mathbf{0.23}\boldsymbol{x_{13}} + \mathbf{2.14}\boldsymbol{x_{15}} + \mathbf{1.48}\boldsymbol{x_{17}} + \mathbf{0.867}\boldsymbol{x_{19}},$$

containing all the true arguments without presence of extra ones.

For the data set with 50 input arguments (25 relevant), the COMBI-GA was most effective when using opCO. The optimal model contained 25 arguments with 20% and 24 with 50% of noise. Table 2 presents the COMBI-GA effectiveness with different genetic operators. Figure 3 shows the convergence of COMBI-GA with opCO.

Table 2. The COMBI-GA efficiency dependent on a used GA operator (noisy data).

Number of input variables	Time to find optimal model, c				
	COMBI	20% of noise		50% of noise	
		COMBI-GA		COMBI-GA	
		oCO	*rCO*	*oCO*	*rCO*
20	85.2	0.158	0.17	0.185	0.2
50	–	0.51	0.7	0.53	0.8

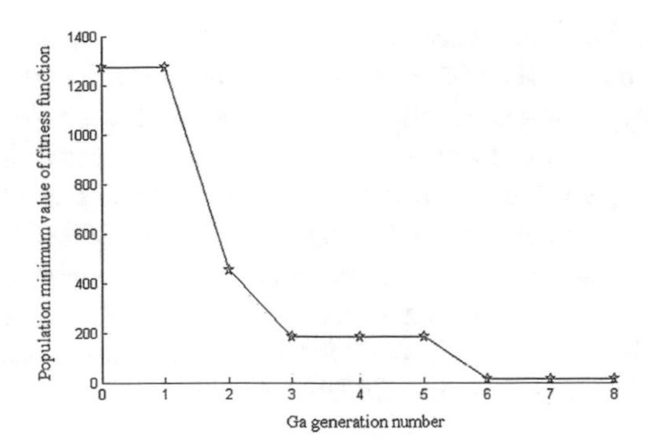

Fig. 3. Convergence of COMBI-GA, 50 arguments and 20% noise, using operators (opCO + 3pMO).

Variant 2. Unlike to the above COMBI-GA research, in this study the initial population of model structures is generated by the binomial distribution with low, average, and high probabilities, and the three mentioned above new versions of mutations are used with these three generation variants of initial population. As a result, we have three cases of experiments with corresponding mutation operators: Case 1 (0.1 + M1), Case 2 (0.5 + M2), Case 3 (0.95 + M3). Level of noise is 20% in all three cases. The Case 3 may be interpreted as an *evolutionary simplification* of models instead of *complication* as in the Case 1. The generation of model structures in the Case 2 is random (irregular).

The optimal model is searched using COMBI-GA in the class of linear functions of all the given arguments, 20 and 1000 respectively. In the case of 20 arguments we can compare the results of modelling by COMBI and COMBI-GA.

As a result of hundred experiments it was found that for 20 input variables the COMBI-GA algorithm works more reliably and accurately in Case 1 unlike to Case 2 and Case 3. The algorithm with models evolutionary simplification (Case 3) performs slowest due to solving high dimension linear systems to estimate model parameters.

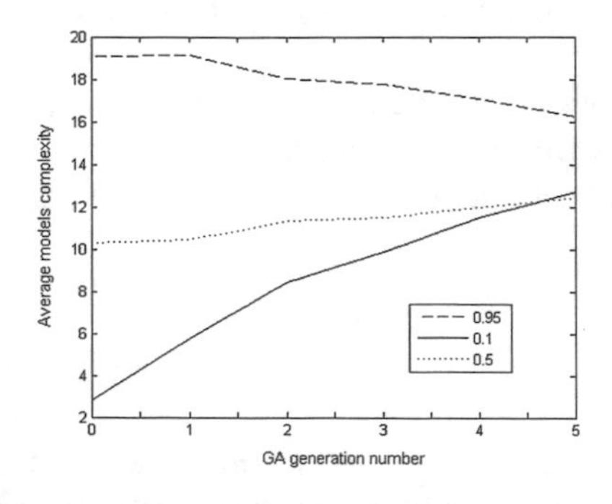

Fig. 4. Changes of average model complexity dependent on number of iterations for three variants of binomial distribution probability.

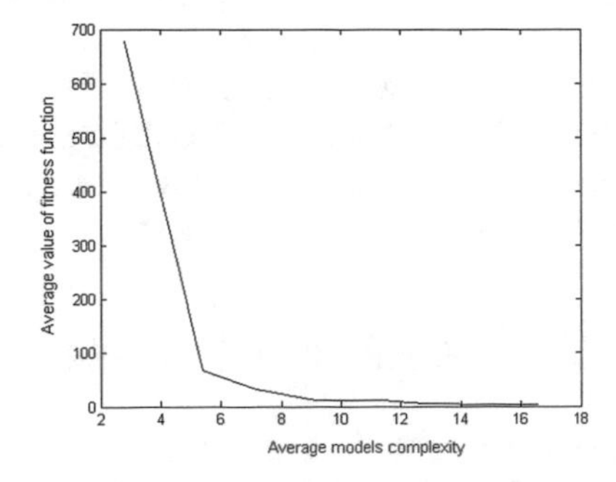

Fig. 5. Changes of average value of fitness function dependent on average models complexity of current population.

Gradual changes in the average model complexity of current population, characterizing performance of COMBI-GA in all three cases, are presented on Fig. 4.

Figure 5 confirms (simulates) the principle of models self-organization: with a successive increasing of the average complexity of model structures of the current population, the average value of the criterion decreases and reaches the minimum.

Optimal model structures obtained using COMBI and COMBI-GA algorithms are the same and consist of 15 arguments including all 10 relevant ones in view of the noise effect, 5 spurious arguments have very small coefficients that illustrate adequate work of COMBI-GA. The developed COMBI-GA algorithm has found the optimal model during 0.3 s. whereas original COMBI for 85.2 s.

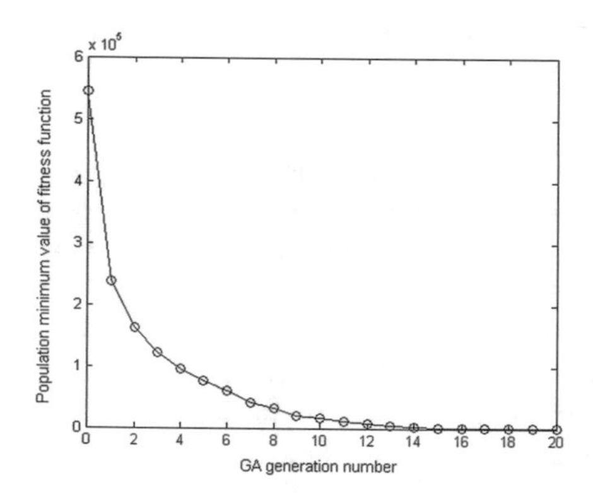

Fig. 6. COMBI-GA convergence for 1000 input variables using models evolutionary complication, noise level 20%.

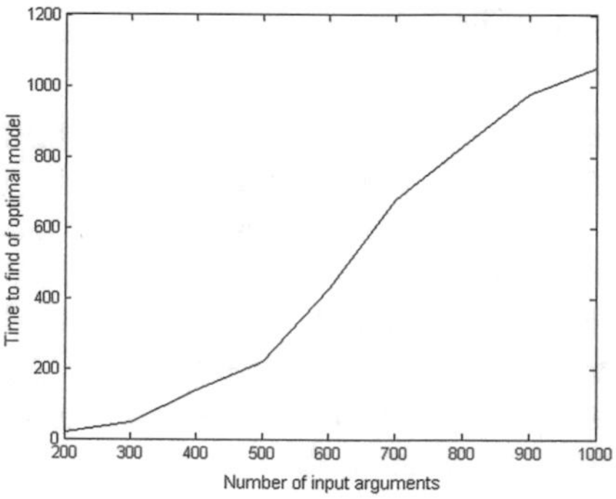

Fig. 7. Time to find the optimal model, s.

Figure 6 shows perfect capability for solving high-dimension problems in Case 1. The COMBI-GA algorithm solves the task with 1000 input variables during 17 s. according to given accuracy. In the other two cases algorithm significantly loses the accuracy and convergence.

Figure 7 shows that complexity of the COMBI-GA algorithm has not exponential character in contrast to the COMBI algorithm, forms S-shaped curve typical for natural processes and illustrate slower increasing the time to find the optimal model depending on the number of input arguments.

7 Real-World Data Example

Experiments with the use of COMBI-GA with opCO in the modification based on the Variant 2 were carried out for solving applied problem based on real-world dataset taken from UCI Machine Learning Repository [15].

Problem of modelling player skills levels. Most studies in the area of cognitive science investigate expertise by comparing experts with novices. In real-time strategy (RTS) games, players develop game pieces called units with the ultimate goal to destroy their opponent's headquarters. RTS play can be considered as an area of expertise; playing well requires a great deal of strategy and knowledge, and these require a great deal of experience.

Data of the analysis of video game telemetry from RTS games are used here to explore the development of expertise. The task deals with two types of variables:

- *Hotkey usage variables.* Players can customize the interface to select and control their units or building more rapidly, thus offloading some aspects of manually clicking on specific units to the game interface.
- *Perception-Action-Cycle variables* (PAC). Each variable pertains to a period of time where players are fixating and acting at a particular location. Many of these variables will therefore reflect both attentional and perceptual processes, and cognitive-motor speed.

There are 18 input variables (both integer and continuous) in this task [16] and the output value as LeagueIndex is coded 1 to 7 of player skills levels (Ordinal): Bronze, Silver, Gold, Platinum, Master, GrandMaster, and Professional leagues. To build model of the skills level, we use only a subset of 55 observations from the complete 3000-instances set.

Optimal model structure built by COMBI algorithm contains only 6 significant arguments out of all 18: x_1, x_7, x_9, x_{11}, x_{16}, x_{17}.

The structure of the best model constructed using both COMBI-GA with uniformly generated initial population [8] and COMBI-GA with binomially generated initial population contains the same set of significant arguments, but it also includes three additional arguments with small parameters. Hybrid COMBI-GA algorithm converges in average during 0.1 s. The COMBI algorithm constructs the model during 31.5 s.

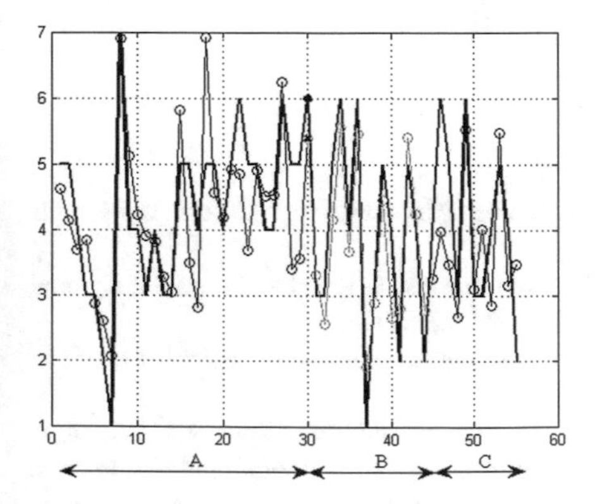

Fig. 8. Approximation of the output in the case of uniformly generated initial population

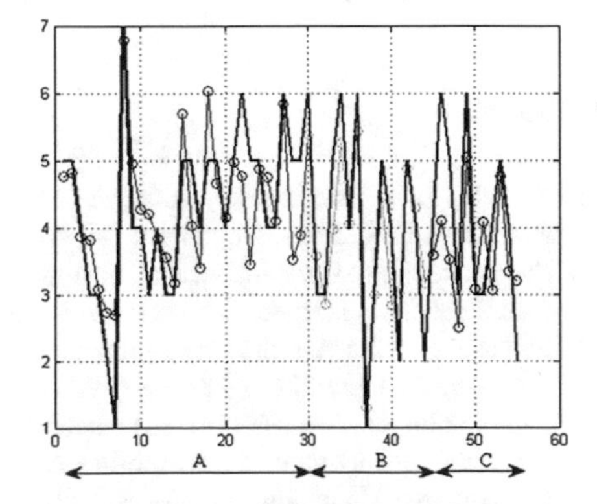

Fig. 9. Approximation of the output value using binomially generated initial population

The COMBI-GA algorithm with binomially generated initial population more accurately approximates the initial value on all three subsamples that is illustrated in Fig. 9 in comparison with similar result obtained using uniformly generated initial population, see Fig. 8. Approximations of the output value are given by grey lines.

The root-mean-square error of the model output on the subsamples A, B and C is 0.14, 0.26 and 0.53 respectively (using uniformly generated initial population) and 0.12, 0.23 and 0.51 (using binomial generated initial population).

8 Conclusion

The proper choice of GA operators essentially depends on the specificity and features of a given task. In particular, on the test tasks, the best results were obtained using a single-point crossover operator in the COMBI-GA algorithm.

Three operation modes of the algorithm were studied using GA initial population generated by binomial random numbers generator with low, average, and high probabilities of unit genes appearance in a chromosome. The three modes were combined with the corresponding mutation operator M1, M2, or M3 (Case 1 (0.1 + M1), Case 2 (0.5 + M2), Case 3 (0.95 + M3)). It was discovered that the Case 1 with low probability in combination with specific mutation operator based on adding units in model structures gives the best results regarding processing speed and accuracy. Such algorithm is helpful when number of relevant arguments is unknown (in real-world problems); it works quickly, reliably and accurately.

Therefore, the use of evolutionary complication of models in COMBI-GA algorithm is very effective to search true model when solving inductive modelling tasks with large number of input variables during really short time.

In the future, it would be reasonable to study the task of setting the GA parameters, particularly concerning how to choose the probability of genetic operators and binomial generator adaptively and automatically. Also it is advisable to find methods for quick solving big systems of linear equations when using GA with evolutionary simplification of models (Case 3). Also the problem of development of hybrid structures of COMBI with other evolutionary algorithms is of real interest.

References

1. Madala, H.R., Ivakhnenko, A.G.: Inductive Learning Algorithms for Complex Systems Modeling. CRC Press, New York (1994)
2. Stepashko, V.S.: Combinatorial algorithm of the group method of data handling with optimal model scanning scheme. Sov. Autom. Contr. **14**(3), 24–28 (1981)
3. Stepashko, V.S.: A finite selection procedure for pruning an exhaustive search of models. Sov. Autom. Contr. **16**(4), 84–88 (1983)
4. Samoilenko, O., Stepashko, V.A.: Method of successive elimination of spurious arguments for effective solution of the search-based modelling tasks. In: 2nd International Conference on Inductive Modelling, pp. 36–39. IRTC ITS NASU, Kyiv (2008)
5. Samoylenko, A.A.: The weight criterion to evaluate the arguments informativeness in the successive selection modeling methods, no. 2, pp. 33–39 (2013). (in Russian)

6. Moroz, O., Stepashko, V.: On the approaches to construction of hybrid GMDH algorithms. In: Proceedings of 6th International Workshop on Inductive Modelling IWIM-2013, pp. 26–30. IRTC ITS NASU, Kyiv (2015)
7. Stepashko, V., Bulgakova, O.: Generalized iterative algorithm GIA GMDH. In: International Conference on Inductive Modelling ICIM-2013, pp. 119–123. IRTC ITS NASU, Kyiv (2013)
8. Stepashko, V., Moroz, O.: Hybrid searching GMDH-GA algorithm for solving inductive modeling tasks. In: 1st International IEEE Conference 2016, Data Stream Mining & Processing, pp. 350–355. IEEE (2016)
9. Srinivas, M., Patnaik, L.M.: Binomially distributed populations for modelling GAs. In: 5th International Conference on Genetic Algorithms 1993, San Francisco, CA, USA, pp. 138–145 (1993)
10. Moroz, O.G., Stepashko, V.S.: Comparative analysis of model structures generators in sorting-out GMDH algorithm. In: Inductive Modeling of Complex Systems, vol. 8, pp. 173–191. IRTC ITS NASU, Kyiv (2016). (in Ukrainian)
11. Moroz, O.: Effectiveness of mutation operators in the genetic search for optimal model in a sorting-out GMDH algorithm. In: 7th International Workshop on Inductive Modeling (IWIM-2016), pp. 14–18. IRTC ITS NASU, Kyiv (2016)
12. Moroz, O.H.: Sorting-out GMDH algorithm with genetic search of optimal model. Contr. Syst. Mach. **6**, 73–79 (2016). (in Russian)
13. Holland, J.: Adaptation in Natural and Artificial Systems: An Introductory Analysis with Application to Biology, Control, and Artificial Intelligence. MIT Press, Cambridge (1975). University of Michigan
14. Kaya, Y., Uyar, M., Tekin, R.: A novel crossover operator for genetic algorithms: ring crossover. Presented at CoRR (2011)
15. http://archive.ics.uci.edu/ml/. Accessed 21 May 2017
16. https://archive.ics.uci.edu/ml/datasets/SkillCraft1+Master+Table+Dataset. Accessed 24 Apr 2017

Development of Combined Information Technology for Time Series Prediction

Oksana Mulesa[1(✉)], Fedir Geche[1], Anatoliy Batyuk[2], and Viktor Buchok[1]

[1] Department of Cybernetics and Applied Mathematics, Uzhhorod National University, Sq. Narodna, 3, Uzhhorod 88000, Ukraine
{oksana.mulesa, fedir.geche}@uzhnu.edu.ua
[2] ACS Department, Lviv Polytechnic National University, Lviv, Ukraine
abatyuk@gmail.com

Abstract. The task of designing information technology for time series forecasting, that bases on fuzzy expert evaluations was considered. A forecasting model, part of which is an expert's unit, were proposed. The algorithm of synthesis predictive scheme based on the basic predictive models was developed. To determine expert evaluation of the forecast value, the task of forecasting was seen as the problem of numerical evaluation of object. The rules for determining the collective numerical evaluations, that are based on fuzzy expert assessments were developed. The developed rules take into account coefficients of experts' competence and also their degree of confidence for their own assessments. The approaches to determining the competence coefficients members of the expert group were systematized. The analysis of features for designing information-analytical system of time series forecasting were done. The structural diagram of the analytical block of information-analytical system for time series prediction, that based on the fuzzy expert estimates, was itemized. The designed information technology should be used for time series forecasting in cases where it is necessary to take account the impact, on the process that is studied, of temporary, informal factors.

Keywords: Information technology · Time series forecasting
Predictive scheme based on the basic predictive models
The task of numeric evaluation of the object · Fuzzy expert evaluations
Competence of expert · System model · Information-analytical system

1 Introduction

The development of information technology (IT) and their integration in all spheres of human activity contributes to the rise efficiency of decision making processes in the production and management [1, 2]. Development such new models, methods and tools as would allow more perfectly tailored to suit each particular subject area contributes to the emergence of new opportunities in the implementation and management of complex systems. Typically, solving the problem of forecasting the main indicators of the system, based on the known retrospective data, is an important aspect in the

N. Shakhovska and V. Stepashko (eds.), *Advances in Intelligent Systems and Computing II*, Advances in Intelligent Systems and Computing 689,
https://doi.org/10.1007/978-3-319-70581-1_26

development of IT. Therefore, the development of new and adaptation to the problem of known methods of time series forecasting is an important step in the process of development IT.

2 Forecasting Problem

The task of time series forecasting for the processes of various nature is arise during the solving of a number of applied problems of health, economics, finance, environment, etc. [3, 4].

Several paradigms for solving the task of time series forecasting are known from the large number of scientific sources. Conventionally, all forecasting methods can be divided into intuitive methods and formal methods. Formal methods are divided into statistical methods among of them distinguish regression methods, autoregressive models, methods of exponential smoothing and others [4, 5]; and structural methods among of them are neural network models, models based on Markov chain, model based on classification-regression trees [3, 4]. To the intuitive methods, include expert methods [6–8], methods of selective pattern matching [9] and others. In practice, also often are used the genetic methods and the neural network methods and others [10, 11].

For the effective solving, a time-series forecasting problem usually is necessary to study the several prediction methods and to select the best of them according to the specified task.

3 The Mathematical Model for the Prediction Problem

The mathematical statement of the problem of time series forecasting is such [5]:

Let $y_1, y_2, \ldots, y_n$ – is the time series. It is necessary to determine the values $y_{n+1}, y_{n+2}, \ldots, y_{n+T}$, where T – step of forecast.

Typically, in time series forecasting, the forecasting model is defined as a functional relationship that with accurate to random component ε_t has such kind [5]:

$$y_t = F_t(y_{t-1}, y_{t-2}, \ldots) + \varepsilon_t \tag{1}$$

In this case, forecasting methods are intended to determine of the most accurate model parameters (1) to better reflection trends in time series. At that, we make the assumption that the value that describe by the time series in the future will be developed by the same laws as the previous time intervals.

However, with such approaches is almost impossible to take into account subjective factors that may have non-systematic or one-time effect on the studied processes. In a number tasks such factors are not permanent or seasonal and often need the forecasting. But they can have a significant impact on predicted value. Such influence impairs the precision of the forecast, which can be obtained by marked or other forecasting methods [7]. Since, as a rule, this effect we can't to forecast and to model, then the using of expert assessments can be useful in getting its numerical characteristics as an additional source of information. Therefore, we will build a model of time series forecasting, component of which is the expert's unit [7, 8]:

Let the experts group with m experts is given: $E = \{E_1, E_2, \ldots, E_m\}$. Each expert E_i, $(i = \overline{1,m})$ assesses the forecast values y_{n+t}, $t = \overline{1,T}$ by some way. It is necessary to determine the total assessment of forecast value y_{n+t}, $t = \overline{1,T}$ based on expert assessments.

We denote the experts' found numerical assessments as $\tilde{y}_{e,n+t}$, $t = \overline{1,T}$. Also, let we have obtained forecasting values for the object of studied using the some method of time series forecasting. We denoted these assessments as the $\tilde{y}_{f,n+t}$, $t = \overline{1,T}$. So the resulting values can be calculated by the formula:

$$y_{n+t} = \alpha \tilde{y}_{e,n+t} + (1 - \alpha)\tilde{y}_{f,n+t}, \tag{2}$$

where $\alpha \in [0; 1]$ – a value that the a person who makes a decision is setting. This value is expressing the degree of influence of expert assessments to the result of prediction.

Thus, the task of time series forecasting is divided into two tasks [12]:

- the task of determining the forecast values using the methods of time series forecasting;
- prediction problem as a problem of numerical assessment of object.

4 An Applying of Predictive Scheme Based on the Basic Forecasting Models for Solving the Problem of Time Series Forecasting

To the models and methods that used effectively in solving the problem of time series forecasting we can include [3, 5]:

- Winters method, which takes into account the seasonal components;
- regression model of forecasting;
- method of autoregression;
- the method of least squares with weights;
- predictive models of Brown and others.

In [13–15], was proved the effectiveness of the predictive scheme based on the basic forecasting models. The algorithm for synthesis of this predictive scheme is given in [13]. According to mentioned algorithm, the synthesis of predictive scheme is as follows.

The functional relationship between the predicted value $\tilde{y}_t$ and the elements of time series we will express this:

$$\tilde{y}_t = f(a_1, \ldots, a_r, y_{t-1}, \ldots, y_{t-k}, t), \tag{3}$$

where $a_1, a_2, \ldots, a_r$ are the parameters of the model f, k is the depth of prehistory.

To find the parameters $a_1, a_2, \ldots, a_r$, usually build the functional

$$L(a_1, \ldots, a_r) = \sum_{t=1}^{n} (y_t - \tilde{y}_t)^2, \quad (4)$$

which should to minimize.

Let $a_1^*, \ldots, a_r^*$ the values of parameters $a_1, a_2, \ldots, a_r$ under which the functiona L takes a minimum value. Then the predictive value $\tilde{y}_{n+\tau}$ by the model f with optimal parameters $a_1^*, \ldots, a_r^*$ can be found as follows:

$$\tilde{y}_{n+\tau} = f\left(a_1^*, \ldots, a_r^*, y_{t-1}, \ldots, y_{t-k}, n+\tau\right), \quad (5)$$

where τ – is the step of prediction.

Depending on the type of functions f with the parameters $a_1^*, \ldots, a_r^*$ we have the different optimal models of time series forecasting.

Let $y_1, y_2, \ldots, y_t, \ldots, y_n$ is the time series. To build a predictive scheme, in the beginning, we consider the method of autoregression by which we define the optimal step of the prehistory k_τ^* for a given time series with fixed prediction step τ. In model of autoregression value of the parameter y_t depends on $y_{t-\tau}, y_{t-\tau-1}, \ldots, y_{t-\tau-k_\tau+1}$, where τ – is the step of prediction, k_τ – the parameter of prehistory with fixed τ. Predictive value of $\tilde{y}_{n+\tau}$, by the method of autoregression, can be found with the following model:

$$\tilde{y}_{n+\tau} = a_1^{(\tau)} y_n + a_2^{(\tau)} y_{n-1} + \ldots + a_{k_\tau}^{(\tau)} y_{n-k_\tau+1}. \quad (6)$$

Let $a_1^{*(\tau)}, \ldots, a_{k_\tau}^{*(\tau)}$ are the optimal parameters of the model (6). Then we have such model of autoregression:

$$\tilde{y}_t = a_1^{*(\tau)} y_{t-\tau} + a_2^{*(\tau)} y_{t-\tau-1} + \ldots + a_{k_\tau}^{*(\tau)} y_{t-\tau-k_\tau+1}, \quad (7)$$

where $t \geq k_\tau + \tau$.

It is obviously, that the value y_t if $\tau = \tau_0$ depends on parameter k_τ $(1 \leq k_\tau \leq n-\tau)$. To determine the optimal value of the parameter of prehistory k_τ if $\tau = \tau_0$, for a given time series y_t, we consider the normalized standard deviations:

$$\delta_1 = \frac{1}{n-\tau} \sum_{t=\tau+1}^{n} \left(y_t - a_1^{*(\tau)} y_{t-\tau}\right)^2,$$

$$\delta_2 = \frac{1}{n-\tau-1} \sum_{t=\tau+2}^{n} \left(v_t - a_1^{*(\tau)} y_{t-\tau} - a_2^{*(\tau)} y_{t-\tau-1}\right)^2,$$

$$\ldots\ldots\ldots\ldots\ldots\ldots\ldots\ldots$$

$$\delta_{n-\tau} = \left(y_n - a_1^{*(\tau)} y_{n-\tau} - \ldots - a_{n-\tau}^{*(\tau)} y_1\right)^2$$

And we find $\min\{\delta_1, \delta_2, \ldots, \delta_{n-\tau}\} = \delta_{k_\tau^*}$. The value k_τ^* determine the optimal value of the parameter of prehistory for the autoregression model with fixed $\tau = \tau_0$.

Once we determined k_τ^* $(\tau = \tau_0)$, we consider the forecast results in different models for time series prediction $M_1, M_2, \ldots M_q$ with the step of prediction τ at the moments of time t: $n - k_\tau^* + 1, n - k_\tau^* + 2, \ldots, n$. Based on forecasts of different models we construct the following table:

Table 1. The predictive values of time series relatively to the basic models

Predicting models	The value of the studied parameters for the period $n - k_\tau^* + 1, n - k_\tau^* + 2, \ldots, n$			
	$y_{n-k_\tau^*+1}$	$y_{n-k_\tau^*+2}$	$\cdots$	y_n
M_1	$\tilde{y}^{(1)}_{n-k_\tau^*+1}$	$\tilde{y}^{(1)}_{n-k_\tau^*+2}$	$\cdots$	$\tilde{y}^{(1)}_n$
M_2	$\tilde{y}^{(2)}_{n-k_\tau^*+1}$	$\tilde{y}^{(2)}_{n-k_\tau^*+2}$	$\cdots$	$\tilde{y}^{(2)}_n$
$\vdots$	$\vdots$	$\vdots$	$\cdots$	$\vdots$
M_q	$\tilde{y}^{(q)}_{n-k_\tau^*+1}$	$\tilde{y}^{(q)}_{n-k_\tau^*+2}$	$\cdots$	$\tilde{y}^{(q)}_n$

In the each column $y_{n-k_\tau^*+1}, y_{n-k_\tau^*+2}, \ldots, y_n$ of Table 1 we find the lowest value of the quadratic deviation between the predictive and the real values of the relevant elements of the time series. Mathematically, this can be written as:

Let

$$j_1 = n - k_\tau^* + 1 \text{ and } \varepsilon_1 = \min\left\{\left(y_{j_1} - \tilde{y}^{(1)}_{j_1}\right)^2, \ldots, \left(y_{j_1} - \tilde{y}^{(q)}_{j_1}\right)^2\right\},$$
$$j_2 = n - k_\tau^* + 2 \text{ and } \varepsilon_2 = \min\left\{\left(y_{j_2} - \tilde{y}^{(1)}_{j_2}\right)^2, \ldots, \left(y_{j_2} - \tilde{y}^{(q)}_{j_2}\right)^2\right\},$$
$$\ldots\ldots$$
$$j_{k_\tau^*} = n \text{ and } \varepsilon_{k_\tau^*} = \min\left\{\left(y_n - \tilde{y}^{(1)}_n\right)^2, \ldots, \left(y_n - \tilde{y}^{(q)}_n\right)^2\right\},$$

Then we define the sets $I_1, I_2, \ldots, I_{k_\tau^*}$ as follows:

$$I_1 = \left\{ i \in \{1, 2, \ldots, q\} \middle| \varepsilon_1 = \left(y_{j_1} - y^{(i)}_{j_1}\right)^2 \right\},$$
$$I_2 = \left\{ i \in \{1, 2, \ldots, q\} \middle| \varepsilon_2 = \left(y_{j_2} - y^{(i)}_{j_2}\right)^2 \right\},$$
$$\ldots\ldots\ldots\ldots\ldots\ldots\ldots\ldots$$
$$I_{k_\tau^*} = \left\{ i \in \{1, 2, \ldots, q\} \middle| \varepsilon_{k_\tau^*} = \left(y_n - y^{(i)}_n\right)^2 \right\},$$

and we construct Table 2

where $a_{ps} = \begin{cases} \beta^{k_\tau^* - s}, & \text{if } s \in I_s, \\ 0, & \text{if } s \notin I_s, \end{cases}$

$$S_p(\beta) = \sum_{j=1}^{k_\tau^*} a_{pj}, \quad 0 < \beta \leq 1, \quad (p = 1, 2, \ldots, q, s = 1, 2, \ldots, k_\tau^*).$$

Table 2. Parameters of the predictive scheme

Predicting models	j_1	j_2	$\cdots$	$j_{k_\tau^*}$	The resulting column
M_1	a_{11}	a_{12}	$\cdots$	$a_{1k_\tau^*}$	$S_1(\beta)$
M_2	a_{21}	a_{22}	$\cdots$	$a_{2k_\tau^*}$	$S_2(\beta)$
$\vdots$	$\vdots$	$\vdots$	$\vdots$	$\vdots$	$\vdots$
M_q	a_{q1}	a_{q2}	$\cdots$	$a_{qk_\tau^*}$	$S_q(\beta)$

Using the $S_p(\beta)$ and $S(\beta) = \sum_{p=1}^{q} S_p(\beta)$ we define weights coefficients of forecasting models M_p $(1 \leq p \leq q)$ which are included in the following predictive scheme

$$\tilde{y}_{n+\tau} = \frac{S_1(\beta)}{S(\beta)} \tilde{y}_{n+\tau}^{(1)} + \frac{S_2(\beta)}{S(\beta)} \tilde{y}_{n+\tau}^{(2)} + \ldots + \frac{S_q(\beta)}{S(\beta)} \tilde{y}_{n+\tau}^{(q)}. \tag{8}$$

Before using the predictive scheme (8) we should conduct a study of this scheme relatively to the β. For it, we build the functional

$$L(\beta) = \sum_{i-1}^{k_\tau^*} \left(y_{j_i} - \frac{S_1(\beta)}{S(\beta)} y_{j_i}^{(1)} - \frac{S_2(\beta)}{S(\beta)} y_{j_i}^{(2)} - \ldots - \frac{S_q(\beta)}{S(\beta)} y_{j_i}^{(q)} \right)^2, \quad (j_i = n - k_\tau^* + i),$$

and we minimize this functional by varying the value of β. For it, the interval (0, 1] is divided into m equal intervals. Then we find the values of $L(\beta_i)$ in the points $\beta_i = i/m,\ (i = 1, 2, \ldots, m)$ on this interval. It is obviously, that the parameter m defines the accuracy of the minimization of functional $L(\beta)$. Let $\beta_m^* = \arg \min_{1 \leq i \leq m} L(\beta_i)$. Then we execute the forecast for time series by the scheme (8), replacing the β on the value β_m^*.

The remark. If $\beta = 1$, then the model (8) does not take into account the distance from the element y_t to predictive value $\tilde{y}_{n+\tau}$.

5 The Way of Use the Fuzzy Models and Methods of Determination the Numerical Evaluations of Object to the Problem of Time Series Forecasting

We consider the problem of determining the numerical evaluation of object for the problem of time series forecasting in such formulation [7]:

Let we have the expert group with m experts: $E = \{E_1, E_2, \ldots, E_m\}$. The competence coefficients of experts are such $\alpha_1, \alpha_2, \ldots, \alpha_m$ $(\alpha_1 \in [0; 1])$ and $\sum_{i=1}^{m} \alpha_i = 1$.

Each of the experts E_i $(i = \overline{1, m})$ for each of the predicted values y_{n+t}, $t = \overline{1, T}$, puts three numeric evaluations: pessimistic, realistic and optimistic. Assign them, accordingly, $y_{i,n+t}^{(p)}$, $y_{i,n+t}^{(r)}$, $y_{i,n+t}^{(o)}$. Without decreasing generality of considerations,

Let us consider that $y_{i,n+t}^{(p)} \le y_{i,n+t}^{(r)} \le y_{i,n+t}^{(o)}$. Let the experts assign the degrees of confidence for each of their assessments. That is $\mu_{i,n+t}^{(p)}$, $\mu_{i,n+t}^{(r)}$, $\mu_{i,n+t}^{(o)}$ for pessimistic, realistic and optimistic assessment, accordingly. It is naturally to consider that $\mu_{i,n+t}^{(r)} > \mu_{i,n+t}^{(p)}$ and $\mu_{i,n+t}^{(r)} > \mu_{i,n+t}^{(o)}$.

In the earlier stages of processing the input data for exception such experts' assessments that have certain features, it is proposed to apply such heuristics [16]:

Heuristic 1. We do the exception of group E those of the experts which are confident in their assessments less then a specified threshold. I.e.

$$E := E \backslash \{E_i\}, \quad \forall i = \overline{1,m}: \quad \overline{\mu}_i \le \Delta, \tag{9}$$

where $\Delta \in (0;1)$ is the specified threshold.

Heuristic 2. We do the exception of group E those of the experts who confident equally in all their numerical assessments. I.e.

$$E := E \backslash \{E_i\}, \ \forall i = \overline{1,m}: \ |\overline{\mu}_i - \min\{\mu_{1i}, \mu_{2i}\}| < \varepsilon, \tag{10}$$

where ε is the specified value of proximity for the degree of confidence.

Suppose that after using the Heuristics 1 and the Heuristics 2 a group of experts is such: $E = \{E_1, E_2, \ldots, E_m\}$.

Then execute the normalization of values of the coefficients competence of experts as follows:

$$\alpha_i := \frac{\alpha_i}{\sum_{i=1}^{n} \alpha_i}. \tag{11}$$

On the next stage in determining the numerical evaluation of object, for each expert $E_i \in E$ $(i = \overline{1,m})$ we build a fuzzy set $A_i = \{(y, \mu_i(y))\}$ [17]. The membership functions of fuzzy sets will be defined by the such rule:

$$\mu_i(y) = \begin{cases} 0, & \text{if } y < y_i^{(p)}; \\ \mu_i^{(p)}, & \text{if } y = y_i^{(p)}; \\ a_{1i}y + b_{1i}, & \text{if } y \in \left[y_i^{(p)}, y_i^{(r)}\right]; \\ \mu_i^{(r)}, & \text{if } y = y_i^{(r)}; \\ a_{2i}y + b_{2i}, & \text{if } y \in \left[y_i^{(r)}, y_i^{(o)}\right]; \\ \mu_i^{(o)}, & \text{if } y = y_i^{(o)}; \\ 0, & \text{if } y > y_i^{(o)}. \end{cases} \tag{12}$$

The coefficients $a_{1i}, a_{2i}, b_{1i}, b_{2i}$ should be determined by solving the corresponding systems of linear algebraic equations.

We offer the following rules for determining the collective numerical evaluation of the object [7, 8]:

Rule 1. Form the fuzzy set of collective numeric object evaluation $C = \{(y, \mathrm{M}(y))\}$ with membership function $\mathrm{M}(y)$, which is calculated in the following way:

$$\mathrm{M}(y) = \sum_{i=1}^{m} \alpha_i \mu_i(y). \tag{13}$$

The algorithm of the calculation of membership function of the resulting fuzzy set should boil down to defining the total of the corresponding membership functions in the intervals of their monotony.

Rule 2. According to the accurate methods of determining collective numeric object evaluation, calculate the total of fuzzy numbers:

$$C = \alpha_1 A_1 + \alpha_2 A_2 + \ldots + \alpha_n A_n$$
$$= \bigcup_{c \in S_C} \max_{\substack{(a_1, a_2, \ldots, a_n): \\ a_i \in S_{A_i}, \sum_i \alpha_i a_i = c}} \min\{\mu_1(a_1), \mu_2(a_2), \ldots, \mu_n(a_n)\}, \tag{14}$$

where S_{A_i} are the carriers of corresponding fuzzy sets A_i, $i = \overline{1, m}$; S_C is the carrier of fuzzy set C.

To determine the collective numerical evaluation of object we propose to use the one of the following relationships:

$$y^* = \frac{1}{l} \sum_{i=1}^{l} y_i, \tag{15}$$

$$y^* = \max_y \left\{ Arg \max_y \{\mathrm{M}(y)\} \right\}, \tag{16}$$

$$y^* = \min_y \left\{ Arg \max_y \{\mathrm{M}(y)\} \right\}, \tag{17}$$

where $y_i \in Arg \max_y \{\mathrm{M}(y)\}$, l is the capacity of set $Arg \max_y \{\mathrm{M}(y)\}$.

Thus obtained the numerical evaluations we will call as the experts' forecasting values.

6 Models and Methods for Determining the Coefficients of Experts' Competence

Coefficients of experts' competence are parts of both conventional and fuzzy rules of the collective numerical evaluation. Therefore, in the context of determining the numeric evaluation of an object, we also have to solve the problem of determining the competence of experts. Mathematically, the formulation of this problem can be written as follows [18]:

let we have the expert group $E = \{E_1, E_2, \ldots, E_m\}$ and a set of input data about this experts Ξ. Based on available data, it is necessary to determine the coefficients of experts' competence $\alpha_1, \alpha_2, \ldots, \alpha_m$.

Depending on the initial conditions of the task, the set Ξ may include:

- experts' personal data such as information about their education, about place of employment, about position, industry experience, and other indicators that can characterize competence of the experts;
- self-evaluations of experts: levels of competence, that experts have identified for themselves;
- mutual evaluations of experts;
- information on the results of previous examinations, which involved experts that are considered and based on which to make conclusions about the competence of experts.

In the depending on the nature of input data about experts, to evaluate their competence, it is possible to use such groups of methods [19, 20]:

Group 1. Documental evaluations: let $D = \{D_1, D_2, \ldots, D_r\}$ – are the parameters of the documental assessments, the characteristics of experts, impact of which on competency can be determined objectively; $L = \{L_1, L_2, \ldots, L_r\}$ – are the corresponding weights of parameters. Then, the coefficients of experts' competency can be determined by the formula:

$$\alpha_i = \sum_{j=1}^{r} L_j D_j^{(i)},$$

where $D_j^{(i)}$ is the value of j-th parameter for the i-th expert.

Group 2. Mutual evaluations: let we have the set $\mathrm{K} = \left\{k_j^{(i)}\right\}, i, j \in \{1, 2, \ldots, m\}$ of the mutual experts' evaluations, such that, if $\exists k_j^{(i)} \in \mathrm{K}$, it means, that expert with number j estimated the competence of the expert with number i with the value $k_j^{(i)}$ (if $i = j$ then $k_j^{(i)}$ is self-evaluation of i-th expert); if $\exists i, j \in \{1, 2, \ldots, m\}$, that $\overline{\exists} k_j^{(i)} \in \mathrm{K}$, it means that j-th expert didn't estimated the competence of i-th expert.

Then, the resulting coefficients competence of experts can be calculated by the formula:

$$\alpha_i = \frac{\sum\limits_{\substack{j = \overline{1, m}: \\ k_j^{(i)} \in \mathrm{K}}} k_j^{(i)}}{\chi_j^{(i)}(\mathrm{K})},$$

where $\chi_j^{(i)}(\mathrm{K}) = \begin{cases} 1, & if\ \exists k_j^{(i)} \in \mathrm{K}; \\ 0, & otherwise. \end{cases}$

Group 3. Aposterior methods – the methods based on information on the results of expert participation in the preliminary expertises. They are divided into statistical and experimental methods. Separately, one can distinguish methods for determining the competence of experts from the axiom of unbiasedness according to which, the majorities' opinion is competent [21].

There are other, more complex methods for determining the competence of experts [22]. However, their practical application requires additional resources and extends the time required to obtain results. Thus, the decision to choose the method for determining the competence of experts should make the person who makes decisions based on how much of it is important to improve the accuracy of the results, and how resources he or she has for solving the problem.

7 Designing of the Information–Analytical System for the Time Series Forecasting

Information-Analytical System, which implemented the developed models and methods based on fuzzy expert assessments for the time series forecasting, along with considered models and methods is an integral part of the respective IT.

The process of developing information-analytical system should be carried out according to the levels of system model [23, 24]:

$$\text{goals} \Rightarrow \text{problem (model)} \Rightarrow \text{methods (algorithms)} \Rightarrow \text{tools}$$

The improving of the efficiency of decision-making processes which related to forecasting of numerical indicators that based on retrospective data, is the primary purpose of designing the information-analytical system.

According to the mentioned goal and the described models and methods, at the design stage the information-analytical system it is necessary to implement the following tasks:

- the determination of the forecast value for predictable valucs;
- the determination of the expert numerical evaluation of forecast value for predictable values;
- the setting of the impact coefficient of expert assessments to the result and calculating the resulting value of forecast.

Next set of mathematical models and methods are the basis of the information-analytical system:

- the models and methods of forecasting based on time series;
- the models and methods of determining the competence of experts;
- fuzzy models and methods for determining the collective numerical assessment.

Thus, the structure of the analytical unit, which is the basis of IAS, can be represented as (Fig. 1):

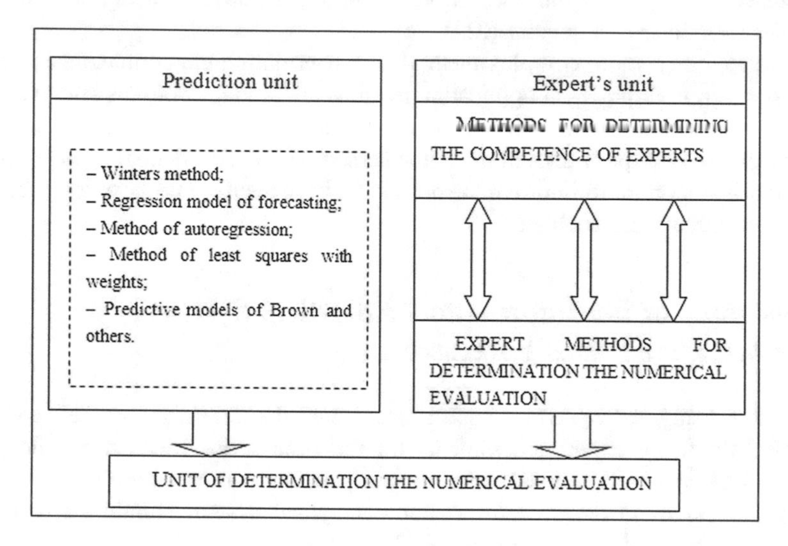

Fig. 1. Unit diagram of the analytical block of IAS

In view of this, we should solve such tasks in the designing of the information-analytical system:

– the task of developing software for primary processing of the input data;
– the task of developing the procedure of synthesis predictive scheme based on the forecasting models;
– the task of developing procedures for processing the results of expert surveys.

8 Summary and Conclusion

The paper contain a description of the study information technology for time series forecasting based on the expert's fuzzy assessments. The mathematical model of determination the predictive values, that based on the results of expert polls, was done. The algorithm of synthesis the predictive scheme that is based on the basic predictive models, which use improves the result of the prediction. There are some heuristics and rules of determining the collective numerical evaluation for the predicted value. Using the proposed information technology for time series prediction allows to takes into account impact of such factors that are temporary and can be displayed in numeric experts' evaluations. The approaches to determining the competence of experts as one of the stages of expertise has been systematized.

The use of the proposed information technology will improve the accuracy of forecasting for some applied tasks.

References

1. Tsmots, I.: Information Technology and Specialized Tools for Signal Processing and Image Processing in Real Time. UAD, Lviv (2005)
2. Mulesa, O., Geche, F., Batyuk, A.: Information technology for determining structure of social group based on fuzzy c-means. In: Xth International Scientific and Technical Conference on Computer Sciences and Information Technologies (CSIT), pp. 60–62 (2015)
3. Kuharev, V.N., Sally, V.N., Erpert, A.M.: Economic-Mathematical Methods and Models in the Planning and Management. Vishcha School, Kiev (1991)
4. Kozadaev, A.S., Arzamasians, A.A.: Prediction of time series with the apparatus of artificial neural networks. The short-term forecast of air temperature. Bull. Univ. Tambov Ser. Nat. Tech. Sci. **11**(3), 299–304 (2006)
5. Snytiuk, V.Y.: Forecasting. Models. Methods. Algorithms: Tutorial. "Maklaut", Kiev (2008)
6. Mendel, A.S.: Method counterparts in predicting short time series: expert-statistical approach. Machine. Telemekh. № 4, pp. 143–152 (2004)
7. Mulesa, O., Geche, F.: Designing fuzzy expert methods of numeric evaluation of an object for the problems of forecasting. Eastern Eur. J. Enterp. Technol. **3**(4(81)), 37–43 (2016). https://doi.org/10.15587/1729-4061.2016.70515
8. Mulesa, O.: Heuristic rules of the collective numerical evaluation of object and their application to the problem of time series prediction. In: Intelligent Decision Support Systems and Computational Intelligence problems, pp. 208–210 (2016)
9. Kuchanky, A., Biloshchytskyi, A.: Selective pattern matching method for time-series forecasting. Eastern Eur. J. Enterp. Technol. **6**(4–78), 13–18 (2015). https://doi.org/10.15587/1729-4061.2015.54812
10. Zaichenko, Y.P., Mohammed, M., Shapovalenko, N.V.: Fuzzy neural networks and genetic algorithms in problems of macroeconomic forecasting. Scientific news "KPI" № 4, pp. 20–30 (2002)
11. Pukach, A., Teslyuk, V., Tkachenko, R., Ivantsiv, R.A.: Implementation of neural networks for fuzzy and semistructured data. In: 11th International Conference the Experience of Designing and Application on CAD Systems in Microelectronics (CADSM), pp. 350–352 (2011)
12. Mulesa, O., Geche, F., Batyuk, A., Buchok, V., Voloshchuk, V.: Information technology for time series forecasting with considering fuzzy expert evaluations
13. Geche, F., Mulesa, O., Geche, S., Vashkeba, M.: Development the method of synthesis of predictive scheme based on the basic forecasting models. Technol. Audit Reserves Prod. **3**(2(23)), 36–41 (2015). https://doi.org/10.15587/2312-8372.2015.44932
14. Geche, F., Mulesa, O., Myronuyk, I., Vashkeba, M.: Prediction the quantitative characteristics of officially registered HIV-infected people in the region. Technol. Audit Reserves Prod. **4**(2(24)), 34–39 (2015). https://doi.org/10.15587/2312-8372.2015.47907
15. Geche, F., Batyk, A., Mulesa, O., Vashkeba, M.: Development of effective time series forecasting model. Int. J. Adv. Res. Comput. Eng. Technol. (IJARCET) **4**(12), 4377–4386 (2015)
16. Mulesa, O.: Methods of considering subjective character of input data for the tasks of voting. Eastern Eur. J. Enterp. Technol. **1**(3(73)), 20–25 (2015). https://doi.org/10.15587/1729-4061.2015.36699
17. Orlovskyi, S.A.: Decision Making with Fuzzy Initial Information. Nauka, Moscow (1981)
18. Gnatienko, G., Snityuk, V.: Experts' Technology of Decision Making. Makaut, Kiev (2008)

19. Korchenko, O.G., Hornitskyy, D.A., Zaharchuk, T.G.: Research priori estimation methods to implement an expert examination in the field of information security. Data Prot. **12** (4(49)) (2010). http://jrnl.nau.edu.ua/index.php/ZI/article/view/1976/1967
20. Kolpakova, T.A.: Determination of the competence of experts in making group decisions. Radioelektronika, computer science, upravlinnya **1**(24), 40–43 (2011)
21. Snityuk, V. E.: Models and methods of determining the competence of experts on the basis of unbiasedness axiom. News CHITI, vol. 4, pp. 121–126 (2000)
22. Shanteau, J.: Competence in experts: the role of task characteristics. Organ. Behav. Hum. Decis. Process. **53**(2), 252–266 (1992)
23. Uyomov, A.: System Approach and General Theory of Systems. Mysl, Moscow (1978)
24. Timchenko, A.A.: Fundamentals of System Design and System Analysis of Complex Objects. Lybed, Kiev (2000)

Keyphrase Extraction Using Extended List of Stop Words with Automated Updating of Stop Words List

Svetlana Popova[1,2] and Gabriella Skitalinskaya[3,4(✉)]

[1] Saint-Petersburg State University, Saint-Petersburg, Russia
svp@list.ru
[2] ITMO University, Saint-Petersburg, Russia
[3] Institute of Technology Tallaght, Dublin, Ireland
gabriellasky@icloud.com
[4] Moscow Institute of Physics and Technology (State University), Moscow, Russia

Abstract. In the paper we consider the problem of keyphrase extraction. Our tasks is to examine additionally the approach to keyphrase extraction, which is based on the use of extended lists of stop words. The second objective of the research is to test the approach for automatic expansion of such extended lists. The obtained results allow to confirm the possibility of improving the quality of algorithms of keyphrase extraction. The results of experiments with the extended lists of stop words show the potential of the proposed approach.

1 Introduction and State-of-the-Art

The problem of keyword extraction has a great number of applications, among which we can distinguish tasks such as topic extraction, data structuring, data clustering and classification, ontology population, search of dependent concepts in large data sets and other tasks. The keyphrase extraction problem expands the problem of keyword extraction and requires not only defining words that are thematically relevant to the document, but also combining these words into agreed phrases reflecting the main topics of the document.

In the field of keyphrase extraction two main approaches can be identified. The first approach [1–4] considers the following steps. At the first step words are selected from the text. Keyphrases are created from these words at the next step. It can be described as follows:

1. Ranking of single words and selecting the best ones.
2. Construction of keyphrases from the selected words.

The reported study was funded by RFBR according to the research project No. 16-37-00430 mol-a and partially supported by the Government of Russian Federation, Grant 074-U01.

N. Shakhovska and V. Stepashko (eds.), *Advances in Intelligent Systems and Computing II*, Advances in Intelligent Systems and Computing 689,
https://doi.org/10.1007/978-3-319-70581-1_27

In the second approach, word sequences are extracted from the text (usually referred to as candidate phrases), and in the next step keyphrases are selected from the extracted set of word sequences. Usually, at this stage, either ranking and selection of top-n phrases, or classification is used. During classification it is determined whether a phrase should be assigned to the class of keyphrases or not. The second approach [5–10] considers the following steps:

1. Construction of candidate keyphrases.
2. (a) Ranking of candidate phrases and selecting the best ones as keyphrases;
   (b) Classification of candidate phrases as keyphrases and not.

Within the framework of the first approach one of the most popular method for ranking words is based on graphs. In this case, text units (for example, words [1], n-gramms or noun phrases [3]) become the vertices of the graph. The arcs show the relationships between the elements forming the vertices. For example, two vertex-words are joined by an arc, if in the text these words occur together in a window of a given size. The weight of the arcs in this case can show how often these words co-occur in the texts. After the graph is constructed, the weight of each of its vertices is estimated. The vertices are ranked according to their weight and the specified percentage of the best of them is selected for further work. For example, let single words be used as vertices [1]. Words selected as the best after the ranking are joined into phrases if in the text these words follow each other. For vertex ranking, authors mostly use various modifications of the PageRank formula [11], for example TextRank and its modifications [1]. In [2], the construction of the graph takes into account the contents of k nearest documents. In [4], information on the semantic proximity between the constructed vertices is taken into account. Usually to calculate the semantic proximity WordNet or Wikipedia are used. The results obtained in [4] are one of the best in the field.

Within the framework of the second approach candidate keyphrase extraction is the first step. To construct the phrases n-grammes, noun phrases, word sequences corresponding to specified patterns, word sequences satisfying restrictions and other are used. For example, following restrictions can be used: absence of stop words in a phrase, limits on word length, minimum frequency in the collection, and other.

At the next stage of the second approach, the problem of selecting parameters for the ranking or classification algorithm is solved. As noted in [9], most of the proposed algorithms use a combination of different word characteristics within an annotated document and the full collection. Examples of such characteristics could be: the semantic similarity, the popularity of the phrase in a set of phrases selected by the expert manually (for example, for a similar collection), lexical and morphological analysis, various heuristics such as the length of the phrase and the position of the phrase in the document (for example, heading, first paragraph, last paragraph, etc.).

In [12], the problem of automatically extracting keyphrases from single tweets is explored. Authors propose a deep recurrent neural network model for the joint processing of the keyword ranking, keyphrase generation, and keyphrase ranking steps. Word embeddings were used as input to the neural network. The word embeddings were pre-trained vectors: part of a Google News dataset and a skip-gram model were used to generate 300-dimensional vectors for 3 million words and phrases. The authors report that the proposed approach can achieve better results than currently presented in the field.

The authors of [13,14] present a research in a field close to extracting keyphrases. The task came from the area of clustering the results of user requests. The idea of this approach is to first extract the key topics presented in the snippets of the search results with further grouping of documents around the topics received. To construct topics, the authors use the most frequent sub-sequences of words extracted using suffix trees and its modifications [13,14]. These sub-sequences that form the topics are, in fact, very close to keyphrases as defined in this paper.

The following articles offer more detailed information on the field of research. The most complete overview of the current state-of-the-art is presented in [15]. Review, results and analysis of the well-known competition in the field "SemEval-2010 Task 5: Automatic Keyphrase Extraction from Scientific Articles" are presented in [9]. The following specialized events should be noted: SemEval 2017 Task 10: Extracting Keyphrases and Relations from Scientific Publications [16], the ACL 2015 Workshop on Novel Computational Approaches to Keyphrase Extraction [17].

In addition, we note some of the important observations in the field. Researchers note that nouns and adjectives are the most significant parts of speech when extracting keyphrases [1,3,18]. In [1,3] it is shown that restrictions on the use of other parts of speech during the construction of phrases can significantly increase the results. The noticeable influence of information on parts of speech on the task of extracting keyphrases was noted also in other works.

In [18], n-grams, NP-chunks, POS Tag Patterns were used to extract keyphrases. The paper notes that the latter method was chosen because when extracting keyphrases only as noun phrases about half of the phrases were not extracted. 56 POS tag patterns were identified. Patterns were built on the basis of manual marking of keyphrases in the training collection, which corresponded to 10 or more phrases. 51 patterns contained one or more nouns. Five most frequent patterns were identified, they consisted of nouns and adjectives. To determine the keyphrases among the extracted sequences a classifier was used. That classifier used 4 main parameters to estimate the word sequence weights. One of the characteristics being the sequence of tags of the candidate phrase. In the work it was shown that the use of only the mentioned feature has a significant effect on the quality of extraction of keyphrases Thus, the information on the sequence of parts of speech in the candidate-phrase is important. The highest recall is achieved when extracting candidate phrases is using POS Tag Patterns. This means that the use of linguistic patterns allows to build many candidate phrases, which will include a large number of true keyphrases. Thus another problem arises, there are too many incorrect phrases in the set of candidate phrases. Therefore, on the one hand, information about the sequences of parts of speech in phrases is important, which justifies the use of patterns, on the other hand, too many candidate phrases are extracted, which complicates the stage of extracting keyphrases from them.

An interesting aspect of the selection and ranking of candidate phrases show the results given in [6]: a large number of generated candidate phrases extremely negatively affects the quality of selection of keyphrases at the ranking stage. The paper also shows that the use of information about the proximity of the phrase to the beginning of the document works well in the case of scientific publications and does not work for literary texts. This is an interesting observation, since one of the main criteria in assessing the weight of a phrase (for scientific texts) in research is the position of the phrase in relation to the beginning of the text.

Another important criterion is the length of the phrase. It is believed that longer phrases should be given higher weight in ranking than shorter phrases, due to the fact that the latter are more rare in the collection. In [19] it was shown that when constructing a set of candidate phrases among single-word phrases, the percentage of phrases that are not keyphrases is higher than for multi-word phrases.

2 Approach

Our approach developed in [19,20] complements and expands the research area and represents a new separate direction. It is based on the use of extended lists of stop words. By stop words we understand words that can not be found in keyphrases and at the same time are separators between phrases. First we note that the task of extracting keyphrases is complicated by the fact that the frequencies of words, which are stop words and which are not, can practically be very similar. Another problem is that the word can be simultaneously a stop word for some texts and a keyword for other texts. Therefore, it is important to consider how the addition of each particular word to the list of stop words is justified by measuring the ratio of the gain in quality (due to the fact that the phrases become more precise) to loss in quality (since part of the correct phrases containing the added word is lost). This ratio justifies the use of different heuristics in the field, which are determined based on the type of processed texts. The latter leads to the assumption that the best approach to selecting stop words is an approach based on the use of a training sample, which nevertheless requires a annotated with keyphrases collection of texts. Although the creation of such a collection can be time and labour-consuming, this approach allows to specify the features and logic of phrases that are supposed to be extracted from the text (for example: should the phrases be long or short, what should they reflect (subject, opinion, location, etc.)).

In this paper, we will further consider the approach based on using extended lists, and also offer an approach for automatically updating such lists. Here we consider only the first stage of keyphrase extraction: the construction of the candidate phrases. We show how the their quality can be improved. The received updated set of phrases can be additionally ranked or classified. But since the ratio of true-positive phrases to false-positive when applying the proposed approach increases, we can expect higher quality when applying methods of ranking and classification.

In the proposed approach candidate phrases are extracted as continuous word sequences of maximum length of given parts of speech. Delimiters between phrases are stop words, punctuation and words of other parts of speech. We use word sequences instead of linguistic patterns. Our assumption is that using word sequences is more effective.

When extracting keyphrases using patterns higher recall and lower precision is achieved. Thus, more phrases are extracted correctly and more phrases extracted are incorrect, which is explained as follows. When using patterns, more phrases are extracted than when using sequences. For example, if the patterns given are: "adjective + noun" and "adjective + adjective + noun", then the phrases "red car" and "good red car" will be extracted from the "good red car" text. At the same time, if the candidate phrases are extracted as the longest sequences of the specified parts of speech (nouns and adjectives), then from the mentioned text only one phrase "good red car" will be retrieved.

In the case of using patterns, their choice is very important. One can, for example, use the most frequent patterns of the gold standard. But our experiments have shown that using the sequences of maximum length allows to get higher quality from the point of view of the F1-score (see Sect. 3 for details) than when using frequent patterns.

When extracting phrases as word sequences of maximum length, we must define the parts of speech that words from the extracted phrases should satisfy. In the field it is shown that the main parts of speech are nouns and adjectives. This is true for a sufficiently large number of different types of texts. In the previous experiments [19,20] we worked with abstracts to scientific publications, where we used sequences of nouns and adjectives. In the current study, we work with messages from Internet forums and accept not only nouns and adjectives, but also verbs, otherwise we get a very low Recall. This can be explained as follows. When working with collections of abstracts of scientific publications, the analysis of the gold standard shows that most of the "good" patterns contain only nouns and adjectives. By saying "good" we mean that the use of this pattern allows to obtain a high ratio of the number of true-positive divided by the number of false-positive. In the case of the collection of posts from Internet car forums with which we work in this study, some of the "best" patterns contain nouns, adjectives and verbs. The latter determines the choice of parts of speech in this work.

3 Quality Evaluation

The task of assessing the quality of keyphrases forms an independent direction of research. In the field researchers tried to use the following methods of quality evaluation, that came from machine translation and summarization evaluations: Bleu, Meteor, Nist, Rouge. Another and one of the most popular measures of quality evaluation was the F1-score measure. This measure allows to integrate information about the precision and recall of the extracted phrases.

$$Precision = \frac{|(C \cap G)|}{|G|} \qquad (1)$$

$$Recall = \frac{|(C \cap G)|}{|C|} \quad (2)$$

$$FScore = \frac{2 \times Precision \times Recall}{Precision + Recall} \quad (3)$$

where $|C \cap G|$ - is the number of correctly extracted phrases when processing all the texts of the collection, $|G|$ - is the total number of phrases automatically extracted by the algorithm from all the texts of the collection, $|C|$ - is the number of all phrases in the "gold standard". The gold standard for each text includes an ideal list keyphrases manually tagged by an expert.

The F1-score method of evaluation does not impose restrictions on the number of phrases to be extracted. In [3], an approach based on the use of R-Precision instead of the F-score is proposed. R-Precision is the Precision value, provided that the number of keyphrases extracted is exactly the same as the number of phrases in the gold standard.

To calculate F1-score and R-Precision, the information about the number of correctly extracted phrases is needed. Let G_t be a set of automatically extracted phrases from text t, and C_T - a set of keyphrases of the gold standard for text. Here the question arises as which of the phrases from G_t belong to $G_t \cap C_t$. In many papers on keyphrase extraction the exact match is used. This means, that the phrase k - "advanced automatic translation" - and the phrase g - "automatic translation" - of the gold standard are recognized as different. Similar conclusions will be made, for example, for the following phrases: "good car" and "good auto". In [3], the following methods of determining the correctness of the extracted phrases are suggested and compared:

- the exact match of two phrases (exact match);
- the coincidence of two phrases in the presence of only a morphological difference (morph);
- the coincidence of two phrases, if the phrase k contains the phrase g (include);
- the coincidence of two phrases, if the phrase g contains the phrase k (part-of).

It is shown that the use of "part-of" is not successful. To evaluate the coincidence of two phrases in [3], "exact match" and "include + morph" were used. In [21], the number of ways to evaluate an extracted keyphrase is expanded. The authors determine a ratio, which best reflects the human judgments of the correctness of a phrase. In the form proposed in [21], the number of words coinciding in the automatically extracted phrase and in the phrase of the gold standard is considered. Then the resulting value is divided by the length of the longest of the two phrases.

However, the question of how to assess the quality of keyphrases remains open. The authors offer new approaches and methods, but the most frequent and popular in use when evaluating exact matches is the F1-score, which we also use for evaluation in this paper.

4 Test Collections

A collection of posts from an Internet car forum related to the purchase or repair of cars was obtained, where keyphrases were manually assigned. The keyphrases should reflect the main content of the text, including: was the car being bought or in service, location, the impression the customer has of the car salon/car service and the reasons for such impression. The constructed collection consists of 120 texts, among which half of the texts contain positive reviews and half are negative. The texts were mixed and divided into two equal collections, hereinafter collection 1 (T1) and collection 2 (T2). Table 1 provides examples of texts and manually assigned keyphrases.

Table 1. Examples of texts and keyphrases of the "gold standard"

Examples of texts	Phrases of the "gold standard"(manually assigned phrases)
Ordered a car - hyundai solaris, they took a prepayment of 50 thousand. rubles., we have been waiting for half a year, they do not really say anything! Disgusting work of the salon, they take not small amounts of money from customers, screw them over, and then offer to re-order the car in another salon!	Ordered a car; took prepayment; waiting for half a year; disgusting work of the salon; offer to re-order the car
заказали у них машину hyundai solaris, взяли предоплату 50тыс. руб., ждем уже пол года, ничего толком не говорят! отвратительная работа салона, берут с клиентов не малыеденьги, прокручивают их, а потом предлагают перезаказать машину в другом автосалоне!	*заказали машину; взяли предоплату; ждем пол года; отвратительная работа салона; предлагают перезаказать машину*
The work took a very long time, not high- quality and expensive. Example: when adjusting the valves, I waited there for 4 hours, but the valves kept knocking as they have knocked before.	A very long time; not high-quality; expensive; valves kept knocking
работу проводят очень долго, не качественно и дорого. пример: при регулировке клапанов я там прождал 4 часа, а клапана как стучали так и стучат.	*очень долго; не качественно; дорого; клапана стучат*
Car service koreana is great. Needed an urgent replacement of the injectors. Called to sign up, was asked to come in an hour. I arrived, gave the car and the injectors - in forty minutes they returned the car (intact and safe). The prices were a pleasant surprise.	Car service koreana; great; urgent replacement of the injectors; prices were a pleasant surprise;
автосервис кореана- зачет. нужна была срочная замена форсунки. позвонил записаться, сказали приехать через часик. приехал, отдал машину и форсунки - через сорок минут отдали готовую(в целости и сохранности). цены приятно удивили.	*автосервис кореана; зачет; срочная замена форсунки; цены приятно удивили*

5 Experiment Description

The goal of the research is to show the effectiveness of the proposed approach. The essence of the approach to keyphrase extraction is to use an extended list of stop words. Before, it was shown the effectiveness of this approach for the English language (for example, by processing scientific publications) [19,20]. In this work, the task was to test a similar approach for the Russian language and texts of another format, such as social media posts. In addition we propose an approach to automatic expanding of existing extended stop lists.

Two groups of experiments were carried out. The first shows that using extended lists of stop words can lead to an improvement in the quality of the extracted keyphrases. The second group shows the potential of updating of the list of stop words with additional words using word2vec models.

The keyphrase extraction process consists in finding sequences of words of given parts of speech, where the sequences are as long as possible. Words of distinct parts of speech, punctuation and stop words are used as separators. This approach showed very good results as described in [20]. In the carried out experiments only the following parts of speech were used: nouns, adjectives and verbs.

5.1 Creation of Extended Lists of Stop Words

Using a training collection, we build an extended list of stop words. Let V be the vocabulary of the training collection. Let S be the set of words included in the extended list of stop words and let $Sbase$ be the set of words included in the standard list of stop words of the selected language. Then $FScore$ evaluates the quality of the algorithm, which uses a list of stop words S. The extended list of stop words is extracted by the Algorithm 1.

Algorithm 1

$S = \emptyset$
for $\forall \nu \in V$: **do**
  if $FScore(\nu \cup Sbase) - FScore(Sbase) > p$ **then** $S = \nu \cup S$
  end if
end for
$S = Sbase \cup S$

Keyphrase extraction algorithm uses the obtained extended list of stop words for processing the test collection.

5.2 Updating the Extended List of Stop Words

Collections used in the work are quite small and the language is quite diverse. This leads to the fact that only part of the stop words for the test collection can be found in the train collection. We assume that this problem can be partially solved with the help of word2vec models [22]. We decided to add to the extended list of stop words - words that are semantically close to the words that already in the list. Our assumption is that by adding words close to the existing stop words, we will be adding words with similar meanings that might be used as stop words in texts of other collections. We have added n ($n = 2, n = 5$) closest words for each stop word to the final list of stop words. For these purposes pre-trained models from the RusVectores project were used [23]. The RusVectores API allows to retrieve up to 10 closest words for each input word. The service allows to work with ready-made models trained on different corpora. In this work we used the ruwikiruscorpora model, which has been trained on the Russian National Corpus and Wikipedia. The model is a continuous bag-of-words model with 300-dimensional vectors.

5.3 Texts Pre-processing

The texts in the training and test collections, as well as their gold standards, underwent the following preprocessing: lemmatization and part-of-speech tagging using MyStem [24] with the some changes. Since the texts often discuss the maintenance of cars such commonly used abbreviations as "To" and "aBTO", which stand for "vehicle inspection" and "car" in the Russian language, are tagged as a nouns during POS-tagging.

6 Experiment Results and Discussion

The experiments have been carried out for following cases:

- where the collection $T1$ was used to create an extended list of stop words, and the collection $T2$ was used to test the obtained lists;
- for the opposite cases, where the collection $T2$ was used to construct the extended list of stop words, and the collection $T1$ was used for testing.

The results are presented in Table 2 using the following denotations: $T1 => T2$ is used for the case when the extended list of stop words was extracted from the collection $T1$, and the quality evaluation of the algorithm was performed on the collection $T2$. The denotation $T2 => T1$ is accepted for the opposite case.

By "standard list of stop words" we mean the case when only the standard stop words of the Russian language are used as stop words. "Extended list of stop words" refers to cases in which extended lists of stop words have been used. The parameter p indicates the value that was used in Algorithm 1 (see Sect. 5.1). The notation "with additional words" is used for cases when words obtained using word2vec models were added to the extended list of stop words. The parameter n determines the number of closest words that was selected for each stop word of the extended list of stop words. The quality of the automatically extracted keyphrases was evaluated by comparing these phrases with phrases in the "gold standard" (manually assigned). For evaluation, the F1-score measure was used for the "exact match" case (for more details, see Sect. 4).

Table 2. Experiment results

Stop words	Standard list of stop words	Extended list of stop words					
		p = 0.005	p = 0.005 with additional words		p = 0.001	p = 0.001 with additional words	
			$n = 2$	$n = 5$		$n = 2$	$n = 5$
T1 =>T2	0.24	0.24	**0.31**	0.25	0.24	0.23	0.24
T2 =>T1	0.24	0.25	0.25	0.26	0.27	**0.28**	0.25

The obtained results show that when using extended lists of stop words, the quality of extracted keyphrases is improved in comparison with the case of using the standard list of stop words. During the experiments no loss in the quality of extraction of keyphrases when using extended lists of stop words has been observed. This allows to assume that there exists some universality of words that are more often not found than are found in keyphrases within a set of similar documents. This means the same extended lists of stop words can be applied to different collections of the same type of texts. The proposed approach works better when mostly identical phrases are used in the train and test collections. The main reason for this we see in the following. If the style of assigning keyphrases is identical in both collections, it starts to follow certain patterns, for example: which words and phrases the expert selects as keyphrases, what information is sought to reflect in the phrases, which words are commonly not used in phrases. The latter is close to the field of authorship and stylometric analysis. Since the proposed approach can identify some of the stylistic features (e.g. untypical words), it introduces an improvement in the quality of the algorithm.

Updating the extended lists of stop words with new words using word2vec models is, by our assumption, expedient. In most cases such an update allows additional improving of the quality of work of the algorithms. Nevertheless, the results obtained show that in some cases such an update may lead to worse results in comparison with using just the extended list of stop words. In our study we found that, when automatically updating the list of stop words, in spite of the fact that most words are very similar in meaning to the original words of the stop words list, there are exceptions. For example, for the word "somnevat'sa" (translated as "to doubt") the following closest words are extracted: to believe, sure, to assure, to verify, to make sure. If, as a result of updating the extended list of stop words, words that are very different from those already in the list are found, adding them to the final list may lead to a list of lower quality. We assume that in this case the problem is caused by the fact that we are using a model trained on texts of a different type (Wikipedia articles instead of texts from Internet car forums). This means that the close words are extracted for the texts on which the model was trained, which may not be similar to texts from car forums. Despite this, the obtained results show the potential of the proposed approach for updating extended lists of stop words. At the next stage of the experiment, we plan to use our own model, which will be trained on texts from Internet car forums.

7 Conclusions

Two tasks have been identified in our research: (1) to further investigate the approach to keyphrase extraction based on the use of extended lists of stop words; (2) to test the approach for automatic updating of such lists.

We have shown that extended lists of stop words created for one collection can be used for other similar collections and can improve the quality of the extracted keyphrases. By similar collections we mean texts that are similar to the texts in the training sample or are obtained from similar sources. It is shown that the creation and use of extended lists of stop words works even for poorly structured texts, such as messages from Internet car forums.

It should be noted, that words that appear in the extended lists of stop words, in general, are not stop words (for example, such as "day", "company", "auto loan"). These words fall into the extended lists of stop words only because their use is not typical in phrases assigned by the expert. It is proposed to update the obtained extended lists of stop words. To do this, words closest to the stop words are found using pre-trained word2vec models and added to the final stop words list.

The positive results obtained during the experiments allow to confirm the possibility of improving the work of algorithms that extract keyphrases using extended lists of stop words obtained from a training collection. The results obtained for the updated extended lists of stop words show the potential of the proposed approach.

References

1. Mihalcea, R., Tarau, P.: TextRank: bringing order into texts. Proc. EMNLP **85**, 404–411 (2004)
2. Wan, X., Xiao, J.: Single document keyphrase extraction using neighborhood knowledge. In: Proceedings of the 23rd National Conference on Artificial Intelligence, vol. 2, pp. 855–860 (2008)
3. Zesch, T., Gurevych, I.: Approximate matching for evaluating keyphrase extraction. In: Proceedings of the 7th International Conference on Recent Advances in Natural Language Processing, pp. 484–489 (2009)
4. Hasan, K.S., Ng, V.: Conundrums in unsupervised keyphrase extraction: making sense of the state-of-the-art. In: Proceedings of the Conference Coling 2010 - 23rd International Conference on Computational Linguistics, pp. 365–373, August 2010
5. Pudota, N., Dattolo, A., Baruzzo, A., Ferrara, F., Tasso, C.: Automatic keyphrase extraction and ontology mining for content-based tag recommendation. Int. J. Intell. Syst. **25**, 1158–1186 (2010)
6. You, W., Fontaine, D., Barthès, J.P.: An automatic keyphrase extraction system for scientific documents. Knowl. Inf. Syst. **34**, 691–724 (2013)
7. El-Beltagy, S.R., Rafea, A.: KP-miner: a keyphrase extraction system for English and Arabic documents. Inf. Syst. **34**, 132–144 (2009)
8. Tsatsaronis, G., Varlamis, I., Nørvåg, K.: SemanticRank: ranking keywords and sentences using semantic graphs. In: Proceedings of the 23rd International Conference on Computational Linguistics, Coling 2010, pp. 1074–1082 (2010)

9. Kim, S.N., Medelyan, O., Kan, M.Y., Baldwin, T.: Automatic keyphrase extraction from scientific articles. Lang. Resour. Eval. **47**, 723–742 (2013)
10. Popova, S., Skitalinskaya, G., Khodyrev, I.: Estimating keyphrases popularity in sampling collections. In: Ciuciu, I., et al. (eds.) On the Move to Meaningful Internet Systems: OTM 2015 Workshops, Lecture Notes in Computer Science, vol. 9416, pp. 481–491 (2015)
11. Page, L., Brin, S., Motwani, R., Winograd, T.: The pagerank citation ranking: bringing order to the web. World Wide Web Internet Web Inf. Syst. **54**, 1–17 (1998)
12. Zhang, Q., Wang, Y., Gong, Y., Huang, X.: Keyphrase extraction using deep recurrent neural networks on Twitter. In: Proceedings of the 2016 Conference on Empirical Methods in Natural Language Processing, EMNLP 2016, pp. 836–845 (2016)
13. Bernardini, A., Carpineto, C., D'Amico, M.: Full-subtopic retrieval with keyphrase-based search results clustering. In: Proceedings - 2009 IEEE/WIC/ACM International Conference on Web Intelligence, WI 2009, vol. 1, pp. 206–213 (2009)
14. Zeng, H.J., He, Q.C., Chen, Z., Ma, W.Y., Ma, J.: Learning to cluster web search results. In: Proceedings of the 27th Annual International ACM SIGIR Conference on Research and Development in Information Retrieval, pp. 210–217 (2004)
15. Hasan, K.S., Ng, V.: Automatic keyphrase extraction: a survey of the state of the art. In: Proceedings of the 52nd Annual Meeting of the Association for Computational Linguistics, 12 p. (2014)
16. Augenstein, I., Das, M., Riedel, S., Vikraman, L., McCallum, A.: SemEval 2017 task 10: Scienceie - extracting keyphrases and relations from scientific publications. CoRR abs/1704.02853 (2017)
17. ACL: The 53rd annual meeting of the association for computational linguistics. In: Proceedings of the ACL 2015 Workshop on Novel Computational Approaches to Keyphrase Extraction (2015)
18. Hulth, A.: Improved automatic keyword extraction given more linguistic knowledge. In: Language, pp. 216–223 (2003)
19. Popova, S., Khodyrev., I.: Ranking in keyphrase extraction problem, is it suitable to use statistics of words occurrences. In: Proceedings of the Institute for System Programming, vol. 26, pp. 123–136 (2014)
20. Popova, S., Kovriguina, L., Mouromtsev, D., Khodyrev, I.: Stop-words in keyphrase extraction problem. In: Conference of Open Innovation Association, FRUCT, pp. 113–121 (2013)
21. Kim, S.N., Baldwin, T., Kan, M.Y.: Evaluating N-gram based evaluation metrics for automatic keyphrase extraction. In: Proceedings of the 23rd International Conference on Computational Linguistics, Coling 2010, pp. 572–580 (2010)
22. Mikolov, T., Corrado, G., Chen, K., Dean, J.: Efficient estimation of word representations in vector space. In: Proceedings of the International Conference on Learning Representations, ICLR 2013, pp. 1–12 (2013)
23. Kutuzov, A., Kuzmenko, E.: WebVectors: A Toolkit for Building Web Interfaces for Vector Semantic Models, pp. 155–161. Springer International Publishing, Cham (2017)
24. Segalovich, I., Titov, V.: MyStem (2011). https://tech.yandex.ru/mystem/. Accessed 19 July 2017

Innovative Concept of the Strict Line Hypergraph as the Basis for Specifying the Duality Relation Between the Vertex Separators and Cuts

Artem Potebnia(✉)

Kyiv National Taras Shevchenko University, Kyiv, Ukraine
potebnia@mail.ua

Abstract. This article presents the original approach for establishing the duality relation between the vertex separators and cuts in hypergraphs. The analysis of the existing mathematical structures providing the edge-focused representation of graphs is followed by the identification of the fundamental drawbacks obstructing the formation of the duality relation on the basis of these structures. With a view to filling this research gap, the article formulates the core concept of the strict line hypergraph and introduces the new notions of the degenerate connected components and vertex separators. The proposed structures underlie the construction of the duality relation between the vertex separators and cuts in the form of theorems accompanied with their rigorous proofs and discussions of the specific situations. Finally, the article considers the application of the established relation for linking the combinatorial optimization problems whose solution spaces are composed of the vertex separators and cuts.

Keywords: Strict line hypergraph · Degenerate vertex separator Degenerate connected component · Vertex separator · Cut · Involution

1 Introduction

The opportunity to present the mathematical object in the different formulations focused on its particular properties and interlinked by the hidden relationship underlies the most fruitful ways of generating the new knowledge in many areas of mathematics. The principle of duality based on translating the primal objects into the dual ones serves as a most widespread and fundamental implementation of such strategy. The discovery and in-depth analysis of the duality relation provides a further comprehensive insight into the nature of the related objects (representing "the two sides of the same coin") and considerably assists in examining their properties. By way of example, the upper bound on the solutions of the maximization optimization problem could be estimated by solving the corresponding dual problem [1].

In 2016 Artem Potebnia graduated from Kyiv National Taras Shevchenko University and currently carries out his research on the personal initiative.

N. Shakhovska and V. Stepashko (eds.), *Advances in Intelligent Systems and Computing II*, Advances in Intelligent Systems and Computing 689,
https://doi.org/10.1007/978-3-319-70581-1_28

The investigation of the duality principle in the combinatorial structures is especially attractive due to the opportunity to illustrate the constructive considerations in the elegant geometrical form along with the great potential for the practical application of the obtained results. In particular, the structural organization of a wide range of complex intelligent systems could be naturally described by the models of the undirected graphs and hypergraphs representing the relationships between the constituent nodes of the system [2]. Therefore, the specification of the duality relations in the graph structures lays the theoretical background required for the development of the new methods for testing and analyzing such models.

In this article, the *undirected graph* $G = (V, E)$ is defined by the vertex set V equipped with the multiset (i.e. set allowing the presence of the repeated elements) E composed of the edges $e \subseteq V$ such that $|e| \leq 2$. In contrast to the undirected graph, the *hypergraph* $\hat{G} = \left(V, \hat{E}\right)$ holds the multiset $\hat{E}$ comprised of the hyperedges $\hat{e} \subseteq V$ having no cardinality limitation. Thus, the undirected graph constitutes the particular case of the hypergraph. Remark that these formulations allow the presence of the parallel edges or hyperedges as well as the loops and even parallel loops. Moreover, such structures also could contain the loose edges or hyperedges that are fully deprived of the incident vertices. Let us denote the classes of all undirected graphs and hypergraphs specified in such manner as Γ and $\hat{\Gamma}$ respectively. Conversely, their subclasses $\Gamma' \subset \Gamma$ and $\hat{\Gamma}' \subset \hat{\Gamma}$ cover only those undirected graphs and hypergraphs that do not contain any loose edge or hyperedge.

The analysis of the potential vulnerabilities to the structural attacks serves as an indispensable preliminary stage in the development of the reliable attack-robust complex intelligent systems [3]. In the graph-based models, the vertex separators and cuts provide a natural representation for the attacks of two fundamental types (those based on removing the entities and those involving the deletion of links between them) leading to the violation of the system integrity. Formally, the set $C_V \subseteq V$ and multiset $C_{\hat{E}} \subseteq \hat{E}$ are considered respectively as a vertex separator and cut in the connected hypergraph $\hat{G}$ only if the exclusion of their constituent elements from $\hat{G}$ results in increasing the number of its connected components. Notice that in the particular case of the connected undirected graph G, any cut takes the form of the multiset $C_E \subseteq E$ satisfying the same requirements [4].

In many contexts, the investigation of the complex systems behavior faces a challenge of the transition from the conventional description focused on the system's entities to the alternative description concentrated on the links between such entities. The main idea of such viewpoint switching lies in representing the links of the analyzed system by the vertices of the graph-based model. In turn, the application of the graph analysis methods to such models could significantly assist in producing the additional information about the system structure (e.g. extract the implicit communities of links) [5]. For example, the interference effects in the ad hoc wireless network could be represented by the graph whose vertices reflect the communication channels established between the network hosts. In addition, each pair of vertices representing the channels that cannot be activated simultaneously due to the interference issues should be connected by an edge. Under such approach, the transmission schedules in the modeled network could be encoded by the vertex colorings of such graph [6].

These considerations contribute to the reasonableness of developing the mathematical apparatus for establishing the relationship between the structures representing the attacks of different types in the graph-based models [7, 8]. Such relationship offers the opportunity for comprehensively analyzing the vulnerabilities of the modeled system as well as its parts separated after the attack implementation. Consequently, ***the ultimate objective of this article*** lays in the construction the duality relation translating the vertex separators into the corresponding “dual” cuts.

In order to ensure the clarity of presentation, the remainder of the article is outlined according to the following structure. Section 2 analyses the concept of the dual graph representing the most valuable related work on establishing the correspondence between the cycles and cuts. Section 3 provides the critical overview of the known theoretical notions intended for converting the node-centric view of the graph into the edge-centric one, discusses the drawbacks of these notions, and detects the fundamental impediments to achieving the stated objective. With a view to overcoming these impediments, this article considers the widest class of hypergraphs with loose hyperedges $\hat{\Gamma}$. In particular, Sect. 4 formulates the requisite concepts of the connected component, vertex separator, and cut in such hypergraphs as well as discusses their specific degenerate forms. Against this background, Sect. 5 introduces the original concept of the strict line hypergraph, which is used in Sect. 6 for establishing the duality relation between the vertex separators and cuts in the $\hat{\Gamma}$ class hypergraphs. Finally, Sect. 7 deals with the concluding remarks.

2 Dual Graphs and Their Application for Establishing the Correspondence Between the Cycles and Cuts

The concept of the *dual graph* represents the most substantial prior advance of the graph theory in the context of establishing the duality relationships. Looking into detail, the operation of constructing the dual graph $D_{\mathcal{E}(G)} \in \Gamma'$ is defined for any undirected connected graph $G = (V, E) \in \Gamma'$ specified together with its particular planar embedding $\mathcal{E}(G)$ that divides the plane into a set of l faces $\Omega(\mathcal{E}(G)) = \{\omega_1, \omega_2, \ldots, \omega_l\}$ (i.e. regions restricted by the graph’s edges). Such partition serves as the basis for generating all nodes v^D and edges e^D of the dual graph $D_{\mathcal{E}(G)}$. In particular, each face $\omega_i \in \Omega(\mathcal{E}(G))$ surrounding t edges of G on both sides is represented in the structure of $D_{\mathcal{E}(G)}$ by the separate vertex v_i^D equipped with t loops corresponding to these edges. In turn, every edge of the initial graph G shared by two distinct faces $\omega_i, \omega_j \in \Omega(\mathcal{E}(G))$ is mapped into the edge of $D_{\mathcal{E}(G)}$ connecting the vertices v_i^D and v_j^D [1, 9].

The construction of the dual graph $D_{\mathcal{E}(G)}$ naturally involves the appropriate transformation of all internal structures contained in the initial graph G. Most importantly, the edges of any simple cycle SC (i.e. closed path without the repetitions of vertices) in G are translated into the edges forming a minimal cut C_E^m (i.e. such cut that does not include any other cut as its proper subset) in $D_{\mathcal{E}(G)}$ and vice versa. In order to demonstrate this relationship, let us consider an arbitrary unicyclic graph containing n edges and having exactly one simple cycle including $m \leq n$ edges. Its dual graph, in turn, is composed of two vertices linked by m edges that constitute a minimal cut corresponding to such cycle, while the remaining $n - m$ edges are transformed into loops. The example of graphs illustrating these considerations is presented in Fig. 1. Apart from that, the concept of the dual graph plays a key role in establishing the relationship between the minimal cuts and shortest paths in the planar networks given by the undirected graphs with specified source and sink nodes lying on the boundary of the infinite face [10].

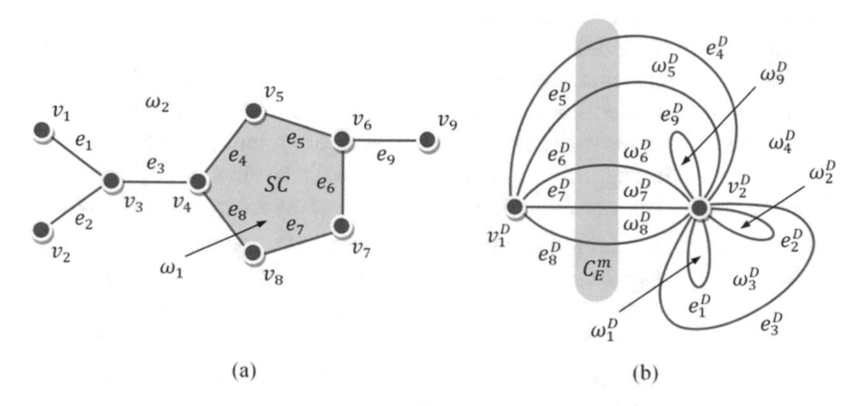

Fig. 1. Example of the planar unicyclic graph (a) having one highlighted simple cycle SC and its dual graph (b) with indicated corresponding minimal cut C_E^m. The dual graph is shown in the special embedding that allows the restoration of the initial graph, while its faces are denoted by ω_i^D.

However, in the general case, the graph G could have multiple possible embeddings $\mathcal{E}_1(G),\ldots,\mathcal{E}_q(G)$ leading to the formation of the corresponding non-isomorphic dual graphs $D_{\mathcal{E}_1(G)},\ldots,D_{\mathcal{E}_q(G)}$. Such extreme sensitivity to the concrete embedding sufficiently complicates the analysis of the dual graphs and constitutes their primary drawback. In particular, the restoration of the initial graph G always could be performed by repetitive constructing the dual graph of $D_{\mathcal{E}(G)}$, but only under the specification of embedding $\mathcal{E}'\left(D_{\mathcal{E}(G)}\right)$ such that $D_{\mathcal{E}'\left(D_{\mathcal{E}(G)}\right)} = G$.

Moreover, the dual graph $D_{\mathcal{E}(G)}$ should be connected even if the graph G is disconnected since under any embedding $\mathcal{E}(G)$ each point inside every internal face could be linked with some point belonging to the external infinite face by a line crossing the graph's edges. This, in turn, allows deducing that the dual graphs are fundamentally unsuitable for recovering the structure of the disconnected initial graphs.

3 Analysis of the Mathematical Structures Providing the Edge-Focused Representation of Graphs

From the first viewpoint, the basic idea of "transforming the graph's edges into nodes" that underlies establishing the duality relation between the vertex separators and cuts is conceptually associated with the known structures of the line and medial graphs.

Definition 1. The *line graph* $L(G) \in \Gamma_S^{'}$ of the undirected graph $G = (V, E) \in \Gamma^{'}$ is represented by a pair (V^L, E^L) composed of the set V^L of vertices $v_i^L \in V^L$ corresponding to the edges $e_i \in E$ of the original graph equipped with the following collection of edges E^L:

$$E^L = \left\{ \left(v_i^L, v_j^L \right) \in \binom{V^L}{2} \,\middle|\, e_i \text{ and } e_j \text{ are adjacent in graph } G \right\}.$$

In other words, the formation of the line graph is conducted through the application of the $L : \Gamma^{'} \to \Gamma_S^{'}$ function (defined as the set of ordered pairs $(G, L(G))$ for all graphs $G \in \Gamma^{'}$) to the specified argument graph. Notice that the resulting line graphs, in contrast to the original ones, are deprived of the multiple edges and loops, thereby, forming the subclass $\Gamma_S^{'} \subset \Gamma^{'}$. In particular, multiset $M_{ij} = \left\{ (v_i, v_j) \middle| (v_i, v_j) \in E \right\}$ composed of the parallel edges incident to the v_i and v_j vertices is reflected in the structure of the corresponding line graph by the clique $C_{ij} \subseteq V^L$ of cardinality $|M_{ij}|$. Conversely, the loops $(v_i) \in E$ receive the same representation in the target graph $L(G)$ as the pendant edges (i.e. edges having one "leaf" endpoint v_l with degree $d(v_l) = 1$). Analogously to the M_{ij} collections, the multisets $M_i = \{(v_i) | (v_i) \in E\}$ comprised of the parallel loops anchored to the same vertex v_i are transformed into the cliques $C_i \subseteq V^L$ including $|M_i|$ nodes [11]. For example, Fig. 2b illustrates the structure of the line graph constructed for the sample instance of the initial graph shown in Fig. 2a.

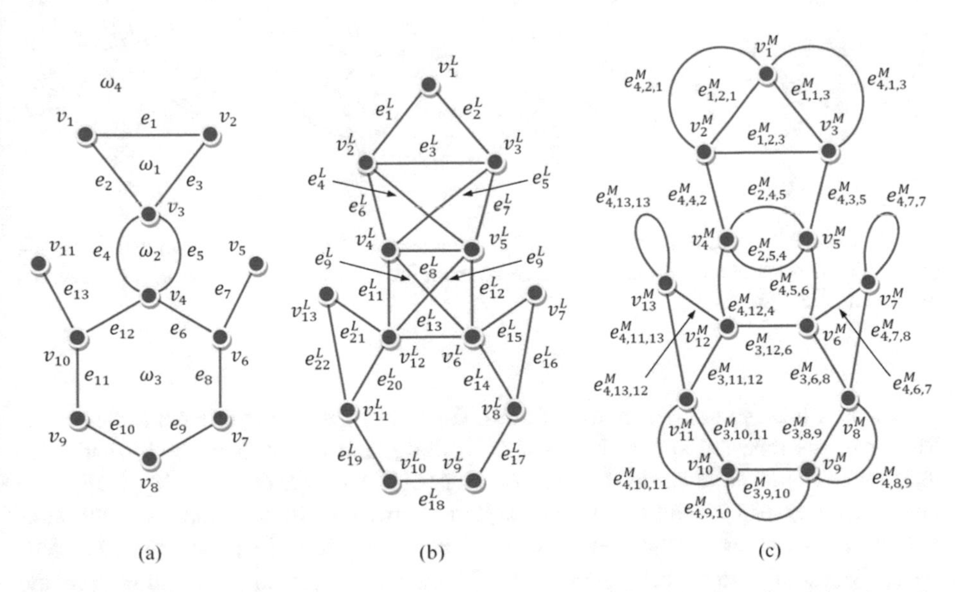

Fig. 2. Example of the initial planar graph (a) and the result of its transformation into the corresponding line (b) and medial (c) graphs. To enhance the traceability of generating the medial graph's edges from the boundary walks of the initial graph, these edges are equipped with three indices representing respectively the face of the initial graph and pair of edges lying on the boundary of this face.

These considerations allow concluding that the function L does not hold the injective property. This, in turn, dramatically restricts the opportunity to completely reconstruct the argument graph G from the image $L(G)$. In order to illustrate the loss of distinctness after applying the function L, let us consider the following list of graphs containing m edges:

1. Multitriangle graphs T_{m_1,m_2,m_3} composed of three vertices joined respectively by m_1, m_2, and m_3 edges, where $m_1 + m_2 + m_3 = m$
2. Multistar graphs $S_n^{m_1,m_2,\ldots,m_n}$ including one root vertex r and $n \geq 1$ leaf nodes. Each leaf node v_i should be incident to m_i edges (r, v_i), while $\sum_{i=1}^{n} m_i = m$.
3. Dipole graphs D_m consisting of two vertices linked by m parallel edges, which constitute the notable particular case of the multistar graphs when $n = 1$.
4. Flower graphs F_m containing only one node r equipped with m parallel loops.

Clearly, all these graphs are indistinguishable in terms of the corresponding line graphs taking the form of the complete graphs K_m. In particular, Fig. 3 demonstrates the sample instances of the listed graphs at $m = 6$. In addition, the application of the L transformation leads to the inevitable loss of all isolated (i.e. having zero degree) vertices. Therefore, the amount of the information encapsulated in the line graphs $L(G)$ is insufficient for recovering the arguments G containing loops, parallel edges, and isolated vertices.

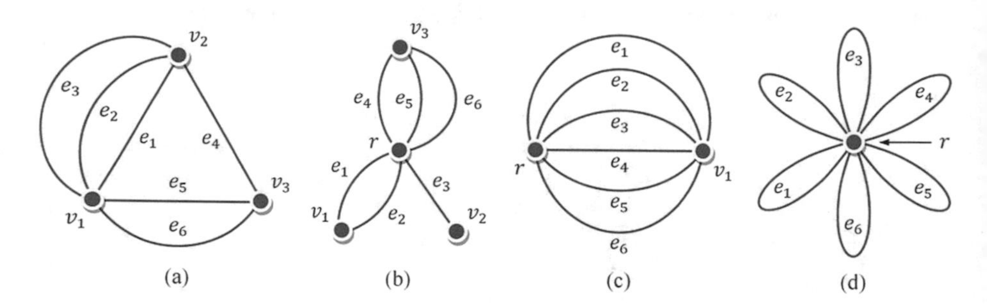

Fig. 3. Instances of the multitriangle $T_{3,1,2}$ (a), multistar $S_3^{2,1,3}$ (b), dipole D_6 (c), and flower F_6 (d) graphs whose structure could not be derived from the corresponding line graphs represented by the complete graphs K_6.

Remark that the non-injectivity of the L function implies that it is not self-inverse. Moreover, its repeated application to some initial graph G results in the formation of the graphs sequence given in the form $L(G), L(L(G)), L(L(L(G))), \ldots$. Typically, the order of all graphs belonging to such sequence grows without bound, i.e. each subsequent item has the greater number of vertices compared to its predecessor. Considering the initial graphs belonging to the $\Gamma_S^{'}$ subclass, the only exceptions are the sequences formed for few specific types of their connected components, such as the cycles, paths (including ones having zero length and, thereby, composed of just one isolated vertex), and graphs $S_3^{1,1,1}$ [12]. These drawbacks clearly demonstrate that the line graphs could not serve as the basis for establishing the duality of the vertex separators and cuts.

On the contrary, the operation of the medial graph formation takes into account the particular embedding $\mathcal{E}(G)$ of the graph G and, thereby, could be viewed as the symbiosis of the ideas underlying the construction of the line and dual graphs. Let us introduce the notations $W(\omega_1), \ldots, W(\omega_l)$ for the closed walks composed of the edges encountered while traversing the boundaries of the faces $\omega_1, \ldots, \omega_l \in \Omega(\mathcal{E}(G))$. Notice that such walks could contain the repeating edges and play a key role in generating the elements of the medial graph according to the next definition:

Definition 2. The *medial graph* $M_{\mathcal{E}(G)} = (V^M, E^M) \in \Gamma^{'}$ of the planar undirected graph $G = (V, E) \in \Gamma^{'}$ under the embedding $\mathcal{E}(G)$ is constructed by mapping the edges $e_i \in E$ into the vertices $v_i^M \in V^M$, while each pair of the subsequent edges e_i and e_j (not necessary distinct) in every walk $W(\omega_k)$ for $k = 1, \ldots, l$ is reflected by the corresponding edge $e_{k,i,j}^M \in E^M$.

The construction of the medial graph is exemplified in Fig. 2c. Notice that in contrast to the line graphs, the medial graphs could contain loops and parallel edges (compare Fig. 2b and c). However, the sensitivity to the particular embedding contributes to the non-uniqueness of such structures, i.e. one initial graph G could be associated with multiple non-isomorphic graphs $M_{\mathcal{E}_1(G)}, \ldots, M_{\mathcal{E}_p(G)}$. Moreover, the operation of the medial graph formation is not self-inverse, leads to the elimination of the isolated vertices and cannot be applied to the non-planar initial graphs G [13, 14]. This list of drawbacks precludes the usage of the medial graph concept for establishing the duality relationship.

In summing up this section, unlike the cycles and cuts, the vertex separators and cuts differ in the basic type of their constituent elements. This difference produces the fundamental impediments to the establishment of the duality relation between such structures in the undirected graphs belonging to the class $\Gamma^{'}$. In such graphs, the removal of vertices automatically implies the exclusion of all their incident edges, while the deletion of any edge does not affect the graph's nodes. Accordingly, the multiset containing all edges of any graph's connected component having two or more nodes serves as the *universal cut*. On the contrary, the conception of the universal vertex separator composed of all nodes contained in the connected component having two or more edges cannot take place in the $\Gamma^{'}$ class graphs. Another challenging issue is associated with the strict limitation imposed on the number of vertices that could be incident to any edge. This stimulates us to consider the widest class of hypergraphs with loose hyperedges $\hat{\Gamma}$ and shows the need for introducing the new mathematical structure that is conceptually similar to the line and medial graphs, but at the same time overcomes their drawbacks discussed in this section.

4 Adaptation of the Connected Component, Cut, and Vertex Separator Concepts for Hypergraphs with Loose Hyperedges

The presence of the loose hyperedges in the $\hat{\Gamma}$ class hypergraphs allows overcoming the fundamental restriction represented by the inevitable elimination of any hyperedge when excluding all its incident vertices. Nevertheless, such interpretation of the hyperedges as the independent structures leads to the existence of the nodeless hypergraphs composed exclusively of the loose hyperedges. This, in turn, underlies the need for introducing the following modified formulation of the connected component:

Definition 3. The *connected component* of the hypergraph $\hat{G} = (V, \hat{E}) \in \hat{\Gamma}$ is represented by the hypergraph $\hat{F} = (V_{\hat{F}}, \hat{E}_{\hat{F}})$ equipped with the vertex set $V_{\hat{F}} \subseteq V$ that is either empty or composed of the nodes connected to each other by paths (i.e. the sequences of hyperedges) and unreachable from any other vertex belonging to $V \backslash V_{\hat{F}}$. In the case $V_{\hat{F}} \neq \emptyset$, the multiset $\hat{E}_{\hat{F}} \subseteq \hat{E}$ comprises all hyperedges that are incident to the nodes of $V_{\hat{F}}$. On the contrary, if $V_{\hat{F}} = \emptyset$, the collection $\hat{E}_{\hat{F}}$ should contain only one loose hyperedge $\{\emptyset\} \in \hat{E}$ having no incident vertices.

Under such approach, each loose hyperedge is covered by the separate connected component represented by a pair $(\emptyset, \{\{\emptyset\}\})$. Notice that such nodeless components are referred to as *degenerate* in the remainder of this article. Conversely, the isolated nodes $v_s \in V$ deprived of the incident hyperedges are placed into the corresponding components given by pairs $(\{v_s\}, \emptyset)$. At the same time, all other components of $\hat{G}$ are required to have both non-empty collections of vertices and hyperedges. With respect to the proposed formulation of the connected component, we can define the vertex separators and cuts in the $\hat{\Gamma}$ class hypergraphs in the following manner:

Definition 4. The non-empty set $C_V \subseteq V$ (multiset $C_{\hat{E}} \subseteq \hat{E}$) belonging to one connected component $\hat{F}$ of the hypergraph $\hat{G} \in \hat{\Gamma}$ is considered as a *vertex separator* (*cut*) in $\hat{G}$ only if the exclusion of its constituent vertices (hyperedges) from $\hat{G}$ results in splitting $\hat{F}$ into two or more disjoint connected components.

Figure 4 illustrates the connected components detected in the particular hypergraph instance along with the few examples of vertex separators and cuts containing only one element. Let us denote the spaces containing all possible cuts $C_{\hat{E}}$ and vertex separators C_V for any component $\hat{F}$ of the hypergraph $\hat{G}$ as $\mathcal{C}(\hat{F})$ and $\mathcal{VS}(\hat{F})$ respectively. Notice that these spaces are closed under union of their elements, i.e. $C^i_{\hat{E}} \bigcup C^j_{\hat{E}} \in \mathcal{C}(\hat{F})$ for any $C^i_{\hat{E}}, C^j_{\hat{E}} \in \mathcal{C}(\hat{F})$ and $C^i_V \bigcup C^j_V \in \mathcal{VS}(\hat{F})$ for any $C^i_V, C^j_V \in \mathcal{VS}(\hat{F})$.

Moreover, the existence of the nodeless connected components allowed by Definition 3 contributes to the necessity for distinguishing the degenerate form of the vertex separators. Formally, the vertex separator C_V could be considered as *degenerate* only if the exclusion of its constituent nodes results in arising less than two non-degenerate (i.e. having at least one vertex) connected components. In contrast to the non-degenerate ones, such separators do not ensure the formation of the disjoint subsets

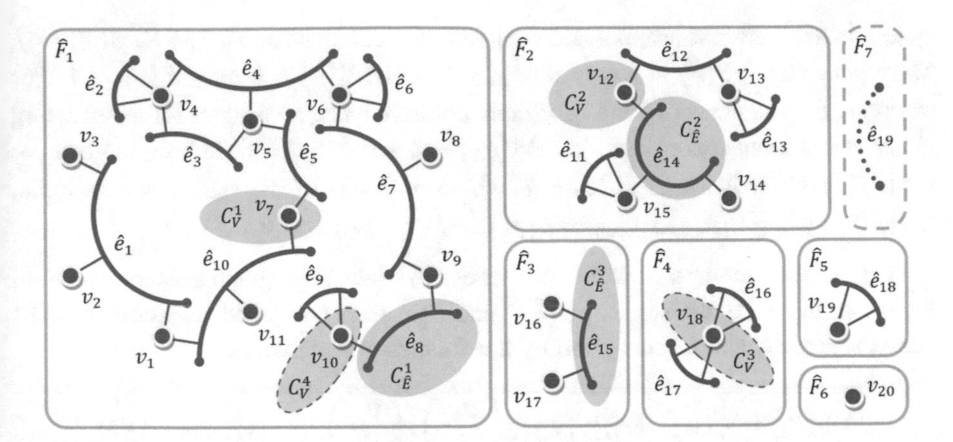

Fig. 4. Example of the $\hat{\Gamma}$ class hypergraph with 7 detected connected components. The degenerate connected component is outlined by the dashed border, while the dotted line depicts the loose hyperedge. The examples of possible vertex separators and cuts are highlighted by the ellipses. Moreover, the ellipses with dashed border indicate the degenerate vertex separators.

of vertices and provide only the segregation of the new loose hyperedges by removing all their incident vertices.

For example, the set $C_V^4 = \{v_{10}\}$ highlighted in Fig. 4 serves as the degenerate vertex separator for the component $\hat{F}_1$. This follows directly from the fact that the removal of the vertex v_{10} produces two connected components given by pairs $\left(V_{\hat{F}_1} \backslash \{v_{10}\}, \hat{E}_{\hat{F}_1} \backslash \{\hat{e}_9\}\right)$ and $(\emptyset, \{\hat{e}_9\})$, while only the first of them is non-degenerate. Moreover, according to the proposed definitions, the entire vertex set $V_{\hat{F}}$ of any connected component $\hat{F}$ for which $\left|V_{\hat{F}}\right| \geq 1$ and $\left|\hat{E}_{\hat{F}}\right| \geq 2$ serves as the *universal degenerate vertex separator* because the removal of its nodes produces $\left|\hat{E}_{\hat{F}}\right|$ loose hyperedges.

Let us introduce the notation $\mathcal{NDVS}(\hat{F}) \subseteq \mathcal{VS}(\hat{F})$ for the subspace containing all possible non-degenerate vertex separators for the connected component $\hat{F}$. Like the overall space $\mathcal{VS}(\hat{F})$, the subspace $\mathcal{NDVS}(\hat{F})$ is closed under union of its elements. At the same time, the division performed by excluding the hyperedges of any cut $C_{\hat{E}}$ always leads to separating only non-degenerate connected components, which clearly shows the fundamental absence of the degenerate form of cuts. In summarizing the above considerations, all possible cuts, vertex separators, and non-degenerate vertex separators for the whole hypergraph $\hat{G}$ having n connected components $\hat{F}_1, \hat{F}_2, \ldots, \hat{F}_n$ are encompassed respectively by the spaces $\mathcal{C}(\hat{G}) = \bigcup_{i=1}^{n} \mathcal{C}(\hat{F}_i)$, $\mathcal{VS}(\hat{G}) = \bigcup_{i=1}^{n} \mathcal{VS}(\hat{F}_i)$, and $\mathcal{NDVS}(\hat{G}) = \bigcup_{i=1}^{n} \mathcal{NDVS}(\hat{F}_i)$.

Clearly, the removal of the isolated vertices or loose hyperedges from the hypergraph $\hat{G}$ is simply equivalent to the elimination of the corresponding connected components. Therefore, such connected components are associated with the empty spaces of the vertex separators and cuts (e.g. the components $\hat{F}_6$ and $\hat{F}_7$ in Fig. 4). More generally, $\mathcal{VS}(\hat{F}) = \emptyset$ only if $|\hat{E}_{\hat{F}}| \leq 1$, while $\mathcal{C}(\hat{F}) = \emptyset$ only if $|V_{\hat{F}}| \leq 1$. For example, the component $\hat{F}_5$ in Fig. 4 does not have enough elements for constructing either vertex separators or cuts, i.e. $\mathcal{VS}(\hat{F}_5) = \emptyset$ and $\mathcal{C}(\hat{F}_5) = \emptyset$. In turn, the components $\hat{F}_3$ and $\hat{F}_4$ illustrate the case in which only one of the spaces is empty, i.e. $\mathcal{VS}(\hat{F}_3) = \emptyset$ and $\mathcal{C}(\hat{F}_4) = \emptyset$, while $\mathcal{C}(\hat{F}_3) = \left\{C^3_{\hat{E}}\right\}$ and $\mathcal{VS}(\hat{F}_4) = \{C^3_V\}$.

Let us take a closer look at the component $\hat{F}_2$ of the hypergraph instance shown in Fig. 4. In this case, the spaces $\mathcal{VS}(\hat{F}_2)$ and $\mathcal{C}(\hat{F}_2)$ are composed respectively of 14 vertex separators and 12 cuts given by the following expressions:

$$\mathcal{VS}(\hat{F}_2) = \left\{\{v_{12}\}, \{v_{13}\}, \{v_{15}\}, \binom{V_{\hat{F}_2}}{2}, \binom{V_{\hat{F}_2}}{3}, \{v_{12}, v_{13}, v_{14}, v_{15}\}\right\};$$

$$\mathcal{C}(\hat{F}_2) = \left\{\{\hat{e}_{12}\}, \{\hat{e}_{14}\}, \binom{\hat{E}_{\hat{F}_2}}{2} \backslash \{\hat{e}_{11}, \hat{e}_{13}\}, \binom{\hat{E}_{\hat{F}_2}}{3}, \{\hat{e}_{11}, \hat{e}_{12}, \hat{e}_{13}, \hat{e}_{14}\}\right\}.$$

At the same time, only 3 of the vertex separators listed above are non-degenerate, i.e.

$$\mathcal{NDVS}(\hat{F}_2) = \{\{v_{12}\}, \{v_{12}, v_{14}\}, \{v_{12}, v_{15}\}\}.$$

In closing this section, we should emphasize that the proposed concept of the degenerate vertex separator plays a key role in expanding the overall space $\mathcal{VS}(\hat{G})$. Such expansion, in turn, lays the groundwork for establishing the duality relation between the vertex separators and cuts.

5 Concept of the Strict Line Hypergraph

The close inspection shows that the restriction of the edges cardinality serves as a principal reason for the shortcomings of the $L(G)$ graphs discussed in Sect. 3. Actually, the vertices that ensure connecting $n \geq 2$ edges are reflected in the structure of the line graph by the collections containing C^2_n edges. In turn, such edges are not accompanied with any additional information allowing to indicate that they were inspired by the same vertex of the original graph. On the contrary, the hyperedges of the $\hat{\Gamma}$ class hypergraphs are exempted from the above-mentioned cardinality limitation. These considerations emphasize the reasonableness of using the hypergraphs as the core of the new mathematical structure formulated below.

Definition 5. The *strict line hypergraph* of the primal hypergraph $\hat{G} = (V, \hat{E}) \in \hat{\Gamma}$ is denoted as $\hat{L}(\hat{G})$ and given by a pair $\left(V^{\hat{L}}, \hat{E}^{\hat{L}}\right) \in \hat{\Gamma}$ including the sets specified as follows:

$$V^{\hat{L}} = \left\{ v_i^{\hat{L}} \,|\, \hat{e}_i \in \hat{E} \right\};\ \hat{E}^{\hat{L}} = \left\{ \hat{e}_i^{\hat{L}} \,|\, v_i \in V \right\};$$
$$\hat{e}_i^{\hat{L}} = \left\{ v_k^{\hat{L}} \,|\, (v_i \in \hat{e}_k) \wedge \left(\hat{e}_k \in \hat{E} \right) \right\}.$$

In contrast to the L function, the transformation $\hat{L} : \hat{\Gamma} \to \hat{\Gamma}$ (specified implicitly as the set of pairs $\left(\hat{G}, \hat{L}(\hat{G})\right)$ for all $\hat{G} \in \hat{\Gamma}$) has both domain and codomain represented by more general class $\hat{\Gamma}$. Notice that each vertex $v_i \in V$ of the primal hypergraph $\hat{G}$ is reflected by only one hyperedge $\hat{e}_i^{\hat{L}} \in \hat{E}^{\hat{L}}$ in the corresponding strict line hypergraph $\hat{L}(\hat{G})$. This underlies the possibility to recover the argument $\hat{G}$ from the image $\hat{L}(\hat{G})$ and constitutes the main difference of the proposed structure from the line graphs [15].

Theorem 1. The function $\hat{L}$ that maps each hypergraph $\hat{G}$ into the corresponding strict line hypergraph $\hat{L}(\hat{G})$ is an involution, i.e. $\hat{L}(\hat{L}(\hat{G})) = \hat{G}$.

▲ Assume that the second application of the $\hat{L}$ function results in obtaining the hypergraph $\hat{L}(\hat{L}(\hat{G})) = \left(V^{\hat{L}\hat{L}}, \hat{E}^{\hat{L}\hat{L}} \right)$. By taking into account that the hypergraph $\hat{L}(\hat{G}) = \left(V^{\hat{L}}, \hat{E}^{\hat{L}} \right)$ serves as the argument at the second application of the $\hat{L}$ function, we can specify the following expressions detailing the structure of $V^{\hat{L}\hat{L}}$ and $\hat{E}^{\hat{L}\hat{L}}$:

$$V^{\hat{L}\hat{L}} = \left\{ v_i^{\hat{L}\hat{L}} \,\middle|\, \hat{e}_i^{\hat{L}} \in \hat{E}^{\hat{L}} \right\};\ \hat{E}^{\hat{L}\hat{L}} = \left\{ \hat{e}_i^{\hat{L}\hat{L}} \,\middle|\, v_i^{\hat{L}} \in V^{\hat{L}} \right\};$$
$$\hat{e}_i^{\hat{L}\hat{L}} = \left\{ v_k^{\hat{L}\hat{L}} \,\middle|\, \left(v_i^{\hat{L}} \in \hat{e}_k^{\hat{L}} \right) \wedge \left(\hat{e}_k^{\hat{L}} \in \hat{E}^{\hat{L}} \right) \right\}.$$

By putting the expressions for the collections $V^{\hat{L}}$ and $\hat{E}^{\hat{L}}$ into the obtained formulas for $V^{\hat{L}\hat{L}}$ and $\hat{E}^{\hat{L}\hat{L}}$, we would have

$$V^{\hat{L}\hat{L}} = \left\{ v_i^{\hat{L}\hat{L}} \,\middle|\, \hat{e}_i^{\hat{L}} \in \left\{ \hat{e}_t^{\hat{L}} \,|\, v_t \in V \right\} \right\};$$
$$\hat{E}^{\hat{L}\hat{L}} = \left\{ \hat{e}_i^{\hat{L}\hat{L}} \,\middle|\, v_i^{\hat{L}} \in \left\{ v_t^{\hat{L}} \,|\, \hat{e}_t \in \hat{E} \right\} \right\}.$$

Obviously, the condition $\hat{e}_i^{\hat{L}} \in \left\{ \hat{e}_t^{\hat{L}} \,|\, v_t \in V \right\}$ is satisfied if there exists such vertex $v_t \in V$ that $t = i$. This allows to present the expression for the set $V^{\hat{L}\hat{L}}$ in the following compact form:

$$V^{\hat{L}\hat{L}} = \left\{ v_i^{\hat{L}\hat{L}} \,|\, v_i \in V \right\}.$$

Therefore, the hypergraph $\hat{L}(\hat{L}(\hat{G}))$ holds the same vertex set as the initial hypergraph $\hat{G}$. In turn, since the condition $v_i^{\hat{L}} \in \left\{ v_t^{\hat{L}} \,|\, \hat{e}_t \in \hat{E} \right\}$ is satisfied at the presence of the hyperedge $\hat{e}_i \in \hat{E}$, we can provide the following simplified form of the expression for the $\hat{E}^{\hat{L}\hat{L}}$ multiset:

$$\hat{E}^{\hat{L}\hat{L}} = \left\{ \hat{e}_i^{\hat{L}\hat{L}} \mid \hat{e}_i \in \hat{E} \right\}.$$

However, this result implies only that the hypergraphs $\hat{G}$ and $\hat{L}(\hat{L}(\hat{G}))$ hold an equal number of hyperedges, i.e. $|\hat{E}| = \left|\hat{E}^{\hat{L}\hat{L}}\right|$. In order to show that the hyperedges belonging to the collections $\hat{E}$ and $\hat{E}^{\hat{L}\hat{L}}$ have the identical structure, we need to perform the transformations of the expression for the hyperedge $e_i^{\hat{L}\hat{L}}$. In particular, using the formulas for $\hat{e}_k^{\hat{L}}$ and $\hat{E}^{\hat{L}}$, this expression could be rewritten as follows:

$$\begin{aligned} \hat{e}_i^{\hat{L}\hat{L}} = \Big\{ v_k^{\hat{L}\hat{L}} \Big| \Big(v_i^{\hat{L}} \in \left\{ v_t^{\hat{L}} \mid (v_k \in \hat{e}_t) \wedge (\hat{e}_t \in \hat{E}) \right\} \Big) \\ \wedge \Big(\left\{ \left\{ v_t^{\hat{L}} \mid (v_k \in \hat{e}_t) \wedge (\hat{e}_t \in \hat{E}) \right\} \in \left\{ \hat{e}_h^{\hat{L}} \mid v_h \in V \right\} \right\} \Big) \Big\}. \end{aligned}$$

Here the condition $v_i^{\hat{L}} \in \left\{ v_t^{\hat{L}} \mid (v_k \in \hat{e}_t) \wedge (\hat{e}_t \in \hat{E}) \right\}$ is satisfied only if $v_k \in \hat{e}_i$. By considering this observation and inserting the formula for the $\hat{e}_h^{\hat{L}}$ component, we would have

$$\begin{aligned} \hat{e}_i^{\hat{L}\hat{L}} = \Big\{ v_k^{\hat{L}\hat{L}} \Big| (v_k \in \hat{e}_i) \wedge \Big(\left\{ v_t^{\hat{L}} \mid (v_k \in \hat{e}_t) \wedge (\hat{e}_t \in \hat{E}) \right\} \\ \in \left\{ \left\{ v_s^{\hat{L}} \mid (v_h \in \hat{e}_s) \wedge (\hat{e}_s \in \hat{E}) \right\} \mid v_h \in V \right\} \Big) \Big\}. \end{aligned}$$

First of all, we need to prove that the second predicate of the conjunction is satisfied for any $v_k \in V$. Note that the expression $\left\{ v_t^{\hat{L}} \mid (v_k \in \hat{e}_t) \wedge (\hat{e}_t \in \hat{E}) \right\}$ represents the set $V^{\hat{L}}(v_k)$ composed of the vertices $v_t^{\hat{L}} \in V^{\hat{L}}$ for which the hyperedges $\hat{e}_t \in \hat{E}$ are incident to the v_k node.

Conversely, the expression $\left\{ \left\{ v_s^{\hat{L}} \mid (v_h \in \hat{e}_s) \wedge (\hat{e}_s \in \hat{E}) \right\} \mid v_h \in V \right\}$ could be rewritten as $\left\{ V^{\hat{L}}(v_h) \mid v_h \in V \right\}$. Thus, the second predicate of the conjunction takes the form of $V^{\hat{L}}(v_k) \in \left\{ V^{\hat{L}}(v_h) \mid v_h \in V \right\}$ and is satisfied for any $v_k \in V$. In turn, as evident from the first predicate, each $\hat{e}_i^{\hat{L}\hat{L}}$ hyperedge of the hypergraph $\hat{L}(\hat{L}(\hat{G}))$ contains the vertices $v_k^{\hat{L}\hat{L}}$ such that $v_k \in \hat{e}_i$ and, thereby, reproduces the structure of the corresponding hyperedge $\hat{e}_i$ of the hypergraph $\hat{G}$. ▼

Figure 5 demonstrates the example of transforming the instance of the primal hypergraph $\hat{G}$ into the instance of the strict line hypergraph $\hat{L}(\hat{G})$. The most natural observation is that the nodes $v_i \in V$ equipped with n neighbor hyperedges of the initial hypergraph are reflected in the resulting structure by the corresponding hyperedges $\hat{e}_i^{\hat{L}} \in \hat{E}^{\hat{L}}$ that are incident to n vertices $v^{\hat{L}} \in V^{\hat{L}}$. Conversely, the hyperedges $\hat{e}_i \in \hat{E}$ covering m nodes are translated into the vertices $v_i^{\hat{L}} \in V^{\hat{L}}$ with degrees $d\left(v_i^{\hat{L}}\right) = m$. In particular, the loops (e.g. the hyperedges $\hat{e}_1$, $\hat{e}_2$, $\hat{e}_4$, and $\hat{e}_7$ in Fig. 5a) are transformed into the "leaf" vertices that are incident to only one hyperedge. The "leaf" nodes (such as the vertices v_2, v_5, v_7, and v_8 in Fig. 5a), for their part, are converted into loops, which is in conformity with Theorem 1. Notice that the isolated vertices $v_s \in V$ are reflected in the strict line hypergraph by the loose hyperedges $\hat{e}_s^{\hat{L}} = \{\emptyset\} \in \hat{E}^{\hat{L}}$ that are fully deprived of the incident nodes and allowed according to Definition 5. The vertex v_6 in Fig. 5a serves as an illustration of such case.

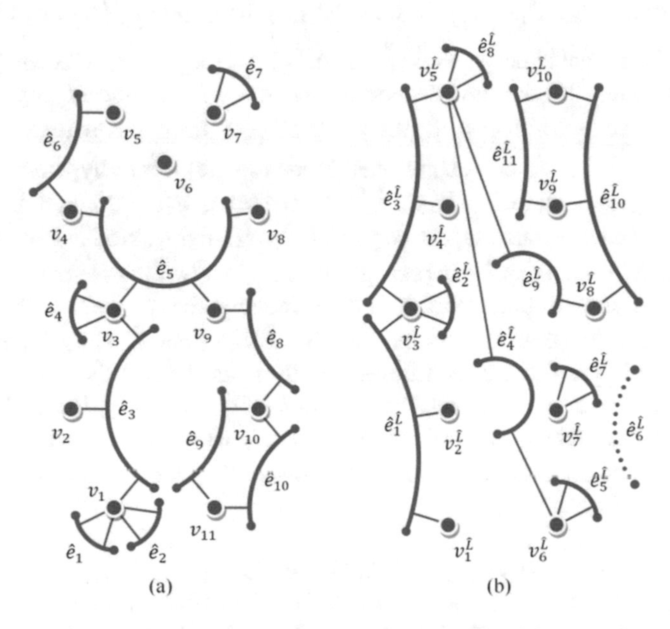

Fig. 5. Example of the primal hypergraph (a) and the result of its transformation into the corresponding strict line hypergraph (b).

Clearly, the strict line hypergraph constructed for the disconnected primal hypergraph is represented by the disjoint union of the strict line hypergraphs that are separately formed for all its connected components. Another important property of the function $\hat{L}$ is associated with the existence of the numerous fixed points represented by the hypergraphs whose structure remains unchanged after applying the function $\hat{L}$. Considering the hypergraph instance shown in Fig. 5a, the connected component containing the vertex v_7 equipped with the loop $\hat{e}_7$ serves as the simplest example of such fixed point. Notice that the hypergraphs whose maximum vertex degree is bounded by 2 are mapped into the strict line hypergraphs belonging to the subclass of undirected graphs $\Gamma \subset \hat{\Gamma}$.

6 Construction of the Duality Relation Between the Vertex Separators and Cuts

Theorem 2. Each vertex separator $C_V \in \mathcal{VS}(\hat{G})$ in the primal hypergraph $\hat{G} = (V, \hat{E}) \in \hat{\Gamma}$ is mapped into the corresponding dual cut $\hat{E}^{\hat{L}}(C_V) = \left\{\hat{e}_i^{\hat{L}} | v_i \in C_V\right\}$ belonging to the space $\mathcal{C}(\hat{L}(\hat{G}))$ of the strict line hypergraph $\hat{L}(\hat{G}) = \left(V^{\hat{L}}, \hat{E}^{\hat{L}}\right)$.

▲ Suppose the primal hypergraph $\hat{G}$ is connected and the set $S \subset V$ is composed of the vertices belonging to the non-degenerate connected component segregated after excluding the nodes of the separator $C_V \in \mathcal{VS}(\hat{G})$ from the overall set V, while $T = V \backslash \{S \bigcup C_V\}$. In this context, we can divide the multiset of hyperedges $\hat{E}$ into the disjoint allowed subcollections $\hat{E}(S)$, $\hat{E}(S, C_V)$, $\hat{E}(C_V)$, $\hat{E}(C_V, T)$, and $\hat{E}(T)$. Assume that their cardinality equals, respectively, t_1, t_2, t_3, t_4, and t_5. Notice that each allowed subcollection $\hat{E}(\bullet)$ contains the hyperedges that are incident to at least one vertex from each subset indicated in brackets and, at the same time, has no links with the nodes of the other subsets. By contrast, the subcollections $\hat{E}(S, T)$ and $\hat{E}(S, C_V, T)$ are forbidden and contain no hyperedges. This follows directly from the fact that the presence of at least one hyperedge that connects the S and T sets bypassing the nodes of C_V is completely inconsistent with considering the set C_V as a vertex separator.

Without losing the generality, we can renumber all hyperedges of the hypergraph $\hat{G}$ in such manner that

$$\hat{E}(S) = \{\hat{e}_1, \ldots, \hat{e}_{\sigma_1}\},\ \hat{E}(S, C_V) = \{\hat{e}_{\sigma_1+1}, \ldots, \hat{e}_{\sigma_2}\},$$
$$\hat{E}(C_V) = \{\hat{e}_{\sigma_2+1}, \ldots, \hat{e}_{\sigma_3}\},\ \hat{E}(C_V, T) = \{\hat{e}_{\sigma_3+1}, \ldots, \hat{e}_{\sigma_4}\},$$
$$\hat{E}(T) = \{\hat{e}_{\sigma_4+1}, \ldots, \hat{e}_m\},$$

where $\sigma_i = \sum_{j=1}^{i} t_j$, while $m \geq 2$ is the cardinality of $\hat{E}$.

By Definition 5, each allowed subcollection $\hat{E}(\bullet)$ is reflected in the structure of the $\hat{L}(\hat{G})$ hypergraph by the corresponding subset of nodes $V^{\hat{L}}(\bullet)$. In turn, the vertex sets S, T, and C_V are transformed respectively into the multisets of hyperedges and $\hat{E}^{\hat{L}}(S)$, $\hat{E}^{\hat{L}}(T)$, and $\hat{E}^{\hat{L}}(C_V)$. Thereby, the remainder of the proof is intended to demonstrate that the multiset $\hat{E}^{\hat{L}}(C_V)$ forms a cut in the dual hypergraph $\hat{L}(\hat{G})$.

Obviously, the vertices of the set C_V could be covered only by the hyperedges belonging to the union $\hat{E}(S, C_V) \bigcup \hat{E}(C_V) \bigcup \hat{E}(C_V, T)$. This allows imposing the restriction on the structure of each hyperedge $\hat{e}_i^{\hat{L}} \in \hat{E}^{\hat{L}}(C_V)$ taking the form of the following superset:

$$\hat{E}^{\hat{L}}(C_V) = \left\{\hat{e}_i^{\hat{L}} | v_i \in C_V\right\},\ \hat{e}_i^{\hat{L}} \subseteq \left\{v_k^{\hat{L}} | k = \sigma_1 + 1, \ldots, \sigma_4\right\}.$$

Conversely, the vertices of the S and T sets could be incident only to the hyperedges belonging to the unions $\hat{E}(S) \bigcup \hat{E}(S, C_V)$ and $\hat{E}(C_V, T) \bigcup \hat{E}(T)$, respectively. Taking this into consideration, we can provide the next expressions limiting the structure of the hyperedges included in the collections $\hat{E}^{\hat{L}}(S)$ and $\hat{E}^{\hat{L}}(T)$:

$$\hat{E}^{\hat{L}}(S) = \left\{\hat{e}_i^{\hat{L}} | v_i \in S\right\}, \hat{e}_i^{\hat{L}} \subseteq \left\{v_k^{\hat{L}} | k = 1, \ldots, \sigma_2\right\};$$

$$\hat{E}^{\hat{L}}(T) = \left\{\hat{e}_i^{\hat{L}} | v_i \in T\right\}, \hat{e}_i^{\hat{L}} \subseteq \left\{v_k^{\hat{L}} | k = \sigma_3 + 1, \ldots, m\right\}.$$

In the case of the non-degenerate vertex separator C_V belonging to the space $\mathcal{NDVS}(\hat{G})$, both S and T sets should be non-empty. Moreover, each of the multisets $\hat{E}(S, C_V)$ and $\hat{E}(C_V, T)$ should contain at least one hyperedge because the set C_V initiates the separation of the non-degenerate connected components and, thereby, is required to be linked with their nodes. This observation produces the limitations $t_2 \geq 1$ and $t_4 \geq 1$ underlying the condition $\sigma_4 - \sigma_2 \geq 1$. Such condition implies that the multisets $\hat{E}^{\hat{L}}(S)$ and $\hat{E}^{\hat{L}}(T)$ are non-empty and contain the hyperedges that are incident to the non-overlapping subsets of vertices. At the same time, since the connectivity of hypergraphs is preserved under the $\hat{L}$ transformation, the hyperedges of the multiset $\hat{E}^{\hat{L}}(C_V)$ play the "bridge" role. As a result, for any C_V belonging to $\mathcal{NDVS}(\hat{G})$, the collection $\hat{E}^{\hat{L}}(C_V)$ represents a cut in the hypergraph $\hat{L}(\hat{G})$. The example illustrating this case is given in Fig. 6.

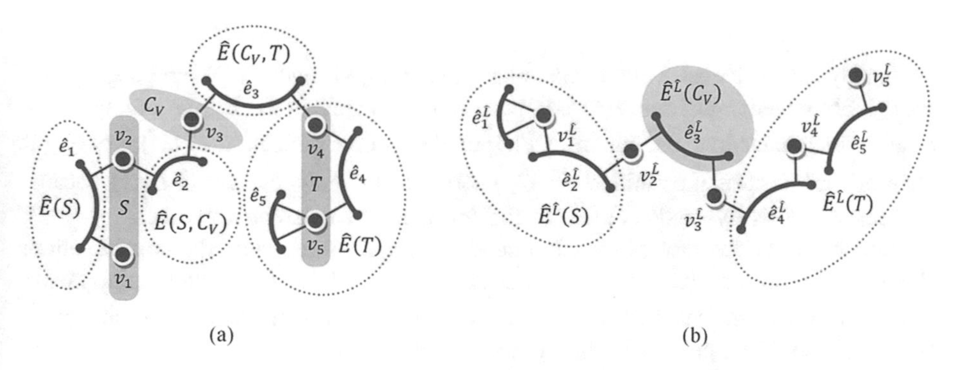

Fig. 6. Example of establishing the duality relation between the non-degenerate vertex separator C_V in the primal hypergraph (a) and the corresponding cut $\hat{E}^{\hat{L}}(C_V)$ in the strict line hypergraph (b).

In order to complete the proof, we need to take into consideration the case of the degenerate vertex separators belonging to the set $\mathcal{VS}(\hat{G}) \backslash \mathcal{NDVS}(\hat{G})$. In particular, regarding the vertex separator C_V that segregates one non-degenerate connected component, only the set S is non-empty, while all other vertices of $V \backslash S$ are included in C_V. As in the previous case, the multiset $\hat{E}(S, C_V)$ should be non-empty. At the same time, the collection $\hat{E}(C_V)$ should contain one or more hyperedges underlying the formation of the degenerate connected components after excluding all vertices of C_V. This contributes to the limitations $t_2 \geq 1$, $t_3 \geq 1$, and $\sigma_3 - \sigma_2 \geq 1$ allowing to conclude that the multiset $\hat{E}^{\hat{L}}(C_V)$ constitutes a cut in $\hat{L}(\hat{G})$ because the exclusion of its hyperedges produces t_3 isolated vertices and the connected component containing $t_1 + t_2$ nodes. Figure 7 illustrates the example of mapping the degenerate vertex separator into the corresponding cut segregating one isolated vertex and one connected component containing three nodes.

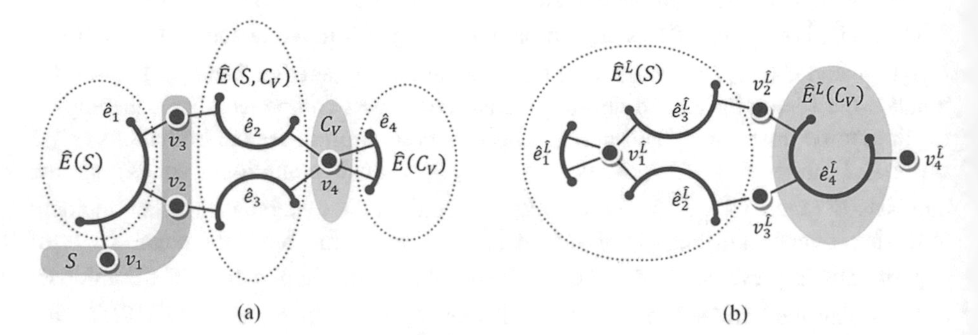

Fig. 7. Example of transforming the degenerate vertex separator C_V in the primal hypergraph (a) into the corresponding cut $\hat{E}^{\hat{L}}(C_V)$ in the strict line hypergraph (b).

Finally, the universal vertex separators containing all nodes of V and segregating exclusively degenerate connected components are characterized by the empty S and T sets, which leads to consolidating all hyperedges in the collection $\hat{E}(C_V)$, i.e. $t_3 \geq 2$. Thereby, in this case, the multiset $\hat{E}^{\hat{L}}(C_V)$ serves as the universal cut in $\hat{L}(\hat{G})$ because the removal of its hyperedges leads to the formation of t_3 isolated vertices. ▼

Notice that in the case of the disconnected primal hypergraph, the considerations given in the above proof should be followed for all its connected components. Moreover, since the function $\hat{L}$ is an involution, the transformation formulated in Theorem 2 could be performed also in the reverse order.

In summarizing the findings presented in this section, the non-degenerate vertex separators in $\hat{G}$ are converted into cuts in $\hat{L}(\hat{G})$ satisfying the restriction that the exclusion of their hyperedges should produce at least two connected components having more than one vertex. However, this conversation is non-exhaustive since it does not put any structures in correspondence with cuts segregating less than two connected components having at least two vertices and, thereby, provides only the partial coverage of the space $\mathcal{C}(\hat{L}(\hat{G}))$. The degenerate vertex separators ensure filling this gap and serve as the dual structures for such cuts.

7 Conclusions

As a corollary of Theorem 2, the established duality relationship is characterized by the cardinality preservation, i.e. the vertex separators in the primal hypergraph $\hat{G}$ and their dual cuts in the corresponding strict line hypergraph $\hat{L}(\hat{G})$ are composed of an equal number of elements. In particular, the articulation points in $\hat{G}$ are converted into the “bridge” connections in $\hat{L}(\hat{G})$. This property underlies the opportunity of setting the linkage between the combinatorial optimization problems whose solution spaces are composed of the vertex separators and cuts. Most importantly, the problem of finding the minimum vertex separator in $\hat{G}$ could be converted into the corresponding problem of calculating the minimum cut in $\hat{L}(\hat{G})$. Consequently, solving both these problems could be reduced to solving only one of them with the subsequent transformation of the received solution into the solution of the linked problem.

At the same time, the exclusion of the degenerate form of the vertex separators from the consideration affects both these problems as well as the linkage between them. In particular, the problem of calculating the minimum non-degenerate vertex separator in $\hat{G}$ could be translated into the restricted problem of finding the minimum cut in $\hat{L}(\hat{G})$. The feasible solutions of such restricted problem are represented by the cuts segregating at least two connected components having more than one vertex.

Therefore, the duality relation formulated in this article is extremely crucial for endowing the complex systems with the enhanced resistance to the structural attacks. This relation allows selecting the most convenient way for searching the vulnerable parts of the modeled system and their representing in terms of the vertex separators and cuts. Such opportunity of the viewpoint switching could serve as the basis for the more comprehensive investigation of the system structure, which contributes to the significance of the obtained results.

References

1. Goh, C.J., Yang, X.Q.: Duality in Optimization and Variational Inequalities. Taylor & Francis, London (2002)
2. Potebnia, A.: Representation of the greedy algorithms applicability for solving the combinatorial optimization problems based on the hypergraph mathematical structure. In: Proceedings of 14th International Conference on the Experience of Designing and Application of CAD Systems in Microelectronics (CADSM), pp. 328–332. IEEE, Lviv (2017). https://doi.org/10.1109/cadsm.2017.7916145
3. Holme, P., Kim, B.J., Yoon, C.N., Han, S.K.: Attack vulnerability of complex networks. Phys. Rev. E **65**(5), 056109 (2002). https://doi.org/10.1103/physreve.65.056109
4. Potebnia, A.: Method for classification of the computational problems on the basis of the multifractal division of the complexity classes. In: Proceedings of 3rd International Scientific-Practical Conference on Problems of Infocommunications, Science and Technology (PIC S&T), pp. 1–4. IEEE, Kharkiv (2016). https://doi.org/10.1109/infocommst.2016.7905318

5. Schaub, M.T., Lehmann, J., Yaliraki, S.N., Barahona, M.: Structure of complex networks: quantifying edge-to-edge relations by failure-induced flow redistribution. Netw. Sci. **2**(1), 66–89 (2014). https://doi.org/10.1017/nws.2014.4
6. Ganesan, A.: Performance of sufficient conditions for distributed quality-of-service support in wireless networks. Wirel. Netw. **20**(6), 1321–1334 (2014). https://doi.org/10.1007/s11276-013-0680-z
7. Lu, Z., Li, X.: Attack vulnerability of network controllability. PLoS ONE **11**(9), e0162289 (2016). https://doi.org/10.1371/journal.pone.0162289
8. Shen, Y., Nguyen, N.P., Xuan, Y., Thai, M.T.: On the discovery of critical links and nodes for assessing network vulnerability. IEEE/ACM Trans. Netw. **21**(3), 963–973 (2013). https://doi.org/10.1109/tnet.2012.2215882
9. Ray, S.S.: Planar and dual graphs. In: Graph Theory with Algorithms and its Applications, pp. 135–158. Springer, India (2013). https://doi.org/10.1007/978-81-322-0750-4_9
10. Ahuja, R.K., Magnanti, T.L., Orlin, J.B.: Network Flows: Theory, Algorithms, and Applications. Prentice-Hall, New Jersey (1993)
11. Xicheng, L., Li, Q.: Line graphs of pseudographs. In: Proceedings of the International Conference on Circuits and Systems, pp. 713–716. IEEE, Shenzhen (1991). https://doi.org/10.1109/ciccas.1991.184458
12. Evans, T., Lambiotte, R.: Line graphs, link partitions, and overlapping communities. Phys. Rev. E **80**(1), 016105 (2009). https://doi.org/10.1103/physreve.80.016105
13. Huggett, S., Moffatt, I.: Bipartite partial duals and circuits in medial graphs. Combinatorica **33**(2), 231–252 (2013). https://doi.org/10.1007/s00493-013-2850-0
14. Brylawski, T., Oxley, J.G.: The Tutte polynomial and its applications. In: White, N. (ed.) Matroid Applications, pp. 123–225. Cambridge University Press, Cambridge (1992)
15. Potebnia, A.: Creation of the mathematical apparatus for establishing the duality relation between the vertex separators and cuts in hypergraphs. In: Proceedings of 12th International Scientific and Technical Conference on Computer Science and Information Technologies (CSIT), pp. 236–239. IEEE, Lviv (2017)

Involutory Parametric Orthogonal Transforms of Cosine-Walsh Type with Application to Data Encryption

Dariusz Puchala(✉)

Institute of Information Technology,
Lodz University of Technology, Lodz, Poland
dariusz.puchala@p.lodz.pl

Abstract. In this chapter we introduce a novel scheme for constructing a class of parametric involutory transforms that are described by the computational lattice structure of fast cosine-Walsh type transform. Further on the practical effectiveness of the proposed class of transforms is evaluated experimentally in the task of data encryption. Finally, the selected aspects of mass-parallel realizations of the proposed class of involutory parametric transforms with usage of modern graphics processing units are considered and discussed.

Keywords: Fast parametric linear transforms · Data encryption

1 Introduction

Modern applications of digital processing and analysis of one- and multidimensional data require highly efficient computational techniques. This is a direct consequence of the systematic increase in the capabilities of data acquisition devices, in particular for recording static images and video sequences. Naturally, this involves a proportional increase in the size of data to be processed or analyzed. Classical methods that are commonly used in practice utilize fast linear transforms which can be characterized by high computational efficiency. Beyond any doubt among these the most popular are: Walsha-Hadamard transform [1, 2], slant transform [3], fast Fourier transform [4], fast trigonometric transforms of Fourier type, i.e. Hartley transform [5], cosine and sine transforms in different variants [6], etc. Fast algorithms for such transforms are constructed using "divide and conquer" strategy and take the form of lattice structures with planar rotation operations realized by fixed and predetermined angles. The natural extension of fast linear transforms are their parametric equivalents. Such transforms can be characterized by high flexibility while retaining the same computational efficiency. The parameterization of a fast linear transforms consists in making the rotation angles of all or only selected base operations dependent on the values of a set of parameters. This allows for adaptation of the basis vectors to the specific aspects of the tasks being performed, as well as the statistical characteristics of input signals. The high increase in interest of researchers in parametric transforms was initiated by the chapter [7]. At the present time, the most well-known classes of parametric transforms include: slant and slant-Haar transforms [8, 9], reciprocal-orthogonal transforms [10, 11], Fourier and

N. Shakhovska and V. Stepashko (eds.), *Advances in Intelligent Systems and Computing II*, Advances in Intelligent Systems and Computing 689,
https://doi.org/10.1007/978-3-319-70581-1_29

Hartley transforms [12], involutory transforms [13], etc. Among these for special attention deserve fast orthogonal parametric involutory transforms. Their additional feature is the property of self-invertibility, which means that the inverse transform matrix U^T is the same as a forward one U, i.e. equality $U^T = U$ takes place. Such a feature is very important primarily from the point of view of hardware dedicated implementations. In such a case the same hardware layout (dedicated for forward transform) is able to perform both forward and inverse transformations which translates directly into the cost reduction.

The areas of application of fast parametric transforms are the same as of fast linear ones. Hence, the examples may include the following tasks: signal denoising [8, 12, 14], encryption of data [11, 13, 15], joint compression and encryption of data [16] and pattern recognition [17]. In this chapter we are interested in the task data encryption. It should be noted that the idea of application of linear transforms to encrypt data is not new and a considerable number of scientific chapters can be indicated were linear parametric transforms are exploited for encryption or joint encryption and compression of multimedia data [15, 18]. It is also obvious that data encryption methods based on linear transformations can be easily cracked with plain text attacks. However, the specificity of multimedia data transmissions (e.g. digital television broadcasts) which can be shortly characterized as: transmission of huge amounts of data of low or moderate value, makes it impossible or economically unprofitable to perform such types of attacks. Moreover the simplicity of implementations both in software and hardware (and also in parallel, mass-parallel, pipelined modes, etc.) combined with high computational effectiveness and the adequate level of security of consumer market data place linear transforms at the forefront of practically usable and desired methods of data encryption.

In this chapter we propose a novel scheme for constructing a class of parametric involutory transforms that are defined by the aptly parametrized lattice structure of fast algorithm for calculation of the discrete parametric cosine-Walsh transform (DCWT) [7]. The necessary conditions imposed on the lattice structure in order to obtain involutory transformations are formulated and given in the form of equations involving all of the free parameters. Further on the usefulness of the proposed class of transforms in the task of data encryption is investigated experimentally operating on model signals. Eventually, in the final sections of the chapter we discuss the practical issues of mass-parallel calculations of the considered variants of DCWT lattice structures with aid of modern graphics processing units (GPUs). It should be noted that mass-parallel realizations of several fast linear transforms, namely: discrete Fourier transform [19] and discrete wavelet transform [20] have proved to be highly efficient.

2 Discrete Parametric Cosine-Walsh Transform

The discrete parametric cosine-Walsh transform introduced in chapter [7] is a parametric transform defined by the Cooley-Tuckey like fast computational lattice structure of the well-known discrete Walsh-Hadamard transform (DWHT) [2] (see Fig. 1). However, in the case of DCWT the base orthogonal operators are parametrized with real-valued parameters representing angles of rotations which can take arbitrary values in contrast to DWHT where all base operators are invariable and represent rotations by constant angle of $\pi/4$ radians. In Fig. 1 the lattice structure of DCWT for the case of $N = 8$ element input data vectors is presented.

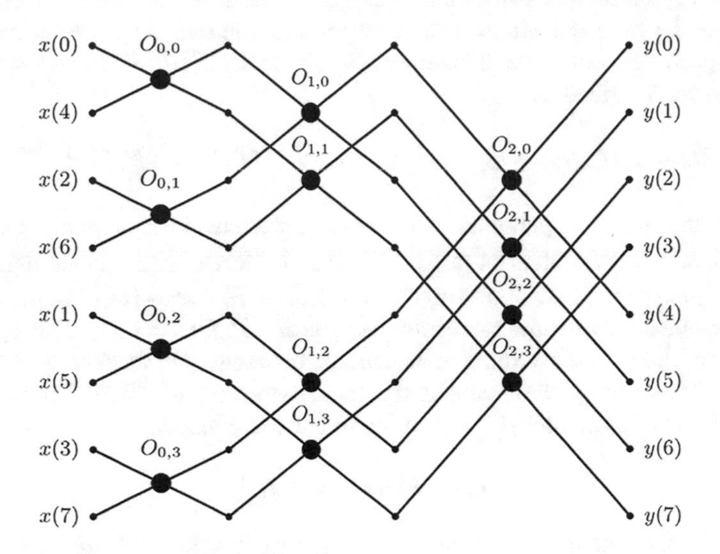

Fig. 1. The computational lattice structure of discrete cosine-Walsh transform for the case of $N = 8$ point transformation.

As the first step the elements $x(n)$ for $n = 0, 1, \ldots, N - 1$ of input vector must be permuted in bit-reversal order (see [2]). Next the permuted vector is propagated forward through $i = 0, 1, \ldots, \log_2 N - 1$ consecutive layers each containing $N/2$ base orthogonal operators $O_{i,j}$ for $j = 0, 1, \ldots, N/2 - 1$ which are represented graphically as "•" and can be defined in the matrix form as:

$$O_{i,j} = \begin{bmatrix} \cos \alpha_{i,j} & \sin \alpha_{i,j} \\ \sin \alpha_{i,j} & -\cos \alpha_{i,j} \end{bmatrix}, \tag{1}$$

where $\alpha_{i,j}$ for $i = 0, 1, \ldots, \log_2 N - 1$ and $j = 0, 1, \ldots, N/2 - 1$ are the free parameters in a number of:

$$L_{PAR} = \frac{N}{2} \log_2 N.$$

The computational lattice structure of DCWT transform is a fast one (understood in the sense of fast linear transforms characterized by the order of $O(N\log_2 N)$ addition/multiplication operations) and can be described in the most general form by a number of:

$$L_{ADD} = N \log_2 N, \; L_{MUL} = 2N \log_2 N \tag{2}$$

real-valued additions and multiplications respectively.

3 Involutory Discrete Cosine-Walsh Transform

The DCWT is an orthogonal transform and as so it can be described in a matrix form as a product of sparse block-diagonal orthogonal matrices and permutation matrices which describe base operations within stages and connections between neighboring stages respectively. Let U_N be a matrix of N-point forward DCWT. Then, according to [21] it can be described as:

$$U_N = I_N(P_{\log_2 N-1}U_{\log_2 N-1}P^T_{\log_2 N-1}) \cdot \ldots \cdot (P_1U_1P^T_1)(P_0U_0P^T_0)P^{br}_N, \tag{3}$$

where P^{br}_N and I_N are the permutation matrices of elements in input and output vectors, I_N is an identity N on N element matrix, P^{br}_N is a bit-reversal permutation matrix, U_i for $i = 0, 1,\ldots, \log_2 N - 1$ are block-diagonal matrices $U_i = \text{diag}\{O_{i,0}, O_{i,1},\ldots, O_{i,N/2-1}\}$ describing operations within consecutive layers, and P_i for $i = 0, 1,\ldots, \log_2 N - 1$ are permutation matrices describing connections between neighboring layers of base operators. Then taking into account the lattice structure of DCWT (c.f. Fig. 1) P_i matrices for $i = 0, 1,\ldots, \log_2 N - 1$ can be defined as follows:

$$P_i = (I_{2^{\log_2 N-i-1}} \otimes P^{eo}_{2^{i+1}}), \tag{4}$$

where "$\otimes$" stands for the Kronecker product of matrices, and P^{eo}_M are the permutation matrices that group elements into the following order: $M/2$ elements with even indices, $M/2$ elements with odd indices, i.e. $P^{eo}_M = [P_{k,l}]^{eo}_M$ for $k, l = 0, 1,\ldots, M - 1$ and:

$$p_{k,l} = \begin{cases} 1 & \text{for } 2k = l \text{ and } 0 \leq l \leq \frac{M}{2} - 1, \text{ or} \\ & \text{for } 2k + 1 = l - \frac{M}{2} \text{ and } \frac{M}{2} \leq l \leq M - 1, \\ 0 & \text{in other cases.} \end{cases}$$

Using introduced P_i permutation matrices the bit-reversal ordering matrix P^{br}_N can be described as a following product of matrices:

$$P^{br}_N = \left(\prod_{i=0}^{\log_2 N-1} P_{2^{i+1}} \right).$$

In Fig. 2 the lattice structure for calculation of $N = 8$ point DCWT in a fashion resulting from the factorization described by formula (3) is presented, wherein the input and output permutation matrices are deliberately omitted.

The involution property requires the transform matrix U_N to be self-invertible, i.e. the following equality $U_N^{-1} = U_N$ must hold. By taking into account the orthogonality of U_N matrix it is straightforward to verify that the quoted equality would take the following form $U_N^T = U_N$. It is also obvious that then we would have $U_NU_N = I_N$. But

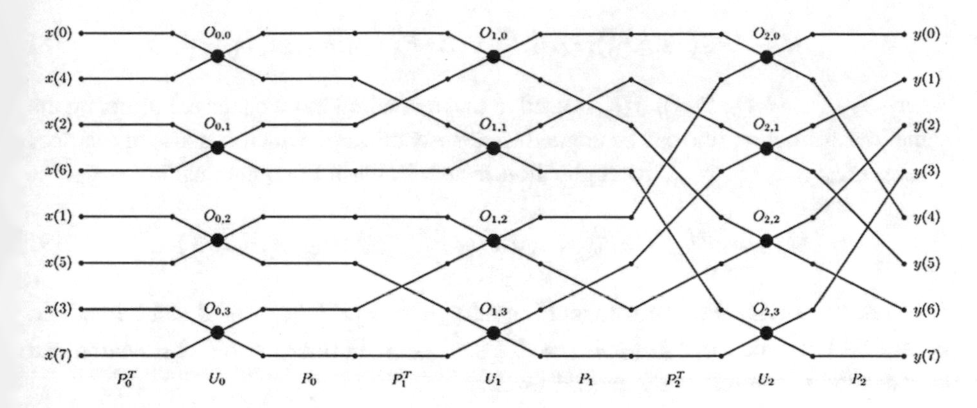

Fig. 2. The lattice structure of discrete cosine-Walsh transform for the case of $N = 8$ point transformation resulting from matrix product factorization (3).

in order to obtain the desired property the additional dependencies between transform parameters are required. In accordance with basic matrix algebra transformations we can define on the basis of Eq. (3) the inverse matrix U_N^T in the form of the following factorization:

$$U_N^T = (P_N^{br})^T (P_0 U_0^T P_0^T)(P_1 U_1^T P_1^T) \cdot \ldots \cdot (P_{\log_2 N-1} U_{\log_2 N-1}^T P_{\log_2 N-1}^T) I_N \tag{5}$$

It should be also noted that $\left(P_N^{br}\right)^T = P_N^{br}$ and in addition for $i = 0, 1, \ldots, \log_2 N - 1$ we have $U_i^T = U_i$ (c.f. Eq. (1)). Then taking into account the observations made we can rewrite formula (5) as:

$$U_N^T = P_N^{br} (P_0 U_0 P_0^T)(P_1 U_1 P_1^T) \cdot \ldots \cdot (P_{\log_2 N-1} U_{\log_2 N-1} P_{\log_2 N-1}^T) I_N. \tag{6}$$

Now comparing Eqs. (3) and (5) it can be conceived that one way to obtain the equality $U_N^T = U_N$ leads by bringing the formula (5) to the form structurally identical to the factorization described by (3). Here, by the structural identity we understand the same structures of matrices used, i.e.: the same permutation matrices P_i at the same positions and in both cases block-diagonal matrices representing base operations.

Let us start the process of refactorization of formula (6) by multiplying its right side by P_N^{br} matrix twice in order to keep the equality undisturbed. We have then: $U_N^T = (\ldots)\left(P_N^{br} P_N^{br}\right)$, which is valid since: $P_N^{br} P_N^{br} = I_N$. Next the following transformation rule is proposed, i.e.:

$$(P_{\log_2 N-1} U_{\log_2 N-1} P_{\log_2 N-1}^T) P_N^{br} = P_N^{br} (P_0 Q_N U_{\log_2 N-1} Q_N^T P_0^T). \tag{7}$$

From Eq. (7) emerges instantly the construction of Q_N which in matrix notation can be written as: $Q_N = P_0^T P_N^{br} P_{\log_2 N-1}$. Then introducing the proposed transformation rule into formula (6) it can be rewritten as follows:

$$U_N^T = P_N^{br}(P_0 U_0 P_0^T)(P_1 U_1 P_1^T) \cdot \ldots \cdot P_N^{br}(P_0 \bar{U}_{\log_2 N-1} P_0^T) P_N^{br}, \tag{8}$$

where $\bar{U}_{\log_2 N-1} = Q_N U_{\log_2 N-1} Q_N^T$ is used to make notation more concise. Further on the same transformation rule can be applied to the second stage which consists of matrices $(P_{\log_2 N-2} U_{\log_2 N-2} P_{\log_2 N-2}^T)$. Its application results in the following equality:

$$(P_{\log_2 N-2} U_{\log_2 N-2} P_{\log_2 N-2}^T) P_N^{br} = P_N^{br}(P_1 Q_N U_{\log_2 N-2} Q_N^T P_1^T). \tag{9}$$

Then by solving Eq. (9) we would obtain $Q_N = P_1^T P_N^{br} P_{\log_2 N-2}$. It can be verified by direct check that: $P_1^T P_N^{br} P_{\log_2 N-2} = P_0^T P_N^{br} P_{\log_2 N-1}$. Further on this observation can be extended to the following general form:

$$Q_N = P_i^T P_N^{br} P_{\log_2 N-i-1} \tag{10}$$

for $i = 0, 1, \ldots, \log_2 N - 1$ which is formulated here without proof. Equation (10) allows to draw an additional conclusion that $Q_N = Q_N^T$. Then on the basis of observation (10) it is clear that the proposed transformation rule can be applied to all remaining layers resulting in the following factorization:

$$\begin{aligned} U_N^T = (P_{\log_2 N-1} \bar{U}_0 P_{\log_2 N-1}^T)(P_{\log_2 N-2} \bar{U}_1 P_{\log_2 N-2}^T) \cdot \ldots \\ \ldots \cdot (P_1 \bar{U}_{\log_2 N-2} P_1^T)(P_0 \bar{U}_{\log_2 N-1} P_0^T) P_N^{br}, \end{aligned} \tag{11}$$

where $\bar{U}_i = Q_N U_i Q_N$ for $i = 0, 1, \ldots, \log_2 N - 1$. Further on it should be noted that Q_N is a permutation matrix which places elements of input vector in bit-reversal order but calculated for $N/2$ elements and applied to the pairs of adjacent elements in input vector. For example for $N = 8$ we would obtain the following matrix:

$$Q_8 = \begin{bmatrix} 1 & 0 & 0 & 0 & 0 & 0 & 0 & 0 \\ 0 & 1 & 0 & 0 & 0 & 0 & 0 & 0 \\ 0 & 0 & 0 & 0 & 1 & 0 & 0 & 0 \\ 0 & 0 & 0 & 0 & 0 & 1 & 0 & 0 \\ 0 & 0 & 1 & 0 & 0 & 0 & 0 & 0 \\ 0 & 0 & 0 & 1 & 0 & 0 & 0 & 0 \\ 0 & 0 & 0 & 0 & 0 & 0 & 1 & 0 \\ 0 & 0 & 0 & 0 & 0 & 0 & 0 & 1 \end{bmatrix}$$

and the following order of elements: [0, 1, 4, 5, 2, 3, 6, 7]. It is obvious that if such permutation is applied to matrix U_i in the form: $Q_N U_i Q_N$, then it would result in the block-diagonal matrix $\bar{U}_i$ with operations $O_{i,j}$ for $j = 0, 1, \ldots, N/2 - 1$ which are permuted in bit-reversal order. Thus, we would have $\bar{U}_i = \mathrm{diag}\{O_{i,p(0)}, O_{i,p(1)}, \ldots, O_{i,p(N/2-1)}\}$, where $p(j)$ for $j = 0, 1, \ldots, N/2 - 1$ stands for bit-reversal order calculated for $N/2$ elements. In Fig. 3 the lattice structure for calculation of inverse DCWT for $N = 8$ points is presented in the form resulting from refactorization described by formula (11).

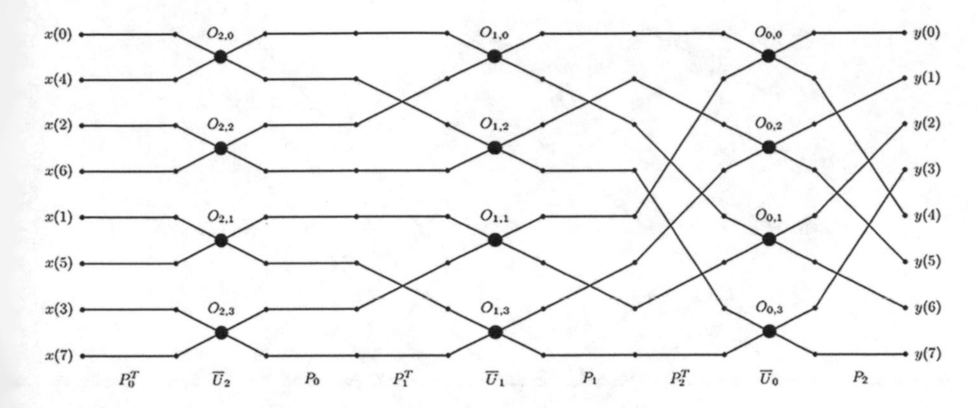

Fig. 3. The lattice structure of inverse discrete cosine-Walsh transform for the case of $N = 8$ point transformation in the form resulting from factorization (11).

Since both factorizations (3) and (11) for U_N, U_N^T transforms respectively are structurally identical (c.f. Figs. 2 and 3) it is now elementary to derive the necessary conditions required in order to achieve the equality $U_N^T = U_N$. The mentioned conditions can be written as: $\bar{U}_i = U_{\log_2 N-i-1}$ for $i = 0, 1,\ldots, \log_2 N - 1$, which further on results in the following dependencies between specific transform parameters:

$$\alpha_{i,p(j)} = \alpha_{\log_2 N-i-1,j} \tag{12}$$

for $i = 0, 1,\ldots,\lceil \log_2 N/2 \rceil - 1$ and $j = 0, 1,\ldots, N/2 - 1$, where $\lceil \cdot \rceil$ is a ceiling operator. The resulting transform constructed in accordance with factorization (3) that takes into account dependencies described by formula (12) must be an involutory parametric orthogonal transform with fast computational lattice structure of cosine-Walsh transform type. In such case a number of free parameters equals:

$$L_{PAR} = \begin{cases} \frac{N}{4}\log_2 N & \text{for even values of } \log_2 N, \\ \frac{N}{4}\log_2 N + \frac{1}{2}\sqrt{\frac{N}{2}} & \text{for odd values of } \log_2 N. \end{cases} \tag{13}$$

In Fig. 4 we show the lattice structure of involutory parametric DCWT transform for the case of $N = 8$ points. It is obvious that the computational complexity of involutory variant of DCWT is the same as in the case of parametric DCWT and hence it can be also described by formulas (2).

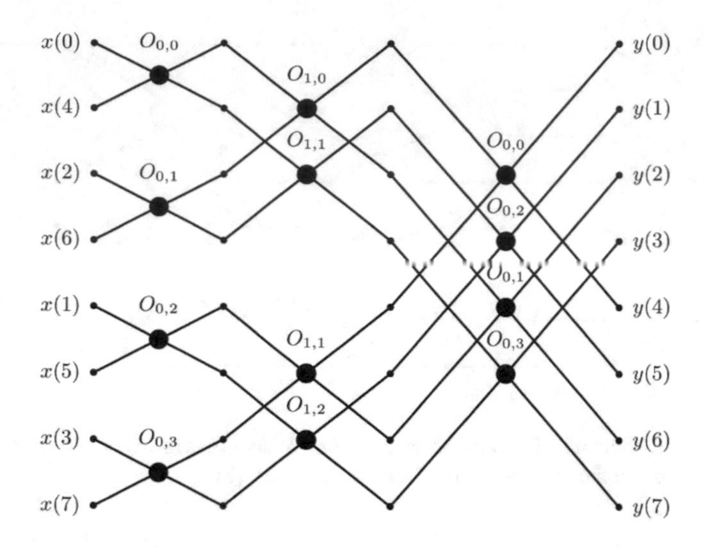

Fig. 4. The lattice structure of involutory parametric cosine-Walsh transform for the case of $N = 8$ point transformation.

4 Experimental Study

As it was previously mentioned the experimental part of the chapter would be focused on the issues of data encryption. The results of similar study but applied to different types of lattice structures were published in chapter [22] (two-stage Beneś network) and also in [23] (the comparison of two-stage Beneś network to Walsh-Hadamard like structure and lattice structure for two-channel bank of orthogonal filters). In this chapter, we adopt similar test procedure which involves the experimental calculation of: (I) the probability distribution of relative mean square error of signal reconstruction with randomly drawn private keys for the decryption step, (II) the expected value of relative mean square error in the function of Hamming distance calculated between private keys for the encryption and decryption steps. In Fig. 5 the flow diagram for the encryption and decryption steps is presented. In the first place an input vector x is encrypted with aid of orthogonal parametric transform U which is parametrized with private key K. It is obvious that the construction of private key should allow to extract the values of all parameters required by the parametric transform. After this step the encrypted (cyphered) vector y is obtained. The decryption process requires knowledge of the form of a private key and involves inverse parametric transform U^T or U in the case of involutory transformations. In Fig. 5 it is assumed that at both stages the same private key K is known. Hence, theoretically (i.e. neglecting computational errors) the input data vector x can be reconstructed without any distortion.

In paper [22] a simple mapping of individual bits of a private key to the values of transform parameters $\alpha_{i,j}$ for $i = 0, 1, \ldots, \log_2 N - 1$ and $j = 0, 1, \ldots, N/2 - 1$ was proposed. It assumes in the first place that the interval $[0, 2\pi)$ of parameters variation is divided into a number of 2^k subintervals of equal lengths $\Delta\alpha = 2\pi/2^k$. Then the discrete

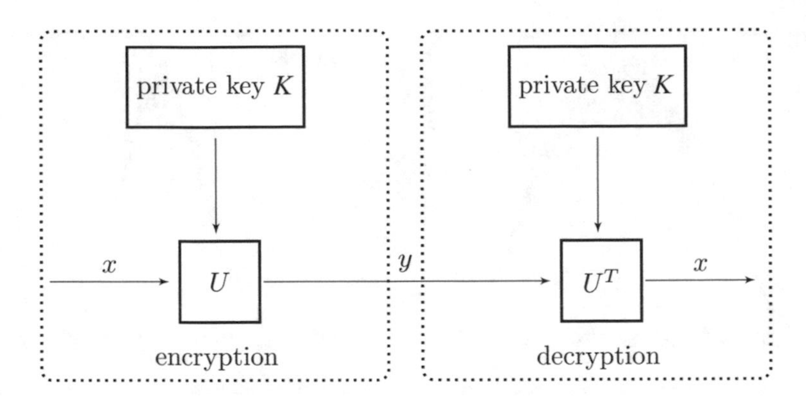

Fig. 5. The diagram of the steps of data encryption and decryption.

values of parameters would be calculated as $\alpha_{i,j} = k_{i,j}\Delta\alpha$, where $k_{i,j}$ denotes the decimal integer number from an interval $[0, 2^k - 1]$, which is encoded, e.g. with natural binary code, with the subsequent bits of a private key assigned to this parameter. Then the size of a private key K can be calculated as $L_K = kL_{PAR}$ bits. It is clear that in the case of involutory transforms the mapping of private key bits to the values of parameters must take into account the necessary dependencies described by formula (12).

In order to verify the effectiveness of the proposed class of parametric transforms a series of experiments involving model first order Markov signals with variance $\sigma^2 = 1$ and the correlation coefficient $\rho = 0.9$ was performed. The process of data encryption and decryption proceeded according to the data flow diagram depicted in Fig. 5. In addition, we assumed $k = 4$ and considered the following lengths of input vectors, i.e. $N = 8, 16, 32$ and 64 points, which correspond to the following lengths of private keys, i.e.: $L_K = 28, 64, 168$ and 384 bits. The obtained results averaged over a number of 10^5 trials are shown in Figs. 6 and 7.

The first part of the study involves the experimental determination of the probability distribution of data reconstruction error during the simulation of brute force attack. As a measure of reconstruction error we adopt the mean square error (MSE) expressed as a percentage of mean signal energy. Such relative MSE can be defined as:

$$\varepsilon_{MSE} = 100 \cdot \left(\frac{1}{M} \sum_{i=0}^{M-1} \frac{\|x_i - z_i\|^2}{\|x_i\|^2} \right) [\%], \tag{14}$$

where x_i and z_i for $i = 0, 1, \ldots, M - 1$ stand for input and reconstructed vectors respectively. In this experiment $x_i = x =$ const. for $i = 0, 1, \ldots, M - 1$ and output vectors are calculated as: $z_i = U_i U x$, where U is an encrypting involutory DCWT parametrized with private key K, and U_i are involutory decrypting DCWTs obtained for randomly guessed keys K_i for $i = 0, 1, \ldots, M - 1$ in a number of $M = 10^5$ trials. The results in the form of probability distribution of discrete values of relative MSE are shown in Fig. 6 for all of the considered transform lengths. It should be noted that the

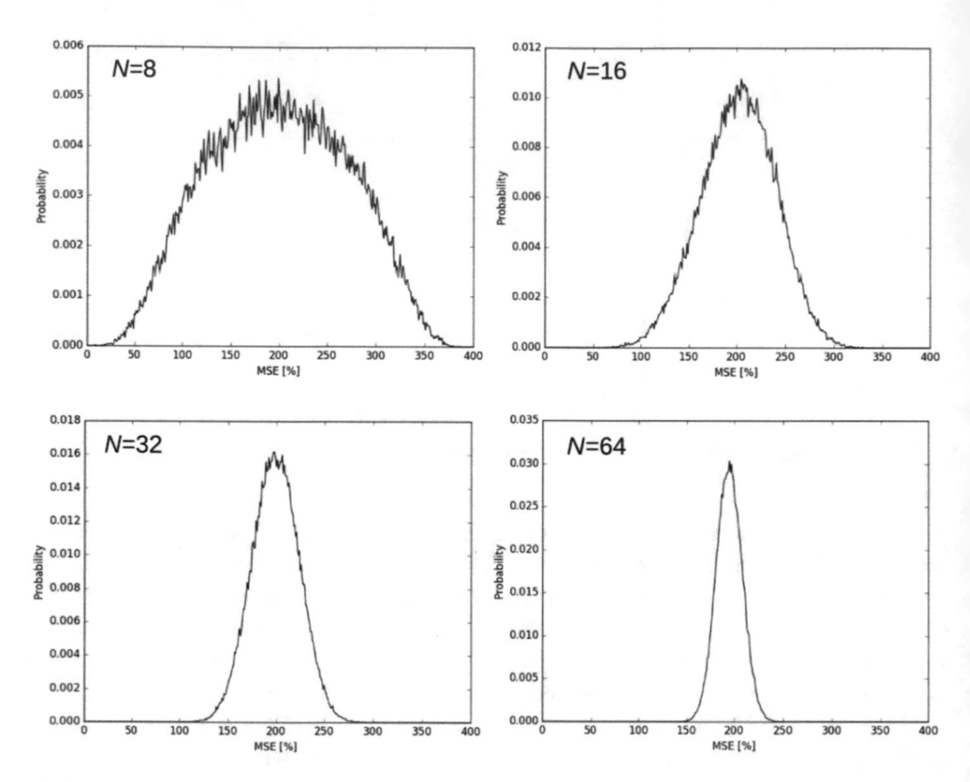

Fig. 6. The probability distribution of relative MSE for the following transform sizes: $N = 8$, 16, 32 and 64 points.

relative MSE have the possible range of variation from 0% to 400%, where 0% means $z = x$ and 400% means $z = -x$. The obtained experimental results show that the most probable value of relative MSE for all transform lengths is close to 200%. Taking into account the shape of probability distribution function it means that the expected value of relative MSE during brute-force attack is close to 200% of energy of input signal, which corresponds to the scenario: $z_i \perp x$. In addition the standard deviation of relative MSE tends to be smaller with an increase of transform size, which clearly results from the increasing lengths of private keys.

The second part of experiments was concentrated on the experimental evaluation of relative MSE in the function of Hamming distance calculated between private keys used at the encryption and decryption stages. The relative MSE results averaged over a number of $M = 10^5$ trials are presented in Fig. 7 for the following transform sizes: $N = 8$, 16, 32 and 64 points. An analysis of experimental results allows to make the following conclusions: (I) the mean value of Hamming distance between randomized private keys equals one half of the key size L_K, (II) the mean value of relative MSE calculated over the possible interval of observation (distances lying outside that interval were simply not observed due to very small probability of occurrence) is close to 200%, what is fully consistent with the results of the first experiment, (III) the width of the observation interval decreases with the growth of transform size which is a direct

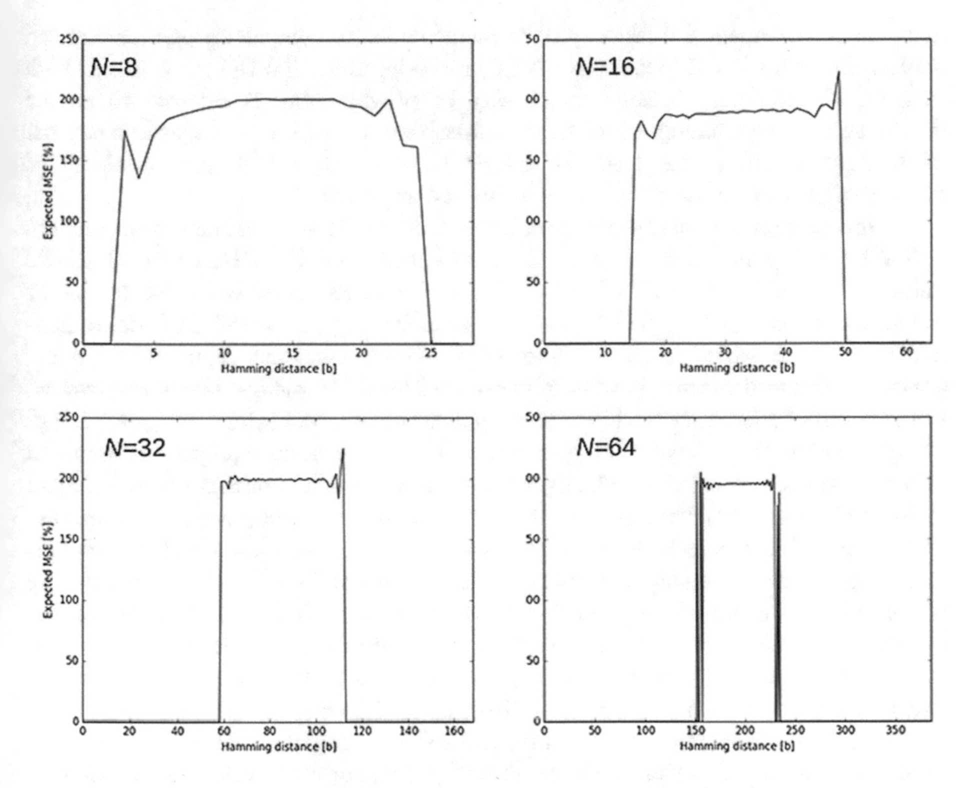

Fig. 7. The mean values of relative MSE in the function of Hamming distance between private keys drawn randomly for the encryption and decryption steps.

consequence of the binomial Bernoulli distribution of the probability of random drawing a pair of keys that differ by a given number of bits (see [22]).

Based on the series of experiments performed we conclude that the proposed class of involutory transforms structurally defined by the fast algorithms for calculation of cosine-Walsh transform can be characterized by high practical effectiveness in the tasks of data encryption, i.e. the high level of data concealment. It should be noted that the considered encryption/decryption process is realized in the direct fashion, i.e. in signal domain, and can be described by the data flow diagram from Fig. 5. Moreover, if we take into account high computational effectiveness and the identity of forward and inverse transformations it can be stated that involutory DCWTs can be applied as effective tools in practical tasks of data encryption, e.g. as a part of the scheme for joint encryption and compression of visual data that was proposed in chapter [17].

5 Selected Aspects of Mass-Parallel Realizations

General purpose scientific and engineering computations realized in parallel mode involving several hundreds or thousands of concurrently operating processors are commonplace nowadays. This is possible by the availability of programming libraries that enable convenient and fast way of implementation of computer programs that can harness the potential of modern graphics processing units (GPUs). Due to the high number of available processors such way of programming is referred to as the mass-parallel programming. In the field of linear transforms we can indicate a number of research chapters (see [19, 20, 24, 25]) which deal with the problems of mass-parallel realizations of well-known linear transforms.

In this section we analyze the selected aspects of DCWT calculation on modern GPUs by taking into account two microarchitectures of NVIDIA cards: (I) G200 microarchitecture (NVS3100M GPU card) which can be treated as an old one to be found in older computers, (II) Maxwell microarchitecture (GTX 960 GPU card), relatively novel one, which establishes a today standard in consumer segment laptop and desktop computers. Since it is well-known that older GPU microarchitectures suffered from the loss of effectiveness while accessing data with improper alignment (coalescing) we try to answer the following two questions: (I) in what measure a proper coalescing of data accesses can increase the effectiveness of DCWT calculation algorithms on older microarchitectures, (II) whether it is still needed to optimize data coalescing in the case of modern GPUs. In order to answers these questions we perform a series of experiments involving two computational structures for calculation of DCWTs. The first structure (Type I) is a standard one of Cooley-Tuckey type shown in Fig. 1. The second one (Type II) is inspired by the lattice structure proposed in monograph [26] for fast calculation of Fourier transformation (see Fig. 8)[1]. Next, regardless of the type of the lattice structure considered we assume the following model of computations, i.e.: (I) we operate on data stored in global memory, (II) a single computational thread is responsible for calculations within a single butterfly operator, (III) calculations between consecutive stages are synchronized (c.f. Fig. 9). Thus, it is clear that the structure of type I can be characterized by a very poor data coalescing at the initial stages which is gradually improving so as to achieve the optimum at the last stage (c.f. the Single Instruction Multiply Threads (SIMT) model of instruction execution within warps of threads [27]). In the case of the lattice structure of type II inputs of butterfly operator at all stages are optimally aligned while outputs refer to the neighboring pairs of data. We try to solve this problem by engaging shared memory for data mixing within warps and thus allowing for improved alignment of data writes into global memory (c.f. Fig. 10). Moreover for both types of structures an additional variant taking advantage of texture memory to store the coefficients of base operations is tested. It is well known that the texture memory although residing in global memory area is provided with additional cache memory which may result in a significant increase of the bandwidth of data transfer. Further on for the simplicity of calculation of indices which are required to access the values of the coefficients for butterfly operators (coefficients are pre-calculated and stored as the pairs of sine/cosine of operators angles) we implement

[1] It should be noted that adaptation of lattice structure presented in monograph [26] for fast calculations of Fourier transform to the field of cosine-Walsh transformation requires proper permutations of butterfly operators within consecutive stages. The required permutations can be described as follows: (I) in the first stage butterfly operators must be permuted in accordance to bit-reversal order, (II) the permutations in the following stages are calculated as even/odd order of operators taken from the directly preceding stage (c.f. Fig. 8).

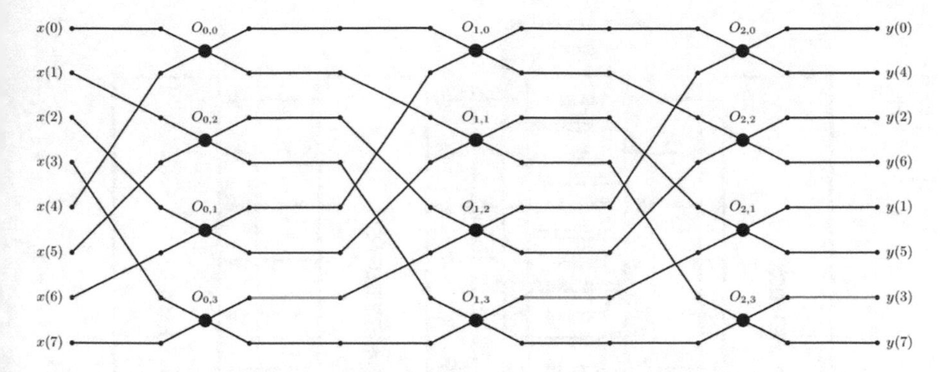

Fig. 8. The lattice structure of type II for fast calculation of DCWT with size of input data $N = 8$ points.

```
__global__ void dcwt_stage(float32 data,int32 sid)
{
        tid    ← calculate thread identifier using grid
                 specific data,
        c,s    ← read coefficients (c-cosα, s-sinα) for base
                 operations from global or texture memory
                 using tid and sid (stage identifier) variables,
        i1,i2 ← on the basis of tid and stage identifier sid
                 calculate input data indices,
        o1,o2 ← accordingly calculate output data indices,
        ; perform calculations within single base operator,
        data[o1] ← c*data[i1] + s*data[i2];
        data[o2] ← s*data[i1] - c*data[i2];
}
```

Fig. 9. The structure of kernel function written in pseudocode.

only the lattice structures for the general case of DCWTs. It is clear that involutory DCWTs can be realized with aid of the same structures if only the proper dependencies regarding the coefficients would be taken into account. Hence, the values of the coefficients of butterfly operators can be always stored in the convenient way in global or texture memory depending on the variant of the experiment. The results obtained with G200 microarchitecture for the following variants of experiments, i.e.: (I) lattice structure of type I, (II) lattice structure of type I using texture memory, (III) lattice structure of type II, (IV) lattice structure of type II with texture memory and (V) lattice structure of type II using texture memory and data mixing are collected in Tables 1 and 2. The representative results obtained for Maxwell microarchitecture can be found in Table 3.

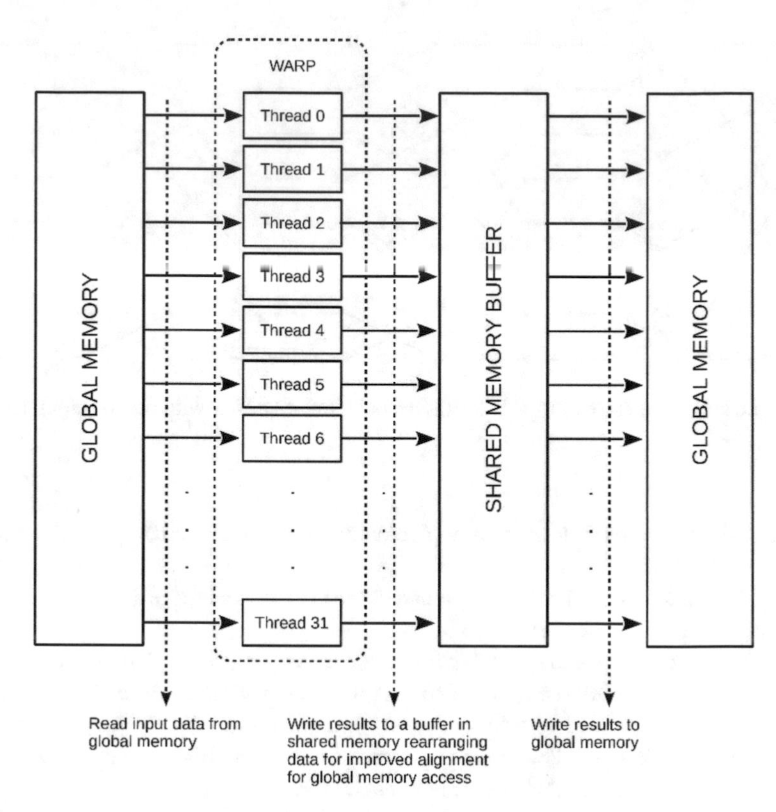

Fig. 10. The applied scheme of data rearrangement using shared memory buffer for better alignment of global memory accesses within warps of threads.

Table 1. The experimental results for DCWT calculated with lattice structure of the first type for G200 microarchitecture.

Basic implementation						
n	**10**	**11**	**12**	**13**	**14**	**15**
t [μs]	56	99	208	340	646	$1.2 \cdot 10^3$
n	**16**	**17**	**18**	**19**	**20**	**21**
t [μs]	$2.6 \cdot 10^3$	$5.3 \cdot 10^3$	$11 \cdot 10^3$	$23 \cdot 10^3$	$49 \cdot 10^3$	$102 \cdot 10^3$
Implementation exploiting texture memory						
n	**10**	**11**	**12**	**13**	**14**	**15**
t [μs]	53	90	198	345	635	$1.1 \cdot 10^3$
n	**16**	**17**	**18**	**19**	**20**	**21**
t [μs]	$2.5 \cdot 10^3$	$5.1 \cdot 10^3$	$10 \cdot 10^3$	$22 \cdot 10^3$	$47 \cdot 10^3$	$98 \cdot 10^3$

Table 2. The experimental results for DCWT calculated with lattice structure of the second type for G200 microarchitecture.

Basic implementation						
n	**10**	**11**	**12**	**13**	**14**	**15**
t [µs]	27	61	114	263	498	994
n	**16**	**17**	**18**	**19**	**20**	**21**
t [µs]	$2.0{\cdot}10^3$	$4.2{\cdot}10^3$	$8.9{\cdot}10^3$	$18{\cdot}10^3$	$39{\cdot}10^3$	$82{\cdot}10^3$
Implementation exploiting texture memory						
n	**10**	**11**	**12**	**13**	**14**	**15**
t [µs]	25	54	97	226	419	820
n	**16**	**17**	**18**	**19**	**20**	**21**
t [µs]	$1.7{\cdot}10^3$	$3.5{\cdot}10^3$	$7.3{\cdot}10^3$	$15{\cdot}10^3$	$32{\cdot}10^3$	$67{\cdot}10^3$
Impl. using texture memory & additional mixing within warps						
n	**10**	**11**	**12**	**13**	**14**	**15**
t [µs]	25	52	90	204	360	681
n	**16**	**17**	**18**	**19**	**20**	**21**
t [µs]	$1.3{\cdot}10^3$	$2.8{\cdot}10^3$	$5.8{\cdot}10^3$	$12{\cdot}10^3$	$25{\cdot}10^3$	$54{\cdot}10^3$

Table 3. The representative results of DCWT calculation for Maxwell microarchitecture.

Representative results						
n	**10**	**11**	**12**	**13**	**14**	**15**
t [µs]	15	17	21	26	35	61
n	**16**	**17**	**18**	**19**	**20**	**21**
t [µs]	97	202	672	1417	3000	6257

An analysis of results collected in Tables 1 and 2 shows that the proper alignment of data is of key importance in the case of older NVIDA GPU microarchitectures. By comparing the results obtained with "type I & texture memory" and "type II & texture memory & data mixing" variants of the structures we would see almost 45% decrease of the time of calculations obtained with the structure of type II. It should be noted that all of the considered improvements gave positive effects. For example an application of the texture memory alone allowed to gain 4% to 18% acceleration. It gives an immediate answer to the first question.

In the case of modern Maxwell architecture the situation looks radically different. Here, in the experiments involving all of the considered variants of lattice structures the obtained results were identical if we only neglect the typical and minor deviations resulting from the different initial states of the computational machine. This is the reason why only one set of the representative results is given in Table 3. Hence, the answer to the second question is: in the case of the considered variants of calculations involving typical lattice structures for the fast computation of Fourier-like transforms the way of data alignment is of marginal practical importance for modern GPU microarchitectures.

6 Summary

In this chapter a novel scheme for constructing fast orthogonal parametric transforms of cosine-Walsh type with an additional involution property is proposed. The involution property means that the inverse transform is obtained simply as: $U^T = U$, which is a desirable feature especially in the case of hardware implementations, where the same electronic circuit can be used for calculation of both transforms, i.e. forward and inverse transformations. In order to obtain involutory transforms in the form of lattice structure of the fast cosine-Walsh type transform the necessary dependencies between transform parameters are formulated and given in the chapter. Further on the proposed class of involutory transforms is tested experimentally in terms of their applicability to data encryption and, moreover, some aspects of their mass-parallel realizations with aid of GPUs are considered and verified experimentally. The obtained experimental results reveal good data concealment properties of the proposed class of transforms which means that such transformations can be used successfully in the practical tasks, e.g. in joint compression and encryption of visual data (see [17]). Moreover the practical tests on mass-parallel realizations of the proposed transforms involving several variants of lattice structures allow to draw the following important conclusions: (I) the proper care of data alignment is crucial in the case of older GPU microarchitectures, (II) modern microarchitectures deal with the problem of data misalignment in the measure which in the case of performed type of calculations did not allow to notice any essential differences.

References

1. Harmuth, H.F.: Transmission of Information by Orthogonal Functions. Springer, Heidelberg (1972)
2. Ahmed, N., Rao, K.R.: Orthogonal Transforms for Digital Signal Processing. Springer, Heidelberg (1975)
3. Fino, B.J., Algazi, V.R.: Slant Haar transform. Proc. IEEE **62**, 653–654 (1974)
4. Cooley, J.W., Tukey, J.W.: An algorithm for the machine calculation of complex Fourier series. Math. Comput. **19**(90), 297–301 (1965)
5. Hartley, R.V.L.: A more symmetrical Fourier analysis applied to transmission problems. Proc. IRE. **30**(3), 144–150 (1942)
6. Rao, K., Yip, P.: Discrete Cosine Transform: Algorithms, Advantages, Applications. Academic Press, New York (1990)
7. Morháč, M., Matoušek, V.: New adaptive cosine-Walsh transform and its application to nuclear data compression. IEEE Trans. Sign. Proc. **48**(9), 2693–2696 (2000)
8. Agaian, S., Tourshan, K., Noonan, J.P.: Parametric slant-Hadamard transforms with applications. IEEE Sign. Process. Lett. **9**(11), 375–377 (2002)
9. Agaian, S., Tourshan, K., Noonan, J.P.: Parameterisation of slant-Haar transforms. IEE Proc. Vis. Image Sign. Process. **150**(5), 306–311 (2003)
10. Bouguezel, S., Ahmad, O., Swamy, M.N.S.: A new class of reciprocal-orthogonal parametric transforms. IEEE Trans. Circ. Syst. I Regul. Pap. **56**(4), 795–805 (2009)

11. Bouguezel, S., Ahmad, O., Swamy, M.N.S.: Image encryption using the reciprocal-orthogonal parametric transforms. In: IEEE Symposium on Circuits and Systems, pp. 2542–2545 (2010)
12. Bouguezel, S., Ahmad, O., Swamy, M.N.S.: New parametric discrete Fourier and Hartley transforms, and algorithms for fast computation. IEEE Trans. Circ. Syst. I Regul. Pap. **58**(3), 562–575 (2011)
13. Bouguezel, S., Ahmad, O., Swamy, M.N.S.: A new involutory parametric transform and its application to image encryption. In: IEEE Symposium on Circuits and System, pp. 2605–2608 (2013)
14. Puchala, D., Yatsymirskyy, M.M.: Fast parametrized biorthogonal transforms. Electr. Rev. **4**, 123–125 (2012)
15. Bhargava, B., Shi, C., Wang, S.Y.: MPEG video encryption algorithms. Multimedia Tools Appl. **24**(1), 57–79 (2004)
16. Puchala, D., Yatsymirskyy, M.M.: Fast parametrized biorthogonal transforms with normalized basis vectors. Electr. Rev. R. **89**(3a), 277–300 (2013)
17. Puchala, D., Yatsymirskyy, M.M.: Joint compression and encryption of visual data using orthogonal parametric transforms. Bull. Polish Acad. Sci. Tech. Sci. **62**(2), 373–382 (2016)
18. Tang, L.: Methods for encrypting and decrypting MPEG video data efficiently. In: ACM Multimedia, pp. 219–229 (1996)
19. Puchala, D., Stokfiszewski, K., Szczepaniak, B., Yatsymirskyy, M.M.: Effectiveness of fast Fourier transform implementations on GPU and CPU. Electr. Rev. **R.92**(7), 69–71 (2016)
20. Yildrin, A., Ozdogan, C.: Parallel wavelet-based clustering algorithm on GPUs using CUDA. Procedia Comput. Sci. **3**, 296–400 (2011)
21. Minasyan, S., Astola, J., Guevorkian, D.: On unified architectures for synthesizing and implementation of fast parametric transforms. In: 5th International Conference, Information Communication and Signal Processing, pp. 710–714 (2005)
22. Puchala, D., Stokfiszewski, K.: Parametrized orthogonal transforms for data encryption. In: Computational Problems of Electrical Engineering, no. 2 (2012)
23. Puchala, D.: Comparison of fast orthogonal parametric transforms in data encryption. J. Appl. Comput. Sci. **23**(2), 55–68 (2015)
24. Moreland, K., Angel, E.: The FFT on a GPU. In: Proceedings of ACM SIGGRAPH/EUROGRAPHICS Conference on Graphics Hardware, pp. 112–119 (2003)
25. Spitzer, J.: Implementing a GPU-efficient FFT. In: SIGGRAPH Course on Interactive Geometric and Scientific Computations with Graphics Hardware (2003)
26. Yatsymirskyy, M.M.: Fast algorithms for orthogonal trigonometric transforms computations. Lviv Academic Express (1997). (in Ukrainian)
27. NVIDIA's Next Generation CUDATM Compute Architecture: FermiTM, NVIDIA. Whitepaper (2009)

On the Asymptotic Methods of the Mathematical Models of Strongly Nonlinear Physical Systems

Petro Pukach[1], Volodymyr Il'kiv[1], Zinovii Nytrebych[1], Myroslava Vovk[1](✉), and Pavlo Pukach[2]

[1] Department of Mathematics,
Lviv Polytechnic National University, Lviv, Ukraine
ppukach@gmail.com, ilkivv@i.ua, znytrebych@gmail.com,
mira.i.kopych@gmail.com
[2] Department of Applied Mathematics,
Lviv Polytechnic National University, Lviv, Ukraine
pavlopukach@gmail.com

Abstract. The approximation methods to investigate the dynamical processes in the strong nonlinear single- or multi-degree physical systems are studied in the chapter. The developed procedure of the investigation of the oscillations in the nonlinear systems with the concentrated mass can solve not only the analysis problems but the problems of the physical oscillation system synthesis on the projective phase choosing such system characteristics enabled the resonance phenomena as well. There is described the real application of the obtained results to analyze the technical systems using to protect the equipment from the vibration load.

Keywords: Mathematical model · Approximation method · Physical system Resonance · Dynamical process

1 Introduction

The mathematical modeling is the important tool to solve the most of the scientific and engineering problems. It also helps to understand the kernel of the processes that are investigating in the modern physics, to interpret the known phenomena and to predict the new ones.

The analysis and optimization of the physical, chemical dynamical processes in the technical systems in the cases of high speed, high pressure and energy demands the investigation of the complicated processes in the nonlinear oscillation theory.

The strong nonlinear oscillation physical systems are widely applied in the engineering. They are well studied in the case while the ordinary linear or with the partial derivatives differential equations are used as the mathematical models.

In the first case these equations describe the processes in the systems with the concentrated masses, in the second one in the system with the distributed parameters.

N. Shakhovska and V. Stepashko (eds.), *Advances in Intelligent Systems and Computing II*, Advances in Intelligent Systems and Computing 689,
https://doi.org/10.1007/978-3-319-70581-1_30

Moreover, for the linear computing models of the oscillations physical systems basing on the superposition principle are developed the analytical methods in virtue of which one can state the reaction on the one- or multi-frequency periodical perturbations. Quasilinear systems i.e. systems with the small maximum values of the nonlinear forces comparing to the linear component of reducing force are similar to the linear systems. The analytical investigating methods for the most of the classes of the quasilinear systems are based on the different modifications of the perturbation methods in particular asymptotic methods of the nonlinear mechanics. Therefore is possible to describe and to state the special attributes of the oscillations in the quasilinear systems that are not proper to the linear systems with the constant in time physical-mechanical characteristics. One can speak firstly about resonance phenomena [1, 2] firstly (main and combinational), process stability [3–5] secondly.

As to the more complicated systems, namely the strong nonlinear systems one can use the analytical investigating methods only in some cases. The nonlinear systems with the reducing force described by the power or close to power nonlinearity can be treated to this case [6–8].

In reference to non-autonomous strong nonlinear systems, i.e. the systems under the external periodic perturbation (even small on value), it is necessary to realize the additional investigation. Their characteristics can be described solely. The general procedure of the investigating some important classes of the nonlinear physical systems with strong nonlinearity, in particular, the procedure to study the resonance phenomena is developed. The restoring forces can't be linearized in such type systems. Single-degree and multi-degree systems are considered. They are studied under perturbation. The perturbation is such that: it can be described by the analytical periodic on time variable functions; and the maximum perturbation value is small comparing to the maximum value of the reducing force. Concerning to the unperturbed systems, the reducing force is described by close to the power dependence and ensures the periodic processes. Such constraint may confine the classes of the physical (mechanic) systems that can be properly analyzed by this procedure. However, for the most cases the nonlinear reducing forces of the autonomous type with the sufficient exactness degree can be approximated by the discussed dependence, thus the procedure can be successfully used for the wide class of the nonlinear systems. This approach is the feasible one (numerical-analytical) to solve the stated problem.

The main issue of the developed in the chapter procedure of the investigating the resonance phenomena in the discussed class of the systems is based on idea to use the special periodic Ateb-functions constructing the solutions of the unperturbed analogue systems and to adopt the general ideas of the perturbation methods – in the case of the non-autonomous systems.

2 The Mathematical Models of the Strong Nonlinear Single-Degree Physical Systems

In this section there is described the procedure of the investigation of the oscillations in the systems with one degree freedom while the reducing force is approximated close to the power law. In this case the differential equation that describes such oscillations can be written as

$$m\ddot{x} + cx^{\nu+1} = \varepsilon f(x, \dot{x}, \theta), \tag{1}$$

where m is the mass of the material point; x is its coordinate at the arbitrary time moment; $F(x) = cx^{\nu+1}$ is function, describing the main part of the reducing force; c, ν are constants, provided $\nu + 1 = \frac{2p+1}{2q+1}$ $(p, q = 0, 1, 2, \ldots)$; $f(x, \dot{x}, \theta)$ is analytical 2π – periodic on $\theta = \mu t$ function, μ is frequency of the external periodic perturbation acting on the system; ε is the small parameter indicating on the small value of the external periodic perturbation and inconsiderable deviation of the restoring force from the power law. The restrictions concerning the parameter value ν are caused by the symmetry with respect to the coordinate origin of the function $F(x) = cx^{\nu+1}$, describing the restoring force. More general representation of the reducing nonlinear force in the form $F(x) = c|x|^p x$ always with the necessary exactness power can be approximated by the upper written dependence, thereby all obtained results, one can transfer also on the systems with such type restoring force. Let's notice, that function $\varepsilon f(x, \dot{x}, \theta)$ can describe the nonlinear forces with the maximum value being the small quantity comparing to the maximum value of the restoring force $F(x) = cx^{\nu+1}$, that is $F(x) \gg \max \varepsilon f(x, \dot{x}, \theta)$. The last inequality can use the general ideas of the perturbation methods to construct the solution of the nonlinear Eq. (1). As per usual, it is necessary to describe the dynamical process of the unperturbed ($\varepsilon = 0$) or generating system, corresponding to (1). The next nonlinear equation is its analogue

$$m\ddot{x} + cx^{\nu+1} = 0. \tag{2}$$

In spite of the simplicity of the solution form, and the fact that the amplitude and the frequency of vibration are exactly obtained, difference between the approximate solution and the exact numerical one may be significant. It was the reason that the exact solution of (2) in the form of cosine Ateb periodic function being an inversion of the incomplete Beta function [9] is introduced. The Ateb-functions are more complex than the trigonometric functions and, at the same moment, are not familiar to engineers and technicians who have to use them [10–14]. Usually, simple approximation of Ateb-functions by elementary functions is proposed, in particular, in [9–14].

The solution of the Eq. (2) is represented via periodic Ateb-functions as

$$x = aca(\nu + 1, 1, \omega(t) + \psi), \tag{3}$$

Where a is the amplitude, $\omega(t) + \psi$ is the oscillation phase of the unperturbed motion, $\omega(a)$ is the oscillation frequency equals

$$\omega(a) = \sqrt{\frac{c(\nu+2)}{2m}} a^{\frac{\nu}{2}}. \tag{4}$$

Hence, even for the unperturbed motion natural frequency of the dynamical process depends on amplitude. This is the first conceptual difference between the strong nonlinear and quasilinear systems. Thus, in case of the unperturbed motion natural frequency $\omega(a)$ depends on the mass of the material body and on the proportionality constant in the restoring force (analogue constant inelasticity), and on the amplitude and the nonlinear parameter ν also. This is the main difference of the considered class of the systems from the linear one. The issue is to construct the solutions of such strong nonlinear systems, taking into account all upper dynamical peculiarities, and to answer on the main questions of the investigating of the resonant oscillations.

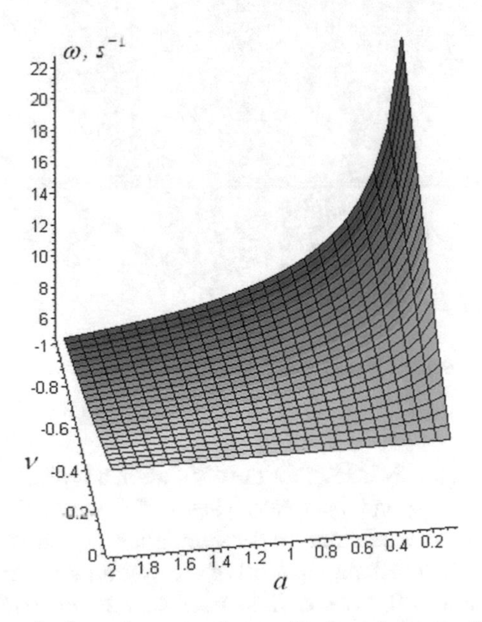

Fig. 1. Natural frequency's dependence on the amplitude and the small nonlinear parameter ν

Figure 1 demonstrates the dependence of the frequency $\omega(a)$ on the parameters defining the dynamical process of the unperturbed motion, Fig. 2 demonstrates the ratio of the periods of natural nonlinear to the linear $\nu = 0$ oscillations on the amplitude and on the parameter ν, i.e. $\eta = \frac{\Pi(1,\nu+1)}{\omega(a)}$. The parameters a and ψ for the unperturbed motion are constants and are defined from the initial conditions $x(0) = x_0$, satisfying the relationships

$$\left(\frac{x_0}{a}\right)^{\nu+2} + \left(\frac{V_0(\nu+2)}{2a\omega(a)}\right)^2 = 1,$$

$$cta(\nu+1, 1, \psi) = \frac{2\omega(a)}{\nu+2}.$$

Figures 1 and 2 demonstrate:

(1) for the systems with small stiffness $-1 < \nu < 0$ the oscillation natural frequency of the physical system for the greater amplitude's values is less; (2) while the nonlinear parameter ν converges to zero process in the system is considered as similar to the isochronous; (3) for the systems of the considerable stiffness $\nu \gg 0$ and for the oscillation amplitude values less than 1 for the greater values of nonlinear parameter

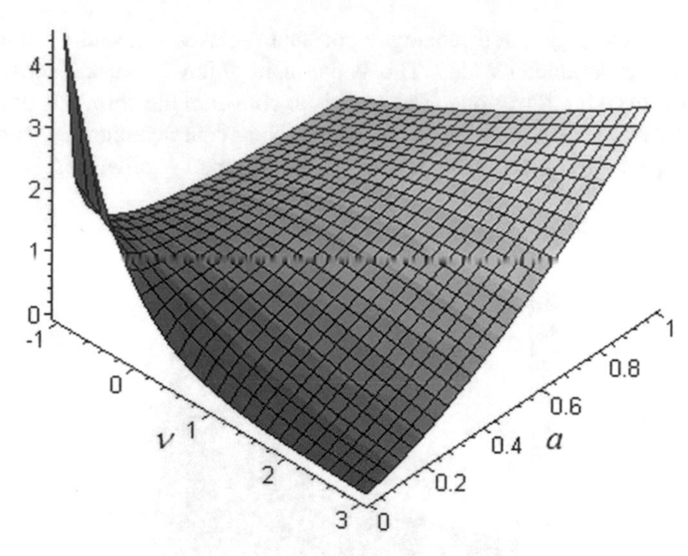

Fig. 2. The ratio of the oscillation period of the nonlinear model to the oscillation period of the linear model

v oscillation natural frequency is less; for the oscillation amplitude values greater than 1 for its greater values correspond the greater values of the natural frequencies; (4) at the equal values of the parameters c and m the oscillation period of the strong nonlinear system is greater than the oscillation period of the system's linear analogue $-1<v<0$ and $a<1$ and less at $v>0$ i $a<1$. It is necessary to take into account all these peculiarities in the resonance phenomena investigation.

3 Resonance Phenomena in the Strong Nonlinear Physical Systems with the Discrete Structure

To develop the investigating methods for the reaction of the strong nonlinear system on the periodic non-autonomous type perturbations it is necessary to study the resonance conditions considering Eq. (2) as the model. The classic definition of the resonance phenomena can be applied for the considered systems noting the difference, that concept "frequency" treating as the meaning of the function $\omega(a)$, is wider, than in the quasilinear systems. At first, the oscillation natural frequency of the system is defined not only by the physical and mechanical characteristics (parameters m, c), but also by the amplitude a; secondly, the periodic process of the unperturbed motion is described by the periodic Ateb-functions. The period 2Π for the last ones by the argument $\psi = \omega(a)t$ depends on the nonlinear parameter v and is defined by the relationship

$$\Pi = \Pi(1, v+1) = \frac{\sqrt{\pi}\Gamma\left(\frac{1}{v+2}\right)}{\Gamma\left(\frac{1}{2} + \frac{1}{v+2}\right)}. \tag{5}$$

On account of this for the considered system class one can accept the derivative from the classic meaning instead of the main resonance, in other words – the natural oscillation period coincides with the perturbation period. Then the resonance condition transforms as

$$\frac{\pi}{\mu} = \frac{\Pi(1, \nu + 1)}{\omega(a)}. \tag{6}$$

Hereafter the parameter value μ means the frequency of the external periodic perturbation (force). Taking into account the function, describing "natural frequency", one can obtain the parameter value a^*, responsible for the resonance phenomena:

$$a^* = \left(\frac{\Pi(1, \nu + 1)}{\pi} \frac{\mu}{k} \sqrt{\frac{2}{\nu + 2}} \right)^{\frac{2}{\nu}}. \tag{7}$$

where $k = \sqrt{\frac{c}{m}}$. Thus, the resonance phenomena in the investigated dynamical systems appears when the oscillation amplitude is close to a^*. If the amplitude is greater than a^*, then existing resistant forces in the real systems produce the decrease of the oscillation amplitude to the value a^*, causing the resonance phenomena in the future.

This explains the second principle difference of the resonance phenomena in the strong nonlinear systems. If there exist one or several stationary oscillation modes in the perturbed strong nonlinear system, then the investigation of the physical system reaction on the periodic perturbation must be considered separately. Figure 3 demonstrates the dependence of the resonance amplitude on the parameters, that characterized the physical and mechanical system properties and forced periodic perturbation frequency.

In particular, from the obtained in the chapter plotted dependences follows, that for the «soft» systems ($-1 < \nu < 0$) the resonance phenomena is possible in the case, when the natural frequency oscillation of the system linear analogue is less than the frequency of the perturbing force; while the parameter ν tends to -1 the resonance amplitude a^* extremely increases. In the case of the «rigid» systems ($\nu > 0$) the resonance phenomena is possible when the frequency of the linear analogue of the system is greater than the frequency of the perturbing force; for the greater values of the nonlinear system parameter and the frequency of the perturbing force the resonance amplitude is greater. Moreover, the results of the chapter allow to state: if in the physical system there are oscillations with the amplitude less than a^*, then the periodic forces of the limited value and arbitrary frequency can't cause the resonance oscillations value and arbitrary frequency can't cause the resonance oscillations in it. Simultaneously it is stated, that for the greater stiffness system the amplitude threshold value is greater. Under the condition that the frequency of the external periodic perturbation $\mu = 8\ \mathrm{s}^{-1}$ the increase of the nonlinear measure ν from $0,4$ to $0,8$ produces the increase of the amplitude threshold from $0,012$ m to $0,04$ m. For the greater frequency values of the perturbation force (with all constant parameters) the amplitude threshold value is greater. While the frequency of the external periodic perturbation

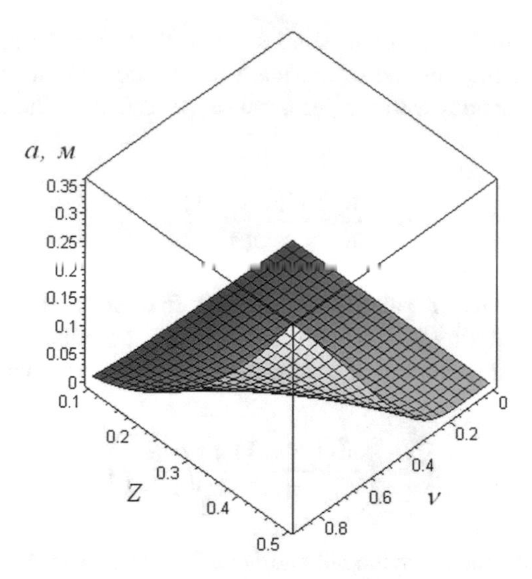

Fig. 3. The dependence of the resonance amplitude on parameters $\nu > 0$ and $Z = \frac{\mu}{k}$

$\mu = 8\ \mathrm{s}^{-1}$ and $\nu = \frac{2}{3}$ the amplitude threshold value equals $0,04$ m, while $\mu = 15\ \mathrm{s}^{-1}$ and $\nu = \frac{2}{3}$ equals $0,1$ m.

4 The Mathematical Models and Resonance Phenomena in Strong Nonlinear Physical Oscillation Multi-degree Systems

The analytical investigation of the dynamical processes in the strong nonlinear multi-degree systems is more complicated, than for their quasilinear analogues or even for the strong nonlinear single-degree systems. It is explained by the fact that the principle of the oscillation superposition can't be realized, hence the analysis of the dynamical process based on the partial solutions or on the application of the numerical methods of the integration of the corresponding differential equations. Besides, the main properties of the generating system and perturbations must be taken into account independently from the investigating methods. As examples of the strong nonlinear systems to applicate the analytical methods for the construction of the solutions corresponding mathematical models can be the systems where the potential energy is defined as follows

$$P(x_1, x_2, \ldots, x_n) = \sum_{i,j=0}^{n} c_{ij}\left(x_i - x_j\right)^{\nu+2}. \tag{8}$$

In the relationship (8) $x_1, x_2, \ldots, x_n$ are the generalized coordinates, c_{ij}, ν are constants, and ν is defined in the same form as upper. Thus, the potential energy in the both cases is the homogenous function on the generalized coordinates. Taking into account the fact, that the kinetic energy as the function of the generalized speeds, is defined by the relationship

$$T(\dot{x}_1, \dot{x}_2, \ldots, \dot{x}_n) = \frac{1}{2}\sum_{i=1}^{n} m_i \dot{x}_i^2, \tag{9}$$

where m_i is the mass of i point, the differential equations describing the dynamical process of the unperturbed motion are the next

$$m_i \ddot{x}_i + (\nu + 1)\sum_{j=0}^{n} c_{ij}\left(x_i - x_j\right)^{\nu+1} = 0. \tag{10}$$

In spite of the strong nonlinearity of the Eq. (10), the dynamical process in the corresponding physical system can be represented via the periodic Ateb-functions. That's why formally suppose, that the phase coordinates are connected by the relations

$$x_i = b_i x_1, \quad i = 2, 3, \ldots, n, \tag{11}$$

where b_i are the unknown constants. The physical explanation of the substitution (11) is the next: in the strong nonlinear system there are oscillations that coincide by the form with the first generalized coordinate.

After the substitution of the variables (11) to describe the "normal" forms of the oscillations one can get the differential equations

$$m_i \ddot{x}_i + (\nu + 1)\sum_{j=0}^{n} c_{ij}\left(b_i - b_j\right)^{\nu+1} = 0,\ i = 2, 3, \ldots, n, \tag{12}$$

where the unknown coefficients b_i are defined from the system of the algebraic equations

$$b_i \sum_{j=0}^{n} c_{ij}\left(1 - b_j\right)^{\nu+1} = \sum_{j=0}^{n} c_{ij}\left(b_i - b_j\right)^{\nu+1}. \tag{13}$$

The solutions of the differential equations (12) can be written via the periodic Ateb-functions in the form

$$x_1 = a ca(\nu + 1, 1, \omega(a)t + \theta),$$

$$x_i = a b_i ca(\nu, 1, \omega(a)t + \theta),\ i = 2, 3, \ldots, n.$$

At the same time the unknown coefficients b_i are related by the system of the nonlinear algebraic equations (13).

5 The Examples of the Application of the Developed Methods for the Investigation of the Mathematical Models of the Strong Nonlinear Vibro-Protection Systems. The Numerical Simulation

A. The resonance phenomena in the quasi-zero stiffness vibro-protection systems, that can be modeled by the ordinary differential equations

In the last decade there is very popular antivibrating procedure, using the systems, known as quasi-zero stiffness vibro-protection systems [15]. In spite of the wide spectrum of such type systems, their mathematical models are described by the ordinary one-type differential equations with strong nonlinearity, exactly

$$m\ddot{x} + \alpha\dot{x} + \beta x + \gamma x^3 = f(t). \tag{14}$$

In the Eq. (14):

- m is the mass of the body under the action of the vibrating load, caused by the motion of the base in accordance with the law $g(t)$. The function $f(t)$ is $f(t) = -m\ddot{g}(t)$;
- α is the constant of proportionality in the resistant force, is taken for the sake of simplicity proportional to the motion speed of the body;
- β, γ are constants defined by stiffness of the system [8];
- the function $g(t)$ and $f(t)$ too, are periodic on time and the maximum value of the periodic perturbation is small comparing to the restoring force, i.e.

$$\max f(t) \ll \max\left(\beta x + \gamma x^3\right).$$

Let note, that the constant γ can be the positive or negative, and can be equal to zero also. The numerical integration and the respective analysis of the results based on it was realized for some cases of the considered equation in [16]. To investigate the dynamical processes of the considered vibro-protection system is effectually to use the results of the Sects. 2 and 3 in this chapter. In accordance with them the first-order approximation of the non-resonant process is described by the relationship

$$x(t) = aca\left(3, 1, \sqrt{\frac{2\beta}{m}}at + \psi\right),$$

where the parameters a and ψ are determined from the system:

$$\dot{a} = -\frac{2\alpha}{m\Pi(1,3)}\frac{\Gamma\left(\frac{3}{2}\right)\Gamma\left(\frac{1}{4}\right)}{\Gamma\left(\frac{7}{4}\right)}, \quad \dot{\psi} = \frac{\gamma}{\sqrt{2m\beta}}\frac{\Gamma\left(\frac{3}{4}\right)}{\sqrt{\pi}\Gamma\left(\frac{1}{4}\right)}\frac{\sqrt{\pi}\Gamma\left(\frac{3}{4}\right)}{\Gamma\left(\frac{5}{4}\right)}a. \tag{15}$$

The resonance oscillations on the frequency μ appear when the amplitude is close to $a^* = 0.425\mu\sqrt{\frac{m}{\beta}}$ and can be described by some system of the differential equations and numerically found.

B. The resonance phenomena in the quasi-zero stiffness vibro-protection systems modeled by the system of two ordinary differential equations

To protect the equipment from different type vibrations one can apply so-called two-degree quasi-zero stiffness systems, their mathematical models are represented as the differential equation system [15] in the form

$$\begin{aligned} \ddot{x}_1 + c_{11}x_1^3 + c_{12}(x_1 - x_2)^3 &= c_{13}x_1 - \alpha_1\dot{x}_1, \\ \ddot{x}_2 + c_{21}x_1^3 + c_{22}(x_1 - x_2)^3 &= c_{23}x_1 - \alpha_2\dot{x}_2. \end{aligned} \tag{16}$$

Let's consider the case, when the both bodies are of the same mass, and the resistant force and the previous deformations of the elastic elements are small. It possible to state, that the maximum values in the right sides of the Eq. (16) are small, and therefore for their investigation one can use the general ideas of the perturbation procedure. So, firstly let's describe the dynamical process of the generating systems as follows

$$\begin{aligned} \ddot{x}_1 + c_{11}x_1^3 + c_{12}(x_1 - x_2)^3 &= 0, \\ \ddot{x}_2 + c_{21}x_1^3 + c_{22}(x_1 - x_2)^3 &= 0. \end{aligned} \tag{17}$$

The aim is to find the solutions (17) in the form

$$\begin{aligned} x_1 &= aca(3, 1, \omega(a)t + \theta), \\ x_2 &= abca(3, 1, \omega(a)t + \theta). \end{aligned} \tag{18}$$

To find the unknown parameter b using the system (17), considering (18), let write the next algebraic equation of the type (13)

$$b^4 + \delta_1 b^3 - \delta_2 b - 1 = 0, \tag{19}$$

where δ_1, δ_2 are known constants expressed via the coefficients of the system (17).

Using (18), one can get the solutions (17).

After defining the normal oscillations forms of the unperturbed system (17), further is proposed the first-order approximation of the perturbed equations system (16). Accordingly to the general method of perturbation theory for the nonlinear oscillation systems, the relationship (18) also can be accepted as the solution of the system (16), where the parameters a and θ are the functions on time. To find them there is the next differential equation system

$$\dot{a} = \frac{\varepsilon a}{2\pi\omega(a)} \int_0^{2\pi} k(1 + b)ca(3, 1, l\psi)sa(1, 3, l\psi)d\psi = 0,$$

$$\dot{\theta} = \frac{\varepsilon a}{2\pi\omega(a)} \int_0^{2\pi} k(1+b)ca^2(3,1,l\psi)d\psi = \frac{\varepsilon 0,457}{\omega(a)} k(1+b),$$

where $k = k_1 = k_2$, $l = \frac{\Gamma(\frac{1}{4})}{\sqrt{\pi}\Gamma(\frac{3}{4})} \approx 1.67$. The amplitudes of the normal modes of oscillations in the considered system in the first approximation of the asymptotic decomposition stay the constants because of conservativity of the system. The frequencies of the perturbed oscillations Ω_s $(s = 1,2,3,4)$ depend on the amplitude and are determined by the relationship $\Omega_s = \omega(a) + \frac{\varepsilon \cdot 0.457}{\omega(a)} k(1+b_s)$, where

$$\omega(a) = \sqrt{\frac{\pi}{2}\left(\delta_3 + \delta_4(1+b_s)^3\right)}a,$$

δ_3, δ_4 are known constants, and b_s are defined according to (19).

Figure 4 demonstrates the dependence of the period of the normal form oscillations on the amplitude at the different values of the system coefficients (16).

From the obtained graphs it follows:

(1) while the amplitude increases the period of all normal forms oscillations of the considered conservative system decreases;
(2) at the same amplitude values the oscillation period is the maximum for the case of the motion in one phase.

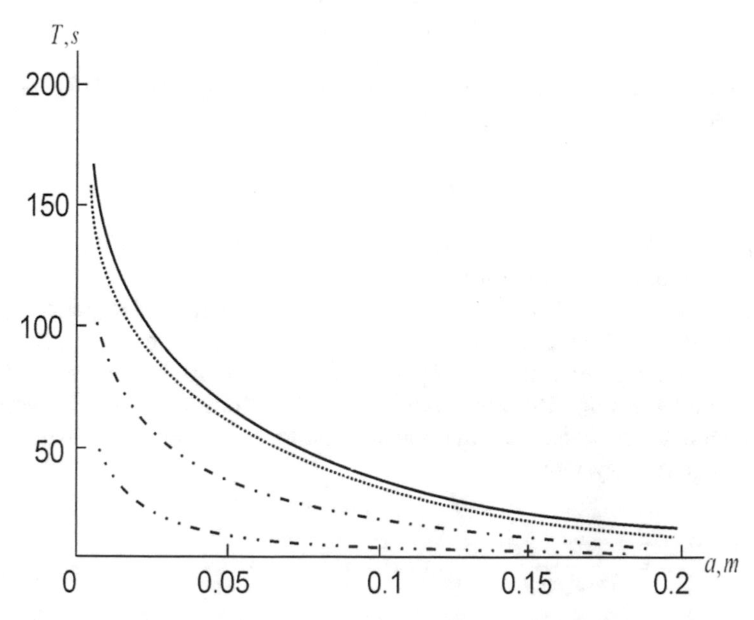

Fig. 4. The graphs of the dependence of the normal form oscillation periods on the amplitude

Figure 4 demonstrates the dependence of the period of the normal form oscillations on the amplitude at the different values of the system coefficients (16).

From the obtained graphs it follows:

(1) while the amplitude increases the period of all normal forms oscillations of the considered conservative system decreases;
(2) at the same amplitude values the oscillation period is the maximum for the case of the motion in one phase.

6 Conclusions

There is developed the procedure to investigate the dynamical processes in the strong nonlinear oscillation single- and multi-degree physical systems. The attribute of the considered systems is the next:

- the oscillation process of the unperturbed analogues can be described via the special periodic Ateb-functions;
- the frequency (period) of this process depends on the amplitude.

The main difficulty of the investigation of the periodic forces action on the process is tied with the last fact. That is why the asymptotic integration procedure (solution construction) of the according strong nonlinear the second-order differential equations of the non-autonomous type is developed. The standard equations describing the changes laws of determining parameters of the dynamical process are obtained for the both non-resonant and resonance cases based on the developed procedure. The next results are obtained for the resonance case.

1. There are built the domains of the resonant (non-resonant) oscillations of the strong nonlinear systems at the arbitrary frequency value of the periodic perturbation.
2. There is stated:
   - in the case, when the resonance amplitude threshold is greater than its initial value, but less than stationary amplitude value, then the system during some time period would be in the resonance zone. The resonance process is very short-time and after it the dynamical process approximates to the stationary dynamical oscillation mode;
   - in the case, when the resonance amplitude threshold is less than its initial value, but greater or equal than stationary amplitude value, then the system during some time period would be in the resonance zone. The resonance process is very short-time After the exit from the resonance zone, due to the strong stability of the stationary oscillation mode once more starts the resonance phenomena. The process repeats until the periodic perturbation is acting.
3. There is developed the procedure of the investigation oscillations of the so-called systems with quasi-zero stiffness, used to protect the equipment from the vibration load. There are obtained the resonance conditions for such type systems and also are described the laws of the resonant and non-resonant amplitude changes.

The proposed method of the investigating the oscillation processes of the strong nonlinear physical discrete structure systems helps to solve not only the analysis problems, but also the important problems of the oscillation systems synthesis on the projective phase, and to choose such dynamical systems characteristics that make the resonance phenomena impossible.

References

1. Bogolyubov, N.N., Mitropol'skii, Y.A.: Asymptotic Methods in the Theory of Nonlinear Oscillations [Asimptoticheskiye metody v teorii nelineynykh kolebaniy]. Nauka, Moscow (1974). (in Russian)
2. Grebennikov, E.A., Ryabov, Y.A.: Constructive Methods for the Analysis of Nonlinear Systems [Konstruktivnyye metody analiza nelineynykh sistem]. Nauka, Moscow (1979). (in Russian)
3. Bellman, R.: A Theory of the Stability of Solutions of Differential Equations [Teoriya ustoychivosti resheniy differentsial'nykh uravneniy]. Foreign Literature Publishing House, Moscow (1954). (in Russian)
4. Perestyuk, M.O., Chernihiv, O.S.: Theory of the Stability [Teoriya stiykosti]. Publishing House "Kyiv", Kyiv (2002). (in Ukrainian)
5. Kyrychenko, Y.O., Samusya, V.I., Kyrychenko, V.Y., Romanyukov, A.V.: Experimental investigation of aero-hydroelastic instability parameters of the deepwater hydrohoist pipeline. Middle East J. Sci. Res. **18**(4), 530–534 (2013)
6. Nayfeh, A.H.: Perturbation Methods [Metody vozmushcheniy]. Mir, Moscow (1976). (in Russian)
7. Pukach, P.Y., Kuzio, I.V.: Nonlinear transverse vibrations of semiinfinite cable with consideration paid to resistance [Nelíníyní poperechní kolivannya napívneobmezhenogo kanata z urakhuvannyam oporu]. Naukovyi Visnyk Natsionalnoho Hirnychoho Universitetu, No. 3, pp. 82–86 (2013). (in Ukrainian)
8. Pukach, P.Y.: Investigation of bending vibrations in Voigt-Kelvin bars with regard for nonlinear resistance forces. J. Math. Sci. **215**(1), 71–78 (2016)
9. Cveticanin, L., Pogány, T.: Oscillator with a sum of noninteger-order nonlinearities. J. Appl. Math. **2012**, 1–20 (2012)
10. Andrianov, I., Awrejcewicz, J.: Asymptotic approaches to strongly non-linear dynamical systems. Syst. Anal. Model. Simul. **43**(3), 255–268 (2003)
11. Cveticanin, L.: Pure odd-order oscillators with constant excitation. J. Sound Vib. **330**(5), 976–986 (2011)
12. Awrejcewicz, J., Andrianov, I.V.: Oscillations of non-linear system with restoring force close to sign(x). J. Sound Vib. **252**(5), 962–966 (2002)
13. Gendelman, O., Vakakis, A.F.: Transitions from localization to nonlocalization in strongly nonlinear damped oscillators. Chaos, Solitons Fractals **11**(10), 1535–1542 (2000)
14. Cveticanin, L., Kovacic, I., Rakaric, Z.: Asymptotic methods for vibrations of the pure non-integer order oscillator. Comput. Math Appl. **60**(9), 2626–2628 (2010)
15. Pukach, P.Y., Kuzio, I.V.: Resonance fenomena in quasi-zero stiffness vibration isolation systems. Naukovyi Visnyk Natsionalnoho Hirnychoho Universitetu, No. 3, pp. 62–67 (2015)
16. Hayashi, T.: Nonlinear Oscillations in Physical Systems [Nelineynyye kolebaniya v fizicheskikh sistemakh]. Mir, Moscow (1968). (in Russian)

Solution of the Discrete Ill-Posed Problem on the Basis of Singular Value Decomposition and Random Projection

Elena G. Revunova(✉)

Department of Neural Information Processing Technologies,
International Research and Training Centre for Information
Technologies and Systems, Kiev, Ukraine
egrevunova@gmail.com

Abstract. A brief overview of our work on the solution of DIP is presented. The stable solution of DIP we obtained, by truncated singular value decomposition and by random projection methods. Analytic and experimental averaging over random matrices for the evaluation of the error of true signal recovery is carried out for the method of solving the discrete ill-posed problems on the basis of random projection. Averaging over random matrices leads to diagonalization of the matrix conditioning both components of the error (deterministic and stochastic). The values of the diagonal elements change monotonically as a function of *k*. This in turn leads to the smoother characteristics and reducing the number of local minima. The results of the experimental study showed the connection of the elements of the diagonalized matrix with the singular values of the original matrix. This provides the basis for investigating the connection of the truncated singular value decomposition method and the random projection method.

Keywords: Ill-posed problem · Singular value decomposition
Random projection · Averaging over random matrices

1 Introduction

Discrete ill-posed problems (DIP) often arise when the signals are reconstructed based on the results of indirect measurements [1–3].

To overcome the instability and hence to improve the accuracy of solutions, regularization is used [3–5]. Regularization imposes some restrictions on the solution that improve its stability - for example, the small l_2-norm of the solution. The classic method of regularization is Tikhonov regularization [4]. Defects inherent in the methods of solving discrete ill-posed inverse problems based on Tikhonov regularization, are the computational complexity and the difficulties in selecting the proper regularization parameter on which largely depends the stability of solutions. Therefore, alternative approaches to the inverse discrete incorrect problem are needed with the accuracy on the level of Tikhonov regularization.

The stable solution of DIP can be obtained, for example, by truncated singular value decomposition (SVD) [5–8] or by random projection (RP) [9–12] methods. RP is a

N. Shakhovska and V. Stepashko (eds.), *Advances in Intelligent Systems and Computing II*, Advances in Intelligent Systems and Computing 689,
https://doi.org/10.1007/978-3-319-70581-1_31

variety of methods for the formation of neural network distributed representations [13–17]. Recently, there has been an increased interest in using RP as a regularizer [18, 19].

It is of interest to study and compare the error of the solution obtained by these two methods. In this paper, the error averaging over an ensemble of random matrices is performed analytically for the method based on random projection. The connection between the error of the methods of truncated singular value decomposition and random projection is investigated. A brief overview of our work on the solution of DIP is also presented.

2 Solution of DIP Using the Methods of Truncated Singular Value Decomposition and Random Projection

The problem of estimating a signal vector $\mathbf{x} \in \Re^n$ for the model $\mathbf{y} = \mathbf{A}\mathbf{x} + \varepsilon$, where the matrix $\mathbf{A} \in \Re^{n \times n}$ and the vector of measurements $\mathbf{y} \in \Re^n$ ($\mathbf{y} = \mathbf{y}_0 + \varepsilon$) are known, if the singular values of the matrix $\mathbf{A}$ gradually decrease to zero, belongs to the class of discrete ill-posed problems [5]. Stable solution of DIP can be obtained by using the truncated SVD [5–8] or the RP method [9–12]. The vector of estimation $\mathbf{x}^*$ based on the truncated SVD is obtained by the following linear model [6]:

$$\mathbf{x}^*_{k\,SVD} = \sum_{i=1}^{k} \mathbf{v}_i s_i^{-1} \mathbf{u}_i^{\mathrm{T}} \mathbf{y}, \tag{1}$$

where $\mathbf{u}_i \in \Re^n$ are the left singular vectors, $\mathbf{v}_i \in \Re^n$ are the right singular vectors and s_i are the singular values of the matrix $\mathbf{A}$. Note that

$$\mathbf{A}_k = \sum_{i=1}^{k} \mathbf{u}_i s_i \mathbf{v}_i^{\mathrm{T}}. \tag{2}$$

For the solution based on random projection both sides of the initial equation are multiplied by a random matrix $\mathbf{R}_k \in \Re^{k \times n}$: $\mathbf{R}_k\mathbf{A}\mathbf{x} = \mathbf{R}_k\mathbf{y}$, and then the estimation vector $\mathbf{x}^*$ is obtained by the following linear model [10]:

$$\mathbf{x}^*_{k\,R} = \sum_{i=1}^{k} \mathbf{h}_i \mathbf{r}_i^{\mathrm{T}} \mathbf{y}, \tag{3}$$

where $\mathbf{r}_i \in \Re^n$ is a column of the matrix $\mathbf{R}_k = [\mathbf{r}_1, \ldots, \mathbf{r}_k]$, elements of this matrix are realizations of a standard Gaussian random variable; $\mathbf{h}_i \in \Re^n$ is a column of the matrix $(\mathbf{R}_k\mathbf{A})^+ = [\mathbf{h}_1, \ldots, \mathbf{h}_k]$.

The expected squared error of the $\mathbf{x}$ recovery is calculated as

$$e(k) = \mathrm{E}\{\|\mathbf{x} - \mathbf{x}_k^*\|^2\}, \tag{4}$$

where E{·} is the expectation with respect to the noise in vector of measurements **y**, and $\mathbf{x}_k^*$ is the recovered signal vector obtained by (1) or (3) with **y** containing the respective noise realization.

In [8, 12], the expression for mean squared error of **x** recovery was obtained for the SVD method as

$$e_{SVD}(k) = \left\|(\mathbf{A}_k^+\mathbf{A}_k - \mathbf{I})\mathbf{x}\right\|^2 + \sigma^2 trace(\mathbf{A}_k^{+\mathrm{T}}\mathbf{A}_k^+), \tag{5}$$

and the error components are

$$e_{SVD\,d}(k) = \left\|(\mathbf{A}_k^+\mathbf{A}_k - \mathbf{I})\mathbf{x}\right\|^2 = \mathbf{x}^\mathrm{T}\mathbf{x} - \mathbf{x}^\mathrm{T}\mathbf{V}_k\mathbf{V}_k^\mathrm{T}\mathbf{x}, \tag{6}$$

$$e_{SVD\,s}(k) = \sigma^2 trace(\mathbf{A}_k^{+\mathrm{T}}\mathbf{A}_k^+) = \sigma^2 trace(\mathbf{U}_k\mathbf{S}_k^{-2}\mathbf{U}_k^\mathrm{T}), \tag{7}$$

where $e_{SVD\,d}$ is the deterministic component of the error, $e_{SVD\,s}$ is the stochastic one, and σ^2 is the variance of the noise.

In [9, 10], the expression for mean squared error of **x** recovery was obtained for RP:

$$e_R(k) = \left\|((\mathbf{R}_k\mathbf{A})^+\mathbf{R}_k\mathbf{A} - \mathbf{I})\mathbf{x}\right\|^2 + \sigma^2 trace((\mathbf{R}_k\mathbf{A})^{+\mathrm{T}}(\mathbf{R}_k\mathbf{A})^+), \tag{8}$$

and its components are

$$e_{R\,d}(k) = \left\|((\mathbf{R}_k\mathbf{A})^+\mathbf{R}_k\mathbf{A} - \mathbf{I})\mathbf{x}\right\|^2, \; e_{R\,s}(k) = \sigma^2 trace((\mathbf{R}_k\mathbf{A})^{+\mathrm{T}}(\mathbf{R}_k\mathbf{A})^+), \tag{9}$$

where $e_{R\,d}$ is the deterministic component of the error and $e_{R\,s}$ is the stochastic one.

Experimental studies [10, 12] have shown that there exists an optimal number of the components of linear models (1) and (3), that minimizes the error (4). The optimum exists because the true signal recovery error can be represented as a sum of two components, one of which (deterministic) decreases with the increasing number of model components (the model dimensionality), and the other (stochastic) grows and is proportional to the noise level in the measurement vector [9–12]:

$$e(k) = e_d(k) + e_s(k) = e_d(k) + \sigma^2 e_g(k), \tag{10}$$

where $e_d(k)$ is the value of the deterministic error component for the model of dimensionality k, $e_s(k)$ is the value of the stochastic error component for the model of dimensionality k, $e_g(k) = e_s(k)/\sigma^2$. Thus, at a certain noise level, the global minimum of the error can be achieved at $1 \leq k \leq n$.

The representation of the error in the form (10), the study of the error components and the development of the model selection criteria (MSC) are the techniques used by the inductive modeling approach [20–22] to find the optimal solution. In practice, it is impossible to calculate the recovery error $e_x(k)$ due to the lack of information about **x**, therefore it is impossible to determine the optimal k directly. For the choice of k close to the optimal, the MSC, that is, a function that would have an extremum at k close to or equal to the optimal one [20, 23–25], is used.

As the basis for the development of MSC for DIP we use the error of $\mathbf{y}_0$ recovery: $e_y(k) = \mathrm{E}\{\|\mathbf{y}_0 - \mathbf{y}_k^*\|^2\}$, where $\mathbf{y}_k^* = \mathbf{A}\mathbf{x}_k^*$. Note that although $\mathbf{y}_0$ is unknown, it can be estimated using a known measurement vector $\mathbf{y}$.

When developing a MSC for the DIP solution, we have to:

- present the error in the form of a sum of two components;
- show the increase and decrease of the corresponding error components with the increase of model dimensionality;
- show that the global minima of the dependencies of the $\mathbf{x}$ recovery error and the $\mathbf{y}_0$ recovery error on k coincide or at least are close to each other;
- obtain the expression for estimation of the recovery error of $\mathbf{y}_0$ using a known measurement vector $\mathbf{y}$.

A tool to show that the global minima of the dependence of $\mathbf{x}$ recovery error and $\mathbf{y}_0$ recovery error on k are the same or close is provided by the function

$$J(k) = \frac{\Delta e_d(k)}{\Delta e_g(k)}. \tag{11}$$

If, for $e(k)$ represented in the form (10), the value of the noise variance is in the range $J(k + 1) < \sigma^2 < J(k)$, then the k dimensional model provides a minimum of the recovery error. If $J(k)$ for the $\mathbf{x}$ recovery error and the $\mathbf{y}_0$ recovery error are the same or close, then the optimal k values for the recovery errors of $\mathbf{x}$ and $\mathbf{y}_0$ are also the same or close.

For solution of DIP by means of truncated SVD, it is possible to obtain analytical expressions for $\Delta e_d(k)$ and $\Delta e_g(k)$ and to demonstrate that $\Delta e_s(k)$ increases and $\Delta e_d(k)$ decreases both for $e_x(k)$ and for $e_y(k)$. A comparison of the positions of the minima $e_x(k)$ and $e_y(k)$ with the use of $J(k)$ then shows that the minima coincide. On the basis of this fact, a MSC minimizing the error of the DIP solution can be constructed. Similar criterion can be developed for the DIP solution by means of RP. Although in this case the minima of $e_x(k)$ and $e_y(k)$ do not necessarily coincide, numerical experiments show that they are close.

Development of the MSC for truncated SVD and RP methods of DIP solution is further elaborated in the following sections.

3 The Model Selection Criterion for the DIP Solution Based on Singular Value Decomposition

The behavior of the components of recovery error for the true signal $\mathbf{x}$ and the noiseless model output $\mathbf{y}_0$ as a function of the number k of truncated singular value decomposition components was investigated in [8, 12], and analytical expressions for $\Delta e_d(k)$ and $\Delta e_g(k)$ were obtained.

For the deterministic error component $e_{x\ d}(k)$, the following expression holds:

$$e_{x\,d}(k) = \mathbf{x}^{\mathrm{T}}\mathbf{x} - \sum_{i=1}^{k} \mathbf{x}^{\mathrm{T}}\mathbf{v}_i\mathbf{v}_i^{\mathrm{T}}\mathbf{x} = e_{x\,d}(k-1) - \mathbf{x}^{\mathrm{T}}\mathbf{v}_k\mathbf{v}_k^{\mathrm{T}}\mathbf{x}. \tag{12}$$

Since $\mathbf{x}^{\mathrm{T}}\mathbf{v}_k\mathbf{v}_k^{\mathrm{T}}\mathbf{x} > 0$, the value of the deterministic component of the error $e_{x\ d}(k)$ decreases with increasing k.

From the recursive expression for the stochastic error component:

$$e_{x\,s}(k) = e_{x\,s}(k-1) + \sigma^2 s_k^{-2} trace(\mathbf{u}_k\mathbf{u}_k^{\mathrm{T}}) = e_{x\,s}(k-1) + \sigma^2 s_k^{-2}. \tag{13}$$

it follows that its value increases with increasing k.

The expression for the deterministic component of the $\mathbf{y}_0$ recovery error has the form:

$$e_{y\,d}(k) = e_{y\,d}(k-1) - \mathbf{y}_0^{\mathrm{T}}\mathbf{u}_k\mathbf{u}_k^{\mathrm{T}}\mathbf{y}_0. \tag{14}$$

Since $\mathbf{y}_0^{\mathrm{T}}\mathbf{u}_k\mathbf{u}_k^{\mathrm{T}}\mathbf{y}_0 > 0$, the value $e_{y\ d}(k)$ decreases with increasing k.

The value of the stochastic component of the $\mathbf{y}_0$ recovery error is:

$$e_{y\,s}(k) = \sigma^2 trace(\sum_{i=1}^{k} \mathbf{v}_k\mathbf{v}_k^{\mathrm{T}}) = \sigma^2 k \tag{15}$$

and it increases with the increasing k.

Given that $\Delta e_{x\ d}(k) = \mathbf{x}^{\mathrm{T}}\mathbf{v}_k\mathbf{v}_k^{\mathrm{T}}\mathbf{x}$ and $\Delta e_{x\ g}(k) = s_k^{-2} trace(\mathbf{u}_k\mathbf{u}_k^{\mathrm{T}}) = s_k^{-2}$,

$$J_x(k) = \frac{\Delta e_{x\,d}(k)}{\Delta e_{x\,g}(k)} = s_k^2\mathbf{x}^{\mathrm{T}}\mathbf{v}_k\mathbf{v}_k^{\mathrm{T}}\mathbf{x}. \tag{16}$$

Let us calculate the ratio $J_y(k)$ for the $\mathbf{y}_0$ recovery error. Since $\Delta e_{y\ d}(k) = s_k^2\mathbf{x}^{\mathrm{T}}\mathbf{v}_k\mathbf{v}_k^{\mathrm{T}}\mathbf{x}$ and $\Delta e_{y\ g}(k) = trace(\mathbf{v}_k\mathbf{v}_k^{\mathrm{T}}) = 1$,

$$J_y(k) = \frac{\Delta e_{y\,d}(k)}{\Delta e_{y\,g}(k)} = s_k^2\mathbf{x}^{\mathrm{T}}\mathbf{v}_k\mathbf{v}_k^{\mathrm{T}}\mathbf{x}. \tag{17}$$

From the expressions (16) and (17) it is clear that $J_x(k)$ and $J_y(k)$ coincide, that is, for the truncated singular value decomposition, the position of the minimum error in recovery of the input $\mathbf{x}$ and the output $\mathbf{y}_0$ always coincide. Therefore, the dependence of the $\mathbf{y}_0$ recovery error on the number of model components can be used to determine the optimal number of said components.

In the work [12], the following expression that approximates the $\mathbf{y}_0$ recovery error was obtained:

$$\begin{aligned} CR_{SVD} = \mathrm{E}\|(\mathbf{A}_k\mathbf{A}_k^+ - \mathbf{I})\mathbf{y}\|^2 - \sigma^2 trace((\mathbf{A}_k\mathbf{A}_k^+ - \mathbf{I})^{\mathrm{T}}(\mathbf{A}_k\mathbf{A}_k^+ - \mathbf{I})) \\ + \sigma^2 trace((\mathbf{A}_k\mathbf{A}_k^+)^{\mathrm{T}}\mathbf{A}_k\mathbf{A}_k^+). \end{aligned} \tag{18}$$

This expression allows to determine the number of components of the truncated singular value decomposition that is close to optimal. To estimate the noise variance σ^2, known methods can be used [26, 27].

Figure 1 shows the plots of the dependence of CR_{SVD} criterion values, average error of the output signal recovery (e_y_mean), and average error of the true input signal recovery (e_x_mean) on the number k of truncated singular value decomposition components for Carasso problem [28]. It can be seen that the dependence curves for the criterion CR_{SVD} and the average recovery error of $\mathbf{y}_0$ are close, and the model selection criterion CR_{SVD} approximates the output recovery error well. The positions of the minima of dependencies of $\mathbf{x}$ recovery error and $\mathbf{y}_0$ recovery error on k coincide.

Further experimental investigation of the developed criterion was carried out in [12].

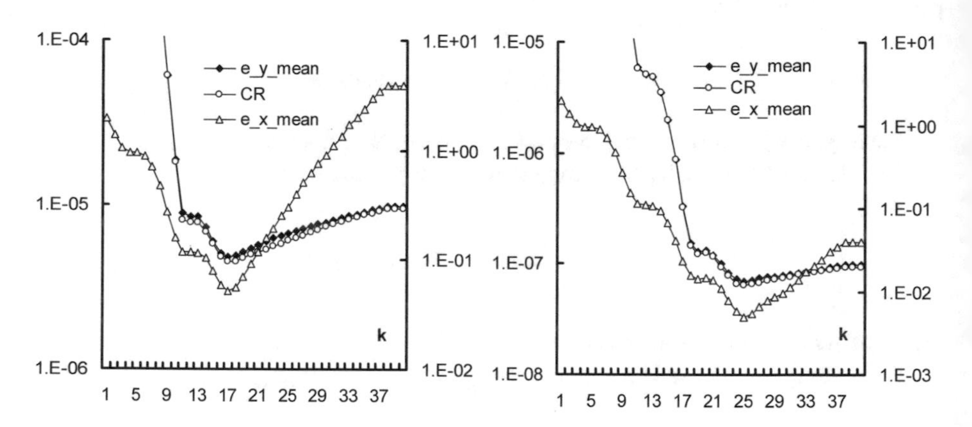

Fig. 1. Dependencies of the CR_{SVD} criterion values and the output recovery error on k. For the noise level of 1e−4 (left) and the noise level of 1e−5 (right).

4 The Model Selection Criterion for the DIP Solution Based on Random Projection

For the DIP solution method based on RP, the increase of stochastic $e_s(k)$ and decrease of deterministic $e_d(k)$ error components with increasing k was shown analytically for $e_x(k)$, and experimentally for $e_y(k)$. Experimental studies have shown the proximity (but not complete coincidence) of the positions of the minima of $e_x(k)$ and $e_y(k)$. The analytical expressions obtained for the random projection method for $\Delta e_d(k)$, $\Delta e_g(k)$, $J_x(k)$, $J_y(k)$, however, do not allow to analytically investigate the closeness of the positions of $e_x(k)$ and $e_y(k)$ minima.

The behavior of the dependence of error components on the number of rows k of the matrix $\mathbf{R}_k$ was analytically investigated in [11]. The study is based on the representation of the matrix $\mathbf{F}_k = \mathbf{R}_k\,\mathbf{A}$ as the sum of the original matrix and the perturbation matrix. The pseudoinverse matrix $\mathbf{F}_k^+ = (\mathbf{R}_k\,\mathbf{A})^+$ (that is used to estimate $\mathbf{x}$) can be represented as a perturbation of the pseudoinverse matrix through the perturbation of the original matrix as proposed in [29].

Based on this representation, recursive expressions for the stochastic and deterministic components of the input signal **x** recovery error were obtained. These expressions provide a tool for the study of the behavior (that is, increase and decrease) of the error components, depending on k. The matrix $\mathbf{F}_k = \mathbf{R}_k \mathbf{A}$ can be represented as

$$\mathbf{F}_k = \begin{bmatrix} \mathbf{F}_{k-1} \\ \mathbf{f}_k \end{bmatrix}, \tag{19}$$

where $\mathbf{F}_{k-1} \in \Re^{k-1 \times N}$, row vector $\mathbf{f}_k = \mathbf{r}_k \mathbf{A}$, $\mathbf{r}_k$ is the k-th row of the matrix $\mathbf{R}_k = [\mathbf{r}_1, \ldots, \mathbf{r}_k]$. As a perturbation of the matrix $\mathbf{F}_{k-1}$, we consider a matrix $\mathbf{E}_k \in \Re^{k \times N}$, containing a single non-zero row $\mathbf{f}_k$ that is added at the k-th step:

$$\mathbf{F}^0_{k-1} + \mathbf{E}_k = \mathbf{F}_k, \tag{20}$$

$$\mathbf{E}_k = \begin{bmatrix} \mathbf{O}_{k-1} \\ \mathbf{f}_k \end{bmatrix}, \ \mathbf{F}^0_{k-1} = \begin{bmatrix} \mathbf{F}_{k-1} \\ \mathbf{O}_1 \end{bmatrix}, \tag{21}$$

where $\mathbf{O}_{k-1}$ is zero submatrix of the size $(k-1) \times N$.

In [11], the recursive expressions for stochastic and deterministic components of the recovery error are formulated. For the stochastic component of the true signal recovery error, the recursive expression has the form:

$$e_{x\,s}(k) = \sigma^2 trace(\mathbf{F}_k^{+\mathrm{T}} \mathbf{F}_k^{+}) = e_{x\,s}(k-1) + \sigma^2 trace(\mathbf{M}_{k-1}^{\mathrm{T}} \mathbf{M}_{k-1}) + \sigma^2 d_k, \tag{22}$$

where $\mathbf{M}_{k-1} = \mathbf{f}_k^{+}\mathbf{f}_k \ \mathbf{F}_{k-1}^{+}$, $d_k = \mathbf{f}_k^{+\mathrm{T}}\mathbf{f}_k^{+}$. If $\mathbf{f}_k$ is nonzero, then $d_k > 0$. For nonzero $\mathbf{M}_{k-1}, trace\left(\mathbf{M}_{k-1}^{\mathrm{T}} \mathbf{M}_{k-1}\right) > 0$. Therefore, the value of the stochastic component of the error increases with increasing k. For the deterministic component of the true signal recovery error, the recursive expression has the form:

$$e_{x\,d}(k) = \mathbf{x}^{\mathrm{T}}\mathbf{x} - \mathbf{x}^{\mathrm{T}}\mathbf{F}_k^{+}\mathbf{F}_k\mathbf{x} = e_{x\,d}(k-1) - \mathbf{x}^{\mathrm{T}}\mathbf{K}_k\mathbf{x}, \tag{23}$$

$$\mathbf{K}_k = \frac{(\mathbf{I} - \mathbf{F}_{k-1}^{+}\mathbf{F}_{k-1})\mathbf{f}_k^{\mathrm{T}}\mathbf{f}_k(\mathbf{I} - \mathbf{F}_{k-1}^{+}\mathbf{F}_{k-1})}{\mathbf{f}_k(\mathbf{I} - \mathbf{F}_{k-1}^{+}\mathbf{F}_{k-1})(\mathbf{I} - \mathbf{F}_{k-1}^{+}\mathbf{F}_{k-1})^{\mathrm{T}}\mathbf{f}_k^{\mathrm{T}}} = \frac{\mathbf{m}_k^{\mathrm{T}}\mathbf{m}_k}{\mathbf{m}_k\mathbf{m}_k^{\mathrm{T}}} = \frac{\mathbf{m}_k^{\mathrm{T}}\mathbf{m}_k}{z_k}. \tag{24}$$

From (23) and (24) it follows that $\mathbf{x}^{\mathrm{T}}\mathbf{K}_k\mathbf{x}$ is a positive number and the value of deterministic component of the true signal recovery error decreases with increasing k.

From the recursive expressions (22) and (23) it follows that for the **x** recovery error

$$\Delta e_{x\,d}(k) = \mathbf{x}^{\mathrm{T}}\mathbf{f}_k^{+}\mathbf{f}_k(\mathbf{I} - \mathbf{F}_{k-1}^{+}\mathbf{F}_{k-1})\mathbf{x}, \ \Delta e_{x\,g}(k) = trace(\mathbf{M}_{k-1}^{\mathrm{T}}\mathbf{M}_{k-1}) + d_k. \tag{25}$$

Respectively,

$$J_x(k) = \frac{\Delta e_{x\,d}(k)}{\Delta e_{x\,g}(k)} = \frac{\mathbf{x}^{\mathrm{T}}\mathbf{f}_k^{+}\mathbf{f}_k(\mathbf{I} - \mathbf{F}_{k-1}^{+}\mathbf{F}_{k-1})\mathbf{x}}{trace(\mathbf{M}_{k-1}^{\mathrm{T}}\mathbf{M}_{k-1}) + d_k}. \tag{26}$$

Unlike in the case of truncated SVD method, for the RP method the forms of $J_x(k)$ and $J_y(k)$ are different, and the optimum k values for the input $\mathbf{x}$ and output $\mathbf{y}_0$ recovery errors do not coincide. The numerical experiments however demonstrate that the difference between $J_x(k)$ and $J_y(k)$ is small. Approximation of the output vector recovery error, obtained as a result of averaging over multiple noise realizations, has the form:

$$\begin{aligned} CR_R = & \mathrm{E}\left\|(\mathbf{A}\mathbf{F}_k^{+}\mathbf{R}_k^{\mathrm{T}} - \mathbf{I})\mathbf{y}\right\|^2 - \sigma^2 trace((\mathbf{A}\mathbf{F}_k^{+}\mathbf{R}_k^{\mathrm{T}} - \mathbf{I})^{\mathrm{T}}(\mathbf{A}\mathbf{F}_k^{+}\mathbf{R}_k^{\mathrm{T}} - \mathbf{I})) \\ & + \sigma^2 trace(\mathbf{F}_k^{+\mathrm{T}}\mathbf{A}^{\mathrm{T}}\mathbf{A}\mathbf{F}_k). \end{aligned} \tag{27}$$

Figure 2 shows the plots of the dependence of CR_R criterion values, average error of the output recovery, and average error of the true signal recovery on k for the Carasso problem.

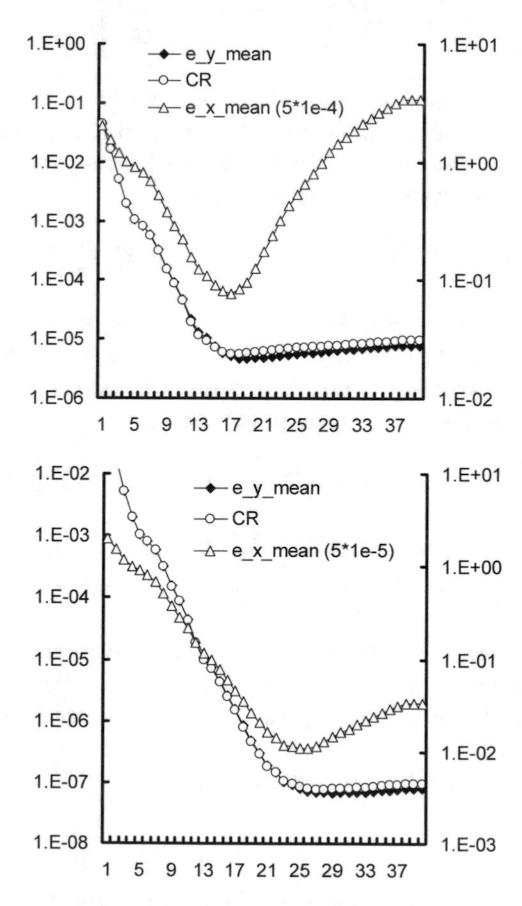

Fig. 2. Dependencies of the CR_R criterion values and the output recovery error on k for the noise levels of 5e−4 and 5e−5.

The dependence curves for the criterion CR_R and the average recovery error of $\mathbf{y}_0$ are close; the model selection criterion CR_R approximates the output recovery error well. The positions of the minima of dependencies of $\mathbf{x}$ recovery error and $\mathbf{y}_0$ recovery error on k are also close. We note that the experimental studies have shown that averaging over random matrices leads to the smoothing of the characteristics $e_x(k)$, $e_y(k)$, $J_x(k)$, $J_y(k)$ and, correspondingly, decrease in the number of local minima (Figs. 3, 4, 5 and 6).

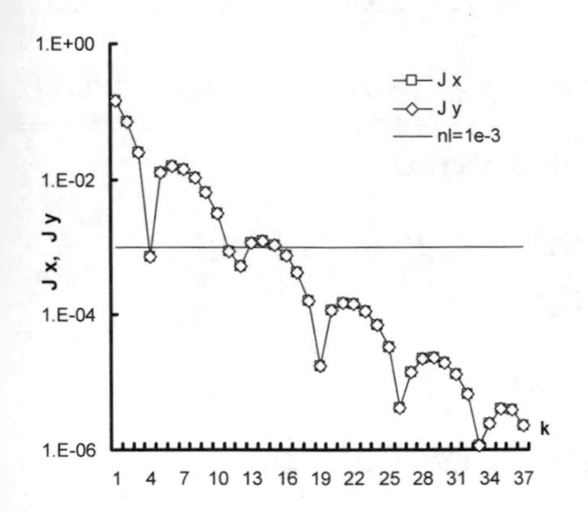

Fig. 3. Dependence of J_x, J_y on k for the Carasso problem.

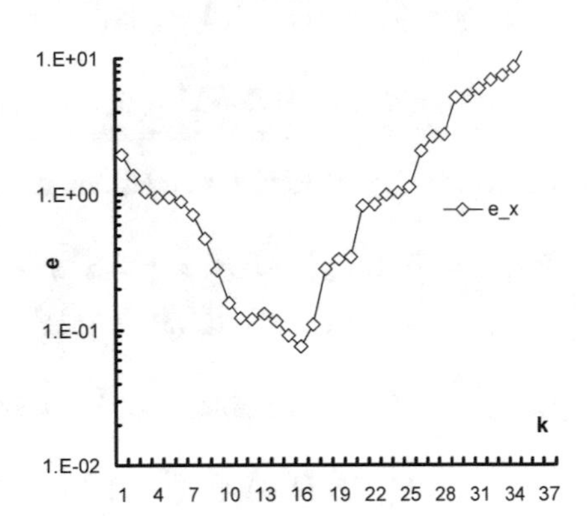

Fig. 4. Dependence of the $\mathbf{x}$ recovery error on k for the Carasso problem.

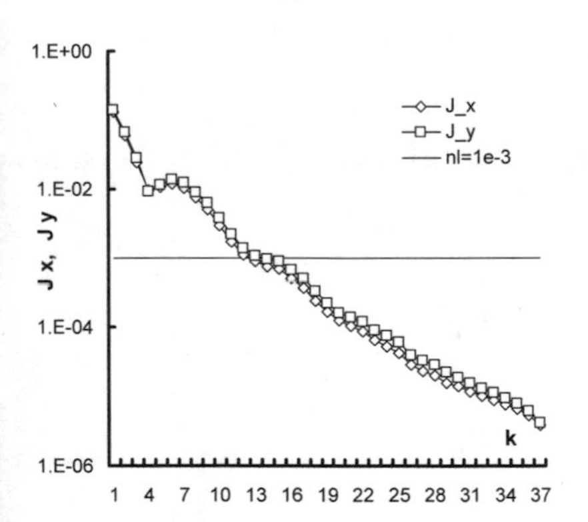

Fig. 5. Dependence of $J_x(k)$, $J_y(k)$ on k for the Carasso problem after averaging over random matrices.

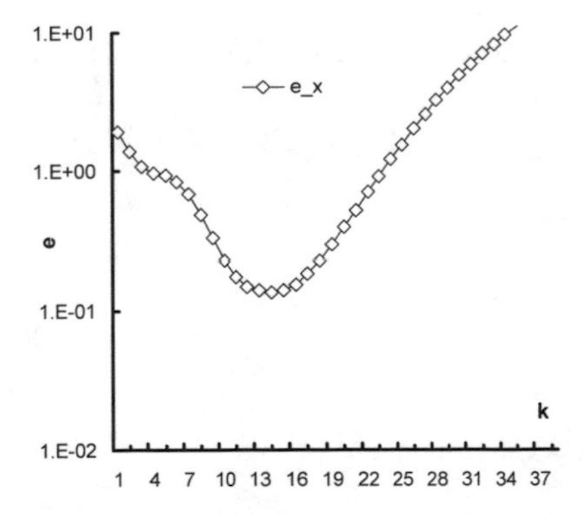

Fig. 6. Dependence of the $\mathbf{x}$ recovery error on k for the Carasso problem after averaging over random matrices.

Therefore, analytical averaging over random matrices can lead to simpler expressions for $e_x(k)$ and $e_y(k)$ and simplify the expressions for $\Delta e_d(k)$, $\Delta e_g(k)$, $J_x(k)$, $J_y(k)$. In addition, numerous experimental studies of the truncated SVD and RP methods have shown the closeness of the $e_x(k)$ dependences for these methods.

To study the connection of $e_{SVD}(k)$ and $e_R(k)$, we need to obtain an expression for the $e_x(k)$ average over an ensemble of random matrices.

5 Error Averaging Over Random Matrices

We transform the expressions for the error components (9) as follows:

$$e_{R\,d}(k) = \left\|((\mathbf{R}_k\mathbf{A})^+\mathbf{R}_k\mathbf{A} - \mathbf{I})\mathbf{x}\right\|^2 = \mathbf{x}^T\mathbf{x} - \mathbf{x}^T\mathbf{A}^T\mathbf{R}_k^T(\mathbf{R}_k\mathbf{A}\mathbf{A}^T\mathbf{R}_k^T)^+\mathbf{R}_k\mathbf{A}\mathbf{x}, \quad (28)$$

$$e_{R\,s}(k) = \sigma^2 trace((\mathbf{R}_k\mathbf{A})^{+T}(\mathbf{R}_k\mathbf{A})^+) = \sigma^2 trace(\mathbf{R}_k^T(\mathbf{R}_k\mathbf{A}\mathbf{A}^T\mathbf{R}_k^T)^+\mathbf{R}_k). \quad (29)$$

Let us average over the realizations of random matrices:

$$\begin{aligned}\mathrm{E}_R\{e\} = \mathrm{E}_R\{e_{R\,d}\} + \mathrm{E}_R\{e_{R\,s}\} &= \mathbf{x}^T\mathbf{x} - \mathbf{x}^T\mathbf{A}^T\mathrm{E}_R\{\mathbf{R}_k^T(\mathbf{R}_k\mathbf{A}\mathbf{A}^T\mathbf{R}_k^T)^+\mathbf{R}_k\}\mathbf{A}\mathbf{x}\\ &+ \sigma^2 trace\mathrm{E}_R\{\mathbf{R}_k^T(\mathbf{R}_k\mathbf{A}\mathbf{A}^T\mathbf{R}_k^T)^+\mathbf{R}_k\}.\end{aligned} \quad (30)$$

Using $\mathbf{A} = \mathbf{U}\mathbf{S}\mathbf{V}^T$ we obtain $\mathbf{A}\mathbf{A}^T = \mathbf{U}\mathbf{S}^2\mathbf{U}^T$ and

$$\mathrm{E}_R\{\mathbf{R}_k^T(\mathbf{R}_k\mathbf{A}\mathbf{A}^T\mathbf{R}_k^T)^+\mathbf{R}_k\} = \mathrm{E}_R\{\mathbf{R}_k^T(\mathbf{R}_k\mathbf{U}\mathbf{S}^2\mathbf{U}^T\mathbf{R}_k^T)^+\mathbf{R}_k\}. \quad (31)$$

By the Lemma 30 in [30]

$$\mathrm{E}_R\{\mathbf{R}_k^T(\mathbf{R}_k\mathbf{U}\mathbf{S}^2\mathbf{U}^T\mathbf{R}_k^T)^+\mathbf{R}_k\} = \mathbf{U}\mathrm{E}_R\{\mathbf{R}_k^T(\mathbf{R}_k\mathbf{S}^2\mathbf{R}_k^T)^+\mathbf{R}_k\}\mathbf{U}^T. \quad (32)$$

In accordance with the expression (45) [30]

$$\mathrm{E}_R\{\mathbf{R}_k^T(\mathbf{R}_k\mathbf{S}^2\mathbf{R}_k^T)^+\mathbf{R}_k\} = diag(\lambda_1, \ldots, \lambda_m, \mu\mathbf{I}_{n-m}) = \mathbf{D}_k, \quad (33)$$

where

$$\lambda_i = \frac{\mu}{1+\mu s_i^2},\ \mu = const,\ n-m \approx \frac{1}{n}\sum_{i=1}^{n}\frac{1}{1+\mu s_i^2}. \quad (34)$$

Consequently

$$\mathrm{E}_R\{\mathbf{R}_k^T(\mathbf{R}_k\mathbf{A}\mathbf{A}^T\mathbf{R}_k^T)^+\mathbf{R}_k\} = \mathbf{U}\mathbf{D}_k\mathbf{U}^T. \quad (35)$$

The resulting expression for the error (8) after averaging over the matrices $\mathbf{R}_k$ takes the form:

$$\mathrm{E}_R\{e\} = \mathbf{x}^T\mathbf{x} - \mathbf{x}^T\mathbf{A}^T\mathbf{U}\mathbf{D}_k\mathbf{U}^T\mathbf{A}\mathbf{x} + \sigma^2 trace(\mathbf{U}\mathbf{D}_k\mathbf{U}^T). \quad (36)$$

By converting $\mathbf{x}^{\mathrm{T}}\mathbf{A}^{\mathrm{T}}\mathbf{U}\mathbf{D}_k\mathbf{U}^{\mathrm{T}}\mathbf{A}\mathbf{x} = \mathbf{x}^{\mathrm{T}}\mathbf{V}\mathbf{S}^{\mathrm{T}}\mathbf{D}_k\mathbf{S}\mathbf{V}^{\mathrm{T}}\mathbf{x} = \mathbf{x}^{\mathrm{T}}\mathbf{V}\mathbf{S}^2\mathbf{D}_k\mathbf{V}^{\mathrm{T}}\mathbf{x}$, we get:

$$\mathrm{E}_R\{e\} = \mathbf{x}^{\mathrm{T}}\mathbf{x} - \mathbf{x}^{\mathrm{T}}\mathbf{V}\mathbf{S}^2\mathbf{D}_k\mathbf{V}^{\mathrm{T}}\mathbf{x} + \sigma^2 trace(\mathbf{U}\mathbf{D}_k\mathbf{U}^{\mathrm{T}}). \tag{37}$$

An experimental study of the fulfillment of expression (33) showed that the sequence $\lambda_1, \ldots, \lambda_m$ is not only upper bounded by the sequence $1/s_1^2, \ldots, 1/s_m^2$, but also several initial values of the diagonal of the matrix $\mathbf{D}_k$ approximate $1/s_i^2$ with great accuracy. For example, for the Carasso problem [28], the values of the diagonal elements of the matrix $\mathbf{D}_k$ at $k = \{2, 3, 6, 9, 13\}$ are shown in Fig. 7, where the values of $1/s_i^2$ and μ are also given.

If it is possible to show analytically that for some $p < m$ we have $d_i = 1/s_i^2$, it will allow us to investigate analytically the connection of the truncated SVD and RP as it is sketched below.

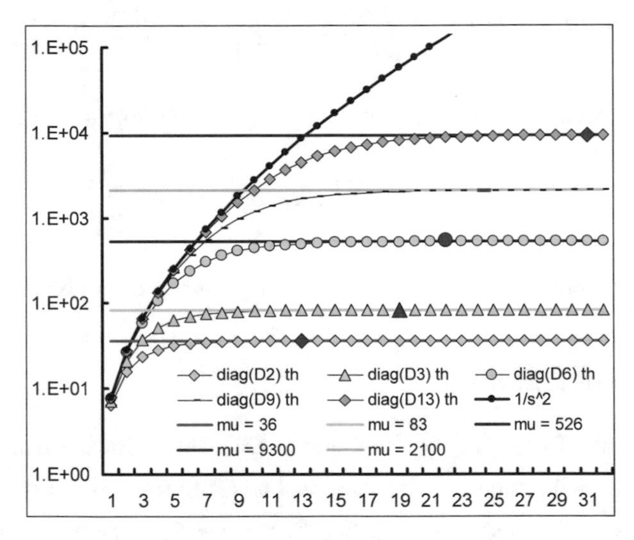

Fig. 7. Elements diag($\mathbf{D}_k$), values of μ and $1/s_i^2$ at $k = \{2, 3, 6, 9, 13\}$ for the Carasso problem.

6 Connection Between the Errors of the Truncated SVD and RP Methods

Using (37), we write the expression for the deterministic error component for the RP method:

$$e_{R\,d} = \mathbf{x}^{\mathrm{T}}\mathbf{x} - \mathbf{x}^{\mathrm{T}} \sum_{i=1}^{n} (\mathbf{v}_i s_i^2 d_i \mathbf{v}_i^{\mathrm{T}})\mathbf{x}, \tag{38}$$

$$e_{R\,d} = \mathbf{x}^{\mathrm{T}}\mathbf{x} - (\mathbf{x}^{\mathrm{T}} \sum_{i=1}^{p} (\mathbf{v}_i s_i^2 \frac{1}{s_i^2} \mathbf{v}_i^{\mathrm{T}})\mathbf{x} + \mathbf{x}^{\mathrm{T}} \sum_{i=p+1}^{n} (\mathbf{v}_i s_i^2 d_i \mathbf{v}_i^{\mathrm{T}})\mathbf{x}). \tag{39}$$

Let us transform this expression so as to isolate the deterministic component e_d^{SVD} of the error for SVD:

$$e_{R\,d} = \mathbf{x}^{\mathrm{T}}\mathbf{x} - \mathbf{x}^{\mathrm{T}}\sum_{i=1}^{p}(\mathbf{v}_i\mathbf{v}_i^{\mathrm{T}})\mathbf{x} - \mathbf{x}^{\mathrm{T}}\sum_{i=p+1}^{n}(\mathbf{v}_i s_i^2 d_i \mathbf{v}_i^{\mathrm{T}})\mathbf{x} = e_{p\,d}^{SVD} - \mathbf{x}^{\mathrm{T}}\sum_{i=p+1}^{n}(\mathbf{v}_i s_i^2 d_i \mathbf{v}_i^{\mathrm{T}})\mathbf{x}. \tag{40}$$

In the expression for the stochastic error component for the RP method

$$e_{R\,s} = \sigma^2 trace(\mathbf{U}\mathbf{D}_k\mathbf{U}^{\mathrm{T}}) = \sigma^2 trace(\mathbf{D}_k\mathbf{U}^{\mathrm{T}}\mathbf{U}) = \sigma^2 trace(\mathbf{D}_k), \tag{41}$$

isolate the stochastic component e_s^{SVD} of the error for SVD

$$e_{R\,s} = \sigma^2 trace(\mathbf{D}_k) = \sigma^2\sum_{i=1}^{p}\frac{1}{s_i^2} + \sigma^2\sum_{i=p+1}^{n} d_i = e_{p\,s}^{SVD} + \sigma^2\sum_{i=p+1}^{n} d_i. \tag{42}$$

Thus, the expression for the error of recovering the true signal by the RP method is:

$$\mathrm{E}_R\{e\} = e_p^{SVD} - \mathbf{x}^{\mathrm{T}}\sum_{i=p+1}^{n}(\mathbf{v}_i s_i^2 d_i \mathbf{v}_i^{\mathrm{T}})\mathbf{x} + \sigma^2\sum_{i=p+1}^{n} d_i, \tag{43}$$

where e_p^{SVD} is the error of recovering the true signal using the p components of SVD.

7 Experimental Study

For the problems of Philips [31] and Carasso [28], the following experiments were carried out. The mathematical expectation $\mathrm{E}_{\mathrm{R}}\left\{\mathbf{R}_k^{\mathrm{T}}\left(\mathbf{R}_k\mathbf{S}^2\mathbf{R}_k^{\mathrm{T}}\right)^{+}\mathbf{R}_k\right\}$ was approximated by averaging over 10^4 random matrices: 10^4 matrices $\mathbf{R}_{ki}$ were generated for each k, 10^4 matrices $\mathbf{M}_{ki} = \mathbf{R}_{ki}^{\mathrm{T}}\left(\mathbf{R}_{ki}\mathbf{S}^2\mathbf{R}_{ki}^{\mathrm{T}}\right)^{+}\mathbf{R}_{ki}$ were calculated, and matrices

$$\mathbf{D}_k = \sum_{i=1}^{10^4}\mathbf{M}_{ki}/10^4, \tag{44}$$

The values of the diagonal elements of the matrix obtained after averaging are shown in Figs. 8 and 9. The theoretical values of the elements of the $\mathbf{D}_k$ matrix diagonal calculated from (33) are shown in Figs. 8 and 9 with the label "*th*".

For the DIP solution by the RP method, the error value and its components (8, 9) were experimentally obtained with averaging over 10^4 matrices (Figs. 10 and 11). The theoretical values of the error and its components calculated using the theoretical values of $diag(\mathbf{D}_k)$ (33) are indicated with the label "*th*".

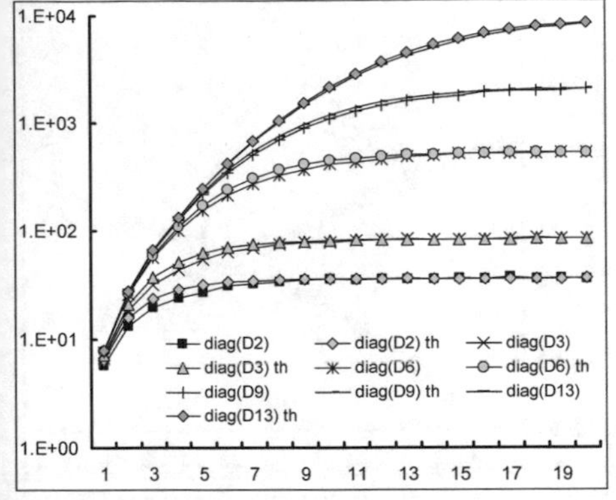

Fig. 8. Elements $diag(\mathbf{D}_k)$ obtained experimentally and theoretically at $k = \{2, 3, 6, 9,13\}$ for the Carasso problem.

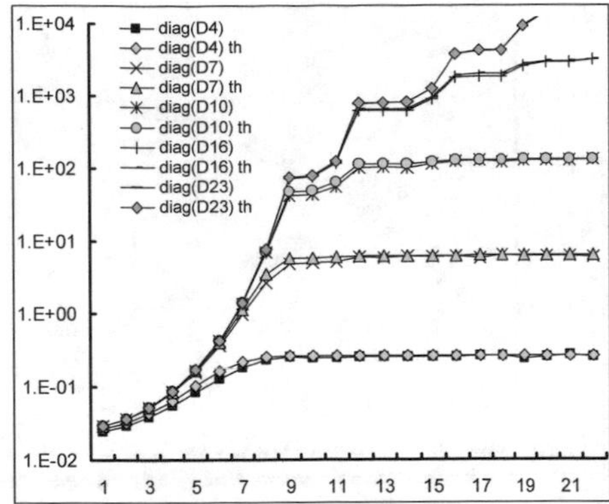

Fig. 9. Elements $diag(\mathbf{D}_k)$ obtained experimentally and theoretically at $k = \{4, 7, 10, 16, 23\}$ for the Phillips problem.

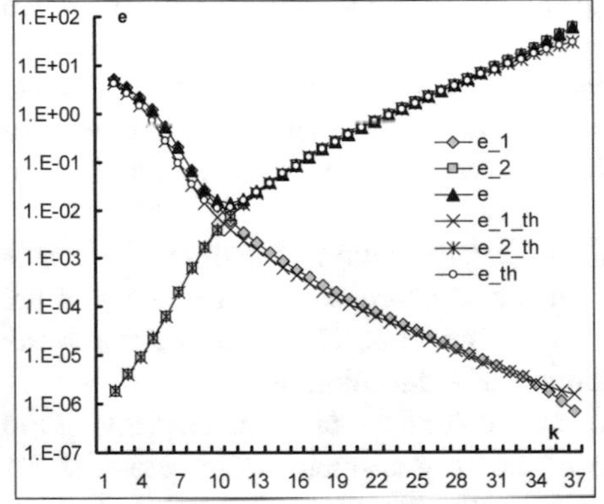

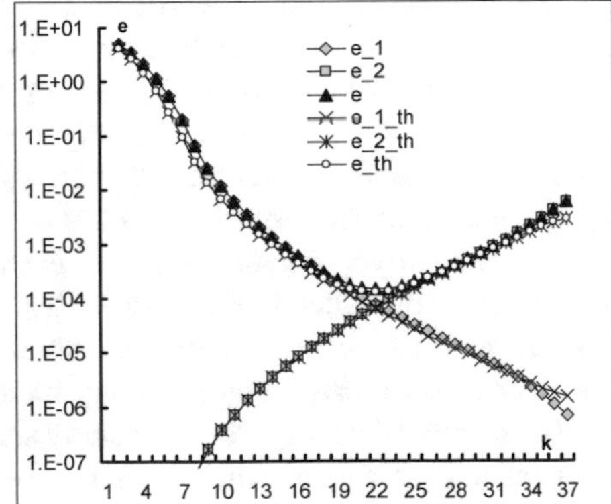

Fig. 10. The problem of Phillips. For the noise level of 1e−2 (left) and noise level of 1e−4 (right).

An example of these dependencies for the discrete ill-posed problems of Phillips (at the noise levels of 1E−2, 1E−3, 1E−4) and Carasso (at the noise levels of 0.3E−3, 1E−4, 1E−5) is shown in Figs. 10 and 11. Here e, e_1, e_2 are the error and its components with averaging over 10^4 matrices, e_{th}, e_{1th}, e_{2th} are the error and its components obtained by formula (36).

When k increases, the deterministic component decreases, and the stochastic component increases, hence the total error norm has a minimum. With increasing noise level the position of error minimum shifts to the lower values of k, and the error value at the minimum increases.

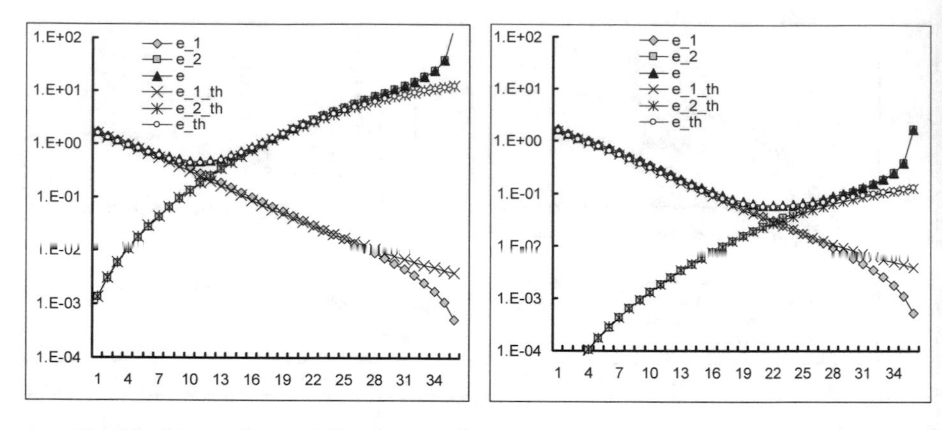

Fig. 11. The problem of Carasso. For the noise level of 0.3e−3 and level of 1e−5.

8 Conclusions

Analytic and experimental averaging over random matrices for evaluation of the error of the true signal recovery is carried out for the method of solving the discrete ill-posed problems on the basis of random projection.

Comparison of the dependencies of the error and its components on model dimensionality k obtained analytically and experimentally after averaging over an ensemble of random matrices demonstrate a good agreement of the theoretical dependences with the experimental ones.

Averaging over random matrices leads to diagonalization of the matrix conditioning both components of the error (deterministic and stochastic). The values of the diagonal elements change monotonically as a function of k. This in turn leads to the smoother characteristics and reducing the number of local minima.

The results of the experimental study showed the connection of the elements of the diagonalized matrix with the singular values of the original matrix. This provides the basis for investigating the connection of the truncated singular value decomposition method and the random projection method.

Acknowledgment. The author is grateful to Dmitri A. Rachkovskij for comprehensive help and support.

References

1. Zabulonov, Y., Korostil, Y., Revunova, E.: Optimization of inverse problem solution to obtain the distribution density function for surface contaminations. Model. Inf. Technol. **39**, 77–83 (2006). (in Russian)
2. Rachkovskij, D., Revunova, E.: Intelligent Gamma-ray data processing for environmental monitoring. In: Intelligent Data Analysis in Global Monitoring for Environment and Security, pp. 124–145. ITHEA, Kiev-Sofia (2009)

3. Starkov, V.: Constructive methods of computational physics in interpretation problems. Naukova Dumka, Kyev (2002). (in Russian)
4. Tikhonov, A., Arsenin, V.: Solution of Ill-Posed Problems. V.H. Winston, Washington (1977)
5. Hansen, P.: Rank-Deficient and Discrete Ill-Posed Problems, Numerical Aspects of Linear Inversion. SIAM, Philadelphia (1998)
6. Hansen, P.: The truncated SVD as a method for regularization. BIT **27**, 534–553 (1987)
7. Revunova, E., Tyshchuk, A.: A model selection criterion for solution of discrete ill-posed problems based on the singular value decomposition. In: 7th International Workshop on Inductive Modelling, IWIM 2015, pp. 43–47. Kyiv-Zhukyn (2015)
8. Revunova, E., Tyshchuk, A.: Model selection criterion for the solution of discrete ill-posed problems based on singular value decomposition. Control Syst. Comput. **6**, 3–11 (2014)
9. Revunova, E., Rachkovskij, D.: Using randomized algorithms for solving discrete ill-posed problems. Int. J. Inf. Theor. Appl. **2**(16), 176–192 (2009)
10. Rachkovskij, D., Revunova, E.: Randomized method for solving discrete ill-posed problems. Cybern. Syst. Anal. **48**(4), 621–635 (2012)
11. Revunova, E.: Analytical study of the error components for the solution of discrete ill-posed problems using random projections. Cybern. Syst. Anal. **51**(6), 978–991 (2015)
12. Revunova, E.: Model selection criteria for a linear model to solve discrete ill-posed problems on the basis of singular decomposition and random projection. Cybern. Syst. Anal. **52**(4), 647–664 (2016)
13. Kussul, E., Baidyk, T., Lukovich, V., Rachkovskij, D.: Adaptive neural network classifier with multifloat input coding. In: Proceedings Neuro-Nimes 1993, pp. 209–216 (1993)
14. Lukovich, V., Goltsev, A., Rachkovskij, D.: Neural network classifiers for micromechanical equipment diagnostics and micromechanical product quality inspection. In: Proceedings EUFIT 1997, vol. 1, pp. 534–536 (1997)
15. Kussul, E., Kasatkina, L., Rachkovskij, D., Wunsch, D.: Application of random threshold neural networks for diagnostics of micro machine tool condition. In: Neural Networks Proceedings, IEEE World Congress on Computational Intelligence, vol. 1, pp. 241–244 (1998)
16. Rachkovskij, D., Slipchenko, S., Kussul, E., Baidyk, T.: Properties of numeric codes for the scheme of random subspaces RSC. Cybern. Syst. Anal. **41**(4), 509–520 (2005)
17. Rachkovskij, D., Misuno, I., Slipchenko, S.: Randomized projective methods for construction of binary sparse vector representations. Cybern. Syst. Anal. **48**(1), 146–156 (2012)
18. Durrant, R., Kaban, A.: Random projections as regularizers: learning a linear discriminant from fewer observations than dimensions. Mach. Learn. **99**(2), 257–286 (2015)
19. Wei, Y., Xie, P., Zhang, L.: Tikhonov regularization and randomized GSVD. SIAM J. Matrix Anal. Appl. **37**, 649–675 (2016)
20. Ivakhnenko, A., Stepashko, V.: Noise-Immunity of Modeling. Naukova dumka, Kiev (1985). (In Russian)
21. Stepashko, V.: Theoretical aspects of GMDH as a method of inductive modeling. Control Syst. Mach. **2**, 31–38 (2003). (In Russian)
22. Stepashko, V.: Method of critical variances as analytical tool of theory of inductive modeling. J. Autom. Inf. Sci. **40**(3), 4–22 (2008)
23. Mallows, C.: Some comments on CP. Technometrics **15**(4), 661–675 (1973)
24. Akaike, H.: A new look at the statistical model identification. IEEE Trans. Autom. Control **19**(6), 716–723 (1974)
25. Hansen, M., Yu, B.: Model selection and minimum description length principle. J. Am. Stat. Assoc. **96**, 746–774 (2001)

26. Bayati, M., Erdogdu, M., Montanari, A.: Estimating LASSO risk and noise level. In: Proceedings of Advances in Neural Information Processing Systems, NIPS (2013)
27. Fan, J., Guo, S., Hao, N.: Variance estimation using refitted cross-validation in ultrahigh dimensional regression. J. Roy. Stat. Soc. Ser. B (Stat. Methodol.) **74**, 1467–9868 (2012)
28. Carasso, A.: Determining surface temperatures from interior observations. SIAM J. Appl. Math. **42**, 558–574 (1982)
29. Stewart, G.: On the perturbation of pseudo-inverses, projections and linear least squares problems. SIAM Rev. **19**(4), 634–662 (1977)
30. Marzetta, T., Tucci, G., Simon, S.: A random matrix-theoretic approach to handling singular covariance estimates. IEEE Trans. Inf. Theor. **57**(9), 6256–6271 (2011)
31. Phillips, D.L.: A technique for the numerical solution of integral equation of the first kind. J. ACM **9**, 84–97 (1962)

Methodology of Research the Library Information Services: The Case of USA University Libraries

Antonii Rzheuskyi[1(✉)], Nataliia Kunanets[1], and Vasil Kut[2]

[1] Information Systems and Networks Department, Lviv Polytechnic National University, Lviv, Ukraine
antonii.v.rzheuskyi@lpnu.ua, nek.lviv@gmail.com
[2] Information Technology and Analytics Department, Augustine Voloshin Carpathian University, Uzhgorod, Ukraine
kytnakryt@mail.ru

Abstract. The article presents the research methodology of the range of information services provided remotely by libraries of USA leading higher educational establishments. The methodology consists of benchmarking method and pairwise comparisons method on the basis of expert estimates. The authors reviewed the existing approaches to carrying out benchmarking researches and proposed their own concept. Innovation is that a comparative analysis of the results obtained in the process of benchmarking studies and the pairwise comparisons method is held. The library remote information services of higher educational establishments of America are analyzed. The study is based on benchmarking and pairwise comparisons methods that allow to identify the group of library-leaders in providing quality information services.

Keywords: Benchmarking · Library · Consumers of information services · Cloud services · Expert evaluation method

1 Introduction

Today, librarians have not solved the problem of developing a single integrated methodology for assessing the quality of information service, although some attempts in this direction were made. In particular, with sociological studies the evaluation of some library services by users was studied. But still the question remains to analyze the parameters of this assessment. Conducted studies are irregular, and to improve the libraries work, there is a demand on a toolkit for systematic analysis of information services provided by libraries. In our previous studies we found out that such toolkit can be considered a benchmarking methodology, which provides an opportunity to search, evaluate and borrow experience from the best algorithms for the organization of effective work. Benchmarking is a part of systematic and structured approach for finding and implementing the best examples of excellence. In librarianship benchmarking can be used for comparative analysis of one service to others. The purpose of the paper is to define USA libraries-leaders of providing high quality remote information service for users.

N. Shakhovska and V. Stepashko (eds.), *Advances in Intelligent Systems and Computing II*, Advances in Intelligent Systems and Computing 689,
https://doi.org/10.1007/978-3-319-70581-1_32

2 Background

Benchmarking is widely used in the economic field. Particularly foreign researchers who turn to this methodology should be called: Wann Jong-Wen, Lu Ta-Jung, Lozada Ina, Cangahuala Guillermo [1], Dalalah Doraid, Al-Rawabdeh Wasfi [2], Franses Philip Hans, de Bruijn Bert [3], Kwon He-Boong, Marvel Jon H., Roh James Jungbae [4]. Among the researchers who using benchmarking in the context of librarianship worth noting: Vivas Salas Leonel Orangel, Briceno Sosa Maria Alejandra, Colls Ojeda Janeyra del Carmen [5], Amos H., Hart S. [6].

Let's consider the main advantages of using the benchmarking methodology in library science:

- Firstly, it contributes to the objective analysis of the strengths and weaknesses of the library work. Such monitoring of strengths and weaknesses is carried out in the context of comparison with the standards.
- Secondly, it allows to study and implement new ideas in the organization of information and library work, marketing services and in the other areas of activity to increase their efficiency.
- Thirdly, it ensures an analysis of the activity of library-leaders for the borrowing of innovative technologies.
- Fourthly, it allows the strategic planning of the library institution activities with the use of advanced technologies, implemented by leaders of the area.
- Fifthly, it helps to conduct regular researches that allows to keep up to date with innovations in librarianship and actively implement them in own institution.
- Sixthly, it allows to overcome the conservative principle – to plan from the achieved, based on the analysis of the libraries activity, which constitute a certain competition [7].

By adapting the methodology of benchmarking to improve the management of social institutions activity, research should be directed to the study, borrowing and implementation of the best technologies, innovative processes and methods of work organization in order to create and disseminate information products among own users.

The use of this method allowed us to develop algorithms for system analysis of the activity of the investigated library and the library-leaders. The advantage of this method is the ability to conduct studies to determine the variability of certain trends in a number of parameters of the library activity for a certain time. The experience has shown that using the method of data comparative analysis allows us to concentrate on studying of one parameter at a certain time.

Taking into account the algorithm, defined by R. Kemp [8], the benchmarking comparisons in the study of the library should be carried out in several stages:

1. Identification and definition of object reference comparisons.
2. Selection of standards and experts of benchmarking stage.
3. Identification of suitable methods of information gathering and data collection.
4. Identification of existing inconsistencies and gaps of the investigated library from libraries-competitors for selected indicators.

5. Establishing the desired results and levels of effectiveness of library activity.
6. Informing representatives of all interested parties about the benchmarking results and obtaining consent for their use.
7. Setting goals and objectives of increasing the efficiency of library activity.
8. The development of an action plan to achieve the objectives.
9. Carrying out planned events and analyzing their results. Integration of experience in the work processes of one's own institution.
10. Re-analysis of selected strategies with using benchmarking.

A simplified version of the use of benchmarking consists of four sequential steps:

1. Awareness and analysis of the details of one's own activities
   In this phase, problem areas of the library activity are set up, which should be investigated with the help of benchmarking. This enables a critical attitude to the library activity as a whole, individual processes or work of departments. At this stage, a decision is made about the type of benchmarking.
   As practice shows, due to the regular compilation of reports on the library work, every section of its activity is thoroughly analyzed, that allows to identify the weaknesses.
2. Analysis of the processes of other libraries
   Having substantiated the goals, it is necessary actively to look for the best libraries that will serve as standards. The potential standards should be the leaders, but also convenient for the most simple comparability. Thus, the dynamics of specific work indicators is analyzed.
3. Comparison of work results of investigated library with the activity results of the library-leaders. At this stage, the collection of additional data of particular value, the analysis of work content, processes or factors associated with their effectiveness is carried out. The analysis in this case means not only finding the similarities and differences, but also to determine causal relationships.
4. Introduction of qualitative or quantitative changes to overcome the lag.

3 An Own Approach to the Benchmarking Research of Library Information Services

In the research process of information library service, benchmarking methodology was used to analyze the service provided with higher educational establishments of USA library websites. We consider that benchmarking investigation of libraries should be held by the following algorithm (Fig. 1).

1. To identify the maximum amount of libraries. For benchmark comparison, the American higher educational establishments scientific libraries were covered. At this stage the problem areas in libraries activity were setting that should be investigated with the benchmarking method. This enables a critical attitude to the library activity in general, individual processes or department units.

2. Determination of the maximum range of services provided by the library institutions. The comparison of the investigated library processes with the results of library-leaders. Substantiated goals it should actively seek the best libraries that act as benchmarks. Potential benchmarks must be leaders, but inherent to the most simple comparability. In this case the dynamic of specific indicators of work is analyzed.
3. At this stage, the collection of extra data, which are especially valuable, content analysis of library processes or factors related to their efficiency is made. To get the information the method of "competitive intelligence" is used. The application of "competitive intelligence" is carrying out a quick search of necessary information

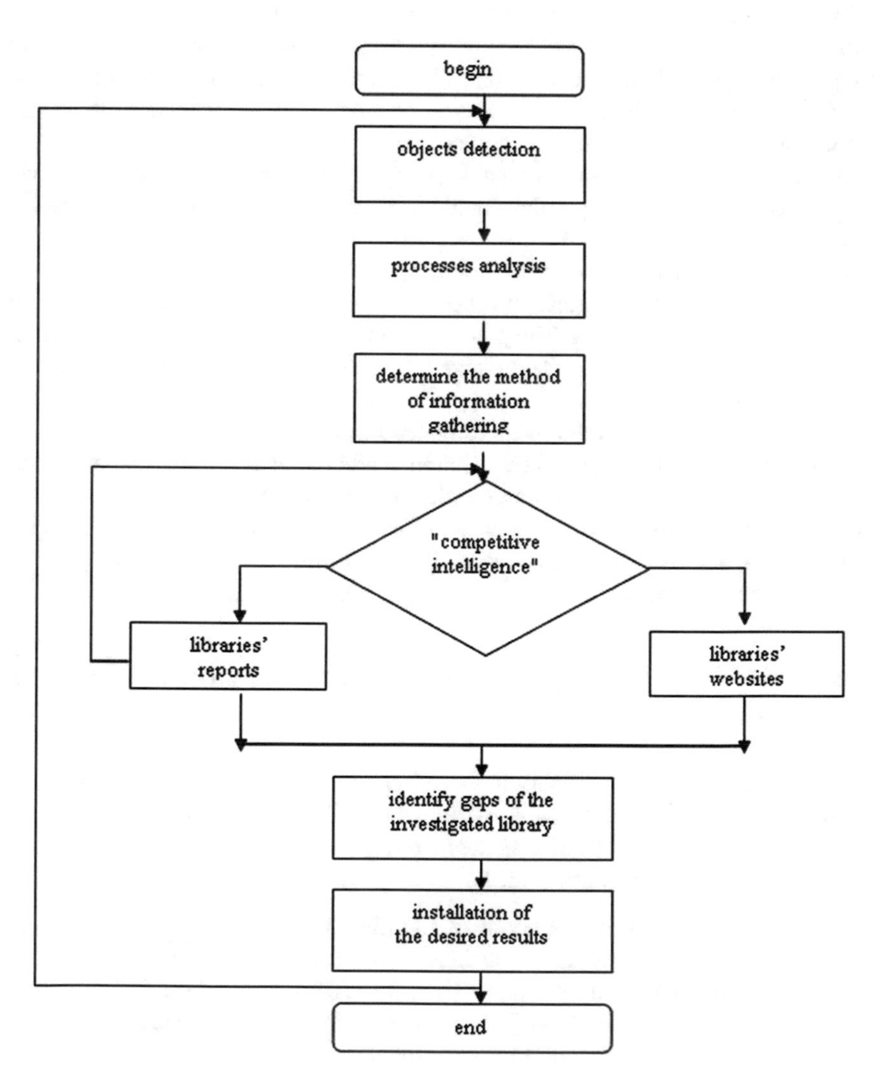

Fig. 1. The algorithm for studying the libraries of USA universities according to benchmarking the methodology.

and proper analysis, using legal methods of collecting and processing information, based on open source to obtain it.

4. Due to reporting about the work of the library, thoroughly every area of its activity has been analyzed, to determine weaknesses. Publicly available library reports serve as a reliable source for the analysis and comparison of library processes.
5. However, not all libraries place free access reports. Besides, these documents contain only statistics. So for a comparative analysis were used library websites as an open source of providing remote service for users.
6. Based on the information services provided by libraries via a web-site, the library-leaders as an example to follow of best practice were defined. The obtained data will overcome the lag in the information service.
7. The introduction of innovative services in the library, to which is used benchmarking methodology, will contribute to the modernization of library information service. The use of benchmarking methodology for effective management of the library should be systematic. Therefore, it is necessary to carry out continuous monitoring of other libraries and information institutions to identify and implement innovative technologies in the work of one's own library.

It was found that USA higher educational establishments scientific libraries provide the following range of remote services:

- electronic catalog;
- institutional repository;
- access to pre-paid databases;
- virtual reference;
- interactive communication services (skype, on-line assistant, etc.);
- electronic delivery of documents (EDD);
- multimedia resources;
- tools for help professionals;
- bibliographic managers;
- cloud services;
- recommended resources (guidance to public information resources in the Internet: databases, on-line journals, web-services).

4 The Research of Remote Information Services of USA University Libraries

Within the research, 19 library websites of higher educational establishments of the United States of America were analyzed. For comparative analysis, the amount of the libraries was divided into several groups. Binghamton University Library took the first place among the libraries in the group A (Fig. 2).

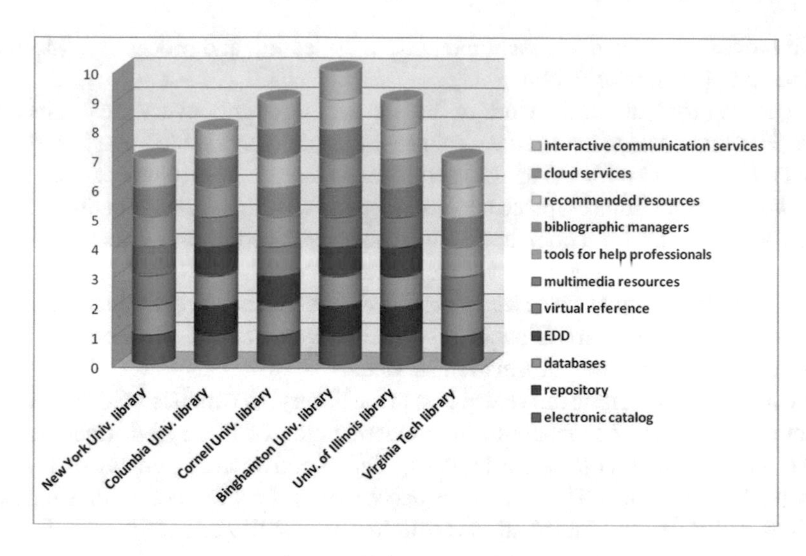

Fig. 2. The comparison of the libraries in the group A.

By the composition of library information services, the championship got Case Western Reserve University, and Brown University of Washington (Fig. 3).

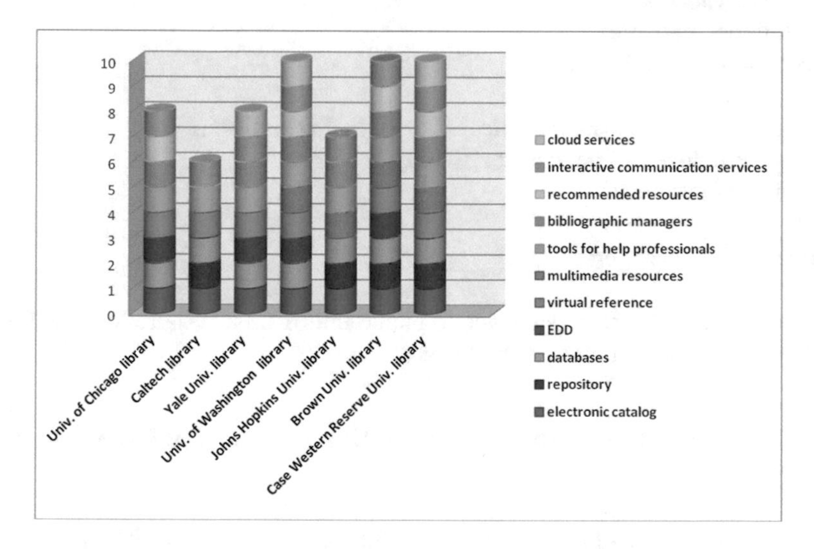

Fig. 3. The comparison of the libraries in the group B.

On this list of the libraries, the leader is Rutgers University Library (Fig. 4).

Benchmarking technology allowed to identify the library-leaders of the United States of America that provide a wide range of the remote library services (Fig. 5).

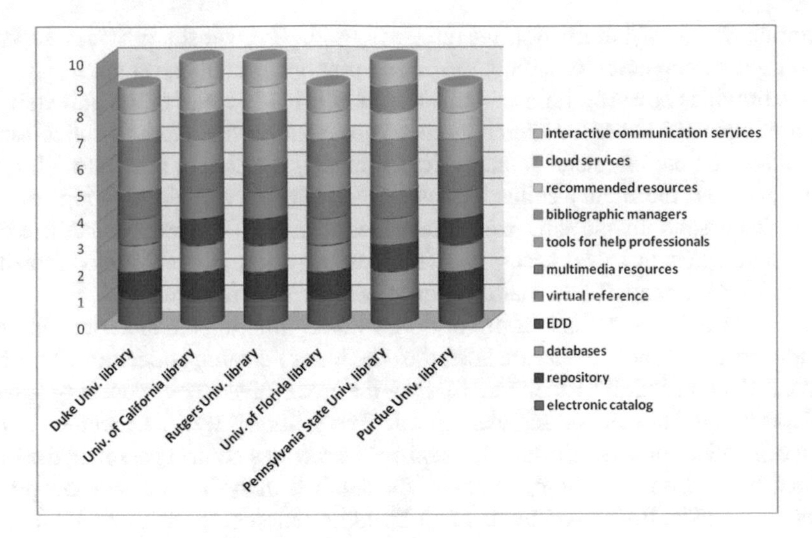

Fig. 4. The comparison of the libraries in the group C.

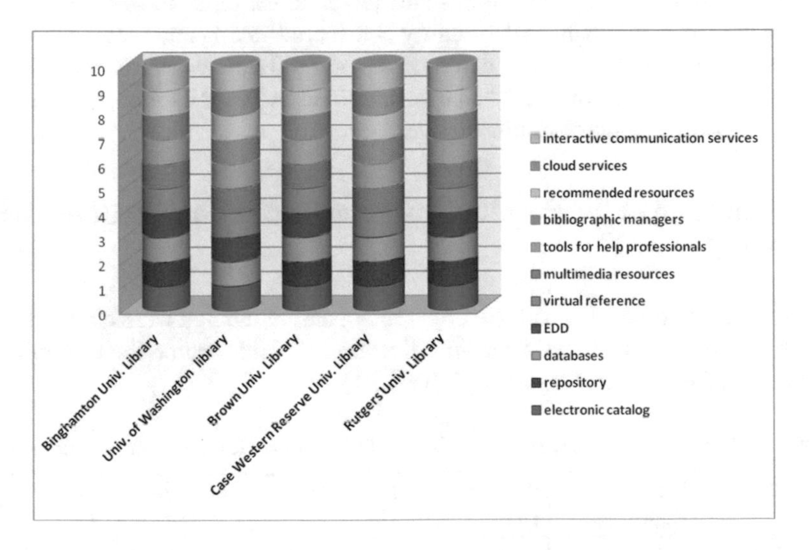

Fig. 5. The comparison of library-leaders.

5 Summarizing the Results of the Research with the Benchmarking Methodology

It was found that USA universities libraries provide users with the ability to use the system of electronic catalogs; repository, which stores theses, articles, abstracts of the university employees; and a wide range of pre-paid and free of charge databases. The decisive fact is that the USA library practice the individual approach to assist the researchers – the curator of a particular area of knowledge, which is guided in an appropriate range of information resources is used. The website provides supervisor (subject guide) contact information: the phone number, e-mail and skype.

This provides powerful information support of scientific staff of the university. The users are suggested to use the open reference resources, on-line journals collections that are available on the Internet. Cloud services include provision of access to full texts of documents from the library collections via OverDrive service for a certain period of time, which essentially plays the role of an electronic loan. It is proposed to use bibliographic management tools, access to which is provided from libraries websites to improve the efficiency of information support of scientific researches.

The weak point is the lack of promotional library multimedia materials. However, the audio and video documents are hosted in the library catalog collections. There is an electronic document delivery service (EDD) – the user is given the opportunity to receive the electronic documents on his own email. Every library website includes a virtual reference service and chat – instant messaging users with a contact person in the library.

Thus, by selecting the library-standard, the detailed analysis of the services provided to users was made. Benchmarking is not a blind imitation of the work of another institution, its main task – to borrow best practices and adaptation in the studied library. It was considered that not all innovations can facilitate the efficient work of one's own library. Each proposal should be thoroughly considered, taking into account the realities of one's own library. Some ideas that at first sight did not deserve of imitation, after some rethinking and improvement had been revalued and their implementation contributed to achieving significant results.

6 Verification of the Results Using the Saati Pairwise Comparisons Method

The results of the research, conducted by benchmarking method were checked with an expert evaluation method. Let's consider the results of this test on the example of the group C with the criterion "repository" (Table 1).

Table 1. The matrix of pairwise comparisons of alternatives for the criterion "repository".

Alternatives	Univ. of California	Rutgers Univ.	Univ. of Florida	Duke Univ.	Pennsylvania State Univ.	Purdue Univ.
Univ. of California	1,00	1,00	1,00	3,00	3,00	5,00
Rutgers Univ.	1,00	1,00	1,00	3,00	3,00	5,00
Univ. of Florida	1,00	1,00	1,00	3,00	3,00	5,00
Duke Univ.	0,33	0,33	0,33	1,00	1,00	3,00
Pennsylvania State Univ.	0,33	0,33	0,33	1,00	1,00	3,00
Purdue Univ.	0,20	0,20	0,20	0,33	0,33	1,00

According to this criterion this library service is compared for determination of the providing of users with remote information resources.

On the basis of an expert assessment method, it was determined the weight coefficients for service "repository" and constructed a matrix.

The results of calculating the weights of alternatives for the criterion "repository" are shown in the Table 1.

Elements of the table columns obtained with normalization of appropriate elements (Table 2).

Table 2. The weight of alternatives according to the criterion "repository".

Alternatives	Univ. of California	Rutgers Univ.	Univ. of Florida	Duke Univ.	Pennsylvania State Univ.	Purdue Univ.	Weight of alternative
Univ. of California	0,259	0,259	0,259	0,265	0,265	0,227	0,2554
Rutgers Univ.	0,259	0,259	0,259	0,265	0,265	0,227	0,2554
Univ. of Florida	0,259	0,259	0,259	0,265	0,265	0,227	0,2554
Duke Univ.	0,086	0,086	0,086	0,088	0,088	0,136	0,0952
Pennsylvania State Univ.	0,086	0,086	0,086	0,088	0,088	0,136	0,0952
Purdue Univ.	0,052	0,052	0,052	0,029	0,029	0,045	0,0432

Thus, by the criterion "repository" may be offered University of California library, Rutgers University library and University of Florida library (Fig. 6), because they have the same and more than the other libraries the weight value – 0.2554.

For the matrix of pairwise comparisons, based on the criterion "repository" the following parameters were calculated:

- evaluation of the largest eigenvalue, which is calculated according to the formula:

$$\lambda_{\max} = \sum_{i=1}^{n} w_i s_i \tag{1}$$

where w_i – weight alternative with a number i, s_i – sum of the elements of column number i matrix of pairwise comparisons, n – number of alternatives.

$$\begin{aligned}\lambda_{\max} &= 0.2554 \cdot 3.87 + 0.2554 \cdot 3.87 + 0.2554 \cdot 3.87 \\ &\quad + 0.0952 \cdot 11.33 + 0.0952 \cdot 11.33 + 0.0432 \cdot 22 = 6.073;\end{aligned} \tag{2}$$

- the index of consistency

$$CI = \frac{\lambda_{\max} - n}{n-1} = \frac{6.073 - 6}{6 - 1} = 0.0146; \tag{3}$$

- the index of the sequence of ratios

$$CR = \frac{CI}{RI} = \frac{0.0146}{1.24} = 0.0118 \tag{4}$$

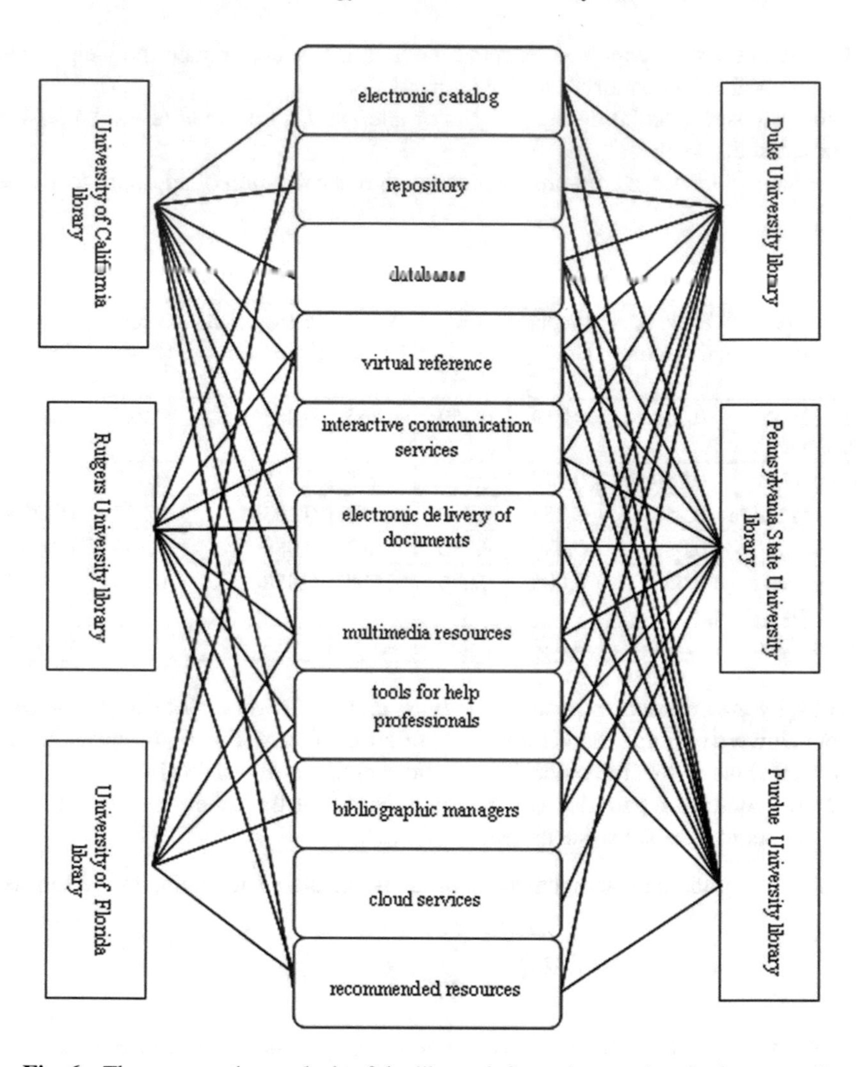

Fig. 6. The comparative analysis of the library information services in the group C.

Here $RI = 1,24$ – random index for $n = 6$, value of which is the same for all further calculations of the weights of alternatives. As $CR = 1.18\ \% < 10\ \%$ the matrix of pairwise comparisons is considered consistent.

7 Conclusions

According to the results of our research, the use of benchmarking methods in the study of libraries contributes to the improvement of the following directions of its activities:

- satisfying the information needs of the library users and the demand on document-information resources of the library;

- socio-communicative relations of the library with the external environment;
- improving quality and expanding the range of services;
- the use of modern technologies;
- traditional library activities in an innovative mode;
- creating a positive image of the institution.

Thus, using the benchmarking and pairwise comparisons methods, among 19 USA universities academic libraries were identified the leaders in providing the most comprehensive list of library and information services. The results, obtained on the basis of the benchmarking libraries, have been confirmed with expert evaluation method, which gives reason to keep using benchmarking method in librarianship studies.

References

1. Wann, J.-W., Lu, T.-J., Lozada, I., Cangahuala, G.: University-based incubators' performance evaluation: a benchmarking approach. Benchmarking Int. J. **24**(1), 34–49 (2017)
2. Dalalah, D., Al-Rawabdeh, W.: Benchmarking the utility theory: a data envelopment approach. Benchmarking Int. J. **24**(2), 318–340 (2017)
3. Franses, P.H., Bruijn, B.: Benchmarking judgmentally adjusted forecasts. Int. J. Fin. Econ. **22**(1), 3–11 (2017)
4. Kwon, H.-B., Marvel, J.H., Roh, J.J.: Three-stage performance modeling using DEA-BPNN for better practice benchmarking. Expert Syst. Appl. **71**(1), 429–441 (2017)
5. Salas, L.O.V., Sosa, M.A.B., del Carmen, J., Ojeda, C.: Benchmarking applied to catalog of library services at the University of Los Andes. Visión Gerencial **1**, 59–72 (2017)
6. Amos, H., Hart, S.: International collaboration for quality: a case study of fostering collaboration across an international network of University Libraries through an activity based benchmarking project. In: 34th Annual IATUL Conference Proceedings, Purdue e-Pubs (2013)
7. Rzheuskiy, A., Kunanets, N.: The concept of benchmarking in the librarianship. Coll. Sci. Works Econ. Sci. **10**, 17–24 (2014)
8. Camp, R.C.: Benchmarking. The Search for Industry Best Practices that Lead to Superior Performance. ASQC Industry Press, Milwaukee (1989)

The Method of Big Data Processing for Distance Educational System

Natalya Shakhovska(✉), Olena Vovk, Roman Hasko, and Yuriy Kryvenchuk

Artificial Intelligence Department, Lviv Polytechnic National University, 12 S. Bandera Street, Lviv 79012, Ukraine
Nataliya.b.shakhovska@lpnu.ua

Abstract. The paper presents the main features of modern educational systems such as huge amount of information and requirements for data processing. The Big data definition and the main characteristics description The model of association between entities and characteristics is constructed. The method of heterogeneous data sharing and bringing to relational data model "entity-characteristic" was created. The testing results of developed methods and algorithms are presented.

Keywords: Big data · XML · NoSQL · Relation database · Distance education

1 Introduction

The distance education is an innovative form of learning that helps overcome barriers to obtaining the necessary knowledge and skills. The distance learning attracted the attention of scientists as a theoretical and trainers, educators and practitioners. Distance education is a promising way for the development of vocational education. Scientists are exploring the development of distance education in the system of higher education institutions. It should be noted that most of the researchers are devoted to a greater extent analysis technology implementation processes of learning at a distance. And almost no issue highlights the specific processes of distance learning of engineering students.

One of the modern forms of education, which was formed on the basis of active the use of new information technologies for distance learning is based on the active use of remote information systems training for engineering students as program-algorithmic and technical complexes that implement a set of information technology for selection, registration, transmission, storage, processing and presentation of educational information, given the specificity of the audience and laboratory work. With stipulates that distance learning emerged as integration complex of modern information technology, telecommunications, and modern educational methods and allows implementation process of remote training of students using interactive educational tools and organize communication students both within the group and with teachers, regardless of where they are geographically. "Thin clients" (laptops, pocket PCs, smart phones, mobile phones and so on) can provide presentation software and information components of distance learning in the languages of HTML, XML, where the interface part provides a

N. Shakhovska and V. Stepashko (eds.), *Advances in Intelligent Systems and Computing II*, Advances in Intelligent Systems and Computing 689,
https://doi.org/10.1007/978-3-319-70581-1_33

web-browser, and significantly improve mobility distance learning. The web technology usage enables mobile organize an effective job of distance learning, which should provide flexible expansion and shift functions to perform the kind of educational tasks solved by remote education specified time. But there is no information about remote laboratory preparation [9].

Modern education systems produce a huge amount of information. The size of individual databases is growing very fast and overcame barrier in PB. Most of the data collected are not currently analyzed, or is only superficial analysis [1].

The main problems that arise in the data processing in education system is the lack of analytical methods suitable for use because of their diverse nature the need for human resources to support the process of data analysis, high computational complexity of existing algorithms for analysis and rapid growth of data collected. They lead to a permanent increase in analysis time, even with regular updating of hardware servers and also arise need to work with distributed database capabilities which most of the existing methods of data analysis is not used effectively. Thus, the challenge is the development of effective data analysis methods that can be applied to distributed databases in different domains. It is therefore advisable to develop methods and tools for data consolidation and use them for analysis [6].

Big Data information technology is the set of methods and means of processing different types of structured and unstructured dynamic large amounts of data for their analysis and use for decision support. There is an alternative to traditional database management systems and solutions class Business Intelligence. This class attribute of parallel data processing (NoSQL, algorithms MapReduce, Hadoop).

The main problems that arise in the data processing is the lack of analytical methods suitable for use because of their diverse nature the need for human resources to support the process of data analysis, high computational complexity of existing algorithms for analysis and rapid growth of data collected. They lead to a permanent increase in analysis time, even with regular updating of hardware servers and also arise need to work with distributed database capabilities which most of the existing methods of data analysis is not used effectively. Thus, the challenge is the development of effective data analysis methods that can be applied to distributed databases in different domains. It is therefore advisable to develop methods and tools for data consolidation and use them for analysis.

This paper is extended version of [10].

2 Analysis of the Information Resources

Big data is a term used to identify data sets that we cannot cope with existing methodologies and software tools because of their large size and complexity. many researchers are trying to develop methods and software tools for data mining or information granules of Big Data [1, 2].

Big Data features are:

- working with unstructured and structured information,
- orientation on the fast data processing

- leads to the fact that traditional query language is ineffective while working with data.

The purpose of the article is to formally describe Big data model and carriers distinguishing and sharing methods.

One of the adapting concepts not only of relational data is NoSQL. The followers of the concept of NoSQL language emphasize that it is not a complete negation of SQL and the relational model, but the project comes from the fact that SQL - is important and very useful tool, that cannot be considered as universal. One problem that point for classical relational database is a problem of dealing with huge data and projects with a high load. The main objective approach is to extend the database if SQL flexible enough, and not displace it wherever it to perform its tasks.

The basis of ideas of the NoSQL are the following [7]:

- non-relational data model,
- distribution,
- open output code,
- good horizontal scalability.

As one of the methodological approach of NoSQL studies used a heuristic principle, a theory known as CAP (Consistence, Availability, Partition tolerance - «consistency, availability, resistance to division»), arguing that in a distributed system cannot simultaneously provide consistency, availability and resistance to splitting distributed system into isolated parts. Thus, if necessary, to achieve high availability and stability of the division is expected not to focus on the means to ensure consistency of data provided by traditional SQL-oriented database with transactional mechanisms on the principles of ACID.

Non-strict proof of the CAP theory is based on a simple reflection. Let the distributed system consisting of N servers, each of which handles the requests of a number of client applications. While processing a request the server must ensure the relevance of the information contained in the response to the request is sent, which previously required synchronizing the contents of his own base with other servers. Thus, the server must wait a full synchronization or generate a response based not synchronized data. However, for any reasons synchronization is only part of the servers. In the first case the requirement of availability is not performed, in the second – consistency and in the third - resistance to division.

The relation model is not suitable for parallel processing. That is why the method of relation data transformation is proposed.

3 The Information Model of Big Data

Formally we can divide all the objects on the following categories [3]:

- e - entities;
- f - characteristics;
- associations between e - entities and f –characteristics.

Therefore, the information model of Big data is triple [4]

$$BigD = \langle e, f, a \rangle,$$

where $e \in E$ is entity, $f \in F$ is characteristic, $a \rightarrow n_{e,f}$ is association between entity e and characteristic f.

From the relational model the model “entity-characteristic” is distinguished by the association with a value ranging from 0 to 1. Opposite to proposed model the relational model is a subspecies of the model “entity-characterization” association with a value equal to 1 for each substance and characteristics associated with it.

For each entity e we calculate importance of association with f. It is necessary to normalize the meaning of importance (weight):

$$V(e,f) = \frac{\left(1 + \log_2\left(n_{e,f}\right)\right)) \cdot \log_2\left(\frac{|E|}{e(f)}\right)}{\sqrt{\sum\left(\left(1 + \log_2\left(n_{e,f}\right)\right) \cdot \left(\frac{|E|}{|e(f)|}\right)\right)^2}}$$

For each entity we calculate weight $V(e,f)$. That’s is why r-statistics [8] between objects E_1 and E_2, is distance between vectors $\left(V(e_1,f), V(e_2,f), \ldots\right)$.

3.1 The Method of Relational Data Conversion in to XML-Database Using Model “Entity-Characteristic”

The main purpose of converting relational data expansion is nearly used Xml-description in XML-enabled fuzzy membership functions. For fuzzy extension we propose to use XML modern architecture MVVM (“View-Model”, “Model-View”) [5]. Its main difference from the well-known architectural software application MCV (Model-View-Controller) is the lack of data binding requirements for their presentation. Under this approach we will show the necessary components designed for automated conversion of a relational database in XML model taking into account the features of the model “entity value” (Fig. 1).

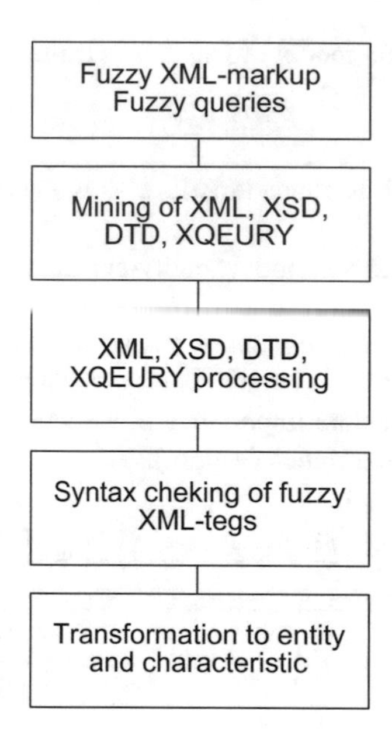

Fig. 1. The scheme of conversion tables in relational XML-database incl model "entity-characteriistic"

The architecture of the applications that are implemented with the structure as shown in Fig. 1, there are several components that require further explanation. One of these components - fuzzy XML-markup. This term means a Xml-description that allows description of fuzzy variables and functions.

Implementation of fuzzy functions provided in the fuzzy XML- document is implemented by means of an external application that has the appropriate computing power.

For the convenience of universal description of fuzzy elements, we used Xsd description, the type of which is given below. The membership function in this example presents Minvalue limited value and Max Value.

```
<>:s : schema id="Fuzzyshema " elementformdefault="F qualified" xmlns
:xs="http://www.w3.org/2001/XMLSchttp">
<xs:complextypename="fuzzy">
<xs:sequence>
<xs:elementname="fuzzy"minoccurs="1"maxoccurs="un-bounded"
type="functions7>
</xs:sequence>
</xs:complextype>
<xs:complextypename="functions">
<xs:sequence>
<xs:elementref="function"minoccurs="1"maxoccurs="unbounded"/>
</xs:sequence>
<xs:attributename="name"type="xs:string"/>
</xs:complextype>
<xs:elementname="function">
<xs:complextype>
<xs:simplecontent>
<xs:extensionbase="xs:string">
<xs:attributename="minvalue"type="xs:string"/>
<xs:attributename="maxvalue"type="xs:string"/>
</xs:extension>
</xs:simplecontent>
</xs:complextype>
</:ts:element>
</xs:schema>
```

The method of converting relational data model to “entity-characteristics” includes the following steps.

Step 1. Create a root element of the XSD schema for the data model FDB.
Step 2. For each entity Efm (i) there is required to create separate element cxcmbi and place it under the root element.
Step 3. For each attribute Attfm (j) for entity Efm (i) there is required to create attribute xs: attribute R XSD, place it inside an appropriate description of the entity and specify its type.
Step 4. For each attribute select and specify the data type and define thresholds.

In the real situation we must filter date. Figure 2 illustrates filtering for data sources in the form of Xml-files from ADM.

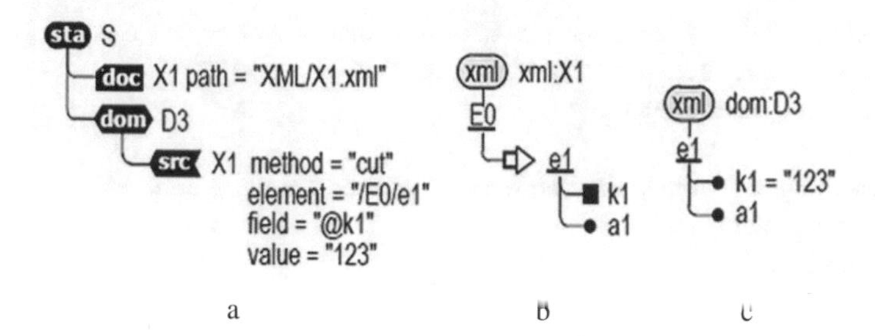

Fig. 2. The example of Dom-object filtering Xml-data: a part of the dynamic model; b - Model Xml-data sources; in - Xml-model data is loaded into the Dom-object

In the dynamic model in Fig. 2, the root condition sta: S Xml-definition sets Document doc: X1, continuing to ADM, and Element Dom-dom: D3, in which the document provided for downloading doc: X1 filtering. Data source src: X1 has an attribute method = "cut", which indicates that the source Xml-document will be "cut" subtree. Additional attributes of «element», «field» and «value» set filtering condition. Attribute «element» contains Xpath-expression that defines a plurality of nodes elements in Xml-Tree, one of which is the root of the downloaded subtrees. Attribute «field» contains Xpath-expression that defines a subtree audited Xml-element or attribute. Attribute «value» includes the required value. In an example we download Dom-object subtree that begins in the element "e1", attribute "k1" which has a value of "123".

The model b in Fig. 2. Xml-Data doc: X1 with XML- root element «E0» can contain child Xml-Elements «e1». Each of these elements contains attributes «k1» (ID) and «a1». Thus, Xpath-expression «/ E0 / e1», specified in the attribute «element», has source src: X1, but addresses all set Xml-items «e1»; attribute "field" addresses the "E1" Xml-Attribute "k1", attribute "of value" asking for a desired value "123".

During processing element dom: D3 interpreter refers to the source src: X1, takes Xml-elements "e1" finding someone who Xml-Attribute "k1" is the desired value "123", and loads the appropriate subtree in Dom-Ob' object. As a result, Dom-object «D3» will be downloaded Xml-data corresponding to the model shown in Fig. 2c.

The streaming algorithm of elements processing of source document starts with initialization tool streaming input Xmlreader.

Next step in the loop is executed streaming node reading from a document source. We check whether the finding element is present in the source attribute, and in this case checked is whether the processed node is desired. The loop is moving names from the list given in the attribute wanted, compare with the processed node name in the case of coincidence checked. At the end of the loop we check the box matches and, if unchecked, the node processing is completed and we are directed to the next processing. If desired processed node is recognized, it is checked for fulfillment of filtering specified attribute cond. These conditions require treatment to other components of the document input. For this item processed (with contents) produced in the cache. Then in a series of "cond" attribute this kind of conditions obtained and tested for processed node is in the cache. If the processed node passed all inspections, he joined as a child element to the target element machined parent document. Further purification is carried captive subsidiary element of internal elements whose names are given in the attribute ignore. Once processed node source document verified attached to the target element and refined, for a worked out recursively internal data sources (Fig. 3).

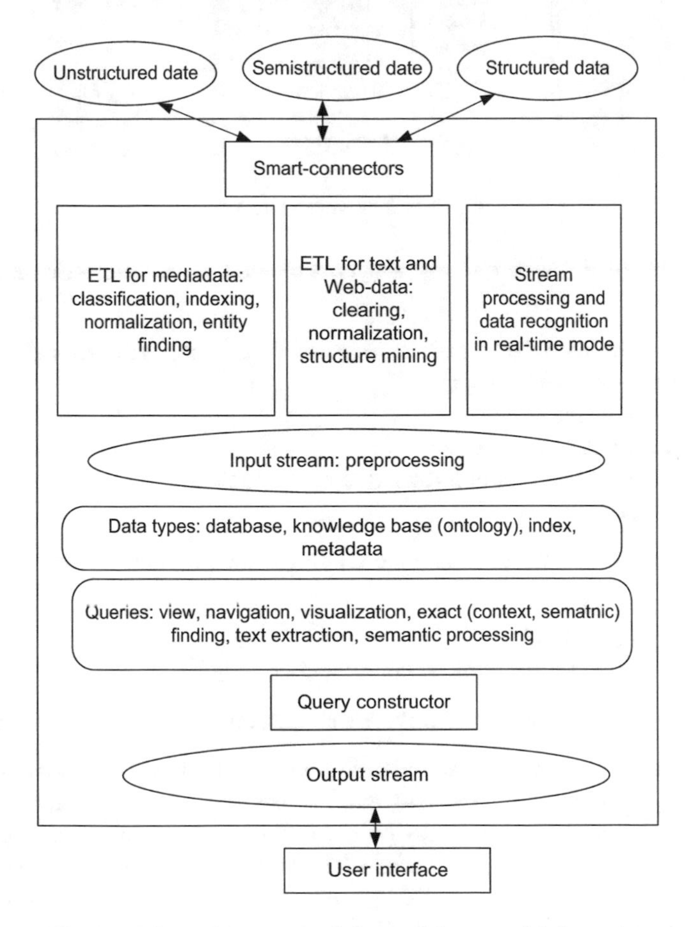

Fig. 3. Universal structure mining multistructural information

3.2 The Structure of the Remote Education Resource Center

The main task of information technology for remote education resource center (RERC) is organizing and conducting a full educational process. Additionally, the system can be used for remote control of knowledge and become a didactic tool for self-training.

The overall structure of the center is shown on Fig. 4.

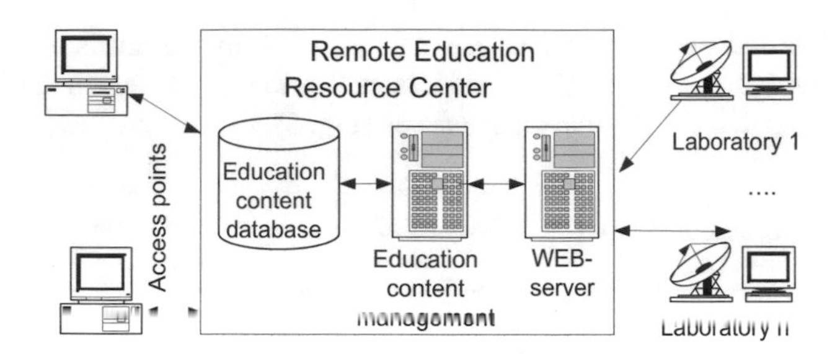

Fig. 4. The overall structure

The system of distance training and consultation center performs the following functions:

- creation of learning objects, which are available as part of educational materials;
- learning Content Management;
- remote student management;
- adaptive Learning Organization;
- testing and remote control of knowledge of the student;
- statistical data analysis.

The laboratory on the Fig. 4 has such technical and program features as:

- thin client as workplace;
- camera for tutor controlling;
- the system OrCAD PSpice for practical WORK.

The following learning modes of the RERC structure in Fig. 4 are available:

1. Testing students' knowledge and learning. For this purpose there is used a database of training materials, training and management server content and Web-server. Student through collective access points (remotely or from the workplace at the university) accesses training materials. The student' testing can be used for assessment of his knowledge (of the interim evaluation) and for selfeducation.
2. Implementation of laboratory and practical work in the laboratories of other universities. In this mode, there are involved laboratories with established hardware and software simulation of analog electronic circuits, a database of training materials, training and management server content and Web-server. With established in the laboratory Web-cameras (one for 2-3 workgroups) teacher can observe the work of the student. This mode of learning is close to the regime of passing certification in IT companies.
3. Implementation of laboratory and practical work in the laboratories of the University. The difference from the previous mode of operation is that the teacher not only oversees the work, but also can adjust it by removing the relevant messages in the system.

3.3 The Results Analysis

The overall structure of the center (Fig. 4) implies the existence of heterogeneous components:

- Relational database (Microsoft SQL Server Database Services, Oracle Database, MySQL, PostgreSQL) 64-bit performance;
- Multidimensional database (Microsoft SQL Server Analysis Services or Hyperion Essbase) 64-bit performance;
- "Hierarchical" database (MongoDB);
- The control system of federated data warehouse, which is a separate program designed specifically to ensure that the data store, and includes file storage.

In order to develop a federal system management data warehouse platform used Microsoft.Net, C# language and development environment Visual Studio. Class Library that comes with .net and high-level language C#, and methodology RAD (rapid application development), which built Visual Studio development environment to quickly build application-oriented database.

In the first study the federated data sources carried. We analyze the relative number of objects or documents available data sources to the total number of objects that hit the federated repository. For the feasibility of different sources, we must have data directory (Data catalog), which is also posted on the Web server. Data catalog is a register of data resources system that contains basic information about each of them: source name, location, size, creation date and owner, etc. Product is infrastructure for most other services RERC dataspace.

Data catalog structure is shown on Fig. 5.

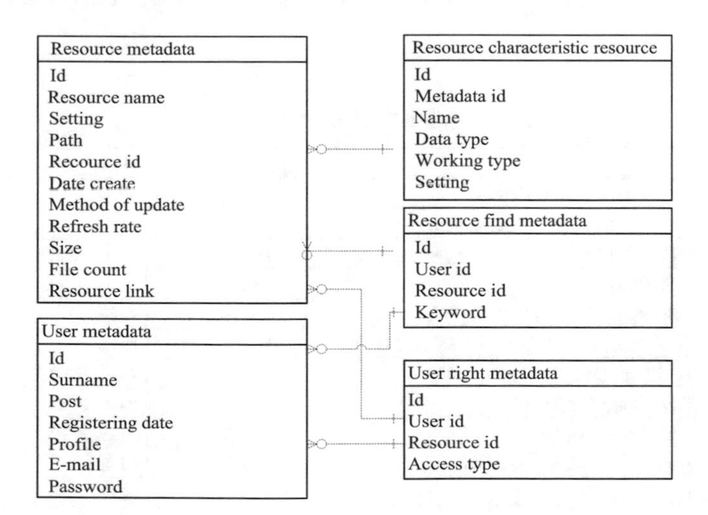

Fig. 5. Data catalog structure

It also indicated the data structuring and placement in storage (serial number field and the type of database).

The next step is design of query processing circuit in the system center (Fig. 6). It consists of the following components:

- CEO server component (Main),
- authentication and authorization component (Auth),
- component to ensure quality of service (Qo),
- component, which provides safety information about the source (Catalog),
- management associations (Class),
- quality control data (Qo).

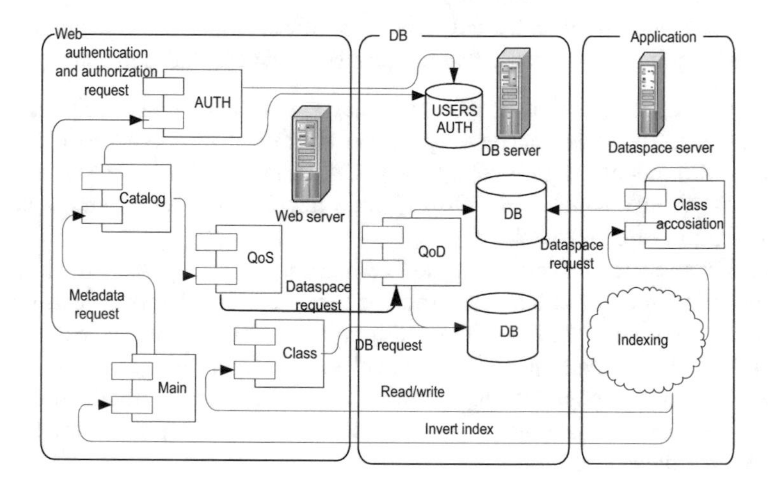

Fig. 6. The scheme of processing queries in Remote education resource center using Big data technology

Table 1. Testing data

Source	Weight	Destination
MFUVUDB.t_aspnet_Roles	0	RDB
MFUVUDB.t_mfuvu_Settlements	1	RDB
MFUVUDB.t_vw_mfuvu_list_sti_correspond	0.666942	RDB, XML
MFUVUDB.t_vw_mfuvu_Users	0.401167	RDB, XML
TransBudgDB.t_VXXDDMMYY	0.961365	RDB
TransBudgDB.t_D_BUDG_LOCAL_DET	0.888913	RDB
TransBudgDB.t_D_ECON_CRED	0.9	RDB, XML
TransBudgDB.t_D_FIN	0.916667	RDB, XML
TransBudgDB.t_DMYYMM	0.533333	RDB, XML
TransBudgDB.t_D_INC_DET	0.903839	RDB
TransBudgDB.t_district	1	RDB
TransBudgDB.t_obl_region	1	RDB
TransBudgDB.t_VW_EXPENSES_DET	0.5	RDB, XML
FUPortalDB.t_People	0.227109	RDB, XML
FUPortalDB.t_Conclusions	1	RDB
FUPortalDB.t_aspnet_Applications	0.666667	RDB, XML

The next step is data source's weight calculation (Table 1).

After that we analyze the quality of data processing method (Fig. 7).

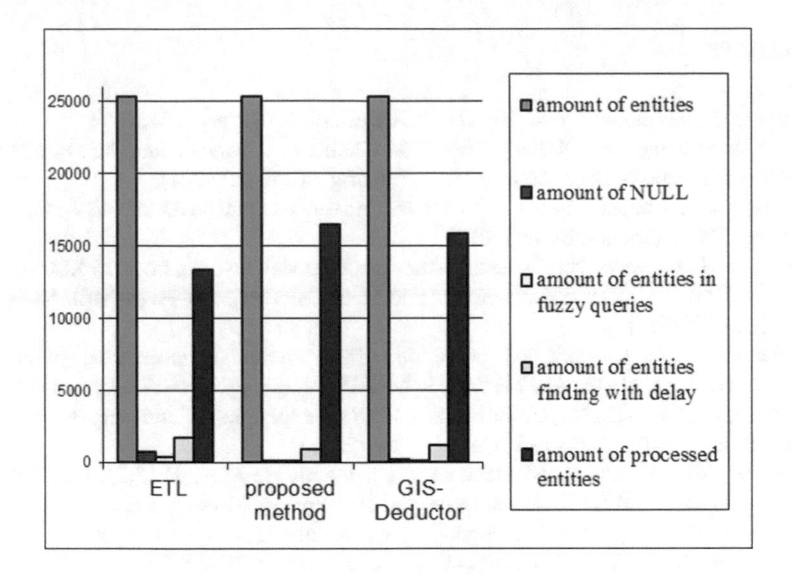

Fig. 7. The quality of data processing

In the submitted work diagram analysis algorithm federal inquiry. The algorithm compared to unmodified integration of the algorithm used in the Oracle Data Integrator. Data in federated repository database come from various institutions known data structures are unknown. The number of input records databases have to get into federated data warehouse - 15000.

4 Conclusions

The paper solved important scientific task of organizing and integrating information resources using Big data technologies for distance learning system.

The main defines of Big data are given and the main characteristics are described. In the first chapter there are analyzed mathematical means submission and processing of Big data and define their limitations. There are posted software for working with large data.

The formal description of Big data is defined. There are posted patterns of associations between entities and characteristics for various categories NoSQL databases. However, the main operation is no consolidation of integration and federalization, allowing capacitive reduce the complexity of requests.

The method of heterogeneous data sharing and bringing to the relational data model "entity-characteristic" is developed. The software created for developed methods and algorithms testing.

To achieve the desired goal, the development of a formal model of the Big Data information technology is made and its structural elements are described.

References

1. Laney, D.: The Importance of 'Big Data': A Definition. Gartner (2012)
2. Beyer, M.: Gartner Says Solving 'Big Data' Challenge Involves More than Just Managing Volumes of Data. Gartner, Archived from the original, 10 July 2011
3. Di Ciaccio, A., Coli, M., Ibanez, J.M.A. (eds.): Advanced Statistical Methods for the Analysis of Large Data. Springer, Berlin (2012)
4. Veres, O., Shakhovska, N.: Elements of the formal model big date. In: 2015 XI International Conference on Perspective Technologies and Methods in MEMS Design (MEMSTECH), pp. 81–83. Lviv (2015)
5. Shakhovska, N.: Consolidated processing for differential information products. In: Perspective Technologies and Methods in MEMS Design, pp. 176–177. Polyana (2011)
6. Fang, L., Sarma, A.D., Yu, C., Bohannon, P.: Rex: explaining relationships between entity pairs. Proc. VLDB Endowment **5**(3), 241–252 (2011)
7. Chen, M., Mao, S., Liu, Y.: Big data: a survey. Mobile Netw. Appl. **19**(2), 171–209 (2014)
8. Van der Waerden, B.L.: Mathematische Statistic. Springer-Verlag, Berlin (1957). English. Translation of 2nd (1965) ed. Springer-Verlag, Berlin and New York (1969)
9. Educational Technology & Society 5(1), pp. 215–221 (2002). ISSN 1436-4522
10. Shakhovska, N.: The method of Big data processing. In: 2017 XII International Conference on Computer sciences and information technlogies (CSIT), pp. 122–125. Lviv (2017)

Developments and Prospects of GMDH-Based Inductive Modeling

Volodymyr Stepashko(✉)

International Research and Training Centre for Information Technologies and Systems of the NAS and MES of Ukraine, Kyiv 03680, Ukraine
stepashko@irtc.org.ua

Abstract. The article provides information on the historical development of the scientific direction of inductive modeling, originated by Ukrainian scholar Professor Oleksiy Ivakhnenko in 1968 with creation of his Group Method of Data Handling, as well as characterizes the basic fundamental, applied and technological achievements. The term inductive modeling can be defined as a self-organizing process of evolutional transition from initial data to mathematical models reflecting some patterns of functioning objects and systems implicitly contained in available experimental, trial or statistical data.

The structured information is presented on the development of GMDH-based inductive modeling in Ukraine and abroad, main fundamental, technological and applied achievements are characterized, as well as the most prospective ways of further research are formulated. The performed survey of the research state in the field of inductive modeling shows that GMDH is one of the most powerful methods of data mining and a promising basis for creating modern information technologies for discovering knowledge from observation data.

Keywords: Inductive modeling · Self-organization · GMDH · Data mining Knowledge discovery · Survey

1 Introduction

The very first article on the group method of data handling (GMDH) was published in 1968 by Professor Oleksiy Ivakhnenko in the Ukrainian journal «Avtomatyka». This article was republished in the USA [1] and followed by next important publications [2, 3] abroad which manifested the creation of new scientific direction called now «inductive modeling of complex systems and processes». Since then, the new modeling method is being recognized by researchers around the world and various GMDH-based techniques and technologies are intensively developing till nowadays.

The author called this scientific direction variously: «heuristic self-organization of models» [4], «self-organization of models from experimental data» [5], «inductive method for self-organizing models» [6], «inductive learning algorithms» [7]. Finally, in 1998, when the Department for information technologies of inductive modeling (ITIM) was established at the IRTC ITS of NAS of Ukraine, the direction obtained its short name «Inductive Modeling» being now generally accepted for scientific forums, articles and books in Ukraine and abroad.

N. Shakhovska and V. Stepashko (eds.), *Advances in Intelligent Systems and Computing II*, Advances in Intelligent Systems and Computing 689,
https://doi.org/10.1007/978-3-319-70581-1_34

The term *inductive modeling* can be defined as a self-organizing process of evolutional transition from initial data to mathematical models reflecting some functioning patterns of the modeled objects and systems implicitly contained in the available experimental, trial or statistical information under the uncertainty conditions. The task of inductive modeling consists in an automated construction of a mathematical model approaching an unknown regularity of functioning the simulated object or process.

An enormous amount of articles has published during the 50-years period in the world on the subject of GMDH-based inductive modeling; some of them are reflected in surveys [8–11]. In this article, a structured information is presented on the development of inductive modeling in Ukraine and abroad, including the ITIM department as the follower of the Ivakhnenko's school; basic fundamental, technological and applied achievements are characterized, as well as the most prospective ways of further research development are formulated.

2 Main Features of the Inductive Approach to Model Building

Inductive modeling consists in construction of models based on empirical data, results of measurements, observations, experiments, when this statistical material is used to form mathematical models, plausibly describing the patterns displayed in the data. This makes it possible to classify GMDH as an effective method of data mining and computational intelligence, since it is aimed at automatic search and description of regularities with the choice of structure and parameters of linear, nonlinear, difference and other models based on data samples under condition of uncertainty and incompleteness of information.

Instead of the traditional *deductive* way of solution of modeling tasks «from general theory – to particular model», Ivakhnenko has formulated the principle of new *inductive* approach based on the intensive use of computers, as way «from particular data – to general model»: a researcher having observation data advances a hypothesis on the class of functions, forms procedure of automatic generation of many thousand variants of models in this class and sets the criterion of choice of the best model among all of generated ones. It gives the most objective result along with maximum activation of creative possibilities of the researcher.

The Ivakhnenko's idea of models self-organization consisted in maximal automation of the whole process of model construction from observation data. The GMDH is the basic instrument of the self-organizing modelling. This is a fundamentally new method of modeling from experimental data.

GMDH differs from other data-driven modeling methods by active application of the following principles:

(1) automatic generation of ever more complex structures of models,
(2) non-final decisions (multiple choice of models) in every stage, and
(3) successive selection of best models by external criteria for the construction of models of optimum complexity.

It possesses an original network-type multilayered procedure of automatic generation of model structures simulating the process of biological selection with the pair-wise account of successive features. For comparison and choice of best models, external criteria based on dividing the sample into two parts are used, thus evaluating the parameters and checking the models quality is executed on different subsamples. The sample division into parts with similar statistical characteristics helps to automatically take into account different types of uncertainties in the modeling process providing a possibility to avoid burdensome a priory assumptions.

3 Brief Historical Information About the Development of Inductive Modeling as a Scientific Area

In the 1970s, the method effectiveness was repeatedly confirmed by solving specific problems of modeling complex objects and processes in the areas of ecology, economy, technology and hydrometeorology [4, 5]. The clearness of the method led to a rapid growth of its popularity among researchers and practitioners around the world.

In the 1980s, intensive research continued on the application of GMDH to solving practical problems of modeling economic and ecological processes and systems [6], including that done by foreign authors [12, 13]. At the same time, research was evolved on the development of algorithms for self-organization in the field adjacent to modeling, namely in problems of classification and pattern recognition [6, 13].

In the early 1980s, A.G. Ivakhnenko established an organic analogy between the problems of constructing a model from noisy experimental data and passing a signal through a channel with noise [14]. This allowed building foundations of the theory of noise-immune modeling [15]. The main result of the theory: the higher is the noise level in the data, the simpler it should be the optimal model complexity. This means that under noise conditions, a reduced "nonphysical" model can have smaller error than the "physical" one.

In the same years, the so-called two-level predictive models were offered [15] for modelling of dynamics of cyclic processes, in particular economic and ecological ones. This was a new method of construction of systems of two-level difference models for cyclicities of the "cycle-time" type. The variables of upper (cycle-averaged) and lower (time-averaged) levels in their balance relationship are thus considered. A correspondent algorithm and its applications were described in [16].

In 1994, the Ivakhnenko's book [7] was published in the USA, containing the main results of his school of the 1980th to early 1990th. These results include such algorithms: objective system analysis (OSA [17]) intended for determination of input and output variables among all measured ones and building the system of difference equations of an object dynamics, and objective computer clusterization (OCC [18]) solving the tasks of clustering with the use of self-organization of the quantity and composition of clusters. Besides that, new nonparametric prediction algorithms based on correlation models, a reverse transformation of the transition probabilities matrix into forecast [19] and the group analogs complexing [20] were introduced.

In 1990th, the classical multilayered GMDH algorithm has commenced to be called as the "polynomial neural network" PNN [21]. The main result in these years was an

Ivakhnenko's idea on the development of GMDH as a "neuronet with active neurons" [22, 23]. The idea was to create a network of neurons in the form of GMDH algorithms, the resulting structure turns into a network with heterogeneous neurons tuned independently. In the same years, GMDH began to be classified as the successful means of intelligent data mining [24].

It is worth to remember other constructive and efficient ideas suggested and evolved by Ivakhnenko regarding various new algorithms of model construction in the following original model classes: harmonic trend models of oscillatory processes in the form of sum of harmonics with aliquant frequencies [25]; nonlinear additive-multiplicative models with automatic determination of non-integer degrees of input variables [26]; models of distributed processes dynamics in the form of difference analogues of partial differential equations [27].

4 Main Results of GMDH Theoretical Studies

In the area of the GMDH theory, algorithms for generating variants of model structures as well as criteria for selection of the best generated models are analyzed first of all. The main properties of the sorting-out GMDH algorithms are described in [15] while of the iterative algorithms in [28].

When studying any of multilayer (iterative) GMDH algorithms, the question of conditions for their convergence should be analyzed in the first line [28]. Historically this problem was examined in 1970–80th [29–31], the most comprehensive results was obtained in [28] and represented in [7]. It should be noted that all the achieved results was obtained only for the case of so-called "internal convergence" of the iterative algorithms, when the sample division is not taken into account and the residual sum of squares RSS is used as the "internal" selection criterion. The problem of "external convergence" of the algorithms for the case of using external criteria based on the sample division is still to be solved.

The task of analyzing the effectiveness of the model selection criteria was considered within the framework of the theory of noise-immune modeling [15]. Essential results are related to analysis of properties of the so-called "ideal" J-criterion being firstly introduced in [32] and investigated in [15, 33, 34]. Selective properties of the predictions balance criterion used for construction of systems of two-level difference models of cyclic processes was studied in [35].

The basic analytical apparatus of the theory of inductive modeling is the method of critical variances for analyzing the regularities of model choice under uncertainty conditions. This instrument, firstly proposed in [32], was applied in [15] and developed in [36–38]. The method is presented systematically in [39].

Asymptotic properties of the external criteria for models selection were studied in [40, 41] under assumption of infinite growth of the sample length. The GMDH theory foundations for objects with multidimensional output are also formed [42, 43].

5 General Characteristics of the Inductive Modeling Task

If a sample $W = [X y]$ of n observations over m input variables X and one output y is given, then the problem of constructing models from experimental data consists generally in finding the minimum of a given external criterion $CR(X, y, f)$ on a discrete set Φ of models of different structures of the form $\hat{y}_f = f(X, \hat{\theta}_f)$, where the parameter estimation $\hat{\theta}_f$ for each $f \in \Phi$ is a solution of additional task of continuous minimization of an internal error criterion $QR(X, y, \theta_f)$, $QR \neq CR$.

In general, the process of solving the problem of structural and parametric identification includes the following main stages: (1) setting a data sample and a priori information; (2) the choice or assignment of a class of basic functions and the respective transformation of data; (3) generation of different model structures in this class; (4) estimation of parameters of generated structures by an internal criterion $QR(\theta_f)$ and formation of the set Φ; (5) minimization of a given external criterion $CR(f)$ and selection of the optimal model f^*; (6) checking the adequacy of the model obtained; (7) completion of the process and/or use the model.

This means that each modeling method including GMDH can be described with help of four main components: (a) model class, (b) structure generator, (c) method for estimating parameters, and (d) criterion for model selection. Typical GMDH applications are modeling of nonlinear systems, forecasting complex processes, function approximation, recognition, classification, clustering and others.

6 Basic Components of the GMDH Algorithms

6.1 Model Classes Used in GMDH Algorithms

In the practice of modeling, the following main types of tasks are occurred: (1) building regression models of static objects; (2) modeling of time series or processes; (3) modeling of dynamic objects, processes and systems.

Models of Static Objects. The term "static object" unites all cases when it is necessary to build a model of the form $y_k = \sum_1^m \theta_j g_j(x_k)$, where $k = 1, \ldots, n$ is the number of the observation point, θ_j are model parameters, $g(x)$ is the given m-dimensional vector function of m_x input variables. The basic functions $g(x)$ in GMDH algorithms are, first of all, polynomials, and each component of $g(x)$ is a monomial.

Time Series Models. To model processes as time series, polynomial, trigonometric, exponential, logarithmic and logistic functions of time are applied in GMDH algorithms, as well as autoregressive models of the form $y_k = \sum_1^{l_y} a_i y_{k-i}$, where l_y is the autoregression order.

Models of Dynamic Objects. The problems of modeling dynamics are solved first of all in the class of linear dynamic models with a vector of external actions x. Then the general form of the linear dynamic model of a multidimensional system is represented by the difference equation $y_k = \sum_1^{l_y} a_\alpha^T y_{k-\alpha} + \sum_1^{l_x} b_v^T x_{k-v+1}$, where l_y, l_x are the numbers of the considered past values (delays, lags) for output and input variables, a_α, b_ν are vectors of unknown parameters.

In each GMDH algorithm there is a data conversion block in accordance with the specified model class, and then the task of model $y = f(x, \hat{\theta}) = x^T\hat{\theta}$ synthesis is solved, where the original task specificity is not taken explicitly into account.

6.2 Basic Generators of Model Structures

Most of the known GMDH generators of model structures are naturally divided into two groups, sorting-out and iterative (like optimization methods) generators differing by ways of forming structures and searching for a minimum of a given criterion.

Sorting-Out Algorithms [15, 44]. They are intended for solving the problem by comparing models from a finite set Φ whose elements can be calculated independently by estimating parameters of all models. Then the problem of discrete programming can be solved by a complete or directed search.

The most known is an exhaustive search implementation in the form of the combinatorial algorithm COMBI [45] generating all possible model structures $y_\nu = X_\nu\hat{\theta}_\nu,\ \nu = 1,\ldots,2^m$. Here, the number of models being compared is 2^m and the exhaustive search is effective only up to approximately $m = 25$.

The purpose of directed search methods is to find the global minimum of the selection criterion, i.e. the result of an exhaustive search, by significantly smaller calculations. They are runnable with hundreds of arguments, e.g. the multistage algorithm MULTI [46] which uses a special step-by-step evolving procedure of the type $\hat{y}^l_s = (X^i_{s-1}|x^j_s)\hat{\theta}_s,\ s = \overline{1,m},\ i,l = \overline{1,F_{s-1}}$, where s is the stage number (and the structure complexity); F_s – number of the best structures (freedom of choice) at the stage s; j is the index of a regressor vector being added to the matrix X^i_{s-1}. In this case, a subset $\Phi_s \subseteq \Phi$ is analyzed, with a high probability containing the result of a complete search. This algorithm has the polynomial complexity of order m^3.

The monograph [44] is devoted to the numerical research and practical application issues of sorting-out GMDH algorithms. A comprehensive survey of the available techniques of exhaustive and partial search of models in such algorithms and their comparative analysis is presented in [47].

Iterative Algorithms [28, 48]. According to the principle of operation, the iterative GMDH algorithms are similar to methods of optimization using successive approaching, but the principle of indecisive solutions (freedom of choice F) is substantially used. Depending on the method of successive approaching, the iterative GMDH algorithms may be divided into two groups: multilayered and relaxational ones.

Multilayered algorithms of the MIA GMDH type are constructed by the analogy with the biological selection of living organisms: the complication of models from r-th layer to $(r + 1)$-th occurs due to the "crossing" of all possible pairs from F best models of the previous layer. Typically, the partial description of MIA is of the form $y^{r+1}_l = f_l\left(y^r_i, y^r_j\right),\ r = 0, 1, \ldots;\ i,j = \overline{1,F};\ l = \overline{1,C^2_F}$, where second order polynomial is usually used, but bilinear or even linear one may be applied. The iterative process of model complication stops after the criterion CR starts to increase.

In algorithms of *relaxational* type called now RIA GMDH, models are complicated on each layer by "crossing" the best models of the previous layer with the initial arguments. In the RIA case, a partial description of the following modification $y_l^{r+1} = f_l(y_i^r, x_k),\ i = \overline{1,F};\ k = \overline{1,m}$ may be used with quadratic, bilinear or linear polynomial. Such kind of descriptions was introduced at first in [48–50] to avoid some known drawbacks of the MIA GMDH: for instance, they help to exclude the possibility of losing relevant arguments.

The monograph [51] is devoted to the description, research and application of high-performance GMDH algorithms of thc relaxational type with partial description $y_l^{r+1} = y_i^r + \phi(x_k),\ i = \overline{1,F};\ k = \overline{1,m}$. Typical architectures of seven different iterative algorithms are presented in the generalized algorithm GIA GMDH [52]. As it was mentioned above, for iterative algorithms, in contrast to sorting-out ones, it becomes necessary to prove the convergence of the iterative process. In general, iterative methods are operable at $m > 1000$, and they allow constructing models even in degenerate problems when $n < m$.

6.3 External Criteria of Models Quality

The criteria are based on dividing the sample into at least two nonoverlapping subsamples A and B, $A \cup B = W$. They are called usually training A, checking (or testing) B and learning (or working) W data sets. An additional subset C called examination (or validation) one may be divided for verification of a model adequacy.

Let us introduce the following denotations: the least-squares estimation of the parameters at some subset G is $\hat{\theta}_G,\quad G = A, B, W$, and the error on a subset Q of a model with parameters estimated on G, is equal to $\Delta(Q|G) = \left\| y_Q - X_Q\hat{\theta}_G \right\|^2$, $Q = A$, B, W. Taking this into account, one can give the following computational formulae for the main external criteria.

Accuracy Criteria. When $Q = G$, the value $\varepsilon_G = \Delta(G|G)$ is equal to the residual sum of squares RSS, and if $G \neq Q$, then it is so called *regularity* criterion: $AR_B = \Delta(B|A)$, $AR_A = \Delta(A|B)$. The symmetric form of the criterion: $AD = \Delta(B|A) + \Delta(A|B)$.

Conformity Criteria. The main one in this group is the *consistency* criterion (or *unbiasedness* criterion): $CB = \left\| X_W\hat{\theta}_A - X_W\hat{\theta}_B \right\|^2 = (\hat{\theta}_A - \hat{\theta}_B)^T X_W^T X_W (\hat{\theta}_A - \hat{\theta}_B)$. This group includes also the so-called *variability* criterion which may be expressed as difference of internal criteria: $CV = (\hat{\theta}_A - \hat{\theta}_W)^T X_W^T X_W (\hat{\theta}_W - \hat{\theta}_B) = (\varepsilon_W - (\varepsilon_A + \varepsilon_B))$.

For more details on the spectrum of various GMDH criteria apply to [15, 28].

7 Up-to-Date Trends in the Development of Algorithms for Self-organization of Models

During the last two decades, the interest to the GMDH-based inductive modelling algorithms is permanently growing which may be attributed both to their good performance in ill-defined modelling tasks and intensive development of new and enhanced algorithms and software tools. These achievements are based on further evolution of the initial Ivakhnenko's ideas on models self-organization performed both by the author with his disciples and a lot of researchers around the world.

GMDH as a Polynomial Neural Network. As it was mentioned above, the multilayered iterative algorithm MIA GMDH is called now as Polynomial Neural Network (PNN) [21, 24]. In this case, the main element of the algorithm, namely the partial description, is considered as an elementary polynomial neuron. The originality and efficiency of the network of such neurons lies in the speed of the process of local training their weights and automatic global optimization (self-organization) of the network structure (numbers of "hidden layers" and "nodes" in each layer). To the explicit advantages of PNN GMDH we can refer the possibility to "fold" the adjusted network directly into an explicit mathematical expression ready to be used for solving tasks of simulation, prediction, control and decision making.

Neuronet with Active Neurons. A typical "GMDH neuron" in the form of quadratic polynomial of two arguments can be called as "passive" because any of the neurons have the same fixed structure, i.e. the PNN GMDH is homogeneous net. In the 1990s, Ivakhnenko proposed a new type of GMDH network with *active* neurons [22, 24] or a *heterogeneous* network in which any of the neurons is in turn also a GMDH algorithm, due to that the structure of the neuron is optimized. As a result, all neurons can get different structures increasing the flexibility of configuring the network to a specific task. Networks of such type are also called as "twice multilayered" ones [24].

GMDH-Like Feedback Neural Network. In [53], a GMDH-like neural network with feedback was proposed: the neuron outputs of each layer are "crossing" with original variables. In this algorithm, the network automatically selects one of three architectures for each neuron: sigmoidal, radial or polynomial transition function. In this case, the structural parameters, e.g. the number of layers, neurons in the hidden layers and input variables, are selected automatically, i.e. self-organized.

Group of Adaptive Models Evolution. An evolutionary algorithm called GAME [54], developed at the Czech Technical University in Prague, is based on the general GMDH architecture. The main modifications of this GMDH-like system with active neurons: the transfer function of a node can be linear, polynomial, logistic, RBF and others; the network structure is heterogeneous; the number of node inputs increases with the depth of the node in the network, and there are inter-row links; there are not all the locations of nodes, but their random subsets; the original GMDH creates one optimal model, and GAME does a group of locally optimal models.

Inductive Algorithms for Classification, Recognition, Clusterization. Essential part of real-world problems requires application of special methods of pattern recognition in wide sense. Among current developments in the inductive modelling field there are many methods and tools dealing with tasks of such type, for instance [55–58].

Hybrid GMDH-Type Algorithms and Neural Networks. New and efficient architectures of neuronets are recently intensively developed on the basis of hybridization of GMDH procedures and various approaches of computational intelligence and nature-inspired solutions, e.g.: particle swarm optimization [59, 60], genetic selection and cloning [61], immune systems [62, 63] etc. A good deal of present hybridization variants is implemented in the aggregated GMDH-based architecture GAME [54].

Fuzzy and Interval Approaches in Inductive Modeling. For real-world tasks with fuzzy variables, there were developed corresponding algorithms: multi-layer hybrid fuzzy PNN [64]; fuzzy GMDH [65] based on the classical structure MIA GMDH; GMDH-like neo-fuzzy and cascade wavelet-neuro-fuzzy networks [66, 67] etc. For case of assumptions on interval-given input data, methods [68, 69] were elaborated.

Algorithms Based on Paralleling Operations. There are several effective realizations of both iterative [70] and combinatorial [71, 72] algorithms with implemented parallel computations.

Automated Data Preprocessing and Ensembling-Based Prediction. An advanced idea to build some kind of automatic means for data preprocessing enabling to enhance the accuracy of classification was suggested in [73]. A non-parametric ensembling procedure [74] for predicting processes is the core idea in the GAME method.

Software Tools Based on GMDH. There are several examples of computer tools for modelling complex systems completely based on GMDH and/or GMDH-type algorithms. The commercial software complex KnowledgeMiner for Mac [75] described in [24] and its recent highly advanced realization [76] contains iterative and several non-parametric GMDH algorithms for fully automatic building and analyzing forecasting models. Another variant of commercial software is GMDHshell [77] for PC based on COMBI and MIA GMDH algorithms. The technology FAKE GAME [54] is intended for constructing, investigation and application of inductive evolutionary algorithms of different architecture.

More detailed information on the developed algorithms, tools and technologies, their research and application can be found on the base site [78], on the site [79] of the ITIM department, as well as on sites [76, 77].

8 Specialization and Results of ITIM Department

The studies of the ITIM department cover a complete cycle of scientific research: *methodology* of the modeling from data samples; *theory* of inductive constructing models of optimal complexity; *algorithmization* of high-performance modeling tools; *intellectualization* of technologies for constructing models; *computer experiments* to evaluate the effectiveness of the developed technologies; *solving* real-world problems of modeling and forecast; testing and *application* of developed tools in monitoring, control and decision support systems.

8.1 Main Scientific Results of ITIM Department

Starting from 1998, there was developed:

- the GMDH-based inductive modeling theory with the use of the method of critical variances [39] which made it possible to explain the nature of the GMDH efficiency as a method for constructing noise-immune models with minimum prediction error variance; the problem of optimizing the data sample partitioning into two parts has been solved [51].

- two-criteria method for extra-determining (additional determination) the model choice using the new errors unbiasedness criterion [80] eliminating the ambiguity in choosing the optimal model [81].
- principles of designing and implementing *high-performance sorting-out* GMDH algorithms based on recurrent calculations [44], parallelling operations [72] and sequential selection of informative variables [82], allowing to enhance the dimensionality of the problems being solved.
- principles of constructing hybrid architectures of iterative GMDH algorithms as a generalization of algorithmic structures of multilayered, relaxational and combinatorial types, based on which a *generalized iterative algorithm* GIA GMDH [52] was developed as a neural network with active neurons in the form of the COMBI algorithm for automatic adjustment of a neuron complexity.
- the *generalized relaxational iterative algorithm* GRIA GMDH based on the use of high-speed recurrent computations and matrices of normal equations, which allows solving inductive modeling problems from high-dimension data [51].
- theoretical foundations of a new class of sorting-out GMDH algorithms with the use of *recurrent-and-parallel computations* on cluster systems [83] as a basis for high-performance intelligent modeling technologies.
- principles of designing technologies for the *intelligent modeling* of complex systems based on the use of knowledge bases, inductive data analysis tools and an intelligent user interface [84]. Such technologies should have three main instrumental levels: *autonomous* modeling from the available database; *embedded* modeling as part of a real-time control system; *combined* modeling of a complex system for identifying optimal operation modes and critical scenarios.
- theoretical principles and tools for predicting interrelated socio-economic processes from statistical data in the class of discrete dynamic models of *vector autoregression* [85].

8.2 Technologies Developed in the Department

ASTRID methodology of designing GMDH-based computer technologies for building models of complex systems from statistical data [86] is a base to develop technologies for discovering regularities, identification, prediction, aimed to informational support of decision-making problems.

Package of software tools [44, 87] is developed for designing, researching and applying modeling methods, conducting experiments on testing modeling methods and their components (model classes, generators of model structures, methods for estimating parameters and models selection criteria).

Integrated environment platform [88] is designed for storing and handling information in tasks of inductive modelling for business intelligence systems.

Cross-platform *software package with an advanced interface* [89] is developed in the Java language for inductive modeling and forecasting of complex objects and large-scale processes on the basis of the fast-operating GMDH algorithm with sequential selection of informative and/or sifting of non-informative arguments [48].

The *software package* ASTRID-GIA [90] for inductive modeling of complex systems based on various iterative GMDH algorithms makes it possible to use the generalized algorithm GIA GMDH and all its special cases [52] in online access mode both over the Internet and in the local network.

Computer system ASPIS [55] for building predictive models on the basis of the high-speed generalized relaxational iteration algorithm GRIA GMDH [91]. It is implemented in C++ and allows solving big data problems.

The *management decisions informational support system* MDISS allows to solve problems of estimation, analysis and forecasting of the state of complex systems of interrelated socio-economic processes with the purpose of making reasonable managerial decisions [92].

8.3 Basic Applied Results

The tasks of modeling of such processes were solved:

- dynamics of the quantity of microorganisms in the soil, depending on environmental factors and the dose of contamination with heavy metals [93];
- dynamics of interdependent indicators of the energy and investment areas in the class of vector autoregression models for a short-term forecast [44, 85];
- dependence of the sputtering (disruption) coefficient of a spacecraft surface under action of ionized gas jets on the physical properties of the surface coating [44];
- analysis of the ecological consequences of trees irrigation with treated wastewater [94].
- quantitative assessment of the impact of sea water pollution with bitumoid substances on the total number of species of benthic organisms in the Sevastopol bays [95].
- predicting the results of testing blood samples with medical products in order to determine the most effective for a particular patient [51];
- classifiers construction for the differential diagnostics of blood diseases for reducing the risks of misdiagnosis [51].

9 Prospects of the Research Development

The most promising areas of research development are:

The methodology of intelligent modeling: new paradigms of neural networks with active neurons and hybrid GMDH-based architectures; theoretical study of modeling efficiency under uncertainty; theory of building models for prediction, classification, recognition, and clustering, including ensemble-based approaches.

The theory and architectures of iterative (neuronet) GMDH algorithms with new properties: increasing the accuracy of solving modeling problems on the basis of generalization of structures of known algorithms; the development of new hybrid architectures using evolutionary and multi-agent approaches; analysis of the "external" convergence of iterative algorithms to the true model.

Novel structures of sorting-out GMDH algorithms: increasing the algorithms efficiency on the basis of recurrent-and-parallel computations; development of effective schemes for sequential selection of the best model variants; construction of hybrid algorithms using the computational intelligence ideas [96].

New algorithms for *inductive solving of classification, recognition and clustering problems*: inductive construction of optimal classification rules for big data problems; increasing the effectiveness of inductive solving the multiple-choice recognition problems based on iterative GMDH algorithms; development of intelligent algorithms for data analysis and automatic clustering to identify patterns.

The intelligent technologies for informational support of decision making based on inductive modeling methods and tools: development of tools in the form of an algorithm construction set as a shell for interactive synthesis of new modeling means; development of *intelligent interface* means for efficient support of user solutions in the GMDH-based inductive modeling process; development of a set of software tools to support the design of intelligent technologies for information support of decisions to control of complex systems of various nature.

Application of the developed methods and tools in applied problems of intelligent decisions support based on inductive modeling: support of operational decisions for managing socio-economic processes at different levels; supporting the solution of medical diagnosis problems and predicting the effectiveness of drugs and procedures; intelligent modeling and control of technological processes and robotic systems; support solutions for control the ecological state of processes based on environmental monitoring data; decisions support in problems of automatic search and analysis of categories and structures of textual information [97]; identification of knowledge based on automated content analysis of messages in social networks in order to support the administrative decision making at different levels.

10 Conclusion

The performed survey of the research state in the field of inductive modeling shows that GMDH is one of the most effective methods of computation intelligence and soft computing as well as a promising basis for creating modern information technologies for data mining, knowledge discovery, and business intelligence. They form a reliable base for promoting intensive current and future development of methodology, theory, algorithms and tools of inductive modelling of complex processes and systems in applied decision making tasks.

During the whole period of GMDH evolution, this method and corresponding software means demonstrated good performance when solving real-world modeling problems of different nature in fields of environment, economy, finance, hydrology, technology, robotics, sociology, biology, medicine, and others. This method represents the original and efficient facility for solving a wide spectrum of artificial intelligence problems including identification and forecast, pattern recognition and clusterization, data mining and search for regularities. This practically boundless area of knowledge is worth of a special targeted survey.

Further development of the ideas of model self-organization assumes the improvement of the inductive modeling theory, the development and implementation of new high-performance algorithms for computer technologies, and the application of these technologies to solve a wide range of real-world applications for tasks of modeling, forecasting, control and decision-making in systems of different nature.

Other kinds of surveys are also expected to be prepared, both general and topical, specialized, because the amount of available GMDH publications is enormous and permanently growing with regard to both theoretical and applied aspects.

Some special attention seems to be done to the relationship between GMDH and the task and methods of the "Deep learning in neural networks" because GMDH is positioned as the very first example of a deep learning architecture [98].

References

1. Ivakhnenko, A.G.: The group method of data handling – a rival of the method of stochastic approximation. Sov. Autom. Control **1**(3), 43–55 (1968)
2. Ivakhnenko, A.G.: Heuristic self-organization in problems of automatic control. Automatica (IFAC) **3**, 207–219 (1970)
3. Ivakhnenko, A.G.: Polynomial theory of complex systems. IEEE Trans. Syst. Man Cybern. **1** (4), 364–378 (1971)
4. Ivakhnenko, A.G.: Systems of Heuristic Self-Organization in Technical Cybernetics. Technika, Kiev (1971). (in Russian)
5. Ivakhnenko, A.G.: Long-Term Forecasting and Control of Complex Systems. Technika, Kiev (1975). (in Russian)
6. Ivakhnenko, A.G.: Inductive Method for Self-Organizing Models of Complex Systems. Naukova Dumka, Kiev (1982). (in Russian)
7. Madala, H.R., Ivakhnenko, A.G.: Inductive Learning Algorithms for Complex Systems Modeling. CRC Press, New York (1994)
8. Ivakhnenko, A.G., Müller, J.-A.: Recent developments of self-organising modeling in prediction and analysis of stock market. Microelectron. Reliab. **37**, 1053–1072 (1997)
9. Anastasakis, L., Mort, N.: The development of self-organization techniques in modelling: a review of the Group Method of Data Handling (GMDH). ACSE research report, vol. 813, 39 p. The University of Sheffield (2001)
10. Snorek, M., Kordik, P.: Inductive modelling world wide the state of the art. In: Proceedings of 2nd International Workshop on Inductive Modelling, pp. 302–304. CTU, Prague (2007)
11. Stepashko, V.: Ideas of Academician O.H. Ivakhnenko in the inductive modelling field from historical perspective. In: Proceedings of the 4th International Conference on Inductive Modelling, ICIM-2013, pp. 30–37. IRTC ITS NASU, Kyiv (2013)
12. Farlow, S.J. (ed.): Self-Organizing Methods in Modeling: GMDH Type Algorithms. Marcel Decker Inc., New York, Basel (1984)
13. Iwachnenko, A.G., Müller, J.-A.: Selbstorganisation von Vorhersagemodellen. VEB Verlag Technik, Berlin (1984)

14. Ivakhnenko, A.G., Karpinsky, A.M.: Computer-aided self-organization of models in terms of the general communication theory (information theory). Sov. Autom. Control **15**(4), 7–15 (1982)
15. Ivakhnenko, A.G., Stepashko, V.S.: Noise Immunity of Modelling. Naukova Dumka, Kiev (1985). (in Russian)
16. Stepashko, V.S., Kostenko, Y.: A GMDH algorithm for two-level modeling of multidimensional cyclic processes. Sov. Autom. Control **20**(4), 49–57 (1987)
17. Ivakhnenko, A.G., Kostenko, Y.: System analysis and long-term prediction on the basis of model self-organisation (OSA algorithm). Sov. Autom. Control **15**(3), 11–17 (1982)
18. Ivakhnenko, A.G.: Objective computer clasterization based on self-organisation theory. Sov. Autom. Control **20**(6), 1–7 (1987)
19. Ivakhnenko, A.G., Osipenko, V.V., Strokova, T.I.: Prediction of two-dimensional physical fields using inverse transition matrix transformation. Sov. Autom. Control **16**(4), 10–15 (1983)
20. Ivakhnenko, A.G.: Inductive sorting method for the forecasting of multidimensional random processes and events with the help of analogs forecast complexing. Pattern Recogn. Image Anal. **1**(1), 99–108 (1991)
21. Oh, S.K., Pedrycz, W.: The design of self-organizing polynomial neural networks. Inf. Sci. **141**, 237–258 (2002)
22. Ivakhnenko, A.G., Ivakhnenko, G.A., Mueller, J.-A.: Self-organization of neuronets with active neurons. Pattern Recogn. Image Anal. **4**(4), 177–188 (1994)
23. Ivakhnenko, A.G., Wunsh, D., Ivakhnenko, G.A.: Inductive sorting-out GMDH algorithms with polynomial complexity for active neurons of neural networks. In: Proceedings of the International Joint Conference on Neural Networks, pp. 1169–1173. IEEE, Piscataway (1999)
24. Muller, J.-A., Lemke, F.: Self-Organizing Data Mining: An Intelligent Approach to Extract Knowledge from Data. Trafford Publishing Press, Berlin (1999)
25. Vysotskiy, V.N., Ivakhnenko, A.G., Cheberkus, V.I.: Long term prediction of oscillatory processes by finding a harmonic trend of optimum complexity by the balance-of-variables criterion. Sov. Autom. Control **8**(1), 18–24 (1975)
26. Ivakhnenko, A.G., Krotov, G.I.: A multiplicative-additive nonlinear GMDH algorithm with optimization of the power of factors. Sov. Autom. Control **17**(3), 10–15 (1984)
27. Ivakhnenko, A.G., Peka, PYu., Vostrov, N.P.: Combined Method for Modeling Water and Oil Fields. Naukova dumka, Kiev (1984). (In Russian)
28. Ivakhnenko, A.G., Yurachkovsky, Y.: Complex Systems Modeling after Experimental Data, Radio and Communication. Radio i Swiaz, Moscow (1987). (in Russian)
29. Ivakhnenko, A.G., Kovalchuk, P.I., Todua, M.M., Shelud'ko, O.I., Dubrovin, O.F.: Unique construction of regression curve using a small number of points. Sov. Autom. Control **6**(5), 29–41 (1973)
30. Yurachkovsky, Y.: Convergence of multilayer algorithms of the group method of data handling. Sov. Autom. Control **14**(3), 29–34 (1981)
31. Kovalchuk, P.L.: Internal convergence of GMDH algorithms. Sov. Autom. Control **16**(2), 88–91 (1983)
32. Stepashko, V.S.: Potential noise stability of modelling using a combinatorial GMDH algorithm without information regarding the noise. Sov. Autom. Control **16**(3), 15–25 (1983)
33. Kocherga, Y.L.: J-optimal reduction of model structure in the Gauss-Markov scheme. Sov. J. Autom. Inf. Sci. **21**(4), 21–23 (1988)
34. Aksenova, T.I., Yurachkovsky, Y.P.: Characterization of unbiased structure and condition of its J-optimality. Sov. J. Autom. Inf. Sci. **21**(4), 24–32 (1988)

35. Stepashko, V.S.: Noise immunity of choice of model using the criterion of balance of predictions. Sov. Autom. Control **17**(5), 27–36 (1984)
36. Stepashko, V.S.: Investigation of the predicting properties of a recurrent structural-parametric identifier. Sov. J. Autom. Inf. Sci. **24**(3), 31–40 (1991)
37. Stepashko, V.S.: Structural identification of predicting models for planned experiment. J. Autom. Inf. Sci. **25**(1), 23–31 (1992)
38. Stepashko, V.S.: Analysis of criteria effectiveness for structural identification of forecasting models. J. Autom. Inf. Sci. **27**(3–4), 13–20 (1994)
39. Stepashko, V.S.: Method of critical variances as analytical tool of theory of inductive modeling. J. Autom. Inf. Sci. **40**(3), 4–22 (2008)
40. Stepashko, V.S.: Asymptotic Properties of External Criteria for Model Selection. Sov. J. Autom. Inf. Sci. **21**(6), 84–92 (1988)
41. Aksenova, T.I.: Sufficient conditions and convergence rate using different criteria for model selection. Syst. Anal. Model. Simul. **20**(1–2), 69–78 (1995)
42. Sarychev, A.P.: System criterion of regularity in the group method of data handling. J. Autom. Inf. Sci. **38**(1), 22–35 (2006)
43. Sarychev, A.P.: Modelling in the class of regression equations systems in conditions of structural uncertainty. In: Proceedings of 2nd International Workshop on Inductive Modelling, IWIM-2007, pp. 193–203. Czech Technical University, Prague (2007)
44. Stepashko, V.S., Yefimenko, S.M., Savchenko, Y.A.: Computerized Experiment in Inductive Modeling. Naukova Dumka, Kyiv (2014). (in Ukrainian)
45. Stepashko, V.S.: A combinatorial algorithm of the group method of data handling with optimal model scanning scheme. Sov. Autom. Control **14**(3), 24–28 (1981)
46. Stepashko, V.S.: A finite selection procedure for pruning an exhaustive search of models. Sov. Autom. Control **16**(4), 84–88 (1983)
47. Moroz, O.G., Stepashko, V.S.: Comparative analysis of model structures generators in sorting-out GMDH algorithm. Inductive Model. Complex Syst. **8**, 173–191 (2016). IRTC ITS NASU, Kyiv (in Ukrainian)
48. Sheludko, O.I.: GMDH algorithm with orthogonalized complete description for synthesis of models by the results of a planned experiment. Sov. Autom. Control **7**(5), 24–33 (1974)
49. Tamura, H., Kondo, T.: Large-spatial pattern identification of air pollution by a combined model of source-receptor matrix and revised GMDH. In: Proceedings of IFAC Symposium on Environmental Systems Planning, Design and Control, pp. 373–380. Elsevier, Oxford (1977)
50. Yurachkovsky, Y.: Restoration of polynomial dependencies using self-organization. Sov. Autom. Control **14**(4), 17–22 (1981)
51. Pavlov, A.V., Stepashko, V.S., Kondrashova, N.V.: Effective Methods of Models Self-Organization. Akademperiodika, Kyiv (2014). (in Russian)
52. Stepashko, V., Bulgakova, O.: Generalized iterative algorithm GIA GMDH. In: Proceedings of the 4th International Conference on Inductive Modelling, ICIM-2013, Kyiv, Ukraine, September 2013, pp. 119–123. IRTC ITS NASU, Kyiv
53. Kondo, T., Ueno, J.: Feedback GMDH-type neural network self-selecting optimum neural network architecture and its application to 3-dimensional medical image recognition of the lungs. In: Proceedings of II International Workshop on Inductive Modelling, September 2007, pp. 63–70. Czech Technical University, Prague. ISBN 978-80-01-03881-9
54. Kordik, P.: Fully automated knowledge extraction using group of adaptive model evolution. Ph.D. thesis, Department of Computer Science and Computers, FEE, CTU in Prague, 150 p. (2006)

55. Sarychev, A.P.: The solution of the discriminant analysis task in conditions of structural uncertainty on basis of the group method of data handling. J. Autom. Inf. Sci. **40**(6), 27–40 (2008)
56. Sarycheva, L.: Quality criteria for GMDH-based clustering. In: Proceedings of the II International Conference on Inductive Modelling, ICIM-2008, pp. 84–90. IRTC ITS NASU, Kyiv (2008)
57. Borisova, I.A.: Calculation of FRiS-function over mixed dataset in the task of generalized classification. In: Proceedings of the III International Conference on Inductive Modelling, ICIM-2010, pp. 44–50. KNTU, Kherson, Ukraine (2010)
58. Cepek, M., Snorek, M., Chudacek, V.: ECG signal classification using GAME neural network and its comparison to other classifiers. In: Proceedings of International Conference on Artificial Neural Networks, ICANN 2008, pp. 768–777. Springer, Heidelberg (2008)
59. Voss, M.S., Feng, X.: A new methodology for emergent system identification using particle swarm optimization (PSO) and the group method of data handling (GMDH). In: Proceedings of the Genetic and Evolutionary Computation Conference, pp. 1227–1232. Morgan Kaufmann Publishers, New York (2002)
60. Onwubolu, G., Sharma, A., Dayal A. et al.: Hybrid particle swarm optimization and group method of data handling for inductive modeling. In: Proceedings of 2nd International Conference on Inductive Modelling, pp. 95–103. IRTC ITS NASU, Kyiv (2008)
61. Jirina, M., Jirina Jr., M.: Genetic selection and cloning in GMDH MIA method. In: Proceedings of the II International Workshop on Inductive Modelling, IWIM 2007, pp. 165–171. CTU, Prague (2007)
62. Lytvynenko, V., Bidyuk, P., Myrgorod, V.: Application of the method and combined algorithm on the basis of immune network and negative selection for identification of turbine engine surging. In: Proceedings of the II International Conference on Inductive Modelling, ICIM-2008, pp. 116–123. IRTC ITS NASU, Kyiv (2008)
63. Lytvynenko, V.: Hybrid GMDH cooperative immune network for time series forecasting. In: Proceedigns of the 4th International Conference on Inductive Modelling, pp. 179–187. IRTC ITS NASU, Kyiv (2013)
64. Oh, S.K., Pedrycz, W., Park, H.S.: Multi-layer hybrid fuzzy polynomial neural networks: a design in the framework of computational intelligence. Neurocomputing **64**, 397–431 (2005)
65. Zaychenko, Y.: The investigations of fuzzy group method of data handling with fuzzy inputs in the problem of forecasting in financial sphere. In: Proceedings of the II International Conference on Inductive Modelling, ICIM-2008, pp. 129–133. IRTC ITS NASU, Kyiv (2008)
66. Bodyanskiy, Y.V., Zaychenko, Y.P., Pavlikovskaya, E.: The neo-fuzzy neural network structure optimization using the GMDH for the solving forecasting and classification problems. In: Proceedings of the 3rd International Workshop on Inductive Modelling, IWIM-2009, Krynica, Poland, pp. 10–17. Czech Technical University, Prague (2009)
67. Bodyanskiy, Y., Vynokurova, O., Teslenko, N.: Cascade GMDH-wavelet-neuro-fuzzy network. In: Proceedings of the IV International Workshop on Inductive Modelling, IWIM-2011, pp. 16–21. IRTC ITS NASU, Kyiv (2011)
68. Dyvak, M., Manzhula, V., Pukas, A., Stakhiv, P.: Structural identification of interval models of the static systems. In: Proceedings of 2nd International Workshop on Inductive Modelling, pp. 132–139. Czech Technical University, Prague (2007)
69. Voytyuk, I., Dyvak, M., Spilchuk, V.: The method of structure identification of macromodels as difference operators based on the analysis of interval data and genetic algorithm. In: Proceedings of the IV International Workshop on Inductive Modelling, IWIM-2011, pp. 114–118. IRTC ITS NASU, Kyiv (2011)

70. Lemke, F.: Parallel self-organizing modeling. In: Proceedings of the II International Conference on Inductive Modelling, ICIM-2008, pp. 176–183. IRTC ITS NASU, Kyiv (2008)
71. Koshulko, O.A., Koshulko, A.I.: Multistage combinatorial GMDH algorithm for parallel processing of high-dimensional data. In: Proceedings of III International Workshop on Inductive Modelling, IWIM-2009, pp. 114–116. CTU, Prague (2009)
72. Stepashko, V., Yefimenko, S.: Parallel algorithms for solving combinatorial macromodelling problems. Przegląd Elektrotechniczny (Electr. Rev.) **85**(4), 98–99 (2009)
73. Čepek, M., Kordík, P., Šnorek, M.: The effect of modelling method to the inductive preprocessing algorithm. In: Proceedings of the III International Conference on Inductive Modelling, ICIM-2010, pp. 131–138. KNTU, Kherson (2010)
74. Kordík, P., Černý, J.: Advanced ensemble strategies for polynomial models. In: Proceedings of the III International Conference on Inductive Modelling, ICIM-2010, pp. 77–82. KNTU, Kherson (2010)
75. http://knowledgeminer.eu. Accessed 17 May 2017
76. Lemke, F.: Insights v.2.0, self-organizing knowledge mining and forecasting tool (2013). http://www.knowledgeminer.eu. Accessed 21 May 2017
77. https://www.gmdhshell.com. Accessed 12 Apr 2017
78. www.gmdh.net. Accessed 15 Apr 2017
79. www.mgua.irtc.org.ua. Accessed 11 May 2017
80. Ivakhnenko, A.G., Ivakhnenko, G.A., Savchenko, E.A.: GMDH algorithm for optimal model choice by the external error criterion with the extension of definition by model bias and its applications to the committees and neural networks. Pattern Recogn. Image Anal. **12**(4), 347–353 (2002)
81. Ivakhnenko, A.G., Savchenko, E.A.: Investigation of efficiency of additional determination method of the model selection in the modeling problems by application of GMDH algorithm. J. Autom. Inf. Sci. **40**(3), 47–58 (2008)
82. Samoilenko, O., Stepashko, V.: A method of successive elimination of spurious arguments for effective solution of the search-based modelling tasks. In: Proceedings of the II International Conference on Inductive Modelling, pp. 36–39. IRTC ITS NASU, Kyiv (2008)
83. Yefimenko, S., Stepashko, V.: Intelligent recurrent-and-parallel computing for solving inductive modeling problems. In: Proceedings of 16th International Conference on Computational Problems of Electrical Engineering, pp. 236–238. LNPU, Lviv, Ukraine (2015)
84. Stepashko, V.S.: Conceptual fundamentals of intelligent modeling. Control Syst. Mach. (USiM) **4**, 3–15 (2016). (in Russian)
85. Yefimenko, S.N.: Construction of systems of predictive models for multidimensional interrelated processes. Control Syst. Mach. (USiM) **4**, 80–86 (2016). (in Russian)
86. Stepashko, V.S.: GMDH algorithms as basis of modeling process automation after experimental data. Sov. J. Autom. Inf. Sci. **21**(4), 33–41 (1988)
87. Yefimenko, S., Stepashko, V.: Technologies of numerical investigation and applying of data-based modeling methods. In: Proceedings of the II International Conference on Inductive Modelling, ICIM-2008, pp. 236–240. IRTC ITS NASU, Kyiv (2008)
88. Shcherbakova, N., Stepashko, V.: Integrated environment for storing and handling information in tasks of inductive modelling for business intelligence systems. In: Setlak, G., Alexandrov, M., Markov, K. (eds.) Artificial Intelligence Methods and Techniques for Business and Engineering Applications, pp. 210–219. ITHEA, Rzeszow, Poland, Sofia, Bulgaria (2012)
89. Samoilenko, O.A.: Designing new GMDH algorithms as basic components of a modeling subsystem. Inductive Model. Complex Syst. **3**, 191–208. IRTC ITS NASU, Kyiv (2011). (in Ukrainian)

90. Bulgakova, O., Zosimov, V., Stepashko, V.: Software package for modeling of complex systems based on iterative GMDH algorithms with the network access capability. Syst. Res. Inf. Technol. **1**, 43–55 (2014). (in Ukrainian)
91. Pavlov, A.: Design patterns of automated structure-parametric identification system. In: Proceedings of 6th International Workshop on Inductive Modelling, pp. 31–35. IRTC ITS NASU, Kyiv (2015)
92. Stepashko, V., Samoilenko, O., Voloschuk, R:. Informational support of managerial decisions as a new kind of business intelligence systems. In: Setlak, G., Markov, K. (eds.) Computational Models for Business and Engineering Domains, pp. 269–279. ITHEA, Rzeszow, Poland, Sofia, Bulgaria (2014)
93. Iutynska, G., Stepashko, V.: Mathematical modeling in the microbial monitoring of heavy metals polluted soils. In: Book of Proceedings of IX ESA Congress, Part 2, pp. 659–660. Institute of Soil Science and Plant Cultivation, Warsaw (2006)
94. Kalavrouziotis, I.K., Vissikirsky, V.A., Stepashko, V.S., Koukoulakis, P.H.: Application of qualitative analysis techniques to the environmental modeling of plant species cultivation. Glob. NEST J. **12**(2), 161–174 (2010)
95. Alyomov, S.V., Bulgakova, O.S., Stepashko, V.S.: Modeling of the Black Sea pollution impact on the total number of benthic organisms species. Collected Art. SNUNE&I Sevastopol **3**(39), 54–62 (2011). (in Ukrainian)
96. Stepashko, V., Moroz, O.: Hybrid searching GMDH-GA algorithm for solving inductive modeling tasks. In: IEEE International Conference on Data Stream Mining & Processing, pp. 350–355, Lviv, Ukraine (2016)
97. Zosimov, V., Stepashko, V. Bulgakova, O.: Inductive building of search results ranking models to enhance the relevance of the text information retrieval. In: Spies, M., et al. (ed.) Proceedings of the 26th International Workshop on Database and Expert Systems Applications, Valencia, Spain, pp. 291–295. IEEE Computer Society, Los Alamitos (2015)
98. Schmidhuber, J.: Deep learning in neural networks: an overview. Neural Netw. **61**, 85–117 (2015)

Construction and Research of the Generalized Iterative GMDH Algorithm with Active Neurons

Volodymyr Stepashko[1], Oleksandra Bulgakova[2(✉)], and Viacheslav Zosimov[2]

[1] Department for Information Technologies of Inductive Modeling, International Research and Training Centre for Information Technologies and Systems, Akademik Glushkov Prospect 40, Kiev 03680, Ukraine
stepashko@irtc.org.ua

[2] Department for Applied Mathematics and Information Computer Technologies, V.O. Sukhomlynsky Mykolaiv National University, Nikolska Street, 24, Mykolaiv 54030, Ukraine
sashabulgakova2@gmail.com, zosimovvvv@gmail.com

Abstract. Architecture of the generalized iterative algorithm GIA GMDH with active neurons is presented based on hybridization of iterative and combinatorial search schemes and the use of interactive technologies. The architecture comprises six standard variants of typical GMDH algorithms. Experiments show that the proposed modifications improve considerably the practical performance of the multilayered GMDH algorithm and accuracy of the simulation results. Solution results are given for modeling Ukraine's Black Sea economic region GRP as dependent from socio-economic indicators that describe the state region development using generalized iterative algorithm GMDH.

Keywords: Inductive modeling · GMDH · Multilayered algorithm Combinatorial algorithm · Generalized iterative algorithm · Active neuron

1 Introduction

Problems of complex systems modeling can be solved using both the logical deductive (theory-driven) and empirical inductive (data-driven) methods. Deductive methods have advantages in the case of simple modeling tasks, if there is known the theory of an object being modeled and therefore it is possible to build a model based on physical principles using knowledge of processes in an object. But these methods cannot give satisfactory results for complex systems. In this case, an approach of knowledge extraction directly from the data based on statistical or experimental measurements has an advantage. Such inductive approach is useful when a priori information about the properties of such objects may be only minimal or even absent.

The world-wide known Group Method of Data Handling (GMDH) [1–3] is one of the most successful methods of inductive modeling. It is used in various tasks of data analysis and knowledge discovery, forecasting and systems modeling, classification and pattern recognition.

N. Shakhovska and V. Stepashko (eds.), *Advances in Intelligent Systems and Computing II*, Advances in Intelligent Systems and Computing 689,
https://doi.org/10.1007/978-3-319-70581-1_35

Main GMDH advantages are as follows [3]: the model structure and its parameters are found automatically; optimal complexity of the model structure is found adequately to the noise level in data sample.

- The method uses information directly from a data sample and minimizes the impact of a priori assumptions of an author on the modeling results;
- Relationships in data are found and informative input variables are selected; The model structure and its parameters are found automatically;
- Optimal complexity of the model structure is found adequately to the noise level in the data sample;
- Any non-linear functions or factors that could influence the output variable are used as input variables (arguments).

As of today, a great variety of GMDH algorithms of the sorting-out and iteration types was developed and explored [4, 5]. The sorting-out algorithms are effective as the tool for structural identification but only for limited number of arguments. Iterative algorithms are capable of working with a lot of arguments but the specific of their architecture do not guarantee constructing the true model structure. Until recently, these two classes of algorithms were developed without combining their strengths.

This chapter considers in some detail the basic structural elements of typical iterative GMDH algorithms, analyzes its advantages and shortcomings, describes the historical ways to overcome these shortcomings, discusses recent trends in GMDH methodology, proposes a new type of a generalized hybrid GMDH algorithm "with active neurons", investigates comparative effectiveness of various algorithms using numerical experiments with artificially generated data, and demonstrates results of applying diverse algorithms to modeling of a real economic process.

2 GMDH as an Inductive Method of Model Building

The development of GMDH theory contributed to appearance of the wide range of algorithms having its own advantages and disadvantages. GMDH algorithms may vary against the type of elementary functions, the way of a model structure formation, the form of external criteria and so on. The choice of an algorithm depends on the level of noise in the data and their sufficiency.

Any GMDH algorithm solves a discrete optimization task to construct the model of optimal complexity by the minimum of a given external criterion based on the data sample division:

$$f^* = \arg \min_{f \in \Phi} CR(f), \tag{1}$$

where CR is a selection criterion as a measure of the quality of a model $f \in \Phi$. A model selection criterion is called "external" if it is based on additional information that is not contained in the data used for calculation of model parameters.

The set Φ of models being compared can be formed by various generators of model structures of diverse complexities. All structure generators developed within the

GMDH framework may be divided into two main groups, sorting-out and iterative ones which differ by techniques of variants generation and organization of search of a given criterion minimum.

To realize the principle of external supplement [1], the parameter estimations (usually by the least squares method LSM) and criteria values are calculated on different parts of the sample $W = [X \vdots y]$, where X and y are matrix and vector of n measurements of m inputs (arguments) and one output respectively.

The simplest case is the sample splitting into two subsets: $W = [W_A^T W_B^T]^T$, consequently $X = [X_A^T X_B^T]^T$, $y = [y_A^T y_B^T]^T$, where A, B, and $W = A \cup B$, $n = n_A + n_B$, are training, checking and learning sets, respectively. The most commonly used among GMDH criteria is the *regularity* criterion calculated for a model $f \in \Phi$ as follows:

$$AR_{B|A}(f) = \|y_B - \hat{y}_{B|A}(f)\|^2 = \left\|y_B - X_{Bf}\hat{\theta}_{Af}\right\|^2, \tag{2}$$

where designation $AR_{B|A}(f)$ means "error on B of a model f with parameters obtained on A", and X_{Af}, X_{Bf} are submatrices of the matrix X containing columns that correspond to a partial model $f \in \Phi$ being considered.

Another type of external criteria is represented by so called *unbiasedness* criterion reflecting the requirement that models obtained on A and B should differ minimally:

$$CB = CB_{W|A,B}(f) = \|\hat{y}_{Wf}(A) - \hat{y}_{Wf}(B)\|^2 = \left\|X_{Wf}\hat{\theta}_{Af} - X_{Wf}\hat{\theta}_{Bf}\right\|^2, \tag{3}$$

where $\hat{\theta}_{Af}$ and $\hat{\theta}_{Bf}$ are LSM parameter estimations on A and B of a model f.

Accuracy (2) and bias (3) are different, not interchangeable, measures of model quality especially when noisy data are used. When the noise is absent, all criteria (internal or external) are equivalent because they all specify the physical model [3]. For more information on the GMDH criteria see [6].

3 Main Historical Types and Modifications of GMDH Iterative Algorithms

Iterative GMDH algorithms solve the task (1) by successive approaching to the criterion minimum using a network-type procedure based on the analogy with the biological selection of living organisms: the complication of models on a layer occurs due to the pairwise "crossing" F best models from the previous layer. The process of complication stops after the criterion starts to increase.

Presently, the classical multilayered iterative algorithm MIA GMDH [1–3] is the most widely known, Fig. 1. However it has some substantial drawbacks: (1) possibility of loss of informative arguments if they were eliminated at the beginning of the selection procedure; (2) inclusion of spurious arguments to the final model if they were included at the beginning of the selection procedure; (3) exponential growth of the polynomial degree (1, 2, 4, 8, …) due to using the non-linear (quadratic) partial description, and others. To enhance the efficiency of the iterative GMDH algorithm, various modifications of MIA have been proposed in different periods of its evolution. Main variants of them are presented below.

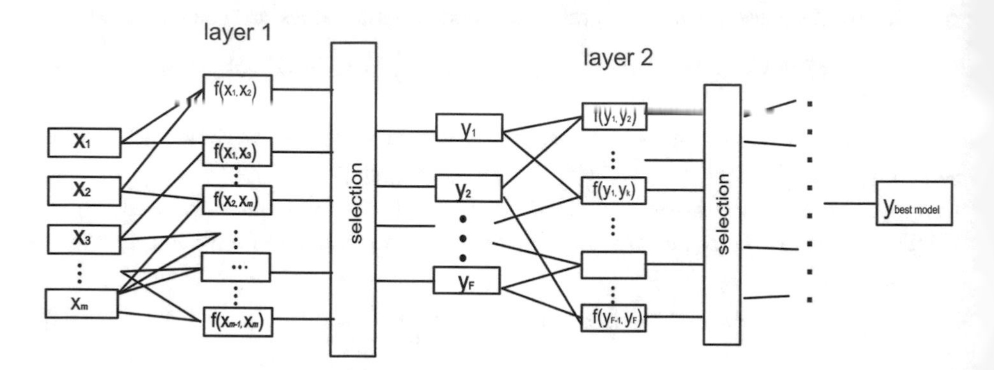

Fig. 1. Classical algorithm MIA GMDH

For example, the mentioned disadvantages of MIA have been partly removed in the following algorithms. The first drawback was eliminated in algorithms with including initial arguments to the selection process at any layer [7]. The second one was partly precluded by adding some kind of sorting variants of partial description structures at any layer [8–10]. The last one may be prevented due to using the second-order polynomial on the first layer and only linear forms for next layers [11] or, on the contrary, by using firstly linear partial descriptions and then, starting from the second layer, quadratic descriptions [12]. In what follows, some of such modified algorithms proposed in 1970–80th are characterized in more details.

So called "Simplified multilayered GMDH algorithm" was developed in 1974 where pairs were formed only from the intermediate and initial arguments to prevent the loss of informative ones [7]. The algorithm was attributed later to the relaxational type of GMDH algorithms; its disadvantage is absence of freedom of choice, $F = 1$.

Structurally similar algorithms were proposed in 1981, also providing the inclusion of initial arguments to the selection process on subsequent layers of the iterative procedure [10]. First one of this kind was CML, multilayered algorithm with linear partial descriptions. This is the only algorithm for which the so called internal convergence was proved. Here the following partial models are considered on a layer r:

$$y_l^r = a^{r-1} y_i^{r-1} + b^{r-1} z_j^{r-1}, \tag{4}$$

where both the intermediate and initial arguments may be used as variable z. Polynomial algorithm with the Gödel numbering GN in which partial models are considered on a layer r based on not two but three inputs:

$$y_l^r = a^{r-1} y_i^{r-1} + b^{r-1} z_j^{r-1} z_k^{r-1}, \tag{5}$$

where similarly the intermediate and/or initial arguments may be used as variable z.

Obviously, the previous algorithm CML is a particular case of GN (with y_k = 1 or y_j = 1). Algorithm of GN type is further developed as GMDH-PNN [13].

Algorithms described above eliminate only the first shortcoming, namely the possibility of informative arguments loss if they were removed or not included at the beginning of the selection procedure. To eliminate the second and third shortcomings (fixing uninformative arguments and growth the degree of polynomial), the use of the full sorting-out of partial model variants was proposed [8, 9].

For example, in so called multilayered-combinatorial algorithm [9] a procedure of multilayered selection was implemented using combinatorial sorting-out of partial models applied to the following partial description: $y = a_1 + a_2 x_i + a_3 x_j + a_4 x_i x_j$. Main steps of the algorithm are as follows:

1. Implementation of multilayered selection of the partial descriptions: generating all possible pairs of arguments for current selection layer.
2. Combinatorial sorting-out of all variants of partial descriptions for a given pair of arguments. Selection of the best partial models by the given criterion.
3. Checking the stop rule and ending the process or switching to the next layer.

But it is obvious that this algorithm, like classical multilayered one, does not still eliminate the first of the disadvantages mentioned above.

4 Current Trends of GMDH Development

At present, many domestic and foreign researchers in Ukraine, Czech Republic, Japan and other countries are actively developing GMDH-like systems based on the classical multilayered algorithm.

GMDH as a polynomial neural network. At early stages of the GMDH theory development, the similarity between neural networks and multilayered GMDH algorithm was observed. O. Ivakhnenko in one of the articles stated that it is acceptable to call GMDH-systems as “perceptron-like systems”.

Starting from 1990th, GMDH algorithm is often called among professionals as Polynomial Neural Network (PNN) [14]. This is because one of the main elements of iterative GMDH algorithms, namely the polynomial partial description, can be considered as an elementary neuron of the GMDH neural network. The network originality with such neurons consists in high speed of the process of local adjustment of neuron weights and automatic global optimization of the network structure (number of units and iterations or hidden layers).

The idea of neural networks with active neurons. In all kinds of standard (passive) neurons, any mechanisms of optimization of the set of input variables are not used, only parametric optimization is performed. The mechanisms are realized in the complex process of self-organization of the whole system of many neurons in general.

A combined method was proposed by prof. Ivakhnenko in [15] extending the theory of self-organization from fixed passive structures to active neural networks. An algorithm known as "neural network with active neurons" is used in the GMDH architecture. Both multilayered and combinatorial GMDH algorithm can be used as active neurons; that leads to increasing the accuracy and reducing the calculation time.

The advantage of GMDH neural networks with active neurons as compared to the conventional neural network with uniform neurons consists in that self-organizing of the network is simplified: each neuron finds necessary connections and its own structure in the process of self-organization. The idea of active neurons served also as the basis for generalization of the previous modifications in order to significantly improve the efficiency of iterative GMDH algorithms.

GMDH-type neural network with feedback. A new multilayered GMDH-type neural network with feedback was proposed in [16] which may automatically choose the optimum neural network architecture using the idea of self-organization. In this algorithm outputs of neurons are connected with the system inputs (feedback) for further calculation. Thus, the complexity of the neural network increases gradually. This GMDH-type neural network algorithm has an ability of self-selecting optimum neural network architecture from three variants such as sigmoid function, radial basis function (RBF) and polynomial neural network. In addition, structural parameters such as number of neurons in hidden layers and input variables are selected automatically by minimizing the Akaike criterion, that is without data division similarly as in [8].

Group of Adaptive Models Evolution. An evolutionary algorithm based on GMDH and called GAME (group of adaptive models evolution) was developed in [17]. Major modifications of this GMDH-type system are [9]: the transfer function of the unit is of several types (linear, polynomial, logistic, etc.); each type of unit has its own learning algorithm for coefficients estimation; choice of the type of units that form a network is determined by the given criterion.

This is so-called heterogeneous network structure: the number of unit inputs increases together with the depth of the unit in the network; there exist interlayer connections in the network; the network construction process does not search all possible layouts of units, it searches just the random subset of these layouts.

The modified GMDH generates a group of models on a single training data set that are locally optimal. Weights and coefficients of units are randomly initialized. Transfer functions of many units types are defined pseudo-randomly when the unit is initialized. Inputs for units are selected pseudo-randomly as well. It results in the fact that the topology of models developed on the same training data set may differ.

Hybrid of Differential Evolution and GMDH for Inductive Modeling. The GMDH and differential evolution (DE) population-based algorithm are two well-known non-linear methods of mathematical modeling. A new design methodology which is a hybrid of GMDH and DE was proposed in [18]. The method constructs a GMDH network model of a population of promising DE solutions.

The modeling methods have many common features, but, unlike the GMDH, DE does not follow a pre-determined path for input data generation. The same input data elements can be included or excluded at any stage in the evolutionary process by virtue of the stochastic nature of the selection process. A DE algorithm can thus be seen as implicitly having the capacity to learn and adapt in the search space and thus allow previously bad elements to be included if they become beneficial in the later stages of the search process.

5 Constructing the Generalized Hybrid GMDH Algorithm

The above analytical results make it possible to advance the following basic ideas in developing new hybrid iterative algorithms methodology for removing all the three disadvantages of MIA GMDH:

- Enabling the use of the initial arguments in each layer to prevent losing the relevant ones in the multilayer self-organizing process;
- Using the idea of active neurons by performing the optimization of every partial model structure by combinatorial algorithm to avoid inclusion of spurious arguments in the final model and overfitting the model complexity;
- Using various partial descriptions in the form of linear, bilinear or nonlinear functions on different layers;
- Enabling to a user applying different modes of the modeling process.

These modifications were applied and studied in [19–21] to construct the so called Generalized Iterative Algorithm GIA GMDH which makes it possible to get the following basic variants of the GMDH algorithms of iterative type:

(1) Classical MIA GMDH algorithm with the use of pairwise combinations of only intermediate arguments;
(2) An algorithm with adding to intermediate arguments only initial ones;
(3) An algorithm with equal usage of both the intermediate and initial arguments in the partial descriptions;
(4) In any of the three variants, the optimization of every partial model structure by combinatorial algorithm may or may not be used;
(5) In the case of using the optimization, this architecture becomes a typical hybrid algorithm with active neurons.

Due to these features the developed hybrid algorithm with quadratic partial description can build linear, bilinear and nonlinear models of complex systems and processes.

5.1 GIA GMDH Algorithm Description

Formally, in general case, a layer of the GIA GMDH may be defined as follows [20]:

(1) The input matrix is $X_{r+1} = (y_1^r, \ldots, y_F^r, x_1, \ldots, x_m)$ for a layer $r + 1$, where $x_1, \ldots, x_m$ are the initial arguments and $y_1^r, \ldots, y_F^r$ are the intermediate ones of the layer r;
(2) The operators of the kind

$$y_l^{r+1} = f(y_i^r, y_j^r), \;\; l = 1,2,\ldots,C_F^2, \quad i,j = \overline{1,F},$$
$$y_l^{r+1} = f(y_i^r, x_j), \; l = 1,2,\ldots,Fm, \; i = \overline{1,F}, \; j = \overline{1,m} \tag{6}$$

may be applied on the layer $r + 1$ to construct linear, bilinear and quadratic partial descriptions:

$$z = f(u, v) = a_0 + a_1 u + a_2 v;$$
$$z = f(u, v) = a_0 + a_1 u + a_2 v + a_3 uv\,; \tag{7}$$
$$z = f(u, v) = a_0 + a_1 u + a_2 v + a_3 uv + a_4 u^2 + a_5 v^2\,.$$

(3) For any description, the optimal structure is searched by combinatorial algorithm; e.g., for the linear partial description the expression holds:

$$f(u, v) = a_0 d_1 + a_1 d_2 u + a_2 d_3 v, \tag{8}$$

where d_k, $k = 1, 2, 3$, $d_k = \{0,\ 1\}$ are elements of the binary structural vector $d = (d_1\, d_2\, d_3)$ where values 1 or 0 mean inclusion or not a relevant argument.

Then the best model will be described as $f(u, v, d_{opt})$, where

$$d_{opt} = \arg \min_{l=\overline{1,q}} CR_l, \quad q = 2^p - 1, \;\; f_{opt}(u, v) = f(u, v,\, d_{opt}). \tag{9}$$

(4) The algorithm stops when the condition $CR^r > CR^{r-1}$ is checked, where CR^r, CR^{r-1} are criterion values for the best models of (r–1)-th and r-th layers respectively. If the condition holds, then stop, otherwise jump to the next layer.

The GIA structure is schematically represented in the Fig. 2.

5.2 Main Particular Cases of the GIA GMDH Architecture

The following six typical algorithms can be generated as particular cases of the generalized architecture:

(a) *Iterative GMDH algorithms*:
   (1) Multilayered Iterative Algorithm MIA in which pairs are only between the intermediate arguments;
   (2) Relaxation Iterative Algorithm RIA in which pairs are between the intermediate and initial arguments;
   (3) Combined Iterative Algorithm CIA in which pairs are possible between both the intermediate and initial arguments;

(b) *Iterative-combinatorial GMDH algorithms with combinatorial optimization of partial descriptions:*
   (4) Multilayered Iterative-Combinational MICA;
   (5) Relaxation Iterative-Combinational RICA;

(6) Combined Iterative-Combinational CICA (being practically the same as GIA).

Any of these variants may be formed and used with the help of the specialized modeling software complex [21] based on the GIA GMDH with implementation of the following features: automatic and interactive options for organization of user interface; administration through the web interface; ensuring multi-access. Best constructed models are presented by system for the graphic and semantic analysis as well as selection of the most informative arguments.

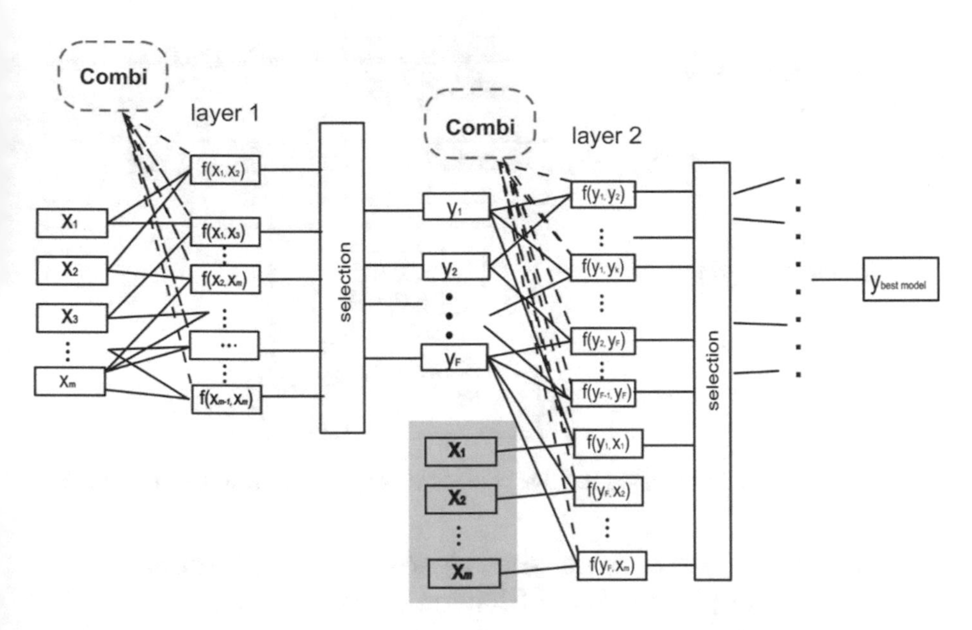

Fig. 2. The generalized architecture of GIA GMDH

5.3 The Ordered Coding of the Iterative GMDH Algorithms Structure

Let us define the GIA GMDH as a set of iterative and iterative-combinatorial algorithms described by the vector of the following three elements: DM (Dialogue Mode), IC (Iterative-Combinatorial), MR (Multilayer-Relaxation).

This means that any iterative algorithm can be defined as a special case of the generalized one of general description GIA(DM,IC,MR). In this case, DM takes three values: 1 – standard automatic mode, 2 – planned automatic mode, 3 – interactive mode; IC takes two values: 1 – iterative, 2 – Iterative-combinatorial algorithms; MR takes three values: 1 – classic iterative, 2 – relaxation, 3 – combined algorithms.

This vector defines any of typical iterative algorithms mentioned in the item 5.2: if DM = 1, we have three standard variants of iterative algorithms MIA = GIA(1,1,1), RIA = GIA(1,1,2), CIA = GIA(1,1,3), as well as three iterative-combinatorial ones:

MICA = GIA(1,2,1); RICA = GIA(1,2,2), CICA = GIA(1,2,3). For DM equal 2 or 3, some new versions of these algorithms can be formed.

A hierarchical representation of the developed GIA GMDH architecture is displayed on Fig. 3, where white blocks correspond to known algorithms and gray ones indicate newly designed structures as special cases of GIA.

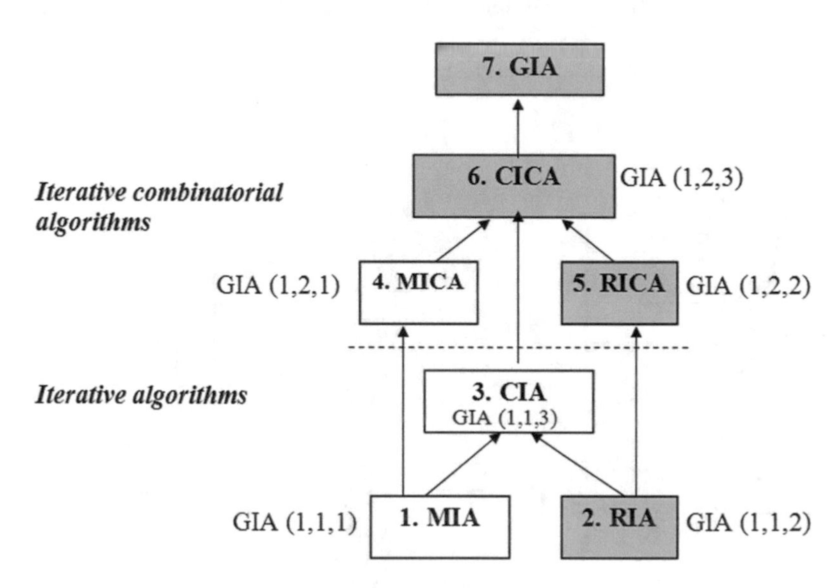

Fig. 3. The hierarchy of typical iterative algorithms as special cases of the GIA GMDH

6 Experimental Research of the Algorithms Effectiveness in the Task of True Model Structure Construction

6.1 Integrated Technique for Analysis of the Effectiveness of Iterative GMDH Algorithms Using Computational Experiments

The aim of computational experiments is investigating the effectiveness of the generalized hybrid algorithm which combines the ideas of initial arguments selection and optimization of the partial models complexity. It also allows performing comparison of various iterative GMDH algorithms as special cases of the generalized one.

Any of the iterative algorithms is characterized by its input (control) parameters as well as intermediate and output indicators of the model building process. The structure of a resulting data table is presented in Table 1.

The input parameters of any algorithm: the number of initial arguments m; number of points n in the sample; procedure to partition the dataset W into the training A and testing B subsets; selection criteria; freedom of choice value F.

The intermediate indicators: plot of minimum and maximum values of the criterion changing by layers; changing of the best model arguments by layers.

Table 1. Data table structure

No	X [n×m]				y [n×1]
	x_1	x_2	…	x_m	
1	x_{11}	x_{12}	…	x_{1m}	y_1
2	x_{21}	x_{22}	…	x_{2m}	y_2
…	…	…	…	…	…
n	x_{n1}	x_{n2}	…	x_{nm}	y_n

The output indicators: the best model and its parameters; values of the selection criteria; composition of arguments of the best model and its proximity to the true one; program run-time (work time of central processor).

Besides studying the impact of key control parameters, in the experiments it is necessary to investigate the effectiveness of the true structure detection. Therefore, to study restoring of the structure by different algorithms we need to generate experiments in which it is necessary to:

- Restore the true structure in case of many redundant arguments from data without noise;
- Restore the true model structure hidden in data and investigate the algorithms effectiveness in tasks with noise. The noise can be generated, for example, as follows:

$$\hat{y} = \mathring{y} + \alpha(2\gamma - 1)\frac{\mathring{y}_{\max} - \mathring{y}_{\min}}{200} \tag{2}$$

where $\mathring{y}$ is the true signal, α is noise percentage, γ is a random value uniformly distributed in the interval [0, 1], $\mathring{y}_{\max}$, $\mathring{y}_{\min}$ are maximum and minimum of the output value respectively.

- Restore the true structure in case of a small number of arguments but with a complex nonlinear dependence;
- Examine the developed algorithms for the existence of so-called "internal convergence" of the iterative process to the true model by structure and parameters.

Based on the proposed methodology of comparative analysis of algorithms effectiveness, one can investigate the properties of iterative GMDH algorithms and the effect of key parameters on modeling performance. It should be noted that in all cases the effect of freedom of models choice F on performance indicators of algorithms was investigated especially as the main of the control parameters.

Experiments described below were based on artificially generated data and carried out to study the effectiveness of various iterative algorithms and the regularity criterion was used in all the experiments. The goal was to compare all the six algorithms introduced above but first of all the three typical ones, namely classical multilayered MIA, advanced combined CIA and the generalized GIA with active neurons.

The comparison was done under various conditions: linear and nonlinear dependences in data; presence of additive noise of different level; testing the internal convergence of the algorithms. In all cases the regularity criterion *AR* (2) was used.

6.2 Effectiveness of Iterative Algorithms in Detecting True Model, the Case of Quadratic Dependence

The experiment goal is to compare the ability of different GMDH algorithms to restore structure of the true nonlinear model hidden in data.

Data characteristics: artificial data, random numbers in the range of [0, 5]; arguments number $m = 200$; sample length $n = 243$; sample division $n_A = 163$, $n_B = 80$; freedom of choice $F = 40$.

Algorithms being compared: (1) multilayered MIA, (2) relaxation RIA, (3) combined iterative CIA, (4) multilayered-combinatorial MICA, (5) relaxation-combinatorial RICA; (6) generalized GIA.

The quadratic dependence is represented by the following formula (196 redundant arguments):

$$\mathring{y} = -3x_1 - 2x_2x_3 + x_{25}^2 + x_{25}x_1 + 12 \quad (3)$$

Table 2 shows that in this case the true structure is restoring by algorithms RICA and GIA but the latter (generalized) enables to recover the true model most precisely by structure and parameters.

Table 2. Comparison of iterative algorithms effectiveness

Algorithm	AR(B)	Model
1	11.309	$\hat{y} = -3x_1 - 1.231x_2x_3 + 2.711x_{25} + 0.998x_{32} + x_{37} - 2.001x_{45} + 12$
2	3.031	$\hat{y} = -3x_1 - 1.231x_2 + 6.700x_{25} - 2.001x_{45} + 11.99$
3	3.011	$\hat{y} = -3x_1 - 1.231x_2x_3 + 6.711x_{25} - 2.001x_{45} + 12$
4	0,462	$\hat{y} = -2.0001x_1x_3 + x_{25}^2 + x_1x_{25} - 2.999x_{45} + 12$
5	$4*10^{-7}$	$\hat{y} = -3.000x_1 - 2.000x_2x_3 + x_{25}^2 + 0.985x_{25}x_1 + 12.000$
6	$3*10^{-8}$	$\hat{y} = -3.000x_1 - 2.000x_2x_3 + x_{25}^2 + x_{25}x_1 + 12.000$

6.3 Effectiveness of Iterative Algorithms Under Conditions of Noisy Data

The experiment goal: to restore the true model structure hidden in the data using various algorithms GMDH and compare effectiveness of them in the case of noisy data.

Data characteristics: artificial data, random numbers in the range of [0, 1]; $m = 40$, $n = 60$, $n_A = 40$, $n_B = 20$; freedom of choice $F = 40$.

Algorithms being compared: (1) multilayered MIA, (2) combined iterative CIA, (3) generalized GIA.

The true dependence is represented by the following formula:

$$\mathring{y} = 0,5 - 1,2x_2 + 5x_{10} - 3,4x_{25}. \quad (4)$$

The executed experiments show that when the output variable is eddied by uniform noise with noise/signal ratio of 10% and 30% then GIA remains to be the best algorithm detecting the true model structure, see Fig. 4.

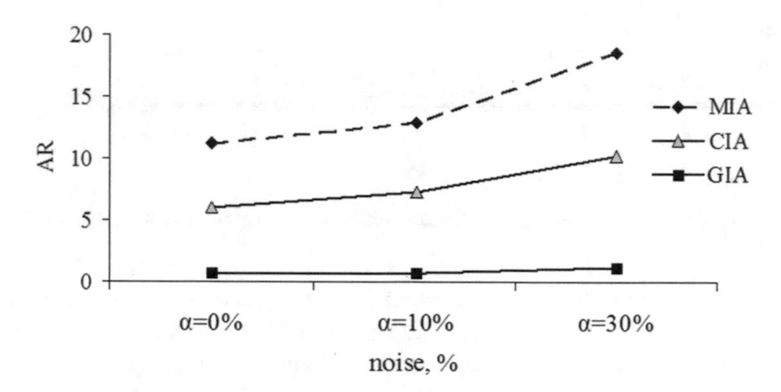

Fig. 4. Comparison of the iterative algorithms effectiveness under conditions of noisy data

6.4 Effectiveness Investigation of Constructing an Essentially Nonlinear Dependence in Case of a Small Number of Arguments

The experiment goal: to study the influence of the freedom of choice F of the best solutions as the key GMDH parameter on the comparative efficiency of algorithms in nonlinear tasks.

The sample parameters: $m = 3$ arguments and $n = 100$ points, the division into sub-samples $n_A = 65$, $n_B = 35$. The effect of the freedom of choice F on the performance of algorithms is investigated.

The true model is the following nonlinear function of three arguments:

$$y = 7 - 6x_1 + 5x_3 - 4x_2^2 + 3x_1x_3 - 2x_1^2x_2 + x_3^3. \quad (13)$$

Figure 5 shows the dependence of minimum values of the criterion AR for different algorithms on the value of freedom of choice of the best F models. The lowest value of the criterion for all F has been achieved for the GIA which detected the true model structure. As it is obvious from the figure, the freedom of choice value $F = 20$ is sufficient for all algorithms.

6.5 Verification of Internal Convergence of Iterative GMDH Algorithms

It was experimentally verified the existence of so-called internal convergence of iterative algorithms theoretically proved earlier in [10].

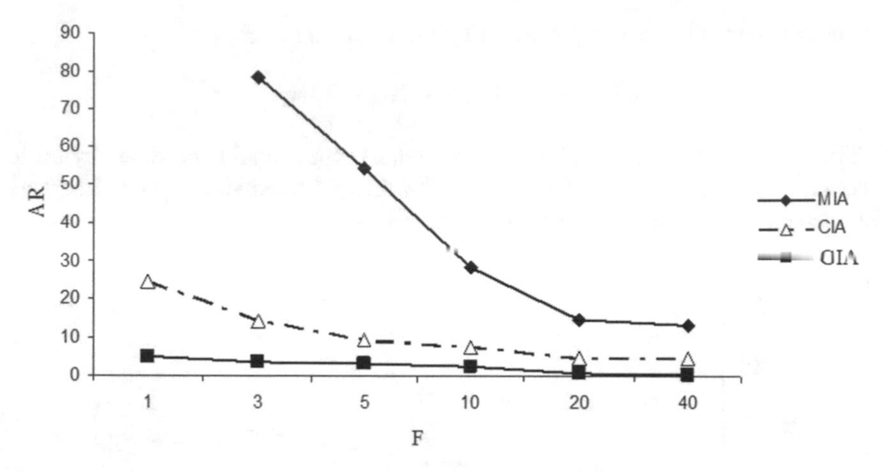

Fig. 5. Changing the selection criterion values while increasing the freedom of models choice F

An iterative algorithm holds the internal convergence if as a result of iteration process with the selection criterion *RSS* being the residual sum of squares on the entire sample *W* (without partition), the estimations of optimal model parameters are converging to the estimates of parameters of the true model by the Least Squares Method.

Data characteristics in this case: $m = 10$, $n = 50$. The internal convergence is studied for the following three typical iterative GMDH algorithms: classical MIA, combined CIA, and generalized GIA [12].

The true model is the following linear function of 10 arguments:

$$\mathring{y} = 3 - 2x_1 + 5x_2 - x_3 + 7x_4 - 2x_5 + 4x_6 - 3x_7 + 2,5x_8 - 1,8x_9 + x_{10}. \tag{14}$$

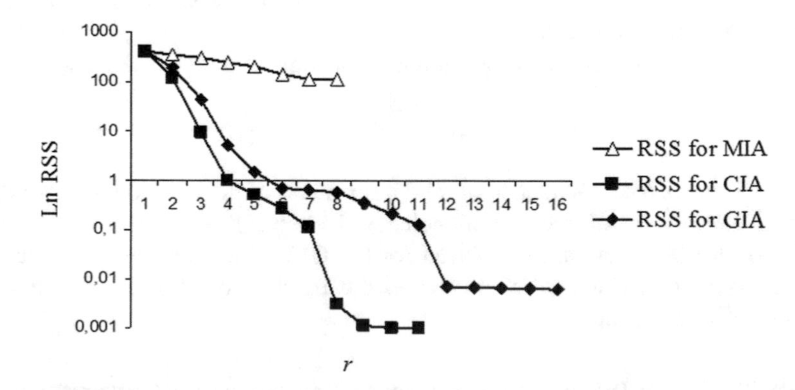

Fig. 6. Change of RSS values for the three iterative GMDH algorithms, $F = 20$ (logarithmic scale)

Figure 6 shows the change of *RSS* values as dependent on the layer number *r* of these algorithms with $F = 20$.

The constructed models are of form [12]:

$$y_{\mathrm{MIA}} = -22.387 + 5.334x_2 - 0.535x_3 + 5.266x_4 + 0.812x_5 \\ + 4.726x_6 + 1.278x_8 - 2.643x_9 + 0.437x_{10} \quad (15)$$

$$y_{\mathrm{CIA}} = 2.913 - 1.978x_1 + 4.999x_2 - 1.015x_3 + 7.004x_4 \\ - 1.988x_5 + 3.997x_6 - 2.994x_7 + 2.514x_8 - 1.808x_9 + 1.003x_{10} \quad (16)$$

$$y_{\mathrm{GIA}} = 3.001 - 1.999x_1 + 4.999x_2 - 1.002x_3 + 7.004x_4 \\ - 1.998x_5 + 4.000x_6 - 2.999x_7 + 2.500x_8 - 1.801x_9 + 1.003x_{10} \quad (17)$$

In the classical multilayered algorithm MIA, the convergence is observed only by criterion but there is no internal convergence with respect to the structure and parameters in view of the loss of true arguments x_1 and x_7, compare (14) and (15).

The Eqs. (16) and (17) generally confirm the internal convergence of the two iterative algorithms because values of the estimated parameters are close to true ones, see (14). The Fig. 6 shows that the generalized algorithm reaches its minimum on the 9th layer and the combined one on the 12th, hence the convergence rate of GIA is much higher than CIA. The accuracy of the obtained models also differs in favor of the generalized: for GIA RSS = 0.001, for CIA RSS = 0.006. This demonstrates the effectiveness of the idea of partial models optimization.

The results obtained above illustrate, first, that the developed architecture of the generalized GMDH algorithm allows investigating various versions of iterative algorithms, and, second, this algorithm demonstrates highest performance in all versions of test problems.

The effectiveness comparison of the generalized hybrid GMDH algorithm with other iterative GMDH algorithms have been made based on study of the effect of key parameters of algorithms on performance indicators of the model construction process. Executed experiments showed the best performance of the generalized algorithm.

7 Modeling the Ukraine Black Sea Economic Region GRP Dependence on Socio-Economic Indicators

The Ukraine's Black Sea economic region includes the territory of Odessa, Mykolaiv, and Kherson oblasts located in the southern and south-western parts of Ukraine.

For this economic region, the gross regional product (GRP) dependence on socio-economic indicators was modeled using the generalized iterative algorithm GMDH. The goal was to identify the dependence of GRP growth (mln. UAH) on the following socio-economic indicators:

Industry: x_1 is index of industrial production, % to the previous year;

Production of major animal products: production of meat x_2, milk x_3, wool x_4, and eggs x_5 in all categories of farms;

Production of main agricultural crops: grains and legumes x_6; sunflower seeds x_7; potatoes x_8; vegetables x_9; fruits and berries x_{10};

Fishing industry: fishery and other aquatic resources x_{11};

Investment and construction activities: commissioning housing x_{12};

Transport: transportation of cargo x_{13} and passengers x_{14},

Foreign trade: export x_{15} and import x_{16} of goods and services;

Internal trade: retail turnover x_{17};

Job Market: unemployment rate x_{18}; average monthly salary x_{19}; wage arrears x_{20}.

Data were taken at the Ukrainian State Statistics Committee.

Statistical sample of quarterly data for Ukraine's Black Sea economic region for the period from 2004 (1st quarter) to 2011 (4th quarter) contains totally 20 variables and 32 data points which are divided into three parts: training A (20 points), testing B (8 points), and examination C (4 points, 2011 year) sub-samples.

Table 3 and Figs. 7, 8 and 9 show the modeling results of the GRP values.

Table 3. Result of modeling the GRP values

Region	*AR*	Model accuracy, %	Model
Mykolaiv region	0.787	99	$y = 1697.10 + 0.0306x_3 - 0.01315x_7 - 0.00549x_{11} + 1.815x_{12} - 0.00023x_{13} + 0.000027x_{16} + 0.00026x_{17} + 9.284x_{19} + + 0.022x_{12}^2 + 24 * 10^{-10}x_{17}^2$
Kherson region	202.01	87	$y = -538.59 + 1.1376x_4 + 0.063x_6 + 8.228x_{19} + 0.000037x_7^2$
Odessa region	189.23	90	$y = -18548.99 - 0.00407x_7 + 0.541x_9 + 17.83x_{15} + 35.868x_{19} - 0.00014x_{15}x_9 - 0.0093x_{15}x_{19} - 0.0006x_{10}^2$

The obtained dependences demonstrate that:

- For Mykolaiv region, only 8 socio-economic indicators among all 20 ones have the most significant effect on GDP: milk production; sunflower seeds; fishing extraction and other aquatic resources; commissioning housing; departure (transportation) of cargo; import goods and services; retail turnover; average monthly salary;
- For Kherson region, only 4 indicators among all 20 ones have significant effect on GDP: wool production; grains and legumes; the average monthly salary; sunflower seeds;
- For Odessa region, only 8 indicators among all 20 ones have the most significant effect on GDP: sunflower seeds; vegetables; export goods and services; average monthly salary; fruits and berries.

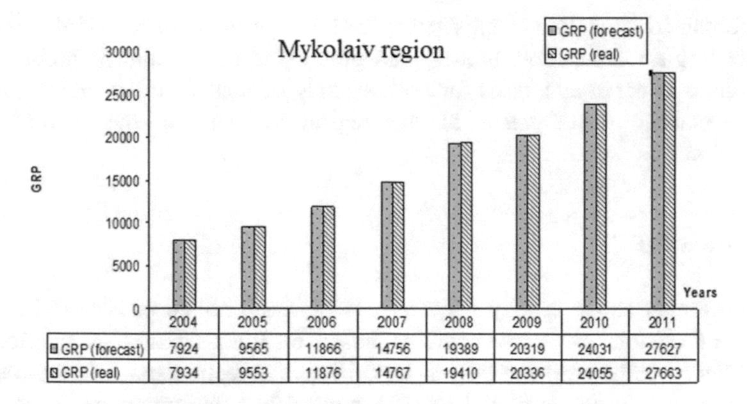

Fig. 7. Result of modeling of the Mykolaiv region GRP values, on years

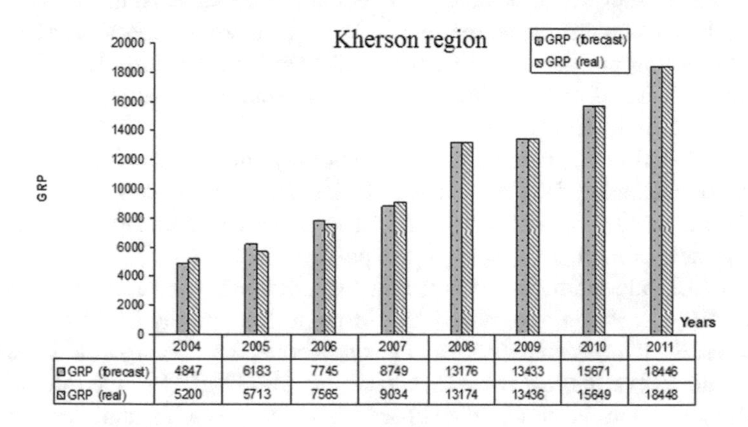

Fig. 8. Result of modeling of the GRP Kherson region values, on years

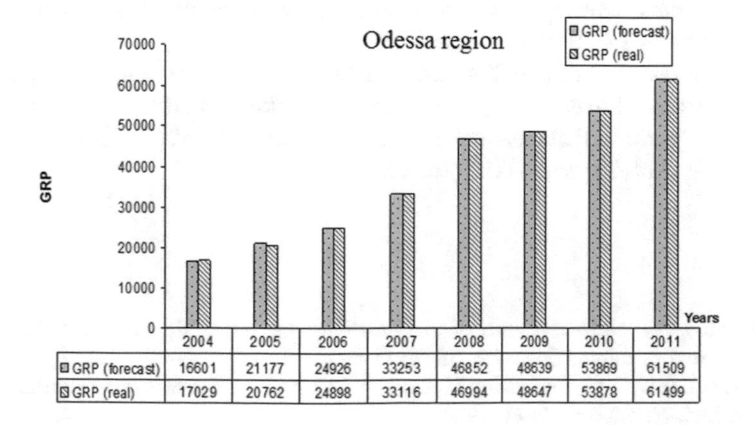

Fig. 9. Result of modeling of the Odessa region GRP values, on years

The obtained modeling results show that the priority sector of the Black Sea Economic Region is the food industry and processing of agricultural products. These conclusions may serve as a basis for further study of reasons of the uneven level and rate of economic development of the region to mitigate the current regional differentiation.

8 Conclusion

Results presented above specify ways to solving the problem of how to improve the efficiency of complex systems modeling based on hybridization of architectures of iterative and combinatorial GMDH algorithms. Effectiveness of the developed improvements has been confirmed by both numerical experiments on artificial examples and solving a modeling problem of a real economic process.

Comparative analysis of advantages and drawbacks of existing iterative GMDH algorithms has resulted in some reasonable ways to increase effectiveness of solving the model building problems. The Generalized Iterative Algorithm GIA GMDH with Active Neurons based on the combinatorial optimization of partial descriptions was introduced and implemented.

The developed computational online-technology for automated solving the modeling tasks on the basis of the generalized GMDH algorithm makes it possible to carry out computations via Internet and local net in automatic and interactive modes for building models of complex system in the polynomial class.

GIA GMDH algorithm is a neural network with active neurons being optimized using COMBI, i.e. it is a generalized hybrid algorithm entirely based on the GMDH architectures. In all the executed numerical experiments, the GIA algorithm was the most effective. The fact of internal convergence of the GIA GMDH to the true polynomial model with respect to the structure and parameters was experimentally confirmed.

The modeling task of the Ukraine Black Sea economic region GRP dependence on socio-economic indicators was solved using the GIA GMDH. Informative indicators of the socio-economic development influencing the dynamics of the Kherson, Mykolaiv and Odessa region's GRP were determined.

The above results illustrate, first, that the developed architecture of the generalized GMDH algorithm allows investigating various versions of iterative algorithms, and, second, this algorithm demonstrates highest performance both in all versions of test tasks and when solving real-world problems.

References

1. Ivakhnenko, A.G.: Group method of data handling as a rival of the stochastic approximation method. Soviet Autom. Control **3**, 58–72 (1968)
2. Farlow, S.J. (ed.): Self-Organizing Methods in Modeling: GMDH Type Algorithms. Marcel Decker Inc., New York, Basel (1984)
3. Madala, H.R.: Inductive Learning Algorithms for Complex Systems Modeling. CRC Press Inc., Boca Raton (1994)

4. Ivakhnenko, A.G., Müller, J.-A.: Recent Developments of Self-Organizing Modeling in Prediction and Analysis of Stock Market. Microelectron. Reliab. **37**, 1053–1072 (1997)
5. Stepashko, V.: Ideas of Academician O.H. Ivakhnenko in the Inductive Modelling Field from Historical Perspective. In: Proceedings of the 4th International Conference on Inductive Modelling ICIM 2013, IRTC ITS NASU, Kyiv, Ukraine, pp. 30–37 (2013)
6. Stepashko, V.S., Kocherga, Y.L.: Classification and analysis of the noise immunity of external criteria for model selection. Soviet Autom. Control **17**(3), 36–47 (1984)
7. Sheludko, O.I.: GMDH algorithm with orthogonalized complete description for synthesis of models after results of a planned experiment. Soviet Autom. Control **7**(5), 46–57 (1974)
8. Tamura H., Kondo T.: Large-Spatial pattern identification of air pollution by a combined model of source-receptor matrix and revised GMDH. In: Proceedings of IFAC Symposium on Environmental Systems Planning, Design and Control, pp. 373–380. Elsevier, Oxford (1977)
9. Ivakhnenko, N.A., Marchev, A.A.: Self-organization of a mathematical model for long-term planning of construction and installation works. Soviet Autom. Control **11**(3), 10–16 (1978)
10. Yurachkovskiy, Y.P.: Recovery of polynomial dependences using self-organization. Soviet Autom. Control **14**(4), 14–19 (1981)
11. Parker R.G.J., Tummala M.: Identification of Volterra systems with a polynomial neural network. In: Proceedings of the 1992 International Conference on Acoustics, Speech and Signal Processing ICASSP 1992, San Francisco, vol. 4, pp. 561–564. IEEE (1992)
12. Hara, K., Yamamoto, T., Terada, K.: Improved dual mode GMDH with automatic switch. Int. J. Syst. Sci. **21**(8), 1553–1565 (1990)
13. Aksyonova, T.I., Volkovich, V.V., Tetko, I.V.: Robust polynomial neural network in quantative-structure activity relationship studies. Syst. Anal. Model. Simul. **43**(10), 1331–1341 (2003)
14. Oh, S.K., Pedrycz, W.: The design of self-organizing polynomial neural networks. Inf. Sci. **141**, 237–258 (2002)
15. Ivakhnenko, A.G., Wunsh, D., Ivakhnenko, G.A.: Inductive sorting-out GMDH algorithms with polynomial complexity for active neurons of neural networks. In: Proceedings of the International Joint Conference on Neural Networks, pp. 1169–1173. IEEE, Piscataway, New Jersey (1999)
16. Kondo T.: GMDH neural network algorithm using the heuristic self-organization method and its application to the pattern identification. In: Proceedings of the 37th SICE Annual Conference SICE 1998, pp. 1143–1148. IEEE, Piscataway, New Jersey (1998)
17. Kordík, P., Náplava, Šnorek, M., Genyk-Berezovskij, P.: The modified GMDH method applied to model complex systems. In: Proceedings of International Conference on Inductive Modelling ICIM 2002, Lviv, Ukraine, pp. 134–138. SRDIII (2002)
18. Onwubolu, G.C.: Design of hybrid differential evolution and group method of data handling for inductive modeling. In: Proceedings of the IInd International Workshop on Inductive Modelling IWIM 2007, CTU in Prague, Czech Republic, pp. 87–95 (2007)
19. Stepashko, V., Bulgakova, O., Zosimov, V.: Experimental verification of internal convergence of iterative GMDH algorithms. In: Proceedings of the Vth International Workshop on Inductive Modelling IWIM 2012, IRTC ITS NASU, Kyiv, pp. 53–56 (2012)
20. Stepashko, V., Bulgakova, O.: Generalized iterative algorithm GIA GMDH. In: Proceedings of the 4th International Conference on Inductive Modelling ICIM 2013, IRTC ITS NASU, Kyiv, pp. 119–123 (2013)
21. Zosimov, V.V., Bulgakova, O.S., Stepashko, V.S.: Software package for complex systems modelling on the basis of iterative GMDH algorithms with the possibility of network access. Syst. Res. Inf. Technol. **1**, 43–55 (2014). (In Ukrainian)

The Fast Fourier Transform Partitioning Scheme for GPU's Computation Effectiveness Improvement

Kamil Stokfiszewski(✉), Kamil Wieloch, and Mykhaylo Yatsymirskyy

Lodz University of Technology University, Lodz, Poland
{kamil.stokfiszewski,mykhaylo.yatsymirskyy}@p.lodz.pl,
kamil.wieloch@dokt.p.lodz.pl

Abstract. In this paper authors present the Fast Fourier Transform (FFT) partitioning scheme aimed at improvement of the effectiveness of the considered transform computation on graphics processing units (i.e. the GPUs). The FFT radix-2 decimation in time (DIT) algorithm is chosen as the base procedure for the FFT calculation which is then partitioned into subtransform blocks of arbitrary sizes enabling for different GPU resources distribution during its computational process and thus resulting in the potential improvement of the overall FFT execution time for chosen consumer segment GPU models. The conducted experiments show that for a chosen GPU architectures running in the single instruction multiple thread (SIMT) mode of operation partitioning of the FFT into 4-point and 8-point subtransforms calculated sequentially within an individual thread, instead of its calculation using standard 2-point butterfly operations, significantly reduces the FFT's computation time. The presented scheme is general and can be used for the partitioning of the FFT into arbitrary size subtransform blocks aimed at the scheme's time effectiveness fine-tuning to the chosen, particular GPU architectures.

Keywords: Fast Fourier Transform · FFT · GPU programming · Parallel computations · Computational effectiveness

1 Introduction

Parallel computations with the aid of graphics processing units, i.e. GPUs, has in recent years been attracting an increasing attention of the scientific community across a broad spectrum of computational research domains [1]. This comes from fact that the GPUs have become an attractive tool for general-purpose industrial and research computational applications, offering a cost-effective hardware solutions, especially within a consumer segment GPUs, coupled with relatively high computational performance, see e.g. [1, 2]. This, on the other hand, has caused an intense interest among researchers in redesigning the implementations of many well-known classical algorithms to meet the requirements of parallel computations, and thus enabling them to build their time-effective variants that are suitable for the GPU realizations - see e.g. [3, 4]. A particular interest has been paid to the discrete Fourier transform (DFT), since it plays a key role

N. Shakhovska and V. Stepashko (eds.), *Advances in Intelligent Systems and Computing II*, Advances in Intelligent Systems and Computing 689,
https://doi.org/10.1007/978-3-319-70581-1_36

in many areas of engineering such as digital signal processing and signal filtering, data compression and encryption, image and pattern and recognition, as well as in a variety of many other data and signal processing applications, e.g. [5, 6] or [7]. A well-known class of DFT computation algorithms is Cooley-Tukey's Fast Fourier Transform algorithm [8, 9], often shortly referenced to as the FFT, which is especially convenient for parallel implementations due to its well-structured computational form. Various FFT realizations were analyzed in literature in terms of the improvement of their parallel GPU implementations' effectiveness, c.f. [10–12].

In this paper the authors analyze the issue of enhancement of time and resource allocation efficiency of parallel GPU FFT realizations by introducing a general FFT partitioning scheme that can be adopted for implementations on different, selected consumer segment GPU models. The scheme makes use of the decimation in time (DIT) FFT algorithm's variant which is partitioned into subtransform blocks of arbitrary sizes for which the computations are performed in a sequential manner. Experimental study reveals that such partitioning ultimately improves the DIT variant of the FFT calculation algorithm's time effectiveness, which is tested for the two selected different GPU architectures. Furthermore the generality of the scheme enables it to be adopted to various, chosen consumer segment GPU architectures performing their computations in the single instruction multiple thread (SIMT) manner.

2 Decimation in Time Fast Fourier Transform Algorithm

The discrete Fourier transform, also often referred to as the DFT, is one most commonly used mathematical tools in the areas of digital signal and data processing and also in many other important engineering fields. For the sake of clarity we'll now introduce its definition. For $k = 0, 1, \ldots, N-1$ the k-th coefficient of the N-point DFT of the complex sequence $x(n), n = 0, 1, \ldots, N-1$ is defined as follows

$$X(k) = DFT_N\{x(n)\} = \sum_{n=0}^{N-1} x(n) \cdot e^{-j2\pi kn/N}, \tag{1}$$

where j is the complex imaginary unit. It can be easily deducted form definition formula (1) that the computational complexity of the DFT calculation of all of its N coefficients is equal to $O(N^2)$. The reduction of the DFT's computational complexity is crucial for practical applications. The historically first class of the computational algorithms which were aimed to achieve the mentioned goal were introduced by Cooley-Tuckey in [8]. In their work Cooley-Tuckey proposed a class of fast computational procedures for the DFT calculation, i.e. Fast Fourier Transforms (FFTs), which require performing only $O(N\,log_2N)$ basic arithmetic operations in order to calculate all of the DFT coefficients, thus reducing its original computational complexity by one order of magnitude, see e.g. [9]. Cooley-Tuckey's original algorithm comes in many possible variants. In our work we have chosen decimation in time (DIT) FFT's version as our departure point. In the DIT FFT approach to DFT calculation the definition formula (1) can be expressed in the following recursive form

$$X(k) = DFT_{N/2}\{x(2n)\} + W_N^k DFT_{N/2}\{x(2n+1)\},$$
$$X(k + N/2) = DFT_{N/2}\{x(2n)\} - W_N^k DFT_{N/2}\{x(2n+1)\}, \quad (2)$$

where $n = 0, 1, \ldots, N-1$, $k = 0, 1, \ldots, N/2 - 1$ for $N = 2^p, p \in \mathbf{N}$ and $W_N^k = e^{-j2\pi kn/N}$. Figure 1 presents the data flow diagram of the 16-point DIT FFT obtained from the successive, recursive application of decomposition Eq. (2).

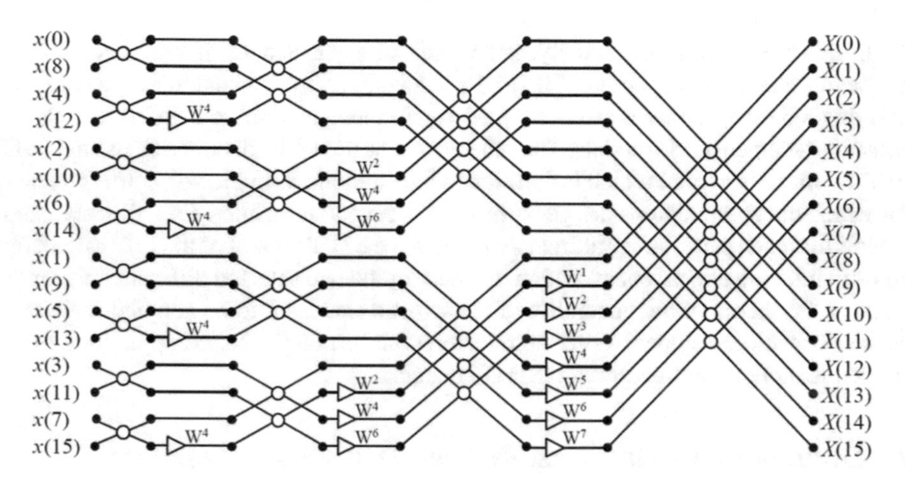

Fig. 1. Data flow diagram of 16-point decimation in time FFT calculation algorithm.

FFT's base operations, so called "butterfly" operations, are defined below in Fig. 2.

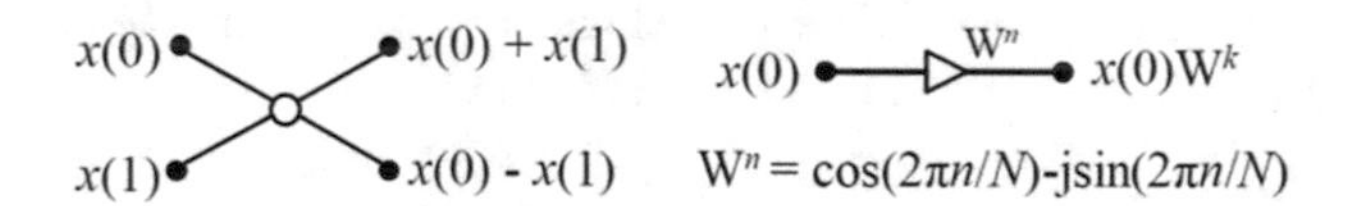

Fig. 2. Base operations of the Cooley-Tuckey FFT calculation algorithm form Fig. 1.

From Eq. (2) it can be easily verified that the computational complexity of FFT calculation algorithm is of $O(N\,log_2 N)$ and it's though justified to call the analyzed procedure a fast one with relation to the original DFT definition (1).

Presented in (2) and depicted in Figs. 1 and 2, DIT FFT algorithm is chosen by the authors as a departure point for the construction of the partitioning scheme for time-effective GPU FFT calculation.

3 Partitioning Scheme for the Effective GPU FFT Computation

Parallel computing with the use of GPUs became very popular in recent years, see e.g. [10–12]. There are many research papers related to this subject, but the majority is still focused on comparing CPU performance against the GPU, see e.g. [11]. The most popular approach in such type of research is based on the use of the maximum possible number of threads working in parallel and the largest defragmentation of given problem in case of GPUs' realizations against standard sequential implementations. During execution of such calculation procedures, basic operations conducted on the CPU and the GPU are the same, so algorithms remain unchanged from the logical point of view.

In this paper our aim is concentrated at attempting to increase the performance of parallel implementation of the FFT algorithm. In our case we we're not interested in comparing CPU and GPU processing times. As mentioned earlier, such study has already been undertaken in many research papers and it is well known that the use of graphics cards may significantly reduce the FFT's execution time, c.f. [13, 14]. In this article we are focusing on the opportunities arising from the possibility of different task-defragmentation strategies of the FFT implementations on chosen GPU architectures to achieve optimal FFT computation time-effectiveness. Such approach demands comparing different parallel implementations of FFT to identify the fastest one in terms of execution times on chosen GPU models. In our research we have used the DIT FFT [15] as the base procedure for further considerations. The most commonly used in practical parallel implementations on GPU devices is the radix-2 DIT FFT algorithm which is schematically illustrated in Fig. 3.

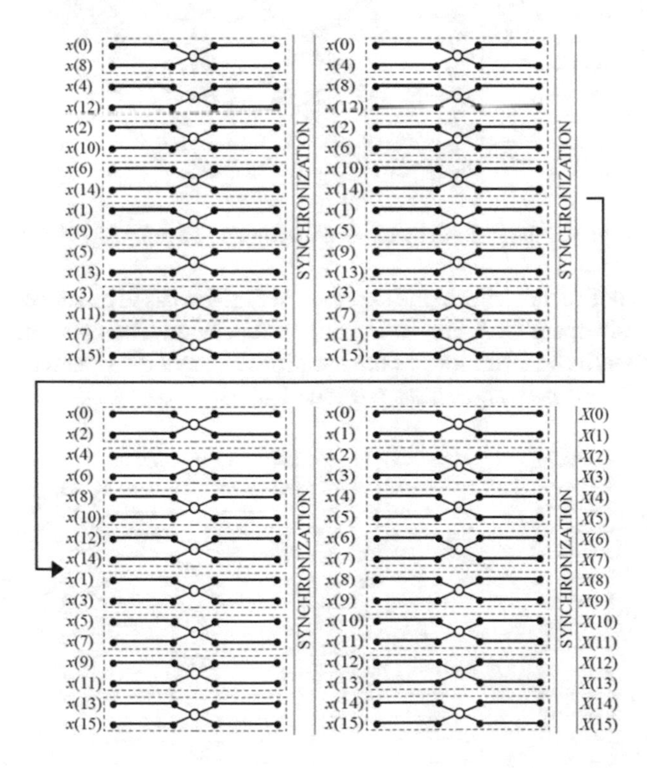

Fig. 3. Graphical interpretation of standard 16-point DIT FFT parallel GPU implementation.

In this implementation a single thread processes only one of the basic butterfly operations depicted in Fig. 2. The advantage of such a solution is the large defragmentation of the problem, which is beneficial for many of the consumer segment GPU devices. Unfortunately, the considered algorithm has also a significant disadvantage which is the necessity of performing synchronizations after each stage to ensure data integrity. Such defragmentation is clearly not the only one possible to conduct. The very structure of the original DIT FFT algorithm [8] allows for the transform's division into customizable blocks of larger sizes than the considered 2-point subtransforms, as depicted in Fig. 3. In Fig. 4 one can see how the original radix-2 DIT FFT algorithm structure could be topologically interpreted using defragmentation into 4-point subtransform blocks.

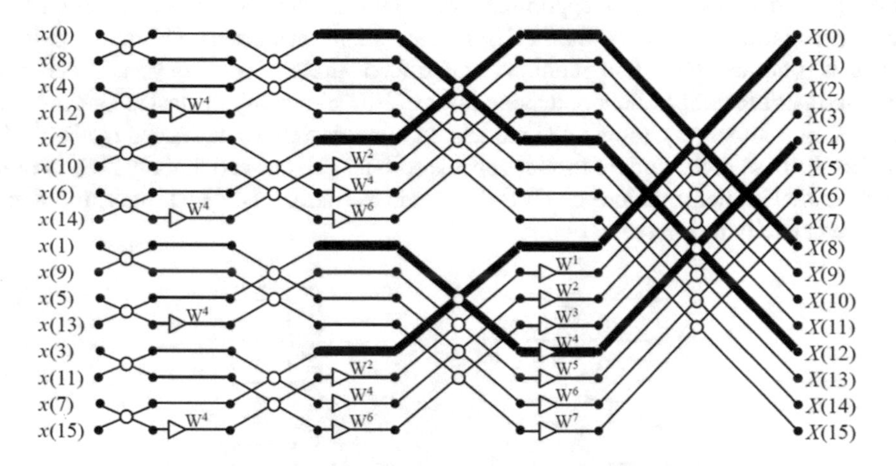

Fig. 4. 4-point subtransform block visualization on 16-point DIT FFT diagram.

It is easy to notice that the scheme depicted in Fig. 4 is based on successive repetition of the same basic operations' graphs in each of its stages. The only difference, which has to be taken into consideration are the locations of input data elements, but from the logical point of view both algorithms are identical.

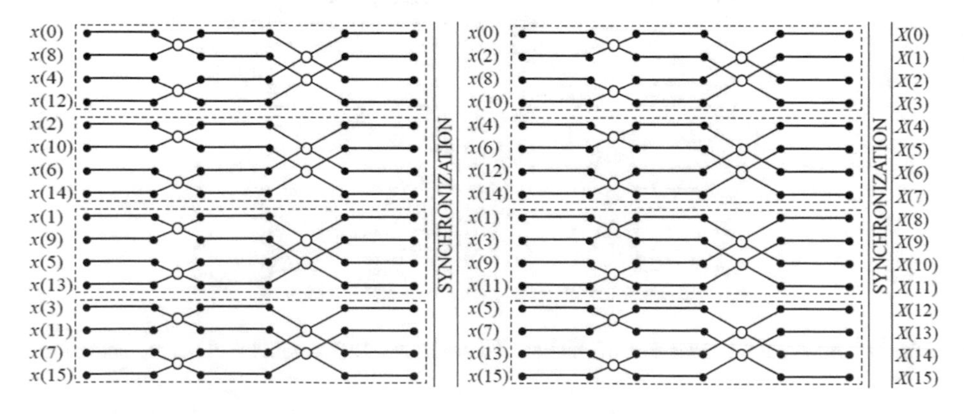

Fig. 5. 16-point FFT realization with the use of 4-point subtransform blocks.

Examples of different defragmentations are presented in Figs. 5 and 6. The following diagrams show two examples of 16-point FFT defragmentation into 4-point and 8-point subtransforms blocks.

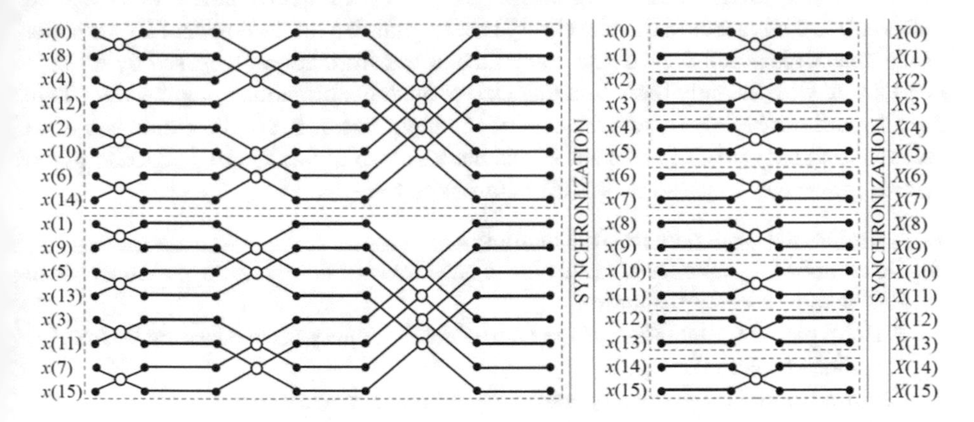

Fig. 6. 16-point FFT realization with the use of 8-point subtransform blocks.

In the case illustrated in Fig. 4, the amount of required synchronizations could be reduced from 4 to 2, with respect to the original FFT algorithm, and every single thread is computing a 4-point subtransform at each computational stage. In case illustrated in Fig. 5 the amount of required synchronizations might also be reduced from 4 to 2 because of the need of computation of 2-point transformations in the algorithm's last stage. Here every single thread is computing 8-point transform in the first stage, and single butterfly (2-point transform) in its final stage. Operations performed by single thread have been marked with the dashed lines in Figs. 4, 5 and 6.

Such grouping, originally proposed in [15], potentially induces significant benefits with respect to parallel GPU implementations of the FFT algorithm see e.g. [3, 14] or [16]. Firstly, the increase in the subtransform size computed by a single thread decreases the overall number of threads in a single stage and, at the same time, provides potential for better memory coalescing for each of the GPU's physical cores. Secondly, the increase in the subtransform size computed by a single thread enables also for the decrease in the number of explicit synchronizations, which are often relatively time consuming operations for many GPU devices. E.g. for an input signal of length 2^N, the classic approach with 2-point transformations partitioning requires N explicit synchronizations, while the same transformation realized with 4-point subtransforms reduces this amount by half, like it's demonstrated in Figs. 3 and 5. Another aspect is the temporal order of memory access operations which, due to suitable subdivision, can be organized more effectively. Unfortunately for specific cases, e.g. like the one presented in Fig. 4, computation of the complete transformation becomes more complicated because it requires the additional maximum subdivision step in its last stage. This becomes obvious, e.g. when looking at the structure presented in Fig. 6, that the calculation of the entire scheme might not be possible with a single decomposition of the FFT to subtransforms of a certain size. Scheme that contains an odd number of stages, cannot be fully realized with the 4-point subtransforms because they them-selves consist of two computational stages.

In order to verify the outlined considerations we have developed a parallel implementations of the discussed FFT algorithms. The research implementation was prepared that is able to compute the FFT of arbitrary input signal size (of the power of 2) providing at the same time possibility to select any size of transformation processed by a single GPU thread with the only limitation that size of the transformation computed by single thread could not be larger than the size of the input vector itself. To create a general solution to the considered problem, it was necessary to divide the FFT algorithm into several elementary steps, so its final structure is as follows:

1. Copy the raw input data to GPU memory.
2. Determine of the number of GPU blocks and threads, taking into account the chosen FFT scheme's defragmentation.
3. Perform parallel calculation of the coefficient dictionary table based on "butterfly" operations depicted in Fig. 2.
4. Perform parallel bit-inverse operation for input array, see e.g. [12].
5. Perform FFT radix-2 diagram realization with the chosen defragmentation strategy:
   - Determine of the number of layers and points processed by a single thread during a single computational stage
   - Launch the appropriate number of parallel threads
   - Determine the number of remaining layers and select the appropriate subdivision strategy
   - Redefine of the number of layers and points processed by a single thread for the remaining stages
   - Launch the appropriate number of parallel threads
6. Copy the data from graphics card memory to RAM
7. Determine the results of time performance statistics

```
SET IteratorUp TO 0;
SET IteratorDown TO subtransform size;
FOR each of subtransform stages DO
{
 FOR each of subtransform points DO
 {
  SET index1 TO (index from thread Id) + IteratorUp;
  SET index2 TO (index from thread Id) % IteratorUp * iteratorDown;
  Perform butterfly operations for real part with use of Index1, Index2;
  Perform butterfly operations for imaginary part with use of Index1, Index2;
 }
 MULTIPLY IteratorUp BY 2;
 DIVIDE IteratorDown BY 2;
 }
}
```

Fig. 7. Pseudocode illustrating the operation performed by a single parallel thread.

The above sequence of operations calculates the FFT for a selected defragmentation strategy and for a single input data *N*-vector. For the purposes of the experiments, the considered sequence was repeated for input data sizes ranging from 2^4 up to 2^{25} elements and defragmentation strategies of subtransform blocks of no greater size than 128-point subtransforms computed by a single thread. Above, in Fig. 7, we also present pseudocode illustrating the operation performed by a single parallel thread used in our implementation. Next section presents the results of the conducted experiments.

4 Results of the Experimental Research

Time effectiveness comparisons of the proposed implementations were performed with input vectors containing random data and also with arbitrarily prepared signals. The hardware used to obtain and collect the results was equipped with Nvidia GeForce GTX 1060 graphics card, built in Pascal architecture with CUDA 6.1 compute capability and 4 GB DRAM memory. Furthermore, to confirm the obtained results, exactly the same computations were performed also on a different architecture graphics card, namely the Nvidia's GeForce GTX 860M, built in Maxwell architecture with compute capability CUDA 5.2, and also equipped with 4 GB DRAM memory. In both cases, the host operating system was Windows 10 with the latest available GPU drivers, and Nvidia CUDA Tookit's 8.0. The research application was written in CUDA C programming language, with the use of the latest available NVCC compiler.

Execution time comparisons of the considered radix-2 FFT GPU implementations with different subtransform block size partitioning are presented in Tables 1 and 2.

Table 1. Time comparison results for GPU model GTX 1060.

Input vector size – 2^N/time in ms	Subtransform block size				
	2	4	8	16	32
4	0.015360	0.012288	0.015360	0.023552	–
5	0.019456	0.017408	0.017408	0.026624	0.052224
6	0.021504	0.018432	0.023552	0.029696	0.056320
7	0.027616	0.021504	0.027648	0.036864	0.059392
8	0.029696	0.023552	0.033792	0.056320	0.068608
9	0.032768	0.028672	0.041920	0.070656	0.093184
10	0.038810	0.032704	0.047104	0.072704	0.156672
11	0.047104	0.043008	0.053216	0.083968	0.158720
12	0.050144	0.080896	0.106496	0.136192	0.186368

The above results include only a part of the overall of the performed tests. Every execution included FFT transform sizes of $2^p, p = 4, 5, \ldots, 25$ points, while Tables 1 and 2 present only the results for FFT sizes of $2^p, p = 4, 5, \ldots, 12$, since the standard approach involving 2-point subtransform block sizes turned out to be the best in terms of time effectiveness for all cases in which $p > 11$. For all the tested input data, FFT's

Table 2. Time comparison results for GPU model GTX 860M.

Input vector size – 2^N/time in ms	Subtransform block size				
	2	4	8	16	32
4	0.022112	0.016096	0.018848	0.026720	–
5	0.026880	0.021216	0.021376	0.032768	0.058368
6	0.033248	0.024416	0.028608	0.037120	0.066240
7	0.038720	0.030624	0.036480	0.045056	0.073536
8	0.044416	0.034464	0.044192	0.071104	0.085664
9	0.050624	0.039552	0.053568	0.090144	0.115712
10	0.054912	0.044672	0.062848	0.094912	0.198784
11	0.072384	0.067296	0.076512	0.105376	0.204576
12	0.079200	0.119168	0.156576	0.197152	0.272704

computations with 2, 4, 8, 16, 32, 64, and 128 subtransform block sizes calculations were performed. The results gathered in Tables 1 and 2 also don't include all the tested cases for the same reason. Table 1 gathers significant tests for GTX 1060.

Table 2 gathers exactly the same tests, but this time performed with the GTX 860 M GPU model.

It turns out that the time-efficiency for subtransform block sizes larger than 16 starts to worsen significantly for both considered GPUs.

In order to illustrate the dependence between the execution times of the FFT's of the selected sizes with respect to their block size partitionings we have prepared three graphs which are presented below.

Figure 8 presents execution times for the FFT of input signal length of 128 points, implemented with partitioning strategies of $2^b, b = 1, 2, \ldots, 7$ subtransform blocks sizes.

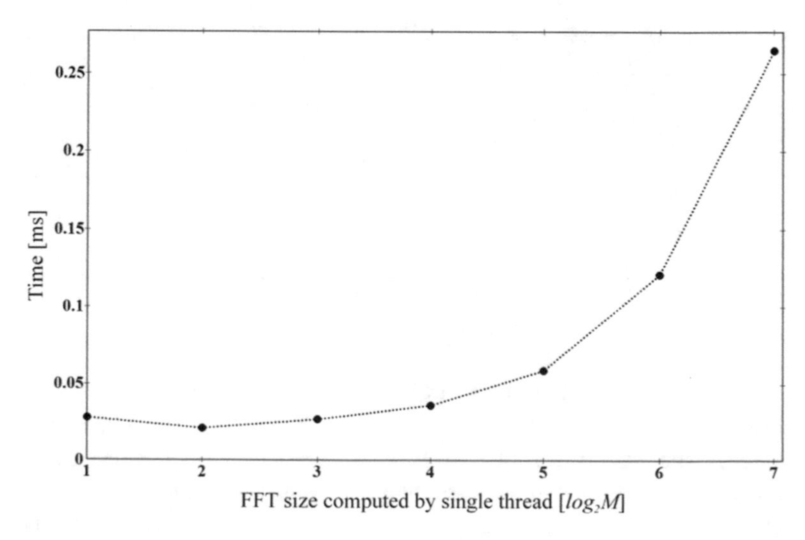

Fig. 8. Execution times for the 128-point FFT with various block sizes defragmentation.

It can be observed that defragmentation into 2-point, 4-point, and 8-point subtransforms performs very similar, although increasing further the size of the subtransform has a negative effect, and this behavior is repeated also for all the other tests.

Figure 9 depicts the same execution times as shown in Fig. 8 but this time of a input signal length of 1024 points and partitioning strategies of $2^b, b = 1, 2, \ldots, 9$ subtransform blocks sizes.

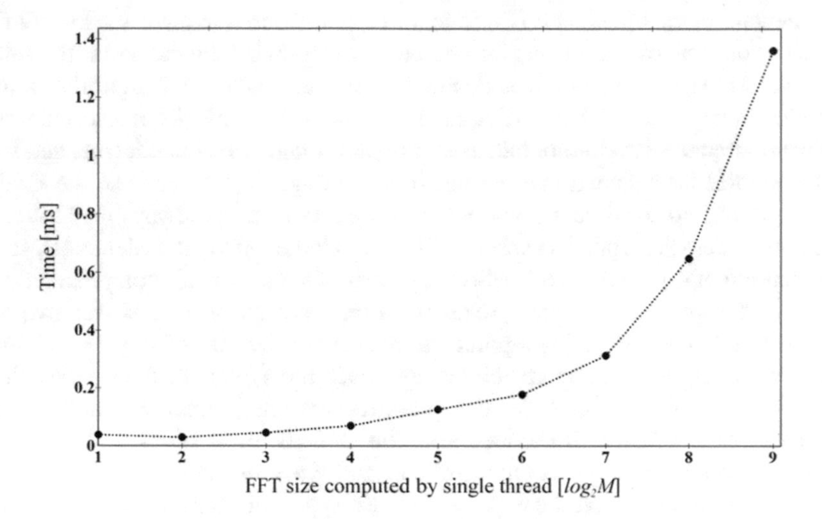

Fig. 9. Execution times for the 1024-point FFT with various block sizes defragmentation.

In Fig. 10 the acceleration ratios between the second fastest, standard 2-point and the most time-effective amongst all of the tested implementations is shown. It can be observed that for transform sizes up to 2^{11} the acceleration of the 4-point subtransform

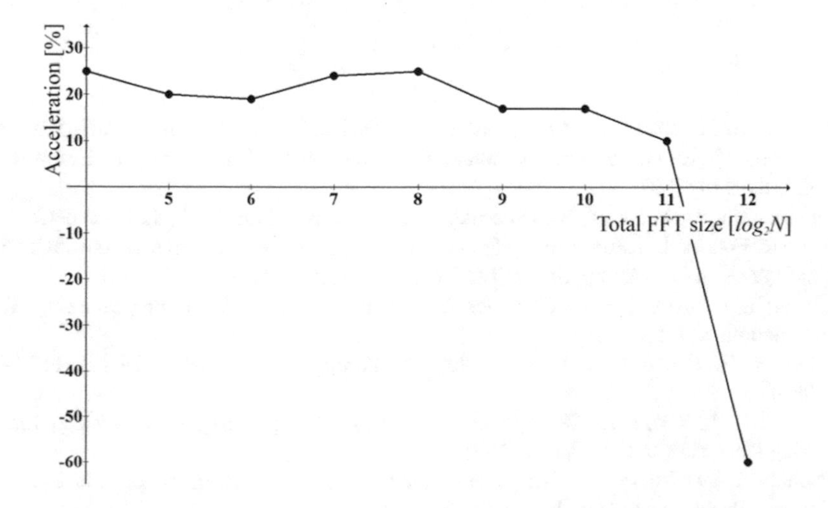

Fig. 10. 4-point subtransforms to 2-point subtransforms defragmentation ratio

implementation with respect to a standard 2-point one achieves levels between 10% up to an almost 30% in particular test cases.

5 Conclusions

In this article authors present a general FFT partitioning scheme that can be adopted for implementations on different GPU architectures whose aim is to lower the overall FFT's the execution time by optimizing computational resources' allocation on the selected GPU models. The scheme utilizes decimation in time (DIT) FFT algorithm's implementation in which the FFT is partitioned into subtransform blocks of selected size and within the subtransforms the computations are performed in a sequential manner. Experimental studies have shown that the maximum fine-grained division of the FFT data flow diagrams into standard 2-point butterfly operations is not always the best choice. In all cases where the input data size is fairly not to large (up to 2048 elements), sequential execution of 4-point transformations by each of single threads comprising the DIT FFT procedure reduces its total processing time even by up to 30% relative to the standard solution of calculating 2-point transformations by each of single GPU threads. However, for larger arrays, more efficient approach turns out to be the smallest defragmentation int standard 2-point butterfly operations. We may conclude that the memory access performed by each GPU core during the FFT calculation does not have a significant effect on its performance. On the other hand synchronization of the GPU threads has significant influence on the execution times, especially for relatively smaller data sizes. If the calculation times are relatively large, even the need for increasing the number of effective synchronizations does not affect the overall execution time results and then the fine-grained FFT division strategy becomes ones again the most beneficial one. This study undoubtedly confirms that defragmentation strategies of the GPU implementations of FFT algorithms might have direct and significant impact on the overall FFTs computational time.

References

1. Owens, J.D., Luebke, D., Govindaraju, N., Harris, M., Kruger, J., Lefohn, A.E., Purcell, T.: A survey of general purpose computation on graphics hardware. Comput. Graph. Forum **26**(1), 80–113 (2007)
2. Govindaraju, N.K., Lloyd, B., Dotsenko, Y., Smith, B., Manferdelli, J.: High performance discrete Fourier transforms on graphics processors. In: Proceedings of the ACM/IEEE Conference on Supercomputing. IEEE Press, Piscataway (2008)
3. Cheng, J., Grossman, M., McKercher, T.: Professional CUDA® C Programming. Wiley, Indianapolis (2014)
4. Hillis, W.D., Steele, G.L.: Data parallel algorithms. Commun. ACM **29**(12), 1170–1183 (1986)
5. Ahmed, U.N., Rao, K.R.: Orthogonal Transforms for Digital Signal Processing. Springer-Verlag New York Inc., Secaucus (1975)
6. Puchala, D., Stokfiszewski, K.: Parametrized orthogonal transforms for data encryption. Comput. Probl. Electr. Eng. J. **3**(1), 93–97 (2013)

7. Yatsymirskyy, M., Stokfiszewski, K., Szczepaniak, P.S.: Image compression using fast transforms realized through neural network learning. Model. Comput. Sci. Technol. Ukrainian Nat. Acad. Sci. **23**, 95–99 (2003)
8. Cooley, J.W., Tukey, J.W.: An algorithm for the machine calculation of complex Fourier series. Math. Comput. **19**(90), 297–301 (1965)
9. Nussbaumer, H.J.: Fast Fourier Transform and Convolution Algorithms. Springer, Heidelberg (1982)
10. Moreland, K., Angel, E.: The FFT on a GPU. In: Proceedings of the ACM Siggraph/Eurographics Conference on Graphics Hardware, pp. 112–119. Eurographics Association Aire-la-Ville, San Diego (2003)
11. Spitzer, J.: Implementing a GPU-efficient FFT. In: Siggraph Course on Interactive Geometric and Scientific Computing with Graphics Hardware (2003)
12. Puchala, D., Stokfiszewski, K., Szczepaniak, B., Yatsymirskyy, M.: Effectiveness of Fast Fourier Transform implementations on GPU and CPU. Przegląd Elektrotechniczny **92**(7), 69–71 (2016)
13. Ambuluri, S.: Implementations of the FFT algorithm on GPU. MSc. thesis, Linköping University, Department of Electrical Engineering (2012)
14. Dotsenko, Y., Baghsorkhi, S.S., Lloyd, B., Govindaraju, N.K.: Auto-tuning of Fast Fourier Transform on Graphics Processors. In: Proceedings of the 16-th ACM Symposium on Principles and Practice of Parallel Programming, San Antonio, TX, USA, pp. 257–266 (2011)
15. Yatsymirskyy, M.: Fast Algorithms for Orthogonal Trigonometric Transforms' Computations (in Ukrainian). Lviv Acad. Express, Lviv (1997)
16. CUDA® Programming Guide. http://docs.nvidia.com/cuda/cuda-c-programming-guide/index.html. Accessed 20 Aug 2016

Consolidation of Virtual Machines Using Stochastic Local Search

Sergii Telenyk, Eduard Zharikov(✉), and Oleksandr Rolik

Department of Automation and Control in Technical Systems, National Technical University of Ukraine "Igor Sikorsky Kyiv Polytechnic Institute", Kiev, Ukraine
telenik@acts.kiev.ua, zharikov.eduard@acts.kpi.ua

Abstract. A modern cloud data center is represented as a complex system where virtual machine consolidation and scheduling influence directly the cloud cost and performance. The virtual machine consolidation is the subject to many constraints originating from multiple domains, such as the resource requirements, user Service Level Agreement (SLA) compliance, security requirements, availability requirements, and other. Properly defined resource management methods and algorithms allow to achieve execution efficiency, SLA compliance, utilization of resources, energy saving, and the increasing profit of cloud providers. In this paper, the authors propose two versions of the Optimization using Simulated Annealing (OSA) algorithm to solve dynamic virtual machine consolidation problem. The virtual machine consolidation problem is considered as a multi-dimensional vector bin-packing problem. The authors take into account that the properties of items can be changed, new items may be requested to be deploy, and existing items may need to be reassigned to bins. Other constraints should be taken into consideration to solve virtual machine consolidation problem such as balanced load of resources of each physical machine, the limitation on maximum number of simultaneous migrations per physical machine, hardware constraints and other. The configuration of the system, the function for obtaining new configuration, the objective function for the optimization problem are determined for the proposed algorithms. The evaluation results show, that using OSA algorithms the simulated data center consumes almost the same amount of energy as while using a not optimized algorithm. On the other hand, the OSA algorithm with constraints allows to decrease overall performance degradation by virtual machines due to migrations, as a result, SLA violation is decreased. Furthermore, both OSA algorithms allow to reserve some resources of physical machine to react to increasing random resource demands in the nearest future.

Keywords: Virtual machine placement · Cloud computing
Energy efficiency · Simulation

1 Introduction

A modern cloud data center is represented as a complex system using virtualization technologies, different cloud service models (IaaS, PaaS, SaaS), software-defined technologies and a large amount of heterogenous resources. In the cloud data center

N. Shakhovska and V. Stepashko (eds.), *Advances in Intelligent Systems and Computing II*, Advances in Intelligent Systems and Computing 689,
https://doi.org/10.1007/978-3-319-70581-1_37

environment, a service provider faces with different problems caused by the expanding set of customer requirements and regulations. On one hand, the cloud service providers should ensure service QoS parameters (such as response time, latency, throughput, space) improving resource utilization and energy efficiency while complying the SLA. On the other hand, there are resource provision, resource allocation, resource mapping, resource scaling, resource estimation, and resource brokering tasks to be performed to achieve a high-quality performance of application services with a minimal amount of resources involved.

The resource allocation problem is an important and an urgent problem in modern cloud data centers now. The IaaS cloud service model enables customers to request dynamically the needed number of virtual machines (VM) based on their business requirements. One of the main tasks of managing resources in IaaS is the VM allocation on the physical machine (PM) of the cloud data center. The VM allocation process must be performed with the minimal number of PMs involved and with decreasing energy consumption. The process of selecting, which VM should be allocated at each PM of a cloud data center, is known as VM consolidation or VM placement problem. The VM placement problem has been extensively studied [1]. The VM placement is the subject to many constraints originating from multiple domains, such as the VM resource requirements, security requirements, availability requirements, and other. Many of the constraints may be defined in an SLA associated with each customer's VM [2].

The virtual machine consolidation problem is considered as a multi-dimensional vector bin-packing problem taking to account that the properties of items can be changed, new items may be requested to be deploy, and existing items may need to be reassigned to bins. The workload of multiple services in a cloud data center can change over time when a set of clients requesting services or jobs spans one or multiple VMs. This requires to solve periodically an optimization problem to deploy new and reallocate existing VMs in the data center. Reassignment of new and existing VMs should be performed periodically in asynchronous mode as needed. Stochastic local search approach is one of the widely used technique to solve such hard combinatorial optimization problem. In this paper, simulated annealing technique is adopted to dynamic VM consolidation problem with uniform loading of CPU, RAM and network interface to achieve maximal productivity and fully utilize data center resources. As shown in the literature [3–5], stochastic local search methods can be applied to cloud management problems finding acceptable solutions in admissible amount of time and resources.

Initial VM allocation calculated by using heuristics such as First Fit Decreasing (FFD) or Best Fit Decreasing (BFD). In this paper, the BFD heuristic with sorting all VMs in decreasing order of their current CPU utilization is used. At the same time, initial VM allocation does not account the usage of each PM's RAM, storage, and network interface utilization. To obtain near optimal VM allocation map with more balanced loading of CPU, RAM and network interface of each PM the optimization process employs simulated annealing technique.

The simulation environments are widely used to conduct repeatable large-scale experiments on different data center resource management algorithms [6]. Many research studies are based on simulation methods, since it is very expensive to conduct experiments on a production data center infrastructure. Thus, the use of simulation tools and frameworks has become a useful and powerful approach in a cloud computing research community. In [7], a CloudSim toolkit is proposed to support system design and modeling of cloud components such as data centers, virtual machines, physical machines and provisioning strategies that can be customized according to researcher's demand.

Thus, it is important to design and develop algorithms and approaches to solve the dynamic VM consolidation problem as a part of the cloud data center management problem.

The remainder of the paper is organized as follows: in Sect. 2 the related work review is presented, Sect. 3 describes the system model of the VM placement problem, in Sect. 4 the problem of VM placement is formulated taking into account new VM deployments and migrations of existing VMs, Sect. 5 describes the modeling process using simulated annealing technique, in Sect. 6 the configuration of simulated environment and modeling results are presented, and Sect. 7 concludes the paper.

2 Related Work

In the recent years, different approaches have been proposed to solve the VM placement problem in the cloud environment [1, 8, 9]. The VM placement problem is an optimization problem with different objective functions. Several researchers have highlighted this problem and proposed various solutions to find the most optimal placement plan for VMs with the objective to minimize number of PM used, with specified SLA. Most of them make use of the ordering algorithms and prediction techniques to generate the VM allocation map.

To solve VM placement problem the meta-heuristics are widely used, such as tabu search [10], evolutionary algorithm [4], ant colony optimization [3], neighborhood search [11], simulated annealing [5] and other. The reason to use the local search algorithms is to improve the results obtained by such heuristics as FFD, BFD and their modifications. There are several objective functions used for VM placement problem including the energy consumption minimization [12], number of PMs minimization [13], network traffic minimization [14], Availability Maximization [15], resource utilization maximization [12] and others.

During the last decade, many approaches to various VM placement problems have been proposed. In [16], the authors propose an efficient multi-start iterated local search metaheuristic for multi-capacity bin packing problem and machine reassignment problem. The proposed solution approach relies on such parameters as the thresholds restart, shaking restart, the size of the shaking operator, and the pruning criteria. The proposed approach produces high-quality solutions in a reasonable computational time for large-scale instances of both problems. In most cases more than 99% of the possible improvement from the initial solution have been achieved by reorganizing the processes with proposed method. However, there are two main drawbacks of the proposed approach. First, all processes migrate (VM migrations performed) simultaneously without any time-processing delay, and second, the proposed approach does not account the limitation on maximum number of simultaneous migrations per physical machine.

To solve the VM consolidation problem, the authors in [10] have proposed two phases VM consolidation algorithm which minimizes the number of required PMs, while also guaranteeing that the migrations performed in the transition to the new mapping to be completed in a specified maximum time. The first phase aims at finding a feasible mapping of VMs to PMs that minimizes the maximum migration time of all virtual machines. The second phase employs tabu search metaheuristic to produce solutions using a smaller number of physical servers, but also respecting a maximum migration time threshold. However, the authors have not determined how to detect overloaded PMs and have not considered other PM resources.

A multi-objective optimization mechanism for VM consolidation have designed and implemented in [11]. The authors also proposed mechanism that allows data center operators to add other optimization objectives. The proposed mechanism allows to reduce computational overheads by classifying PMs into a relatively small number of equivalent sets on the basis of the status of each one. It also realizes neighborhood search to find the lowest cost VM placement to serve VM deployment request from a user. But the proposed mechanism does not consider existing VMs and constraints on the number of VM migrations per PM. Other limitation is that the VM placement is not adapted at run-time.

In [3], the authors propose a multi-objective ant colony system algorithm for the virtual machine placement problem to efficiently obtain a set of non-dominated solutions that simultaneously minimize total resource wastage and power consumption. However, the constraints on the number of VM migrations per PM are not considered and to characterize a VM and a PM only CPU and memory resources are used. The proposed VM placement algorithm also does not allow dynamically consolidate existing VMs.

In this paper, the approach to solving VM consolidation problem is based on Power Aware Best Fit Decreasing (PABFD) heuristic [17] but, at the same time, balanced load of each physical machine resources is optimized using simulated annealing technique while taking into account the limitation on maximum number of simultaneous migrations per physical machine.

3 The System Model

The system model is shown in Fig. 1 [18]. The structure of the control object is represented by a set of PMs, each of which is characterized by the heterogeneous hardware and operating platform. Besides, each PM has a fixed capacity of resources. A PM enables a multiple virtual machine allocation. The PMs are connected using some form of topology and constitute the data center. In production data center, the number of VMs always changes. It is necessary to consider that in dynamic environment some VMs determined for migration may cease to exist during control.

The Global Manager (GM) is the main module of data center centralized management and is implemented as a failover and high-availability cluster of the PMs [19]. The GM coordinates data center virtual and physical resource management and allows to select a variety of resource and virtual machine management policies in order to adapt to the impact of external factors. The GM performs VM consolidation, PM and

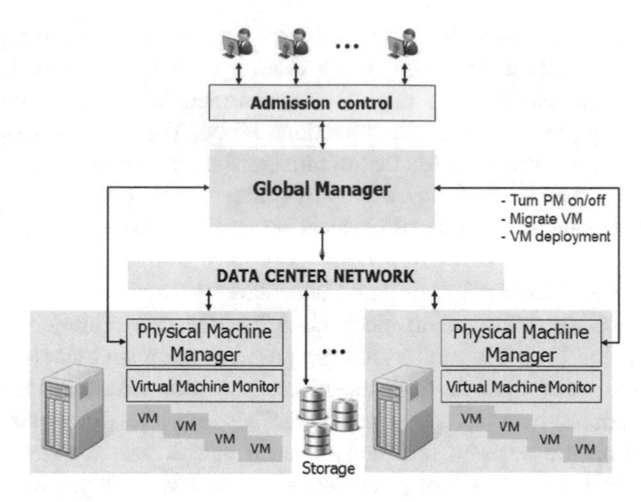

Fig. 1. The system model.

VM state management, VM scheduling, and new VM placement. The GM decides in which physical machines should VMs be allocated.

On the application layer, there is a set of clients requesting services or jobs spanning one or multiple VMs. Each VM is associated with specific performance goals specified in the SLA. The admission control module is responsible for handling user requests so that SLAs of admitted requests are met. A single VM can process one task at a time.

4 Problem Formulation

In a cloud data center, within a certain time there are M PMs, with heterogeneous configuration. Each PM is represented by multi-dimensional resources such as CPU, memory, storage and network interface bandwidth. All VMs on a physical machine share CPU, memory, storage and network bandwidth resources provided by the PM. The VM placement problem arises when one needs to determine which VM should be placed on which PM. This problem is to be solved.

Virtual machine placement problem consists of two parts. The first part of VM placement problem is an initial allocation of new N VMs with different resource demand to appropriate physical machines with specific resource capacity. The second part of VM placement problem is a reallocation of some VMs to other PMs due to demand and constant workload changes. Both parts of VM placement problem can be viewed as an extension of the bin-packing problem, where N numbers of items with different sizes (properties) to be placed on M numbers of bins, with an objective to minimize the number of bins used. The VM placement problem differ from classical bin-packing problem, as the properties of items can be changed, and there are other constraints while mapping virtual machines to physical machines [20]. For example, to provide VM live migration without downtime it is recommended to limit maximum number of concurrent migrations per PM. The VM placement problem can be considered also as a multi-dimensional vector bin-packing problem because each item has more than two properties.

It is important to note that to provide a near 100% guarantee of no SLA violation during VM migrations hypervisors limit maximum number of simultaneous migrations per PM, per network resource, and per datastore [21]. On some workloads that wildly use RAM or on older hardware, the big number of simultaneous migrations (more than recommended number or practically determined number) could saturate a system resource and result in downtime.

The problem of dynamic VM consolidation consists of four parts [17]: (1) determining an overloaded PM and the migration of one or more VMs from this PM is required; (2) determining an underloaded PM and the migration of all VMs from this PM is required to switch it to the sleep mode; (3) determining VMs that should be migrated from an overloaded PM; and (4) finding a new PMs for the VMs selected for migration from the overloaded and underloaded PMs. As a result of solving part (3) of the problem, GM determines a list of VMs from overloaded and underloaded PMs to be placed.

As shown in [22], cloud data centers continuously receive an unpredictable number of VM deploy requests over time. In [23], to account admission of new requests for VM provisioning the high-level algorithm of new and existing VMs placement is proposed. According to the high-level algorithm [23], Global Manager runs a separate instance of the control process for each request from the PM or for a new VM provisioning. It is proposed initially to serve a request from the PM overloaded, after that the request for new VMs creation, and then the request from underloaded PM. As a result of high-level algorithm, the GM obtains a set G^{VM} of virtual machines to be placed in the data center. GM also determines a set G^{PM}. The G^{PM} is the set of suitable PMs as expectants for hosting new and migrating VMs.

Thus, it is necessary to develop an algorithm of allocation VMs from the G^{VM} set using minimum PMs with uniform loading of CPU, RAM and network interface to achieve maximal productivity and fully utilize data center resources.

5 VM Placement Modeling

According to the problem formulation, the dynamic allocation of VMs based on simulated annealing algorithm is presented in this section. The problem of VM consolidation can be seen as a bin packing problem with variable bin sizes and prices. To solve it a modification of the simulated annealing algorithm is applied.

Simulated annealing is a general purpose combinatorial optimization technique [24]. It uses the Metropolis Criterion [25] and Boltzmann distribution to accept a new solution. Using this technique, the solution space of the VM consolidation problem is optimized. It is explored in a controlled fashion using a temperature T as a control parameter so that allocation map of VMs with successively better measures for the objective function is obtained. To adopt simulated annealing, it is necessary to determine a configuration of the system, a function for obtaining new configuration, the objective function for the optimization problem, and maximum number of allowed consecutive iterations that yield no improvement relatively to current configuration.

The allocation map of VMs MAP_{best}, the set of VMs G^{VM}, and the set of PMs G^{PM} constitute the configuration of the system for simulated annealing. The aim of the algorithm is to obtain better VM allocation map in term of objective function. At every evolution step of the algorithm, the allocation map of VMs is changed into a new neighborhood state by using search function. Each new allocation map of VMs is estimated by comparison of its acceptance indicator with the acceptance indicator of current allocation map of VMs to take a decision whether or not to accept a new allocation map as a current allocation map. In this paper, for VM placement problem the maximum number of allowed consecutive iterations, that yield no improvement relatively to current configuration, is accepted being equal to 100.

The pseudo-code for the algorithm of Optimization using Simulated Annealing is presented in Algorithm 1. As a result of Algorithm 1 close-to-optimal solutions of VM placement can be obtained. The primary objective of the OSA algorithm is to calculate VM allocation map and pass it to GM for migration process. Next run of the OSA algorithm is performed after completing of all VM migrations prescribed by previous run. The solution has to be an allocation map with minimum PMs used and with uniform CPU, RAM and network interface utilization of each PM to achieve maximal productivity and utilize different resources fully.

The new neighborhood state is determined by using of two types of search functions. First, the simple search function (simple OSA) is developed and estimated. A simple strategy is to swap two different randomly picking VMs changing their assignment even if the VM was rearranged at previous steps of the algorithm. Second strategy (OSA with constrains) is to limit number of migrations per PM. In this case there are two constraints taking into account: (1) particular VM can be reassigned only once during the optimization and (2) the number of VMs migrating to and from each PM is limited.

As an input the algorithm gets the set of VMs G^{VM} to be allocated, the set of PMs G^{PM}, temperature *T*, and maximum number of allowed consecutive iterations that yield no improvement *MaxCount*. The allocation map *MAP* stores not only mappings of VMs to PMs calculated by Algorithm 1 but also a state of VMs and PMs. For the OSA with constrains, each PM can be limited to perform VM migrations, so the limit on maximum number of concurrent migrations per PM can be reached. In lines 2 and 3 the initial state of each PM (*MigrationCount*) is set to zero. Each VM can be in migration state (*VM.isInMigrationSatate* = true) or not (*VM.isInMigrationSatate* = false). In lines 4 and 5 the initial state of each VM is set to false.

Algorithm 1. VM allocation Optimization using Simulated Annealing
Input: G^{VM}, G^{PM}, *T*, *MaxCount*
Output: A near-optimal allocation map MAP_{best}

1 $iter \leftarrow 0$
2 **for each** *PM* **from** G^{PM} **do**
3 $\quad PM.MigrationCount \leftarrow 0$
4 **for each** *VM* **from** G^{VM} **do**
5 $\quad VM.InMigrationSatate \leftarrow$ false
6 $MAP_{best} \leftarrow getInitialSolution\left(G^{VM}, G^{PM}\right)$
7 $MAP_{cur} \leftarrow MAP_{best}$
8 **while** $iter < MaxCount$ **do**
9 $\quad MAP_{new} \leftarrow swap(MAP_{cur})$
10 $\quad IB_{new} \leftarrow getEstimation(MAP_{new})$
11 $\quad IB_{cur} \leftarrow getEstimation(MAP_{cur})$
12 $\quad dev \leftarrow IB_{cur} - IB_{new}$
13 $\quad$ **if** $(T > 10^{-5}$ and $e^{dev/T} >$ random()) OR
14 $\quad\quad (T < 10^{-5}$ and $dev > 0)$ **then**
15 $\quad\quad MAP_{cur} \leftarrow MAP_{new}$
16 $\quad\quad IB_{cur} \leftarrow IB_{new}$
17 $\quad$ **end if**
18 $\quad$ **if** $IB_{new} < IB_{best}$ **then**
19 $\quad\quad MAP_{best} \leftarrow MAP_{new}$
20 $\quad\quad IB_{best} \leftarrow IB_{new}$
21 $\quad\quad iter \leftarrow 0$
22 $\quad$ **end if**
23 $\quad iter{++}$
24 $\quad T \leftarrow 0.99T$
25 **end while**

The function *getInitialSolution*() returns initial VM allocation calculated by using heuristics such as BFD or FFD. In our experiments, the BFD heuristic with sorting all VMs in decreasing order of their current CPU utilization is used. The initial VM allocation is calculated according to one of the defined strategy. For the OSA with constrains the maximum number of concurrent migrations per PM is constrained. At the same time, initial VM allocation does not account the usage of each PM's RAM and network interface utilization.

To find more balanced and productive solution the function *swap*() is called. The pseudo-code of the function *swap*() for the simple OSA algorithm is presented in Algorithm 2. The pseudo-code for the algorithm of the function *swap*() with migration constraints is presented in Algorithm 3. As a result of *swap*() function, a new VM allocation map is obtained. A new VM allocation map is acquired by picking two different VMs and changing their PMs randomly by swapping their assignment. This rearrangement of VMs results in creation of new configuration and reflects the basic operations of doing bin-packing to load a PM more uniformly. The main benefit of Algorithm 2 is its simplicity for obtaining new allocation map. As a result, more states can be explored during the optimization process. However, the main drawback of Algorithm 2 is that it does not account production requirements and generates too many VM migrations for each optimization cycle.

As the Algorithm 1 visits more solutions, its temperature T drops. In high temperature states, the algorithm is allowed to choose a new allocation map even if it is worse than the one currently existing.

In this paper, the objective function being optimized is being minimized, and when we say that some allocation map of VMs results in a better objective function than some other allocation map, what we mean is, that allocation map of VMs has a smaller value for the objective function. The function *getEstimation*() is used to calculate the objective function.

The objective function IB for VM allocation optimization can be represented as follows (1):

$$IB = \sum_{i=1}^{K} IB_i \tag{1}$$

$$IB_i = \frac{1}{CPU_i RAM_i NET_i} \tag{2}$$

where IB_i is an imbalance indicator of i-th PM, CPU_i, RAM_i, NET_i are utilization of CPU, memory and network interface of i-th PM respectively, K is a number of PMs used in initial solution.

High values of IB_i refer to unbalanced utilization of CPU, RAM and network interface while a value close to 1 indicates a good level of balancing. The values very close to 1 are not recommended because resource demand in the nearest future can increase, so it is preferable to reserve some resources to prevent SLA violations. Three kinds of resources (CPU, RAM, and network interface) are considered in the objective function. However, the approach can be extended to any number of resources supported by simulation environment.

Algorithm 2. *swap*() function (for simple OSA)
Input: *MAP*
Output: A new allocation map *MAP*

1 **for each** VM_{first} **from** *MAP* **do**
2 $ipm = getIndexOfPM(MAP, VM_{first})$
3 **for each random** *PM* **from** *MAP* **do**
4 $jpm = getIndexOfPM(MAP, PM)$
5 **if** $ipm <> jpm$ **then**
6 **for each** VM_{second} **from** PM_{jpm} **do**
7 **if** $PM_{jpm}.isSuitableForVM(VM_{first})$ and $PM_{ipm}.isSuitableForVM(VM_{second})$ **then**
9 $MAP.swapped(VM_{first}, VM_{second})$
10 **return** *MAP*
11 **end if**
12 **end for**
13 **end if**
14 **end for**
15 **end for**

Algorithm 3. *swap*() function (with constrains on concurrent migrations per PM)
Input: *MAP*
Output: A new allocation map *MAP*

1 **for each random** VM_{first} **from** *MAP* **do**
2 $ipm = getIndexOfPM(MAP, VM_{first})$
3 **if** not $PM_{ipm}.MaxMigrationLimit$ and not $VM_{first}.isInMigrationSatate$ **then**
4 **for each random** *PM* **from** *MAP* **do**
5 $jpm = getIndexOfPM(MAP, PM)$
6 **if** $ipm <> jpm$ and not $PM_{jpm}.MaxMigrationLimit$ **then**
7 **for each** VM_{second} **from** PM_{jpm} **do**
8 **if** $PM_{jpm}.isSuitableForVM(VM_{first})$ and
9 $PM_{ipm}.isSuitableForVM(VM_{second})$ and
10 not $VM_{second}.isInMigrationSatate$ **then**
11 $VM_{first}.InMigrationSatate \leftarrow$ true
12 $VM_{second}.InMigrationSatate \leftarrow$ true
13 $PM_{ipm}.MigrationCountInc$
14 $PM_{jpm}.MigrationCountInc$
15 $MAP.swapped(VM_{first}, VM_{second})$
16 **return** *MAP*
17 **end if**
18 **end for**
19 **end if**
20 **end for**
21 **end if**
22 **end for**

The authors also consider the case, when there are no overloaded PMs detected. In some cases, there are only underloaded PMs during some long period of time and there are no new VM deployed. If the number of unused resources on a PM is greater than the threshold, then that PM is added to a consolidation list G^{PM}. The algorithm restarts if the total number of unused resources on the PMs, included in the consolidation list, exceeds the resources of one PM. All migrations initiated by the OSA at the previous stage must be completed. The selection of the threshold value is not considered in this paper.

6 Evaluation

To conduct repeatable large-scale experiments to evaluate the proposed algorithms on a data center infrastructure is very expensive. To evaluate the proposed OSA algorithms, a CloudSim toolkit [7] is used. It is a modular and extensible open source toolkit which has built-in capability to implement and compare management algorithms for different cloud environments and workloads. The extended version of CloudSim is used [26] to enable energy-aware simulations.

A data center that comprises 800 heterogeneous PMs with two cores was used for simulation [17]. One half of PMs are HP ProLiant ML110 G4 servers with 1860 Million Instructions Per Second (MIPS), and the other half consists of HP ProLiant ML110 G5 servers with 2660 MIPS. Each PM is modeled to have 1 GB/s network bandwidth. VMs characteristics correspond to Amazon EC2 instance types but with one core because of the nature of workload data. The workload traces from a real system were used as a workload for simulations. The workload traces are obtained as a part of the CoMon project, a monitoring infrastructure for PlanetLab. The interval of utilization measurements is 5 min [27].

The amount of RAM for VMs is as follows: High-CPU Medium Instance (2500 MIPS, 0.85 GB); Extra Large Instance (2000 MIPS, 3.75 GB); Small Instance (1000 MIPS, 1.7 GB); and Micro Instance (500 MIPS, 613 MB). Initially the VMs are allocated according to the resource requirements defined by the VM types [12].

Following methods [26] are modified in order to realize OSA algorithm: getNewVmPlacement, getNewVmPlacementFromUnderUtilizedHost, and findHostForVm from the package org.cloudbus.cloudsim.power.

The simulation results have been obtained from a set of experiments. To compare the efficiency of the proposed algorithm with CloudSim built-in ones the metrics from [17] are used as follows. In [17], two metrics were proposed for measuring the level of SLA violations in an IaaS cloud: the percentage of time, during which active hosts have experienced the CPU utilization of 100%, SLA violation Time per Active Host (*SLATAH*) (3); and the overall performance degradation by VMs due to migrations, Performance Degradation due to Migrations (*PDM*) (4). The authors in [17] concluded that if a PM serving applications is experiencing 100% utilization, the performance of the hosted applications is bounded by the PM capacity, therefore, VMs experience performance degradation.

$$SLATAH = \frac{1}{M}\sum_{i=1}^{M}\frac{T_{S_i}}{T_{a_i}} \quad (3)$$

$$PDM = \frac{1}{N}\sum_{j=1}^{N}\frac{C_{d_j}}{C_{r_j}} \quad (4)$$

where M is the number of PMs; T_{S_i} is the total time during which the host i has experienced the utilization of 100% leading to an SLA violation; T_{a_i} is the total time of the host i being in the active state; N is the number of VMs; C_{d_j} is the estimate of the performance degradation of the VM j caused by migrations; C_{r_j} is the total CPU capacity requested by the VM j during its lifetime. C_{d_j} is estimated as 10% of the CPU utilization in MIPS during all migrations of the VM j [17].

Another metric proposed in [17] is a combined metric SLA Violation (*SLAV*) that encompasses both performance degradation due to PM overloading and due to VM migrations (5). The combined metric that captures both energy consumption and the level of SLA violations is denoted Energy and SLA Violations (*ESV*) (6).

$$SLAV = SLATAH \cdot PDM \quad (5)$$

$$ESV = E \cdot SLAV \quad (6)$$

Each experiment has been repeated ten times for each day of the workload traces. The final results are averaged over these experiments. Simulation results for THR, IRQ, MAD, LRR, and LR algorithms are obtained from [17]. The metrics of both OSA algorithms (3)–(6) are calculated and presented in Table 1.

Table 1. Simulation results of the algorithms in [17] and both OSA algorithms (median values).

Algorithm	ESV ($\times 10^{-3}$)	E (kWh)	SLAV ($\times 10^{-5}$)	SLATAH %	PDM %
THR-MMT-0.8	4.19	89.92	4.57	4.61	0.10
IQR-MMT-1.5	4.00	90.13	4.51	4.64	0.10
MAD-MMT-2.5	3.94	87.67	4.48	4.65	0.10
LRR-MMT-1.2	2.43	87.93	2.77	3.98	0.07
LR-MMT-1.2	1.98	88.17	2.33	3.63	0.06
OSA algorithm (simple)	1.81	88.24	2.05	3.41	0.06
OSA algorithm (with constraints)	1.71	88.4	1.93	3.85	0.05

The objective of the experimental phase has been to evaluate the quality indicators of OSA algorithm under different load conditions and to compare them with results obtained in [17]. Results shown in Table 1 indicate that using both OSA algorithms simulated data center consumes almost the same amount of energy as LR-MMT-1.2 because the OSA algorithm uses LR and MMT policies but only performs VM placement according to simulated annealing optimization technique. Using simple OSA algorithm the *PDM* metric is not changed comparing to LR-MMT-1.2 algorithm because more uniform PM loading does not influence the number of VM migrations directly. As a result of both OSA algorithms, it should be noted that the more uniform PM loading allows to decrease *SLATAH* metric. Using the OSA algorithm with migration constraints it is possible to decrease the number of migrations and, as a result, to decrease *PDM* metric. The uniform PM loading has the effect which reduces SLA violations due to reservation of some PM resources to react to increasing random resource demands in the nearest future.

7 Conclusion

To meet cloud data center management requirements, in this paper, the virtual machine consolidation problem is solved with uniform loading of CPU, RAM and network interface of PM. That has been done to achieve maximal productivity and fully utilize data center resources.

The VM consolidation problem is considered as a multi-dimensional vector bin-packing problem taking to account that the properties of items can be changed, and there are other constraints while mapping virtual machines to physical machines.

The authors propose two variants of Optimization using Simulated Annealing algorithms to solve virtual machine consolidation problem for hosting new and already deployed VMs. The configuration of the system, the function for obtaining new configuration, the objective function for the optimization problem are determined for the proposed simulated annealing algorithm.

To decrease the number of SLA violation during VM migrations the constraint on the number of simultaneous migrations per PM is considered in the OSA algorithm with constraints. The evaluation results show, that using the OSA algorithms simulated data center consumes almost the same amount of energy as while using a not optimized algorithm. On the other hand, the OSA algorithm with constraints allows to decrease overall performance degradation by VMs due to migrations, as a result, SLA violation is decreased. Furthermore, both OSA algorithms allow to reserve some resources of physical machine to react to increasing random resource demands in the nearest future.

References

1. Pires, F.L., Barán, B.: A virtual machine placement taxonomy. In: 15th IEEE/ACM International Symposium on Cluster, Cloud and Grid Computing (CCGrid), pp. 159–168 (2015)
2. Calcavecchia, N., Biran, O., Hadad, E., Moatti, Y.: VM placement strategies for cloud scenarios. In: 5th IEEE International Conference on Cloud Computing CLOUD, pp. 852–859 (2012)

3. Gao, Y., Guan, H., Qi, Z., Hou, Y., Liu, L.: A multi-objective ant colony system algorithm for virtual machine placement in cloud computing. J. Comput. Syst. Sci. **79**(8), 1230–1242 (2013)
4. Mark, C.C., Niyato, D., Chen-Khong, T.: Evolutionary optimal virtual machine placement and demand forecaster for cloud computing. In: IEEE International Conference on Advanced Information Networking and Applications (AINA), pp. 348–355 (2011)
5. Wu, Y., Tang, M., Fraser, W.: A simulated annealing algorithm for energy efficient virtual machine placement. In: IEEE International Conference on Systems, Man, and Cybernetics (SMC), pp. 1245–1250 (2012)
6. Kaleem, M.A., Khan, P.M.: Commonly used simulation tools for cloud computing research. In: 2nd International Conference on Computing for Sustainable Global Development (INDIACom), pp. 1104–1111 (2015)
7. Calheiros, R.N., Ranjan, R., Beloglazov, A., De Rose, C.A., Buyya, R.: CloudSim: a toolkit for modeling and simulation of cloud computing environments and evaluation of resource provisioning algorithms. Softw. Pract. Experience **41**(1), 23–50 (2011)
8. Salimian, L., Safi, F.: Survey of energy efficient data centers in cloud computing. In: 2013 IEEE/ACM 6th International Conference on Utility and Cloud Computing, pp. 369–374. IEEE Computer Society (2013)
9. Mills, K., Filliben, J., Dabrowski, C.: Comparing VM-placement algorithms for on-demand clouds. In: IEEE Third International Conference on Cloud Computing Technology and Science (CloudCom), pp. 91–98 (2011)
10. Ferreto, T., De Rose, C., Heiss, H.U.: Maximum migration time guarantees in dynamic server consolidation for virtualized data centers. In: Euro-Par 2011 Parallel Processing, pp. 443–454. Springer (2011)
11. Shigeta, S., Yamashima, H., Doi, T., Kawai, T., Fukui, K.: Design and implementation of a multi-objective optimization mechanism for virtual machine placement in cloud computing data center. In: Cloud Computing, pp. 21–31. Springer (2013)
12. Cao, Z., Dong, S.: An energy-aware heuristic framework for virtual machine consolidation in cloud computing. J. Supercomput. **69**, 1–23 (2014)
13. Sun, M., Gu, W., Zhang, X., Shi, H., Zhang, W.: A matrix transformation algorithm for virtual machine placement in cloud. In: 12th IEEE International Conference on Trust, Security and Privacy in Computing and Communications (TrustCom), pp. 1778–1783 (2013)
14. Pires, F.L., Barán, B.: Multi-objective virtual machine placement with service level agreement: a memetic algorithm approach. In: 2013 IEEE/ACM 6th International Conference on Utility and Cloud Computing, pp. 203–210. IEEE Computer Society (2013)
15. Wang, W., Chen, H., Chen, X.: An availability-aware virtual machine placement approach for dynamic scaling of cloud applications. In: 9th International Conference on Ubiquitous Intelligence and Computing and 9th International Conference on Autonomic and Trusted Computing (UIC/ATC), pp. 509–516 (2012)
16. Masson, R., Vidal, T., Michallet, J., Penna, P.H.V., Petrucci, V., Subramanian, A., Dubedout, H.: An iterated local search heuristic for multi-capacity bin packing and machine reassignment problems. Expert Syst. Appl. **40**(13), 5266–5275 (2013)
17. Beloglazov, A., Buyya, R.: Optimal online deterministic algorithms and adaptive heuristics for energy and performance efficient dynamic consolidation of virtual machines in cloud data centers. Concurrency Comput. Pract. Experience **24**(13), 1397–1420 (2012)
18. Telenyk, S., Zharikov, E., Rolik, O.: An approach to software defined cloud infrastructure management. In: XI International Scientific and Technical Conference on Computer Science and Information Technologies Congress on Information Technology (CSIT 2016), pp. 21–26 (2016)

19. Telenyk, S., Zharikov, E., Rolik, O.: Architecture and conceptual bases of cloud IT infrastructure management. In: Advances in Intelligent Systems and Computing, vol. 512, pp. 41–62. Springer (2017)
20. Li, X., Qian, Z., Chi, R., Zhang, B., Lu, S.: Balancing resource utilization for continuous virtual machine requests in clouds. In: 6th IEEE International Conference on Innovative Mobile and Internet Services in Ubiquitous Computing, IMIS, pp. 266–273 (2012)
21. Limits on Simultaneous Migrations. https://docs.vmware.com/en/VMware-vSphere/6.0/com.vmware.vsphere.vcenterhost.doc/GUID-25EA5833-03B5-4EDD-A167-87578B8009B3.html. Accessed 10 July 2017
22. Amazon Usage Estimates. http://blog.rightscale.com/2009/10/05/amazon-usage-estimates/. Accessed 10 July 2017
23. Telenyk, S., Zharikov, E., Rolik, O.: An approach to virtual machine placement in cloud data centers. In: International Conference Radio Electronics and Info Communications (UkrMiCo), pp. 1–6 (2016)
24. Kirkpatrick, S., Gelatt, C.D., Vecchi, M.P.: Optimization by simulated annealing. Science **220**, 671–680 (1983)
25. Metropolis, N., Rosenbluth, A.W., Rosenbluth, M.N., Teller, A.H., Teller, E.: Equation of state calculations by fast computing machines. J. Chem. Phys. **21**, 1087–1092 (1953)
26. CloudSim: A Framework For Modeling And Simulation Of Cloud Computing Infrastructures And Services. https://github.com/Cloudslab/cloudsim. Accessed 10 July 2017
27. Park, K., Pai, V.S.: CoMon: a mostly-scalable monitoring system for PlanetLab. ACM SIGOPS Oper. Syst. Rev. **40**(1), 65–74 (2006)

Architecture and Models for System-Level Computer-Aided Design of the Management System of Energy Efficiency of Technological Processes at the Enterprise

Taras Teslyuk(✉), Ivan Tsmots(✉), Vasyl Teslyuk(✉), Mykola Medykovskyy(✉), and Yuriy Opotyak(✉)

Department of Automated Control Systems, Lviv Polytechnic National University, S. Bandera Street, 12, Lviv, Ukraine
taras.teslyuk@gmail.com, ivan.tsmots@gmail.com, vasyl.m.teslyuk@lpnu.ua, medyk@lp.edu.ua, yuvoua@ukr.net

Abstract. Multilevel system architecture of energy efficiency management in the region is developed in this work. The architecture model consists of three levels, namely, level of data collection and management of executive mechanisms, level of control and management of technological process, and level of operator control and formation of administrative decisions. Hardware and software of each hierarchical level are developed. Furthermore, models of system-level computer-aided design, which are based on a Petri net theory and allow the dynamics of the system operation to be examined, are developed. The results of a research of the system using constructed models are represented by a reachability graph for states of the system.

Keywords: Multi-level management system · Energy efficiency
Microcontrollers · Architecture · Hardware and software · Petri net models
Reachability graph for states of the system

1 Introduction

The development of enterprises characterized with the intensive introduction of information technologies that provide automation of processes and the accumulation of large amounts of information about their dynamics. Energy efficiency management at the level of technological processes requires the creation of a common information space with accurate, complete and operative information about flow of technological processes at the enterprise. Approaching of processing means (microcontroller systems) to the sources of information (sensors) and executive mechanisms is one of the ways to reduce the volume of information. Accumulated data are processed using computational intelligence technologies. The received results are used to form management solutions. Enhancement of the management of energy efficiency of technological processes at the enterprise can be achieved by developing multi-level management system of energy efficiency of technological processes (MMSEETP), which should integrate all the functions of monitoring and management in a single system. In

N. Shakhovska and V. Stepashko (eds.), *Advances in Intelligent Systems and Computing II*, Advances in Intelligent Systems and Computing 689,
https://doi.org/10.1007/978-3-319-70581-1_38

developing of MMSEETP it is advisable to focus on the widespread use of telecommunication and Web technology, database tools for data collection, evaluation, operative analytical and intellectual data In developing of MMSEETP it is advisable to focus on the widespread use of telecommunication and Web technology, database tools for data collection, evaluation, operative analytical and intellectual data visualization of their processing and decision making, visualization of processing results and decision making.

In works [1, 2] the hardware and software means of MMSEETP are analyzed, modern information technology and expediency of their use for the development of automated process control systems are considered. The industrial control systems, sensors and executive mechanisms are also considered in details.

In the next works [1] the modern microcontrollers and micro controlled systems, programmable logic controllers, sensors, wired and wireless communication facilities are considered. The issues of collection, storage and processing of data are reviewed in details [3–7]. According to the analysis of publications [3] the formation of effective administrative decisions in MMSEETP is based on the results of the analytical and intellectual processing of collected data.

However, it is paid insufficient attention to an integrated approach to the development of software and hardware means and synthesis MMSEETP in the above works.

The aim is to develop architecture MMSEETP focused on maximum use of ready-made components and Internet technologies.

To achieve this goal, it is necessary to solve the following tasks: to determine a list of problems to solve MMSEETP; to form MMSEETP requirements; to MMSEETP design principles; to elaborate MMSEETP structure; to identify problems and tools designed for every level of the system.

2 Model of MMSEETP System Architecture

2.1 Tasks of MMSEETP

To ensure management of energy efficiency on the level of technological processes MMSEETP should provide resolving of the following tasks:

- real-time collection, storage and pre-processing of data on the manufacturing processes;
- forecasting, management of process control and executive mechanisms;
- setting parameters of technical means depending on environmental conditions;
- drafting and analysis of the energy balances of production and consumption of energy carriers;
- generation of reports of energy efficiency and the creation of their templates;
- integration of various data with databases and web servers;
- data protection in the system from unauthorized access;
- modeling of processes and intelligent processing of collected data;
- analysis of collected data and determining ways of reducing technical and non-productive losses of energy resources;

- visualization of multidimensional data on energy efficiency and presenting the results of processing data in graphs and charts;
- formation and control of administrative decisions.

2.2 Requirements and Design Principles of MMSEETP

Structure of MMSEETP should focus on collecting data, creating a database with operational, reliable, comprehensive information, processing it, preparation of management solutions for operators and formation, if necessary, executive mechanisms control signals. The peculiarity of MMSEETP work should be: high flexibility of configuration, support of remote control and using wired and wireless interfaces for communication between system components.

Computer components used for the synthesis MMSEETP should ensure solving problems in real time. In particular, this concerns problem of filtering the previous data processing and control actuators. Application of computer components directly from the sensors and executive mechanisms imposes severe restrictions on their weight and size characteristics. At the same time, the following components are put forward strict requirements for power consumption, affecting the power supply size and heat dissipation means. In addition, these components meet high requirements for survivability and reliability [8]. Computer Components of MMSEETP must provide verification of performance and rapid localization problems.

It's proposed to develop MMSEETP with using a systematic approach that covers all levels of process integration, hardware and software. Synthesis of MMSEETP is based on the finished hardware and software, because the development and production of new ones demands considerable time and money. When selecting components it is necessary to consider set of factors, namely: information about ready hardware and software components and their specifications, compliance with standards interfaces, the possibility of their purchase and more.

2.3 Development of MMSEETP Structure

Modern MMSEETP are a handy tool to support decision-making at all levels of government. Managers who use such systems do not focus on the problems of integration and data processing, and they focus directly to the energy problems by creating effective management decisions.

To ensure these requirements MMSEETP designed structure, which is given in Fig. 1, where PC - personal computer; PLC - programmable logic controller; MK - microcontroller; EM - executive mechanism, SCADA - (Supervisory Control And Data Acquisition) supervisory control and data collection, PLC - (Programmable logic controller) DCS - (Distributed Control System) distributed control system.

Developed MMSEETP basic structure consists of three levels:

- Data collection and control executive mechanisms;
- monitoring and process control;
- operator control and formation of management decisions.

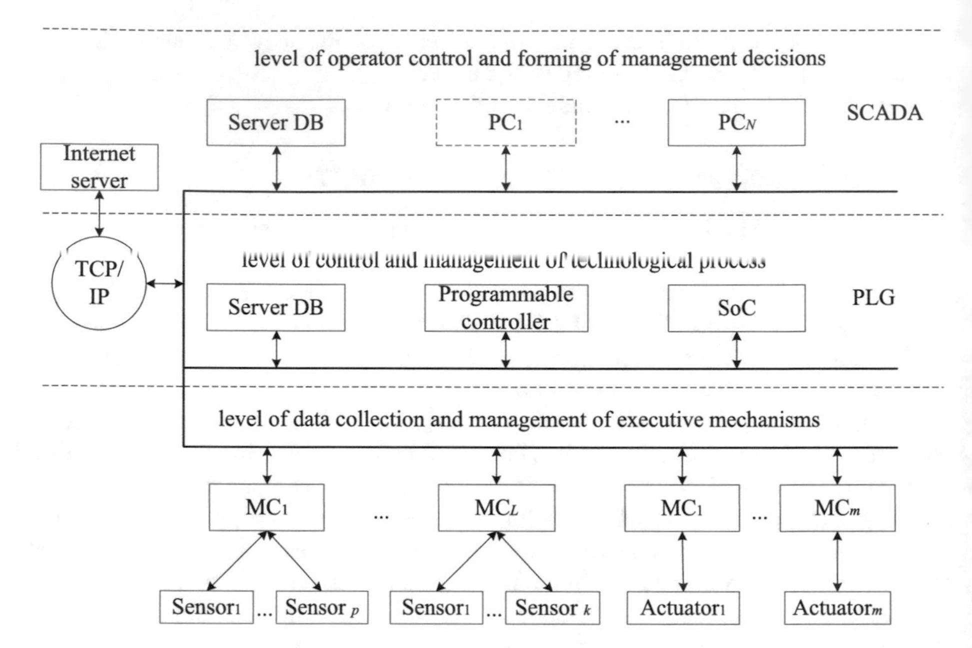

Fig. 1. Basic structure of MMSEETP

The specifics of each level process control system are defined by hardware and software components [4]. At each hierarchical level management the tasks the appropriate level of difficulty are solved.

An operation algorithm for each level of the system being designed is shown in Fig. 2 through Fig. 4. In particular, the system at the first level carries out the following functions: data collection from sensors; impact on the state of actuators; independent operation; communication with higher levels; execution of higher-level commands; data preprocessing, etc.

The first-level elements are usually single threaded and run a main program in an infinite loop. The main tasks of the first level of the system to be designed is to execute commands from outside and implement internal logic. For this purpose, it is necessary to consider the ability to switch between the major commands.

That is why the invented algorithm includes the next features such as a cyclical inquiry about available input commands and their execution, or execution of an internal cycle of the system's operation.

Step 1: While the system is started, the initialization begins, which involves reading performance configuration, initialization of ports and communication channels, etc.
Step 2: Upon system initialization, input communication channels are scanned so as to make sure that external commands are received. If the input command is received, advance to step 10. If no input command is received, proceed to step 3.
Step 3: The device must perform commands according to the internal logic of the system. For the analysis of the environment state, it is necessary to consistently enquire its state and data received from the sensors.

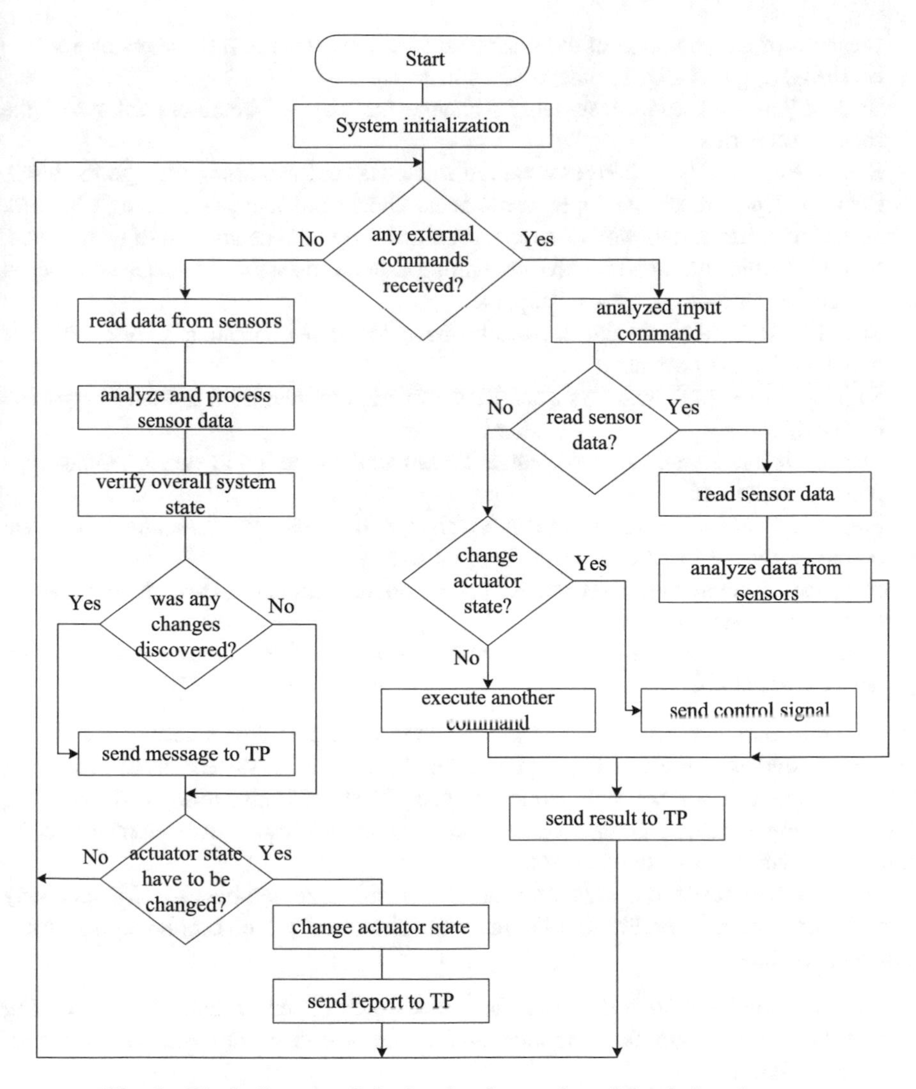

Fig. 2. Block diagram of the first-level operation of the designed system

Step 4: Based on data from the sensors, a preliminary analysis and transformation of data is carried out, which may presuppose normalization, scaling and other simple manipulations.
Step 5: Changes in the state of environment should be verified. The comparison with previously recorded values of the state of the system is made for this. If the state undergoes a change – go to step 6. If no changes occur – advance to step 7.
Step 6: A change in the state has to be notified to the higher levels. For this an incoming message is generated and sent to the level of TP.
Step 7: As first-level elements are able to perform simple logical operations and operate automatically, it is necessary to check the correspondence between internal

scenarios of the operation of the system and the current state. If the actuator needs to be changed, go to step 8, otherwise choose step 2.
Step 8: Transmit control signals to change the state of actuators following the internal scenarios.
Step 9: Report TP on changes occurred in the state of actuators, then go to step 2.
Step 10: Upon receiving input commands, check and analyse a string obtained. A type of command as well as support command options needs also to be received.
Step 11: If this command involves obtaining data on the state of the sensor, proceed to step 12. Otherwise, choose step 14.
Step 12: The state and current results have to be read from a sensor which is described by the command.
Step 13: Carry out a preanalysis and transformation of data from the sensor, proceed to step 17.
Step 14: If this command is to change the actuator's state, go to step 15. Otherwise, proceed to step 16.
Step 15: Transmit control signals to change the state of actuators which are described by the input command. Proceed to step 17.
Step 16: Execute another external command according to the advanced serial communications protocols.
Step 17: Make and send a report about execution of the input command, then advance to step 2.

At the second level the system comprises: storage of data on the current state of the TP system; analysis of lower-level input data; checking a TP system state; generating control commands for the stabilization of the TP state; transmitting and enquiring devices of lower levels; control of connection with the lower-level elements, communication with the mentioned ones.

At the second level of design, one can implement more complex logic, which may comprise execution of periodic tasks, multithreading and more complex communication between levels.

Step 1: While the system is started, the initialization begins, which involves reading performance configuration, initialization of ports and communication channels, kernel initialization.
Step 2: Upon system initialization, input communication channels are scanned so as to make sure that external commands are received. If an input command is received, go to step 15. If there isn't any input command, proceed to step 3.
Step 3: The status of periodic tasks has to be checked. If it is time for accomplishing the task, go to step 4. Otherwise, proceed to step 8.
Step 4: Each periodic task is assigned a list of subtasks for execution. The list of subtasks is being made.
Step 5: The subtasks are transformed into the commands for the system and the elements of the lower level in order to read data, change the state of actuators, etc.
Step 6: A queue of commands is formed to send them to the lower-level elements.
Step 7: The commands are consistently sent to the lower-level elements, proceed to step 2.

Step 8: Launch an internal logic core. Update the value of environment state and running actuators.
Step 9: Check the performance of the system following internal logic of the TP operation. It is necessary to verify whether the system parameters are not beyond the limits.
Step 10: Report about changes in the state of TP to the level of operator control (OC).
Step 11: If the output of the system parameters is detected, go to step 12. Otherwise, go back to step 2.
Step 12: Generate a list of necessary tasks for the stabilization of the system.
Step 13: Transform the tasks intended to stabilize the state of the system into the commands for the elements of microcontroller (MC).
Step 14: Form a queue of commands for the lower-level elements. Transfer the commands consistently, then go back to step 2.
Step 15: Check the recipient of the command. If the command was received from OK, move to step 16. Otherwise, go to step 18.
Step 16: Check and analyze the command given from OC. Receiving a type of command as well as support command options is also necessary.
Step 17: Change the operating parameters of TP following the command from OC. Report about the command execution, go back to step 8.
Step 18: Check and analyze the command or report from the MC. A type of message and its support parameters have to be obtained.
Step 19: Update the value of the system state, go back to step 8.

At the third level of design, the following functions are fulfilled: storage of a large amount of data about the system's state over time; an in-depth analysis of data about the system; presentation of the system's state in a more understandable form to the operator (SCADA systems); user interactions; generating commands for the lower-level elements; control of connection with the elements of the lower levels.

At the third level of design, communication with an operator should be implemented. It is therefore necessary to consider the task showing the state and a two-way interaction with the operator.

Step 1: While the system is started, the initialization begins, which involves reading performance configuration, the initialization of ports and communication channels, the kernel initialization and initialization of a user interface.
Step 2: Upon system initialization, input communication channels are scanned so as to make sure that external commands are given from TP. If an input command is received, advance to step 10. If there isn't any input command, proceed to step 3.
Step 3: If the command is received form the operator, go to step 4. Otherwise, go back to step 2.
Step 4: Check the type of input command given from the operator. If this is the command that forecasts the state of TP, go to step 5. Otherwise, move to step 7.
Step 5: Launch the kernel of data analysis of the system. Configure the kernel, specify search parameters and make prediction. Make calculations of predictions.
Step 6: Representing the prediction results for the user. Return to step 2.

Step 7: If this is a command that changes the state of the TP, go to step 8. Otherwise, go back to step 2.
Step 8: Generate a list of necessary changes and transfer these tasks to a set of commands for TP.
Step 9: Send commands to the level of TP. Return to step 2.
Step 10: Analyze the incoming report given from TP. Selecting and excluding the type of report and a set of input data.
Step 11: Update the value of database system according to the report obtained.
Step 12: Display the changes in the state of TP on the user interface. Return to step 2.

3 Levels Design of MMSEETP

3.1 The Level of Data Collection and Management of Executive Mechanisms

At this level, the primary information that is pre-processed, collected and comes to the controls is formed. With the use of the information generated signals to manage executive mechanisms and process are formed. Keep in mind that the companies built in recent years are provided with means of collecting technological information, and the selection and integration process of collecting data is already in the design phase. In this case, it's used staff means of mentioned systems, to solve problems of MMSEETP (Fig. 3).

A more difficult situation is observed in companies that are already in operation. In these companies, especially for small and medium businesses deployment of systems for collecting and processing are usually not planned due to their high cost in the past. The performed analysis of typical manufacturing processes of current production makes it possible to formulate a number of features that must be considered when creating MMSEETP:

- monitoring of process parameters is desirable to use the technologies for efficiency and simplicity of deployment;
- the number of points of measurement parameters can be changed, and therefore, the system should have an open architecture with the possibility of scaling;
- hardware system should be developed on the base of modern standard solutions to ensure simplicity and low cost;
- a broad application of basic telecommunications protocols is necessary.

To pre-processing of data from sensors and management of executive mechanisms it is appropriate to use the most simple hardware in the form of ready industrial components. To create this level of MMSEETP today it's necessary to focus on single-chip microcontrollers, System on Chip (SoC) and Programmable logic integrated circuits (PLICs) [9].

The class of single-chip microcontrollers, that are made on the basis of available modules for a wide range of systems, is presented today by means of companies ST Microelectronics, Atmel, STCmicro, Microchip, TI and others. It is usually 8 bit

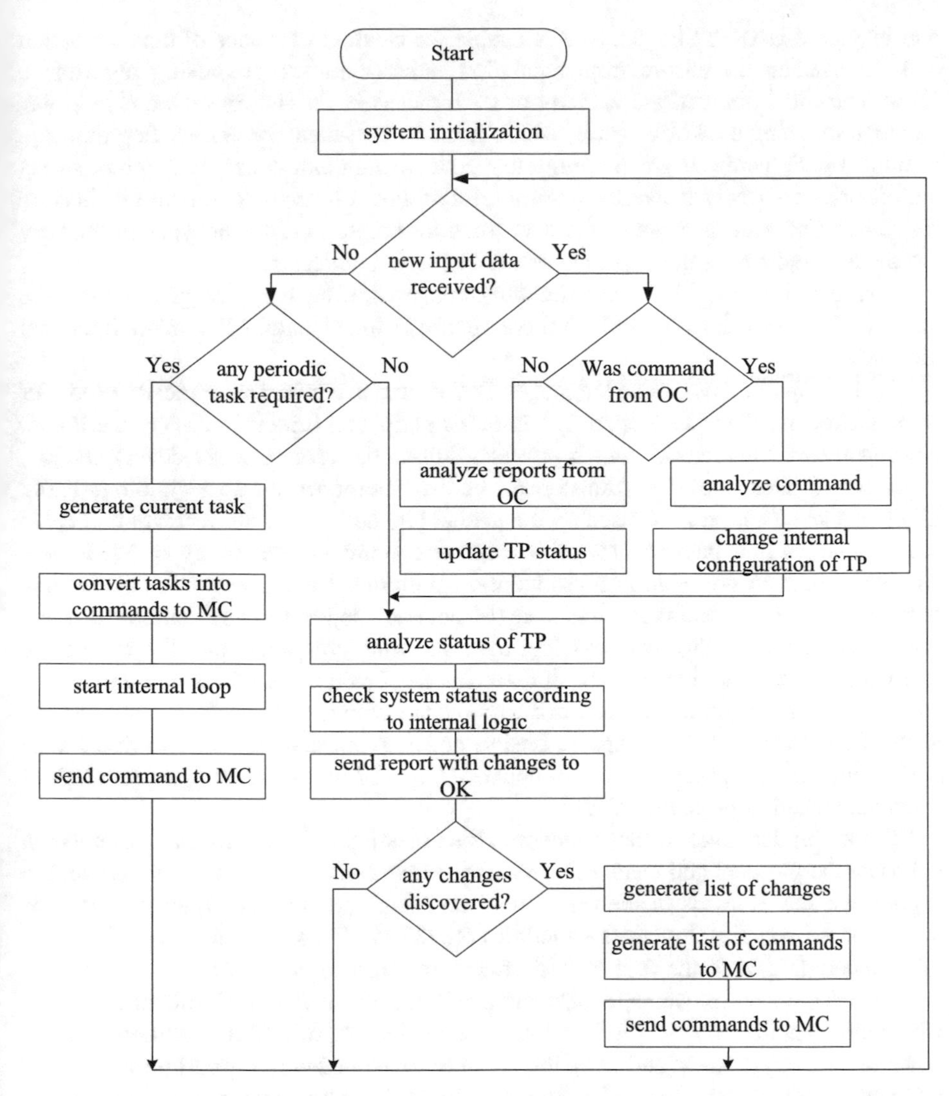

Fig. 3. Block diagram of the second-level operation of the designed system

programmable microcontroller with architecture RISC or CISC and clock speed of tens MHz, built-in RAM, Flash-memory, Eeprom and peripherals that supports interfaces UART, SPI, CAN, I2C. The amount of internal memory for storing data and programs is small and ensures the simple functions of measurement data to the appropriate interface and basic management. The advantages of these microcontrollers include, above all, low cost of finished industrial modules based on them, debugging boards, programmer and debugger.

Recently, 32 bit microcontrollers and ready modules based on them, for example, based on ARM architecture company ST Microelectronics appeared on the market [10]. Clock frequency of hundreds of MHz RAM at hundreds kB, embedded Flash-memory and interfaces UART, SPI, I2C, CAN enable the creation of means of data collection and management to ensure implementation rather complex processing algorithms. From a practical point of view, using of such means in the first place we should pay attention to a single address space, which hosted Flash-memory, RAM, Eeprom, registers and peripherals. It greatly simplifies code writing and codes of different stacks and libraries are easily converted as primarily designed for von Neumann architecture (in meaning of address space). However, tires for access to different types of memory are divided, indicating the presence of the Harvard architecture.

Considered 8 and 32 bit microcontrollers provide an easy integration into the system of various sensors and controls, because they contain the main industrial interfaces.

To develop microprograms languages C, C++ and assembler are typically used. For some microcontrollers developed free libraries peripheral drivers and software framework (software framework), which provides substantial ease software development.

Recently, means of data transferring at the enterprises create technology-based Ethernet and Wi-Fi, that is caused by the availability and a wide nomenclature of cable systems, active and passive network equipment. Wireless technology of Wi-Fi networks are most comfortable in a production environment where requirements include portability, ease of installation and use (of course, subject to consideration of electromagnetic compatibility with existing manufacturing equipment and the absence of interference from its side and the impact on the functioning of the network). The advantage is the flexibility of network architecture with the ability to change dynamically the network topology, speed, design and implementation, that is critical with strict requirements for the duration of network, no need in cabling, support for stack of telecommunication protocol TCP/IP.

To transfer data over wireless networks it's necessary to have a special radiomodem to ensure the physical and channel layers of the OSI model in wireless network and to support the senior levels of the OSI model on a high performance microprocessor is need. Today specialized devices - modules RS232-Wi-Fi are produced, but their cost may be comparable to the cost of basic microcontroller system [11].

However, in recent years, a new class of specialized microcontrollers appeared, which provides full support for Wi-Fi and protocol stack TCP/IP. In addition, a rather powerful processor, RAM and peripherals rather developed with support protocols SPI, I2C and UART are placed on the chip. On the basis of this device, you can create a complete wireless device that can receive data from the sensor, perform pre-processing and transmit the information via telecommunication protocols on a local or remote server. These tools gave a new boost to the development of Internet technology things. In addition, on the basis of these specialized microcontrollers ready industrial modules are available that can be used in the implementation of tasks MMSEETP lower level. Among these tools should indicate microcontroller NL6621 of Nufront firm, RTL8710 of Realtek firm, ESP8266 of company Espressif. For example ESP8266 provides a full support of Wi-Fi functions (host mode and access points with providing authentication protocols WEP and WPA/WPA2) and protocol stack TCP/IP. ESP8266 has an integrated 32 bit RISC processor Tensilica Xtensa LX106 with a clock speed of 80 MHz, 64 KB RAM and 96 KB command RAM data supports external Flash-memory storage

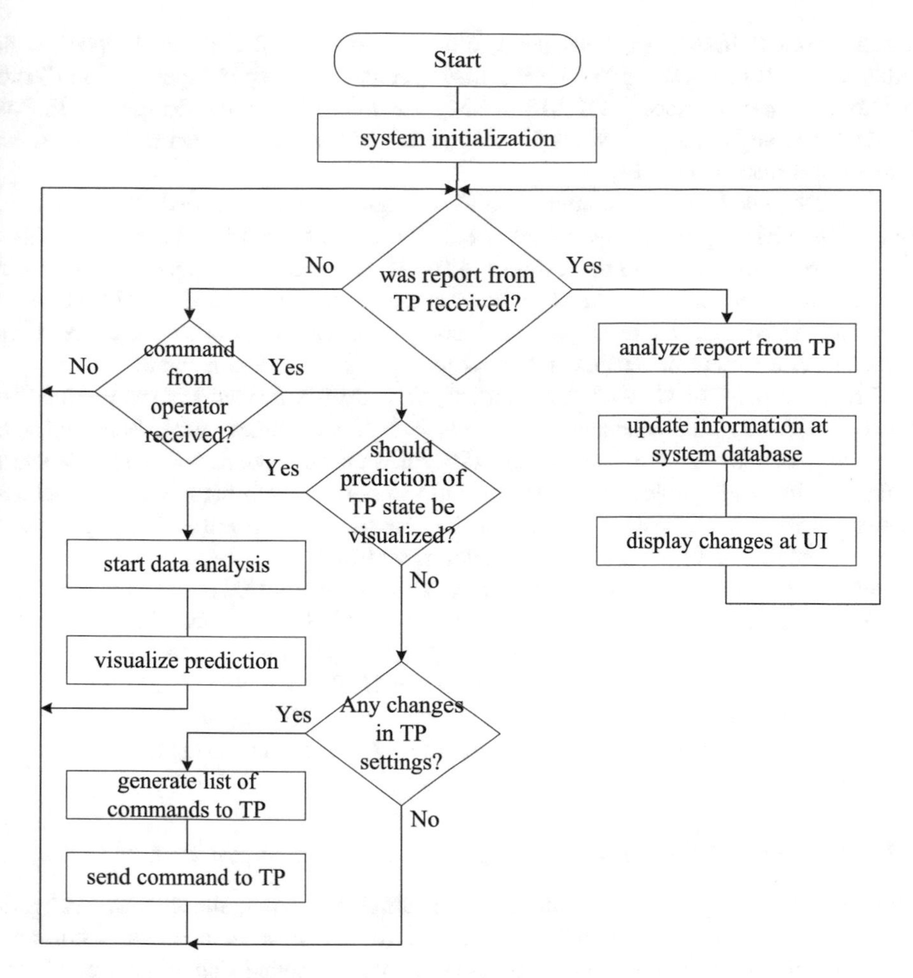

Fig. 4. Block diagram of the third-level operation of the designed system

of firmware from 512 KB to 16 MB. The chip provides support interfaces UART, SPI, I2C and includes single-channel 10-bit ADC [12].

To create software, software framework, libraries of a wide range of sensors supporting are accessible. For development C language can be used, in some cases development can be simplified because of use by the use LUA interpreter.

To implement on the lower level complex data processing algorithms single-board computers based on specialized processor SoC can be used. Single-board microcomputer is a promising platform for automation systems, and it has an open architecture, low price and uses the operating system Linux. One of the first in this class, the Raspberry Pi microcomputer appeared in 2012 and built on the system (SoC) Broadcom BCM2835, which includes ARM processor with a clock speed of 700 MHz GPU VideoCore IV, and 512 or 256 MB RAM, SD-card is used as an additional memory card. Microcomputer is produced in several versions: junior (A) (700 MHts clock speed, 256 MB RAM, one USB port), older (B) (up to 1.2 GHz clock speed with Ethernet, 1 GB of RAM, up to 4 USB ports), recently appeared cheaper version - Zero (up to 1 GHz clock speed, 512 MB RAM, one microUSB port). Raspberry Pi has GPIO ports, supporting interfaces UART, SPI, I2C, which you can use to connect sensors and manage [13, 14].

Recently single-board computer on SoC significantly expanded, for example, BeagleBone Black processor AM3359 ARM Cortex-A8 firms Texas Instruments; Intel Edison processor Intel Atom; pcDuino A10 ARM processor company AllWinner, Cubieboard2 A20 A20 with ARM processor, a line OrangePI processor H3, H2 + of company AllWinner. These single-board computers have developed periphery, they can connect a variety of sensors and provide work in a wireless network.

From the point of view of development, they are full-fledged Linux systems. To implement the algorithms languages C, C++, Python, sometimes - assembler are used. The problem may be the lack of specific software framework for a single-board computer, however, project Armbian can help to solve the problem, which develops Linux distributions on basis of Debian for SoC based on ARM, and each computer has already installed and has ready to use development tools.

For direct management of executive mechanisms software DCS and PLC are used.

At this level of MMSEETP it's advisable to use FPGA as finished industrial units FPGA/CPLD firm Altera - MAXII EPM240, Altera Cyslone ll EP2C5T144, Altera Cyclone IV EP4CE6, firm Xilinx - CoolRunner-II FPGA CPLD XC2C64A, FPGA Spartan-3E XC3S250E and other more productive to solve problems that require high flow performance. Developing for entry-level FPGA can be done using free versions of the software.

3.2 The Level of Control and Management of Technological Process

This level of management is assumed to be sufficiently autonomous, that can work independently for a long time without loss of information in the absence of communication with the upper level. At this level an above-mentioned single-board computer on the basis of SoC, and, for example, programmable logic controllers Mitsubishi Melsec FX3U, the standard means of visual inspection and process control can be used.

Mitsubishi FX3U programmable controllers are the most powerful and high-performance controllers in the line of controllers MELSECFX (FX1S, FX3S, FX1 N, FX3G). Architecture of controller is two-tired that increases its capacity. Mitsubishi programmable controller provides connectivity expansion modules as the previous generation and new generation units of FX3U series. When connecting modules of FX3U controller automatically switches to its high switching bus mode and data exchange takes place at high speed. Modules FX0 N, FX2 N controller works at normal speed. Thus, the use of expansion modules provides increased number of inputs/outputs up to 256 (with direct address), and a station of the decentralized input/output to 384.

In addition, Mitsubishi programmable controller provides high-speed connectivity adapter modules FX3U-XXX-ADP, that extend the capabilities of the controller when dealing with analog signals and increase the number of additional communication interfaces (RS232/422/485).

Over the last decade there have been several PLC programming languages, for which the standard IEC 61131 is developed. This standard specifies changes in the area of programming languages for the process control systems. It covers requirements for hardware, installation, testing, documentation, communication and PLC programming. The standard defines five programming languages, which are divided into textual and graphical. The textual PLC programming languages include the Instruction List (IL), the language of instructions; Structured Text (ST), the language of structured text.

The IL language is a typical assembler of batteries and transition tags. A set of instructions is standardized, and it is independent from a specific target platform. This language provides work with any type of data; calls functions and function blocks implemented in other languages. With IL the algorithms of any complexity are easily implemented.

The ST language is a high-level language, which is syntactically similar to Pascal. Instead of Pascal procedures, in ST the IEC-defined software components are used.

The graphical PLC languages include: Sequential Function Chart (SFC) is a language of consistent function charts; Function Block Diagram (FBD) means a language of functional and block diagrams; Ladder Diagrams (LD) denotes a language of relay-switching circuits.

The SFC language is a high-level graphical tool, in which a graphic chart consists of steps and transitions between them. Allowance for the transition is determined by the condition while with the step specific actions are associated.

The FBD language is oriented towards a description of schematic circuits of a chip-based electronic device. The conductors in FBD are used for transmission of any type of signals (logic, analog, time signal, etc.). The signals from the outputs of the blocks can be issued to the inputs of other units or directly to the PLC outputs. The blocks themselves, shown in the diagram as "black boxes", can fulfil any function. By using FBD, it is provided a description of interconnections between the inputs and outputs of the diagram. If the algorithm is well described in terms of signals, then its FBD representation is always more vivid than in textual languages.

The LD language is designed to implement the structure of electric circuits. Graphically the LD chart is shown as two vertical power rails. Between them are located the circuits formed by coupling the contacts. A relay serves as the load on each circuit. Each relay has contacts that can be used in other circuits. Sequential (I), parallel (OR) connection of contacts and inversion (NOT) constitute a Boolean basis. As a result, LD is ideal not only for developing relay-controlled devices, but also for software implementation of combinational logic circuits. Due to the ability of LD to use functions and functional blocks written in other languages, the sphere of its application is increasing.

The languages of IEC 61131 are developed on the basis of the most popular programming languages for modern controllers. The programs written for modern controllers can be transferred into the environment IEC 61131-3 at minimum expense. The feature of IEC 61131 is that there is an ability to create hardware-independent libraries for the implementation of regulators, filters, motor control, fuzzy logic modules, etc.

For programming controllers with the IEC 61131-3 defined languages, it is developed a programming environment CoDeSys (Controllers Development System) whose editors and debugging tools are based on the principles of a popular professional programming environment (Visual C++, etc.). The feature of the programming environment CoDeSys [https://www.codesys.com/] is that it's not restricted to a specific hardware platform. A modification is made to different PLC programming environments by adapting the program to low-level resources such as memory allocation, communication interfaces, and input-output interfaces. The CoDeSys environment provides: direct machine code generation; the full implementation of standardized IEC 61131-3 languages; language intelligent editors that correct mistakes common to the beginners; an embedded controller emulator that allows debugging a project without additional hardware; built-in visualization elements providing a creation of a model of a control object and debugging a project without making simulation tools; the use of ready-made libraries and service functions.

3.3 The Level of Operator Control and Forming of Management Decisions

This level is represented by an automated operator workplace. Hardware that can be used at this level is determined by the intensity of flow data, processing algorithms complexity and reliability requirements that apply to the control system [16, 17]. As hardware can be used for operator workstations RISC- or Intel-platforms and industrial computers such as Mitsubishi Electric, BECKHOFF, Eaton, AXIOMTEK, that are corresponding to industrial conditions with high requirements for durability and reliability. Tasks at this level are: collecting data from peripheral controllers and microcontroller systems; data storage; data processing; processing video streams, image recognition and scenes in vision systems; synchronization of distributed subsystems; visualization and mapping of the implementation process; formation of management decisions.

To solve such problems the system SCADA is used, its main function is to create an operator interface and the data collection process. To control and manage technological process, tool software is used, which is a combination of hardware channels and algorithmic and software means.

Features of the tasks solved at the level of operator control and formation of management decisions are:

- large volume and diversity of data;
- inconsistency and incompleteness of data;
- consistency and high intensity of incoming data;
- a large amount of computation with a predominance of logical computing operations in processing of video streams, image recognition and scenes in vision systems;
- constant complication of algorithms and increasing requirements for accuracy of results;
- a possibility of parallelization of data processing both in time and in space.

In consideration of the widespread development and implementation of wireless means of data exchange and the increasing integration of different systems from the Internet in recent years, there is the task of creating a universal platform that would support as a function ensure guaranteed data exchange with peripherals and functions to create universal secure storage, access, in accordance with their authority in any part of the world. Thus, with the implementation of MMSEETP arises another task (level) - the formation of an integrated system of sharing and storing information from distributed systems.

Herewith, the aim of this system is to provide an association between sensors, microcontroller systems and executive mechanisms with leading, wireless channels, and arrangement of coordination with the Internet. An open platform of data interchange Device Hive can be used for this. This technology is flexible, scalable and easy to use and provides communication between hardware components on the basis of M2M. Application of Device Hive provides the formation of the communication environment, program control and the use of multiplatform libraries for development of remote management and monitoring, telemetry, remote management and monitoring. With technology Device Hive it's able to work using a wide range of technologies, such as "embedded Linux", Python, libraries of C++, the protocol JSON, or connect the AVR, Microchip microcontrollers [10, 15]. The peculiarity of working with Device Hive is an organization above all access to the network, and then - programming of specific applications, reducing design time transfer methods.

The solutions of implementation of distributed data exchange microcontroller devices and systems are offered by leading companies as Google, Oracle, Amazon, and a number of lesser-known providers of this service.

4 Development of Petri Net Models

For the research and analysis of operation in a system-level design of MMSEETP, a Petri net model is employed [18]. Accordingly, the Fig. 5 shows a developed structured model based on the theory of Petri nets, which is designed to analyse the dynamics of MMSEETP's operation and generate a reachability graph for states of the system.

The model based on the Petri nets can be described using the following expression:

$$Pn = \{PS, TrS, ArcS, BBPoSMark\}, \tag{1}$$

where $PS = \{p1, p2, \ldots, pn\}$ – a set of position (states); $TrS = \{t1, t2, \ldots, tn\}$ – a set of transitions; $ArcS$ – a set of curves, input and output arcs according to the transitions; $BBPoSMark$ – set, which sets the initial marking of Petri nets.

An example of segments of the reachability graph for parts of system are shown in Figs. 6 and 7. These results make it possible to argue that the system is alive, the required states are reachable and deadlocks are absent.

The models based on the theory of coloured Petri nets offer wider possibilities for analyzing the operation of a system designed. In this case a model can be described using the following expression [19]:

$$Pn\_col = \{PS, TrS, ArcS, BBPoSMark, TpS, PTpS, ArcStpS, CnD\}, \tag{2}$$

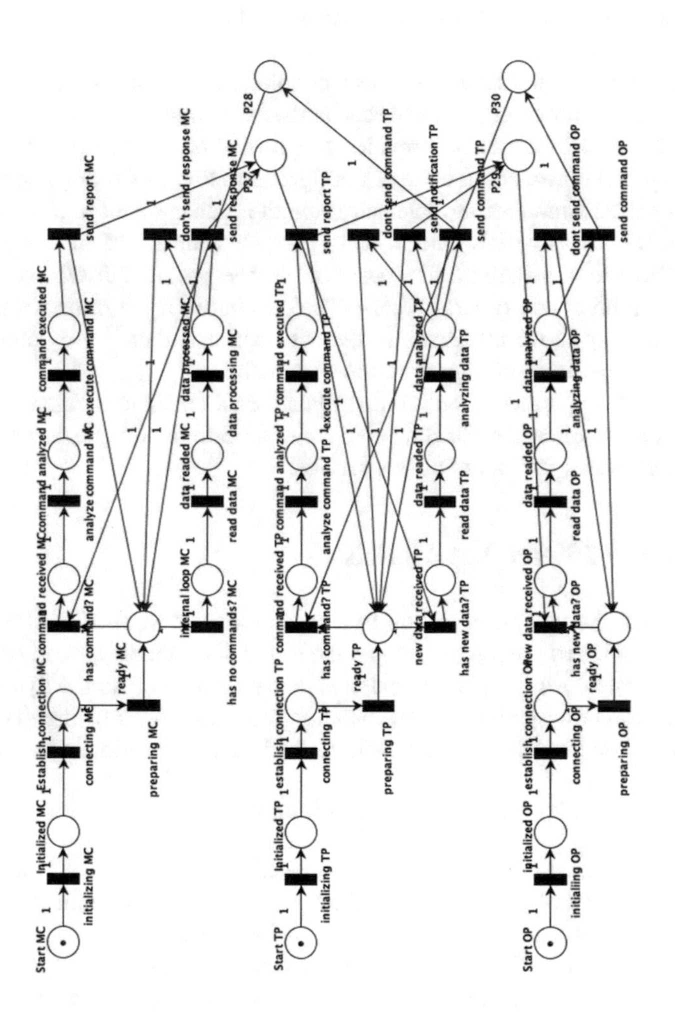

Fig. 5. Block diagram model of a Petri net system

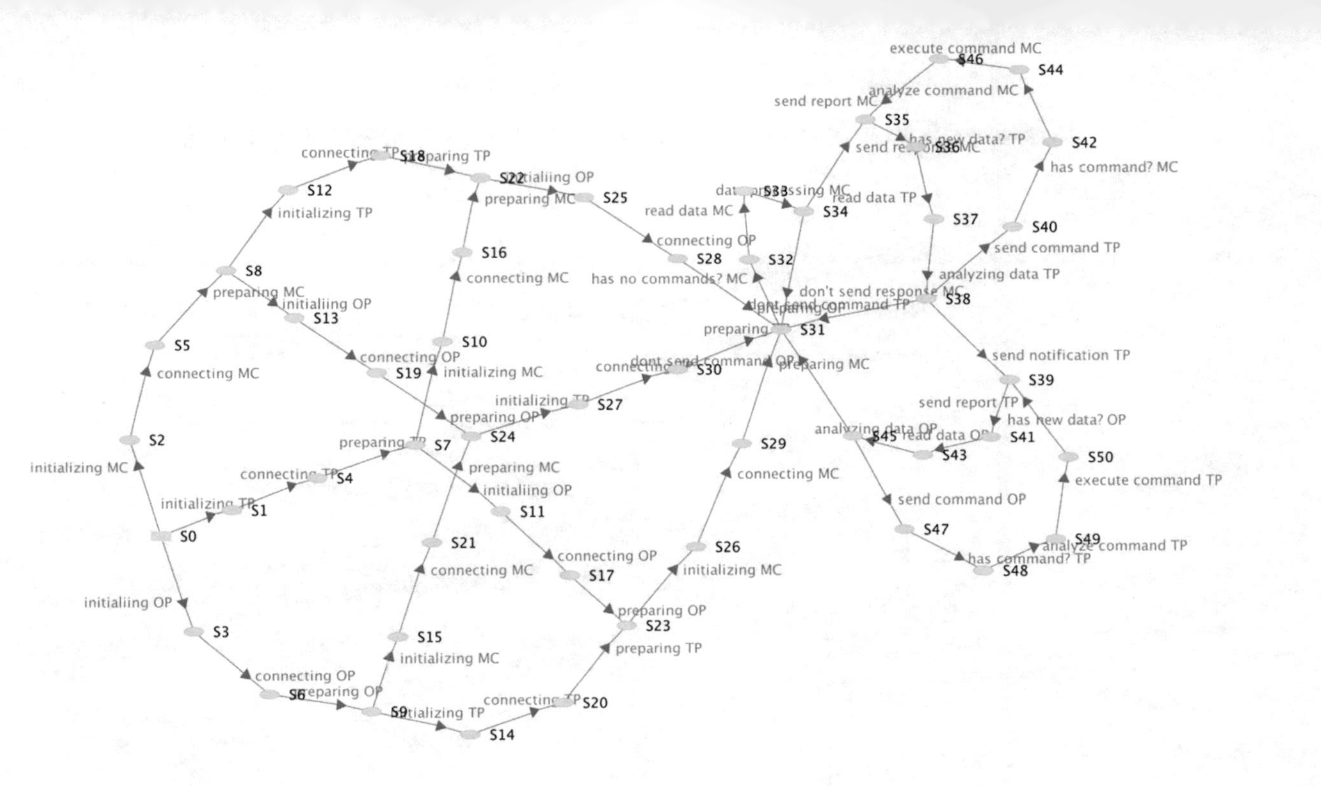

Fig. 6. Research results in a form of the reachability graph for states of the system designed

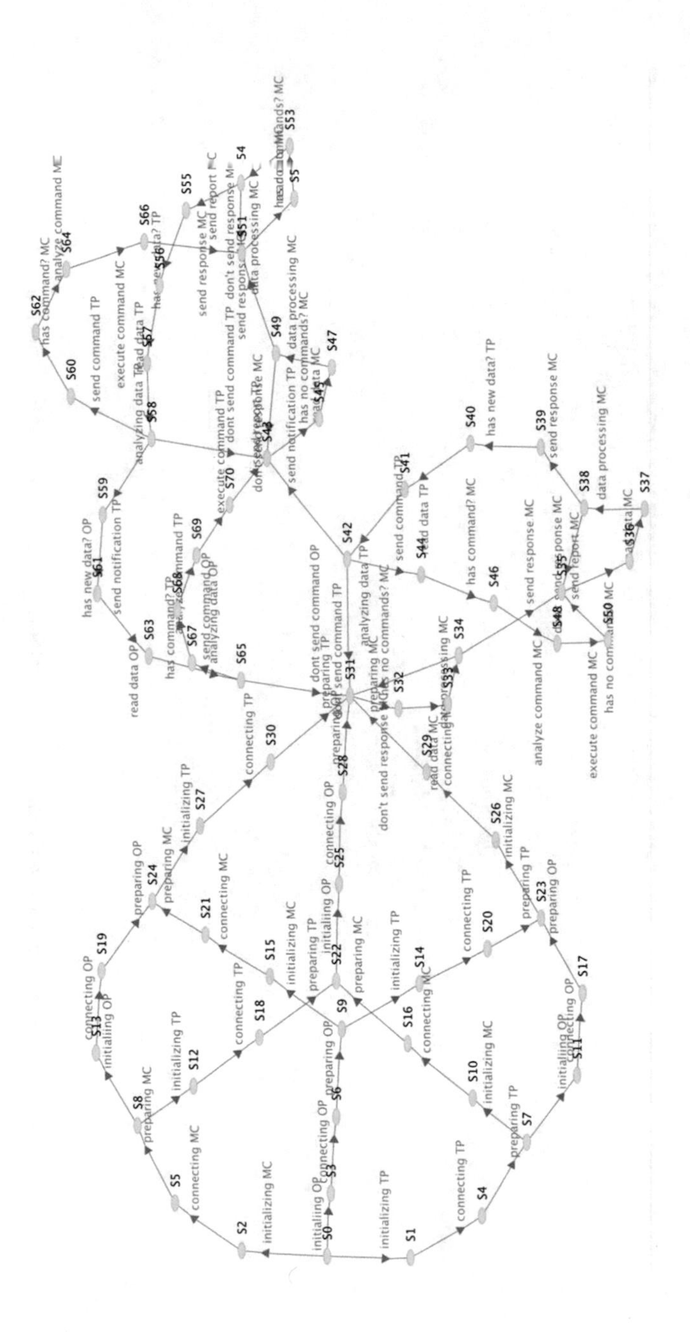

Fig. 7. Research results in a form of the reachability graph for states of the system designed

where *TpS* – a set of types; *PTpS* – a set that reflects the set of available positions in the network; *ArcSTpS* – a set of the markers that stimulate the transition, or indicate what types of tokens need to be generated by the transition; *CnD* – a set of the markers that need to be stimulated by the transition, or indicate what types of tokens need to be generated; *m* – conversions in colored Petri nets.

5 Conclusions

1. The architecture of BSUETP, which is based on the principles such as system integration, use of pre-built components and basic design decisions, modularity, openness and interoperability, has been developed.
2. The computer components of BSUETP should provide data processing in real time, considering limitations on size, power consumption and cost.
3. The high-speed, small-sized and with low power consumption hardware components of BSUETP are developed on the basis of the fixed-point microcontrollers, a small word length, a truncated interface and a shortened system of commands.
4. For programming microcontrollers, it is advisable to use the programming environment CoDeSys (Controllers Development System) whose editors and debugging tools are based on the principles of professional programming environment (Visual C++, etc).
5. It is suggested to combine sensors, microcontroller systems and actuators with each other through wired and wireless channels and connect them to the Internet via the Device Hive technology.
6. The algorithm for the system operation has been developed and the models based on the Petri net theory have been built, which allows the dynamics of the system operation to be examined in the system level design. The results that are acquired from examining the dynamics of the system appear to argue that the system operates properly and correctly.

References

1. Medykovskyi, M.O., Tsmots, I.G., Tsymbal, Y.V.: Information analytical system for energy efficiency management at enterprises in the city of Lviv (Ukraine). Actual Probl. Ekonomics **1**(175), 379–384 (2016)
2. Medykovskyi, M.O., Tsmots, I.G., Tsymbal, Y.V.: Intelligent data processing tools in the systems of energy efficiency management for regional economy. Actual Probl. Ekonomics **12**(150), 271–277 (2013)
3. Аракелов, В.Е., Кремер, А.И.: Методические вопросы экономии энергоресурсов. М: Энергоатомиздат. 188 с (1990)
4. Ганжа, В.Л.: Основы эффективного использования энергоресурсов. Мн.: Белоруская наука. 452 с (2007)
5. Минаев, И.Г., Самойленко, В.В.: Программируемые логические контроллеры: практическое руководство для начинающего инженера. Ставрополь: АГРУС, 100 с (2009)

6. Яковлев, О.Г.: Программирование ПЛК. Москва (2001)
7. Программируемые логические контроллеры Mitsubishi Electric. http://www.sovras.com/fx.php
8. Sydor, A.R., Teslyuk, V.M., Denysyuk, P.Y.: Recurrent expressions for reliability indicators of compound electropower systems. Tekhnichna elektrodynamika **4**, 47–49 (2014)
9. Patti, R.S.: Three-dimensional integrated circuits and the future of system-on-chip designs. Proc. IEEE **94**(6), 1214–1224 (2006)
10. Teslyuk, T., Denysyuk, P., Kernytskyy, A., Teslyuk, V.: Automated control system for arduino and android based intelligent greenhouse. In: Proceeding of the XIth International Conference Perspective Technologies and Methods in MEMS Design, Polyana-Lviv, Ukraine, pp. 7–10 (2015)
11. Parkhomenko, A., Gladkova, O., Kurson, S., Sokolyanskii, A., Ivanov, E.: Internet-based technologies for design of embedded systems. In: Proceedings of 13th International Conference: The Experience of Designing and Application of CAD Systems in Microelectronics, CADSM (2015)
12. Thaker, T.: ESP8266 based implementation of wireless sensor network with Linux based web-server. In: Symposium on Colossal Data Analysis and Networking (CDAN) (2016)
13. Kochlan, M., Hodon, M., Cechovic, L., Kapitulik, J., Jurecka, M.: WSN for traffic monitoring using Raspberry Pi board. In: Federated Conference on Computer Science and Information Systems (FedCSIS), 7–10 September 2014, pp. 1023–1026 (2014)
14. Leccese, F., Cagnetti, M., Trinca, D.: A smart city application: a fully controlled street lighting system isle based on Raspberry-Pi Card, ZigBee sensor network and WiMAX. Sensors **14**(12), 24408–24424 (2014). Special Issue on "Sensors and Smart Cities"
15. Jidin, A.Z., Yusof, N.M., Sutikno, T.: Arduino based paperless queue management system. TELKOMNIKA (Telecommun. Comput. Electron. Contr.) **14**(3), 839–845 (2016)
16. Stefanovych, T., Shcherbovskykh, S., Drożdziel, P.: The reliability model for failure cause analysis of pressure vessel protective fittings with taking into account load-sharing effect between valves. Diagnostyka **16**(4), 17–24 (2015)
17. Mulyak, A., Yakovyna, V., Volochiy, B.: Influence of software reliability models on reliability measures of software and hardware. East. Eur. J. Enterp. Technol. **4**, 9(76), 53–57 (2015)
18. Teslyuk, V.M., Beregovskyi, V.V., Pukach, A.I.: Development of smart house system model based on colored Petri nets. In: Proceedings of International Seminar/Workshop on Direct and Inverse Problems of Electromagnetic and Acoustic Wave Theory, DIPED 2013, Lviv, Ukraine, pp. 205–208, September 2013
19. Jensen, K., Kristensen, L.M.: Coloured Petri Nets: Modelling and Validation of Concurrent Systems, 1st edn. Springer, Heidelberg (2009). 395

Basic Components of Neuronetworks with Parallel Vertical Group Data Real-Time Processing

Ivan Tsmots[1](✉), Vasyl Teslyuk[1](✉), Taras Teslyuk[1](✉), and Ihor Ihnatyev[2](✉)

[1] Lviv Polytechnic National University, Lviv, Ukraine
ivan.tsmots@gmail.com, vasyl.m.teslyuk@lpnu.ua, taras.teslyuk@gmail.com
[2] Ternopil National Economic Universiity, Ternopil, Ukraine
iiv@tneu.edu.ua

Abstract. Neuroalgorithms and neuronetwork structures were analyzed, basic components of neuronetworks were defined and the principles of their development were chosen. It was shown that using the method of parallel vertical group data processing for the implementation of the neuronetworks basic components provides speed increase, reduce of hardware costs and increasing of the equipment use efficiency. Parallel vertical group codes converter, which provides time alignment of data receipt processes and bit sections formation, was developed. The methods and the structures of the components with parallel vertical group data processing for definition of maximum and minimum numbers in the arrays, calculation of the sum of differences squares and scalar product, which due to parallel processing of bit sections groups, provide speed increase, were developed. It was shown that use of the developed basic components for neuronetworks synthesis will provide reduction of time and development cost.

Keywords: Neuroalgorithms · Real-time processing
Neural oriented computer system · An integrated approach
Principles · Structure · Specialized module algorithm

1 Formulation of the Problem

When we use real-time neurotechnologies in industry [1] (control of technological processes and complex objects), electric power industry [2] (load optimization in the electric networks), military industry [3, 4] (technical vision, motion control of mobile robot), transport (motion and engine control), medicine [5–7] (diagnostics of diseases) and instrument making [8, 9] (image recognition and management optimization) we require processing of intensive data streams with neuronetwork, which meet limitations concerning dimensions, weight, energy consumption etc. In order to provide a wide range of applications, such neuronetwork means should simply adapt to the requirements, and costs and time of thir development should be minimal. Existing hardware neuronetwork means can't meet such requirements, since thay have structural

N. Shakhovska and V. Stepashko (eds.), *Advances in Intelligent Systems and Computing II*, Advances in Intelligent Systems and Computing 689,
https://doi.org/10.1007/978-3-319-70581-1_39

organization of the universal type, which functionally and structurally is redundant and does not take into account the requirements of specific applications.

Development of neuronetwork hardware, which meets listed limitations, requires wide use of modern elements base (half-ordered and ordered super integrated circuits (VLSI), single-crystal neuroprocessors) and development of new methods, algorithms and specialized VLSI-structures for the implementation of basic neurooperations.

In this regard, the problem of basic components (ПЕ) development of real-time hardware neuronetworks, which simply adapt to the requirements of specific applications, provide high efficiency of the equipment use and are oriented on synthesis of wide range of neuronetworks, becomes especially actual.

2 Analysis of Recent Research and Publications

Analysis of tasks and fields of application of real-time neuronetworks shows that they have the following features [10–16]:

- high intensity and stability of input data streams;
- permanent complication of processing algorithms and increasing the requirements for the accuracy results;
- possibility of parallel processing both in time and in space;
- ability for generalization and abstraction;
- ability for learning, self learning and self-organization under the influence of the environment.

Existing neuronetworks and means of their implementation were analyzed in the works [10–12]. The analysis shows that the majority of neuronetworks are implemented using software. Such neuronetworks have relatively low performance and need significant hardware costs for their implementation.

It becomes evident from the analysis of works [10, 11, 14–16], that in order to ensure high performance of real-time neuronetworks, it is necessary to use hardware implementation with conveyorization and spatial parallelism.

The analysis of neuronetwork algorithms shows that their operational basis consists of three groups of basic operations: (1) pre-processing; (2) processor operations; (3) calculation of transfer functions.

The group of pre-processing operations provides conversion of initial data for their better perception by neuronetwork. This group includes the following operations:

- input data normalization, during which input data are reduced to one range;
- quantization is performed with continuous quantities, for which finite set of discrete values is defined;
- filtration is performed for «noisy» data and is designed to discard values, which, most likely, are incorrect.

Normalization of input data is the procedure of input data preprocessing (educational, test and work samples), during which the parameter, forming input vector, is reduced to some given range. After such normalization all values of input parameters will be reduced to some narrow range (usually, [0, 1] or [−1, 1].

There are many methods of input parameters normalization. The simplest one, but in most cases effective, is linear normalization. If initial data should be reduced to the range [0, 1], it is executed in this way:

$$X_j^* = \frac{X_j - X_{\min}}{X_{\max} - {\min}}, \tag{1}$$

where $j = 1,\ldots,N$, N – input data number.

In order to reduce initial data to the range [− 1, 1], linear normalization is carried out as follows:

$$X_j^* = \frac{X_j}{X_{\max}}. \tag{2}$$

It is necessary to develop the methods and structures for calculating such operations for hardware implementation of such normalization:

- definition of maximum number from the group of numbers;
- dividing.

After data normalization, depending of the network type, other procedures of data pre-processing can be used. In particular, it is necessary to perform calculation of the Euclidean distance from each input vector to all others after data normalization in RBF- and GRNN-networks. Basic operation for calculation of the sum of differences squares is used for this calculation:

$$Y = \|X_j^e - X_j^b\|^2 = \left(X_1^e - X_1^b\right)^2 + \left(X_1^e - X_2^b\right)^2 + \ldots + \left(X_N^e - X_N^b\right)^2. \tag{3}$$

Another types of data pre-processing can be used for other types of neuronetworks. For instance, for neuro- indefinite networks input data transformations using indefinite sets and the rules of fuzzy logic, having good approximation capabilities, are carried out. Wavelet-transformations, for instance, Haar wavelet, or Daubeşi wavelets [17–20] are used for the analysis of different frequency components of input data in wavelet neural networks.

The group of processor multi-operand operations consists of such operations:

- calculation of scalar product;
- group adding.

The group of operations, realizing transfer functions. This group of operations uses mainly such functions: rigid threshold function, sigmoid or hyperbolic tangent.

In threshold function, for neuron output definition, total sum is compared with some threshold. If the sum is larger than the threshold parameter, processing element generates a signal, otherwise the signal is not generated or braking signal is generated.

Sigmoid or *S-shaped curve* approximates minimum and maximum value in asymptotes within the range [0, 1]. Hyperbolic tangent approximates minimum and maximum value within the range [−1, 1]. The important feature of these curves is continuity of functions and their derivatives.

Comparison schemes are used for hardware implementation of threshold function, and tabular realization is used for calculation of sigmoid and hyperbolic tangent [21].

Disadvantage of existing hardware for implementation of basic neurooperations is complexity of provision of data input intensity receipts with computational intensity of such hardware.

3 Purpose and Tasks of the Research

The purpose of work – development of methods and structures for hardware implementation of basic components of real-time neuronetworks with parallel vertical group processing of data.

In order to achieve the set goal, it is necessary to solve the following tasks:

- to formulate requirements and to choose the principles of basic components development of real-time neuronetworks with parallel vertical group processing of data;
- to develop parallel vertical group converter of codes;
- to develop method and structure of the component of maximum and minimum values calculation;
- to develop method and structure of the component for calculation of the sum of differences squares;
- to develop method and structure of the component for calculation of scalar product.

4 The Main Results of the Research

The Requirements and Principles of Development of Hardware Neuronetworks Basic Components. The structure of hardware neuronetworks basic components should provide implementation of such basic operations: data pre-processing, processor operations and calculation of transfer functions. Hardware basic components of real-time neuronetworks should meet such requirements:

- high efficiency of equipment use;
- efficient use of opportunities and benefits of VLSI-technologies;
- small terms of development and small price;
- coordination of data input intensity with calculation intensity of basic components;
- taking into account the requirements for specific applications;
- reduction of number of interface output and interneural connections;
- real-time work.

In oder to decrease the number of interface outputs and hardware costs, it is offered that data should come parallel, by bit sections groups (vertically), a processing should be performed by with parallel vertical group method using multi-operand approach. Parallel vertical group method of data processing provides that weighting coefficients W_j and input data X_j come parallel by sections from k bits, and for that they are written as follows:

$$W_j = \sum_{i=1}^{n} 2^{-(i-1)} w_{ji} = \sum_{g=1}^{m} 2^{-(g-1)k} \left(w_{j[(g-1)k+1]} + 2^{-1} w_j[(g-1)k+2] + \ldots + 2^{-(k-1)} w_{j[(g-1)k+k]} \right), \quad (4)$$

$$X_j = \sum_{i=1}^{n} 2^{-(i-1)} x_{ji} = \sum_{g=1}^{m} 2^{-(g-1)k} (x_{j[(g-1)k+1]} + 2^{-1} x_{j[(g-1)k+2]} + \ldots + 2^{-(k-1)} x_{j[(g-1)k+k]}) , \quad (5)$$

where w_{ji}, x_{ji} – parameter of i-bits of weighting coefficients and input data; n – bit capacity of weighting coefficients and input data, $m = \lceil \frac{n}{k} \rceil$, k - bits number in the group, in which weighting coefficients W_j and input data X_j are separated.

The following principles are chosen to develop basic components of hardware real-time neuronetworks with parallel vertical group processing method:

- use of the basis of elementary arithmetic operations for implementation of algorithms of performed basic neurooperations;
- spatial and time parallelization of calculation algorithms of basic neurooperations;
- implementation of algorithms of basic neurooperations calculation as a single macrooperation;
- ensuring of regularity, modularity and wide use of standard elements;
- localization and reduction of connections number between the elements of basic components;
- specialization and adaptation of basic components to the processing algorithms structure;
- programmability of architecture through use of reprogrammed logic integrated circuits.

Development of Basic Components. The main purpose of development of basic components of hardware real-time neuronetworks is obtaining high performance, modular and regular VLSI-structure with calculation intensity, coordinated with data input intensity. Output information for development of highly efficient basic components are:

- algorithms of basic neurooperations implementation;
- quantity and number of operands;
- intensity of receipt of input data $P_d = F_d m n_\kappa$, where F_d, m i n_κ – respectively, the frequency of data receipt, number and bit capacity of channels;
- requirements to the interface of basic components;
- calculation accuracy;
- technical and economic requirements and limitations.

Structural organization of basic components is defined with the set of features, the main of which are the following: number of operands, which are processed simultaneously; operating modes; the way of organization of connections between processor elements (PE). Basic components can be devided into synchronous and asynchronous according to the operating modes. In the last case such basic components are called single-cycle, since input data processing is performed without intermediate memorization. Speed of single-cycle basic components is defined with response time of PE, located on the longest path of data passing. Single-cycle basic components are consequential in terms of implementation of data processing algorithm. It is expedient to use synchronous basic components, processing of which is conducted according to the conveyor principle, for processing of intensive data streams. Conveyor basic components are devided into steps with the buffer memory. Basic components should implement as simple as possible operations with approximately the same execution time to provide high speed and efficiency of using the equipment of ПЕ steps. Recording of the results of operations fulfillment in the buffer memory is conducted during timing pulses application in the conveyor. The frequency of such timing pulses F_{TI} is equal to:

$$F_{TI} = \frac{1}{t_{БП} + t_{ОП}}, \tag{6}$$

where $t_{БП}$ – time of record into buffer memory; $t_{ОП}$ – time of operation fulfillment.

In general, the task of development of basic components of hardware real-time neuronetworks can be formulated as follows:

- to define composition, properties and quantity of each PE type;
- to define buffer memory parameters;
- to define necessary connections between PE and exchange methods;
- to synthesize control devices;
- to evaluate main parameters of basic components.

The process of synthesis development of basic components of hardware real-time neuronetworks can be conducted using the next steps:

- development of parallel vertical-group algorithm of basic neurooperations fulfillment and its presentation as specific coordinated stream graph;
- synthesis of basic components structure, connections topology and synchronization functions of change between PE;
- development of basic components interface;
- synthesis of PE structure due to given set of operations considering the element base features and manufacturing technology.

The final results of the first three design stages are set of structures, adequate to algorithm, which consist of buffer memory, limited set of PE, combined with a switching system and provide technical requirements towards basic components.

High efficiency of equipment use is achieved with hardware costs minimization at real time provision of hardware real-time neuronetworks in the basic components. Transition from the algorithm of basic operation implementation to the basic components structure formally presumes minimization of hardware costs

$$W_{БК} = W_{IC} + W_{БП} + W_{БУ} + \sum_{i=1}^{p} W_{ПЕ_i} s_i, \tag{7}$$

providing such condition:

$$\frac{Nn}{F_{д} m n_{к}} \geq t_{БО}, \tag{8}$$

where $W_{БК}$, $W_{БУ}$, W_{I}, $W_{БП}$,$W_{ПЕ}$ - hardware costs for basic components, control block, interface circuits, buffer memory, processor element, s_i – number of ПЕ of i-type, N– number of input data, n – bit rate of input data, $t_{БО}$ – time of calculation of basic operation accordingly.

The main ways of costs minimization of the equipment at development of basic components of hardware real-time neuronetworks are choosing of algorithms of basic neurooperations fulfillment and coordination of intensity of data receipt with calculation intensity of basic components, through change of conveyor time period $t_{к} = t_{БП} + t_{ОП}$ or number m and bit capacity $n_{к}$ of data entry channels.

In order to select a structure option of basic components of hardware real-time neuronetworks, we use the criterion of efficiency of equipment use E, taking into account the number of interface outputs, structure homogeneity, number and location of connections, connect efficiency with equipment costs and evaluate the elements by the efficiency. Quantitative value of efficiency of equipment use for basic components is defined as follows:

$$E = \frac{Rmn_{к}}{t_{к} Nn(k_1 \sum_{i=1}^{s} W_{ПЕ_i} s_i + k_2 Q + k_3 Y)} \tag{9}$$

where R - complexity of the algorithm of basic neuroperation implementation; $t_{к}$ - conveyor time; $W_{ПЕi}$ – equipment costs for the implementation of PE_i, s_i – number of i-type PE, k_1 – coefficient, taking into account homogeneity $k_1 = f(s)$, k_2 – coefficient, taking into account connections regularity $k_2 = f(\Delta j)$, Δj - distance between communication lines, Q – number of communication lines, k_3 – coefficient, taking into accountof number of communication interface outputs $k_3 = f(Y)$, Y - number of communication interface outputs.

5 Methods and Structures of Basic Components Implementation

Parallel Vertical Group Converter of Codes. Converter of codes should provide parallel consecutive group transformation, that is forming of group bit sections. Parallel consecutive group transformation is carried out by consecutive reception of N numbers with further forming of bit sections from k bits for all N numbers in each cycle. Such transformation is used for download of hardware neuronetworks component with parallel vertical group processing of input data. At processing of data streams in each *g*-work cycle on the input of the component it is necessary to form such bit sections with *k* bits for all *N* numbers. In order to provide processing of continuous data streams, a parallel vertical group converter of codes, the structure of which is shown on Fig. 1, was developed, where *БлРг* – registers block, *Рг* –register, *V* – operation mode of converter, D_j – input data.

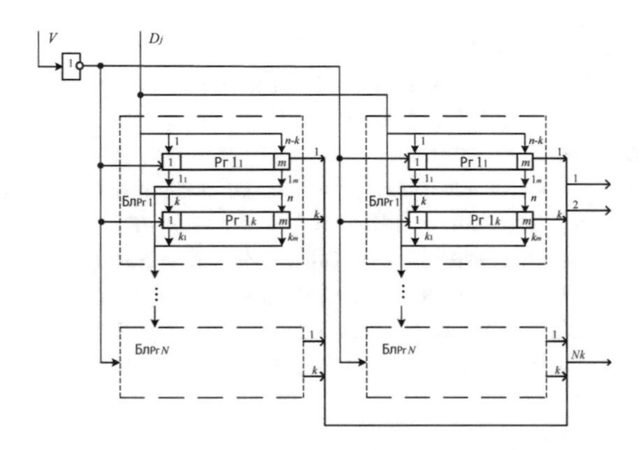

Fig. 1. Developed structure of parallel vertical group converter of codes

Parallel vertical group converter of codes consists of two groups of register blocks, one of which works for parallel record of input data, and another one for bitwise output of previously accumulated data. The groups of registers work in two operating modes: $V = 0$ - parallel record; $V = 1$ – bitwise data output. At parallel recording sequential outputs of registers are in the third state.

6 Calculation Component of Maximum and Minimum Numbers

Parallel vertical group method of calculation maximum D_{max} i minimum D_{min} numbers in one-dimensional $\{D_k\}_{h=1}^{N}$ and two-dimensional $\{D_{kj}\}_{h=1;j=1}^{N;M}$ arrays provides in each *g*–cycle ($g = 1,\ldots,m$ where $m = \lceil \frac{n}{k} \rceil$, *k* – bits number in the group, *n* - bit rate numbers), parallel receipt of *N* numbers with older bits before section with *k* bits [12]. Calculation of maximum D_{max} i minimum $D_{\min}$ numbers in one-dimensional array $\{D_k\}_{h=1}^{N}$ by this method is based on the implementation for each *r*-bit section ($r = 1,\ldots, k$) of the same type basic макрооperations, which are performed on the basis of three simple operations. At calculation of maximum number $D_{\max}$ in one-dimensional array $\{D_k\}_{h=1}^{N}$ such operations are used:

(1) forming of value of r-bit section Pr by the formula:

$$P_r = \bigvee_{h=1}^{N} D_{rh} \wedge y_{rh}, \tag{10}$$

where Drh – the value of r-cycle of h-number of array, yrh – the value of h –cycle of r-word of control, the value of 1-word of control is equal to y11 = y12 = … = y1 N = 1;

(2) definition of r-cycle of maximum number Dmaxr by the expression:

$$D_{\max r} = \begin{cases} 0, & \text{when} \quad P_r = 0 \\ 1, & \text{when} \quad P_r = 1 \end{cases}; \tag{11}$$

(3) forming of h bits (r + 1)-word of control by the formula:

$$y_{(r+1)h} = \begin{cases} 0, & \text{when} \quad P_r = 1, \quad D_{hr} \neq y_{hr} \\ 1, & \text{when} \quad P_r = D_{hr} = y_{hr} = 1 \\ y_{hr}, & \text{when} \quad P_r = 0 \end{cases}. \tag{12}$$

At calculation of minimum number of Dmin in one-dimensional array $\{D_k\}_{h=1}^{N}$ basic macrooperation is realized on such operations:

(1) forming the value r-o of bit section Pr, which is performed by the formula:

$$P_r = \bigvee_{h=1}^{N} \bar{D}_{rh} \wedge y_{rh}, \tag{13}$$

where $\bar{D}_{rh}$ – inverse value of r- cycle of h-number array, yrh – the value of h-cycle of r-word of control, the value of 1-word of control is equal to y11 = y12 = … = y1 N = 1;

(2) definition of r-cycle minimum number Dminr by the expression:

$$D_{\min r} = \begin{cases} 0, & \text{when} \quad P_r = 1 \\ 1, & \text{when} \quad P_r = 0 \end{cases}; \tag{14}$$

(3) forming of h bits (r + 1)-word of control, which is performed by the expression:

$$y_{(r+1)h} = \begin{cases} 0, \text{ when } P_r = 1, \ \overline{D}_{khr} \neq y_{rh} \\ 1, \text{ when } P_r = \overline{D}_{hr} = y_{hr} = 1 \\ y_{hr}, \text{ when } P_r = 0 \end{cases} \quad (15)$$

The peculiarity of the considered parallel vertical-group method of maximum (minimum) number calculation is that in each g-work cycle k bits of maximum Dmax (minimum Dmin) number are defined.

Definition of maximum D_{max} and minimum D_{min} numbers in a two-dimensional array $\{D_{hj}\}_{h=1;j=1}^{N;M}$ is founded on basic macrooperations, which are performed by the formulas accordingly (10)–(12) and (13)–(15). Difference of definition of maximum D_{max} (minimum D_{min}) number in a two-dimensional array $\{D_{kj}\}_{k=1;j=1}^{N;M}$ is that after execution of each m cycles with one-dimensional array from (N + 1) numbers maximum (minimum) number of new j- one-dimensional arrays is used in calculation and units record in all bits of control register is conducted. Calculation maximum (minimum) number in a two-dimensional array $\{D_{kj}\}_{k=1;j=1}^{N;M}$ requires fulfillment of $M Ч m$ basic operations.

The peculiarity of the parallel vertical-group method of maximum and minimum numbers calculation of arrays numbers is:

- use of one basic macro-operation;
- ability to use parallelization and calculation conveyorization;
- ability of simultaneous processing of N- bits sections;
- time of calculation is mainly defined with bits number in the group k, as well as with bit capacity of numbers n, and not with their number N.

Structure of the component calculation of maximum and minimum numbers. Calculation of the component of maximum and minimum numbers is realized on the basis of the same PE type. Each PE mechanically conducts k basic macrooperations of maximum and minimum numbers calculation.

Built structure of the component of maximum and minimum numbers calculation in one-dimensional array $\{D_k\}_{h=1}^{N}$ with parallel vertical group method is shown on Fig. 2, where *TI* – timing pulses, *ПУ* – initial setting, Tr – trigger, Pr – register, $D_{h1} - D_{hk}$ - h-input of the group with k bits, $D_{h1min} - D_{hkmin}$ and $D_{h1max} - D_{hkmax}$ D_{imin} – output from the groups with k bits of maximum and minimum numbers respectively.

Number of PE, connected with the common bus of results, at simultaneous calculation of maximum and minimum numbers for one-dimensional array $\{D_k\}_{k=1}^{N}$ is defined according to its size. Use of common buses of the results provides parallelization of the process of bit section processing, processing time of which defines work cycle of the device. Calculation of maximum and minimum numbers with parallel

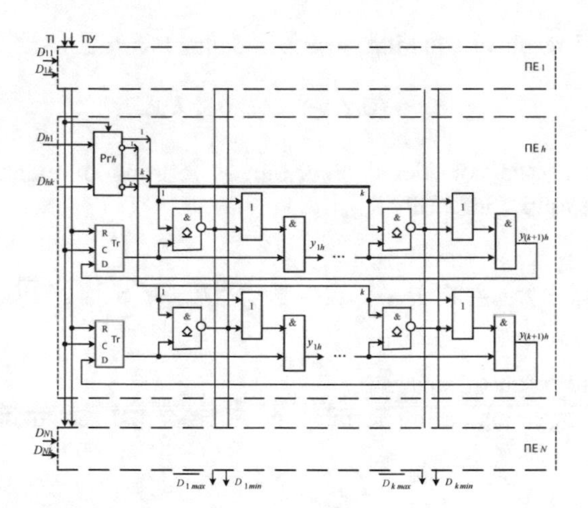

Fig. 2. Structure of the calculation component of maximum and minimum numbers with parallel vertical group method

vertical group method in such device is made during the period, which is defined as follows:

$$t_{Обч}=m(t_{Tг}+3kt_I), \tag{16}$$

where $t_{Tг}$ *and* t_I response time trigger and logical elements of the type OR, AND, AND-NOT accordingly, k – bits number in the group.

Calculation Component of the Sum of Differences Squares. *Parallel vertical group method of the sum of differences squares calculation* requires, that each operand should be presented in the form of groups with k bits. As such the operands are recorded as follows:

$$X_j = \sum_{i=1}^{n} 2^{-(i-1)} x_{ji} = \sum_{g=1}^{m} 2^{-(g-1)k} (x_{j[(g-1)k+1]} + 2^{-1} x_{j[(g-1)k+2]} + \ldots + 2^{-(k-1)} x_{j[(g-1)k+k]}), \tag{17}$$

where x_{ji} –value of i-cycle of j- operand; n – bit rate of operand, m - number of groups, into which the operand is devided.

Squaring is the main operation of calculation of differences squares sum. We'll use a vertical algorithm for execution of such operation:

$$\begin{aligned} X^2 &= (0.01) \wedge x_1 + 2^{-1}(0.x_1 01) \wedge x_2 + 2^{-2}(0.x_1 x_2 01) \\ &\quad \wedge x_3 + \ldots + 2^{-(n-1)}(0.x_1 x_2 \ldots x_{n-1} 01) \wedge x_n \\ &= \sum_{i=1}^{n} 2^{-(i-1)} R_i, \end{aligned} \tag{18}$$

where R_i – partial result of squaring, which is defined as follows:

$$R_i = (0.x_1 x_2 \ldots x_{i-1} 01) \wedge x_i. \tag{19}$$

Development of the considered algorithm is forming of macropartial result of squaring for the group with k bits R_{Mg}:

$$R_{Mg} = R_{g1} + 2^{-1} R_{g2} + \ldots + 2^{-(k-1)} R_{gk} = \sum_{r=1}^{k} 2^{-(r-1)} R_{gr}, \tag{20}$$

where R_{gr} - partial result of squaring.

Algorithm of squaring, using forming of macropartial results $R_{Mg,}$ is written as follows:

$$X^2 = \sum_{g=1}^{m} 2^{-(g-1)k} R_{Mg}. \tag{21}$$

We will calculate differences squares sum on the basis of multi-operand approach, which consists of simultaneous processing of all operands and forming of macropartial result of the sum of differences squares. We will perform calculation of sum of differences squares, using parallel vertical group method, which is written as follows:

$$\begin{aligned} y &= (X_1^e - X_1^b)^2 + (X_2^e - X_2^b)^2 + \ldots + (X_N^e - X_N^b)^2 = \Delta X_2^2 + \ldots + \Delta X_N^2 \\ &= \sum_{g=1}^{m} 2^{-(g-1)k} R_{1Mg} + \ldots + \sum_{g=1}^{m} 2^{-(g-1)k} R_{NMg} = \sum_{j=1}^{N} \sum_{g=1}^{m} 2^{(g-1)k} R_{jMg} = \sum_{g=1}^{m} 2^{(g-1)k} R_{jmg} = \sum_{j=1}^{N} R_{jMg} = \sum_{g=1}^{m} 2^{(g-1)k} p_{Mg}, \end{aligned} \tag{22}$$

where P_{Mg} – g-macropartial result of the sum of differences squares.

The main stages of parallel vertical-group method of calculation of the sum of differences squares is:

- simultaneous consecutive group input of operands X_i^e, X_i^b and calculation of module ΔX_i;
- forming of macropartial results of squaring R_{Mg};
- forming of macropartial result of the sum of differences squares P_{Mg} through adding of macropartial results of squaring R_{Mg};
- obtaining the result of sum of differences squares using adding of macropartial results of calculation P_{Mg} with the right shift on k bits.

Structure of the compinent for parallel vertical group calculation of differences squares sum. Dependning on the method of forming and adding of macropartial results of the sum of differences squares P_{Mg}, the following variants of implementation of the component of differences squares sum calculation are possible:

- with sequential forming and adding of P_{Mg};
- with parallel forming and sequential adding of P_{Mg};
- with parallel forming and adding of P_{Mg}.

Developed structure of the component of calculation of differences squares sum with parallel forming and sequential adding of P_{Mg} is shown on Fig. 3, where Pг – register, БСм – multichannel adder, См – adder, PE – processor element.

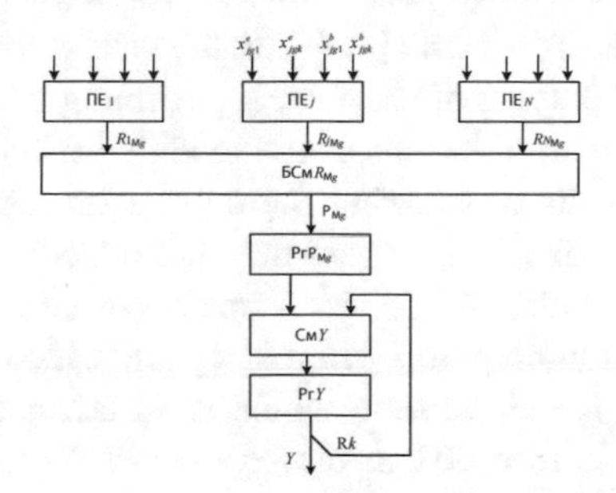

Fig. 3. Structure of the component of calculation of differences squares sum with parallel forming and sequential adding

The main elements of this structure are: PE$_j$, designed for forming of macropartial results of squaring R_{jMg}; БСмR_{jMg}, which with the help of parallel adding of R_{jMg} forms macropartial result of sum of differences squares P_{Mg}; Pг P_{Mg}, СмY and PгY, which provide sequential calculation of the sum of differences squares by the formula:

$$Y_g = 2^{-k} Z_{g-1} + P_{Mg}, \tag{23}$$

where $Y_0 = 0$.

Built structure of PE$_j$ is shown on Fig. 4, where Від – deductor, Tг – trigger, ПК – converter of parallel code into vertical group, $|\Delta X_j|$ - shaper of difference module, R_{rg} – shaper of partial result of squaring.

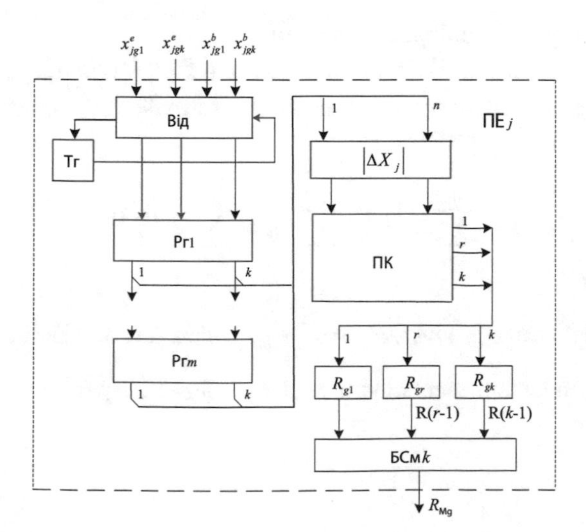

Fig. 4. Developed *ПЕ* structure

The operands X_j^e nad X_j^b entry to the input ПЕ$_j$ is carried out sequentially by the groups with k bits, beginning with younger bits. In each ПЕ$_j$ with the help of deductor Від during m cycles the difference ΔX_j is calculated and recorded in the registers Рг1, …,Ргm. The calculated difference ΔX_i comes to the inputs of shaper $|\Delta X_j|$, on the output of which we receive its module $|\Delta X_j|$. In the next work cycles we receive partial results of squaring on the outputs of shapers R_{rg}. Forming of partial results of squaring R_{rg} is realized beginning with older bits of the module $|\Delta X_i|$ according to the formula (19). Formed k partial results of squaring R_{rg} come with the right shift on (r – 1) bit inputs of multichannel adder БСмk, where they are added. The sum, obtained on the output of multichannel adder БСмk is a macropartial result of squaring R_{jMg}. Макроpartial results of squaring R_{1Mg},…, R_{NMg} are added with the help of multichannel adder БСм R_{Mg}. Received sum, which is a macropartial result of the sum of differences squares P_{Mg}, is recorded in the register РгP_{Mg}. Adding of data from the output of the register РгP_{Mg} to the previously accumulated amount from the register РгY with the right shift on k bits is performed on the adder СмY in each cycle.

The feature of work of this component is time alignment of data input processes of one array and calculation of another one. Using such combination of the sum of differences squares calculation in this component will be made for m cycles.

Calculation Component of Scalar Product. *Parallel vertical group method of calculation of scalar product* is based on elementary arithmetic operations, focused on HBIC-implementation, and provides decrease of work cycles number, and calculation time accordingly. Calculation of scalar product by this method presumes forming and adding of partial multiplication product in accordance with the following formula:

$$Z = \sum_{j=1}^{N} W_j X_j = \sum_{j=1}^{N}\sum_{g=1}^{m} 2^{-(g-1)k}\left(W_j X_{j[(g-1)k+1]} + 2^{-1} W_j X_{j[(g-1)k+2]} + \ldots + 2^{-(k-1)} W_j X_{j[(g-1)k+k]}\right)$$
$$= \sum_{j=1}^{N}\sum_{g=1}^{m} 2^{-(g-1)k} P_{jg}, \qquad (24)$$

where P_{jg} - group partial multiplication result,

$P_{jg} = W_j X_{j[(g-1)k+1]} + 2^{-1} W_j X_{j[(g-1)k+2]} + \ldots + 2^{-(k-1)} W_j X_{j[(g-1)k+k]}$.

After making all necessary changes in the formula (24), calculation of scalar product can be written as follows:

$$Z = \sum_{g=1}^{m} 2^{-(g-1)k} \sum_{j=1}^{N} P_{jg} = \sum_{g=1}^{m} 2^{-(g-1)k} R_g, \qquad (25)$$

where R_g - partial scalar product, $R_g = \sum_{j=1}^{N} P_{jg}$, which can be calculated both parallel and sequential. Sequential calculation of scalar product R_g is performed by the formula:

$$R_g = \sum_{g=1}^{m} 2^{-(g-1k)} \sum_{j=1}^{N} \left(w_{j[(g-1)k+1]} x_{j[(g-1)k+1]} + 2^{-1} w_{j[(g-1)k+2]} x_{j[(g-1)k+2]} + \ldots + 2^{k-1} w_{j[(g-1)k+k]} x_{j[(g-1)k+k]} \right)). \quad (26)$$

It becomes evident from the formula (25), that in roder to calculate scalar product it is necessary to perform m cycles, and each of them includes such operations:

- forming of partial multiplication product in accordance with the formula $P_{j[(g-1)k+r]} = W_j X_{j([(g-1)k+r]}$, where r = 1,…, for each j- couple of operands k;
- calculation of group partial multiplication product P_{jg} for j-couple of operands in accordance with the formula $P_{jg} = \sum_{r=1}^{k} 2^{-(r-1)} W_j X_{j[(g-1)k+r]}$;
- calculation of partial result of scalar product R_g using adding of групових partial multiplication product P_{jg} in accordance with the formula $R_g = \sum_{j=1}^{N} P_{jg}$;
- adding of partial results of scalar product R_g according to the expression $Z_g = 2^{-k} Z_{g-1} + R_g$, where $Z_0 = 0$.

The structure of the component parallel vertical group calculation of scalar product. Depending on the ways of operands coming, number of used buses for operands entering, forming and adding of partial final result of scalar product R_g, conducting of entering processes and calculation of scalar product in time, the following variants of implementation of the component of scalar product calculation are possible:

- with parallel vertical group input of data by younger (older) bits forward;
- with individual or multiplexed buses for input of data X_j and weighting coefficients W_j;
- with parallel or sequential forming of partial result of scalar product;
- with parallel or sequential adding of м partial results scalar product;
- with separation or combination of input processes and calculation of scalar product.

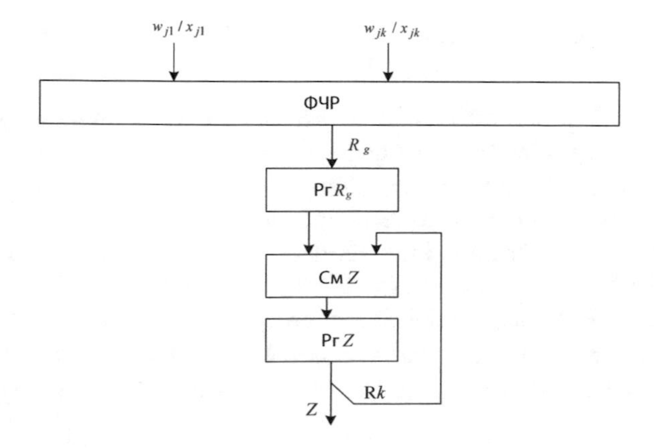

Fig. 5. The structure of the component with parallel vertical group calculation of scalar product

The most efficient structure for VLSI-implementation is the following: with data input by younger bits forward; use of multiplexed buses for operands entry; parallel forming of partial result of scalar product R_g; sequential adding of partial results of scalar product R_g; combining of input processes and scalar product calculation. Така Structure is shown on Fig. 5, where ФЧР – shaper of partial results of scalar product.

The main element of this structure is ФЧР, which can perform both parallel and sequential forming of partial result of scalar product R_g. The structure of ФЧР with parallel forming of partial result of scalar product R_g is shown on the Fig. 6, where $w_{j1}/x_{j1}\ldots w_{jk}/x_{jk}$ multiplexed single-bit information inputs; Pr – register, ПК – converter of bits sequential groups into parallel code; *k*См, *N*См – *N* i *k* input adders.

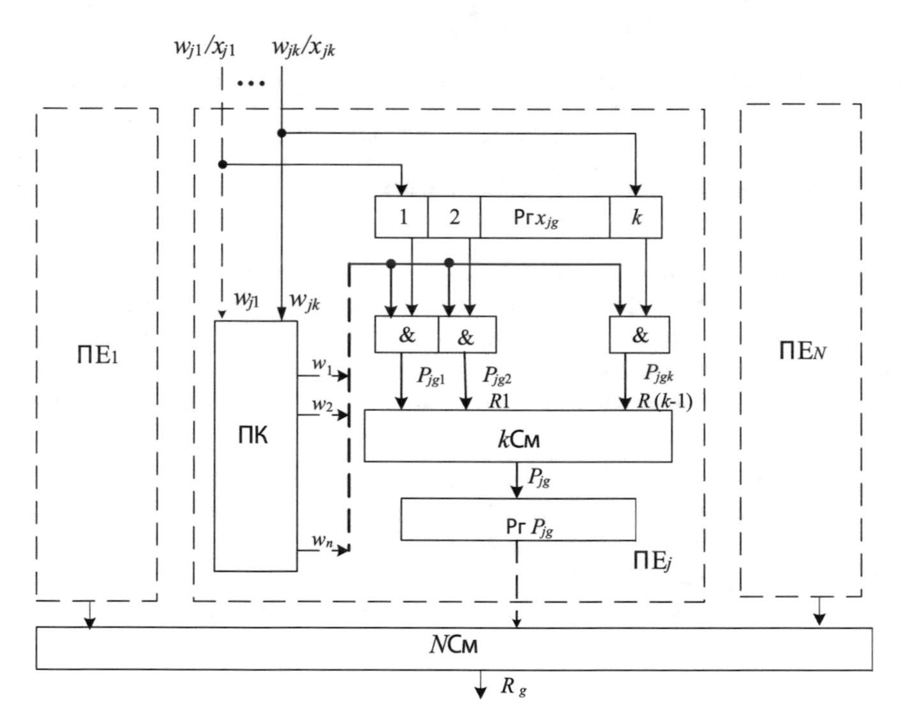

Fig. 6. The structure of the shaper of partial results of scalar product

Parallel vertical group calculation of scalar product in this component can be devided into two stages, and each of them is performed for *m* cycles.

k bits of multiplier W_j enter to the *j*- information entry on the first stage in each *g*-cycle, beginning with younger bits. Accumulation of sequential groups bits of multiplied W_j and their transformation into parallel code is performed in ПК.

On the second stage in each *g* in the work cycle in ΠE_j for the group of multiplier bits $X_{j_{g1}}X_{j_{g2}}\ldots X_{j_{gk}}$ *k* partial multiplication product in accordance with the formula $P_{jgr} = W_jX_{jgr}$ is formed. Formed partial multiplication products come to the input of *k*-input adder, at that *r*-й ($r = 1,\ldots,k$) partial multiplication product W_jX_{jgr} is shifted with regard to (*r*-1)- partial multiplication product $W_j X_{jg(r-1)}$ in one bit to the right. Using adding of partial multiplication product on the output of *k*-input adder, we obtain a group with partial multiplication result P_{jg}, which is recorded in the register PrP_{jg}. Group partial multiplication products P_{jg} from outputs ПЕ come to the inputs of adder *N*См, where they are added. On the output of adder *N*См we obtain *g*-partial result of scalar product R_g, which is recorded in the register PrR_g. Adding of partial results of scalar product R_g, according to the expression $Z_g = 2^{-k} Z_{g-1} + R_g$, where $Z_0 = 0$, is realized on the adder СмZ.

Developed component for scalar product calculation works by the conveyor principle and is oriented for processing of continuous data streams. Conveyor work cycle of such device is defined as follows:

$$T_k = t_{P2} + t_{NCM},$$

where t_{P2} – response time of register, t_{NCM} – time of *N* numbers adding. Calculation of scalar product is realized for *m* conveyor cycles.

7 Conclusions

1. Neuroalgorithms and structures of neuronetworks were analyzed, the followwing basic components were defined: converter of codes, calculation of maximum and minimum numbers, calculation of the sum of differences squares, calculation of scalar products.
2. It was offered to conduct implementation of basic components according to such principles: modularity, coordination of intensity of data receipt with computing power of the components, conveyorization and spatial parallelism, localization and simplifying of the links between the elements, specialization and adaptation of hardware and software to the requirements of specific application.
3. Use of parallel vertical group method of data processing and basis of elementary arithmetic operations for hardware implementation of neuronetworks basic operations provides performance increase, decrease of hardware costs and their orientation for VLSI implementation.
4. Parallel vertical group converter of codes, which due to time alignment of sequential recording processes of bits groups and forming of bit section groups for data array provides faster performance, was developed, using dual memory.
5. Parallel vertical group method of calculation of maximum and minimum numbers in arrays, which due to parallel section processing from the group of bits of all numbers provides time reduction of calculation, was developed.
6. Parallel vertical group method of calculation of the sum of differences squares, which is founded on the calculation operations of differences modules, forming of partial results of squaring and group adding, was developed. This method, due to parallel processing of section from the group of bits of all numbers, provides faster performance.

7. Parallel vertical group method for scalar product calculation, which in comparison with wellknown, due to increasing of bits channels of multipliers input and number analysis of bits for forming partial multiplication result, provides time reduction for calculation of scalar product, was developed.

References

1. Mohamad, H.: Hassoun Fundamentals of Artificial Neural Networks, p. 511. MIT Press, Cambridge (1995)
2. Teich, T., Roessler, F., Kretz, D., Frank, S.: Design of a prototype neural network for smart homes and energy efficiency. In: Proceedings of 24th DAAAM International Symposium on Intelligent Manufacturing and Automation, pp. 603–608. Zwickau, Germany (2013)
3. Цюй Дуньюэ. Управление мобильным роботом на основе нечетких моделей / Цюй Дуньюэ // Современные проблемы науки и образования. – № 6. – С, pp. 115–121 (2012)
4. Matviichuk, K., Teslyuk, V., Teslyuk, T.: Vision system model for mobile robotic systems. In: Proceeding of the XIIh International Conference "Perspective Technologies and Methods in MEMS Design", MEMSTECH 2016, 20–24 April 2016, pp. 104–106. Polyana, Lviv, Ukraine (2016)
5. Cagnoni, S., Coppini, G., Rucci, M., et al.: Neural network segmentation of magnetic resonance spin echo images of the brain. J. Biomed. Eng. **15**(5), 355–362 (1993)
6. Fujita, H., Katafuchi, T., Uehara, T., et al.: Application of artificial neural network to computer-aided diagnosis of coronary artery disease in myocardial SPECT bull's-eye images. J. Nucl. Med. **33**(2), 272–276 (1992)
7. Astion, M.L., Wener, M.H., Thomas, R.G., Hunder, G.G., Bloch, D.A.: Application of neural networks to the classification of giant cell arteritis. Arthritis Reum. **37**(5), 760–770 (1994)
8. Peleshko, D., Ivanov, Y., Sharov, B., Izonin, I., Borzov, Y.: Design and implementation of visitors queue density analysis and registration method for retail videosurveillance purposes. In: 2016 IEEE First International Conference on Data Stream Mining & Processing (DSMP), pp. 159–162. Lviv, Ukraine (2016)
9. Badlani, A., Bhano, S.: Smart home system design based on artificial neural networks. In: Proceedings of the World Congress on Engineering and Computer Science 2011, pp. 106–111. San Francisco, USA, 19–21 October 2011
10. Pukach, A.I., Teslyuk, V.M., Tkachenko, R.O., Ivantsiv, R.-A.D.: Implementation of neural networks for fuzzy and semistructured data. In: Proceedings of 11-th International Conference on the Experience of Designing and Application of CAD Systems in Microelectronics, CADSM 2011, pp. 350–352. Lviv, Polyana, Ukraine, 23–25 February 2011
11. Осовский С. Нейронные сети для обработки информации / Пер. с польского. – М.: Финансы и статистика, 344 с (2009)
12. Патент №101922 Україна, G06F 7/38. Пристрій для обчислення скалярного добутку/ Цмоць І.Г., Скорохода О.В., Теслюк В.М. Бюл. №9 (2013)
13. Патент №110187 Україна, G06F 7/38. Пристрій для визначення максимального числа з групи чисел/ Цмоць І.Г., Скорохода О.В., Медиковський М.О., Антонів В.Я. Бюл. №22 (2015)

14. Цмоць І.Г. Модифікований метод та НВІС - структура пристрою групового підсумовування для нейроелемента. / І.Г. Цмоць, О.В. Скорохода, Б.І. Балич // Вісник НУ « Львівська політехніка » – Львів. – № 732: « Комп'ютерні науки та інформаційні технології » . – С, pp. 51–57 (2012)
15. Tsmots, I., Skorokhoda, O., Rabyk, V.: Structure software model of a parallel-vertical multi-input adder for FPGA implementation. In: Proceedings of XIth International Scientific and Technical Conference CSIT 2016, pp. 158–160. Lviv, Ukraine, 6–10 September 2016
16. Цмоць І.Г., Скорохода О.В., Ігнатєв І.В. Синтез компонентів паралельних нейромереж вертикально-групового типу. Вісник НУ "Львівська політехніка" "Комп'ютерні науки та інформаційні технології" № 826 Львів. С. pp. 69–79 (2015)
17. Bodyanskiy, Y., Dolotov, A., Vynokurova, O.: Evolving spiking wavelet-neuro-fuzzy self-learning system. Appl. Soft Comput. **14**, 252–258 (2014)
18. Bodyanskiy, Y., Vynokurova, O., Pliss, I., Peleshko, D., Rashkevych, Y.: Hybrid generalized additive wavelet-neuro-fuzzy-system and its adaptive learning. In: Zamojski, W., Mazurkiewicz, J., Sugier, J., Walkowiak, T., Kacprzyk, J. (eds.) Dependability Engineering and Complex Systems: Proceedings of the Eleventh International Conference on Dependability and Complex Systems DepCoS-RELCOMEX. 27 June 2016–1 July 2016, pp. 51–61. Brunow, Poland (2016)
19. Haar, A.: Zur Theorie der orthogonalen Funktionensysteme. Math. Ann. **69**(3), 331–371 (1910)
20. Daubechies, I.: Ten Lectures on Wavelets, SIAM, p. 194 (1992)
21. Galushkin, A.I.: Neurocomputers. Book 3 – M.: IPRZR, p. 528 (2000)

Recommendation Systems as an Information and Technology Tool for Virtual Research Teams

Nataliia Veretennikova(✉) and Nataliia Kunanets(✉)

Information Systems and Networks Department,
Lviv Polytechnic National University, Lviv, Ukraine
Nataver19@gmail.com, nek.lviv@gmail.com

Abstract. The paper analyzes options of concept interpretation of "e-Science" and its author's definition as a systemically integrated complex of computer information telecommunication and social and communicative technologies, which ensure the fulfillment of functions and the solution of the actual tasks of science in the information society. The basic principles of the modern concept of a "virtual scientific team" are considered. It is analyzed the creation problems of effective information and technological tools for the comfortable and qualitative implementation of information and communication processes in the social and communication environments of virtual scientific teams that conduct research on the platform of e-Science. The main tendencies of using recommender systems are highlighted for an effective implementation of information and communication processes in virtual scientific teams. A description of the functional capabilities and the basic algorithms of work of the recommender system "Information assistant of a scientist" is presented, which supports the implementation of processes for effective information support of researchers in the virtual scientific teams.

Keywords: E-science · Information support · Virtual research team · Consolidated resources · Recommender system

1 Introduction

In the information society, the development of e-science involves the development of a large-scale distributed computing, communication and information infrastructure, middleware and the formation of virtual research teams. This concept is related to the emergence of inter-organizational and interdisciplinary research, based on innovative approaches in the dissemination, use and analysis of data.

A successful solution to such problems requires the creation of effective tools for the implementation of information and communication processes among such creative scientific communities.

N. Shakhovska and V. Stepashko (eds.), *Advances in Intelligent Systems and Computing II*, Advances in Intelligent Systems and Computing 689,
https://doi.org/10.1007/978-3-319-70581-1_40

2 Analysis of Recent Researches and Publications

The methodological principles of the construction of e-science system is devoted the work by J. Taylor [1], the questions of planning scientific tasks in the field of e-science are researched in the work of E. Deelman, D. Gannon, M. Shields [2], the problems of management of scientific projects are analyzed by D. Spencer, A Zimmerman, D. Abramson [3], A. Warr and others [4]. The combination of WEB- and GRID-technologies in the formation of the infrastructure of e-science is described in the works of M.Z. Zgurovskyi and A.I. Petrenko [5].

The problems of developing a new innovative scientific and technological area, which is e-science, are discussed in the articles "Communication and Collaboration in e-Science Projects" [6], "Supporting Scientific Collaboration: Methods, Tools and Concepts" [7] and "Collaborative e-Science Experiments and Scientific Workflows" [8]. D. Sonnenwald analyzes the concepts of transformations of organizational structures in the form of virtual organizations [9] and the peculiarities of information exchange technologies in them [10]. A number of publications analyze the ways to establish communication in the environment of participants in a particular scientific project [11]. At the same time, we should note that there are practically no works in which the tasks of developing a comprehensive technological tool for providing information and communication processes of virtual research teams are analyzed.

The problem of formation of consolidated information resources that is relevant to the profile of researches of virtual creative teams remains not completely examined, as well as their technological support with the use of conceptual approaches of data warehouses, data spaces and large data. A separate actual informational and technological problem in the field of e-science is the lack of tools for providing the user with personalized advice taking into account the personal information needs of each member of the virtual research team in the context of implementing a joint project.

3 E-Science and Virtual Scientific Teams

The actual scientific research in many technologically advanced countries undergoes dramatic changes in connection with the introduction of an innovative approach called e-science. This term was first used by John Taylor, a chairman of the UK Department of Education and Technology in 1999. According to Taylor, the term e-Science means global collaboration in key areas of science and innovative infrastructure development that will enable such collaboration [12]. It should be noted that the researcher did not emphasize the need for the development of information and communication infrastructure, which became especially relevant. M. Riedel defines e-Science as a new area of research that is based on collaboration in key areas of science and uses innovative infrastructure to expand the capacity of scientific calculations [13]. A. Bosin, N. Dessi, B. Pes understand e-Science as a certain type of large-scale interaction within the typical scenario of an experiment that requires multiple use of data (preliminary processing, calculations, post processing and data storing) [14]. The above definitions are based on the assertions that many of the complex and topical scientific problems of our time are solved by large groups of globally distributed researchers who need an access to modern computer infrastructure for the complete, timely and qualitative support them by necessary data, computing and network resources. Today, the key areas of the use of e-Science concept, according to leading experts, are physics, astronomy, biology, medicine, computer science, etc.

In its turn, a number of domestic scientists define e-Science as a system-integrated complex of computer information, telecommunication and social and communicative technologies, which ensure the fulfillment of functions and the solution of the actual tasks of science in the information society [15].

There are several definitions of the concept of a virtual scientific team, which were formed as a result of durable professional discussions [16, 17]:

(1) The virtual team is a temporarily formed group of representatives from various scientific institutions that through the systematic use of modern information and communication technologies virtually unite for joint research and implementation of complex scientific projects. Typically, they integrate vertically in order to combine the key competencies of individual participants in such virtual geographically distributed team and organize work while solving problems as a single organizational research unit.
(2) Virtual team is an identified group of people or organizations that, based on the widespread use of information and communication technologies, integrate into joint business or research.

The virtual team usually creates a separate virtual organization to achieve a certain goal. A virtual organization is a continuously evolving entity, redefines the direction of its work to achieve the practical goals of a particular business, combining geographically distant employees [16].

The main task of a virtual team (organization) is to provide system-integrated, purposeful management of its functioning, which is aimed at achieving the goal. The directions of activity of virtual organizations can be the most diverse: production, marketing, research and development, etc. In this paper, a virtual organization is considered that deals with scientific research, and it requires for organizing its work to create a qualitative social and communicative environment based on information, communication and computer resources that are geographically distributed and have a diverse organizational and administrative subordination.

One of the essential components of the successful implementation of projects based on the e-Science infrastructure and the achievement of the set goals in the integrated scientific projects is an effective management of e-science teams and the quality implementation of communication processes in the research teams. The task of the leader of the e-Science project is to support synchronous work of a creative team, which is to provide, develop, implement and use large-scale technological systems and relevant infrastructures, which include numerous teams of highly specialized researchers, long-term cooperation in a dynamic environment [3], and ensure an effective implementation of large-scale and correct information and communication processes.

The daily interaction of all participants in large, distributed multidisciplinary projects implemented on the e-Science platform faces with a number of problems, the main reason is the lack of effective means and tools that would ensure communication in the team and quick access to geographically distributed expensive information, communication and computer resources. One of the possible tools that would implement this feature is a recommender system developed by the authors.

4 Information Assistant of a Scientist

4.1 Recommender Systems

Recommender system is an intelligent information system to form recommendations for the sequence and the list of possible user actions in the process of solving a particular problem [18].

These systems provide a solution to a number of problems such as:

- ensuring quick access to information resources and information and analytical services;
- support and formation of thematic databases.

The main task of recommender system is to provide users with personalized information online products or online recommendation services from the growing information flow and the improvement of data management. The researchers acknowledge that the recommender systems are inquiring in the fields of business, administration, education and others. This is confirmed by a number of recent successful recommender systems that are used in a wide range of real-world tasks.

"Information assistant of a scientist" [19], that belongs to the class of recommender systems, allows participants in the virtual scientific team without resorting to additional programming to perform the following tasks: editing the page content, including add or delete graphics; adding new pages; changing the structure of the document and various data; setting up registration forms; managing the surveys, polls and forums; statistics of visits; distributing the human resource management among users. A system with such functional capabilities is best suited for providing information support of a virtual creative team.

The main functionality of the recommender system can be formulated as the following:

- to create user profile in the database;
- to fill the consolidated database with documents that are relevant to the subject area of a project;
- to form metadata and abstracts;
- to search documents in the database according to different criteria;
- to generate recommendations according to the rating of information needs.

4.2 Consolidated Information Resources of the Scientific Project

The first step in the functioning of a recommender system of a virtual creative team is the formation of a problem-oriented consolidated information resource. Members of virtual scientific teams need quick information obtaining in a convenient form of submission and place where the researcher is staying at the moment. This functionality can be provided by ensuring each researcher with on-line access to large amounts of data that must be properly processed in advance, which is achieved through the formation of consolidated information resources.

According to the definition, the concept of "consolidated information" is interpreted as received from different sources and systematically integrated multi-type information resources, which together have features of completeness, integrity, consistency and constitute an adequate information model of the problem area for its analysis, processing and effective use in the decision-making processes [20]. The formation of a consolidated information resource for a virtual research team is the basis for effective research. The formation of a consolidated information resource has several stages: the search and data collection; preliminary processing and structuring of data (data transformation into information); information analysis and synthesis – transforming it into knowledge; consolidation of information resources; creation of a series of information products to meet the information needs of potential consumers.

The information arrays proposed to a user-scientist should be accompanied by a system of user search and provide an interactive study of ontologies to facilitate refinement of search queries.

Storing the required information, including scientific articles on relevant topics conducted by a system administrator or registered users (Fig. 1). Each user can fill up the database with publications and sources that are relevant to this subject area. The efficiency of this system is determined by saving not own materials but their attributes.

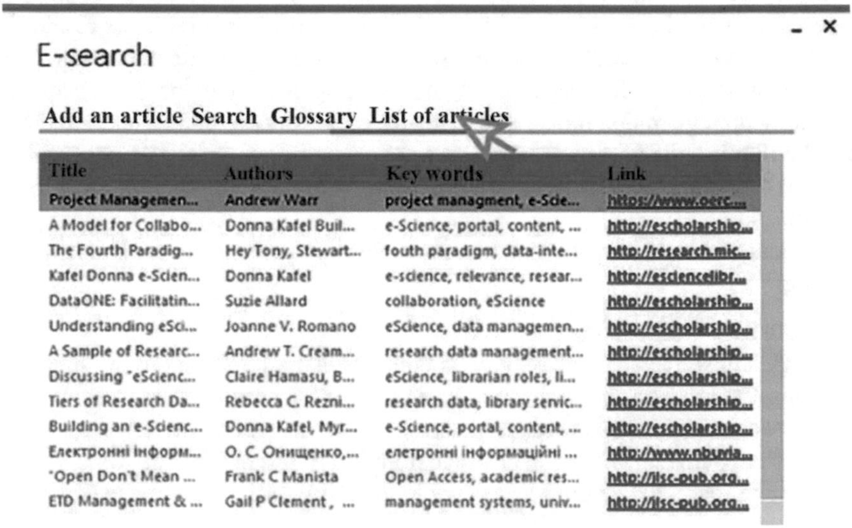

E-search

Add an article Search Glossary List of articles

Title	Authors	Key words	Link
Project Managemen...	Andrew Warr	project managment, e-Scie...	https://www.oerc...
A Model for Collabo...	Donna Kafel Buil...	e-Science, portal, content, ...	http://escholarship...
The Fourth Paradig...	Hey Tony, Stewart...	fouth paradigm, data-inte...	http://research.mic...
Kafel Donna e-Scien...	Donna Kafel	e-science, relevance, resear...	http://esciencelibr...
DataONE: Facilitatin...	Suzie Allard	collaboration, eScience	http://escholarship...
Understanding eSci...	Joanne V. Romano	eScience, data managemen...	http://escholarship...
A Sample of Researc...	Andrew T. Cream...	research data management...	http://escholarship...
Discussing "eScienc...	Claire Hamasu, B...	eScience, librarian roles, li...	http://escholarship...
Tiers of Research Da...	Rebecca C. Rezni...	research data, library servic...	http://escholarship...
Building an e-Scienc...	Donna Kafel, Myr...	e-Science, portal, content, ...	http://escholarship...
Електронні інформ...	О. С. Онищенко,...	елетронні інформаційні ...	http://www.nbuv.gov.ua...
"Open Don't Mean ...	Frank C Manista	Open Access, academic res...	http://lisc-pub.org...
ETD Management & ...	Gail P Clement , ...	management systems, univ...	http://lisc-pub.org...

Fig. 1. List of articles

The system provides the formation of a database that stores document metadata: title, authors, abstract, keywords, and hyperlinks to the full text document in free access. It ensures minimization of the resources needed for the system work, and enables the avoidance of misunderstandings in reference to copyright represented in search results of articles. It ensures minimization of the resources required for the system work, and enables the avoidance of potential misunderstandings in reference to copyright in the articles included in the lists generated by the recommender system. It facilitates the processing of multilingual sources of use of diverse dictionaries.

In addition to the database, which stores articles that are relevant to the users queries, the system contains an interpretative (in Ukrainian and English) dictionary of the terms in the e-science field, the definitions of them are automatically provided to the user when the term is included in the keyword field to search. Users are also given the opportunity to supplement the system with new terms with author's interpretation.

4.3 Formation of User Profiles

The next step in the functioning of the recommender system involves the formation of user profiles.

The rapid growth of volumes and speed of formation of information flows set new requirements for technological instruments that provide the implementation of information communication among members of virtual scientific teams. The wide technological capabilities of the worldwide web along with the implementation of functions for the real transformation of information resources into a digital form, the rapid development of intellectual capabilities of search engines, affect in general the essential nature of the information needs of scientists and the level of technology for their satisfaction. In the recommender systems, the member of a virtual team will register by entering an email address and a convenient password (Fig. 2). The registration procedure involves submitting information about the user that will be used during generating one or another recommendation. One of the key elements of the registration procedure is to identify

Fig. 2. Interface for user registration

the scope of the interests of a scientist and the subject area of his research, the system of concepts and the relevant keywords. This ensures to avoid the problem of "cold start" by a recommender system, as it guarantees at least one rating of the document for the formation of system recommendations the rated lists [21]. In this case, the authors take into account the possibility of data processing in situations that are referred to as the appearance of an "unusual user". Its essence is based on the user inherent of interdisciplinary information needs that do not allow unambiguously classifying his needs in generating relevant recommendations [22].

4.4 Generation of Recommendations

The next step is to generate recommendations, which experts often call the step of similarity determination.

The system proceeds in its actions on the similarity (distinction) of the information needs of users and the consolidated information resources available in the database of metadata documents. In the system, they differ in the profiles of information needs of users. Search of information resources at the user request involves paralleling the process of searching for metadata and full-text documents. The process for generating recommendations for user queries is shown in Fig. 3. The ranking and prioritization of the generation of the information presentation sequence on a user's request is determined by the data contained in the user's information profile. In this case, the search processes in the database and the generation of recommendations do not affect the procedures for updating the database.

The recommender system developed by the authors ensures the implementation of the function of information filtering and building ratings of information objects offered to the user. The system generates a database containing one-way metadata, obtained from some members of the research team and forms recommendations on its basis to other members of the team using the so-called graph of their interests. Unlike search algorithms, the recommender system generates personalized metadata lists that are relevant to information requests.

The recommender system "Information assistant of a scientist" provides collaborative filtration, in particular in accordance with the classical model of user-item, based on the user's behavior, embodied in his information profile.

The search follows the key words in the abstract, the article title and the article text presented in open sources and based on the database of "Information Assistant of a Scientist". Search results are displayed in the program window. The recommendations generated by the system are sent to the email address of registered user.

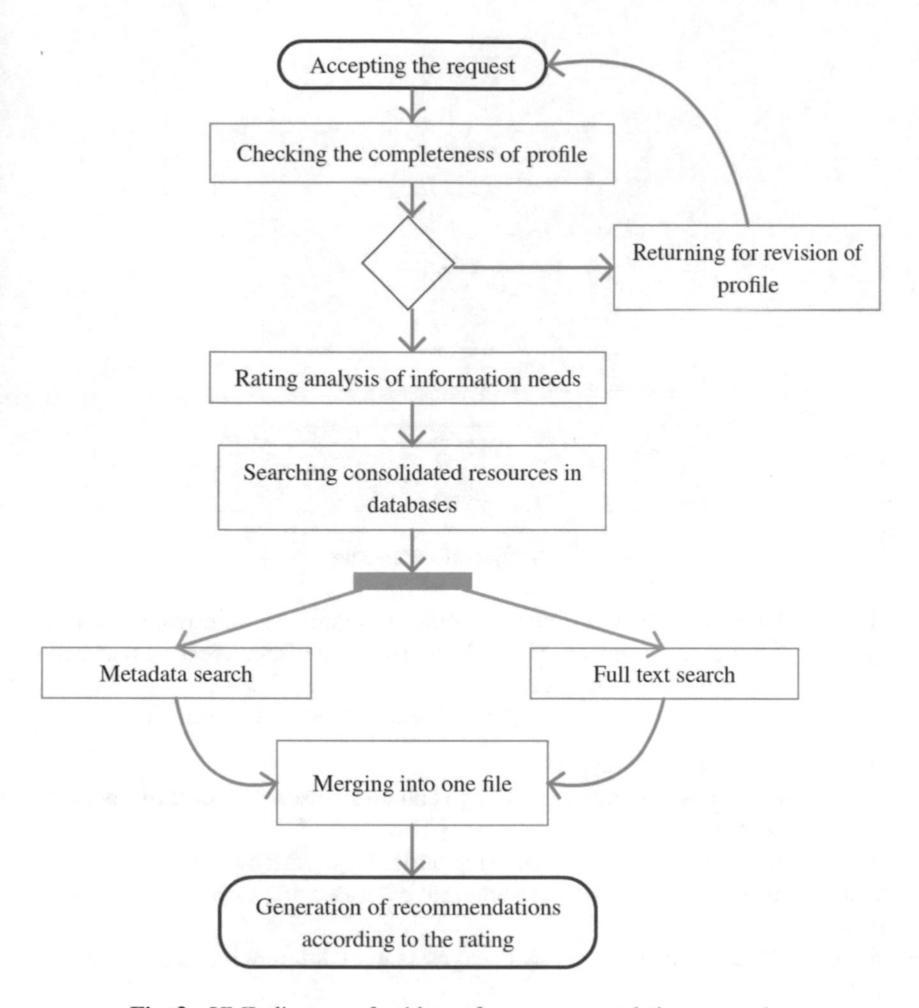

Fig. 3. UML diagram of guidance from recommendation generation

4.5 The Main Characteristics of the System

It is used an object oriented programming language C # to develop an "Information assistant of a scientist". It is chosen an integrated development environment of software systems – MS Visual Studio to implement the software solution, as this product allows through the use of web applications with a graphical interface to create the convenient user interface. The software system is developed using three-tier client-server architecture (Fig. 4). It is used the technology WCF to implement the service part. When designing the user interface technology, it is used Windows Forms. To work with the databases, it is used MS SQL Server and language T-SQL.

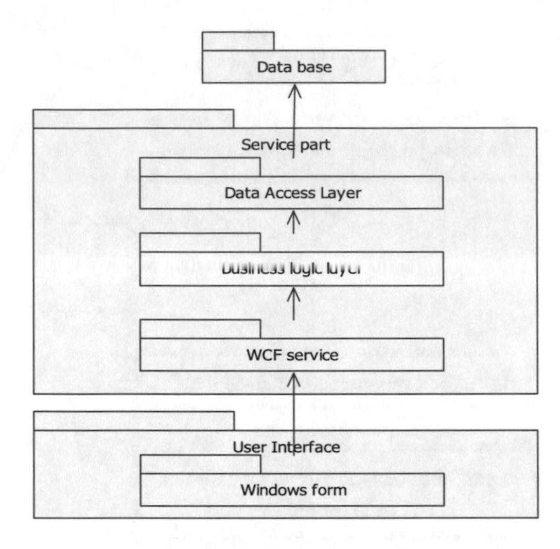

Fig. 4. System architecture

The developed software is designed for efficient search and information storage for the purpose of information support of members in the virtual creative team that conducts research on the platform of e-science.

The main advantages of "information assistant of a scientist" are:

- simplicity and convenience of use;
- system specialization, focused on storage and information retrieval for members of virtual creative team;
- speed, which is provided by search using attributes described above;
- avoiding the misunderstanding of copyright by preserving only links to documents in the system;
- simple and accessible interface, aimed at users with different levels of technological skills;
- the possibility of using by a wide range of users maintaining a number of dictionaries;
- improving performance and reliability through the development of common infrastructure that is provided by technology WCF.

5 Conclusion

The paper presents the main problem tasks concerning the information communications of the participants of scientific virtual teams that conduct research on the platform of e-science. An analysis of the use of recommender systems in the formation of information and communication processes in virtual creative teams is carried out. The main approaches to the formation of recommendations are investigated as well as the use of the collaborative filtration method for the generation of recommendations by the intellectual system. It is developed the recommender system "Information Assistant of a Scientist" which implements processes of effective information support of members in virtual scientific teams. The system functionality is analyzed, the algorithms of its work and the processes of recommendation generation are given. The recommender system "Information Assistant of a Scientist" is implemented in the social and communicative environment of an active virtual research laboratory.

References

1. Taylor, I.J., Deelman, E., Gannon, D.B., Shields, M.: Workflows for e-Science, p. 526. Springer-Verlag, London (2007)
2. Deelman, E.: Pegasus: a framework for mapping complex scientific workflows onto distributed systems. Sci. Program. J. **13**(3), 219–237 (2005)
3. Spencer, D., Zimmerman, A., Abramson, D.: Special theme: project management in e-science: challenges and opportunities. Comput. Support. Coop. Work **20**(3), 155–163 (2011)
4. Warr, A.: Project Management in e-Science. A report from the 'Embedding e-Science Applications: Designing and Managing for Usability' project (EPSRC Grant No: EP/D049733/1). https://www.oerc.ox.ac.uk/sites/default/files/uploads/ProjectFiles/FLESSR/HiPerDNO/embedding/Project%20Management%20Report.pdf
5. Zhurovskii, M., Petrenko, A.: E-science towards a semantic grid. Part 1: Combining WEB and GRID technologies. System research and information technologies, no. 1, pp. 26–38 (2010)
6. Darch, P., Turilli, M., Lloyd, S., Jirotka, M., de la Flor, G.: Communication and Collaboration in e-Science Projects. https://www.oerc.ox.ac.uk/sites/default/files/uploads/ProjectFiles/FLESSR/HiPerDNO/Comm_-_collaboration%2030%20June.pdf
7. Jirotka, M., Lee, C.P., Olson, G.M.: Supporting scientific collaboration: methods, tools and concepts. Comput. Support. Coop. Work **22**(4–6), 667–715 (2013)
8. Belloum, A.: Collaborative e-science experiments and scientific workflows. IEEE Internet Comput. **15**(4), 39–47 (2011)
9. Sonnenwald, D.H., Whitton, M.C., Maglaughlin, K.L.: Evaluating a scientific collaboratory: results of a controlled experiment. ACM Trans. Comput. Human Interact. (TOCHI) **10**(2), 150–176 (2003)
10. Sonnenwald, D.H., Whitton, M.C., Maglaughlin, K.L.: Designing to support situation awareness across distances: an example from a scientific collaboratory. Inf. Proc. Manag. **40**(6), 989–1011 (2004)
11. Darch, P.: Shared Understandings in e-Science Projects. Technical report, Oxford e-Research Centre, Oxford University. https://www.oerc.ox.ac.uk/sites/default/files/uploads/ProjectFiles/FLESSR/HiPerDNO/embedding/Shared_Understanding%2030%20June.pdf
12. Defining e-Science. http://www.nesc.ac.uk/nesc/define.html
13. Riedel, M.: Classification of different approaches for e-science applications in next generation computing infrastructures. In: Proceedings of the e-Science Conference, Indianapolis, Indiana, USA (2008)
14. Bosin, A., Dessì, N., Pes, B.: Extending the SOA paradigm to e-Science environments. Future Gener. Comput. Syst. **27**, 20–31 (2011)
15. Veretennikova, N., Pasichnyk, V., Kunanets, N., Gats, B.: E-Science: new paradigms, system integration and scientific research organization. In: Computer Science and Information Technologies: Proceedings of the Xth International Scientific and technical Conference CSIT, pp. 76–81. Lviv Polytechnic Publishing House, Lviv (2015)

16. Serrano-Guerrero, J., Herrera-Viedma, E., Olivas, J.A., Cerezo, A., Romero, F.P.: A google wave-based fuzzy recommender system to disseminate information in university digital libraries 2.0. Inf. Sci. **181**, 1503–1516 (2011)
17. Creamer, A.T., Morales, M.E., Kafel, D., Crespo, J., Martin, E.R.: A sample of research data curation and management courses. J. eSci. Librarianship (2012). http://escholarship.umassmed.edu/jeslib/vol1/iss2/4/
18. Ricci, F., Rokach, L., Shapira, B., Kantor, P.B.: Recommender Systems Handbook, p. 875. Springer, Boston (2011)
19. Certificate of registration of copyright № 68261. Computer program "Information Assistant of a scientist" (in Ukrainian)
20. Kunanets, N.E., Pasichnyk, V.V.: Introduction to the specialty: "Consolidated Information", p. 196. Lviv (2010)
21. Schein, A.I., Popescul, A., Ungar, L.H., Pennock, D.M.: Methods and metrics for cold-start recommendations. In: Proceedings of the 25-th Annual International ACM SIGIR Conference on Research and Development in Information Retrieval, pp. 253–260. ACM Press, New York, Finland (2002)
22. Vozalis, E., Margaritis, K.G.: Analysis of Recommender Systems (2003). http://lsa-svd-applicationfor-analysis.googlecode.com/svnhistry/r72/trunk/LSA/Other/LsaToRead/hercma2003.pdf

Analytical Model for Availability Assessment of IoT Service Data Transmission Subsystem

Bogdan Volochiy(✉), Vitaliy Yakovyna(✉), and Oleksandr Mulyak(✉)

Lviv Polytechnic National University, Lviv, Ukraine
bvolochiy@ukr.net, vitaliy.s.yakovyna@lpnu.ua,
mulyak.oleksandr@gmail.com

Abstract. This paper discusses the role of safety, availability, and dependability of Internet of Things (IoT) data services dedicated to the monitoring and control of objects in a physical world. These services are designed to be available to devices and users on request at any time and at any location. The Internet of Things differs from today's global Internet in a number of ways. For instance, the networks are typically unmanaged, most of IoT services are safety critical, the link layers are optimized for low power usage, and most nodes have to be implementable in a lightweight manner.

The design of IoT data services should take into account that this type of system has huge scalability (hundreds, thousands of clients). On the one hand this leads to significant financial costs for maintenance the IoT infrastructure, on the other hand, is necessary to provide the appropriated availability and safety level. In addition, today most of IoT services are operated in a critical system such as oil and gas industry, smart cities, medicine, the financial sector. According to this point an actual question of availability and safety estimations IoT services appears, because the fault of IoT system part can lead to huge financial losses or to observed objects death.

In this paper, the model of IoT services as queue network is proposed that allows estimating the availability and safety of these systems.

Keywords: Internet of Things (IoT) · Queue networks · Discrete-continuous stochastic system · Reliability behavior · Availability · Safety

1 Introduction

Nowadays the developments of IoT services [1–6] are a part of automotive, medical, smart cities, energy and other critical systems. One of the main tasks is to provide requirements of reliability, availability and functional safety. Thus the two types of possible risks related to the assessment of risk, and to ensure their safety and security.

The next wave in the era of computing will be outside the realm of the traditional desktop. In the Internet of Things (IoT) paradigm, many of the objects that surround us will be on the network in one form or another. Generally speaking, IoT refers to the networked interconnection of everyday objects, which are often equipped with ubiquitous intelligence. The term Internet of Things was first coined by Kevin Ashton in 1999

N. Shakhovska and V. Stepashko (eds.), *Advances in Intelligent Systems and Computing II*, Advances in Intelligent Systems and Computing 689,
https://doi.org/10.1007/978-3-319-70581-1_41

in the context of supply chain management [7]. However, in the past decade, the definition has been more inclusive covering a wide range of applications like healthcare, utilities, transport, etc. [8]. Although the definition of 'Things' has changed as technology evolved, the main goal of making a computer sense information without the aid of human intervention remains the same. IoT increases the ubiquity of the Internet by integrating every object for interaction via embedded systems, which leads to a highly distributed network of devices communicating with human beings as well as other devices. Thanks to rapid advances in underlying technologies, IoT is opening tremendous opportunities for a large number of novel applications that promise to improve the quality of our lives. In recent years, IoT has gained much attention from researchers and practitioners from around the world [9, 10]. Actually, many challenging issues still need to be addressed and both technological, as well as social knots, have to be untied before the IoT idea being widely accepted. Central issues are making a full interoperability of interconnected devices possible, providing them with an always higher degree of smartness by enabling their adaptation and autonomous behavior, while guaranteeing trust, privacy, and security [10].

The goal of this paper is to suggest a technique to develop a Markovian chain for critical IoT system using the proposed formal procedure and tool. The main idea is to decrease risks of errors during development of Markovian chain (MC) for systems with very large (tens and hundreds) number of states. We propose a special notation which allows supporting development chain step by step and designing final MC using software tools. The paper is structured in the following way. The aim of this research is calculating the availability function of critical IoT system. To achieve this goal we propose a newly designed reliability model of critical IoT system. As an example, a critical IoT system is researched (Fig. 1).

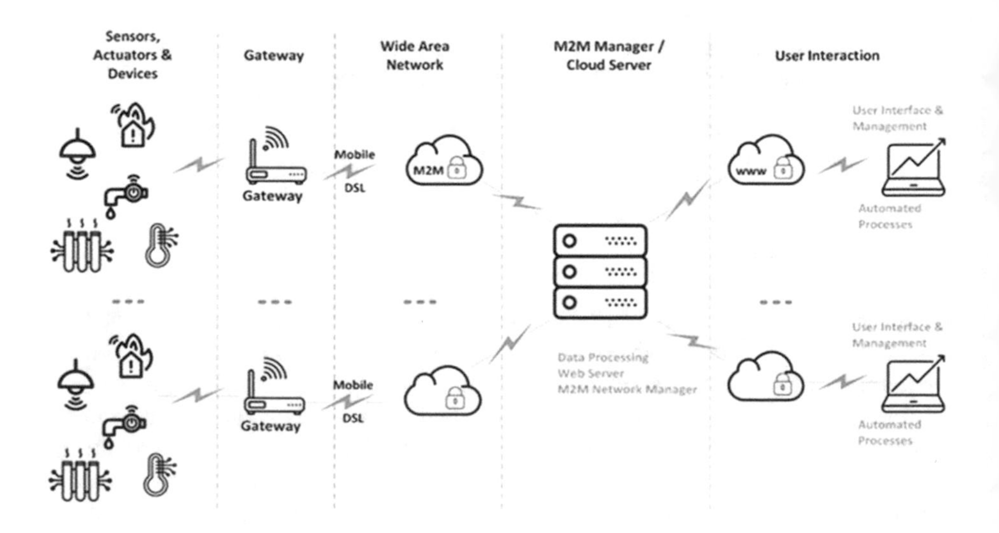

Fig. 1. Typical structure of IoT service with n-clients

2 Typical IoT Systems and Their Usage

The Internet of Things is a concept in which the virtual world of information technology integrates seamlessly with the real world of things. The real world becomes more accessible through computers and networked devices in business as well as everyday scenarios. With access to fine-grained information, management can start to move freely from macro to micro levels and will be able to measure, plan and act accordingly. However, the Internet of Things is more than a business tool for managing business processes more efficiently and more effectively – it will also enable a more convenient way of life [11]. Last years the IoT systems market is greatly growing and used in many sectors of people life.

IoT for manufacturing: Predictive maintenance brings Sandvik to the cutting edge of digital manufacturing [12], Predictive analytics help optimize Rolls-Royce airplane engine performance [13], IoT solutions fuel oil and gas industry success at Rockwell Automation [14], Jabil uses Microsoft IoT to create a digital, intelligent, and predictive factory [15].

IoT for smart cities: Johnson Controls creates connected buildings for a greener planet [16]; BCA partners Microsoft to leverage IoT, data analytics and the cloud for next-generation Green Mark buildings [17]; Ecolab solves global water challenges with the help of Microsoft IoT [18].

IoT for transportation: Fathom's IoT-enabled Weather Cloud enhances driver safety during inclement weather [19]; Revolutionizing air-traffic control at NAV CANADA using IoT [20].

IoT for retail: Coop Italia develops the supermarket of the future with Microsoft IoT [21]; Mondelez builds an intelligent, interactive snack food experience [22]; Hershey creates IoT solutions for sweeter success [23].

IoT for health care: Weka builds a smart fridge to help save lives [24]; Advancing hospital hand hygiene practices through IoT solutions [25]; Liebherr unveils intelligent pharmaceutical fridges with IoT technology [26].

This list of usage the IoT system confirms that these systems are critical and have a huge impact on people living and perform more critical and responsible functions. All of this is increase requirements to IoT system dependability and safety. Here we provide the typical structure (Fig. 1) of researched IoT system, that lets to manage and monitoring the numbers of clients on the one platform. In general is possible to divide IoT system to few levels, such as:

Level 1. It is the lowest level which includes the devices (sensors, actuators, and other). These devices are operated in autonomous mode and receiving the information about the condition of monitoring object. All devices which operated in this level are hardware/ software system that usually is unmaintained and without software updates. However is possible devices restart to return it for operational conditions after temporary software failure which instigated by hardware failure. In some cases is possible to replace devices but this lead to increase the total system cost, therefore this device designed in a fault-tolerant manner to achieve huge reliability level. Usually, devices on this level operated in 24/7 mode and communicate with gateway by Wi-Fi channel.

Level 2. Hardware/software gateway level which provides the communications of all first levels devices on the one network, managements (hub function) and communicates the first levels devices with a cloud server. Communications with cloud server can occur by mobile (2G, 3G, 4G, 5G), DSL and other connections. This gateway is usually hardware/software devices with the possibility of software restart and updates. In addition, the gateway should process the incoming data stream from first levels devices and reliable and safety transfer this data to the cloud server.

Level 3. Cloud Server which provides receiving, processing and manages data from all clients. In the design of IoT system should consider that system would operate when all clients would have a reliable and stable connection to the cloud server. To ensure a high level of reliability in some cases is required to use two or more cloud servers. The cloud server is a combination of hardware, software parts, and communications network. At this point of reliability, the cloud servers have structural redundancy which allows a high level of reliability provided. The software in cloud servers has a large variety: networks firmware, operational systems, and services software with the possibility of version redundancy. In such systems is possible to repair elements of services without it operational losses through redundancy, also software restart and updates.

Level 4. End user level (Desktop and mobile app.). This level is hardware/software platform with a weak spot in software. This software is constantly changing according to each IoT services and client requirements.

One of based technology which was used in solutions of this class is M2M (Machine-to-Machine) [27]. M2M, the acronym for Machine-to-Machine Applications (or more aptly, Mobile to Machine and Machine to Mobile communications) is an emerging area in the field of telecom technologies. This technology allows simple, reliable and profitable the data transfer between different devices provides. M2M allows a wide variety of machines to become nodes of personal wireless networks and provide to develop monitoring and remote control applications. This will decrease costs for involved human resources and will make machines more intelligent and autonomous. The M2M can operate by stationary and mobile mode. The stationary mode was used for manage the technology process, security monitoring, payment terminals, meters, vending machines and other. Mobile mode of M2M lets you moving objects management (vehicle), onboard devices, navigations, diagnostics, positions, and protection. Using such system allows collecting data about monitoring objects, conduct an analysis of this data and provide business recommendations regarding operation or modernization of objects.

3 An Approach to Development of the Availability Model for IoT Services

To solve this problem one should develop a mathematical representation of IoT system that will allow evaluating the safety, reliability and determining the requirements of each of its components at the design phase.

As was mentioned in the description of the typical IoT system structure, important tasks that arise when addressing the design and analysis of safety and reliability of IoT system are:

1. Reliability synthesis of the system and its components using structural-automaton model (SAM).
2. Considering the performance reliability of software development at SAM.
3. Considering the processes of reloading and updating software in the reliable behavior of IoT system. Herewith, to improve the authenticity of reliability indicators, in reliability model you should take into account the phenomenon of an imperfect fix of software defects after replacing it with the new version.
4. In the evaluation of the availability and safety of IoT systems, you should take into account the ability of routers and cloud services to handle the flow of applications from the device of the IoT first level architecture (sensors and actuators). To solve this problem, you can use the formalism of the queuing theory.
5. Evaluation of the safety and availability of IoT systems cannot be authentic without consideration of communication channels reliability.

Figure 2 shows a model of the overall IoT structure as a queuing system, with multiple queues and system maintenance. Specifically, the inputs of service 1 … n (Service Facility 1 … n) (software and hardware router) via a wireless communication channel receives messages from first level devices with the intensity of Arrival rate 1 … m. In service systems, the adaptive service discipline with assigning priorities to critical messages for a maximum speed of transmission is implemented. The next step

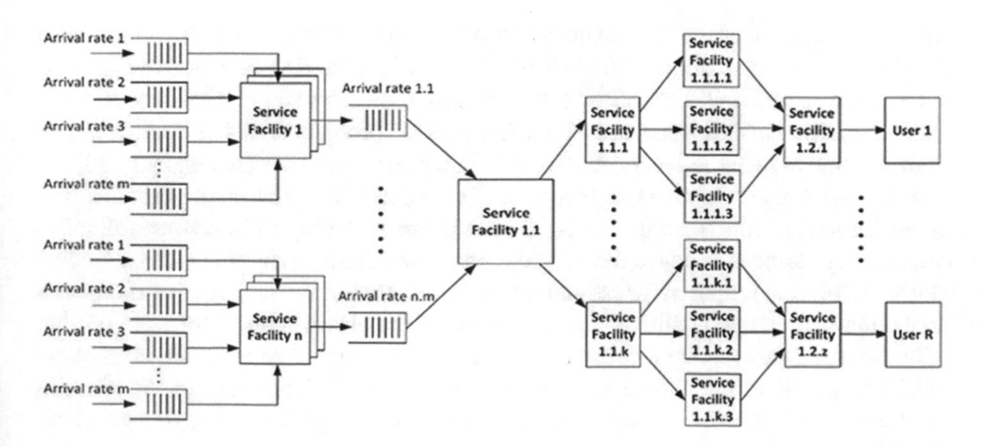

Fig. 2. Queue networks of the IoT service

is forming queues messages that represent communication links between routers and cloud server. These message queues with the intensity of Arrival rate 1.1…n.m are transmitted via the communication channel (Mobile (2G, 3G, 4G), DSL) to the service system 1.1 (Service Facility 1.1) (cloud service), in which they are pretreated. The user through the service system 1.1.1 … 1.2.z gets the access to the cloud service. This access provides sophisticated telecommunications network.

Suggested mathematical model of the IoT service should provide answers to the following questions:

1. What is the availability coefficient of the system? The cost of maintenance and development of the system will depend on the required level of availability.
2. Which frequency of messages should be from the first level devices, to not to "overwhelm" a communication channel with the router and at the same time to react in time on an event that actually leads to the need to solve the optimization problem?
3. Which communication channel (in the case of mobile communication, what traffic is optimal?) has to be chosen between the second and third levels of the system to provide the reliable (lossless) and timely data transmission.
4. How should the cloud server system be built in order to provide a given level of reliability?
5. What should be the permissible intensity of software failures (defects) on the fourth level of service to ensure the required level of reliability of the system as a whole?

Thus, further research will be focused on building reliability models of the IoT system and its components. In these models, the following aspects should be considered: software reliability indices, processes of its restart and update, the data flow processing by routers and cloud services, the reliability of data transmission channel. The aim of this research is the evaluation of safety and availability level of the IoT system.

4 Forecasting of Software Failure and Failure Rate for IoT Transmission Subsystem

All IoT systems consist of two major components: hardware and software (firmware). Software reliability is different from hardware reliability in a sense that software does not wear out or burn out. The software itself does not fail unless flaws within the software result in a failure in its dependent system. The greatest problem facing the industry today is how to assess quantitatively software reliability characteristics (see e.g. [29, 30] and others). This is also of high importance to the future networks, in particular to both cloud-based services and the software running on IoT devices. Research on software reliability engineering has been conducted during the past three decades and numerous statistical models have been proposed for estimating software reliability. Most existing models for predicting software reliability are based purely on the observation of software product failures where they require a considerable amount of failure data to obtain an accurate reliability prediction. Some other research efforts recently have developed reliability models addressing fault coverage, testing coverage, and imperfect debugging processes [28–30].

This paper outlines the software that is operated on IoT devices (sensors, actuators) and gateways. These two types of software have the difference between functionality and structural complexity, and software updates capabilities.

The IoT devices firmware is less complicated than gateways firmware and usually is not updated. The main task of estimating the IoT devices firmware is to calculate the residual numbers of firmware bugs based on testing results.

For solving this task, we propose to use the nonhomogeneous Poisson process (NHPP) in a software reliability growth model with an index of complexity [31].

The group of models based on NHPP have analytical results for descriptions of the software failure behaviors during the testing. The main problem of this sort of software reliability models is the determining of software failures mean value function. A lot of models belonging to this group have been developed for the latest four decades including exponential Musa model [32], Goel and Okumoto NHPP reliability growth model [33], Yamada and Ohba S-shaped reliability growth model [34] etc. The nonhomogeneous Poisson distribution works well in different areas where the subject of interest is the quantities of independent random events [35].

The main assumption of the NHPP software reliability growth models is that $M(t)$, failures count during the time interval $(0, t]$, belongs to the Poisson distribution, and $\{M(t), t \geq 0\}$ corresponds to nonhomogeneous Poisson process. Let $\mu(t) = E[M(t)]$ is a mathematical expectation of $M(t)$ or a mean value of Poisson process. Different models of this group differs by the $\mu(t)$ function expression. Wide NHPP software reliability models exploitation, among others, is caused by the ability to calculate the expected amount of software failures at any time point due to explicit definition of the mean value function; ease to obtain point estimations of model parameters using maximum likelihood or mean squares methods.

The NHPP reliability models with the following software failure rate form were developed in [31]:

$$\lambda(t) = \alpha\beta^{s+1}t^{s}\exp(-\beta t), \tag{1}$$

here α is the coefficient, which defines the expected number of failures to be observed eventually, β is the fault detection rate per fault ($\beta > 0$), s is the dynamic software complexity index.

With given failure rate function (1) the expected number of failures observed by time t has the following form:

$$\mu(t) = \int_{o}^{t} \lambda(\tau)d\tau = \alpha\left[-\beta^{s}t^{t}e^{-\beta t} + s\Gamma_{\beta t}(s)\right], \tag{2}$$

here, $\Gamma_{z}(\mathrm{p}) = \int_{0}^{Z} t^{p-1}e^{-t}dt$, ($\mathrm{Re}\, p > 0$), is the incomplete gamma-function.

The expected number of faults to be eventually detected is defined as the expected number of failures to be observed when $t \to \infty$, thus it has the form:

$$\mu(\infty) = \alpha s\Gamma(s), \tag{3}$$

where $\Gamma(s)$ is the gamma-function.

To use the model for the reliability behavior description of concrete software package using Eqs. (1) and (2) one has to obtain the point estimations of the model parameters. Maximum likelihood method was used in [31] to perform this task.

Using the described model for software reliability indices assessment allows one significantly improve the accuracy of the software failures rate evaluation comparing to other widely used models keeping the expected number of faults value evaluation practically identical [36].

Thus, based on software failures count at the testing stage, using the Eqs. (1) and (3) one can estimate the values of software failures rate and the number of residual faults within the sensor or actuator firmware, which is used in IoT system studied.

The software part of IoT system gateways unlike the device's software usually is more complex in a functional sense, possesses higher values of complexity metrics, and usually is updated by the vendor e.g. by firmware over the air update. After an update, new firmware version will have different values of failure rate and a number of residual defects. The reliability model of the IoT system should include the mentioned facts that in turn will affect estimated risk value and functional model of the system.

Implementing the software complexity index s into the software reliability model (1) and estimation of its value ranges [37] allows to apply this model for gateways firmware failure rate and residual faults assessment. This results from the fact that the typical gateway firmware complexity does not exceed the limits of the model adequacy, and furthermore it is usually no need to consider software structure and different cases usage in such devices as IoT gateways. The latter would result in the need to apply an architecture software reliability model [38] instead.

5 Availability Model for IoT Services

To find an answer to the first questions: what is the availability coefficient of the system, we proposed the availability model of IoT system. The method of automated development of the Availability Model in Markovian chain form of the researched IoT systems is described in the monograph [39]. It involves a formalized representation of the study object as a "structural-automated model". To develop this model of the IoT systems one needs to perform the following tasks: develop a verbal description of the research object (Fig. 1); define the basic events; define the components of vector states that can be described as a state of random time; define the parameters for the object of research that should be in the model; and shape the tree of the modification of the rules and component of the vector of states.

5.1 A Set of Events for the IoT Services

As a result of structure analysis, ten basic events, in particular, were determined: **Event 1** – "Hardware/Firmware failure of sensor/actuator/device"; **Event 2** – "Hardware/ Firmware failure of gateway"; **Event 3 –** "Hardware failure of cloud server"; **Event 4** – "Software failure of cloud server"; **Event 5 –** "Hardware failure of user's device";

Event 6 – "Repair of sensor, actuator or device"; **Event 7 –** "Repair of gateway"; **Event 8 –** "Repair of hardware failure on cloud server"; **Event 9 –** "Fix software failure of cloud server"; **Event 10 –** "Repair hardware failure of user's device.

5.2 Components of the State's Vector for the IoT Services

To describe the state of the system, nine components are used: V1 – displays the current number of sensors/actuators/devices (the initial value of components V1 is equal to n); V2 – displays the current number of gateways (the initial value of components V2 is equal to m); V3 – displays the number of user's devices (the initial value of components V3 is equal to k); V4 – displays the number of cloud servers (the initial value of components V4 is equal to 1); V5 – displays HW fault on cloud server; V6 – displays SW fault on cloud server; V7 – displays the number of non-operational units due to fault of sensor, actuator or device; V8 – displays the number of non-operational units due to fault of gateway; V9 – displays the number of non-operational units due to HW fault of user's device.

5.3 Sets of Parameters Which Describe the IoT Services

Developing the Markov model of the IoT Services, its composition and separate components should be set to relevant parameters in particular: n – number of sensors/actuators/devices; m – number of gateways; k – number of user's devices; λ_{hw0} – the failure rate of sensor, actuator or device; $\lambda_{firmware0}$ – the firmware failure rate of sensor, actuator or device; λ_{hw1} – the failure rate of gateway; $\lambda_{firmwarer1}$ – the firmware failure rate of gateway; λ_{hw2} – the failure rate of cloud server hardware; λ_{hw3} – the failure rate of user's device hardware; λ_{sw} – the failure rate of cloud server software; T_{rep0} – the duration of sensor, actuator or device repair; T_{rep1} – the duration of gateway repair; T_{rep2} – the duration of cloud server hardware repair; T_{rep3} – the duration of user's device hardware repair; T_{sw} – the duration of software failure fix.

5.4 Structural-Automated Model of the IoT Services for the Markovian Chain Development

According to the technology of analytical modeling, the discrete-continuous stochastic systems [39] based on certain events using the component vector state and the parameters that describe IoT Services, and model of the IoT Services for the automated development of the Markovian chains are presented in Table 1.

Proposed structural-automated model is the input sets for cutting-age problem-oriented software tool ASNA. Using this technology we automatically build Markovian Chain for a different structure of IoT system [38, 39]. We can easily change the number of sensors, gateways, users devices and maintenance parameters. Based on a Markovian chain we shape the mathematical model in Chapmen-Kolmogorov differential equation form. A solution of this equation gives answers to the following questions: What is the Probability of Faultless? What is the mean time between failures? What is the mean time between repairs? Which fault-tolerance configurations of cloud server are better?

Table 1. Structural-automated model of the IoT System for the automated development of the Markovian chains

Terms and conditions	Formula used for the intensity of the events	Rule of modification component for the state vector
Event 1. Hardware/Firmware failure of sensor/actuator/device		
(V1 > 0) AND (V2 = 1) AND (V4 = 1) AND (V3 = 1) AND (V7 < n)	$V1 \cdot \lambda_{hw0} \cdot \lambda_{firmware0}$	V1 := V1 – 1; V2 := 1; V3 := 1; V4 := 1; V7 := V7 + 1
Event 2. Hardware/Firmware failure of gateway		
(V2 = 1) AND (V1 > 0) AND (V4 = 1) AND (V3 = 1) AND (V8 = 0)	$V2 \cdot \lambda_{hw1}$	V2 := 0; V1 := V1; V3 := 1; V4 := 1; V8 := 1
Event 3. Hardware failure of cloud server		
(V4 = 1) AND (V1 > 0) AND (V2 = 1) AND (V3 = 1) AND (V5 = 0)	$V4 \cdot \lambda_{hw2}$	V4 := 0; V1 := 0; V2 := 0; V3 = 0; V5 := 1; V7 := n; V8 := 1; V9 := 1
Event 4. Software failure of cloud server		
(V4 = 1) AND (V1 > 0) AND (V2 = 1) AND (V3 = 1) AND (V6 = 0)	$V4 \cdot \lambda_{sw}$	V4 := 0; V1 := 0; V2 := 0; V3 = 0; V5 := 1; V6 := 1; V7 := n; V8 := 1; V9 := 1
Event 5. Hardware failure of user's device		
(V3 = 1) AND (V1 > 0) AND (V2 = 1) AND (V4 = 1) AND (V9 = 0)	$V3 \cdot \lambda_{hw3}$	V3 := 0; V1 := V1; V2 := 1; V4 := 1; V9 := 1
Event 6. Repair of sensor, actuator or device		
(V1 < n) AND (V2 = 1) AND (V4 = 1) AND (V3 = 1) AND (V7 > 0)	$1/T_{rep0}$	V1 := V1 + 1; V2 := 1; V3 := 1; V4 := 1; V7 := V7 – 1
Event 7. Repair of gateway		
(V2 = 0) AND (V1 > 0) AND (V4 = 1) AND (V3 = 1) AND (V8 = 1)	$1/T_{rep1}$	V2 := 1; V1 := V1; V3 := 1; V4 := 1; V8 := 0
Event 8. Repair of hardware failure on cloud server		
(V4 = 0) AND (V1 = 0) AND (V2 = 0) AND (V3 = 0) AND (V5 = 1) AND (V7 = n) AND (V8 = 1) AND (V9 = 1)	$1/T_{rep2}$	V4 := 1; V1 := n; V2 := 1; V3 := 1; V5 := 0; V7 := 0; V8 := 0; V9 := 0
Event 9. Fix software failure of cloud server		
(V4 = 0) AND (V1 = 0) AND (V2 = 0) AND (V3 = 0) AND (V5 = 1) AND (V6 = 1) AND (V7 = n) AND (V8 = 1) AND (V9 = 1)	$1/T_{sw}$	V4 := 1; V1 := n; V2 := 1; V3 := 1; V5 := 0; V6 := 0; V7 := 0; V8 := 0; V9 := 0
Event 10. Repair hardware failure of user's device		
(V3 = 0) AND (V1 > 0) AND (V2 = 1) AND (V4 = 1) AND (V9 = 1)	$1/T_{rep3}$	V3 := 1; V1 := V1; V2 := 1; V4 := 1; V9 := 0

This mathematical model gives us the opportunity to solve tasks of parametric synthesis of IoT system components.

6 Conclusion

This paper presents a motivation to research the availability and safety of IoT service. The queue networks for estimating the availability and safety of IoT service and tasks for next research were presented. The availability model was developed using the structural-automated model of studying the discrete-continuous stochastic systems. The technique of IoT systems software reliability indices assessment has been described that allows increasing the adequacy of IoT system reliability assessment. The reliability model of IoT system in the form of SAM was developed. The developed model allows carrying out the parametric synthesis of IoT system components, select an optimal maintenance strategy depending on given exploitation environments.

References

1. Atzori, L., Iera, A., Morabito, G.: The Internet of Things: a survey. Comput. Netw. **54**, 2787–2805 (2010)
2. Ma, H.-D.: Internet of things: objectives and scientific challenge. J. Comput. Sci. Technol. **26**, 919–924 (2011)
3. Miorandi, D., Sicari, S., Pellegrini, F.D., Chlamtac, I.: Internet of things: vision applications and research challenges. Ad Hoc Netw. **10**, 1497–1516 (2012)
4. Roman, R., Alcaraz, C., Lopez, J., Sklavos, N.: Key management systems for sensor networks in the context of the Internet of Things. Comput. Electr. Eng. **37**, 147–159 (2011)
5. Silva, M.D., Leandro, R., Batista, D.M., Guedes, L.A.: A dependability evaluation tool for the Internet of Things. Comput. Electr. Eng. **39**, 2005–2018 (2013)
6. Zin, T.T., Tin, P., Hama, H.: Reliability and availability measures for Internet of Things consumer world perspectives. In: 5th IEEE Global Conference on Consumer Electronics, pp. 1–2 (2016)
7. Ashton, K.: That 'Internet of Things' thing. RFID J., 1–2 (2009)
8. Sundmaeker, H., Guillemin, P., Friess, P., Woelfflé, S.: Vision and challenges for realising the Internet of Things. In: Cluster of European Research Projects on the Internet of Things —CERP IoT (2010)
9. Xia, F., Yang, L.T., Wang, L., Vinel, A.: Internet of Things. Int. J. Commun Syst **25**, 1101–1102 (2012)
10. Gubbi, J., Buyya, R., Marusic, S., Palaniswami, M.: Internet of Things (IoT): a vision, architectural elements, and future directions. Future Gener. Comput. Syst. **29**, 1645–1660 (2013)
11. Uckelmann, D., Harrison, M., Michahelles, F.: An architectural approach towards the future Internet of Things. In: Uckelmann, D., Harrison, M., Michahelles, F. (eds.) Architecting the Internet of Things, pp. 1–24. Springer, Heidelberg (2011)
12. Azure IoT Suite helps Sandvik Coromant stay on cutting edge within "digital manufacturing". https://blogs.microsoft.com/iot/2016/09/12/azure-iot-suite-helps-sandvik-coromant-stay-on-cutting-edge-within-digital-manufacturing/#08ckShkjVuPGPOBP.99
13. Rolls-Royce and Microsoft collaborate to create new digital capabilities. https://customers.microsoft.com/en-US/story/rollsroycestory
14. Fueling the oil and gas industry with IoT. https://customers.microsoft.com/en-US/story/fueling-the-oil-and-gas-industry-with-iot-1

15. How manufacturers are creating the digital, intelligent and predictive factory. https://blogs.microsoft.com/transform/2016/04/24/how-manufacturers-are-creating-the-digital-intelligent-and-predictive-factory/#sm.0000ns1xd9hpqcolwy71tyhiubvkz
16. Connecting Buildings to the Cloud for a Greener Planet. https://customers.microsoft.com/en-US/story/connecting-buildings-to-the-cloud-for-a-greener-planet
17. BCA partners Microsoft to leverage IoT, data analytics and the cloud for next-generation Green Mark buildings. https://news.microsoft.com/en-sg/2016/09/07/bca-partners-microsoft-to-leverage-iot-data-analytics-and-the-cloud-for-next-generation-green-mark-buildings/#jwEJJSbDCDrtL6fK.99
18. Ecolab uses cloud computing to save fresh water on the ground. https://blogs.microsoft.com/iot/2016/04/05/ecolab-uses-cloud-computing-to-save-freshwater-on-the-ground/#XMQgmhjSRmdtDQ7B.99
19. Fathym's IoT-enabled Weather Cloud enhances driver safety during inclement weather. https://blogs.microsoft.com/iot/2016/12/09/fathyms-iot-enabled-weathercloud-enhances-driver-safety-during-inclement-weather/#wpbM9QQvdxROoDx8.99
20. Azure IoT Technology helps NAV CANADA revolutionize air-traffic control. https://blogs.microsoft.com/iot/2016/03/17/azure-iot-technology-helps-nav-canada-revolutionize-air-traffic-control/#JvuzE3WFYvjuqU6h.99
21. Italian grocery co-op develops supermarket of the future. https://blogs.microsoft.com/iot/2016/04/08/italian-grocery-co-op-develops-supermarket-of-the-future/#4vi5fAJ5xT87wAjZ.99/
22. Immersive, interactive, intelligent: New retail experiences on display at NRF. https://blogs.microsoft.com/iot/2016/01/19/immersive-interactive-intelligent-new-retail-experiences-on-display-at-nrf/
23. Hershey enhances global brands and productivity with cloud technology. https://customers.microsoft.com/en-US/story/hershey-office365
24. IoT-enabled Smart Fridge helps manage vaccines and saves lives. https://blogs.microsoft.com/iot/2016/08/16/iot-enabled-smart-fridge-helps-manage-vaccines-and-saves-lives/
25. Advancing hospital hand hygiene practices through IoT solutions. https://www.microsoft.com/en-us/internet-of-things/customer-stories#healthcare&gojoindustries
26. Liebherr Domestic Appliances collaborates with Microsoft to build new smart fridge for medicine. https://blogs.microsoft.com/transform/2016/04/24/liebherr-domestic-appliances-collaborates-with-microsoft-to-build-new-smart-fridge-for-medicine/#sm.0000ns1xd9hpqcolwy71tyhiubvkz
27. Krishnan, V., Braswar, S.: M2M Technology: Challenges and Opportunities. Tech Mahindra (2010)
28. Pham, H.: System Software Reliability. Springer Series in Reliability Engineering. Springer, London (2006)
29. Trivedi, K.S., Bobbio, A., Muppala, J.K.: Greenbook: Reliability and Availability Engineering: Modeling, Analysis, and Applications. Cambridge University. (2017, in Press)
30. Maevsky, D.A.: A new approach to software reliability. In: Gorbenko, A., Romanovsky, A., Kharchenko, V. (eds.) SERENE 2013. LNCS, vol. 8166, pp. 156–168. Springer, Heidelberg (2013)
31. Chabanyuk, Y.M., Yakovyna, V.S., Fedasyuk, D.V., Seniv, M.M., Khimka, U.T.: Development and study the software reliability model with project size index. Software Eng., 24–29 (2010). (in Ukrainian)
32. Musa, J.D.: A theory of software reliability and its application. IEEE Trans. Software Eng. **SE-1**(3), 312–327 (1975)

33. Goel, A.L., Okumoto, K.: Time-dependent error-detection rate model for software and other performance measures. IEEE Trans. Reliab. **R-28**, 206–211 (1979)
34. Yamada, S., Ohba, M., Osaki, S.: S-shaped reliability growth modeling for software error detection. IEEE Trans. Reliab. **R-32**, 475–478 (1983)
35. Goel, A.L.: Software reliability models: assumptions, limitations, and applicability. IEEE Trans. Software Eng. **SE-11**, 1411–1423 (1985)
36. Mulyak, A., Yakovyna, V., Volochiy, B.: Influence of software reliability models on reliability measures of software and hardware systems. Eastern Eur. J. Enterp. Technol. **4**, 53–57 (2015)
37. Bobalo, Yu., Volochiy, B., Lozynskyi, O., Mandziy, B., Ozirkovskii, L., Fedasyuk, D., Shcherbovskyh, S., Yakovyna, V.: Mathematical Models and Methods for Reliability Analysis of Electrical, Electronics and Software Systems. Lviv Polytechnic Publishing House, Lviv (2013). (in Ukrainian)
38. Yakovyna, V., Nytrebych, O.: Discrete and continuous time high-order Markov models for software reliability assessment. In 11th International Conference ICTERI 2015, Lviv, Ukraine, 14–16 May 2015. CEUR-WS.org, CEUR-WS.org/Vol-1356/paper_62.pdf
39. Volochiy, B.: Technology of Modelling of Algorithms of Behavior of Information Systems. Lviv Polytechnic Publishing House, Lviv (2004). (in Ukrainian)

Building Vector Autoregressive Models Using COMBI GMDH with Recurrent-and-Parallel Computations

Serhiy Yefimenko(✉)

Department for Information Technologies of Inductive Modelling, International Research and Training Center for Information Technologies and Systems, Ave Glushkov, 40, Kyiv 03680, Ukraine
syefim@ukr.net

Abstract. The paper presents theoretical grounds of recurrent-and-parallel computing applying for modelling and prediction of complex multidimensional interrelated processes in the class of vector autoregressive models. The combinatorial GMDH algorithm is used for vector autoregressive (VAR) modelling by exhaustive search of all possible variants and finding the best model for every time series. The algorithm with selection a few best models for every process is used. The procedure of structural and parametric identification of VAR models is proposed. It is applied the combining all possible variants of systems from the selected best models and choosing the best system model according to an additional criterion. The test experiment on solving the problem of structural and parametric identification for experimental testing of algorithm efficiency was carried out. The effectiveness of constructed algorithm is demonstrated by prediction of interrelated processes in the field of Ukraine energy sphere with the purpose of effective managerial decision making.

Keywords: Multidimensional time series · VAR · Inductive modelling GMDH · Combinatorial algorithm · Recurrent-and-parallel computing

1 Introduction

The problem of mathematical modelling and prediction of complex multidimensional interrelated time series finds the application foremost in economic, ecological, sociological spheres [1–4]. In the case of prediction of a vector process presented as a set of time series (or multidimensional time series), it is advisable to use such class of models as vector autoregression [5, 6]. Granger causality has been applied to reveal interdependence structure in such kind of time series [7–9].

An approach to the problem of structural and parametric identification of this process, when parameters for every model are estimated independently, is considered. The drawback of the approach is that the parameters of separate models for interrelated processes are interdependent. The paper uses an algorithm with selection not one but a few best models for every process to eliminate the shortcoming. It is applied the combining all possible variants of systems from the selected best models and choosing the best system model according to an additional criterion.

N. Shakhovska and V. Stepashko (eds.), *Advances in Intelligent Systems and Computing II*, Advances in Intelligent Systems and Computing 689,
https://doi.org/10.1007/978-3-319-70581-1_42

We use the Group Method of Data Handling (GMDH) [10] as one of the most effective inductive modelling methods. It is widely used for solving broad spectrum problems of artificial intelligence such as identification, forecasting, recognition, clustering, and macromodelling.

When solving the problem of vector autoregressive modelling with the use of inductive modelling methods, the run-time is one of the most important criteria of efficiency. The most effective ways to achieve the high performance are recurrent parameter estimation and parallel computing.

High-performance algorithms and software tools on the basis of both recurrent and parallel computing were already developed and proved their efficiency [11–13].

The purpose of this work is to apply the technology, combining both these approaches, for effective solving the problem of vector autoregressive modelling using GMDH combinatorial algorithm.

2 Vector Autoregressive Modelling

The vector autoregressive (VAR) model generalizes the model of autoregression to multidimensional case. It is built by the stationary time series. It is the system of equations, in which every variable (component of multidimensional time series) is linear combination of all variables in the previous time points. The order of such model is determined by the order of the lags. For the simplest case of two time series with one lag the model of VAR will be as follows:

$$\begin{aligned} x_1(t) &= \theta_{11}x_1(t-1) + \theta_{12}x_2(t-1); \\ x_2(t) &= \theta_{21}x_1(t-1) + \theta_{22}x_2(t-1), \end{aligned} \tag{1}$$

where $\theta_{ij},\ i,j = 1,2$ – model parameters.

In the general case for m time series and k lags the model will be the system of m equations:

$$\begin{aligned} x_1(t) &= \theta_{11}x_1(t-1) + \ldots + \theta_{1k}x_1(t-k) + \theta_{1,k+1}x_2 \\ &\quad \times (t-1) + \ldots + \theta_{1,2k}x_2(t-k) + \ldots + \theta_{1,mk}x_m(t-k); \\ \ldots \\ x_m(t) &= \theta_{m1}x_1(t-1) + \ldots + \theta_{mk}x_1(t-k) + \theta_{m,k+1}x_2 \\ &\quad \times (t-1) + \cdots + \theta_{m,2k}x_2(t-k) + \ldots + \theta_{m,mk}x_m(t-k), \end{aligned} \tag{2}$$

or alternatively in matrix form:

$$X(t) = \sum_{j=1}^{k} \Theta_j X(t-j), \tag{3}$$

where $\Theta_j,\ j = \overline{1,k}$ – matrices of model parameters (3) of size $m \times m$.

3 Procedure of Structural and Parametric Identification of VAR Models

The general models structure in the form of a system of m difference equations is determined as a result of the sequence of such operations initially developed in [14]:

1. Data array of $m \cdot k$ arguments is composed under the number of interrelated processes m and lags k.
2. Maximal complexity for restricted search is defined [14]. COMBI algorithm with sequentially complicated structures of models on the basis of recurrent-and-parallel computing is used for modelling (see next section). For every time series the best F (by the value of the regularity criterion [10]) models are selected. Overall $F \cdot m$ models are passed to the next step.
3. The sorting of $G = F^m$ possibilities of systems of models is carried out. The best model (by the value of the systemic integral criterion of vector models quality) is selected. The value of the criterion is calculated on the given part of initial data set in the prediction mode of the process for the given steps number n_S:

$$B = \sum_{i=1}^{n_S} \sum_{j=1}^{m} (x_{ij} - x_{ij}^*)^2, \tag{4}$$

where x_{ij}^* is the result of step-by-step integration (recursive computations) of the system of m equations.

Acceptable modelling time should be taken into account when assignment the value of degree of freedom F (number of the best models being passed to the next step). Figure 1 shows experimental results for the test problem with m = 11 time series and k = 2 lags. The time of sorting of systems (and computing the value of the criterion (4)) grows exponentially.

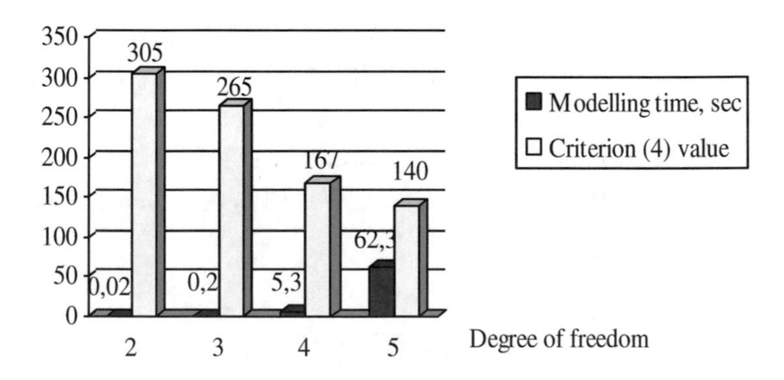

Fig. 1. Dependence of modelling time and criterion (4) value on degree of freedom

4 Recurrent-and-Parallel Computing

The exponential dependence of number of system models G on the number of interrelated processes m results in huge amount possibilities sorting and requires improvement of modelling aids. Therefore both paralleling of computing and recurrent algorithm of parameters estimation are used here. Recurrent computing increase modelling efficiency (in terms of modelling time) nearly five times [16].

According to theoretical study, efficiency of paralleling the COMBI algorithm computations is approximate to maximum possible 100% (i.e., paralleling on l processors reduces the modelling time nearly l times). The results of testing the efficiency of parallel computing are given below (Sect. 8).

Figure 2 shows approximate modeling time of exhaustive search for 40 arguments with the use of COMBI algorithm for 4 cases:

- without recurrent-and-parallel computing (I);
- with recurrent parameters estimation (II);
- with parallel computing, 100 processors (III);
- with recurrent-and-parallel computing, 100 processors (IV).

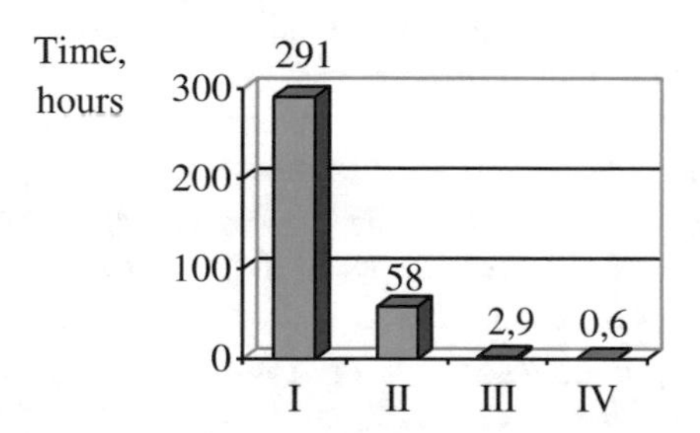

Fig. 2. Approximate modeling time of exhaustive search

5 Combinatorial GMDH Algorithm with Recurrent Parameters Estimation

The combinatorial algorithm is used for VAR modelling by exhaustive search of all possible variants and finding the best model for every time series containing the most informative subset of input arguments (regressors). It consists of such main units:

- data conversion according to a basic class of models (linear in parameters);
- forming models of different complexity;
- calculation values of external quality criteria for all models being formed;
- selection of the best models.

For linear object with m input, all possible models are compared in the process of exhaustive search. Total quantity of all generated models of the type

$$\widehat{y}_v = X_v \widehat{\theta}_v, \ v = 1, \ldots, 2^m - 1 \quad (5)$$

is 2^m–1. Decimal number v corresponds to binary number d_v in (5). Unit elements of d_v indicate inclusion regressors with corresponding numbers in the model, whereas zero elements signify exclusion.

Due to the exponential growth of 2^m as function of arguments amount, it is advisable to use algorithms recurrent in the number of parameters in structural identification problems for the parameters estimation of model structures being sequentially complicated.

Efficient recurrent modifications of classic Gauss and Gramm-Schmidt algorithms were offered in [12]. As far as the recurrent variant of Gauss method is useful for combinatorial algorithm paralleling, its short-form description is done below.

The modification, in a nutshell, is as follows. The matrix $H_s = X_s^T X_s$ of the size $s \times s$ is reduced to superdiagonal form by computing only elements $h_{i\,s}^s, \quad i = \overline{2,\ s-1}, h_{s\,i}^s, \quad i = \overline{2,\ s}$, and $g_s = X_s^T y$ at every step $s, \quad s = \overline{1,\ m}$ during the direct motion. The elements of the nested matrix H_{s-1} of size (s–1) × (s–1) (reduced to superdiagonal form on the previous step) remain changeless. So only "bordering elements" (bold fonts) are computed on step s:

$$\left[\begin{array}{ccccccccc|c}
h_{11} & h_{12} & h_{13} & \ldots & h_{1s\text{-}1} & \boldsymbol{h_{1s}} & \ldots & h_{1m} & & g_1 \\
h_{21} & h_{22} & h_{23} & \ldots & h_{2s\text{-}1} & \boldsymbol{h_{2s}} & \ldots & h_{2m} & & g_2 \\
h_{31} & h_{32} & h_{33} & \ldots & h_{3s\text{-}1} & \boldsymbol{h_{3s}} & \ldots & h_{3m} & & g_3 \\
\ldots & \ldots & \ldots & \ldots & \ldots & \ldots & \ldots & \ldots & & \ldots \\
h_{s\text{-}1,1} & h_{s\text{-}1,2} & h_{s\text{-}1,3} & \ldots & h_{s\text{-}1,s\text{-}1} & \boldsymbol{h_{s\text{-}1,s}} & \ldots & h_{s\text{-}1,m} & & g_{s\text{-}1} \\
\boldsymbol{h_{s1}} & \boldsymbol{h_{s2}} & \boldsymbol{h_{s3}} & \boldsymbol{\ldots} & \boldsymbol{h_{s,s\text{-}1}} & \boldsymbol{h_{ss}} & \ldots & h_{sm} & & \boldsymbol{g_s} \\
\ldots & \ldots & \ldots & \ldots & \ldots & \ldots & \ldots & \ldots & & \ldots \\
h_{m1} & h_{m2} & h_{m3} & \ldots & h_{m,s\text{-}1} & h_{ms} & \ldots & h_{mm} & & g_m
\end{array}\right]$$

6 Paralleling of COMBI Algorithm with Standard Binary Counter

The scheme of COMBI paralleling based on the modified recurrent Gauss algorithm with standard binary counter is described in [15].

The sequence of all possible combinations for models comprising e.g. m = 3 arguments will be as follows (with corresponding binary structural vector):

$$
\begin{array}{ll}
y_1 = a_1x_1 & \{1,0,0\} \\
y_2 = a_2x_2 & \{0,1,0\} \\
y_3 = a_1x_1 + a_2x_2 & \{1,1,0\} \\
y_4 = a_3x_3 & \{0,0,1\} \\
y_5 = a_1x_1 + a_3x_3 & \{1,0,1\} \\
y_6 = a_2x_2 + a_3x_3 & \{0,1,1\} \\
y_7 = a_1x_1 + a_2x_2 + a_3x_3 & \{1,1,1\}
\end{array}
$$

Table 1 represents approximate dependence of modelling time on arguments number and used processors for constructed algorithm. Already for more than 50 arguments, an exhaustive search (in acceptable modelling time) becomes impossible even for cluster system containing one hundred processors. Any effective reducing of exhaustive search is impossible due to the feature of the standard binary generator: complexity of structural vectors changes inconsequentially.

Table 1. Approximate time of exhaustive search

Arguments	Models	Time	
		1 processor	100 processors
20	1048575	1 s	0,01 s
21	2097151	2 s	0,02 s
…	…	…	…
40	1,1E + 12	~12 days	~3 h
…	…	…	…
50	1,1E + 15	~12 years	~124 days

Another scheme of COMBI paralleling is used for this case.

7 Paralleling of COMBI Algorithm with Sequential Binary Counter

This scheme uses such sequence of binary numbers generation when all combinations with one unit in structural vector appears first of all (totally $C_m^1 = m$ possible variants is generating), then with two units ($C_m^2 = \frac{m(m-1)}{2}$ possible variants), and so on to complete model $\left(C_m^m = 1\right)$ comprising all arguments.

The sequence of all possible combinations for models comprising three arguments will be the following:

$$
\begin{array}{ll}
y_1 = a_1x_1 & \{1,0,0\} \\
y_2 = a_2x_2 & \{0,1,0\} \\
y_3 = a_3x_3 & \{0,0,1\} \\
y_4 = a_1x_1 + a_2x_2 & \{1,1,0\} \\
y_5 = a_1x_1 + a_3x_3 & \{1,\ 0,\ 1\} \\
y_6 = a_2x_2 + a_3x_3 & \{0,1,1\} \\
y_7 = a_1x_1 + a_2x_2 + a_3x_3 & \{1,1,1\}
\end{array}
$$

The scheme can be easily used for COMBI paralleling on the given amount of processors [14].

It allows to partially solve the problem of exhaustive search when arguments number exceeds capability of the algorithm with a standard binary generator. In this case it is advisable to execute an exhaustive search not among all possible models but only for models of the restricted complexity.

8 Testing the Efficiency of the Algorithm Paralleling

The test experiment on solving the problem of structural and parametric identification for experimental testing of algorithm efficiency was carried out. The system model for $m = 11$ time series with $k = 2$ lags was built. The computing was distributed on $l = 5$ threads and carried out sequentially on PC with processor Intel Pentium M (CPU clock 1.73 GHz). Hence we have a result close to theoretical due to avoiding interprocessor interaction.

The best $F = 5$ models were selected for every time series by the value of the regularity criterion. The run-time of every thread and run-time of the whole program (without paralleling) were measured and compared. Figure 3 represents result of the experiment as a run-time diagram.

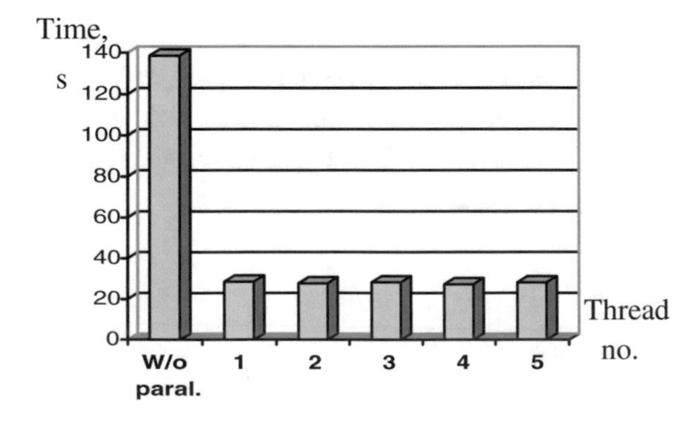

Fig. 3. Program run-time diagram

The results of the experiment may be used for calculation the efficiency of paralleling:

$$E = \frac{T_1}{l \times T_{l\max}} \times 100\% \tag{6}$$

and the uniformity of the computer load:

$$P = \left(1 - \frac{T_{l\max} - T_{l\min}}{T_{l\max}}\right) \times 100\%, \tag{7}$$

where T_1 is the program run-time for 1 thread (i.e. without paralleling), T_{lmax} is the run-time for $l = 5$ threads (maximal run-time from 5 threads), T_{lmin} is the minimal run-time.

Table 2 shows numerical values of efficiency indices for this experiment.

Table 2. Efficiency indices

Efficiency of paralleling, %	97
Uniformity of the load, %	96

Thus, according to the test experiment, paralleling the COMBI algorithm computations on l processors reduces the modelling time nearly l times.

9 Modelling and Prediction of Indices of Ukraine Energy Sphere

As an example of applying the constructed technology, an economic problem is solved with the use of vector autoregressive modelling.

In total, 4 algorithms (the classical COMBI algorithm, its recurrent modification, parallel implementation, and recurrent-and-parallel algorithm as their synergism) are compared. The main criterion is the modelling time. And these variants (in represented order) reduce the criterion value.

Ukrainian Ministry of Economy data for 11 energy sphere indices for years 1996 to 2005 (n = 10 records) are used for modelling. In addition it is known value of first 9 indices in 2006 year. The indicators are as follows:

x_1 – share of domestic supply in energy balance, %;
x_2 – share of dominant fuel resources in energy balance, %;
x_3 – share of fuel import from one country (company) in its total volume, %;
x_4 – tearing of fixed assets of fuel-energy complex (FEC) enterprises, %;
x_5 – ratio of investments amount in FEC enterprises to GDP;
x_6 – energy intensity of GDP;
x_7 – volume of coal production, million tons;

x_8 – oil transit, million tons;
x_9 – gas transit, billion cubic metres;
x_{10} – natural gas production, billion cubic metres;
x_{11} – oil and gas condensate production, million tons.

Two approaches is used for modelling: (1) traditional, when every index (i.e. single time series) is modelling independently in the class of autoregressive models; (2) modelling the vector of indices (i.e. interdependent time series) in the class of discrete dynamic VAR models.

Autoregressive modelling. Figures 4 and 5 presents modelling results and real values for x_3 and x_8. All other results have similar character. It may be concluded that autonomous autoregressive models are inappropriate for the description of experimental data (have a character of a simple shifting the curve) and cannot be used for reliable prediction.

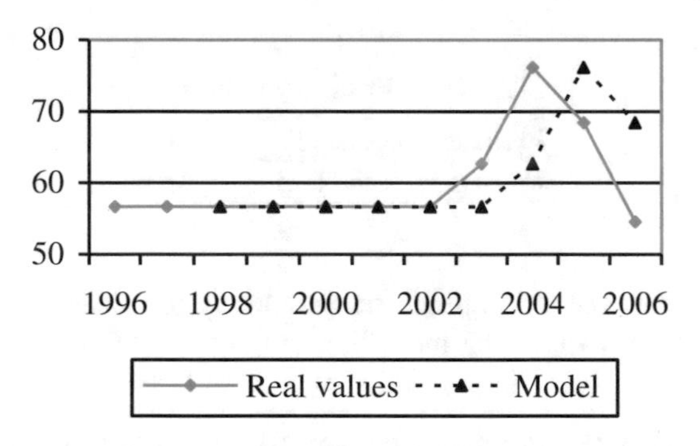

Fig. 4. Modelling results for index x_3

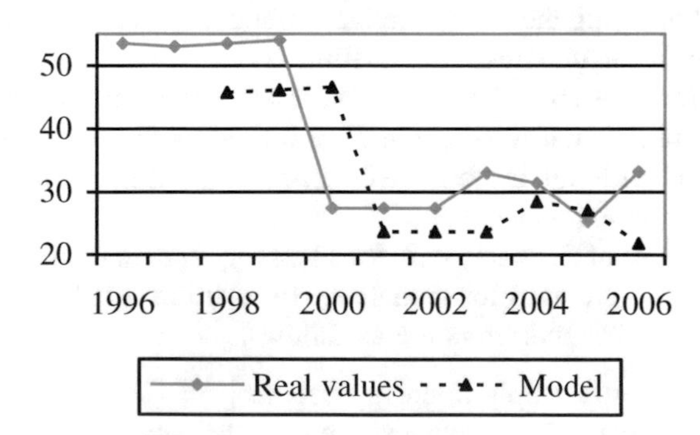

Fig. 5. Modelling results for index x_8

Vector autoregressive modelling. The best 5 models (system of linear difference equations) were selected for every index to build VAR model. Such value of the degree of freedom was chosen to have acceptable modelling time.

After the selection of 55 best models, sorting of $G = 5^{11}$ possible systems of models were carried out. The best system model was selected by the value of systemic integral criterion (5). It was calculated in the prediction mode of the process for 8 steps.

A built VAR model is as follows:

$$x_1(t) = 0.74x_3(t-2) + 42.93x_6(t-2) - 0.06x_9(t-2) + 7.96x_{11}(t-1) - 15.16x_{11}(t-2)$$
$$x_2(t) = 0.27x_2(t-1) - 0.93x_3(t-2) - 0.59x_9(t-1) - 0.44x_9(t-2) + 57.04x_{11}(t-2)$$
$$x_3(t) = -5x_1(t-1) + 4.28x_1(t-2) - 0.29x_7(t-2) - 0.5x_9(t-2) + 9.98x_{10}(t-2)$$
$$x_4(t) = 1.46x_4(t-2) + 0.08x_7(t-2) + 0.14x_8(t-2) - 0.04x_9(t-2) - 8.68x_{11}(t-2)$$
$$x_5(t) = 0.03x_2(t-2) - 0.003x_7(t-1) - 0.05x_8(t-1) + 0.01x_9(t-1) + 0.01x_9(t-2)$$
$$x_6(t) = 0.02x_1(t-1) - 0.03x_4(t-1) + 0.75x_6(t-2) + 0.004x_{10}(t-1) + 0.15x_{11}(t-1)$$
$$x_7(t) = -1.04x_2(t-1) + 0.78x_4(t-2) + 288.53x_6(t-1) - 249.19x_6(t-2) + 9.33x_{11}(t-2)$$
$$x_8(t) = 1.81x_2(t-1) - 4.71x_5(t-1) - 1.96x_9(t-1) - 1.86x_9(t-2) + 118.35x_{11}(t-2)$$
$$x_9(t) = 7.38x_1(t-2) - 13.94x_4(t-2) - 1.46x_8(t-2) - 0.6x_9(t-2) + 41.94x_{10}(t-2)$$
$$x_{10}(t) = 0.32x_4(t-2) - 11.34x_6(t-2) + 0.04x_8(t-2) - 0.04x_9(t-2) + 3.35x_{11}(t-2)$$
$$x_{11}(t) = 0.08x_4(t-2) - 0.21x_5(t-1) - 2.06x_6(t-2) + 0.01x_7(t-1) + 0.27x_{11}(t-1)$$

Modelling time for this enough complex problem was about 5 min.

Table 3 shows ratio errors for first 9 indicators.

Table 3. Model accuracy

	x_1	x_2	x_3	x_4	x_5	x_6	x_7	x_8	x_9
Ratio error, %	9.3	11	38.9	5.7	6.2	4.3	22	4.3	30.6

Figures 6 and 7 present modelling results and real values for indicators x_3 and x_8 with the worst and the best model accuracy on the testing data set, respectively.

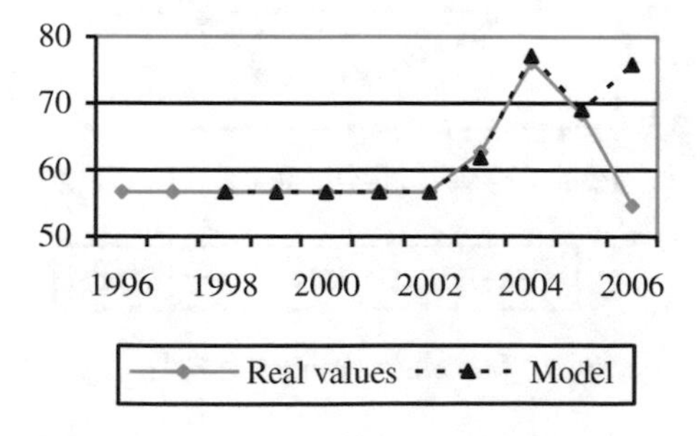

Fig. 6. Modelling results for index x_3

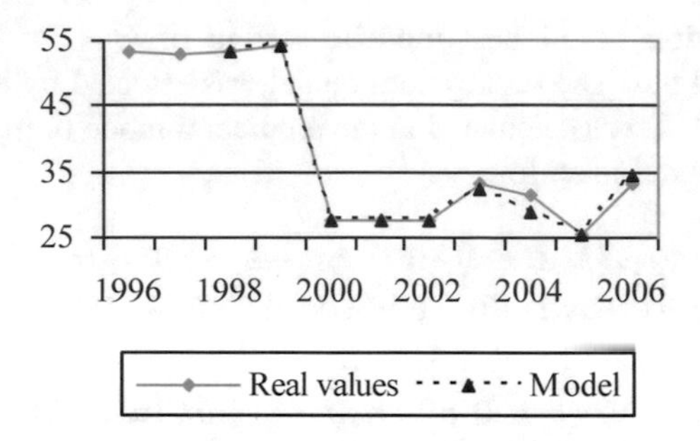

Fig. 7. Modelling results for index x_8

Figures 8 and 9 show modelling results for indices x_{10} and x_{11} with pure prediction on 2006 year. It is significant to note that modelling values for every index were calculated by integrating the process from initial conditions (first two records).

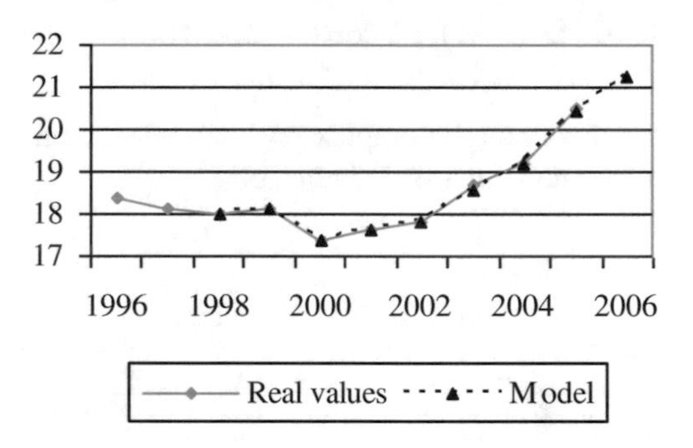

Fig. 8. Modelling results for index x_{10}

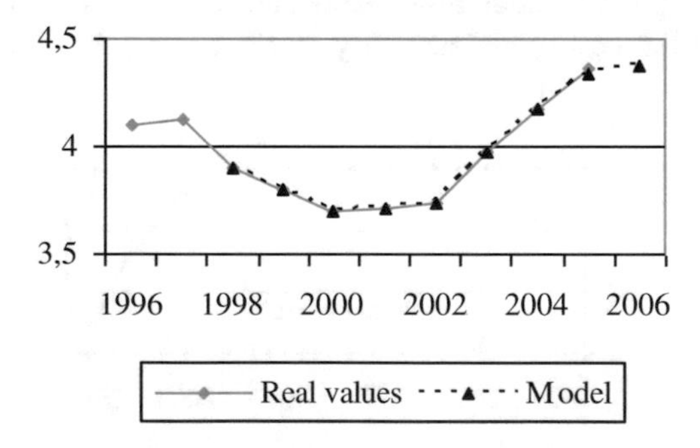

Fig. 9. Modelling results for index x_{11}

Constructed VAR model is accurate on experimental data for all indices of Ukraine energy sphere. And it has acceptable prediction capabilities for most of them. Divergence between real and model values for some indices may be caused by too short data set. Another possible reason consists in neglecting some economic or other factors.

Hence, the constructed system model can be used for intermediate-term forecasting – integration of the optimal system has high accuracy on 8 points (and for most of indicators – even on 9 points).

10 Conclusion

The technique of recurrent-and-parallel computing in COMBI GMDH algorithm for building discrete prediction models of complex multidimensional interrelated processes is proposed.

Software tools for modelling and prediction of complex multidimensional interrelated processes on the basis of high-performance combinatorial algorithm in the class of discrete dynamic VAR models are developed.

The tools were used for modelling and prediction of interrelated processes of Ukraine energy sphere with the purpose of effective managerial decision making in this field.

References

1. Neusser, K.: Time Series Econometrics. Springer International Publishing, Switzerland (2016)
2. Lui, G.C.S., Li, W.K.: Modelling algal blooms using vector autoregressive model with exogenous variables and long memory filter. Ecol. Model. **200**, 130–138 (2007)
3. Primiceri, G.E.: Time varying structural vector autoregressions and monetary policy. Rev. Econ. Stud. **72**, 821–852 (2005)
4. Brandt, P.T., Williams, J.T.: Multiple Time Series Models. Sage (2007)
5. Canova, F.: VAR models: specification, estimation, inference and forecasting. In: Pesaran, H., Wickens, V. (eds.) Handbook of Applied Econometrics. Blackwell, Basil (1994)
6. Enders, W.: Applied Econometric Time Series, New York (1992)
7. Granger, J.: Investigating causal relations by econometric models and cross-spectral methods. Acta Phys. Pol. B **37**, 424–438 (1969)
8. Lutkepohl, H.: New Introduction to Multiple Time Series Analysis. Springer, Heidelberg (2005)
9. Hoover, K.: Causality in Economics and Econometrics. The New Palgrave Dictionary of Economics, 2nd edn. Palgrave Macmillan, London (2008)
10. Madala, H.R., Ivakhnenko, A.G.: Inductive Learning Algorithms for Complex Systems Modeling. CRC Press, New York (1994)
11. Stepashko, V.S.: A combinatorial algorithm of the group method of data handling with optimal model scanning scheme. Soviet Autom. Control **14**(3), 24–28 (1981)
12. Stepashko, V.S., Efimenko, S.N.: Sequential estimation of the parameters of regression model. Cybern. Syst. Anal. **41**(4), 631–634 (2005)
13. Stepashko, V., Yefimenko, S.: Parallel algorithms for solving combinatorial macromodelling problems. Przegląd Elektrotechniczny (Electrical Review) **85**(4), 98–99 (2009)
14. Yefimenko, S., Stepashko ,V.: Intelligent recurrent-and-parallel computing for solving inductive modeling problems. In: 16th IEEE International Conference on Computational Problems of Electrical Engineering (CPEE), pp. 236–238. LNPU, Lviv, Ukraine (2015)

15. Stepashko, V.S., Kostenko, Y.V.: A GMDH algorithm for two-level modeling of multidimensional cyclic processes. Sov. J. Autom. Inform. Sci. **20**(4), 49–57 (1987)
16. Yefimenko, S.: Comparative effectiveness of parallel and recurrent calculations in combinatorial algorithms of inductive modelling. In: Proceedings of the 4th International Conference on Inductive Modelling ICIM 2013, pp. 231–234. IRTC ITS NASU, Kyiv (2013)

Adaptive Enhancement of Monochrome Images with Low Contrast and Non-uniform Illumination

Elena Yelmanova[1(✉)] and Yuriy Romanyshyn[1,2]

[1] Lviv Polytechnic National University, Lviv, Ukraine
yelmanova@lp.edu.ua, yuriy.romanyshynl@gmail.com
[2] University of Warmia and Mazury, Olsztyn, Poland

Abstract. The problem of adaptive enhancement for complex images with low contrast and non-uniform illumination is considered. The histogram-based method of image contrast enhancement in automatic mode on the basis of the estimations of parameters of distribution of brightness values at the boundaries of objects and background for the various known definitions of contrast kernels is proposed. The research of the effectiveness of the proposed and well-known methods of image contrast enhancement is carried out using the known no-reference metrics of generalized contrast of image.

Keywords: Image processing · Contrast enhancement · Low-contrast Non-uniform illumination · Small-sized objects · Contrast kernel

1 Introduction

Operative (in real-time) images quality enhancement in automatic mode is extremely relevant for the vast majority of practical applications in imaging, image processing and analysis [1].

Wide applying of modern technologies of imaging and image processing makes the automatic quality enhancement for the formed images more relevant than ever [1].

The widespread use of various types of mobile gadgets (mobile phones, tablets, laptops, car dash cams, etc.) equipped with video image sensors (high-resolution photo and video cameras), of the systems of remote sensing, imagery intelligence, surveillance and reconnaissance of various destination and of basing, of the remotely piloted vehicles and remote-controlled robotic systems (UAVs, copters, robotic cars), etc., requires addressing the problem of real-time images enhancement in automatic mode.

The real-time image enhancement in automatic mode is currently one of the most pressing and difficult problems in imaging and image pre-processing [2].

The objective quality of digital images is characterized by several basic parameters [3]. Contrast is the main quantitative characteristic that determines the objective quality of the image [3].

Therefore, the development of new effective techniques for real-time image contrast enhancement in automatic mode (with acceptable level of computational costs) is especially relevant at present [1]. Particularly relevant is the addressing of the problem

N. Shakhovska and V. Stepashko (eds.), *Advances in Intelligent Systems and Computing II*, Advances in Intelligent Systems and Computing 689,
https://doi.org/10.1007/978-3-319-70581-1_43

of effective contrast enhancement of complex images with low-contrast and small-sized objects and non-uniform illumination.

The problem of developing of histogram-based methods of contrast enhancement in automatic mode for complex low-contrast images with small-sized objects and non-uniform illumination is considered (Sect. 2). The object of study is the process of improving the image quality. The problem of contrast enhancement in automatic mode for complex monochrome images with low-contrast and small-sized objects and non-uniform illumination is considered. The purpose of the work is to improve the effectiveness of contrast enhancement of small-sized low-contrast objects on the complex multi-element monochrome images with non-uniform illumination. The subject of the study is histogram-based methods of adaptive contrast enhancement of complex monochrome images in automatic mode.

In this paper the new histogram-based method for adaptive enhancement in automatic mode of the contrast of monochrome low-contrast images with small-sized objects and non-uniform illumination on the basis of the estimation of parameters of contrast distribution at the boundaries of image elements (objects and background) is proposed (Sect. 3).

The proposed method of adaptive contrast enhancement provides an efficient redistribution of the contrast of image objects by maximizing the contrast for poorly distinguishable low-contrast objects with the proviso that the contrast of high-contrast objects is guaranteed to remain at a level which ensures their reliable detection and recognition. Research of the effectiveness of the proposed and well-known histogram-based methods of image contrast enhancement in automatic mode was carried out using known no-reference metrics of generalized contrast for four groups of low-contrast monochrome images with a complex structure and non-uniform illumination (Sects. 4 and 5).

2 Image Contrast Enhancement

Various approaches for improving image quality are known at present [1, 2].

One of the most effective methods of quality improvement for image is enhancement of its contrast (of its generalized contrast).

Generalized contrast of image is the most important quantitative characteristic, which to a large extent determines the objective quality of the image, as well as the efficiency, reliability and accuracy of its subsequent analysis and interpretation [2, 4].

The generalized contrast of complex (multi-element) subject images is usually determined on the basis of the analysis of contrast values for all individual pairs of image elements (objects and background) [5]. The contrast of a pair of image elements characterizes the quantitative difference between the brightness values of two image areas (of two objects or an object and a background) [5]. Increasing the generalized contrast allows to effectively improve the quality of the initial image [3].

There are various approaches to enhance the contrast of monochrome images in automatic mode. The known techniques of image contrast enhancement in a spatial domain are most often subdivided into two main classes: global and local methods.

Currently, the histogram-based techniques of image enhancement in the spatial domain, such as [1, 2, 6, 7]:

- adaptive linear stretching,
- methods of nonlinear stretching (logarithmic, exponential and power transformations and their modifications, etc.),
- histogram specification, histogram alignment and its modifications, and others, are of the greatest interest for images contrast enhancement in real time.

A main shortcoming for the majority of known histogram-based techniques of contrast enhancement is a possible reduction of contrast and the disappearance of small-sized objects with low contrast which is unacceptable to image processing in automatic mode [2].

The effective contrast enhancement of the small-size objects on the complex image is possible by the analysis the contrast distribution at the boundaries of image elements (objects and background).

To address these shortcomings, the histograms-based method of image contrast enhancement on the basis of the estimations of contrast values at the boundaries of objects and background is proposed.

3 Proposed Method

In this paper we propose the histogram-based method of contrast enhancement of monochrome images on the basis of the estimations of parameters of distribution of brightness values at the boundaries of objects and background for the various definitions of contrast kernels.

The proposed method is designed to enhance the contrast for complex low-contrast images with small-sized objects and non-uniform illumination in automatic mode.

The proposed contrast enhancement method is designed to redistribute the contrast of objects on the image for maximizing the contrast of poorly discernible low-contrast objects, with the proviso that the contrast of high-contrast objects is guaranteed to remain at a level which ensured their reliable detection and recognition.

To process the initial image, a known transformation is used [1]:

$$L_i^* = L_{\min}^* + \left(L_{\max}^* - L_{\min}^*\right) \cdot f_{tr}(L_i), \tag{1}$$

where L_i^* - brightness value of i-th pixel of the transformed image, $L_{\min}^*$ and $L_{\max}^*$ - minimum and maximum brightness values of the transformed image, L_i - brightness value of i-th pixel of the initial image, $f_{tr}(L_i)$ - transformation function of brightness of initial image.

As a rule, the transformation function $f_{tr}(L_i)$ is a monotonically increasing function and is normalized to the range [0, 1].

In the proposed method, the transformation function $f_{tr}(L_i)$ for the initial image is defined as:

$$f_{tr}(L_i) = M \cdot \int_0^{L_i} \mu_{con}(L_j)\, dL_j, \quad (2)$$

$$M = \left[\int_0^1 \mu_{con}(L_k)\, dL_k \right]^{-1}, \quad (3)$$

where $\mu_{con}(L_j)$ - distribution function of contrast for the initial image, M - normalizing coefficient, constant for current image.

The distribution function $\mu_{con}(L_i)$ of contrast is the assessment of averaged contrast of all image elements (objects and background) relative to the value L_i of preset adaptation level.

In this case, the distribution function $\mu_{con}(L_i)$ of contrast for the initial image is defined as:

$$\mu_{con}(L_i) = \int_0^1 \varphi(C(L_i, L_j)) \cdot \rho_{bou}(L_i, L_j)\, dL_j, \quad (4)$$

where $C(L_i, L_j)$ - contrast value for two image elements i and j, $\varphi(\bullet)$ - correction function for contrast values of image elements, $\rho_{bou}(L_i, L_j)$ - two-dimensional distribution of brightness on the boundaries of image elements.

Expressions (1)–(4) describe the proposed method of image contrast enhancement.

However, it should be noted that for the practical implementation of the proposed method, it is necessary to solve a number of rather complicated problems.

In particular, it is necessary to solve the problems of choosing the contrast definition for two image elements (the contrast kernel), of choosing of the weighting function for contrast values and also of choosing of the method to estimate the parameters of the two-dimensional distribution of brightness values at the boundaries of image elements.

To demonstrate the possibilities and limitations of the proposed method of image enhancement, let us consider a concrete example of its implementation.

In expression (4), the contrast of a pair of image elements (object and background) was defined in a generalized form.

Various approaches to measuring the contrast value for the two image elements are known at present.

To demonstrate the implementation of the proposed method, it is proposed to consider the three most well-known contrast definitions for two elements of the image, namely:

(1) the weighted contrast [1]:

$$C^{wei}(L_i, L_j) = \frac{L_i - L_j}{L_i + L_j}, \quad (5)$$

(2) the relative contrast [5]:

$$C^{rel}(L_i, L_j) = \frac{L_i - L_j}{\max(L_i, L_j)}, \quad (6)$$

(3) the absolute contrast, invariant to linear transformations [8]:

$$C^{abs}(L_i, L_j) = \frac{L_i - L_j}{L_{\max} - L_{\min}}, \quad (7)$$

where $L_{\min}, L_{\max}$ - minimum and maximum brightness value of the initial image.

The definitions (5)–(7) of the contrast of the image elements are called the contrast kernels [5] and are the basis for calculating the distribution function of image contrast.

The correction function $\varphi(C)$ provides a redistribution of contrast between low-contrast and high-contrast objects in the image by increasing of contrast values for low-contrast objects and limiting the contrast values for objects with excessively high contrast.

The correction function $\varphi(C)$ provides a predominant increasing of contrast for low-contrast objects as compared to objects with high contrast when forming a distribution function of image contrast $\mu_{\text{con}}(L)$.

To demonstrate the capabilities of the proposed method, the correction function of contrast values $\varphi(C)$ can be defined as:

$$\varphi(C(L_i, L_j)) = \min(|C(L_i, L_j)|^{\gamma}, \delta), \quad (8)$$

where γ - exponent, parameter; δ - threshold value, parameter.

To estimate the distribution parameters of brightness on the boundaries of image elements it is necessary to solve the problem of finding the boundaries of contiguous image elements, which in itself is quite a complex and resource intensive task.

For the case where image elements are independent events in relation to each other, it can be suggested that the distribution of brightness values at the boundaries of image elements has the form:

$$\rho_{bou}(L_i, L_j) = N \cdot p(L_i)^{\beta} \cdot p(L_j)^{\beta}, \quad (9)$$

$$N = \left(\int_0^1 p(L_k)^{\beta} dL_k \right)^{-2}, \quad (10)$$

where $p\ (L)$ - probability density function of brightness of initial image, β - exponent, parameter, N - normalizing coefficient, constant for current image.

The proposed method of image contrast enhancement is defined in accordance with (1)–(4).

Research of the effectiveness of the proposed method was carried out through comparison with well-known histogram-based methods of contrast enhancement. Research of the effectiveness of the proposed and known methods were carried out by measuring the contrast using known no-reference metrics of generalized contrast for four groups of low-contrast monochrome images with a complex structure and non-uniform illumination.

Comparative analysis of the effectiveness of contrast enhancement for the well-known and proposed histogram-based method of contrast enhancement was carried out in Sects. 4 and 5.

4 Research

Research was carried out through a comparative analysis of the proposed method and four well-known histogram-based methods of contrast enhancement, namely:

(1) method of linear stretching, $\alpha = 0.01$ [1];
(2) method of global histogram equalization (GHE) [2];
(3) BBHE method [6];
(4) DSIHE method [7];
(5) proposed method using weighted contrast (1)–(5), (8) and (9) with $\gamma = 1.0$, $\delta = 1.0$, $\beta = 0.5$;
(6) proposed method using relative contrast (1)–(4), (6), (8) and (9) with $\gamma = 1.0$, $\delta = 1.0$, $\beta = 0.5$.

Research of the effectiveness of the proposed and well-known methods of contrast enhancement was carried out by measuring of contrast using known no-reference histogram-based metrics of generalized contrast.

To measure the contrast of the processed image by its histogram, the known no-reference metrics of generalized contrast were used, namely:

(1) generalized contrast on the basis of definition of weighted contrast [5]:

$$C_{gen}^{wei} = \int_{0}^{1} \frac{|L - \bar{L}|}{L + \bar{L}} \cdot p(L)\, dL, \qquad (11)$$

where $\bar{L}$ - mean value of image brightness.

(2) generalized contrast on the basis of definition of relative contrast [5]:

$$C_{gen}^{rel} = \int_{0}^{1} \frac{|L - \bar{L}|}{\max(L, \bar{L})} \cdot p(L)\, dL, \tag{12}$$

(3) generalized contrast on the basis of definition of absolute contrast [5]:

$$C_{gen}^{abs} = \int_{0}^{1} \left| \frac{(L - \bar{L})}{LMAX} + \frac{1}{2} - \left| \frac{(L - \bar{L})}{LMAX} - \frac{1}{2} \right| \right| \cdot p(L) dL, \tag{13}$$

where $LMAX$ - maximum possible brightness value,

(4) averaged contrast on the basis of definition of weighted contrast [8]:

$$C_{ave}^{wei} = \int_{0}^{1} \int_{0}^{1} \frac{|L_i - L_j|}{L_i - L_j} \cdot p(L_i) \cdot p(L_j)\, dL_i dL_j, \tag{14}$$

(5) averaged contrast on the basis of definition of relative contrast [8]:

$$C_{ave}^{rel} = \int_{0}^{1} \int_{0}^{1} \frac{|L_i - L_j|}{\max(L_i, L_j)} \cdot p(L_i) \cdot p(L_j)\, dL_i dL_j, \tag{15}$$

(6) averaged contrast on the basis of definition of absolute contrast [8]:

$$C_{ave}^{abs} = \int_{0}^{1} \int_{0}^{1} \frac{|L_i - L_j|}{LMAX} \cdot p(L_i) \cdot p(L_j)\, dL_i dL_j. \tag{16}$$

Research of the effectiveness of the proposed and known methods were carried out for four groups of test monochrome images with a complex structure and non-uniform illumination.

Four complex monochrome images with small-sized objects and non-uniform illumination were used as the initial images.

Appearance of the four initial images and their histograms is shown in Figs. 1, 2, 3 and 4.

Each group of test images consisted of the initial image and of the results of its processing with the use of various (the most well-known and proposed) contrast enhancement methods by histogram transformation.

The results of processing the four initial images (Figs. 1, 2, 3 and 4) with using earlier considered methods are shown in Figs. 5, 6, 7 and 8.

Fig. 1. The first test image and its histogram

Fig. 2. The second test image and its histogram

Fig. 3. The third test image and its histogram

Fig. 4. The fourth test image and its histogram

a) method of linear stretching
b) histogram equalization method
d) BBHE method
e) DSIHE method
f) proposed method using weighted contrast
g) proposed method using relative contrast

Fig. 5. The results of processing for the first initial image (Fig. 1)

a) method of linear stretching
b) histogram equalization method
d) BBHE method
e) DSIHE method
f) proposed method using weighted contrast
g) proposed method using relative contrast

Fig. 6. The results of processing for the second initial image (Fig. 2)

a) method of linear stretching
b) histogram equalization method
d) BBHE method
e) DSIHE method
f) proposed method using weighted contrast
g) proposed method using relative contrast

Fig. 7. The results of processing for the third initial image (Fig. 3)

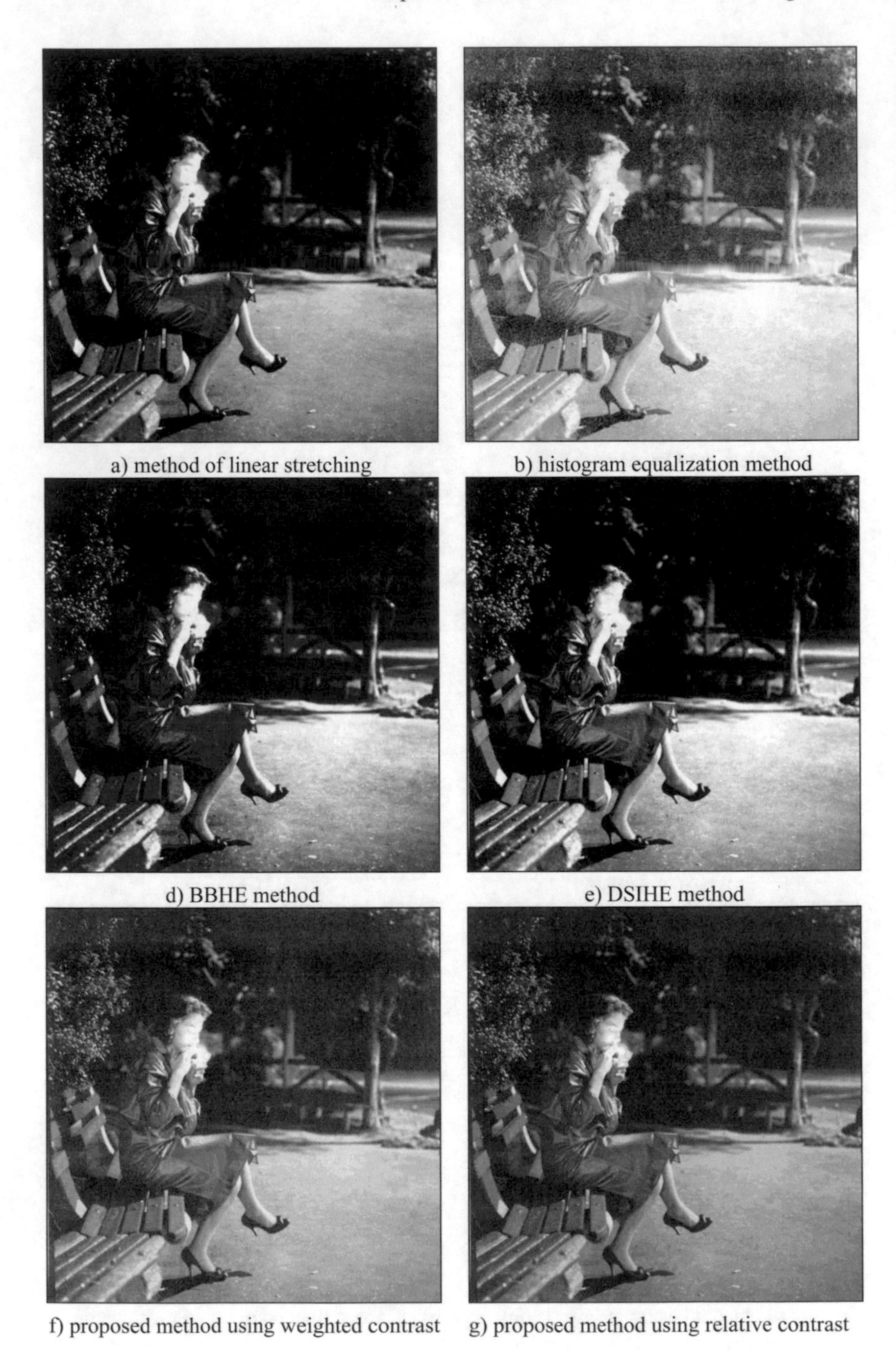

a) method of linear stretching
b) histogram equalization method
d) BBHE method
e) DSIHE method
f) proposed method using weighted contrast
g) proposed method using relative contrast

Fig. 8. The results of processing for the fourth initial image (Fig. 4)

Table 1. The results of measurements of generalized contrast for four groups of test images

	C_{gen}^{wei}	C_{gen}^{rel}	C_{gen}^{abs}	C_{ave}^{wei}	C_{ave}^{rel}	C_{ave}^{abs}
1	0.247	0.370	0.103	0.264	0.371	0.074
5.a	0.345	0.471	0.128	0.383	0.492	0.095
5.b	0.288	0.403	0.500	0.387	0.500	0.334
5.d	0.496	0.636	0.267	0.469	0.574	0.176
5.e	0.560	0.675	0.518	0.578	0.669	0.328
5.f	0.193	0.294	0.235	0.273	0.391	0.173
5.g	0.202	0.305	0.242	0.284	0.403	0.178
2	0.231	0.350	0.118	0.271	0.386	0.083
6.a	0.309	0.434	0.197	0.369	0.487	0.143
6.b	0.288	0.404	0.502	0.387	0.500	0.334
6.d	0.505	0.645	0.358	0.501	0.607	0.232
6.e	0.523	0.646	0.517	0.555	0.654	0.329
6.f	0.214	0.323	0.275	0.293	0.415	0.195
6.g	0.224	0.336	0.284	0.305	0.428	0.201
3	0.284	0.414	0.216	0.328	0.451	0.156
7.a	0.388	0.515	0.226	0.457	0.576	0.165
7.b	0.289	0.405	0.502	0.388	0.502	0.335
7.d	0.432	0.568	0.388	0.484	0.597	0.259
7.e	0.469	0.600	0.517	0.527	0.634	0.333
7.f	0.240	0.354	0.327	0.326	0.450	0.233
7.g	0.253	0.369	0.337	0.339	0.464	0.239
4	0.383	0.533	0.223	0.395	0.511	0.137
8.a	0.517	0.646	0.430	0.542	0.644	0.267
8.b	0.289	0.405	0.502	0.388	0.500	0.334
8.d	0.473	0.613	0.442	0.500	0.609	0.285
8.e	0.535	0.656	0.525	0.563	0.659	0.334
8.f	0.291	0.423	0.407	0.363	0.488	0.265
8.g	0.308	0.442	0.420	0.378	0.503	0.272

The results of contrast measurements for four groups of test images using known no-reference metrics of generalized contrast (11)–(16) are shown in Table 1.

The results of the contrast measurements (Table 1) for each of the four groups of test images are also shown in the form of graphs in Figs. 9, 10, 11 and 12.

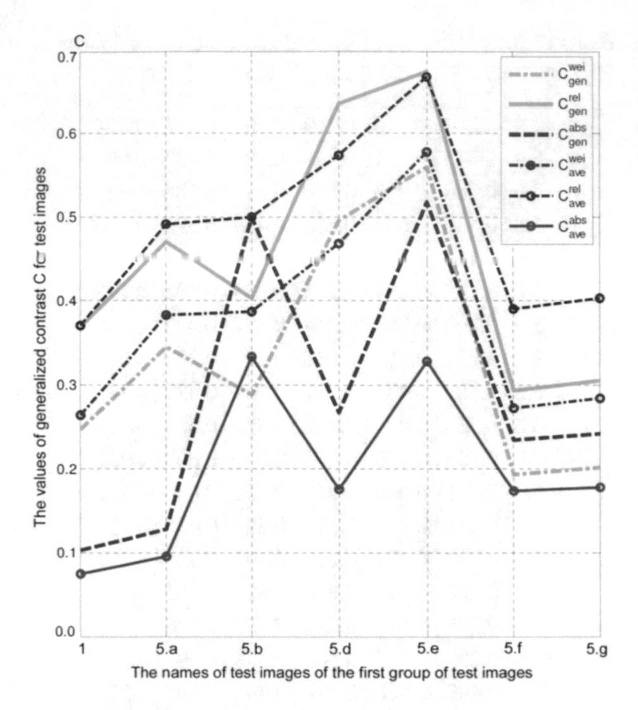

Fig. 9. The values of contrast for the first group of test images (Figs. 1 and 5)

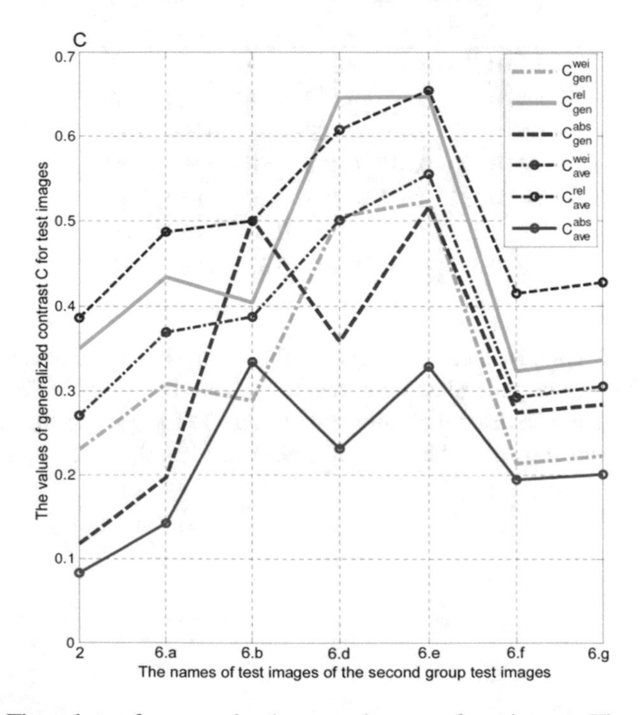

Fig. 10. The values of contrast for the second group of test images (Figs. 2 and 6)

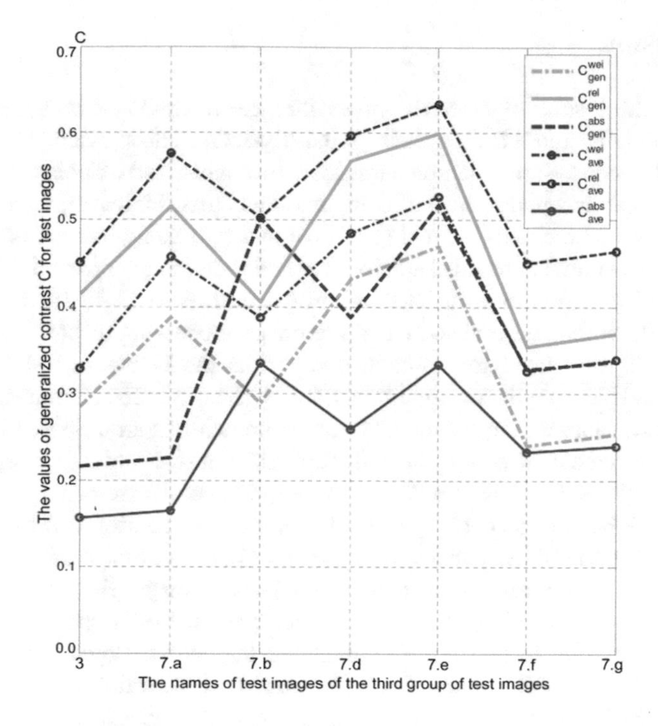

Fig. 11. The values of contrast for the third group of test images (Figs. 3 and 7)

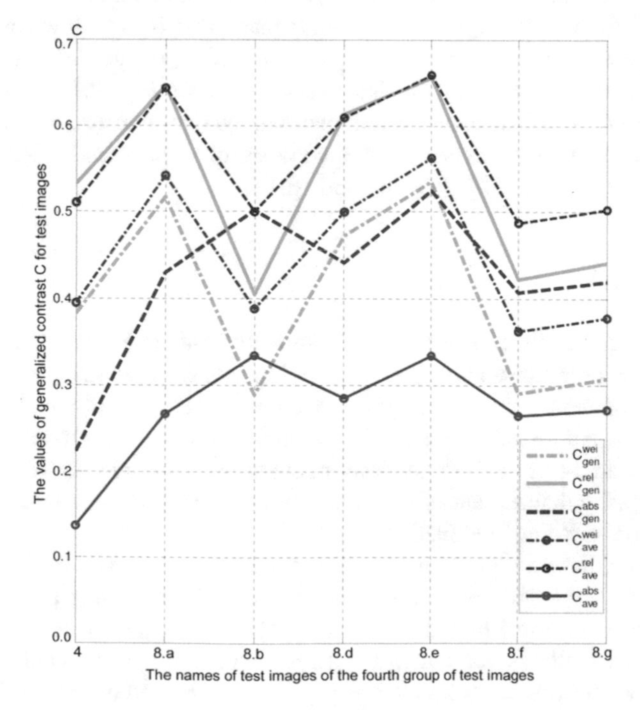

Fig. 12. The values of contrast for the fourth group of test images (Figs. 4 and 8)

5 Discussion

Analysis of the results of research shows that the methods of image processing in automatic mode by modification their global histogram allow essentially enhance the contrast of low-contrast monochrome images with small-sized objects and non-uniform illumination with acceptable level of computational costs. However, it should be noted that the considered metrics (11) and (12), (14) and (15) using kernels of weighted (5) and relative (6) contrast are substantially dependent on the average brightness level of the image and give greatly overstated values of contrast (Figs. 9, 10, 11 and 12).

The results of the research show that the image processing techniques on the basis of the technology of histogram equalization and its modifications (GHE, BBHE and DSIHE methods) ensure the maximum increase of the generalized contrast of the initial image. A main shortcoming for the majority of known histogram-based techniques of contrast enhancement is a possible reduction of contrast and the disappearance of small-sized objects [2]. The results of the experimental research have shown that methods on the basis of technology of the histogram equalization and its modifications (GHE, BBHE and DSIHE methods) in some cases can lead to a significant decrease of contrast or to the disappearance of small-size objects in image (Figs. 5(b), (d), (e), 6(d), (e), 7(b), (d), 8(b), (d) and (e). Possible significant reduction of contrast of small-sized objects and excessive increasing of contrast for large-sized objects in image significantly limits the use the technologies of histogram equalization (in particular, GHE, BBHE and DSIHE methods) for processing of complex images in automatic mode. The effectiveness of methods of linear stretching and its modifications essentially depends on the dynamic range of brightness of initial image and is very low for the complex images with non-uniform illumination (Figs. 5(a), 6(a), 7(a) and 8(a). The results of the experimental research have shown that proposed method pro-vides the increase of efficiency of contrast enhancement in automatic mode (by an average of 51%–98%) for all initial images without reducing the contrast of small-sized objects on image (Figs. 5(f), (g), 6(f), (g), 7(f), (g), 8(f) and (g)).

6 Conclusion

Operative images quality enhancement in automatic mode is extremely relevant for the vast majority of practical applications in imaging, image processing and analysis.

The development of new effective techniques for real-time image contrast enhancement in automatic mode (with acceptable level of computational costs) is especially relevant at present. Particularly relevant is the addressing of the problem of effective contrast enhancement of complex images with low-contrast and small-sized objects and non-uniform illumination.

In this paper, the problem of development of histogram-based methods of contrast enhancement in automatic mode for complex monochrome images with small-sized objects and non-uniform illumination was considered. In this paper it is shown that histogram-based methods for contrast enhancement allow significantly increase the contrast of low-contrast monochrome images with small-sized objects and non-uniform illumination with an acceptable level of computational costs.

However, the known technologies for image contrast enhancement by histogram equalization have several significant disadvantages, which significantly limit their practical use. A main shortcoming for the majority of known histogram-based techniques of contrast enhancement is a possible reduction of contrast and the disappearance of small-sized objects with low contrast which is unacceptable to image processing in automatic mode.

The histogram-based method of image contrast enhancement in automatic mode on the basis of the estimations of parameters of distribution of brightness values at the boundaries of objects and background for the various definitions of contrast kernels was proposed to address these shortcomings. The proposed histogram-based method of adaptive contrast enhancement provides an efficient redistribution of the contrast of image objects by maximizing the contrast for poorly distinguishable low-contrast objects with the proviso that the contrast of high-contrast objects is guaranteed to remain at a level which ensures their reliable detection and recognition.

Research of the effectiveness of the proposed and several well-known histogram-based methods of contrast enhancement was carried out using the known no-reference metrics of generalized contrast for four groups of complex monochrome images with small-sized objects and non-uniform illumination.

The results of research confirm the effectiveness of the proposed method for contrast enhancement in automatic mode for low-contrast monochrome images with small-sized objects and non-uniform illumination.

References

1. Gonzalez, R.C., Woods, R.E.: Digital Image Processing, 2nd edn. Prentice Hall, New Jersey (2002)
2. Pratt, W.K.: Digital Image Processing: PIKS Inside, 3rd edn. Wiley, New York (2001)
3. Wang, Z., Bovik, A.C.: Modern image quality assessment. In: Synthesis Lectures on Image, Video, and Multimedia Processing, Vol. 2, no. 1, pp. 1–156. Morgan and Claypool Publishers, New York (2006). https://doi.org/10.2200/S00010ED1V01Y200508IVM003
4. Kosarevych, R.J., Rusyn, B.P., Korniy, V.V., Kerod, T.I.: image segmentation based on the evaluation of the tendency of image elements to form clusters with the help of point field characteristics. Cybern. Syst. Anal. **1**, 704–713 (2015)
5. Vorobel, R.A.: Logarithmic Image Processing. Naukova Dumka, Kyiv, Ukraine (2012). (In Ukrainian)
6. Kim, Y.T.: Contrast enhancement using brightness preserving bi-histogram equalization. IEEE Trans. Consum. Electron. **43**(1), 1–8 (1997)
7. Wang, Y., Chen, Q., Zhang, B.: Image enhancement based on equal area dualistic sub-image histogram equalization method. IEEE Trans. Consum. Electron. **45**(1), 68–75 (1999)
8. Yelmanova, E., Romanyshyn, Y.: No-reference contrast assessment by image histogram. In: Proceedings of 14th International Conference CADSM 2017, pp. 148–152. IEEE, Lviv, Ukraine (2017)

Secure Routing in Reliable Networks: Proactive and Reactive Approach

Oleksandra Yeremenko(✉), Oleksandr Lemeshko, and Anatoliy Persikov

Kharkiv National University of Radio Electronics, 14 Nauka Ave., Kharkiv, Ukraine
oleksandra.yeremenko.ua@ieee.org,
oleksandr.lemeshko@nure.ua,
persikovanatoliy@gmail.com

Abstract. In this paper, the approach to providing a given level of information security for multipath routing of confidential messages in a network is considered. A method for providing secure routing over overlapping paths is developed and belongs to the class of proactive solutions for ensuring a given level of information security. The analysis has shown that using the proposed method within the presented calculated examples can improve the probability of compromising transmitted messages at average from 5–10% to 25–50% due to the possibility of using composite paths that are one of the subclasses of overlapping paths. A method of Secure Fast ReRouting (S-FRR) of messages in the network has been synthesized, the novelty of which lies in the fact that it focuses on the implementation of both proactive and reactive secure routing confidential messages. In this case, the proactive nature of the solutions is conditioned by the calculation of the set of primary composite paths forming the primary multipath, along which parts of the confidential message are transmitted. However, in the case of violation of the information security requirements in the network caused by the increased probability of compromising one or multiple composite paths constituent the primary multipath, the messages will be transmitted over the calculated set of the backup composite paths determining the backup multipath. Within the framework of the proposed S-FRR method, it is possible to protect both the primary multipath as a whole and one or several precomputed composite paths included in this primary multipath. The developed methods of secure routing can be used as the basis for new network protocols for routing and fast rerouting for multipath transmission of parts of a confidential message with specified requirements regarding the probability of its compromise in the network.

Keywords: Information security · Reliable network · Secure routing
Probability of compromise · Link · Multipath · Composite path
Secure fast rerouting

1 Introduction

As shown by the analysis, one of the most important tasks, which is regulated by standards of next generation networks (NGN) construction, is the problem of the implementing information security functions. In accordance with the requirements of

N. Shakhovska and V. Stepashko (eds.), *Advances in Intelligent Systems and Computing II*, Advances in Intelligent Systems and Computing 689,
https://doi.org/10.1007/978-3-319-70581-1_44

the International Telecommunication Union (ITU) standards, provision of information security is carried out within three levels: security of infrastructure, security of services and security of applications [1]. At the same time, the effectiveness of the top two levels is entirely determined by the efficiency of functioning of the infrastructure level security means, the main tasks of which are: ensuring security at the level of network elements (switches, routers, servers), links, and routes consisting of links in general.

As a rule, the security level of network elements is evaluated using such an important indicator as the probability of compromise, where compromising means the fact of unauthorized access to protected information, as well as the implementation of a suspicion of such access. At the same time, the security services should be provided by appropriate levels of OSI model protocols [1–3]. In turn, the security at the network level should be maintained and provided by routing protocols. With the help of routing tools, as shown in [4–8], secure delivery of various confidential information, in particular session keys, information on authentication, critical user data, etc., can be carried out.

To ensure a given level of information security in practice, depending on the time of reaction to a possible compromise of communication links and network fragments, both proactive and reactive means should be complementarily applied, including those related to routing solutions. Proactive means are used, as a rule, at the stage of preventing the compromise of transmitted messages or minimizing the probability of its occurrence [9]. Reactive means are used in those cases when the security of the transmitted data is violated and, for example, it is important for the means of routing to quickly restore the required level of security.

An example of a proactive approach can be a solution related to the provision of a given level of security. In this case, transmission of messages divided into parts according to the Shamir's scheme from the source to the destination is organized using the multipath routing with balancing the number of parts over non-overlapping paths [10]. When changing the network state that causes a violation of the security level of the transmitted messages, it is important to determine the operational order of changing the set of paths used to transfer parts of confidential messages. Therefore, fast re-routing solutions can be considered as elements of the reactive approach while ensuring secure routing.

It is known that the effectiveness of protocol solutions is largely determined by the quality of the mathematical models and methods for calculating the desired paths, which they are based. The use of theoretically grounded solutions allows to significantly enhance the functionality of the routing protocol in relation to the network tasks assigned to it. Therefore, we will propose theoretical solutions presented by mathematical models and methods of secure routing and combine the capabilities of both proactive and reactive approaches when implementing secure routing of messages in the network.

2 Method of Secure Routing of Messages Using Non-overlapping Paths: Proactive Approach

As the conducted analysis has shown [10–12], the possibility of calculating probability of compromising the message being transmitted in the network is mostly determined by the features of the structural network construction and types of paths used. It is common knowledge that a set of paths in the network could be divided into two subsets: a subset of non-overlapping paths and a subset that allows nodal or link overlapping [13]. Moreover, non-overlapping paths mean only those routes with common source-destination nodes. If the paths contain at least one common node and/or link, then they are called overlapping paths. In addition, if the paths have common nodes, they are called as the paths that are overlapping over nodes. If they have common links, they are called as the paths overlapping over links.

One of the directions for ensuring a given level of information security in communication networks is the implementation of a mechanism based on multipath routing of the transmitted message previously divided into parts in accordance with the Shamir's scheme [10]. As a result of using such a scheme, it is possible to reduce the probability of compromising the transmitted message, because an attacker in order to compromise the message, must compromise all paths, usually non-overlapping, over which parts of the divided message are transmitted.

Currently there are known analytical expressions for calculating the probability of compromising the message transmitted in parts over a set of non-overlapping paths [10–12]. In addition, it is assumed that the following initial data are known:

Constants

S_{msg} and D_{msg}	source and destination nodes for the transmitted message
M	number of the used non-overlapping paths during routing of message parts
M_i	number of links in the i th path, which can be compromised $(i=\overline{1,M})$
p_i^j	probability of compromising the j th link of the i th path $(i=\overline{1,M},\ j=\overline{1,M_i})$
(T,N)	Shamir`s scheme parameters, where $N$ – total number of message parts, obtained by applying the Shamir`s scheme; T – minimum number of parts $(T \le N)$ needed for the message reconstruction
γ_P	allowable probability of message compromise in the network

Indices

p_i	probability of compromising the i th path $(i=\overline{1,M})$
P_{msg}	probability of compromise for the whole message during its transmission in parts over the network

Variables

n_i	integer variable, which is the number of message parts, transmitted over the i th path $(i=\overline{1,M})$

In this case, during the multipath routing and balancing the number of message parts over paths, it is necessary to ensure a specified level of information security, provided by, for example, probability of compromising the message P_{msg} being transmitted:

$$P_{msg} \leq \gamma_P. \tag{1}$$

The following considerations assume that the source and the destination are secure, i.e. the probabilities of compromising the source node and the destination node are equal to zero. It is assumed in [10, 12] that if an element (node, link) of the path has been compromised, then all the fragments sent over this element will also be compromised. Then the probability of compromising the i th path consisting of M_i elements can be calculated using the expression

$$p_i = 1 - (1 - p_i^1)(1 - p_i^2)\ldots(1 - p_i^{M_i}) = 1 - \prod_{j=1}^{M_i}(1 - p_i^j). \tag{2}$$

Besides, during the calculation of the control variables n_i $(i = \overline{1, M})$ regulating the allocation of the message parts over the non-overlapping paths the following condition [10, 12] must be met:

$$\sum_{i=1}^{M} n_i = N. \tag{3}$$

In the case of Shamir's scheme with redundancy when $T < N$ the condition below must be satisfied

$$N - n_i < T, \quad \left(i = \overline{1, M}\right), \tag{4}$$

while when $T = N$ the following conditions must be met in the non-redundant sharing scheme

$$1 \leq n_i \leq T - 1, \quad \left(i = \overline{1, M}\right). \tag{5}$$

The condition (4) ensures that in the case of compromising all the paths except the i th path, an adversary cannot reconstruct the whole message.

One of the main conditions that must be fulfilled in secure routing is that the probability of message compromising, when it is transmitted over the network, should not exceed the set allowable value (1). Then, for instance, the probability of compromising the message divided into N parts according the Shamir's scheme (N, N) and transmitted over M paths, is determined by the expression [10].

$$P_{msg} = \prod_{i=1}^{M} p_i. \tag{6}$$

The advantages of the described method include the fact that using a set of non-overlapping paths when transferring parts of a confidential message greatly simplifies the process of calculating the probability of its compromising in the network using expressions (2)–(6). In this case, the condition (1) is completely determined by the parameters of the non-overlapping paths used, and the task of balancing the parts of the message along these paths is already secondary and lies in fulfilling the conditions (3)–(5). Therefore, if the use of an accessible set of non-overlapping paths does not allow satisfying the requirement (1), then the task of ensuring the given level of information security remains unsolved.

It can be intuitively assumed that the use of overlapping paths on the same network topology could lead to an improvement in the probability of compromising the transmitted message and, as a result, to the successful solution of the task. However, as the analysis has shown, in the case of using overlapping paths in the network, the procedure for calculating the probability of compromising a message is significantly complicated, and sometimes becomes impossible (in the analytical form) [10–12]. In this regard, the relevant task seems to be finding a trade-off solution concerned with the definition of such a class of overlapping routes, for which it is possible to analytically calculate and hence to control the probability of compromising the message being transmitted.

3 Method of Secure Routing of Messages Using Overlapping Paths: Proactive Approach

The use of expressions (2) and (6) yields adequate results for the case of nodal overlapping of paths, but only if there is a true hypothesis that the probabilities of compromising all network nodes are equal to zero, i.e. only links can be compromised. This is true for some class of wireless networks. In a more general case, any network node subject to network attacks and other impacts can also be always represented by a conditional communication link, where the probability of its compromise is equal to the probability of compromising the simulated node. The method of calculating paths with a nodal overlapping is proposed, for example, in [13].

In the paper [14], an attempt has been made to expand the class of overlapping paths, in which it is still possible to analytically estimate the probability of compromising the transmitted message. This will allow to create conditions for monitoring compliance with the requirements for the level of information security (1) under implementation of overlapping paths.

In this context, we must additionally enter two more types of paths: simple and composite. A simple path is always formed by the series connection of communication links of the network, and the probability of its compromising is calculated using the expression (2). In turn, composite paths represent more complex structural forms including the overlapping of simple paths. In this regard, we refine the previously introduced notations and introduce some additional ones:

Constants

$\widetilde{M}$	number of the non-overlapping composite paths that could be used during routing of message parts
$\widetilde{M}_i$	number of fragments in the i th composite path, which can be compromised ($i = \overline{1, \widetilde{M}}$)
M_i	number of links in the i th composite path, which can be compromised ($i = \overline{1, \widetilde{M}}$)
p_i^j	probability of compromising the j th link of the i th composite path ($i = \overline{1, \widetilde{M}}$, $j = \overline{1, M_i}$)

Indices

$\widetilde{p}_i^j$	probability of compromising the j th fragment of the i th composite path ($i = \overline{1, \widetilde{M}}$, $j = \overline{1, \widetilde{M}_i}$)
$\widetilde{p}_i$	probability of compromising the i th composite path ($i = \overline{1, \widetilde{M}}$)
$\widetilde{P}_{msg}$	probability of compromising for the whole message during its transmission in parts over composite paths

Variables

n_i	integer variable, which is the number of message parts, transmitted over the i th composite path ($i = \overline{1, \widetilde{M}}$)

In order to provide the formulation of the expression for calculating the probability of compromising a composite path in the analytical form during secure routing, the path must contain two types of fragments consisting of a series (Fig. 1(a)) or parallel (Fig. 1(b)) connection of links. Figure 1(c) presents an example of a composite path with a series connection of two network fragments. The first fragment is represented by the parallel connection of links, and the second is a series one.

Figure 1(c) shows the structure of the network containing one composite path, which includes links of two overlapping simple paths. The first simple path is represented by the nodes 1 → 2 → 3 → 4, the second one – by the nodes 1 → 3 → 4. On the other hand, this composite path consists of two subsequently connected fragments. The first fragment includes a parallel connected link 1 → 3 and a sequence of links 1 → 2 and 2 → 3, whereas the second fragment is represented by the link 3 → 4.

Then the probability of compromise of the composite path (Fig. 1(c)) is calculated according to the expression

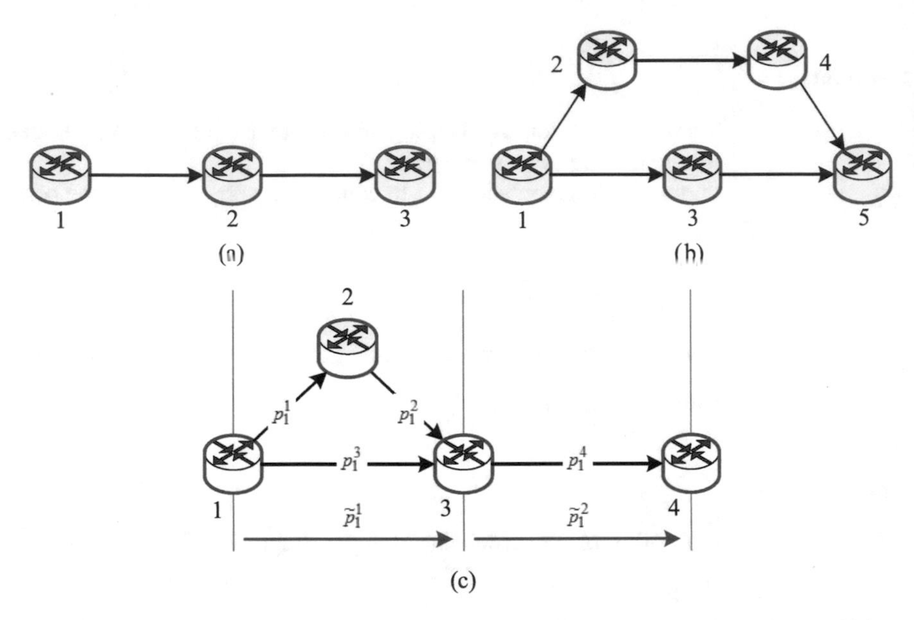

Fig. 1. The examples of fragment types and composite path: (a) series, (b) parallel, and (c) composite path.

$$\tilde{p}_1 = 1 - (1 - \tilde{p}_1^1)(1 - \tilde{p}_1^2), \tag{7}$$

where the probabilities of compromising the first and second fragments are determined via the probabilities of compromise of their links:

$$\tilde{p}_1^1 = \left[1 - (1 - p_1^1)(1 - p_1^2)\right] p_1^3,\ \tilde{p}_1^2 = p_1^4.$$

Thus, in general, the probability of compromising the *i* th composite path consisting of $\tilde{M}_i$ fragments can be calculated according to the following expression:

$$\tilde{p}_i = 1 - \prod_{j=1}^{\tilde{M}_i} (1 - \tilde{p}_i^j). \tag{8}$$

In case if a single composite path is used to deliver the message, the probability of compromising this message is determined by the probability of compromising this composite path. In a more general case, when parts of a message are transmitted over a set of non-overlapping composite paths, the following expression must be used to calculate the probability of compromising a message:

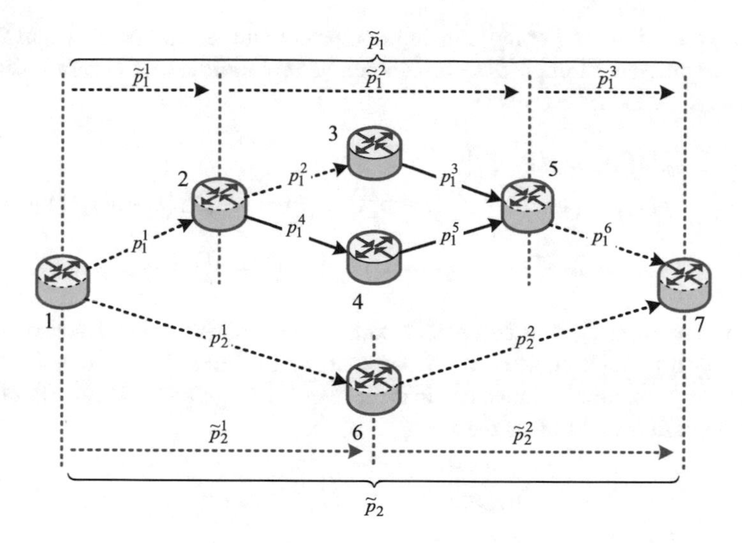

Fig. 2. The example of using two paths: the composite and the simple ones.

$$\tilde{P}_{msg} = \prod_{i=1}^{\tilde{M}} \tilde{p}_i, \tag{9}$$

which is a modification of the formula (6).

The example of such a case is given in Fig. 2, when in order to transmit message parts two non-overlapping paths are used:

- the first path is composite and it includes the following communication links $1 \to 2$, $2 \to 3$, $2 \to 4$, $3 \to 5$, $4 \to 5$, $5 \to 7$;
- the second path is simple and it contains the communication links $1 \to 6$, $6 \to 7$.

In its turn, the first (composite) path includes three series-connected network fragments:

- the first fragment is presented by the link $1 \to 2$;
- the second fragment is based on parallel connection of the links $2 \to 3$, $3 \to 5$ and $2 \to 4$, $4 \to 5$;
- the third fragment is presented by the link $5 \to 7$.

Then for the considered network structure (Fig. 2), the probability of compromising the message when using two different above-described paths will be determined as follows:

$$P_{msg} = \tilde{p}_1 \cdot \tilde{p}_2, \tag{10}$$

where the probabilities of compromising composite and simple paths 1 and 2 respectively are expressed through the corresponding probabilities of compromising their fragments and links as

$$\begin{aligned}\tilde{p}_1 &= 1 - \left(1 - \tilde{p}_1^1\right)\left(1 - \tilde{p}_1^2\right)\left(1 - \tilde{p}_1^3\right) \\ &= 1 - \left(1 - p_1^1\right)\left(1 - \left[1 - \left(1 - p_1^2\right)\left(1 - p_1^3\right)\right] \times \left[1 - \left(1 - p_1^4\right)\left(1 - p_1^5\right)\right]\right)\left(1 - p_1^6\right);\end{aligned} \tag{11}$$

$$\tilde{p}_2 = 1 - (1 - \tilde{p}_2^1)(1 - \tilde{p}_2^2) = 1 - (1 - p_2^1)(1 - p_2^2). \tag{12}$$

In the general case, one composite path may comprise several series-connected fragments with parallel connection of communication links. Let us denote the maximum number of parallel connected links h_i over all fragments of the i th composite path. Then condition (5) takes the form

$$h_i \le n_i \le T - 1, \quad \left(i = \overline{1, \tilde{M}}\right), \tag{13}$$

and its fulfillment will thus allow to distribute parts of the message over parallel links of network fragments of composite paths so that in each of them a nonzero number of such parts of the message is transmitted and expressions (8) and (9) are valid.

In addition, the condition (4) taking into account the composite nature of the used paths will take the form:

$$N - n_i < T, \quad \left(i = \overline{1, \tilde{M}}\right). \tag{14}$$

In this regard, the basis of the proposed method of secure routing for the parts of the transmitted message over a set of overlapping paths can be the solution of the optimization problem associated with the use of the optimality criterion

$$\min_{n_i} \prod_{i=1}^{\tilde{M}} \tilde{p}_i(n_i), \tag{15}$$

which ensures the minimization of the probability of compromising the transmitted message. In addition, the constraints (8), (13) or (14) are imposed on the control variables depending on the Shamir's scheme used, as well as the analog of condition (3) represented by the equality

$$\sum_{i=1}^{\tilde{M}} n_i = N. \tag{16}$$

The formulated optimization problem belongs to the class of Nonlinear Integer Programming problems. The variables n_i to be calculated are integer, and the optimality criterion (15) is nonlinear.

The proposed method for secure routing of messages across multiple overlapping paths is a tool for the Proactive Approach to improve the level of information security. This is determined by the fact that on the basis of constant analysis of the network state, its structure and parameters of security of the communication links, and also during the optimal balancing of the confidential message parts over the overlapping paths, all available capabilities are realized in order to minimize the probability of compromising transmitted data.

4 Numerical Analysis of the Proposed Method for Secure Routing of Messages Using Overlapping Paths

4.1 Numerical Study of Using the Unique Composite Path

Using the proposed approach, we will analyze the effect of the security parameters of individual links and network fragments influence on the probability of compromising the message. In addition, we will estimate the gain on compromise probability obtained using the approach proposed in Sect. 3 in comparison to the known one described in Sect. 2. First, the specific features of calculating the probability of compromising a message will be demonstrated for the network structure shown in Fig. 1(c). The values given in Table 1 have been used as the initial data. The last row in Table 1 shows the result of the solution of the stated optimization problem associated with minimizing the expression (15) under the constraints (8), (13) or (14), (16) when implementing the Shamir's scheme (10, 10) and $h_1 = 2$.

Table 1. Initial research data for the case of using the unique composite path.

Link	$1 \to 2$	$2 \to 3$	$1 \to 3$	$3 \to 4$
Link number in path	1	2	3	4
Probability of compromising the link	0.1	0.2	$0 \div 1$	$0 \div 1$
Number of message parts	5	5	5	10

When transmitting the message from the first to the fourth node, its parts are directed along the two overlapping routes: $1 \to 2 \to 3 \to 4$ and $1 \to 3 \to 4$, i.e. the $3 \to 4$ link has been common to them. In the course of the study, it was assumed that the probabilities of compromising the first and second links have been fixed and amounted to 0.1 and 0.2, respectively; the probabilities of compromising the third and fourth links varied from 0 to 1.

The probability of compromise has been calculated for two cases:

- in the first case, to calculate the probability of message compromising ($\tilde{P}_{msg}$) we used the approach proposed in Sect. 3 based on expressions (7)–(9), (14);
- in the second case, for calculations we used the approach (2)–(6) given in Sect. 2, which assumes using of only non-overlapping paths. In respect to Fig. 1(c) this assumes using either the path $1 \to 3 \to 4$, to which the probability of compromise P^1_{msg} corresponded to, or the path $1 \to 2 \to 3 \to 4$, the compromise of which was estimated by the probability P^2_{msg}.

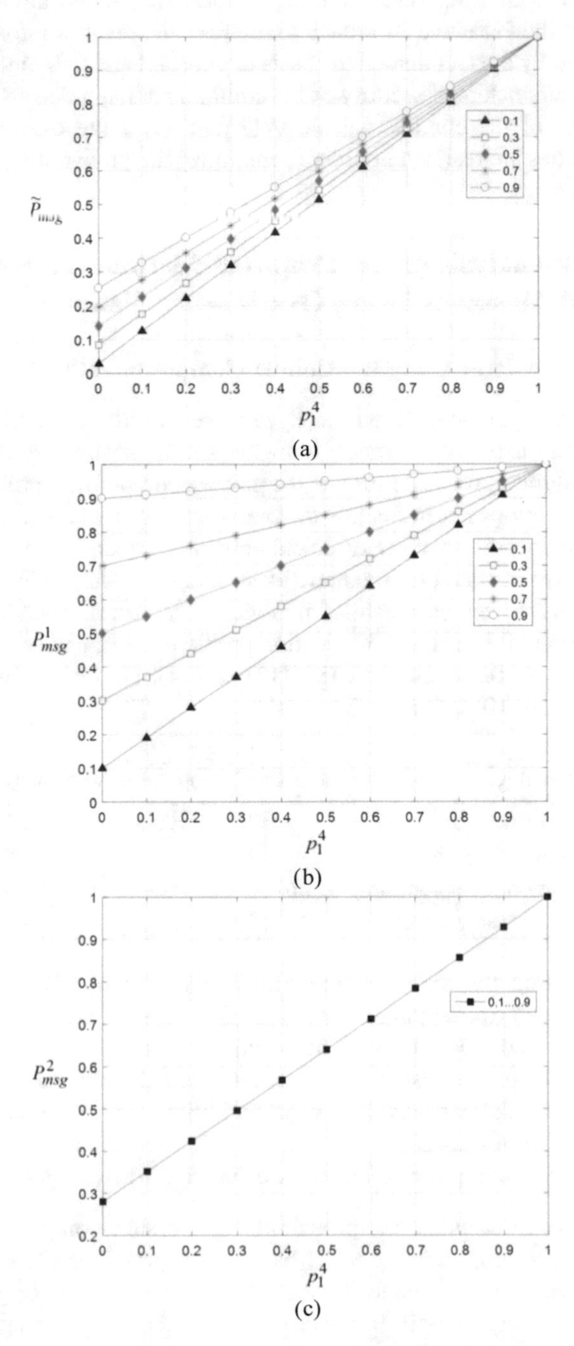

Fig. 3. Dependence of the probability of compromising the message transmitted along different types of paths for the network structure, as shown in Fig. 1(c).

Then, Fig. 3 shows the nature of the dependence of the probability of compromising a message transmitted along different types of paths for the network structure given in Fig. 1(c) on the values of the probability of compromising the fourth link p_1^4 (plotted along the abscissa axis). Each of the set of lines corresponded to its value of the probability of compromising the third link (p_1^3). As shown in Fig. 3, with increasing p_1^3 and p_1^4 the probability of compromising the transmitted message using the composite path and simple path $1 \to 3 \to 4$ has always increased, but the nature of the dependence when using paths of different types (overlapping and non-overlapping) was significantly different. Given that the simple path $1 \to 2 \to 3 \to 4$ did not contain the third link (Fig. 1(c)), the probability of its compromise depended only on p_1^4 and did not depend on p_1^3 (Fig. 3(c)).

To quantify the gains from the probability of compromising messages due to the implementation of the proposed method based on the use of overlapping paths, in comparison with earlier known solutions, the following expressions were used:

$$\Delta_1 = \frac{P^1_{msg} - \tilde{P}_{msg}}{P^1_{msg}} \cdot 100\%, \tag{17}$$

and

$$\Delta_2 = \frac{P^2_{msg} - \tilde{P}_{msg}}{P^2_{msg}} \cdot 100\%. \tag{18}$$

In accordance with these expressions, the graphs presented in Fig. 4 have been obtained.

Based on the results presented in Fig. 4, we can conclude that the use of the proposed method for secure routing of message parts along two overlapping simple paths, united in a single composite path, has led to an improvement in the probability of compromising the transmitted message:

- in comparison to implementation of one simple path $1 \to 3 \to 4$ at average by 20–55% under $p_1^4 = 0.1 \div 0.3$; and by 5–20% under $p_1^4 = 0.5 \div 0.9$ (Fig. 4(a));
- in comparison to implementation of one simple path $1 \to 2 \to 3 \to 4$ at average by 5–50% under $p_1^4 = 0.1 \div 0.3$; and by 3–15% under $p_1^4 = 0.5 \div 0.9$ (Fig. 4(b)).

The gain on the probability of compromise of the transmitted message decreased at $p_1^4 \to 1$ because all the considered routes, both simple and composite, passed through this link.

4.2 Numerical Study of Using the Two Different Types of Non-overlapping Paths

Similarly, we perform a comparative analysis of the effectiveness of the proposed method (see Sect. 3) and the previously known solutions (see Sect. 2) for the network structure presented in Fig. 2. Using the method suggested in the paper, the probability

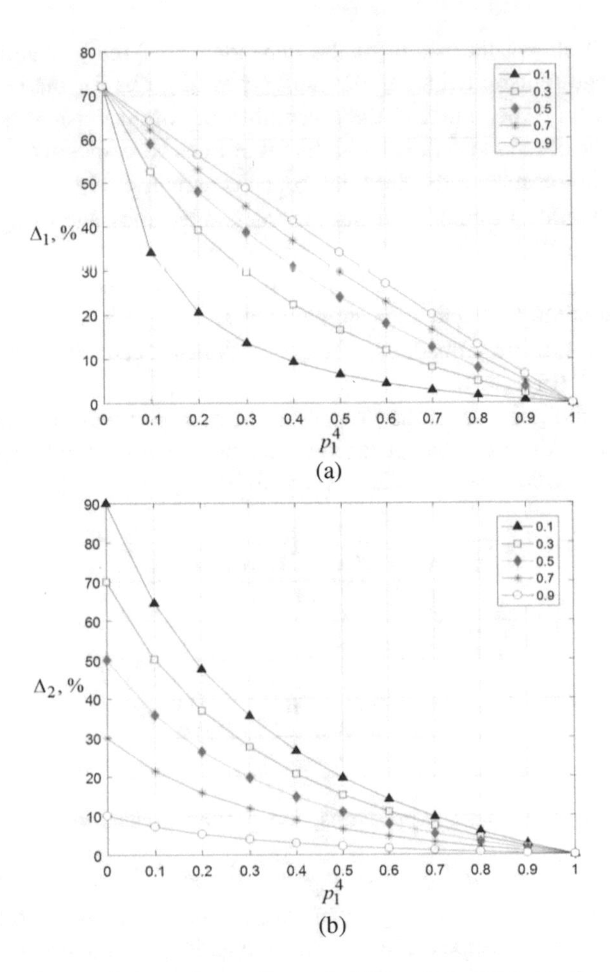

Fig. 4. The dependence of the gain on the probability of compromise from the use of the proposed method in comparison with the previously known solutions for the network structure given in Fig. 1(c).

of message compromise ($\tilde{P}_{msg}$) transmitted using all available communication links included into one composite and one simple path was estimated.

Using the previously known method (2)–(6), the probability of compromising the message was estimated, the parts of which were transmitted using two non-overlapping simple paths. Two possible cases of a combination of the choice of such paths were considered. In the first case, the paths $1 \rightarrow 2 \rightarrow 4 \rightarrow 5 \rightarrow 7$ and $1 \rightarrow 6 \rightarrow 7$ were used, to which the probability of compromising the message P^1_{msg} corresponded. In the second case, parts of the message were transmitted over another pair of non-overlapping paths: $1 \rightarrow 2 \rightarrow 3 \rightarrow 5 \rightarrow 7$ and $1 \rightarrow 6 \rightarrow 7$, to which the probability of compromise P^2_{msg} corresponded.

The indicator of the effectiveness of secure routing was again the probability of compromising transmitted message. In the course of the research, the impact of the probabilities of compromising the links 2 → 4 and 1 → 6, varying from 0 to 1, on the effectiveness were analyzed. The link 2 → 4 was a part of the composite path under the number four, and the link 1–6 had the first number in the structure of the simple path (Table 2). Table 2 also shows the probabilities of compromising all the links included in these two paths.

Table 2. Initial research data for the case of using two different types of non-overlapping paths.

Path number	1 (Composite)						2 (Simple)	
Link	1 → 2	2 → 3	3 → 5	2 → 4	4 → 5	5 → 7	1 → 6	6 → 7
Link number in path	1	2	3	4	5	6	1	2
Probability of compromising the link	0.2	0.1	0.1	0 ÷ 1	0.1	0.2	0 ÷ 1	0.2
Number of message parts	5	3	3	2	2	5	5	5

The last line in Table 2 shows the result of the solution for the stated optimization problem associated with minimizing the expression (15) under the constraints (8), (13) or (14), (16) in the Shamir's scheme (10, 10) and $h_1 = 2$, $h_2 = 1$. On the first (composite) path and the second (simple) path, 5 parts of the original message were transmitted, i.e. $n_1 = n_2 = 5$.

Figure 5 shows the nature of the dependence of the probability of compromising the message transmitted along different types of paths for the network structure, as shown in Fig. 2, on the values of the probability of compromising the fourth link in the composite path p_1^4 (plotted along the abscissa axis). Each of the set of lines in Fig. 5 corresponded to its value of the probability of compromising the first link of the simple path (p_2^1). As shown in Fig. 5(a) and (b), with increasing p_2^1 and p_1^4 the probability of compromising the transmitted message using the composite path and the simple path 1 → 2 → 4 → 5 → 7 has always increased. Given that the simple paths 1 → 2 → 3 → 5 → 7 and 1 → 6 → 7 did not contain the link 2 → 4 (Fig. 2), the probability of their compromising depended only on p_2^1 and did not depend on p_1^4 (Fig. 5(c)).

A quantitative analysis of the gain in the probability of compromising messages (Fig. 6) from the implementation of the proposed method based on the use of overlapping paths is performed in comparison with known solutions using expressions (17) and (18).

As shown in Fig. 6, the gain according to the formulas (17) and (18) on the probability of compromising the message transmitted along the paths of various types depends only on the security parameters of the links entering the composite path. In this case, this is the link 2 → 4, which is the fourth link of the first (composite) path with the probability of compromise p_1^4. The gain (17) and (18) did not depend (Fig. 6) on the value of the probability of compromising the link 1 → 6 ($p_2^1 = 0.1 \div 0.9$), which is the first link of the second (simple) path.

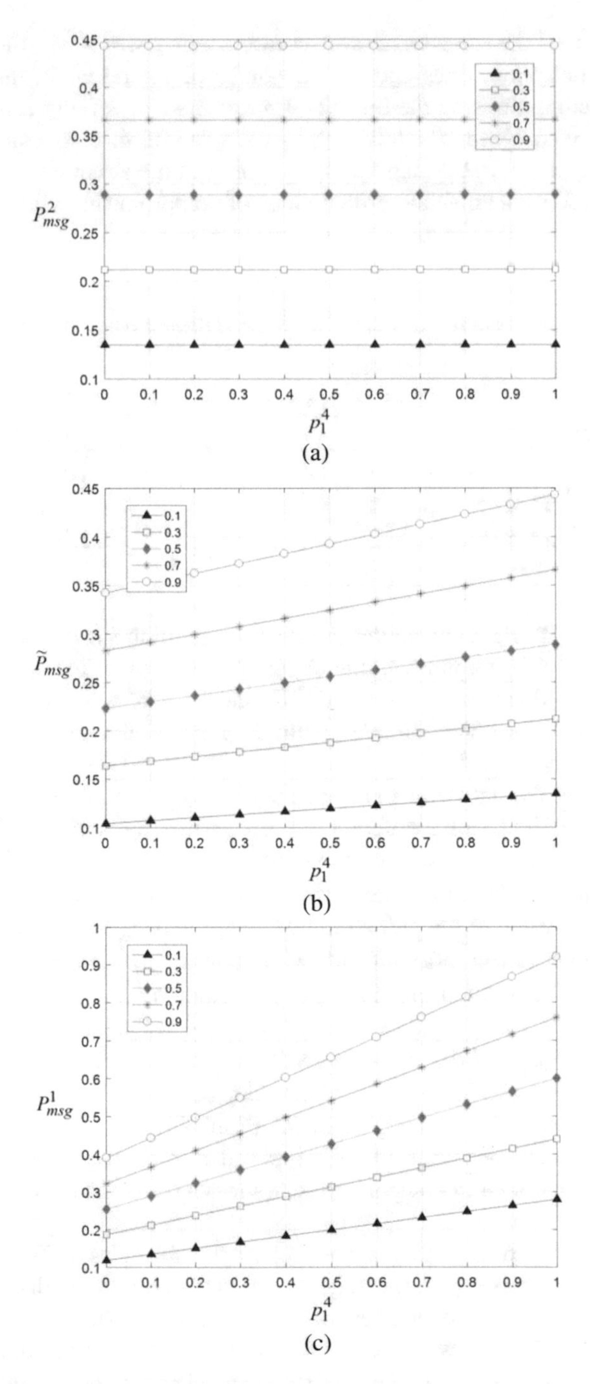

Fig. 5. Dependence of the probability of compromising the message transmitted along different types of paths for the network structure, as shown in Fig. 2.

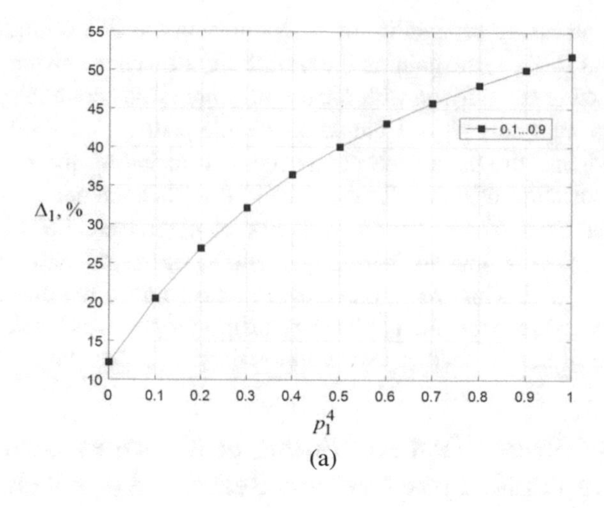

(a)

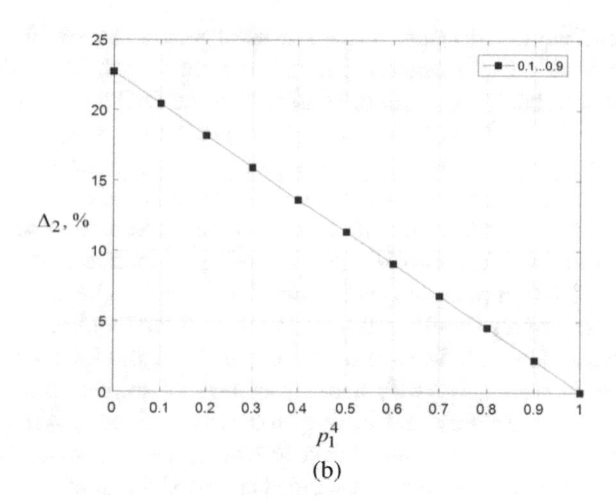

(b)

Fig. 6. The dependence of the gain on the probability of compromise from the use of the proposed method in comparison with the previously known solutions for the network structure given in Fig. 2.

Based on the results presented in Fig. 6, it can also be concluded that the use of the proposed method of secure routing has led to an improvement in the probability of compromising the transmitted message:

- in comparison to implementation of two non-overlapping simple paths $1 \to 2$ $4 \to 5 \to 7$ and $1 \to 6 \to 7$ at average by 20−33% under $p_1^4 = 0.1 \div 0.3$; and by 40–50% under $p_1^4 = 0.5 \div 0.9$ (Fig. 6(a));
- in comparison to implementation of two non-overlapping simple paths $1 \to 2$ $3 \to 5 \to 7$ and $1 \to 6 \to 7$ at average by 16–20% under $p_1^4 = 0.1 \div 0.3$; and by 3–12% under $p_1^4 = 0.5 \div 0.9$ (Fig. 6(b)).

Thus, with increasing probability of compromising the link included in the composite path, i.e. at $p_1^4 \to 1$, the gain on the probability of compromising the transmitted message increased in comparison with the use of simple paths containing the same link (Fig. 6(a)). After all, under $p_1^4 = 1$ the whole simple path $1 \to 2 \to 4 \to 5 \to 7$ will be compromised, and the use of the composite path including the network fragment with parallel connection of links allows to avoid this phenomenon.

On the other hand, if the link $2 \to 4$ was compromised, i.e. $p_1^4 = 1$, then the composite path actually lost its advantage, turning de facto into the simple path $1 \to 2 \to 3 \to 5 \to 7$. This led to a decrease in the gain on the probability of compromising the message from the application of the proposed method in comparison with the method of secure routing over non-overlapping paths (Fig. 6(b)).

5 Method of Secure Fast ReRouting of Messages Over Composite Paths: Proactive and Reactive Approaches

In order to extend the functionality of the means of secure routing, it is important for the proposed method to implement the principles of not only the proactive but also the reactive approach. In other words, in the structure of the method of secure routing it is important to provide procedures for prompt response to possible violations of the information security level. Currently, routing protocols react to possible changes in the network state in the time scale of tens of seconds, which is not always acceptable both in terms of the required level of quality of service and information security.

Therefore, methods and protocols of fast rerouting are increasingly used in practice, during which two types of paths are precomputed: primary and backup. In this case, the use of each type separately should lead to satisfaction of the requirements regarding the level of information security. Then, if the main path fails, the transmitted data will be routed almost instantaneously (with a delay of tens of milliseconds) using backup routes. Of course, the primary and backup routes should not overlap on the failed network elements (routers, communication links or routes in general) [15–17]. The causes of denial of service can be both overload and breakdown of network equipment, and the consequences of network attacks and the impact of malicious software (viruses).

Then, within the framework of Secure Fast ReRouting (S-FRR), the use of multiple primary paths refers to Proactive Approach solutions for providing a given level of information security, and the application of backup paths meets the requirements of Reactive Approach. At the same time, in the framework of the proposed method, the calculation of the set of primary and backup paths should be carried out as consistently as possible in order to improve the efficiency of the final solutions.

Division of paths into primary and backup means that parts of the message will not be transmitted over all accessible composite and simple paths, only in their limited number, but with the fulfillment of the requirements for the probability of compromise (1).

Considering the fact that in order to increase the level of information security of transmitted messages, it is necessary to implement multipath routing of their parts, primary and backup paths will be represented by not individual composite or simple paths, but by the multipaths formed by them. In this case, the composition of both the primary and backup multipath can include several composite and (or) simple paths.

In the calculation of the backup multipath, it is proposed to implement the following two protection schemes for the primary multipath:

- protection scheme for the primary multipath as a whole, in which the primary and backup multipaths do not overlap either by nodes or by communication links;
- protection scheme for a single path (composite or simple) of the primary multipath where the backup multipath should not contain the protected path of the primary multipath.

The implementation of each of the protection schemes is aimed at restoring a given level of information security by eliminating the primary multipath and moving to the use of a backup multipath. In this regard, we refine the previously set notations and introduce some additional ones:

Indices

$\tilde{p}_i^{pr}$	probability of compromising the i th composite or simple path of primary multipath, ($i = \overline{1,\widetilde{M}}$)
$\tilde{p}_i^{b}$	probability of compromising the i th composite or simple path of backup multipath, ($i = \overline{1,\widetilde{M}}$)
$\tilde{P}_{msg}^{pr}$	probability of compromise for the whole message during its transmission in parts over composite or simple paths of primary multipath
$\tilde{P}_{msg}^{b}$	probability of compromise for the whole message during its transmission in parts over composite or simple paths of backup multipath

Variables

n_i	integer variable, which is the number of message parts, transmitted over the i th composite or simple path, included into the primary multipath, ($i = \overline{1,\widetilde{M}}$)
$\bar{n}_i$	integer variable, which is the number of message parts, transmitted over the i th composite or simple path, included into the backup multipath, ($i = \overline{1,\widetilde{M}}$)

In accordance with the above notations, in order to calculate the probability of compromising the message transmitted in parts over a set of composite paths, it is necessary, by analogy with formulas (1) and (9), to use expressions

$$\tilde{P}^{pr}_{msg} = \prod_{i=1}^{\tilde{M}} \tilde{p}^{pr}_i \text{ and } \tilde{P}^{b}_{msg} = \prod_{i=1}^{\tilde{M}} \tilde{p}^{b}_i. \tag{19}$$

It should be noted that the probabilities of compromising network fragments $\tilde{p}^{pr}_i$ and $\tilde{p}^{b}_i$ are the function of the number of message parts transmitted in them, i.e. from n_i and $\bar{n}_i$. Then, taking into account (8), we have the conditions

$$\tilde{p}^{pr}_i = \begin{cases} 1 - \prod\limits_{j=1}^{\tilde{M}_i} (1 - \tilde{p}^j_i),\ n_i > 0; \\ 1,\ n_i = 0, \end{cases} \text{ and } \tilde{p}^{b}_i = \begin{cases} 1 - \prod\limits_{j=1}^{\tilde{M}_i} (1 - \tilde{p}^j_i),\ \bar{n}_i > 0; \\ 1,\ \bar{n}_i = 0. \end{cases} \tag{20}$$

The systems (20) can be rewritten as follows:

$$\tilde{p}^{pr}_i = 1 - H_0(n_i) \prod_{j=1}^{\tilde{M}_i} (1 - \tilde{p}^j_i) \text{ and } \tilde{p}^{b}_i = 1 - H_0(\bar{n}_i) \prod_{j=1}^{\tilde{M}_i} (1 - \tilde{p}^j_i), \tag{21}$$

where H_0 is the Heaviside function, which, taking into account expression (20), is calculated as follows

$$H_0(n) = \begin{cases} 0,\ n = 0; \\ 1,\ n > 0. \end{cases}$$

The conditions (16) due to realization of S-FRR are added by the expression

$$\sum_{i=1}^{\tilde{M}} \bar{n}_i = N. \tag{22}$$

In turn, to protect the main multipath, by analogy with [15, 16], it is necessary to ensure the following condition:

$$\sum_{i=1}^{\tilde{M}} n_i \bar{n}_i = 0. \tag{23}$$

If it is necessary to protect the individual i th composite path, it is important to ensure fulfillment of the condition

$$n_i \bar{n}_i = 0, \tag{24}$$

which is also nonlinear (bilinear).

In order to meet the requirements regarding the probability of compromising the messages transmitted using both the primary and backup multipath, the following conditions are introduced by analogy with (1):

$$P_{msg}^{pr} \leq \gamma_P \text{ and } P_{msg}^{b} \leq \gamma_P. \tag{25}$$

Therefore, the basis of the developed S-FRR method can be the solution of the optimization problem of Nonlinear Integer Programming with the optimality criterion

$$J = \sum_{i=1}^{\tilde{M}} \tilde{p}_i n_i + \sum_{i=1}^{\tilde{M}} \tilde{p}_i \bar{n}_i, \tag{26}$$

and the constraints represented by the conditions (13), (14), (16), (22)–(25). In this case, the constraints (23)–(25) are nonlinear, and the calculated variables n_i and $\bar{n}_i$ are integer. In the criterion (26), the values $\tilde{p}_i$ calculated in accordance with expressions (8) are the cost weight coefficients. This ensures secure routing over the network when the maximum number of message parts will be sent over the path with the minimum probability of compromise. Conversely, the minimum number of message parts will be transmitted over the path with the highest probability of compromise or none of them will be transmitted.

6 Numerical Study of Secure Fast ReRouting

Let us demonstrate the features of the proposed mechanism of Secure Fast ReRouting. The initial structure of the network is shown in Fig. 7, and the corresponding probabilities of compromising the communication links are indicated in Table 3.

The source of the message is the first node, and the destination is the seventeenth node. The solid lines in Fig. 7 show the communication links used to form the primary and backup multipaths for message transmission.

Suppose that with the secure fast rerouting, the Shamir's scheme (10, 10) is realized and according to the structure of the paths shown in Fig. 7, $h_1 = 1$, $h_2 = 2$, $h_3 = 2$ and $h_4 = 1$, and the allowable value of the probability of compromising the transmitted message, determined by the parameter γ_P, is 0.3. Then, in the course of the study, two cases were considered demonstrating the features of the implementation of the protection schemes described in Sect. 5:

- the first case is associated with the implementation of the protection scheme of the second (composite) path;
- the second case describes the protection scheme for the primary multipath as a whole.

Let us consider Case 1 in more detail. Thus, according to the data in Table 3, based on the calculation method proposed in Sect. 5, the primary multipath includes two composite paths: the second and the third, with the smallest probabilities of compromising: 0.5339 and 0.4061, respectively. The parameters of these paths and the order of balancing the parts of the transmitted message by the paths are presented in Table 4. On the third (composite) path, 8 parts of the message were transmitted, because the probability of compromising this path is minimum and equal to 0.4061. On the second (composite) path two parts of the message were transmitted as its probability of compromising was already 0.5339, but the lower threshold n_2 under the conditions (13) was $h_2 = 2$. The use of these two paths as the primary multipath in accordance with expression (19) provides the probability of the message compromise equal to 0.2168, which satisfies the requirements (0.3).

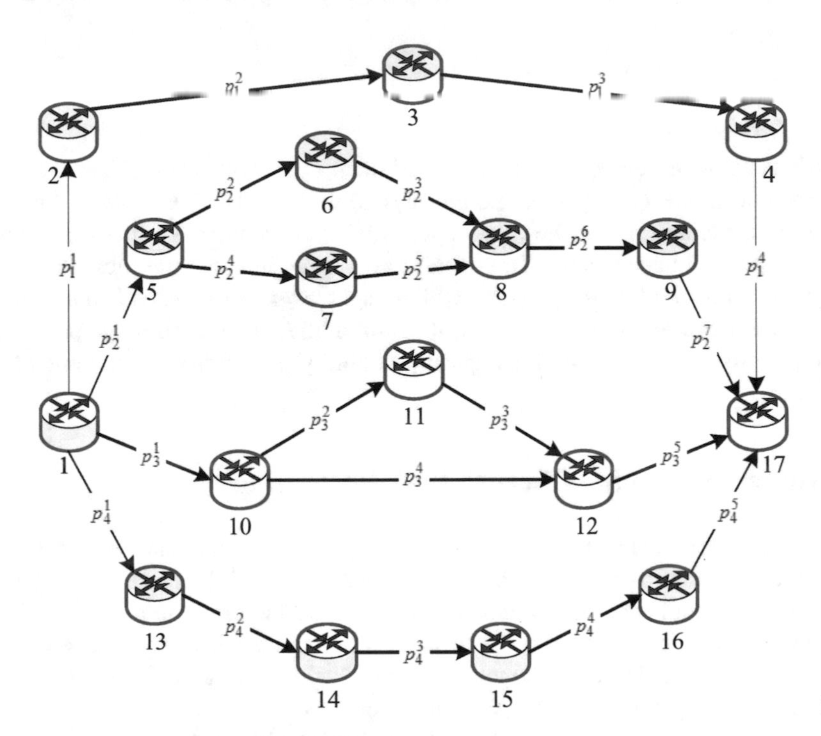

Fig. 7. Initial network structure.

Table 3. Initial research data for the Secure Fast ReRouting.

Path number	1 (Simple)						
Link	1→2	2→3	3→4	4→17			
Link number in path	1	2	3	4			
Probability of compromising the link	0.15	0.19	0.17	0.2			
Path number	2 (Composite)						
Link	1→5	5→6	6→8	5→7	7→8	8→9	9→17
Link number in path	1	2	3	4	5	6	7
Probability of compromising the link	0.2	0.2	0.1	0.2	0.2	0.19	0.2
Path number	3 (Composite)						
Link	1→10	10→11	11→12	10→12	12→17		
Link number in path	1	2	3	4	5		
Probability of compromising the link	0.2	0.2	0.2	0.2	0.2		
Path number	4 (Simple)						
Link	1→13	13→14	14→15	15→16	16→17		
Link number in path	1	2	3	4	5		
Probability of compromising the link	0.2	0.18	0.1	0.15	0.1		

Table 4. Parameters of the primary multipath.

Path number	2 (Composite)						
Probability of compromising the path	0.5339						
Number of message parts in path	2						
Link	1→5	5→6	6→8	5→7	7→8	8→9	9→17
Link number in path	1	2	3	4	5	6	7
Number of message parts in link	2	1	1	1	1	2	2
Path number	3 (Composite)						
Probability of compromising the path	0.4061						
Number of message parts in path	8						
Link	1→10	10→11	11→12	10→12	12→17		
Link number in path	1	2	3	4	5		
Number of message parts in link	8	4	4	4	8		

Table 5. Parameters of the backup multipath (Case 1).

Path number	1 (Simple)				
Probability of compromising the path	0.5428				
Number of message parts in path	1				
Link	1→2	2→3	3→4	4→17	
Link number in path	1	2	3	4	
Number of message parts in link	1	1	1	1	
Path number	3 (Composite)				
Probability of compromising the path	0.4061				
Number of message parts in path	9				
Link	1→10	10→11	11→12	10→12	12→17
Link number in path	1	2	3	4	5
Number of message parts in link	9	5	5	4	9

When protecting the second (composite) path of the primary multipath, the first (simple) and third (composite) paths, the parameters of which are presented in Table 5, will already be a part of the calculated backup multipath. The third (composite) path will transmit 9 parts of the message, and the first one (simple) will transmit only 1 part, because the probability of compromising the first path was 0.5428 at $h_1 = 1$. The use of a backup multipath also allowed to satisfy the requirements for the level of information security, because the probability of compromising the transmitted message was 0.2204.

Let us consider Case 2 in more detail, in which it was necessary to protect the primary multipath as a whole. The application of the proposed S-FRR method has left the basic multipath (Table 4) unchanged. Then, according to the initial data in Table 3, the backup multipath includes two simple paths: the first and the fourth (Table 6).

The probability of compromising the fourth (simple) path was 0.5483. Then the use of a backup multipath allowed to ensure the probability of compromising the transmitted message to 0.2977 under the requirements $\gamma_P = 0.3$. In the first path, 9 parts of

Table 6. Parameters of the backup multipath (Case 2).

Path number	1 (Simple)				
Probability of compromising the path	0.5428				
Number of message parts in path	9				
Link	1→2	2→3	3→4	4→17	
Link number in path	1	2	3	4	
Number of message parts in link	9	9	9	9	
Path number	4 (Simple)				
Probability of compromising the path	0.5483				
Number of message parts in path	1				
Link	1→13	13→14	14→15	15→16	16→17
Link number in path	1	2	3	4	5
Number of message parts in link	1	1	1	1	1

the message were transmitted, because the probability of compromise was lower than that of the fourth path, over which one part of the message was transmitted under $h_4 = 1$.

7 Conclusion

1. The paper considers the approach to providing a given level of information security by means of multipath routing of confidential messages in the network. The proposed solutions are the further development of secure routing methods for non-overlapping paths proposed in [10–12], when the probability of compromising transmitted messages was the main indicator of information security.
2. The method has been developed for the secure routing of messages over overlapping paths, which belongs to the class of proactive solutions to ensure a given level of information security. The novelty of the method is that it is fair, also when using a special class of overlapping paths, which form the basis of the so-called composite paths containing network fragments with series and/or parallel connection of communication links of the network. The method is based on optimizing the process of selecting a set of composite paths and balancing the parts of the transmitted message along them with the provision of specified values of its probability of compromise. The formulated optimization problem belonged to the class of Nonlinear Integer Programming problems. This is due to the fact that the set of variables to be calculated characterizing the number of transmitted message parts in the composite path were integer, and the constraints (1), (9) as well as the criterion (15) associated with calculating the probability of compromising the message were nonlinear.
3. As the analysis has shown, the use of the proposed method within the framework of the presented calculated examples (Sect. 4) makes it possible to improve the probability of compromise of transmitted messages at average from 5–10% to 25–50% (Figs. 4 and 6) due to the possibility of using composite paths that are one of the subclasses of overlapping paths.

4. The method of Secure Fast ReRouting of messages in the network has been synthesized. The novelty of the method is that it focuses on the implementation of both proactive and reactive secure routing of confidential messages. In this case, the proactive nature of the solutions is conditioned by the calculation of the set of primary composite paths that form the primary multipath, along which parts of the confidential message are transmitted. In the case of violation of the information security requirements in the network caused by the increased probability of compromising one or multiple composite paths entering the primary multipath, the messages will be transmitted according to the precomputed set of backup composite paths determining the backup multipath. Based on the use of backup paths, such a solution refers to the reactive approach while ensuring a given level of information security.
5. Moreover, within the proposed S-FRR method, the possibility of protecting both the primary multipath as a whole and one or several precomputed composite paths included in this primary multipath is provided. The application of the developed S-FRR method allows real-time provision of specified values of such an important information security indicator as the probability of compromising transmitted messages even in conditions of dynamic network state change (probability of link and path compromise) due to the calculation and operative switching to the backup composite paths (backup multipath) in the multipath transmission of parts of the confidential message.
6. The developed methods of secure routing can be used as a basis for new network routing protocols and fast rerouting for multipath transmission of parts of a confidential message with specified requirements regarding the probability of its compromise in the network.

References

1. ITU-T X-805. Security architecture for systems providing end-to-end communications (2003)
2. ISO 7498–2:1989 Information processing systems – Open Systems Interconnection – Basic Reference Model – Part 2: Security Architecture (1989)
3. ITU-T X-800. Security architecture for Open Systems Interconnection for CCITT applications (1991)
4. Stallings, W.: Cryptography and Network Security: Principles and Practice, 7th edn. Pearson, London (2016)
5. Schneier, B.: Data and Goliath: The Hidden Battles to Collect Your Data and Control Your World, 1st edn. WW Norton & Company, New York (2015)
6. Cisco Networking Academy (ed.): Routing Protocols Companion Guide, 1st edn. Cisco Press (2014)
7. Santos, O., Kampanakis, P., Woland, A.: Cisco Next-Generation Security Solutions: All-in-one Cisco ASA Firepower Services, NGIPS, and AMP, 1st edn. Cisco Press (2016)
8. Wang, M., Liu, J., Mao, J., Cheng, H., Chen, J.: NSV-GUARD: constructing secure routing paths in software defined networking. In: Proceedings of the 2016 IEEE International Conferences on Big Data and Cloud Computing (BDCloud), Social Computing and Networking (SocialCom), Sustainable Computing and Communications (SustainCom) (BDCloud-SocialCom-SustainCom), pp. 293–300 (2016)

9. Almerhag, I.A., Almarimi, A.A., Goweder, A.M., Elbekai, A.A.: Network security for QoS routing metrics. In: Proceedings of the 2010 International Conference on Computer and Communication Engineering (ICCCE), pp. 1–6 (2010)
10. Lou, W., Liu, W., Zhang, Y., Fang, Y.: SPREAD: improving network security by multipath routing in mobile ad hoc networks. Wirel. Netw. **15**(3), 279–294 (2009)
11. Alouneh, S., Agarwal, A., En-Nouaary, A.: A novel path protection scheme for MPLS networks using multi-path routing. Comput. Netw. **53**(9), 1530–1545 (2009)
12. Yeremenko, O.S., Ali, A.S.: Secure multipath routing algorithm with optimal balancing message fragments in MANET. Radioelectron. Inform. **1**(68), 26–29 (2015)
13. Yeremenko, O.: Enhanced flow-based model of multipath routing with overlapping by nodes paths. In: Proceedings of the 2015 Second International Scientific-Practical Conference Problems of Infocommunications Science and Technology (PIC S&T), pp. 42–45 (2015)
14. Yeremenko, O., Lemeshko, O., Persikov, A.: Enhanced method of calculating the probability of message compromising using overlapping routes in communication network. In: Proceedings of the 2017 XIIth International Scientific and Technical Conference Computer Sciences and Information Technologies (CSIT), pp. 87–90 (2017)
15. Lemeshko, O., Romanyuk, A., Kozlova, H.: Design schemes for MPLS fast reroute. In: Proceedings of the 2013 12th International Conference on the Experience of Designing and Application of CAD Systems in Microelectronics (CADSM), pp. 202–203 (2013)
16. Lemeshko, O.V., Yeremenko, O.S., Tariki, N., Hailan, A.M.: Fault-tolerance improvement for core and edge of IP network. In: Proceedings of the 2016 XIth International Scientific and Technical Conference Computer Sciences and Information Technologies (CSIT), pp. 161–164 (2016)
17. Lemeshko, O., Yeremenko, O., Nevzorova, O.: Hierarchical method of inter-area fast rerouting. Transp. Telecommun. J. **18**(2), 155–167 (2017)

Linguistic Comparison Quality Evaluation of Web-Site Content with Tourism Documentation Objects

Pavlo Zhezhnych[1(✉)] and Oksana Markiv[2]

[1] Department of Social Communication and Information Activities, Lviv Polytechnic National University, Lviv, Ukraine
pavlo.i.zhezhnych@lpnu.ua

[2] Department of Information Systems and Networks, Lviv Polytechnic National University, Lviv, Ukraine
oksanasoprunyuk@gmail.com

Abstract. Information support for tourism service consumers is the important element of tourism activity. Open websites are the main sources of such support. Widespread approach to the analysis of tourism services quality is the analysis of feedbacks from consumers based on the results of tourism services provision. In this work, it is proposed to evaluate the information provision of tourism services consumer formed in the form of tourism documentation. This approach allows even before the provision of tourism services to influence the choice of the consumer and satisfaction with tourism services. The developed method for evaluating the quality of tourism documentation generated on the basis of the website content is based on ISO/IEC-25010 quality indices that are suitable for assessing the quality of the tourism documentation from the point of view of the consumer of its content. These indices include such sub-characteristics of Operability, as Appropriateness recognizability, Helpfulness, Attractiveness.

Keywords: Tourism · Information quality · Documentation · Web-site Content

1 Introduction

Nowadays activities in providing a wide range of tourism services (TS) are practically impossible without proper information support through the World Wide Web (WWW). Obviously, the main source of information for tourism services support are open web sites. According to Maurice de Kunder and his colleagues calculations, about 4.66 billion web pages are available in the WWW via search engines (as of March 2016) [18]. In modern conditions, the content of a large number of web pages is generated by users of web sites (bloggers). According to the blogging industry, about 11% of blogs cover such topics as "Travel" and "Food & Beverage" [14], that is, they directly concern tourism problems. Therefore, even on the basis of the data on the subject of blogs it is easy to predict that millions of web pages that are related to the tourism industry are accessible in the WWW. Qualitative processing, consolidation and

N. Shakhovska and V. Stepashko (eds.), *Advances in Intelligent Systems and Computing II*, Advances in Intelligent Systems and Computing 689,
https://doi.org/10.1007/978-3-319-70581-1_45

presentation of information from such an array is one of the important tasks of effective tourism activity.

This paper dwells upon the presentation of information for tourism services provision in the form of relevant documentation. That is, tourism documentation (TD) is technical documentation for TS, which allows to provide the consumer with the necessary information about these TS. Consumer orientation allows to consider TD as information product, on the quality of which depends the satisfaction of the TD consumers needs. Inadequate TD may result in the poor provision of tourism services because of misinformation and dissatisfaction of tourists, whose negative experience extends and weakens the image of tourism institutions. The construction of high quality TD requires constant monitoring of its quality, its analysis and formation of modern quality requirements, so that TD will be available for use and as informative as possible. Formation of quality TD is related to the achievement of such goals in the interests of the consumer:

- increase of the TS attractiveness in the eyes of the consumer (potential tourist), that can increase the competitiveness of TS;
- support of the information that is needed in order to understand the features of TS provision, that allows the TS generator to be protected from possible claims that may arise during inappropriate consumption of TS.

The basis of the TD formation is the linguistic analysis of the information content of websites with open access in the WWW. The application of the appropriate method of TD formation allows automated processing of large amounts of information from open web resources, as well as its consolidation, classification, structuring and further use.

The main result of this work is the development of method for assessing the quality of TD, formed on the basis of websites content. This method is based on quality indicators of ISO/IEC-25010 [15], which are suitable for assessing the quality of the TD from the point of view of the consumer of its content.

2 State-of-the-Art

The construction of various information systems to support tourism activity is a widespread trend in scientific research. The main emphasis of these studies is on the effective presentation of tourism information for the end user. In particular, [2, 13] discuss general approaches to the construction of tourism support systems and services, and [6] highlights the problems of effective selection of information for its collection in such systems. Personalization of tourism information depending on the needs of the TS consumer is important for planning travel through e-Tourist services [3]. In this case, not only the causes that influence the choice of travel, but also the consequences of this choice are important [16]. An important consequence of the choice of travel is the ability to predict the happiness of the consumer of tourism services, which allows adapting tourism services to a particular consumer [11]. An important element of information systems of tourism activity support is the visualization of information related to geodata [8]. In this case, useful information source of such information is the user generated content [9], which is often associated with a GPS location [7]. User generated content is also used to generate new knowledge based on the common tourism key phrases and text corpus [12].

Key element of tourism services provision is the assessment of their quality, since it directly affects the frequency and duration of tourist services consumption [10]. The main method of such evaluation is consumer feedback analysis [5]. At the same time, typical methods of detecting information about the impressions of tourism services consumer are based on the use of fuzzy evaluation [21] with regard to the level of trust to the consumer [4].

This paper dwells upon the issues of assessing the quality of tourism services not from the point of view of their provision, but from the point of view of their information support. Actually, the initial information provided to the consumer largely influences his choice, as well as his satisfaction with tourism services and feedback. Therefore, this information needs to be properly structured in the form of tourism documentation.

The quality of tourism documentation is a category that reflects the degree of dependence of its characteristics and properties conditioned or pre-seen needs and requirements of the tourism product consumer [20]. TD in terms of quality should have the following main properties:

- provide a set of structural elements that are filled with information about tourism activities;
- be clear to the consumer;
- simple and easy to use;
- contain no gaps in the filling of structural elements;
- contain actual information.

To assess the quality of TD, the ISO/IEC-25010 quality standard is acceptable. In this case, among all the indicators of quality applicable are those sub-characteristics that meet the requirements of TD construction. These are the following sub-characteristics of Operability [16]:

- Appropriateness recognizability;
- Helpfulness;
- Attractiveness.

3 Tourism Documentation Structure

3.1 General Structure of Tourism Documentation

General structure of tourism documentation is an extension of simplified tourism documentation structure described in [19]. The tourism documentation consists of the set of TS and sets of TS consumer experiences (Fig. 1). The structure of the TS is similar to that of TS consumer experiences (TE) and consists of the set of tourism objects (TO) and tourism objects actions (TOA). The structure of the TO is similar to the structure of TOA and consists of the sets of facts and events. Facts and events are elementary (indivisible) informational units in the structure of TD. They describe TO and TOA both in the structure of TS and in the structure of TE.

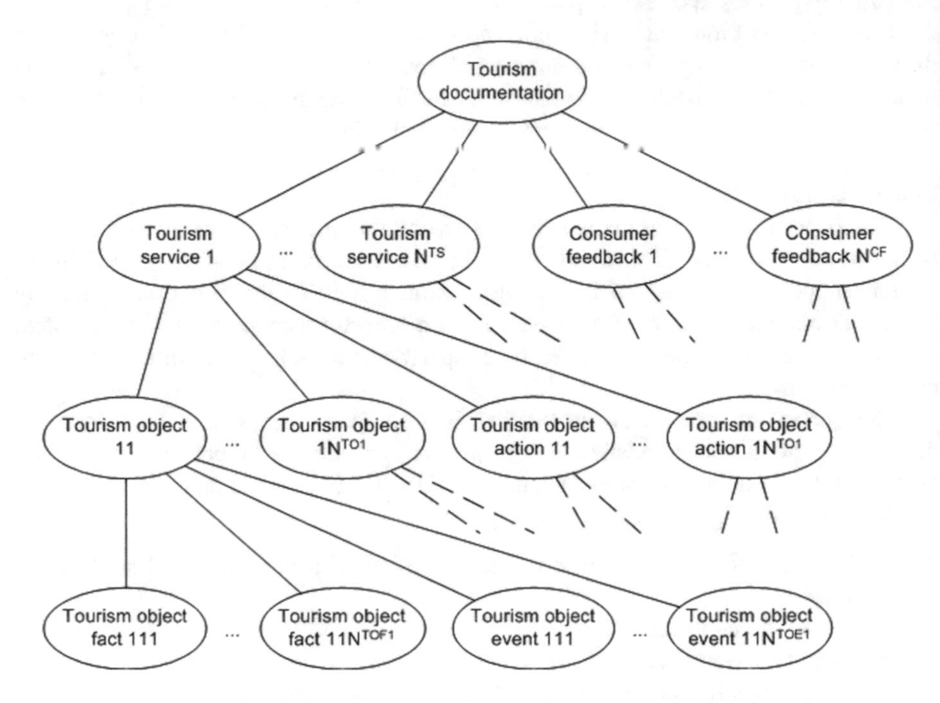

Fig. 1. General structure of tourism documentation.

The same TO and TOA can occur in different TS and TE. That is, in fact, TD has the form of network, and not a tree, where each TO and TOA should be associated with all TS and TE. However, such more complex structure of TD is needed first of all for the construction of a specific IS intended for the formation, processing and use of TD. The simplification of the network model to the model in the form of a tree is used in the methods of reduction of graphs [17], and this approach is used in the construction of a formal TD model, which is sufficient for the application of linguistic approaches to the automated formation of TD [19].

3.2 Eligibility of Tourism Documentation Content

The eligibility of the content of TD depends on the requirements that are necessary to meet the needs of the user. In particular, the following requirements are included:

- Lack of content gaps;
- Information authenticity;
- Information actuality.

Gaps in TD content arise because of the lack of information on the facts and events associated with certain structural elements of TD as TO and TOA. In general, the presence of content gaps shows that for the structural element of the TD there are not all necessary facts and events associated with it.

If $Element_i$ is TO or TOA, $ElementFact^{(TD)}(Element_i)$ is a set of all facts of this TO or TOA in TD, $ElementFact^{(Expected)}(Element_i)$ is an expected set of all facts of TO or TOA, then there is a fact content gap, if:

$$\begin{aligned} & ElementFact^{(Gap)}(Element_i) \\ & = ElementFact^{(Expected)}(Element_i) \backslash ElementFact^{(TD)}(Element_i) \neq \varnothing \end{aligned} \quad (1)$$

If $ElementEvent^{(TD)}(Element_i)$ is a set of all events of TO or TOA $Element_i$ in TD, $ElementEvent^{(Expected)}(Element_i)$ – expected set of all events of TO or TOA, then there is event content gap, if:

$$\begin{aligned} & ElementEvent^{(Gap)}(Element_i) \\ & = ElementEvent^{(Expected)}(Element_i) \backslash ElementEvent^{(TD)}(Element_i) \neq \varnothing \end{aligned} \quad (2)$$

In general, the authenticity of the information provided on the website depends directly on the level of trust to its author [1]. The credibility of the author, in particular, is determined by the adequacy of the data provided in the relevant user profile of the website [4]. Each fact or event contained in TD must be associated with separate author or related website. If the author of the information submitted on the website is unknown, then the owner of this website should be considered as the author of such information. Thus, the authenticity of the facts and events of TD is entirely determined by the level of trust to the author of the relevant information from the website.

The actuality of the information is determined by the time of its post publishing. If T_0 – the starting time point of the actuality index (that is, information that was published earlier then T_0 is considered irrelevant), T_{now} – the current time moment. Then the actuality of the information in Post, which was published at the time $PostTime(Post)$, is:

$$Actuality(Post) = \begin{cases} \frac{(T_{now} - T_0) - (T_{now} - PostTime(Post))}{T_{now} - T_0} & PostTime(Post) \geq T_0 \\ 0 & PostTime(Post) < T_0 \end{cases} \quad (3)$$

4 Tourism Documentation Quality Indices

4.1 Appropriateness Recognisability

Appropriateness recognisability is a measure of the user's ability to recognize whether the TD meets his/her needs. Appropriateness recognizability will is determined by the full description of the elements of TD:

$$Quality^{(O.AR)} = \frac{A^{(O.AR)}}{B^{(O.AR)}} \quad (4)$$

$A^{(O.AR)}$ displays the number of described elements of TD, $B^{(O.AR)}$ – the total quantity of TD elements. If $Element^{(\Pr esent)}$ is a set of available structural elements of TD, $Element^{(ContentGap)}$ is a set of structural elements of TD, which contains content gaps in the content defined by (1) or (2), then for all structural elements of TD (TO and TOA) associated with facts and events:

- $A^{(O.AR)} = \left|Element^{(\Pr esent)}\right| - \left|Element^{(ContentGap)}\right|$, $B^{(O.AR)} = \left|Element^{(\Pr esent)}\right|$

Thus, the appropriateness recognisability in accordance with the recognition indicates the level of presence of content gaps in the TD according to TO, TOA.

Also, this quality index is modified to display content gaps in the authenticity and actuality of facts and events (the facts of definition are not considered here). $Element^{(TrustGap)}$ – set of elements, containing facts or events whose authors have level of confidence lower than admitted, $Element^{(ActualityGap)}$ – set of elements containing non-actual facts or events in accordance with (3). Then the quality of the TD in the fullness of authentic and actual descriptions is determined by (5), if:

- $A^{(O.AR)} = \left|Element^{(\Pr esent)}\right| - \left|Element^{(TrustGap)}\right|$, $B^{(O.AR)} = \left|Element^{(\Pr esent)}\right|$ – for authenticity;
- $A^{(O.AR)} = \left|Element^{(\Pr esent)}\right| - \left|Element^{(ActualityGap)}\right|$, $B^{(O.AR)} = \left|Element^{(\Pr esent)}\right|$ – for actuality.

The authenticity varies in time, since the credibility of the post author, on the basis of which the fact or event are generated in TD, may change depending on its behavior in time.

4.2 Helpfulness

Helpfulness is a measure of user support by the help through the use of information from TD. Helpfulness on the availability of assistance is defined as follows:

$$Quality^{(O.H)} = \frac{A^{(O.H)}}{B^{(O.H)}} \quad (5)$$

$A^{(O.H)}$ – quantity of TO or TOA, that have TE in TD, $B^{(O.H)}$ – general quantity of TO or TOA. If $Element^{(Experience)}$ is a set of TO or TOA, for which TE are present, $Element^{(ExperienceContent)}$ is TO or TOA, for which TE are present and which do not have content gaps, according to (1) and (2), then:

- $A^{(O.H)} = \left|Element^{(ExperienceContent)}\right|$, $B^{(O.H)} = \left|Element^{(Experience)}\right|$.

Thus, helpfullness according to trust level shows the proportion of TO or TOA that contain content in the form of facts and events that cause interest of TS consumers displayed by their feedbacks.

4.3 Attractiveness

Attractiveness is the level of influence (impressions) of the TD on the user. Attractiveness is defined as follows:

$$Quality^{(O.A)} = \frac{A^{(O.A)}}{B^{(O.A)}} \tag{6}$$

$A^{(O.A)}$ – quantity of TO and TOA, which are in description part of TD and for which TE are present, $B^{(O.A)}$ – general quantity of TO and TOA, which are in description part of TD. If $Element^{(Present)}$ is a set of presented TO or TOA in TD and $Element^{(Experience)}$ is a set of TO or TOA, for which TE are present, then:

- $A^{(O.A)} = \left|Element^{(Experience)}\right|$, $B^{(O.A)} = \left|Element^{(Present)}\right|$.

In this way, the suitability for use for attractiveness shows the proportion of TO or TOA that have raised interest in TS consumers as reflected in their feedbacks.

5 Tourism Documentation Quality Monitoring

Monitoring of the TD quality on the indicators of suitability for use is made during the formation of TD on two tourism companies.

The first tourism company 1 provides services of the organization of tourism trips and recreation. The formation of the company TD through linguistic comparison of the website content with tourism documentation objects has begun in September 2014 and lasted 25 months. At the end of the experimental term, TD contained:

- 235 tourism services (dynamics of TS quantity change is presented in Fig. 2);
- 1545 tourism objects (dynamics of TO quantity change is presented in Fig. 3).

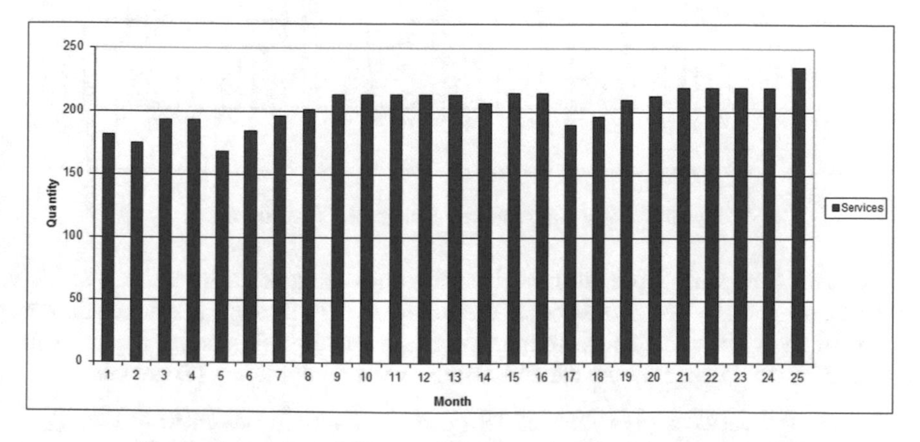

Fig. 2. Dynamics of TS quantity change in TD of the company 1.

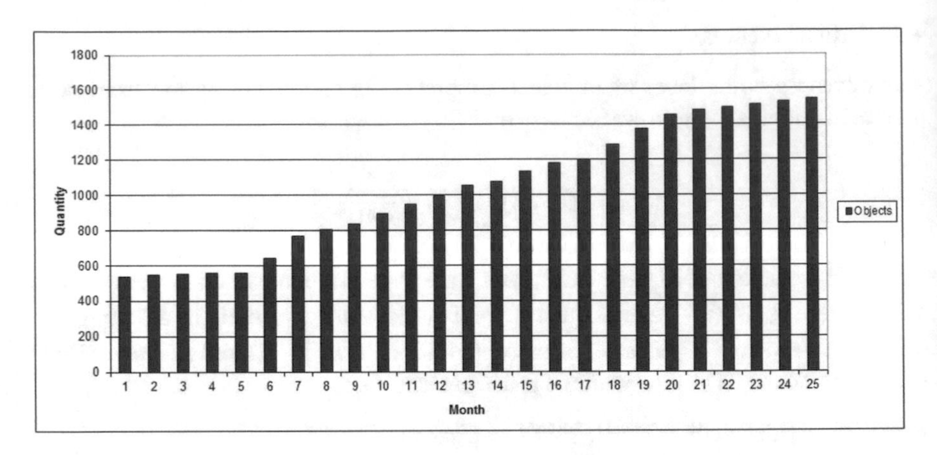

Fig. 3. Dynamics of TO quantity change in TD of the company 1.

Quality monitoring of TD has been made by the appropriateness recognisability indices according to the fullness of TO description (Fig. 4).

Figure 4 shows that in the company 1 after the first year of formation and support of TD in the current state it is possible to fully stabilize the level of TD quality at the sufficiently high level.

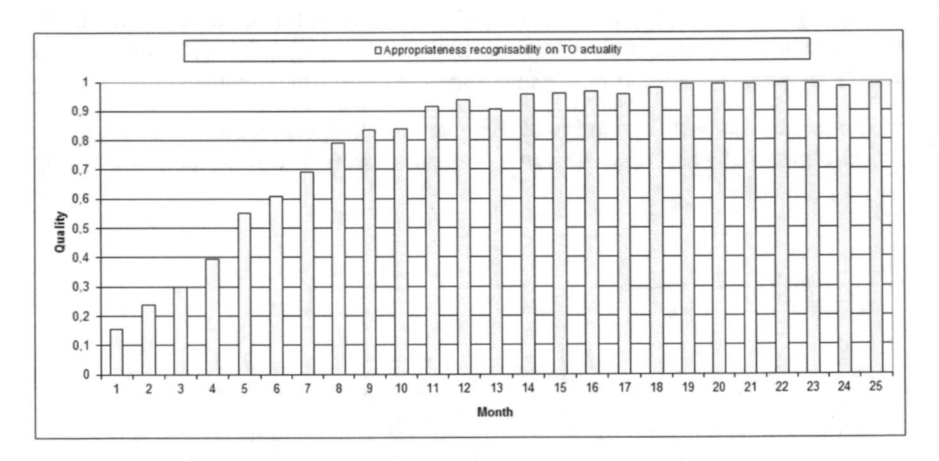

Fig. 4. Dynamics of quality indices change of TD of the company 1.

Tourism Company 2 provides services for organizing tours in the city of Lviv, Ukraine and abroad. The formation of the company TD through the linguistic comparison of the website content with tourism documentation objects has begun in July 2015 and lasted 15 months. At the end of the experimental term, TD contained:

- 210 tourism services (dynamics of TS quantity change is presented in Fig. 5);
- 210 tourism experiences (dynamics of TE quantity change is presented in Fig. 6);

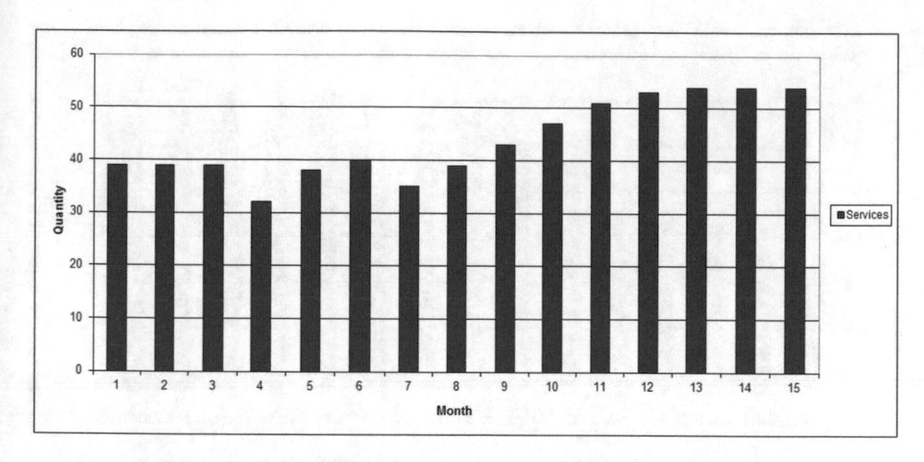

Fig. 5. Dynamics of TS quantity change in TD of the company 2.

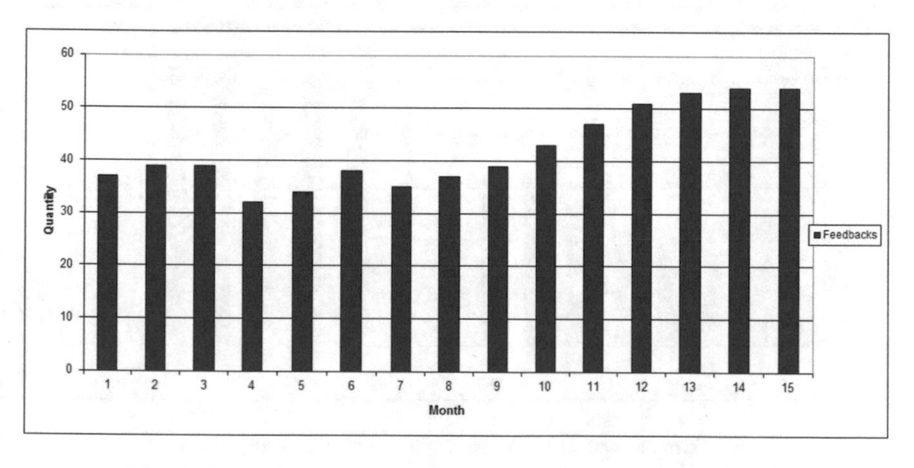

Fig. 6. Dynamics of TE quantity change in TD of the company 2.

- 1248 tourism objects (dynamics of TO quantity change is presented in Fig. 7);
- 2053 tourism objects actions (dynamics of TOA quantity change is presented in Fig. 8).

Quality monitoring of TD has been made by such indices (Fig. 9):

- appropriateness recognisability by the fullness of TO and TOA description;
- level of helpfulness by availability of TO and TOA;
- attractiveness of TO and TOA.

Figure 9 shows that in the company 2 after the first year of formation and support of TD in the current state it is possible to fully stabilize the level of TD quality at a sufficiently high level.

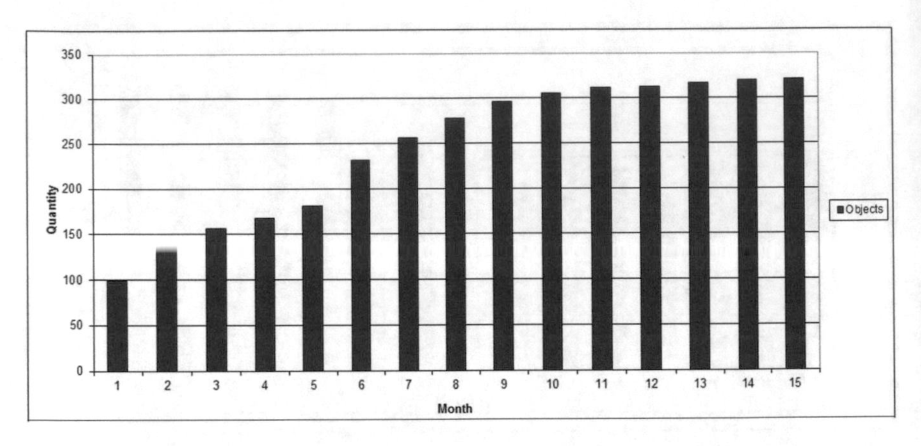

Fig. 7. Dynamics of TO quantity change in TD of the company 2.

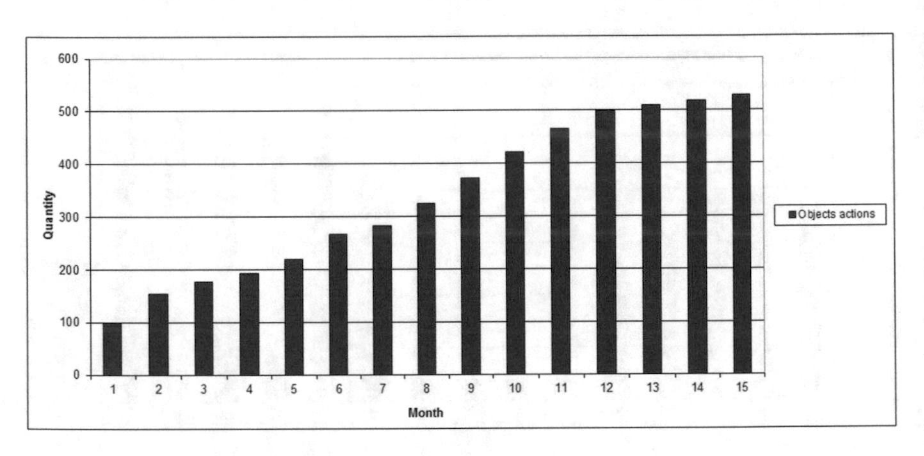

Fig. 8. Dynamics of TOA quantity change in TD of the company 2.

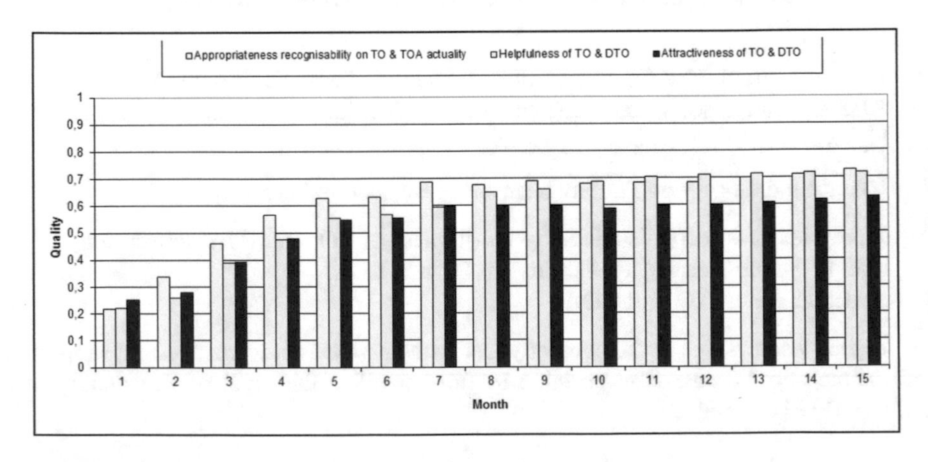

Fig. 9. Dynamics of quality indices change of TD of the company 2.

6 Conclusions

This paper dwells upon the method of assessing the TD quality, formed by linguistic analysis of the open websites content in the WWW. The basis of this method is ISO/IEC-25010 quality standards, which, unlike the common methods of analysis of feedbacks from tourism services users, is oriented towards the assessment of primary information collected in the form of tourism documentation and is provided to the consumer. This approach allows even before tourism services provision to influence the choice of the consumer and satisfaction with tourism services and feedbacks.

The developed method is based on the sub-characteristics of Operability of the ISO/IEC-25010 quality standard such as Appropriateness Recognition, Helpfulness, Attractiveness. These indices allow to control the ability of the TD to meet the needs of the TS consumer in terms of providing by the necessary information.

The monitoring of the TD quality, which is automatically formed with the help of information systems, was carried out in two travel companies within 25 and 15 months. Quality indices have been determined on a monthly basis. The dynamics of TD quality indices changes of these companies has showed that after the first year of formation and support of the TD in the current state, it was possible to fully stabilize the level of TD quality at the level of 0,6–0,9.

References

1. Berezko, O.: Using Web badges as personifying inserts in web-oriented public actions. In: Proceedings of the 5th International Scientific and Technical Conference "Computer science and information Technology" (CSIT 2010), Lviv, Publishing house "Vezha and Co", p. 55 (2010)
2. Chu, Y., Wang, H., Zheng, L., Wang, Z., Tan, K.-L.: TRSO: a tourism recommender system based on ontology. In: Lehner, F., Fteimi, N. (eds.) KSEM 2016. LNCS, vol. 9983, pp. 567–579. Springer, Cham (2016)
3. Cvetkovic, B., et al.: e-Turist: an intelligent personalised trip guide. Informatica J. Comput. Inform. **40**(4), 447–455 (2016)
4. Fedushko, S., Syerov, Y., Peleschyshyn, A., Korzh, R.: Determination of the account personal data adequacy of web-community member. Int. J. Comput. Sci. Bus. Inform. **15**(1), 1–12 (2015)
5. Fux, M., Noti, M., Myrach, T.: Quality of feedback to E-Mail requests - an explorative study in alpine tourism destinations. In: Hitz, M., Sigala, M., Murphy, J. (eds.) Information and Communication Technologies in Tourism, p. 370. Springer, Vienna (2006)
6. Gregorash, B.J.: Restaurant revenue management: apply reservation management. Inform. Technol. Tourism **16**(4), 331–346 (2016). Springer, Berlin, Heidelberg
7. Hendrikx, J., Johnson, J., Shelly, C.: Using GPS tracking to explore terrain preferences of heli-ski guides. J. Outdoor Recreation Tourism-Research Plann. Manage. **13**, 34–43 (2016)
8. Hirakawa, G., Nagatsuji, R., Shibata, Y.: A Collection and delivery method of contents in tourism with location information. In: Proceedings of 2016 19th International Conference on Network-Based Information Systems, pp. 393–396 (2016)

9. Li, Q.S., Wu, Y.D., Wang, S., Lin, M.S., Feng, X.M., Wang, H.Y.: VisTravel: visualizing tourism network opinion from the user generated content. J. Vis. **19**(3), 489–502 (2016)
10. Neal, J.D., Sirgy, M.J., Uysal, M.: Measuring the effect of tourism services on travelers' quality of life: further validation. Soc. Indic. Res. **69**(3), 243–277 (2004)
11. Neidhardt, J., Rummele, N., Werthner, H.: Predicting happiness: user interactions and sentiment analysis in an online travel forum. Inform. Technol. Tourism **17**(1), 101–119 (2017). Springer, Berlin, Heidelberg
12. Paramonov, I., Lagutina, K., Mamedov, E., Lagutina, N.: Thesaurus-based method of increasing text-via-keyphrase graph connectivity during keyphrase extraction for e-Tourism applications. In: Ngonga Ngomo, A.-C., Křemen, P. (eds.) KESW 2016. Communications in Computer and Information Science, vol. 649, pp. 129–141. Springer, Cham (2016)
13. Smirnov, A., Ponomarev, A., Kashevnik, A.: Tourist attraction recommendation service: an approach, architecture and case study. In: Proceedings of the 18th International Conference on Enterprise Information Systems, vol 2, pp. 251–261 (2016)
14. State of the Blogging Industry, ConvertKit. https://convertkit.com/reports/blogging/. Accessed 30 June 2017
15. Systems and software engineering – Systems and software Quality Requirements and Evaluation (SQuaRE) – System and software quality models, ISO/IEC 25010.2:2008. http://sa.inceptum.eu/sites/sa.inceptum.eu/files/Content/ISO_25010.pdf. Accessed 21 June 2017
16. Tanti, A., Buhalis, D.: The influences and consequences of being digitally connected and/or disconnected to travelers. Inform. Technol. Tourism **17**(1), 121–141 (2017). Springer, Berlin, Heidelberg
17. Tkachenko, S., Soprunyuk, O., Tkachenko, V., Solomko, I.: Efficiency enhancement of optimal reduction method by strengthening parallelism of structural models formation. Mach. Dyn. Res. Poland **3**(2), 85–90 (2013)
18. Van den Bosch, A., Bogers, T., de Kunder, M.: Estimating search engine index size variability: a 9-year longitudinal study. Scientometrics, An International Journal for all Quantitative Aspects of the Science of Science, Communication in Science and Science Policy, vol. 106, #2 (2016). http://www.dekunder.nl/Media/10.1007_s11192-016-1863-z.pdf
19. Zhezhnych, P., Markiv, O.: A linguistic method of web-site content comparison with tourism documentation objects. In: Proceedings of 12th International Scientific and Technical Conference Computer Science and Information Technologies (CSIT 2017), Lviv, Ukraine (2017)
20. Zhezhnych, P., Soprunyuk, O.: Analysis of the tourism documentation quality improvement. In: Proceedings of the 7th International Scientific and Technical Conference "Computer science and information technology" (CSIT 2012), pp. 34–36. Lviv, Publishing house "Vezha and Co" (2012)
21. Zhou, R., Lin, S., Lin, M.: Design of evaluation system of residents' tourism quality of tourism area based on fuzzy evaluation. In: Cao, B.-Y., Ma, S.-Q., Cao, H.-h. (eds.) Ecosystem Assessment and Fuzzy Systems Management. AISC, vol. 254, pp. 189–197. Springer, Cham (2014)

Correction to: Optimizing Wind Farm Structure Control

Vitalii Kravchyshyn, Mykola Medykovskyy, Roman Melnyk, and Marianna Dilai

Correction to: Chapter 23 in: N. Shakhovska and V. Stepashko (eds.), *Advances in Intelligent Systems and Computing II*, Advances in Intelligent Systems and Computing 689, https://doi.org/10.1007/978-3-319-70581-1_23

The reference number [9] and its citation were incorrect in the original version of this chapter. The correct version is given below:

Kuchansky, O., & Biloshchytskyi, A. (2015). Selective pattern matching method for time-series forecasting. Eastern-European Journal of Enterprise Technologies, 6(4(78), 13–18. https://doi.org/10.15587/1729-4061.2015.54812 The correction chapter and the book have been updated with the change.

The updated version of this chapter can be found at https://doi.org/10.1007/978-3-319-70581-1_23

N. Shakhovska and V. Stepashko (eds.), *Advances in Intelligent Systems and Computing II*, Advances in Intelligent Systems and Computing 689,
https://doi.org/10.1007/978-3-319-70581-1_46

Author Index

N. Shakhovska and V. Stepashko (eds.), *Advances in Intelligent Systems and Computing II*, Advances in Intelligent Systems and Computing 689,
https://doi.org/10.1007/978-3-319-70581-1

Printed in the United States
by Baker & Taylor Publisher Services